Spatial Ecology via Reaction-Diffusion Equations

WILEY SERIES IN MATHEMATICAL AND COMPUTATIONAL BIOLOGY

BÜRGER—The Mathematical Theory of Selection, Recombination, and Mutation

CHAPLAIN/SINGH/McLACHLAN—On Growth and Form: Spatio-temporal Pattern Formation in Biology

CHRISTIANSEN—Population Genetics of Multiple Loci

CLOTE/BACKOFEN—Computational Molecular Biology: An Introduction

DIEKMANN/HEESTERBEEK—Mathematical Epidemiology of Infectious Diseases: Model Building, Analysis and Interpretation

CANTRELL/COSNER–Spatial Ecology via Reaction-Diffusion Equations

Reflecting the rapidly gorwing interest and research in the field of mathematical biology, this outstanding new book series examines the integration of mathematical and computational methods into biological work. It also encourages the advancement of theoretical and quantitative approaches to biology, and the development of biological organization and function.

The scope of the series is broad, ranging from molecular structure and processes to the dynamics of ecosystems and the biosphere, but unified through evolutionary and physical principles, and the interplay of processes across scales of biological organization.

Topics to be covered in the series include:

- Cell and molecular biology
- Functional morphology and physiology
- Neurobiology and higher function
- Genetics
- Immunology
- Epidemiology
- Ecological and evolutionary dynamics of interacting populations

A fundamental research tool, the *Wiley Series in Mathematical and Computational Biology* provides essential and invaluable reading for biomathematicians and development biologists, as well as graduate students and researchers in mathematical biology and epidemiology.

Spatial Ecology via Reaction-Diffusion Equations

ROBERT STEPHEN CANTRELL and CHRIS COSNER
Department of Mathematics, University of Miami, U.S.A

WILEY

Telephone (+44) 1243 779777

Email (for orders and customer service enquiries): cs-books@wiley.co.uk
Visit our Home Page on www.wileyeurope.com or www.wiley.com

Other Wiley Editorial Offices

John Wiley & Sons Inc., 111 River Street, Hoboken, NJ 07030, USA

Jossey-Bass, 989 Market Street, San Francisco, CA 94103-1741, USA

Wiley-VCH Verlag GmbH, Boschstr. 12, D-69469 Weinheim, Germany

John Wiley & Sons Australia Ltd, 33 Park Road, Milton, Queensland 4064, Australia

John Wiley & Sons (Asia) Pte Ltd, 2 Clementi Loop #02-01, Jin Xing Distripark, Singapore 129809

John Wiley & Sons Canada Ltd, 22 Worcester Road, Etobicoke, Ontario, Canada M9W 1L1

Wiley also publishes its books in a variety of electronic formats. Some content that appears in print may not be available in electronic books.

Library of Congress Cataloging-in-Publication Data

Cantrell, Robert Stephen.
Spatial ecology via reaction-diffusion equations/Robert Stephen Cantrell and Chris Cosner.
p. cm. – (Wiley series in mathematical and computational biology)
Includes bibliographical references (p.).
ISBN 0-471-49301-5 (alk. paper)
1. Spatial ecology–Mathematical models. 2. Reaction-diffustion equations. I. Cosner, Chris. II. Title. III. Series.

QH541.15.S62C36 2003
577'.015'1–dc21

2003053780

British Library Cataloguing in Publication Data

A catalogue record for this book is available from the British Library

ISBN 0-471-49301-5

Typeset in 10/12pt Times from LaTeX files supplied by the author, processed by Laserwords Private Limited, Chennai, India
Printed and bound by CPI Antony Rowe, Eastbourne
This book is printed on acid-free paper responsibly manufactured from sustainable forestry in which at least two trees are planted for each one used for paper production.

For our families and friends

Contents

Preface

The "origin of this species" lies in the pages of the journal *Biometrika* and precedes the birth of either of the authors. There, in his remarkable landmark 1951 paper "Random dispersal in theoretical populations," J.G. Skellam made a number of observations that have profoundly affected the study of spatial ecology. First, he made the connection between random walks as a description of movement at the scale of individual members of some theoretical biological species and the diffusion equation as a description of dispersal of the organism at the scale of the species' population density, and demonstrated the plausibility of the connection in the case of small animals using field data for the spread of the muskrat in central Europe. Secondly, he combined the diffusive description of dispersal with population dynamics, effectively introducing reaction-diffusion equations into theoretical ecology, paralleling Fisher's earlier contribution to genetics. Thirdly, Skellam in particular examined reaction-diffusion models for the population density of a species in a bounded habitat, employing both linear (Malthusian) and logistic population growth rate terms, one- and two-dimensional habitat geometries, and various assumptions regarding the interface between the habitat and the landscape surrounding it. His examinations lead him to conclude that "[just] as the area/volume ratio is an important concept in connection with continuance of metabolic processes in small organisms, so is the perimeter/area concept (or some equivalent relationship) important in connection with the survival of a community of mobile individuals. Though little is known from the study of field data concerning the laws which connect the distribution in space of the density of an annual population with its powers of dispersal, rates of growth and the habitat conditions, it is possible to conjecture the nature of this relationship in simple cases. The treatment shows that if an isolated terrestrial habitat is less than a certain critical size the population cannot survive. If the habitat is slightly greater than this the surface which expresses the density at all points is roughly dome-shaped, and for very large habitats this surface has the form of a plateau."

The most general equation for a population density u mentioned in Skellam's paper has the form

$$\frac{\partial u}{\partial t} = d\nabla^2 u + c_1(x, y)u - c_2(x, y)u^2.$$

Writing in 1951, Skellam observed that "orthodox analytic methods appear in adequate" to treat the equation, even in the special case of a one-dimensional habitat. The succeeding half-century since Skellam's paper has seen phenomenal advances in many areas of mathematics, including partial differential equations, functional analysis, dynamical systems, and singular perturbation theory. That which Skellam conjectured regarding reaction-diffusion models (and indeed much more) is now rigorously understood mathematically and has been employed to provide new ecological insight into the interactions of populations and communities of populations in bounded terrestrial (and, for that matter, marine) habitats. Heretofore, the combined story of the mathematical development of reaction-diffusion theory and its application to the study of populations and communities of populations in bounded habitats has not been told in book form, and telling said story is the purpose of this work. Such is certainly not to suggest in anyway that this is the first book on the mathematical development of reaction-diffusion theory or its applications to ecology, just the first combining a (mostly) self-contained development of the theory with the particular application at hand. There are two other principal uses of reaction-diffusion theory in ecology, namely in the study of ecological invasions (dating from the work of Fisher in the 1930s) and in the study of pattern formation (dating from the work of Turing in the 1950s). It is fair to say that both these other applications have been more widely treated than has the focus of this work. (We discuss this issue further at an appropriate point in Chapter 1 and list some specific references.)

The book is structured as follows. In Chapter 1, we are primarily concerned with introducing our subject matter so as to provide a suitable context–both ecologically and mathematically–for understanding the material that follows. To this end, we begin with an overview of ecological modeling in general followed by an examination of spatial models. So doing enables us then to focus on reaction-diffusion models in more particular terms–how they may be derived, what sorts of ecological questions they may answer, and how we intend to use them to examine species' populations and communities of such populations on isolated bounded habitats. We follow our discussion of reaction-diffusion models as models with a (hopefully) self-contained compilation of the mathematical results that are needed for the analyses of subsequent chapters. For the most part, these results are well-known, so we mainly refer the reader interested in their proofs to appropriate sources. However, our analyses will draw on the theories of partial differential equations, functional and nonlinear analysis, and dynamical systems, and there is quite simply no single source available which contains all the results we draw upon. Consequently, we believe that the inclusion of this material is not merely warranted, but rather essential to the self-containment and readability of the remainder of the book. In Chapter 2, we consider linear reaction-diffusion models for a single species in an isolated bounded habitat and argue that the notion of principal eigenvalue for a linear elliptic operator is the means for measuring average population growth of a species over a bounded habitat which Skellam anticipated in his phrase "perimeter/area concept (or some equivalent relationship)." As with all subsequent chapters, our approach here is a blend of ecological examples, perspective, and applications with model development and analysis. The results in Chapter 2 enable us to turn in Chapter 3 to density dependent reaction-diffusion models for a single species in a bounded habitat. The predictions of such models *viz-a-viz* persistence versus extinction of the species in question may be described rather precisely by employing the notion of a positive (or negative) principal eigenvalue. Frequently, a prediction of persistence corresponds to the existence of a globally attracting positive equilibrium to the model. When we turn to the corresponding models for interacting populations in Chapter 4, the notion of a principal eigenvalue as a measure of average population growth retains its importance. However, the predictive outcomes of such models are not usually so tidily described as in the case of single species models. Frequently, a prediction of persistence cannot be expected to correspond to a componentwise positive globally attracting equilibrium. Instead, one needs to employ the more general notion of a globally attracting set of configurations of positive species' densities. Such configurations include a globally attracting equilibrium as a special case. This notion has come to be called permanence, and Chapter 4 is devoted to the development and application of this concept, followed in Chapter 5 with discussion of notions of persistence beyond permanence. The material in Chapters 4 and 5 is then applied in Chapter 6 to models for two competing species in an isolated bounded habitat and finally in Chapter 7 to nonmonotone models such as models for predation and food chain models.

Many people have offered us encouragement during the preparation of this work and we thank all of them. However, there are a number of individuals whose contributions we would like to mention explicitly. First of all, we are forever indebted to our thesis advisors, Murray Protter and Klaus Schmitt. We are very grateful to Simon Levin for the suggestion that we write this book. Vivian Hutson, Bill Fagan, Lou Gross and Peter Kareiva all made very significant contributions to the development of the research that led to this work or to the research itself, and again we are very grateful. We also want to thank the staff of the Department of Mathematics at the University of Miami, most especially Lourdes Robles for her able job in word processing the manuscript, Rob Calver, our editor at Wiley, the National Science Foundation for its support via the grants DMS99-73017 and DMS02-11367, and the late Jennifer Guilford for her kindness in reviewing our contract. Finally, there is one individual who is most responsible for our having begun research in a direction that made the book possible, and for that and many other kindnesses through the years, we gratefully acknowledge our colleague Alan Lazer.

Series Preface

Theoretical biology is an old subject, tracing back centuries. At times, theoretical developments have represented little more than mathematical exercises, making scant contact with reality. At the other extreme have been those works, such as the writings of Charles Darwin, or the models of Watson and Crick, in which theory and fact are intertwined, mutually nourishing one another in inseparable symbiosis. Indeed, one of the most exciting developments in biology within the last quarter-century has been the integration of mathematical and theoretical reasoning into all branches of biology, from the molecule to the ecosystem. It is such a unified theoretical biology, blending theory and empiricism seamlessly, that has inspired the development of this series.

This series seeks to encourage the advancement of theoretical and quantitative approaches to biology, and to the development of unifying principles of biological organization and function, through the publication of significant monographs, textbooks and synthetic compendia in mathematical and computational biology. The scope of the series is broad, ranging from molecular structure and processes to the dynamics of ecosystems and the biosphere, but it is unified through evolutionary and physical principles, and the interplay of processes across scales of biological organization.

The principal criteria for publication, beyond the intrinsic quality of the work, are substantive biological content and import, and innovative development or application of mathematical or computational methods. Topics will include, but not be limited to, cell and molecular biology, functional morphology and physiology, neurobiology and higher function, immunology and epidemiology, and the ecological and evolutionary dynamics of interacting populations. The most successful contributions, however, will not be so easily categorized, crossing boundaries and providing integrative perspectives that unify diverse approaches; the study of infectious diseases, for example, ranges from the molecule to the ecosystem, involving mechanistic investigations at the level of the cell and the immune system, evolutionary perspectives as viewed through sequence analysis and population genetics, and demographic and epidemiological aspects at the level of the ecological community.

The objective of the series is the integration of mathematical and computational methods into biological work; hence the volumes published should be of interest both to fundamental biologists and to computational and mathematical scientists, as well as to the broad spectrum of interdisciplinary researchers that comprise the continuum connecting these diverse disciplines.

Simon Levin

1

Introduction

1.1 Introductory Remarks

A fundamental goal of theoretical ecology is to understand how the interactions of individual organisms with each other and with the environment determine the distribution of populations and the structure of communities. Empirical evidence suggests that the spatial scale and structure of environments can influence population interactions (Gause, 1935; Huffaker, 1958) and the composition of communities (MacArthur and Wilson, 1967). In recent decades the role of spatial effects in maintaining biodiversity has received a great deal of attention in the literature on conservation; see for example Soulè (1986) or Kareiva et al. (1993). One of the most common ways that human activities alter environments is by fragmenting habitats and creating edges. Some habitat fragments may be designated as nature reserves, but they are fragments nonetheless.

One way to try to understand how spatial effects such as habitat fragmentation influence populations and communities is by using mathematical models; see Tilman and Kareiva (1997), Tilman (1994), Molofsky (1994), Holmes et al. (1994), Goldwasser et al. (1994). In this book we will examine how one class of spatial population models, namely reaction-diffusion equations, can be formulated and analyzed. Our focus will be primarily on models for populations or communities which occupy an isolated habitat fragment. There are several other types of spatial population models, including cellular automata, interacting particle systems, metapopulation models, the ideal free distribution, and dispersal models based on integral kernels. Each type of model is based on some set of hypotheses about the scale and structure of the spatial environment and the way that organisms disperse through it. We describe some of these types of models a bit later in our discussion of model formulation; see also Tilman and Kareiva (1997). Some of the ideas used in analyzing reaction-diffusion systems also can be applied to these other types of spatial models. We also describe a few of the connections between different types of models and some unifying principles in their analysis.

Reaction-diffusion models provide a way to translate local assumptions or data about the movement, mortality, and reproduction of individuals into global conclusions about the persistence or extinction of populations and the coexistence of interacting species. They can be derived mechanistically via rescaling from models of individual movement which are based on random walks; see Turchin (1998) or Durrett and Levin (1994). Reaction-diffusion models are spatially explicit and typically incorporate quantities such as dispersal rates, local growth rates, and carrying capacities as parameters which may vary with location or time.

Spatial Ecology via Reaction-Diffusion Equations R.S. Cantrell and C. Cosner
 ISBN: 0-471-49301-5

Thus, they provide a good framework for studying questions about the ways that habitat geometry and the size or variation in vital parameters influence population dynamics.

The theoretical advances in nonlinear analysis and the theory of dynamical systems which have occurred in the last thirty years make it possible to give a reasonably complete analysis of many reaction-diffusion models. Those advances include developments in bifurcation theory (Rabinowitz 1971, 1973; Crandall and Rabinowitz 1971, 1973), the formulation of reaction-diffusion models as dynamical systems (Henry 1981), the creation of mathematical theories of persistence or permanence in dynamical systems (Hofbauer and Sigmund 1988, Hutson and Schmitt 1992), and the systematic incorporation of ideas based on monotonicity into the theory of dynamical systems (Hirsch 1982, 1985, 1988a,b, 1989, 1990, 1991; Hess 1991; Smith 1995). One of the goals of this book is to show how modern analytical approaches can be used to gain insight into the behavior of reaction-diffusion models.

There are many contexts in which reaction-diffusion systems arise as models, many phenomena that they support, and many ways to approach their analysis. Existing books on reaction-diffusion models reflect that diversity to some extent but do not exhaust it. There are three major phenomena supported by reaction-diffusion models which are of interest in ecology: the propagation of wavefronts, the formation of patterns in homogeneous space, and the existence of a minimal domain size that will support positive species density profiles. In this book we will focus our attention on topics related to the third of those three phenomena. Specifically, we will discuss in detail the ways in which the size and structure of habitats influence the persistence, coexistence, or extinction of populations. Some other treatments of reaction-diffusion models overlap with ours to some extent, but none combines a specific focus on issues of persistence in ecological models with the viewpoint of modern nonlinear analysis and the theory of dynamical systems. The books by Fife (1979) and Smoller (1982) are standard references for the general theory of reaction-diffusion systems. Both give detailed treatments of wave-propagation, but neither includes recent mathematical developments. Waves and pattern formation are treated systematically by Grindrod (1996) and Murray (1993). Murray (1993) discusses the construction of models in considerable detail, but in the broader context of mathematical biology rather than the specific context of ecology. Okubo (1980) and Turchin (1998) address the issues of formulating reaction-diffusion models in ecology and calibrating them with empirical data, but do not discuss analytic methods based on modern nonlinear analysis. Hess (1991) uses modern methods to treat certain reaction-diffusion models from ecology, but the focus of his book is mainly on the mathematics and he considers only single equations and Lotka-Volterra systems for two interacting species. The book by Hess (1991) is distinguished from other treatments of reaction-diffusion theory by being set completely in the context of time-periodic equations and systems. The books by Henry (1981) and Smith (1995) give treatments of reaction-diffusion models as dynamical systems, but are primarily mathematical in their approach and use specific models from ecology or other applied areas mainly as examples to illustrate the mathematical theory. Smith (1995) and Hess (1991) use ideas from the theory of monotone dynamical systems extensively. An older approach based on monotonicity and related ideas is the method of monotone iteration. That method and other methods based on sub- and supersolutions are discussed by Leung (1989) and Pao (1992) in great detail. However, Leung (1989) and Pao (1992) treat reaction-diffusion models in general without a strong focus on ecology, and they do not discuss ideas and methods that do not involve sub- and supersolutions in much depth. One such idea, the notion of permanence/uniform persistence, is discussed by Hofbauer and Sigmund (1988, 1998) and in the survey paper by Hutson and Schmitt (1992). We will use that idea fairly extensively but our treatment differs from those given by Hofbauer and Sigmund (1988, 1998) and Hutson and Schmitt (1992) because we examine the specific applications of permanence/uniform persistence to reaction-diffusion systems in more depth, and we also use other analytic methods. Finally, there are some books on spatial ecology which include discussions of reaction-diffusion models as well as other approaches. Those include the volumes on spatial ecology by Tilman and Kareiva (1997) and on biological invasions by Kawasaki and Shigesada (1997). However, those books do not go very far with the mathematical analysis of reaction-diffusion models on bounded spatial domains.

We hope that the present volume will be interesting and useful to readers whose backgrounds range from theoretical ecology to pure mathematics, but different readers may want to read it in different ways. We have tried to structure the book to make that possible. Specifically, we have tried to begin each chapter with a relatively nontechnical discussion of the ecological issues and mathematical ideas, and we have deferred the most complicated mathematical analyses to Appendices which are attached to the ends of chapters. Most chapters include a mixture of mathematical theorems and ecological examples and applications. Readers interested primarily in mathematical analysis may want to skip the examples, and the readers interested primarily in ecology may want to skip the proofs. We hope that at least some readers will be sufficiently interested in both the mathematics and the ecology to read both.

To read this book effectively a reader should have some background in both mathematics and ecology. The minimal background needed to make sense of the book is a knowledge of ordinary and partial differential equations at the undergraduate level and some experience with mathematical models in ecology. A standard introductory course in ordinary differential equations, a course in partial differential equations from a book such as Strauss (1992), and some familiarity with the ecological models discussed by Yodzis (1989) or a similar text on theoretical ecology would suffice. Alternatively, most of the essential prerequisites with the exception of a few points about partial differential equations can be gleaned from the discussions in Murray (1993). Readers with the sort of background described above should be able to understand the statements of theorems and to follow the discussion of the ecological examples and applications.

To follow the derivation of the mathematical results or to understand why the examples and applications are of interest in ecology requires some additional background. To be able to follow the mathematical analysis, a reader should have some knowledge of the theory of functions of a real variable, for example as discussed by Royden (1968) or Rudin (1966, 1976), and some familiarity with the modern theory of elliptic and parabolic partial differential equations, as discussed by Gilbarg and Trudinger (1977) and Friedman (1976), and dynamical systems as discussed by Hale and Koçak (1991). To understand the ecological issues behind the models, a reader should have some familiarity with the ideas discussed by Tilman and Kareiva (1997), Soulè (1986), Soulè and Terborgh (1989), and/or Kareiva et al. (1993). The survey articles by Tilman (1994), Holmes et al. (1994), Molofsky (1994), and Goldwasser et al. (1994) are also useful in that regard. For somewhat broader treatments of ecology and mathematical biology respectively, Roughgarden et al. (1989) and Levin (1994) are good sources.

1.2 Nonspatial Models for a Single Species

The first serious attempt to model population dynamics is often credited to Malthus (1798). Malthus hypothesized that human populations can be expected to increase geometrically with time but the amount of arable land available to support them can only be expected to increase at most arithmetically, and drew grim conclusions from that hypothesis. In modern terminology the Malthusian model for population growth would be called a density independent model or a linear growth model. In nonspatial models we can describe populations in terms of either the total population or the population density since the total population will just be the density times the area of the region the population inhabits. We will typically think of the models as describing population densities since that viewpoint still makes sense in the context of spatial models. Let $P(t)$ denote the density of some population at time t. A density independent population model for $P(t)$ in continuous time would have the form

$$\frac{dP}{dt} = r(t)P(t); \tag{1.1}$$

in discrete time the form would be

$$P(t+1) = R(t)P(t). \tag{1.2}$$

These sorts of models are linear in the terminology of differential or difference equations, which is why they are also called linear growth models. In the discrete time case we must have $R(t) \geq 0$ for the model to make sense. If r is constant in (1.1) we have $P(t) = e^{rt}P(0)$; if $R(t)$ is constant in (1.2) we have $P(t) = R^t P(0)$. In either case, the models predict exponential growth or decay for the population. To translate between the models in such a way that the predicted population growth rate remains the same we would use $R = e^r$ or $r = \ln R$.

The second major contribution to population modeling was the introduction of population self-regulation in the logistic equation of Verhulst (1838). The key element introduced by Verhulst was the notion of density dependence, that is, the idea that the density of a population should affect its growth rate. Specifically, the logistic equation arises from the assumption that as population density increases the effects of crowding and resource depletion cause the birth rate to decrease and the death rate to increase. To derive the logistic model we hypothesize that the birthrate for our population is given by $b(t)-a(t, P)$ and the death rate by $d(t)+e(t, P)$ where b, a, d, and e are nonnegative and a and e are increasing in P. The simplest forms for a and e are $a = a_0(t)P$ and $e = e_0(t)P$ with $a_0, e_0 \geq 0$. The net rate of growth for a population at density P is then given by

$$\begin{aligned} \frac{dP}{dt} &= ([b(t) - a_0(t)P] - [d(t) + e_0(t)P])P \\ &= ([b(t) - d(t)] - [a_0(t) + e_0(t)]P)P \\ &= (r(t) - c(t)P)P, \end{aligned} \tag{1.3}$$

where $r(t) = b(t) - d(t)$ may change sign but $c(t) = a_0(t) + e_0(t)$ is always nonnegative. We will almost always assume $c(t) \geq c_0 > 0$. If r and c are constant we can introduce the new variable $K = r/c$ and write (1.3) as

$$\frac{dP}{dt} = r\left(1 - \left[\frac{P}{K}\right]\right)P. \tag{1.4}$$

Equation (1.4) is the standard form used in the biology literature for the logistic equation. The parameter r is often called the intrinsic population growth rate, while K is called the carrying capacity. If $r(t) > 0$ then equation (1.3) can be written in the form (1.4) with K positive. However, if K is a positive constant then letting $r = r(t)$ in (1.4) with $r(t)$ negative some of the time leads to a version of (1.3) with $c(t) < 0$ sometimes, which contradicts the underlying assumptions of the model. We will use the form (1.4) for the logistic equation in cases where the coefficients are constant, but since we will often want to consider situations where the intrinsic population growth rate r changes sign (perhaps with respect to time, or in spatial models with respect to location) we will usually use the form (1.3). Note that by letting $p = P/K$ and $\tau = rt$ we can rescale (1.4) to the form $dp/d\tau = p(1-p)$. We sometimes assume that (1.4) has been rescaled in this way. A derivation along the lines shown above is given by Enright (1976). The specific forms $a = a_0(t)P$, $e = e_0(t)P$ are certainly not the only possibilities. In fact, the assumptions that increases in population density lead to decreases in the birth rate and increases in the death rate may not always be valid. Allee (1931) observed that many animals engage in social behavior such as cooperative hunting or group defense which can cause their birth rate to increase or their death rate to decrease with population density, at least at some densities. Also, the rate of predation may decrease with prey density in some cases, as discussed by Ludwig et al. (1978). In the presence of such effects, which are typically known as Allee effects, the model (1.3) will take a more general form

$$\frac{dP}{dt} = g(t, P)P \tag{1.5}$$

where g may be increasing for some values of P and decreasing for others. A simple case of a model with an Allee effect is

$$\frac{dP}{dt} = r(P-\alpha)(K-P)P \tag{1.6}$$

where $r > 0$ and $0 < \alpha < K$. The model (1.6) implies that P will decrease if $0 < P < \alpha$ or $P > K$ but increase if $\alpha < P < K$.

The behavior of (1.4) is quite simple. Positive solutions approach the equilibrium K monotonically as $t \to \infty$ at a rate that depends on r, so that the equilibrium $P = 0$ is unstable and $P = K$ is stable. The behavior of (1.6) is slightly more complicated. Solutions which start with $0 < P < \alpha$ will approach 0 as $t \to \infty$; solutions starting with $P > \alpha$ will approach K monotonically as $t \to \infty$. Thus, the equilibrium $P = \alpha$ is unstable but $P = 0$ and $P = K$ are stable.

There are various ways that a logistic equation can be formulated in discrete time. The solution to (1.4) can be written as $P(t) = e^{rt}P/(1 + [(e^{rt} - 1)/K]P)$. If we evaluate $P(t)$ at time $t = 1$ we get $P(1) = e^r P(0)/(1 + [(e^r - 1)/K]P(0))$; by iterating we obtain the discrete time model

$$P(t+1) = e^r P(t)/(1 + [(e^r - 1)/K]P(t)). \tag{1.7}$$

The model (1.7) is a version of the Beverton-Holt model for populations in discrete time (see Murray (1993), Cosner (1996)). A different formulation can be obtained by integrating the equation $dP/dt = r[1-(P(0)/K)]P(t)$; that yields $P(1) = \exp(r[1-(P(0)/K)])P(0)$ and induces an iteration

$$P(t+1) = \exp(r[1 - (P(t)/K)])P(t). \tag{1.8}$$

This is a version of the Ricker model (see Murray (1993), Cosner (1996)). The difference in the assumptions behind (1.7) and (1.8) is that in (1.7) intraspecific competition is assumed to occur throughout the time interval (t, t_1) while in (1.8) the competitive effect is only based on conditions at time t. The behaviors of the models (1.7) and (1.8) are quite different. Model (1.7) behaves much like the logistic model (1.4) in continuous time. Solutions that are initially positive converge to the equilibrium $P = K$ monotonically (see Cosner (1996)). On the other hand, (1.8) may have various types of dynamics, including chaos, depending on the parameters (see Murray (1993)). In most of what follows we will study continuous time models which combine local population dynamics with dispersal through space, and we will describe dispersal via diffusion. Some of the ideas and results we will discuss can be extended to models in discrete time, but the examples (1.7), (1.8) show that models in discrete time may or may not behave in ways that are similar to their continuous time analogues, so some care is required in going from continuous to discrete time.

In many populations individuals are subject to different levels of mortality and have different rates of reproduction at different ages or stages in their lives. Models which account for these effects typically classify the population by developmental stage, age, or size and specify the rates at which individuals move from one stage to another, what fraction survive each stage of their life history, and the rates at which individuals at each of the stages produce offspring. The type of models which have been used most frequently to describe age or stage structured populations are discrete time matrix models of the sort introduced by Leslie (1948) and treated in detail by Caswell (1989). These models divide a population into n classes, with the population in each class denoted by P_i. Usually the class P_0 represents eggs, seeds, or recently born juveniles. The total population is then given by $\sum_{i=0}^{n} P_i$. The models typically specify the fraction S_i of individuals in class i that survive and enter class $i+1$ at each time step, the fraction S_{n+1} that survive and remain in class n, and the number of offspring R_i of class $i = 0$ produced in each time step by an individual of class i. The models then take the form

$$\vec{P}(t) = M\vec{P}(t) \tag{1.9}$$

where $\vec{P} = (P_0, \dots, P_n)$ and M is the matrix

$$M = \begin{pmatrix} R_0 & R_1 & R_2 & \dots & R_n \\ S_1 & 0 & 0 & \dots & 0 \\ 0 & S_2 & \ddots & \ddots & \dots \\ \vdots & \ddots & \ddots & 0 & 0 \\ 0 & \dots & 0 & S_n & S_{n+1} \end{pmatrix}.$$

Models of the form (1.9) are discussed at length by Caswell (1989). In general the entries in the matrix M may depend on $\vec{P}$ in various ways. A key property of matrices of the form shown for M with constant positive entries is that M^n has all its entries positive. It follows from the theory of nonnegative matrices that M has a positive eigenvalue λ_1 whose corresponding eigenvector $\vec{v}$ is componentwise positive. (This is a consequence of the Perron-Frobenius theorem. See Caswell (1989), Berman and Plemmons (1979), or the discussion of positivity in Chapter 2.) The eigenvalue λ_1 is called the principal eigenvalue of M, and it turns out that λ_1 is larger than the real part of any other eigenvalue of M. If $\lambda_1 > 1$ then the population will increase roughly exponentially; specifically, if $\vec{v}$ is the componentwise positive eigenvector of unit length corresponding to λ_1 we will have $P(t) \approx \lambda_1^t(\vec{P}(0) \cdot \vec{v})\vec{v}$ for t large. (See Caswell (1989).) Similarly, if $\lambda_1 < 1$ then the population will decline roughly exponentially. Thus, λ_1 plays the same role as R plays in (1.2). If we viewed λ_1 as giving an overall growth rate for the entire population $\sum_{i=0}^{n} P_i$, which is reasonable in view of the asymptotic behavior of (1.9), we would use $r = \ln \lambda_1$ in the corresponding continuous model. In this case $r > 0$ if and only if $\lambda_1 > 1$. Because they break down the life history of an organism into simpler steps, models of the form (1.9) are useful in deriving population growth rates from empirical data on survivorship and fecundity; again, see Caswell (1989). The principal eigenvalue of M in effect averages population growth rates over the age or stage classes of a structured population. The use of eigenvalues to obtain something like an average growth rate for a structured population will be a recurring theme in this book. However, the populations we consider will usually be structured by spatial distribution rather than age, and the eigenvalues will generally correspond to differential operators rather than matrices. If the entries in the matrix M depend on $\vec{P}$ then the model (1.9) can display the same types of behavior as (1.7) and (1.8). See Caswell (1989) or Cosner (1996) for additional discussion of density dependent models of the form (1.9).

It is also possible to formulate age structured population models in continuous time. The simplest formulation of such models describes a population in terms of $P(a, t)$ where a is a continuous variable representing age, so that the number of individuals in the population at time t whose ages are between a_1 and a_2 is given by $\int_{a_1}^{a_2} P(a, t)da$. The basic form of a linear (or density independent) model for a population with a continuous age structure consists of an equation describing how individuals age and experience mortality, and another equation describing the rate at which new individuals are born. The equation describing how individuals age is the McKendrick-Von Foerster equation

$$\frac{\partial P}{\partial t} + \frac{\partial P}{\partial a} = -d(a)P \tag{1.10}$$

where $d(a)$ is a age-specific death rate. The equation describing births is the birth law

$$P(0, t) = \int_0^\infty b(a)P(a, t)da \tag{1.11}$$

where $b(a)$ is an age dependent birth rate. Density dependent models arise if d or b depends on P. Age structured models based on generalizations of (1.10) and (1.11) are discussed in detail by Webb (1985).

Our main goal is to understand spatial effects, so we will usually assume that the population dynamics of a given species at a given place and time are governed by a simple continuous time model of the form (1.5). We will often consider situations where the population dynamics vary with location, and we will typically model dispersal via diffusion. Before we discuss spatial models, however, we describe some models for interacting populations which are formulated in continuous time via systems of equations analogous to (1.5).

The population models described above are all deterministic, and all of them can be interpreted as giving descriptions of how populations behave as time goes toward infinity.

It is also possible to construct models based on the assumption that changes in population are stochastic. Typically such models predict that populations will become extinct in finite time, and often the main issue in the analysis of such models is in determining the expected time to extinction. We shall not pursue that modeling approach further. A reference is Mangel and Tier (1993).

1.3 Nonspatial Models For Interacting Species

1.3.1 Mass-Action and Lotka-Volterra Models

The first models for interacting species were introduced in the work of Lotka (1925) and Volterra (1931). Those models have the general form

$$\frac{dP_i}{dt} = \left(a_i + \sum_{j=1}^{n} b_{ij} P_j \right) P_i, \quad i = 1, \ldots, n, \tag{1.12}$$

where P_i denotes the population density of the ith species. The coefficients a_i are analogous to the linear growth rate $r(t)$ in the logistic model (1.3). The coefficients b_{ii} represent intraspecies density dependence, in analogy with the term $c(t)P$ in (1.3), so we have $b_{ii} \leq 0$ for all i. The coefficients b_{ij}, $i \neq j$, describe interactions between different species. The nature of the interaction–competition, mutualism, or predator-prey interaction–determines the signs of the coefficients b_{ij}. If species i and j compete then b_{ij}, $b_{ji} < 0$. If species i preys upon species j, then $b_{ij} > 0$ but $b_{ji} < 0$. If species i and j are mutualists, then b_{ij}, $b_{ji} > 0$. (In the case of mutualism Lotka-Volterra models may sometimes predict that populations will become infinite in finite time, so the models are probably less suitable for that situation than for competition or predator-prey interactions.) Usually Lotka-Volterra competition models embody the assumption that $b_{ii} < 0$ for each i, so the density of each species satisfies a logistic equation in the absence of competitors. In the case where species i preys on species j, it is often assumed that $b_{jj} < 0$ (so the prey species satisfies a logistic equation in the absence of predation), but that $b_{ii} = 0$ while $a_i < 0$. Under those assumptions the predator population will decline exponentially in the absence of prey (because $a_i < 0$), but the only mechanism regulating the predator population is the availability of prey (because $b_{ii} = 0$, implying that the growth rate of the predator population does not depend on predator density). If the predator species is territorial or is limited by the availability of resources other than prey, it may be appropriate to take $b_{ii} < 0$. Lotka-Volterra models are treated in some detail by Freedman (1980), Yodzis (1989), and Murray (1993).

The interaction terms in Lotka-Volterra models have the form $b_{ij} P_i P_j$. If species i and species j are competitors then the equations relating P_i and P_j in the absence of other species are

$$\begin{aligned} \frac{dP_i}{dt} &= (a_i - b_{ij} P_j - b_{ii} P_i) P_i \\ \frac{dP_j}{dt} &= (a_j - b_{ji} P_i - b_{jj} P_j) P_j. \end{aligned} \tag{1.13}$$

In the context of competition, the interaction terms appear in the same way as the self-regulation terms in the logistic equation. Thus, if b_{ii} is interpreted as measuring the extent to which members of species i deplete resources needed by that species and thus reduce the net population growth rate for species i, then b_{ij} can be interpreted as measuring the extent to which members of species j deplete the same resources. This interpretation can be used to study the amount of similarity in resource utilization which is compatible with coexistence; see MacArthur (1972) or Yodzis (1989). The interpretation in the context of predator-prey interaction is more complicated. The interaction rate $b_{ij}P_iP_j$ can be interpreted as a mass-action law, analogous to mass-action principles in chemistry. The essential idea is that if individual predators and prey are homogeneously distributed within some region, then the rate at which an individual predator searching randomly for prey will encounter prey individuals should be proportional to the density of prey, but predators will search individually, so that the number of encounters will be proportional to the prey density times the predator density. Another assumption of the Lotka-Volterra model is that the birth rate of predators is proportional to the rate at which they consume prey, which in turn is directly proportional to prey density. Both of these assumptions are probably oversimplifications in some cases.

1.3.2 Beyond Mass-Action: The Functional Response

A problem with the mass-action formulation is that it implies the rate of prey consumption by each predator will become arbitrarily large if the prey density is sufficiently high. In practice the rate at which a predator can consume prey is limited by factors such as the time required to handle each prey item. This observation leads to the notion of a functional response, as discussed by Holling (1959). Another problem is that predators and prey may not be uniformly distributed. If predators search in a group then the rates at which different individual predators encounter prey will not be independent of each other. Finally, predators may spend time interacting with each other while searching for prey or may interfere with each other, so that the rate at which predators encounter prey is affected by predator density. These effects can also be incorporated into predator-prey models via the functional response.

We shall not give an extensive treatment of the derivation of functional response terms, but we shall sketch how functional responses can be derived from considerations of how individuals utilize time and space. We begin with a derivation based on time utilization, following the ideas of Holling (1959) and Beddington et al. (1975). Suppose a predator can spend a small period of time ΔT searching for prey, or consuming captured prey, or interacting with other predators. (The period of time ΔT should be short in the sense that the predator and prey densities remain roughly constant over ΔT.) Let P_1 denote the predator density and P_2 the prey density. Let ΔT_s denote the part of ΔT that the predator spends searching for prey. Let ΔT_1 denote the part of ΔT the predator spends interacting with other predators and let ΔT_2 denote the part of ΔT the predator spends handling prey. We have $\Delta T = \Delta T_s + \Delta T_1 + \Delta T_2$, but ΔT_1 and ΔT_2 depend on the rates at which the predator encounters other predators and prey, and on how long it takes for each interaction. Suppose that during the time it spends searching each individual predator encounters prey and other predators at rates proportional to the prey and predator densities, respectively (i.e. according to mass action laws.) The number of prey encountered in the time interval ΔT will then be given by $e_2P_2\Delta T_s$, while the number of predators encountered will be $e_1P_1\Delta T_s$, where e_1 and e_2 are rate constants that would depend on factors such as the predator's movement rate while searching or its ability to detect prey or other predators. If h_1 is the length of time required for each interaction between predators and h_2 is the length of time required for each interaction between a predator and a prey item, then $\Delta T_1 = e_1h_1P_1\Delta T_s$ and $\Delta T_2 = e_2h_2P_2\Delta T_s$. Using the relation $\Delta T = \Delta T_s + \Delta T_1 + \Delta T_2$, we have $\Delta T = (1 + e_1 h_1 P_1 + e_2 h_2 P_2)\Delta T_s$. Also, the predator encounters $e_2 P_2 \Delta T_s$ prey items during the period ΔT, so the overall rate of encounters with prey over the time interval ΔT is given by

$$\begin{aligned} e_2 P_2 \Delta T_s / \Delta T &= e_2 P_2 \Delta T_s / (1 + e_1 h_1 P_1 + e_2 h_2 P_2)\Delta T_s \\ &= e_2 P_2 / (1 + e_1 h_1 P_1 + e_2 h_2 P_2). \end{aligned} \tag{1.14}$$

The expression $g(P_1, P_2) = e_2 P_2/(1 + e_1 h_1 P_1 + e_2 h_2 P_2)$ is a type of functional response, introduced by Beddington et al. (1975) and DeAngelis et al. (1975). Under the assumption that predators do not interact with each other, so that $h_1 = 0$, it reduces to a form derived by Holling (1959), known as the Holling type 2 functional response. If we maintain the assumption that the rate at which new predators are produced is proportional to the per capita rate of prey consumed by each predator, and assume the prey population grows logistically in the absence of predators, the resulting model for the predator-prey interaction is

$$\begin{aligned} \frac{dP_1}{dt} &= \frac{a e_1 P_2 P_1}{1 + e_1 h_1 P_1 + e_2 h_2 P_2} - dP_1 \\ \frac{dP_2}{dt} &= r\left(1 - \left[\frac{P_2}{K}\right]\right) P_2 - \frac{e_2 P_1 P_2}{1 + e_1 h_1 P_1 + e_2 h_2 P_2}. \end{aligned} \tag{1.15}$$

(The coefficient a represents the predator's efficiency at converting consumed prey into new predators, while d represents the predator death rate in the absence of prey.) Note that if the prey density is held fixed at the level P_2^* the predator equation takes the form

$$\frac{dP_1}{dt} = \frac{AP_1}{B + CP_1} - dP_1 = \left[\frac{A}{B + CP_1} - d\right] P_1, \tag{1.16}$$

where $A = ae_2 P_2^*$, $B = 1 + e_2 h_2 P_2^*$, and $C = e_1 h_1$. If $A/B > d$ and $C > 0$ the function $[A/(B+CP_1)] - d$ is positive when $P_1 > 0$ is small but negative when P_1 is large. Thus, the model (1.16) behaves like the logistic equation in the sense that it includes self-regulation.

The derivation of the Beddington-DeAngelis (and Holling type 2) functional response in the preceding paragraph from considerations of time utilization retained the assumption that the total rate of encounters between searching predators and items of prey should follow a mass-action law. Other types of encounter rates can arise if predators or prey are not homogeneously distributed. This point is discussed in some detail by Cosner et al. (1999). Here we will just analyze one example of how spatial effects can influence the functional response and then describe the results of other scenarios. Let E denote the total rate of encounters between predators and prey per unit of search time. The rate at which prey are encountered by an individual predator will then be proportional to E/P_1 where P_1 is the predator density. The *per capita* encounter rate E/P_1 reduces to $e_2 P_2$ if $E = e_2 P_1 P_2$, as in the case of mass action. Substituting the form $E/P_1 = e_2 P_2$ into the derivation given in the preceding paragraph yields the Holling type 2 functional response if we assume that predators do not interact with each other. However, the mass action hypothesis $E = e_2 P_1 P_2$ is based on the assumption that predators and prey are homogeneously distributed in space. Suppose instead that the predators do not search for prey independently but form a group in a single location and then search as a group. In that case, increasing the number of predators in the system will not increase the area searched per unit time and thus the number of encounters with prey will not depend on predator density. (This assumes that adding more predators to the group does not significantly increase the distance at which predators can sense prey or otherwise increase the searching efficiency of the predators.) In that case we would still expect the rate of encounters to depend on prey density, so that $E = e^* P_2$. Since E represents the total encounter rate between all predators and all prey, the per capita rate at which each individual predator encounters prey will be given by $e_2^* P_2/P_1$. (We are assuming that predators and prey inhabit a finite spatial region so that the numbers of predators and prey are proportional to their densities.) Since we are assuming that all the predators are in a single group, they will not encounter any other predators while searching for prey. Using the *per capita* encounter rate with prey $e_2^* P_2/P_1$ instead of $e_2 P_2$ in the derivation of (1.14) leads to

$$(e_2^* P_2/P_1)/(1 + e_2^* h_2(P_2/P_1)) = e^* P_2/(P_1 + e_2^* h P_2). \tag{1.17}$$

The corresponding predator-prey model is

$$\begin{aligned} \frac{dP_1}{dt} &= \frac{a e_2^*(P_2/P_1)P_1}{1 + e_2^* h_2(P_2/P_1)} - dP_1 = \left[\frac{a e_2^* P_2}{P_1 + e_2^* h_2 P_2} - d\right] P_1 \\ \frac{dP_2}{dt} &= r\left(1 - \left[\frac{P_2}{K}\right]\right) P_2 - \frac{e_2^* P_1 P_2}{P_1 + e_2^* h_2 P_2}. \end{aligned} \tag{1.18}$$

The model (1.18) is said to be ratio-dependent, because the functional response depends on the ratio P_2/P_1. Other types of functional responses arise from other assumptions about the spatial grouping of predators. These include the Hassell-Varley form $eP_2/(P_1^\gamma + ehP_2)$ where $\gamma \in (0, 1)$, among others; see Cosner et al. (1999). In the ratio-dependent model (1.18) the functional response is not smooth at the origin. For that reason the model can display dynamics which do not occur in predator-prey models of the form

$$\begin{aligned} \frac{dP_1}{dt} &= \left[ag(P_1, P_2) - d\right] P_1 \\ \frac{dP_2}{dt} &= r\left(1 - \left[\frac{P_2}{K}\right]\right) P_2 - g(P_1, P_2)P_1 \end{aligned} \tag{1.19}$$

with $g(P_1, P_2)$ smooth and $g(P_1, 0) = 0$. In particular, the ratio-dependent model (1.18) may predict that both predators and prey will become extinct for certain initial densities; see Kuang and Beretta (1998).

There are several other forms of functional response which occur fairly often in predator-prey models. Some of those arise from assumptions about the behavior or perceptions of predators. An example, and the last type of functional response we will discuss in detail, is the Holling type 3 functional response $g(P_2) = eP_2^2/(1 + fP_2^2)$. The key assumption leading to this form of functional response is that when the prey density becomes low the efficiency of predators in searching for prey is reduced. This could occur in vertebrate predators that have a "search image" which is reinforced by frequent contact with prey, or that use learned skills in searching or in handling prey which deteriorate with lack of practice; i.e. when prey become scarce. It will turn out that the fact that the Holling type 3 functional response tends toward zero quadratically rather than linearly as $P_2 \to 0$ can sometimes be a significant factor in determining the effects of predator-prey interactions.

There are many other forms of functional response terms that have been used in predator-prey models. Some discussion and references are given in Getz (1994) and Cosner et al. (1999). The various specific forms discussed here (Holling type 2 and type 3, Beddington-DeAngelis, Hassell-Varley, etc.) are sometimes classified as prey dependent ($g = g(P_2)$ in our notation), ratio-dependent ($g = g(P_2/P_1)$) and predator dependent ($g = g(P_1, P_2)$). There has been some controversy about the use of ratio-dependent forms of the functional response; see Abrams and Ginzburg (2000) for discussion and references. In a recent study of various data sets, Skalski and Gilliam (2001) found evidence for some type of predator dependence in many cases. In what follows we will often use Lotka-Volterra models for predator-prey interactions, but we will sometimes use models with Holling type 2 or 3 functional response or with Beddington-DeAngelis functional response, depending on the context. Our main focus will generally be on understanding spatial effects, rather than exhaustively exploring the detailed dynamics corresponding to each type of functional response, and the forms listed above represent most of the relevant qualitative features that occur in standard forms for the functional response. We will not consider the ratio dependent case. That case is interesting and worthy of study, but it presents some extra technical problems, and it turns out that at least some of the scaling arguments which lead to diffusion models can destroy ratio dependence.

1.4 Spatial Models: A General Overview

The simple models we have described so far assume that all individuals experience the same homogeneous environment. In reality, individual organisms are distributed in space and typically interact with the physical environment and other organisms in their spatial neighborhood. The most extreme version of local interaction occurs among plants or sessile animals that are fixed in one location. Even highly mobile organisms encounter only those parts of the environment through which they move. Many physical aspects of the environment such as climate, chemical composition, or physical structure can vary from place to place. In a homogeneous environment any finite number of individuals will necessarily occupy some places and not others. The underlying theoretical distribution of individuals may be uniform, but each realization of a uniform distribution for a finite population will involve some specific and nonuniform placement of individuals. These observations would not be of any great interest in ecology if there were no empirical reasons to believe that spatial effects influence population dynamics or if simple models which assume that each individual interacts with the average environment and the average densities of other organisms adequately accounted for the observed behavior of populations and structure of communities. However, there is considerable evidence that space can affect the dynamics of populations and the structure of communities. An early hint about the importance came in the work of Gause (1935). Gause conducted laboratory experiments with *paramecium* and *didnium* and found that they generally led to extinction of one or both populations, even though the same species appear to coexist in nature. In a later set of experiments Huffaker (1958) found that a predator-prey system consisting of two species of mites could collapse to extinction quickly in small homogeneous environments, but would persist longer in environments that were subdivided by barriers to dispersal. Another type of empirical evidence for the significance of spatial effects comes from observations of natural systems on islands and other sorts of isolated patches of favorable habitat in a hostile landscape. There are many data sets which show larger numbers of species on larger islands and smaller numbers of species on smaller islands. These form the basis for the theory of island biogeography introduced by MacArthur and Wilson (1967); see also Williamson (1981) or Cantrell and Cosner (1994). A different sort of empirical evidence for the importance of space is that simple nonspatial models for resource competition indicate that in competition for a single limiting resource the strongest competitor should exclude all others (MacArthur, 1972; Yodzis, 1989), but in natural systems many competitors coexist (see Hutchinson (1961)). This point has been studied systematically from both empirical and theoretical viewpoints by Tilman (1994, 1982). Finally, there are some biological phenomena, such as invasions by exotic species, which are intrinsically spatial in nature and thus require models that involve space. In recent years the amount of attention given to issues of biological conservation has greatly increased. A major reason why many species are threatened or endangered is the destruction or fragmentation of their habitats; see for example Pimm and Gilpin (1989), Quinn and Karr (1986), McKelvey et al. (1986), Groom and Schumaker (1990). The goal of understanding how patterns of habitat destruction and fragmentation affect the persistence of populations provides a strong motivation to develop models for spatial effects.

There are many ways that space and the organisms inhabiting it can be represented in models. Some models treat space explicitly, that is, they incorporate something analogous to a map of a spatial region and they give some sort of description of what is happening at each spatial location at any given time. Other models treat space implicitly, perhaps by incorporating parameters that correlate with spatial scale or by describing what fraction of an environment is occupied by some species without specifying how that fraction of the environment is actually arranged in physical space. Among the models that treat space explicitly, some treat space as a continuum and others treat space as a discrete collection of patches. Similarly, some models explicitly keep track of individuals; others represent populations in terms of densities, and others describe only the probability that a given location is inhabited by a given species. Finally, some models treat the birth, movement, and/or death of individuals as stochastic phenomena, while others are completely deterministic. (Some models can be viewed as giving deterministic predictions for the mean, expected value, or some other attribute of a random variable.) We shall briefly describe a number of representative approaches to spatial modeling before we narrow our focus to the reaction-diffusion models that are the main topic of this book. Our goal is not to give a systematic survey of spatial modeling, but rather to place reaction-diffusion models in a broader context of spatial models and to delineate to some extent the circumstances under which reaction-diffusion models provide an appropriate modeling approach. Other discussions of similarities, differences, and relationships among different types of spatial models are given by Durrett and Levin (1994) and Tilman et al. (1997).

Models that treat both space and population dynamics implicitly include the MacArthur-Wilson (1967) models for island biogeography and the classical metapopulation model of Levins (1969). Both models describe populations strictly in terms of their presence or absence and account for patterns of occupancy in terms of a balance between colonizations and extinctions, which are assumed to occur stochastically. The MacArthur-Wilson model in its simplest form envisions a single island and a collection of species which may colonize the island or, if already present, may become extinct. Let S_0 denote the total number of species that might colonize the island and let S be the number present on the island. If species not already on the island immigrate to it at rate I and species inhabiting the island experience local extinctions at rate E, then an equilibrium value for S is obtained by balancing immigrations and extinctions so that $I(S_0 - S) = ES$. This leads to the formula $S = IS_0/(I + E)$ (see MacArthur and Wilson 1967). If I and E are assumed to depend on the area, location, and other attributes of the island then the model can yield species-area relationships. The model of Levins (1969) describes a species which inhabits an environment consisting of discrete sites, and which may colonize empty sites or experience local extinctions in occupied sites. Let p represent the fraction of sites which are occupied (so $0 \leq p \leq 1$). Let c be the rate at which colonists are produced if all sites are occupied, so if a fraction p of sites are occupied then the rate at which colonists are produced is cp. All sites are assumed to be equally accessible to colonists, so the fraction of sites which are unoccupied when colonists reach them is $1 - p$; thus the total rate of colonization of empty sites is $cp(1-p)$. Let e be the rate of local extinctions on occupied sites. The model for the fraction of sites occupied is then $dp/dt = cp(1 - p) - ep = (c - e)p - cp^2$. This model behaves just like a logistic equation, predicting that the fraction of sites occupied will approach zero as $t \to \infty$ if $c \leq e$ and will approach the equilibrium $1 - (e/c)$ as $t \to \infty$ if $c > e$. (See Levins (1969), Tilman et al. (1997), Tilman (1994).) The model can be extended to multispecies systems; see Tilman (1994).

The Levins model can be modified to treat space in a more explicit way. That is an essential theme in recent work by Hanski and his colleagues (Hanski, 1997, 1999; Hanski and Ovaskainen, 2000, 2001). A key idea is to think of the quantity p as the probability that a patch is occupied rather than the fraction of patches that are occupied. This interpretation makes sense because if a patch is chosen at random from a collection of patches where a fraction p are occupied then the probability of selecting an occupied patch is equal to p. The important thing about interpreting p as a probability of occupancy is that it can be allowed to vary from patch to patch, along with probabilities of colonization or extinction. Thus, Hanski's formulation of metapopulation models envisions a collection of n patches and describes the probability p_i that each patch is occupied in terms of the occupancy of other patches. Assume that a patch can only be colonized if it is empty, that when patch j is occupied colonists from patch j arrive at patch i at a rate c_{ij}, and that when patch i is occupied the population inhabiting it experiences a local extinction in unit time with probability e_i. The formulation of metapopulation models given by Hanski (1997, 1999) then describes the probability p_i that the ith patch is occupied by the equation

$$\frac{dp_i}{dt} = \sum_{\substack{j=1 \\ j \neq i}}^{n} c_{ij} p_j (1 - p_i) - e_i p_i. \tag{1.20}$$

The terms c_{ij} and e_i can be used to incorporate some aspects of the spatial structure of the patch network into the model. Specifically, Hanski and his colleagues use $c_{ij} = cA_j \exp(-\alpha d_{ij})$ and $e_i = e/A_i$, where A_i is the area of the ith patch, d_{ij} is the distance between patch i and patch j, exp denotes the natural exponential, α is a parameter describing the hostility of landscape between patches (which is sometimes called the matrix between patches), and c and e are parameters describing properties of the species which are related to the likelihood of colonizations or extinctions.

The model (1.20) incorporates some aspects of space explicitly in the sense that spatial attributes of the environment appear in the model, but it is not spatially explicit in the sense of keeping track of the locations of individuals as they move through space or of population density as a function of spatial location. Following the terminology of fluid mechanics, models that track the location of individuals within some explicit representation of a spatial region are sometimes called Lagrangian, while models that describe the variations of population density over some explicit representation of a spatial region are sometimes called Eulerian. Lagrangian models are often called individual based because they keep track of individuals. In practice, individual based models have been used primarily in computer simulations. They can capture enough details of behavior and life history to make predictions about the behavior of natural systems, but they do not seem to be amenable to mathematical analysis via existing analytic methods. For a discussion of individual based models and modeling, see DeAngelis et al. (1994). It is sometimes possible to calibrate Eulerian models precisely enough to make useful predictions, but a major reason for using them is that they often can be analyzed mathematically in ways that lead to broad insights about general systems, as opposed to precise predictions about specific systems. Both specific prediction and general understanding are worthy goals, but it is not always feasible to achieve both with the same model. Our discussion in most of this book will focus on reaction-diffusion models, which constitute a particular type of Eulerian model, but for now we will describe some other types of Eulerian models, and in the next section we will explore connections between different types of models.

Eulerian models for the dynamics of spatially distributed populations can incorporate various assumptions about the structure of space, the measurement of time, and the dispersal of organisms through space over time. Specifically, they can treat space and time as continuous or discrete, and they can treat dispersal and population dynamics as stochastic or deterministic.

The simplest widely used population model that can incorporate space explicitly is probably the ideal free distribution of Fretwell (1972); see also Fretwell and Lucas (1970). In its original form, the model simply describes the equilibrium distribution of a population of fixed size dispersing through a spatially discrete environment in a deterministic way. The essential idea is that in an environment which is spatially heterogeneous, individuals will position themselves in the most favorable locations, but the favorability of any location is reduced by crowding. To be more specific, the model envisions an environment divided into n habitats, with P_i denoting the population in the nth habitat. Let $P = \sum_{i=1}^{n} P_i$ denote the total population. Each habitat is assumed to have an intrinsic "fitness" a_i which is reduced logistically by crowding to $a_i - b_i P_i$. The habitats are arranged in order of their intrinsic fitness, so that $a_1 > a_2 > a_3 \cdots > a_n$. The model is given by the rule that if $a_1 - b_1 P > a_2$, then $P_1 = P$, $P_i = 0$ for $i = 2, \ldots, n$. When P is large enough that $a_1 - b_1 P \leq a_2$, individuals distribute themselves so that $a_1 - b_1 P_1 = a_2 - b_2 P_2$ with $P_1 + P_2 = P$, and with $P_i = 0$ for $i = 3, \ldots, n$ as long as $a_1 - b_1 P_1 = a_2 - B_2 P_2 > a_3$, etc. In words, individuals distribute themselves deterministically so that each individual's fitness is maximized. The theory simply describes the equilibrium distribution that develops when the population follows that rule. The model assumes that individuals can assess environmental quality and move deterministically in response to it. A version of the ideal free distribution that treats space as a continuum is formulated in Kshatriya and Cosner (2002).

A class of models that assume space to be discrete and movement to be deterministic (at least at the population level) but which include population dynamics are known as discrete diffusion or island chain models. These models treat space as a discrete set of patches and describe how the population (or density) P_i on each patch varies with time. These models can be set in either discrete or continuous time. Suppose that in each patch the population grows or declines according to a population dynamical equation $dP_i/dt = f_i(P_i)$, and that individuals disperse from patch i at a rate $D_i \geq 0$ and arrive at patch j at a rate $d_{ji} \geq 0$ with $\sum_{\substack{j=1 \\ j \neq i}}^{n} d_{ji} \leq D_i$. (If there is no mortality in transit, $\sum_{\substack{j=1 \\ j \neq i}}^{n} d_{ji} = D_i$.) The model then takes the form

$$\frac{dP_i}{dt} = \sum_{\substack{j=1 \\ j \neq i}}^{n} d_{ij} P_j - D_i P_i + f(P_i), \quad i = 1, \ldots, n. \tag{1.21}$$

Models of the form (1.21) can be extended to include density-dependent dispersal and multispecies interactions. If Q_i denotes the population (or density) of another species on patch i, an extension of (1.21) which allows density-dependent dispersal and describes interspecific interactions is

$$\frac{dP_i}{dt} = \sum_{\substack{j=1\\ j\neq i}}^{n} d_{ij}(P_i, P_j, Q_i, Q_j)P_j - D_i(P_i, Q_i)P_i + f_i(P_i, Q_i)$$

$$\frac{dQ_i}{dt} = \sum_{\substack{j=1\\ j\neq i}}^{n} \tilde{d}_{ij}(P_i, P_j, Q_i, Q_j)Q_j - \tilde{D}_i(P_i, Q_i)Q_i + \tilde{f}_i(P_i, Q_i), \qquad (1.22)$$

$$i = 1, \ldots, n.$$

An analogous formulation can be given in discrete time. To see how such a model should be formulated, consider a two-patch dispersal model

$$\frac{dP_1}{dt} = D_2P_2 - D_1P_1, \quad \frac{dP_2}{dt} = D_1P_1 - D_2P_2.$$

Adding the two equations shows that for $P = P_1 + P_2$ we have $dP/dt = 0$ so $P_1 + P_2 = P = P(0)$ so that $P_2 = P(0) - P_1$; then $dP_1/dt = D_2(P(0) - P_1) - D_1P_1$. This is a simple linear first order equation. Solving for P_1 and then P_2 yields

$$P_1(t) = [(D_2 + D_1e^{-Dt})/D]P_1(0) + [D_2(1 - e^{-Dt})/D]P_2(0)$$
$$P_2(t) = [D_1(1 - e^{-Dt})/D]P_1(0) + [(D_1 + D_2e^{-Dt})/D]P_2(0)$$

where $D = D_1 + D_2$. If we set $t = 1$, we arrive at a model of the form

$$P_1(1) = d_{11}P_1(0) + d_{12}P_2(0)$$
$$P_2(1) = d_{21}P_1(0) + d_{22}P_2(0)$$

where $d_{11} + d_{21} = d_{12} + d_{22} = 1$ and $0 \leq d_{ij} \leq 1$ for all i, j. This last model can be interpreted as saying that at each time step a fraction d_{ii} of the population of patch i remains there and a fraction d_{ji} move to the other patch. Combining the dispersal models with population dynamics gives

$$P_1(t+1) = f_1(d_{11}P_1(t) + d_{12}P_2(t))$$
$$P_2(t+1) = f_2(d_{21}P_1(t) + d_{22}P_2(t))$$

(if we assume that at each time step dispersal occurs first, then population dynamics take effect) or as

$$P_1(t+1) = d_{11}f_1(P_1(t)) + d_{12}f_2(P_2(t))$$
$$P_2(t+1) = d_{21}f_1(P_1(t)) + d_{22}f_2(P_2(t))$$

(if we assume population dynamics act first and dispersal follows). Models of this type can be extended to systems involving several patches and several interacting species.

Two other types of spatially explicit models which treat space as a discrete grid are interacting particle systems (Durrett and Levin, 1994) and cellular automata (Comins et al. 1992; Hassell et al., 1994); see also Tilman and Kareiva (1997). These sorts of models keep track of what happens at each point of the spatial grid. Typically, the state space for such models consists of integer valued functions (or vectors of such functions, if the model describes more than one species) defined on the grid points which give the population at each grid point at any given time. In some cases the function may take on just the values zero and one, indicating whether a given site is empty or occupied. A key feature of cellular automata and interacting particle systems is that the transitions between states at a given grid point are not described by deterministic equations but by rules which may include logical alternatives or may be stochastic. This feature allows for a relatively high degree of realism but can make the mathematical analysis of such models difficult. Interacting particle systems have stochastic rules for transitions between states and usually are set in continuous time; cellular automata may have deterministic or stochastic rules for transitions and often are set in discrete time. We will not try to describe these types of models further here, but discuss interacting particle systems in more detail in the next section, when we explore the connections between reaction-diffusion models and other spatial models.

The last major class of spatial models are those that treat space as a continuum and describe the distribution of populations in terms of densities that vary deterministically in time (but which may sometimes have close connections to stochastic processes). These models include the reaction-diffusion models which are the main subject of this book, along with more general types of models based on partial differential equations and discrete-time models based on integral kernels. We shall discuss the derivation of such models in more detail in the next section, but the essential idea is to envision individuals dispersing via random walks, so that at large spatial scales a collection of dispersing individuals will behave analogously to a collection of particles diffusing under the action of Brownian motion. For simplicity, suppose the spatial environment is one dimensional. If we ignore population dynamics, we can describe the density of a population dispersing via diffusion as $u(x, t)$ where

$$\frac{\partial u}{\partial t} = d\frac{\partial^2 u}{\partial x^2}. \tag{1.23}$$

(In (1.23) the coefficient d describes the rate of movement.) To get a full model we augment (1.23) with population dynamical terms:

$$\frac{\partial u}{\partial t} = d\frac{\partial^2 u}{\partial x^2} + f(x, t, u). \tag{1.24}$$

This type of model describes local population interactions in the same way as the nonspatial models treated in sections 1.2 and 1.3. Models of the form (1.24) were introduced into ecology by Skellam (1951) and Kierstead and Slobodkin (1953). They extend readily to more space dimensions and to several interacting species. They can also be extended to account for dispersal behavior that is more complex than simple diffusion, specifically advection or taxis, and dispersal in response to densities of conspecifics, prey, or predators; see for example Belgacem and Cosner (1995), Kareiva and Odell (1987) or the discussion of chemotaxis and cross-diffusion in Murray (1993). In principle, models such as (1.24) can be deduced from assumptions about the local dispersal behavior and life history of individuals and can be analyzed to give insights about the dynamics of a population at a larger scale. If we start with the dispersal model (1.23) and allow it to act on an initial density $u(x, 0)$ for unit time, we obtain

$$u(x, 1) = \int_{-\infty}^{\infty} K(x - y, 1)u(y, 0)dy \qquad (1.25)$$

where $K(x, t)$ is the integral kernel giving the fundamental solution of the diffusion equation (1.23): $K(x, t) = \exp(-x^2/4dt)/\sqrt{4\pi dt}$ (Strauss, 1992). If we assume that a population engages in reproduction and then disperses via diffusion for a unit time, we can construct a discrete-time population model

$$u(x, t + 1) = \int_{-\infty}^{\infty} K(x - y, 1)f(u(y, t))dy. \qquad (1.26)$$

Other forms of the integral kernel $K(x, t)$ can be used to describe models of dispersal other than simple diffusion. This approach is discussed by Lewis (1997); see also Van Kirk and Lewis (1997), and Hardin et al. (1988b, 1990).

The sorts of models we have described all combine some description of dispersal with some type of population dynamics (perhaps only a specification of the probability of local extinctions) so that they can be used directly to address questions about the persistence of populations. There are also models that only describe movement, specifically the formation of schools or swarms. We shall not discuss those further (see Grünbaum (1994, 1999), Flierl et al. (1999)).

We have described a variety of types of spatial models, but how can we decide which to use in a given situation? The key factors in choosing a type of model are the biology of the organisms being modeled and the structure of the spatial environment they inhabit, the goal of the modeling effort, and the spatial scale of the system. For organisms which have nonoverlapping generations (such as animals that breed once a year) or which disperse only as seeds or juveniles and then remain in one place, so that dispersal occurs via reproduction, models that operate on short to moderate time scales should generally be cast in discrete time. Over longer time scales and for large populations, even these types of organisms can often be adequately described via continuous time models. Different types of models make different assumptions about dispersal. The ideal free distribution assumes that individuals assess environmental quality and locate themselves deterministically to maximize their fitness. Thus, it is suitable as a model only for fairly complex organisms that can monitor environmental quality and move in response to it. In contrast, reaction-diffusion models and interacting particle systems, and many models based on integral kernels, assume that dispersal has a random component. (These types of models can incorporate some directed movement along with random dispersal.) Individual based models can incorporate essentially any type of dispersal, so they may be needed for organisms with highly complex dispersal behavior. At the other extreme, metapopulation models do not describe dispersing individuals at all, only the probability that a patch will be colonized given the current pattern occupancy of patches. Since plants occupy fixed sites and disperse by colonizing empty sites, metapopulation models may be especially suitable for plant populations; see Tilman (1994).

A single patch of habitat, possibly with some internal heterogeneity, can often be viewed as a continuum, so that it is appropriate to use reaction-diffusion models or models based on integral kernels to describe the density of a population inhabiting it. A landscape viewed at a moderately large scale may also be a continuum, or it may be better described as a network of discrete habitat patches within a (possibly hostile) matrix of habitat of other types. In the first case, a reaction-diffusion or integral kernel model might be appropriate, but in the second a patch model such as (1.21) or some type of metapopulation model would probably be more suitable. If a network of patches is viewed on a sufficiently large scale, patch models such as (1.21) may be well approximated by reaction-diffusion models, so those may again be appropriate at large spatial scales.

The scale of the underlying spatial environment can affect the choice of models. So can scaling by the level of detail a model should capture. Interacting particle systems can capture a large amount of detail but are difficult to analyze. Moving to a larger spatial scale reduces the resolution of the models, but in some cases interacting particle systems can be rescaled into reaction-diffusion models which are easier to analyze.

Usually models that incorporate a significant level of detail or account for many factors that might affect population dynamics are difficult to analyze mathematically, although they can often be used in computer simulations. Thus we encounter a trade-off between the resolution of models in making specific predictions and their amenability to analytic approaches which can lead to general insights. Individual based models, cellular automata, and interacting particle systems provide a significant level of detail but are hard to analyze. Thus, they are good choices for doing computer experiments. Metapopulation models, reaction-diffusion models, discrete diffusion or patch models, and models based on integral kernels all provide less detail but are easier to analyze, so they are good candidates for mathematical analysis aimed at gaining general insights. Finally, there is the issue of robustness of conclusions. If similar conclusions follow from diverse models, we can be somewhat confident that those conclusions describe some actual phenomenon that might occur in biological systems. (Clearly, deciding whether or not phenomena that *might* occur actually *do* occur and determining how important they are require data as well as models.) On the other hand, if different models lead to different results, that suggests that the phenomena they propose to describe are not adequately understood and that new empirical data or a different conceptual framework will be needed for the theory to progress further.

1.5 Reaction-Diffusion Models

1.5.1 Deriving Diffusion Models

Diffusion models can be derived as the large scale limits of dispersal models based on random walks. Such derivations are discussed by Okubo (1980) and Turchin (1998). They can also be derived from Fick's law (which describes the flux of a diffusing substance in terms of its gradient), as discussed by Okubo (1980) and Murray (1993), or from stochastic differential equations, as discussed by Gardiner (1985). Finally, they can be derived from interacting particle systems. We will describe the derivation from interacting particle systems in some detail later in this section, but first we sketch some of the other derivations. As we shall see, scaling turns out to be a crucial issue in the derivation of diffusion models.

Suppose we think of an individual (organism or particle) that moves along a line in discrete time steps by jumping one spatial step to the right with probability α or one step to the left with probability $(1-\alpha)$. Let Δt denote the time step and Δx denote the space step, and let $p(x,t)$ denote the probability that the individual is at location x at time t. The probability $p(x, t+\Delta t)$ that the individual is at location x at time $t+\Delta t$ can be computed by observing that to get to position x at time $t+\Delta t$ the individual must either be at position $x-\Delta x$ at time t and move to the right or be at position $x+\Delta x$ at time t and move to the left. Thus, we have

$$p(x,t+\Delta t) = \alpha p(x-\Delta x,t) + (1-\alpha)p(x+\Delta x,t) \tag{1.27}$$

which we can also write as

$$\begin{aligned}\frac{p(x,t+\Delta t)-p(x,t)}{\Delta t} &= \frac{1}{2\Delta t}[p(x+\Delta x,t)-2p(x,t)+p(x-\Delta x,t)] \\ &+ \frac{\beta}{\Delta t}[p(x+\Delta x,t)-p(x-\Delta x,t)]\end{aligned} \tag{1.28}$$

where $\beta = (1/2) - \alpha$. To obtain a diffusion equation from (1.28) we must relate $(1/2\Delta t)$ and $(\beta/\Delta t)$ to Δx via scaling. If we let $d = (\Delta x)^2/2\Delta t$ and let $v = -2\beta\Delta x/\Delta t$ then $1/2\Delta t = d/(\Delta x)^2$ and $\beta/\Delta t = -v/2\Delta x$ so that (1.28) becomes

$$\begin{aligned}\frac{p(x,t+\Delta t)-p(x,t)}{\Delta t} &= d\left[\frac{p(x+\Delta x,t)-2p(x,t)+p(x-\Delta x,t)}{(\Delta x)^2}\right] \\ &- v\left[\frac{p(x+\Delta x,t)-p(x-\Delta x,t)}{2\Delta x}\right].\end{aligned} \tag{1.29}$$

We can now pass to the limit as Δt and Δx approach zero as in (Okubo, 1980, p. 68) to obtain

$$\frac{\partial p}{\partial t} = d\frac{\partial^2 p}{\partial x^2} - v\frac{\partial p}{\partial x}. \tag{1.30}$$

If there is no preferred direction of motion then $\beta = 0$, so $v = 0$. Thus, the term $d\partial^2 p/\partial x^2$ describes the aspect of movement coming from symmetric random displacements, i.e. the aspect due to diffusion. If we return to equation (1.27) and let $\alpha = 1$ we obtain $p(x,t+\Delta t) = p(x-\Delta x,t)$ which reflects a deterministic movement to the right. We can rewrite this last relation as $\dfrac{p(x,t+\Delta t)-p(x,t)}{\Delta t} = \dfrac{p(x-\Delta x,t)-p(x,t)}{\Delta t}$. When $\alpha = 1$, $\beta = -1/2$, so the scaling $-2\beta\Delta x/\Delta t = v$ becomes $v = \Delta x/\Delta t$, so that (1.28) becomes $\left[\dfrac{p(x,t+\Delta x)-p(x,t)}{\Delta t}\right] = -v\left[\dfrac{p(x,t)-p(x-\Delta x,t)}{\Delta x}\right]$ which has the limiting equation

$$\frac{\partial p}{\partial t} = -v\frac{\partial p}{\partial x}. \tag{1.31}$$

The equation (1.31) has solution $p(x,t) = p_0(x-vt)$ where $p(x,0) = p_0(x)$. Thus, it indeed describes motion to the right with velocity $\Delta x/\Delta t = v$. The interpretation of the coefficient d in (1.30) is more subtle, but the relation $(\Delta x)^2/\Delta t = 2d$ suggests that d can be viewed as being half of the square of the distance that is traversed by an individual in unit time by symmetric random movements to the left or right.

The random walk described above can also be analyzed in terms of probability distributions. For simplicity, assume that an individual starts at $x = 0$ and at each time step Δt the individual moves a distance Δx to the right with probability $1/2$ or moves to the left with a distance Δx with probability $1/2$. After n time steps, the probability that the object has moved to the right r times (and thus to the left $n-r$ times) is $(1/2)^n n!/r!(n-r)!$. Assume, again for simplicity, that $-n \le m \le n$ and that m and n are both even or both odd. To arrive at position $m\Delta x$ at time $n\Delta t$, the individual must move so that $r-(n-r) = m$, that is, the difference in the numbers of steps to the right and to the left must be m. Thus, $r = (n+m)/2$ so $n - r = (n-m)/2$, so the probability $p(m\Delta x, n\Delta t)$ that the object is at position $m\Delta x$ at time $n\Delta t$ is

$$p(m\Delta x, n\Delta t) = (1/2)^n n!/[(n+m)/2]![(n-m)/2]!. \tag{1.32}$$

The distribution in (1.32) is a type of binomial distribution. This distribution can be well approximated by the normal (i.e. Gaussian) distribution for n and m large; see for example Dwass (1970). To determine the coefficients in the limiting Gaussian distribution for (1.32), we observe that the position of the individual at time $n\Delta t$ is the sum of n jumps to the right or left, each with probability 1/2. Thus, we can describe the position of the individual at time $n\Delta t$ as arising from the sum $X_1 + \cdots + X_n$ of n independent random variables, each having the value $-\Delta x$ with probability 1/2 or Δx with probability 1/2. Thus, X_k has mean 0 and variance $\sigma^2 = (1/2)(-\Delta x)^2 + (1/2)(\Delta x)^2 = (\Delta x)^2$. Let $Y_n = (X_1 + \cdots + X_n)/\sigma\sqrt{n} = (X_1 + \cdots + X_n)/\Delta x\sqrt{n}$, and let $P(X \le z)$ denote the probability that $X \le z$, where X is any random variable. By the central limit theorem (Dwass 1970) (and using the fact that the mean of X_k is 0 for each k) we have

$$\lim_{n\to\infty} P(Y_n \le z) = \frac{1}{\sqrt{2\pi}} \int_{-\infty}^{z} e^{-r^2/2} dr. \tag{1.33}$$

Since $t = n\Delta t$ is the total time for n steps,

$$P(X_1 + \cdots + X_n \le x) = P(Y_n \le x/\Delta x\sqrt{t/\Delta t}) = P(Y_n \le x/\sqrt{2dt}) \tag{1.34}$$

where we have used the scaling $(\Delta x)^2/\Delta t = 2d$, as before. Thus, we obtain from (1.33) and (1.34),

$$\lim_{n\to\infty} P(X_1 + \cdots + X_n \le x) = \frac{1}{\sqrt{2\pi}} \int_{-\infty}^{x/\sqrt{2dt}} e^{-r^2/2} dr. \tag{1.35}$$

Making the substitution $y = (\sqrt{2dt})r$ in (1.35) yields

$$\lim_{n\to\infty} P(X_1 + \cdots + X_n \le x) = \frac{1}{\sqrt{4\pi dt}} \int_{-\infty}^{x} e^{-y^2/4dt} dy. \tag{1.36}$$

The expression $(1/\sqrt{4\pi dt})e^{-y^2/4dt}$ on the right side of (1.36) is the fundamental solution of the diffusion equation (1.30) in the case $v = 0$, i.e. in the case where there is no bias in the direction of motion; see Strauss (1992). If we use that as a model for the probability distribution $p(y,t)$ for the position of an individual at time t, then $p(y,0) = \delta(y)$, that is, $p(y,0)$ is a point-mass or delta distribution at zero. Furthermore, $\partial p/\partial t = d\partial^2 p/\partial x^2$. If a single particle starts at a point z then the distribution after time t should be $(1/\sqrt{4\pi dt})e^{-(y-z)^2/4dt}$. If we start with a collection of particles with density at time zero given by $u_0(x)$ then at time t the density should be given by

$$u(x,t) = \frac{1}{\sqrt{4\pi dt}} \int_{-\infty}^{\infty} e^{-(y-x)^2/4dt} u_0(y) dy, \tag{1.37}$$

which is equivalent to

$$\frac{\partial u}{\partial t} = d\frac{\partial^2 u}{\partial x^2}, \quad u(x,0) = u_0(x). \tag{1.38}$$

The notion of a random walk can be extended to more space dimensions and to more complex types of movement. We shall not explore those extensions further here, except to note that if there is no bias in the direction of a random walk and no correlation between successive steps, the distribution for an individual starting at the origin in $I\!R^n$ after time t is well approximated by $(1/(4\pi dt)^{n/2})e^{-r^2/4dt}$, where $r = \sqrt{x_1^2 + \cdots + x_n^2}$; this form is the fundamental solution to the equation

$$\frac{\partial u}{\partial t} = d\left(\frac{\partial^2 u}{\partial x_1^2} + \cdots + \frac{\partial^2 u}{\partial x_n^2}\right) = d\nabla \cdot \nabla u \tag{1.39}$$

which is the n-dimensional diffusion equation. In any dimension, $d = \sigma^2/2$ where σ^2 is the variance of the distribution $(1/(4\pi dt)^{n/2})e^{-r^2/4dt}$ when $t = 1$. Since the distribution at $t = 1$ describes the probability that an individual has moved a distance r from its starting point at time 1, the quantity σ^2 can be calculated from data obtained, e.g., via mark-recapture experiments. See Okubo et al. (1989), Andow et al. (1990) and Turchin (1998) for more discussion of how to calibrate diffusion models from data.

Random walks can be described in other ways. One approach is to describe movement in terms of stochastic differential equations, where the position $x(t)$ of an individual is determined by

$$dx = b(x,t)dt + \sigma(x,t)dW \tag{1.40}$$

where dx and dt can be viewed as ordinary differentials and $dW = \xi(t)dt$, where $\xi(t)$ is a random variable describing white noise. One typically requires $\xi(t)$ to have mean zero and $\xi(t_1)$, $\xi(t_2)$ to have covariance $\delta(t_1 - t_2)$. The diffusion equation can be obtained as a partial differential equation for the distribution of x at time t, known as the Fokker-Plank equation. See Gardiner (1985), Okubo (1980) or Belgacem (1997) for some additional discussion of stochastic differential equations. (Note that the stochastic differential equation is a Lagrangian description of movement; the Fokker-Plank equation translates it into an Eulerian form.) A reason to mention stochastic differential equations is that they make explicit the notion that steps are not correlated. If one assumes that the probability of moving a given direction at a certain time step is correlated with the direction moved in the previous time step, derivations similar to those given above lead to the equation

$$a\frac{\partial^2 u}{\partial t^2} + b\frac{\partial u}{\partial t} = c\frac{\partial^2 u}{\partial x^2} \tag{1.41}$$

where a, b, and c are positive constants related to the spatial and temporal scales of the random walk; see Okubo (1980). Equation (1.41) is called the telegraph equation. It has some features that differ from those of the diffusion equation, but it turns out that for our purposes the differences usually will not be too important. We return to that point later.

A completely different approach to deriving diffusion equations is based on Fick's law and the notion of flux. This approach is analogous to some standard derivations of the heat equation. Fick's law, in the one-dimensional case, is the empirically derived hypothesis that diffusion transports particles or individuals across a specified point at a rate which is proportional to the spatial derivative of the concentration or density at that point, and in the direction of decreasing concentration. In higher space dimensions the situation is slightly more complicated, because the transport rate across a surface element is proportional to the directional derivative of the concentration or density in the direction normal to the surface, that is, the component of the gradient of the concentration or density in the normal direction. Again, the direction of transport is in the direction of decreasing concentration, so the constant of proportionality is negative. This leads to the formulation of the diffusive flux as $\vec{J}_D = -d\nabla u$, where u represents a density or concentration. If S is a flat surface element, such as a line segment in the plane or a finite subset of a plane in three-dimensional space, and $\vec{N}$ is a unit normal vector to S, the total rate of diffusive transport across S is $|S|(\vec{J}_D \cdot \vec{N})$ where $|S|$ is the size (length, area, etc. depending on dimension) of S. If Ω is a region with boundary $\partial\Omega$ and $\vec{n}$ denotes the outer unit normal vector to $\partial\Omega$, the total rate of transport *into* Ω by diffusion is given by the surface integral

$$\int_{\partial\Omega} [(-\vec{n}) \cdot \vec{J}_D] dS = \int_{\partial\Omega} d\nabla u \cdot \vec{n} dS.$$

The rate of transport into Ω equals the rate of change in the total number of individuals (or amount of diffusing substance) in Ω, so we have

$$\frac{\partial}{\partial t} \int_{\Omega} u dx = \int_{\partial\Omega} d\nabla u \cdot \vec{n} dS. \tag{1.42}$$

Assume that u and $\partial\Omega$ are smooth enough that we can bring the derivative inside the integral on the left side of (1.42) and apply the divergence theorem to the right side; doing those things and dividing by $|\Omega|$ (the volume of Ω) we obtain

$$\frac{1}{|\Omega|} \int_{\Omega} \left(\frac{\partial u}{\partial t} \right) dx = \frac{1}{|\Omega|} \int_{\Omega} (\nabla \cdot d\nabla u) dx. \tag{1.43}$$

Equation (1.43) says that the average of $\partial u/\partial t$ over any set Ω is the same as the average of $\nabla \cdot d\nabla u$, which we often denote as $d\nabla^2 u$ if d is constant. Since Ω could be any set, we can try to take the limit of those averages as $|\Omega| \to 0$, and if $\partial u/\partial t$ and $d\nabla^2 u$ are continuous we obtain

$$\frac{\partial u}{\partial t} = d\nabla^2 u, \tag{1.44}$$

which is the diffusion equation. The notion of flux extends to advective transport or directed motion; the flux $\vec{J}_{\vec{v}}$ arising from advective or directed motion with velocity $\vec{v}$ is simply $\vec{J}_{\vec{v}} = \vec{v}u$, where u is the concentration or density. If we let $\vec{J} = \vec{J}_D + \vec{J}_{\vec{v}}$, and allow d and $\vec{v}$ to depend on x and repeat the calculations (1.42), (1.43) leading to (1.44) we obtain

$$\frac{\partial u}{\partial t} = \nabla \cdot [d(x)\nabla u - \vec{v}u] = \nabla \cdot d(x)\nabla u - \nabla \cdot (\vec{v}u). \tag{1.45}$$

(We could also allow d and $\vec{v}$ to depend on t, but then we would have to go through a similar calculation based on taking integrals over $\Omega \times (t, t + \Delta t)$.)

Finally, diffusion equations can be derived via scaling of deterministic spatially discrete models. If we begin with the model

$$\frac{dU_i}{dt} = D(U_{i+1} - 2U_i + U_{i-1}), \tag{1.46}$$

use $D = d/(\Delta x)^2$, let $x = m\Delta x$, and make the identification $u(x, t) = u(m\Delta x, t) = U_m(t)$, then taking the limit as $\Delta x \to 0$ of (1.46) yields the diffusion equation (1.38). The same is true if we identify $u(m\Delta x,\ n\Delta t) = U_{m,n}$ and let $\Delta x,\ \Delta t \to 0$ with $(\Delta x)^2/\Delta t = s < 1/2$ in

$$\frac{U_{m,n+1} - U_{m,n}}{\Delta t} = d\left[\frac{U_{m+1,n} - 2U_{m,n} + U_{m-1,n}}{(\Delta x)^2}\right]. \tag{1.47}$$

(See Strauss (1992) and John (1982)). The same sort of results apply to systems in more space dimensions. If the plane is divided into a square grid and the density at point $(i\Delta x,\ j\Delta y)$ is denoted U_{ij}, the system

$$\frac{\partial U_{ij}}{\partial t} = D\left(U_{i+1j} + U_{i-1j} + U_{ij+1} + U_{ij-1} - 4U_{ij}\right) \tag{1.48}$$

will approximate the diffusion equation (1.44) under an appropriate scaling as the mesh size $\Delta x, \Delta y \to 0$. If (1.46) is modified to $dU_i/dt = D_1 U_{1+i} - (D_1 + D_2)u_i + D_2 u_{i-1}$ with $D_1 \neq D_2$, the limiting equation will have a drift term as in (1.30) or (1.45).

It follows from the above discussion that island chain models such as (1.21) where the dispersal coefficients d_{ij} are zero unless patches i and j are nearest neighbors will behave approximately like reaction-diffusion models if viewed on a sufficiently large spatial scale. The same turns out to be true for some interacting particle systems, but for those systems the scaling can also affect the interaction terms. We discuss that topic in detail next.

1.5.2 Diffusion Models Via Interacting Particle Systems: The Importance of Being Smooth

Interacting particle systems treat space as a discrete set of points upon which particles move and interact stochastically. Typically they keep track of the presence or absence of a particle at each point, or the number of particles at each point. The formulation we present here is essentially the one given by Durrett and Levin (1994); various mathematical treatments are given by Liggett (1985), Spohn (1991), DeMasi and Presutti (1991), and Kipnis and Landim (1999). In the systems we consider, space is identified with $\mathbb{Z}^2$, i.e. the set of points in the plane with integer coordinates. The state space for the model is the set of nonnegative integer valued functions on $\mathbb{Z}^2$, so the states specify the number of individuals at each point on the lattice. The dynamics of the model are specified by the rates at which individuals move and die or reproduce. (To say that something happens at a given rate R in this context typically means that it occurs at times which are determined by an exponentially distributed random variable with mean $1/R$.) The rates at which an individual gives birth or dies may be affected by the presence of other individuals at or near its location. If we rescale the grid $\mathbb{Z}^2$ by ϵ (so that each grid point has the form $(m\epsilon, n\epsilon)$ for integers m and n) and rescale the movement rate appropriately, solutions to the interacting particle system can sometimes be shown to converge to solutions to a related reaction-diffusion model via a process known as taking hydrodynamic limits (Durrett and Levin, 1994; DeMasi and Presutti, 1991; Spohn, 1991; Levin and Pacala, 1997; Kipnis and Landim, 1999). There are many technical difficulties connected with proving that interacting particle systems do indeed converge to reaction-diffusion models, but since we are primarily interested in reaction-diffusion models we limit our discussion to calculations of what the hydrodynamic limits would be for certain systems if the limits do indeed exist.

To specify an interacting particle system model we need to describe dispersal, interactions between individuals at a given location, and interactions between individuals in a local neighborhood. The distinction between location and neighborhood makes it possible to account for interactions between individuals at neighboring sites. The nearest neighbors of a point (m, n) are $(m+1, n)$, $(m-1, n)$, $(m, n+1)$, and $(m, n-1)$. A local neighborhood would consist of all points $(m, n) + (p, q)$ where (p, q) belongs to some neighborhood $\mathcal{N}$ of $(0, 0)$. The neighborhood $\mathcal{N}$ might consist only of $(0, 0)$, or of $(0, 0)$ and its nearest neighbors, or might be larger. We denote the number of grid points in $\mathcal{N}$ by $|\mathcal{N}|$. Suppose that individuals

(i) move to a randomly chosen nearest neighbor of their location at rate D, and

(ii) reproduce or die at rates which depend on the number of individuals at the same location, or in the local neighborhood of that location.

To give a specific example for ii), let $\eta_t(x)$ denote the number of individuals at point $x = (m, n)$ at time t. Suppose that for each individual the birth rate is decreased (or death rate increased) logistically by other individuals at the same location; then the *per capita* birth/death rate is given by $a - b(\eta_t(x) - 1)$. (The fact that b multiplies $\eta - 1$ instead of η reflects the idea that individuals do not compete with themselves.) The overall birth/death rate at location x and time t is then $\eta_t(x)(a - b(\eta_t(x) - 1))$. The hydrodynamic limit for the system defined above arises by scaling the grid as $(\epsilon m, \epsilon n)$ and scaling $D = 4d/\epsilon^2$. (The 4 enters the scaling because each site has 4 nearest neighbors, so that the rate of dispersal into a given neighboring site is $D/4$.) Notice that $1/D$ represents a time scale and that $1/D$ scales as ϵ^2, where ϵ represents the spatial scaling. Thus, once again, the diffusion model arises from a scaling where Δt is proportional to $(\Delta x)^2$. Let $\eta_t^\epsilon(x)$ denote the number of individuals at point x at time t in the scaled system, and let $u^\epsilon(x, t) = E(\eta_t^\epsilon(x))$, that is, let $u^\epsilon(x, t)$ be the mean of $\eta_t^\epsilon(x)$. If the hydrodynamic limit can indeed be taken, then $u^\epsilon(x, t) \to u(x, t)$ where the dispersal equation for $u(x, t)$ is

$$\frac{\partial u}{\partial t} = d\nabla^2 u. \tag{1.49}$$

Under some technical assumptions (see Durrett and Levin (1994), Spohn (1991), DeMasi and Presutti (1991), Kipnis and Landim (1999)), the joint distribution for the number of individuals at any finite set of sites converges to independent Poisson distributed random variables at each point, each with mean $u(x, t)$. Thus, to compute the reaction terms, we would calculate the expected birth or death rate by computing $E([a - b(U - 1)]U)$ where U is a Poisson distributed random variable with mean u. The Poisson distribution with mean u has $P(U = k) = e^{-u}u^k/k!$ so that $E(f(U)) = \sum_{k=0}^{\infty} e^{-u} f(k)u^k/k!$. If we compute this expression for aU we simply recover au since $u = E(U)$ is the mean of U. For $U(U - 1)$ we have

$$\begin{aligned} E(U(U-1)) &= e^{-u}\sum_{k=0}^{\infty} k(k-1)u^k/k! = e^{-u}\sum_{k=2}^{\infty} u^k/(k-2)! \\ &= u^2 e^{-u}\sum_{k=2}^{\infty} u^{k-2}/(k-2)! = u^2 e^{-u}\sum_{j=0}^{\infty} u^j/j! = u^2. \end{aligned} \tag{1.50}$$

Thus, $E(aU - bU(U - 1)) = au - bu^2$, so the hydrodynamic limit for this system is the diffusive logistic model

$$\frac{\partial u}{\partial t} = d\nabla^2 u + (a - bu)u. \tag{1.51}$$

This hydrodynamic limit can be derived rigorously; see DeMasi and Presutti (1991).

If we wanted to consider interactions between two species and to allow interactions to occur in the local neighborhood of each point, we would need to introduce variables $\xi_t(x)$ and $\hat{\xi}_t(x)$, where $\xi_t(x)$ denotes the number of individuals of the second species at location x at time t, and $\hat{\xi}_t(x)$ denotes the numbers of individuals in the local neighborhood of x at time t. In the hydrodynamic limit, these quantities would converge to independent Poisson random variables with the means corresponding to $\xi, \hat{\eta}$, and $\hat{\xi}$ being $v, |\mathcal{N}|u, |\mathcal{N}|v$ respectively, where v represents a density of the second species and $|\mathcal{N}|$ the size of the interaction neighborhood. If a birth (or death) rate is given by $f(\xi, \eta, \hat{\xi}, \hat{\eta})$ in the interacting particle system, the corresponding term in the reaction-diffusion model arising as the hydrodynamic limit should be $E(f(U, V, \hat{U}, \hat{V}))$ where $U, V, \hat{U}$, and $\hat{V}$ are independent Poisson random variables with means $u, v, |\mathcal{N}|u$, and $|\mathcal{N}|v$. In principle this expectation can be calculated via computations along the lines of (1.50); however, in practice, the result can be represented in terms of elementary functions only for certain fairly simple forms of f. A consequence of the independence of the random variables is that, for example, $E(\hat{U}V) = E(\hat{U})E(V) = |\mathcal{N}|uv$. It follows that scaling by hydrodynamic limits has no qualitative effect on the interaction terms in Lotka-Volterra models. It turns out that if $f(U) = U/(U + 1)$, as might be encountered in a model with a Holling type 2 functional response, $E(f(U)) = 1 - [(1 - e^{-u})/u]$. On the other hand, $E(U/(U + 2)) = 1 - 2[u - 1 + e^{-u}]/u^2$ and $E(U/(U + \sqrt{2}))$ does not appear to have a simple representation. Some properties can be deduced from the general form $E(f(U)) = \sum_{k=0}^{\infty} e^{-u} f(k)u^k/k!$. Clearly if $|f(k)| \leq M^k$ for some M then $E(f(U))$ is an analytic function, since the power series for $E(f(U))$ will converge. It is less obvious but still true that if $f(U)$ is increasing then so is $E(f(U))$ as a function of u. These and other properties are derived in Cantrell and Cosner (in press). Because most of the models we will consider involve Lotka-Volterra interactions, or other smooth interaction terms such as the Holling type 2 response, we usually simply augment the original interaction terms with diffusion rather than attempting to calculate hydrodynamic limits. This is certainly justified in the Lotka-Volterra case, since in that case the hydrodynamic limit is still Lotka-Volterra. In most of the other models we will consider, we will be concerned with qualitative effects that depend more on general properties of interaction terms such as monotonicity rather than on details of their algebraic form. Thus, using the original form of a Holling type 2 functional response rather than the form arising from hydrodynamic limits (which may not have a representation in terms of elementary functions) usually would not affect the qualitative properties of a model very much. There are, however, some important classes of models where the smoothing property of hydrodynamic limits, i.e. the fact that $E(f(U))$ is an analytic function as long as $f(U)$ does not grow faster than exponentially in U, has significant qualitative effects. These include ratio-dependent models and the hawk-dove game discussed by Durrett and Levin (1994). We will discuss the hawk-dove game in some detail and then describe briefly the analogous effects in ratio-dependent models.

The hawk-dove game is a model for evolution of a pair of strategies that individuals might use for interacting with conspecifics, in a context where the results of using a given strategy depend on the fraction of the local population that is using each of the two strategies. The payoff in this case is interpreted as a birth/death rate. The interactions between individuals using different strategies are assumed to occur in the local neighborhood of each location. Recall that we denoted the number of individuals of each type at x by $\eta_t(x)$ and $\xi_t(x)$, and the numbers in the local neighborhood of x by $\hat{\eta}_t(x)$ and $\hat{\xi}_t(x)$. For the moment we refer to the first type as "hawks" and the second as "doves". Let $\hat{p}_t(x) = \hat{\eta}_t(x)/(\hat{\eta}_t(x)+\hat{\xi}_t(x))$, that is, $\hat{p}_t(x)$ is the fraction of hawks among individuals in the local neighborhood of x at time t. Suppose that hawks at the location x have the birth (or death) rate $a\hat{p}_t(x) + b(1 - \hat{p}_t(x))$; similarly, suppose that doves at x have birth (death) rate $c\hat{p}_t(x) + d(1 - \hat{p}_t(x))$. Finally, suppose that all individuals at x have a logistic death rate proportional to the total number of individuals at x; that is, the death rate is $k(\eta_t(x) + \xi_t(x))$. If we simply treat η, $\hat{\eta}$, ξ, and $\hat{\xi}$ as densities u, v then a nonspatial model with interaction terms corresponding to the rates shown above would be

$$\begin{aligned} \frac{du}{dt} &= \left[\frac{au}{u+v} + \frac{bv}{u+v} - k(u+v)\right]u \\ \frac{dv}{dt} &= \left[\frac{cu}{u+v} + \frac{dv}{u+v} - k(u+v)\right]v. \end{aligned} \tag{1.52}$$

The traditional way to obtain a spatial model from (1.52) would be to simply add diffusion to obtain

$$\begin{aligned} \frac{\partial u}{\partial t} &= d\nabla^2 u + \left[\frac{au}{u+v} + \frac{bv}{u+v} - k(u+v)\right]u \\ \frac{\partial v}{\partial t} &= d\nabla^2 v + \left[\frac{cu}{u+v} + \frac{dv}{u+v} - k(u+v)\right]u. \end{aligned} \tag{1.53}$$

In the case $a = -0.6$, $b = 0.9$, $c = -0.9$, $d = 0.7$, both (1.52) and (1.53) can be shown to predict that $(u, v) \to (0, 0)$ as $t \to \infty$; see Durrett and Levin (1994). On the other hand, the interacting particle system itself predicts that the hawks and doves will coexist (Durrett and Levin, 1994). Computing the hydrodynamic limit for the interacting particle system via computations analogous to (1.50) yields the system

$$\begin{aligned} \frac{\partial u}{\partial t} &= d\nabla^2 u + \left[a\left(h + (1-h)\frac{u}{u+v}\right) + b(1-h)\frac{u}{u+v} - k(1+u+v)\right]u \\ \frac{\partial u}{\partial t} &= d\nabla^2 v + \left[c\left(h + (1-h)\frac{u}{u+v}\right) + d(1-h)\frac{v}{u+v} - k(u+v+1)\right]v \end{aligned} \tag{1.54}$$

where $h(u, v) = [1 - e^{-|\mathcal{N}|(u+v)}]/|\mathcal{N}|(u + v)$; recall that $|\mathcal{N}|$ is the size of the local neighborhood. See Durrett and Levin, (1994), and Perrut (2000). (If we were to assume that individuals do not interact with themselves, then the rate terms in the original interacting particle system would be modified and the form of the hydrodynamic limit would be slightly different, but the general qualitative features of the models would be very similar.) It turns out that (1.54) has an equilibrium with u and v both positive which is globally attracting for positive solutions (Durrett and Levin, 1994). Thus, in this case, taking the hydrodynamic limit of the interacting particle system yields a prediction which is opposite to what is obtained by just using the rates in that system directly, with or without diffusion. This is quite different form the cases of logistic or Lotka-Volterra models. For those, taking hydrodynamic limits typically has no effect at all on the form of the interaction terms, and the only quantitative effect is that some terms may be multiplied by $|\mathcal{N}|$ (Cantrell and Cosner in press.) What is the difference between these cases? The key observation is that the terms $u^2/(u+v)$, $uv/(u+v)$, etc. in (1.52) and (1.53) are not smooth at $(0,0)$; that is, they have partial derivatives which are not continuous at $(0,0)$. The corresponding terms in the hydrodynamic limit are smooth. (This can be seen by expanding $h(u,v)$ in a power series or by recalling that all hydrodynamic limits arise from convergent power series as in (1.50), and that functions with convergent power series representations are smooth.) When the interaction terms are not smooth, pairs of ordinary differential equations can have equilibria of types that are different than the usual stable and unstable nodes, saddles, spiral points, etc. which occur in smooth systems. This point has been explored in some detail in the context of ratio-dependent predator-prey models (Kuang and Berreta, 1998). In smooth systems with two components, there is only a single direction along which a trajectory can approach a saddle point. In predator-prey models the origin is often a saddle point, and in smooth systems the only direction from which trajectories approach the origin is generally along the predator axis. In similar models with ratio-dependent interaction terms there can be a whole sector, that is, a range of values of the ratio of the two components, within which trajectories can enter the origin. Taking the hydrodynamic limit of such a ratio-dependent system generally leads to a smooth system where the region is again an ordinary saddle point. It is not surprising that the model (1.52) supports phenomena similar to those seen in ratio-dependent predator-prey models, because the terms $u/(u+v)$ and $v/(u+v)$ are ratio dependent. If we derive a reaction-diffusion model for a system where the local interaction rates are given by functions that are not smooth, then the results obtained by simply adding diffusion may not accurately reflect the behavior of the interacting particle system. Deriving the reaction-diffusion model via hydrodynamic limits smooths out the interaction terms and thus seems to capture more accurately the behavior of the original system in some cases.

Many of the systems we will study are based on Lotka-Volterra models, and essentially all of them are based on smooth interaction terms whose key properties (e.g. monotonicity) are preserved by hydrodynamic limits. Thus, we usually just add diffusion to the terms describing local interactions, because in the cases we will consider taking hydrodynamic limits has few if any qualitative effects on the structure of the model. This approach would not be appropriate for deriving diffusion models for systems where the local interactions are not smooth, such as ratio-dependent models or variations on the hawk-dove game. For those systems it may be necessary to use hydrodynamic limits if the resulting reaction-diffusion model is supposed to display the same behavior as the underlying interacting particle system.

1.5.3 What Can Reaction-Diffusion Models Tell Us?

Reaction-diffusion models can explain three types of spatial phenomena that are relevant in ecology: waves of invasion by exotic species, the formation of patterns in homogeneous space, and the effects of the size, shape, and heterogeneity of the spatial environment on the persistence of species and the structure of communities. These ideas were introduced in four classic papers on diffusion theory. The idea that reaction-diffusion models can support traveling waves was introduced by Fisher (1937) in the context of models for the spatial spread of an advantageous gene. The idea that adding diffusion to a nonspatial model (with two or more components) can destabilize spatially homogenous equilibria and lead to the formation of patterns was introduced by Turing (1952) in the context of models for morphogenesis. The idea that reaction-diffusion models predict the minimal patch size needed to sustain a population was introduced by Skellam (1951) and Kierstead and Slobodkin (1953), specifically in the context of spatial ecology. (Skellam also extended Fisher's idea of a traveling wave to the spread of populations, as opposed to genes within a population.) In what follows we focus our attention almost exclusively on the effects of habitat geometry and heterogeneity on the persistence, coexistence, and extinction of species in finite habitats. We have chosen to pursue that topic in this book (and the research leading to it) in part because some good treatments of traveling waves (and invasions in general) and pattern formation are already available. Traveling waves in reaction-diffusion models are discussed from a mathematical viewpoint by Fife (1979), Smoller (1982), and Grindrod (1996). They are discussed from the viewpoint of biological applications by Murray (1993). Models for biological invasions, including but not limited to reaction-diffusion models, are discussed by Kawasaki and Shigesada (1997). Pattern formation is discussed by Grindrod (1996) and, again in the biological context, by Murray (1993). There are some general treatments of reaction-diffusion systems in bounded spatial domains, including Lotka-Volterra models with diffusion, for example Leung (1989) and Pao (1992), and in the time periodic case (Hess, 1991), but those treatments are essentially mathematical in nature and generally do not attempt to make close connections with specific applications in ecology. Also, the material we present includes a number of methods and applications which to our knowledge have only appeared in journal articles.

The phenomena that can be described via reaction-diffusion models can often be treated via other types of models. If highly detailed specific predictions are required, it is probably best to use simulations, perhaps via individual based models, cellular automata, or interacting particle systems. Some of these sorts of approaches are discussed by Tilman et al. (1997). A limitation of simulation models is that it is usually difficult to analyze them mathematically and extract general properties which can provide insights into the mechanisms underlying their predictions. However, they can be used in numerical experiments to construct artificial data sets from which general properties can be inferred. In particular, cellular automata models have been observed to generate spatial patterns analogous to those produced by reaction-diffusion models (Comins et al., 1992; Hassell et al., 1994). It is sometimes possible to obtain information about the rate at which a population expands its range from interacting particle systems; see Ellner and et al. (1998). Traveling waves can be shown to exist in island chain models; see Zinner (1991,1992). A limitation of reaction-diffusion models for the propagation of traveling waves is that diffusion equations on unbounded domains predict that an initial density which is zero except on some bounded set will be positive everywhere for all positive times. This seems to be at odds with the notion that organisms move with finite speed. That could be resolved by replacing reaction-diffusion models with models based on the telegraph equation $\epsilon^2\partial^2 u/\partial t^2 + \partial u/\partial t = d\partial^2 u/\partial x^2 + f(u)$. However, it turns out that for parameter values that occur in natural systems, the predictions of the telegraph equation are very close to those of the corresponding reaction diffusion model (Holmes, 1993). A more serious problem is that diffusion models do not account for long-distance movement, e.g. for the movement of an insect that "hitch-hikes" on a car or truck instead of crawling on its own. More generally, diffusion predicts that a population which is initially concentrated at a single point will develop a normal (i.e. Gaussian) distribution in space as time passes. Other patterns are certainly possible, and these can be examined by using models based on integral kernels. It turns out that the details of how the kernel decays at infinity can have profound effects on wave propagation; see Lewis (1997). Thus, there are sometimes good reasons to use such models instead of reaction-diffusion models in the study of biological invasions.

However, in a finite habitat patch the issue of long distance dispersal is much less important, especially if the primary goal is to understand the long term effects of local dispersal and habitat geometry on population dynamics. Thus, while it is possible to use integral kernels to study long term persistence in habitat patches (Hardin et al., 1988a,b 1990; VanKirk and Lewis, 1997, 1999), it is also reasonable to use reaction-diffusion models. Metapopulation models, especially as formulated by Hanski and his co-workers (1997, 1999) and Tilman (1994) can address the issue of persistence in finite habitats, but those models treat networks of patches and treat local population dynamics implicitly, in terms of presence or absence of populations. Thus, they are typically appropriate models for spatial effects on a different set of spatial scales than reaction-diffusion models. Discrete diffusion models, i.e. island chain models, can also be used to model patch networks. To describe systems where different species operate on different spatial scales, it may be necessary to combine reaction-diffusion models and patch network models. An example is discussed in Cantrell and Cosner (1996).

The phenomena of traveling waves and pattern formation differ from that of minimal patch in a fundamental way: they can occur in homogeneous space, while the very notion of "patch" requires at least enough spatial heterogeneity to distinguish the patch from its surroundings. A defining feature of any finite habitat is that it has a boundary, or edge. Edges can mediate numerous effects in population dynamics (Fagan et al., 1999). Habitat edges can be created by physical features such as rivers, roads, or (for aquatic systems) shorelines; they can also arise from interfaces between different types of ecological communities such as forests and grasslands. Edges can influence population dynamics in various ways. They can affect movement patterns, act as a source of mortality or resource subsidy, or function as a unique environment with its own rules for population interactions (Fagan et al., 1999). Edges can have different effects on different species; for example, a road may act as a barrier for some species and a source of mortality for others. Thus, because edges can exert different effects on different species, the presence of edges can influence community structure in ways that are not completely obvious from the ways in which they affect each species. Reaction-diffusion models provide a natural framework for the study of edge effects, because to correctly formulate a reaction-diffusion model in a finite patch it is necessary to specify boundary conditions. In other words, we must describe not only how individuals disperse throughout a patch, but also what they do when they reach the edge of the patch. An advantage of reaction-diffusion models is that they can readily incorporate simple rules about the effects of edges. They can also incorporate effects of internal heterogeneity within a patch. We will use those features of reaction-diffusion models to study how environmental heterogeneity affects populations.

1.5.4 Edges, Boundary Conditions, and Environmental Heterogeneity

The simplest way to formulate boundary conditions for reaction-diffusion models is probably via Fick's law. Recall that Fick's law is based on the idea that the rate of diffusion across an interface is given by $\vec{J} \cdot \vec{n}$ where $\vec{n}$ is the unit normal vector to the interface. If we are describing the density u of a population of individuals that diffuse at a rate $d(x)$ and are advected or engage in directed movement with velocity $\vec{v}(x)$, then $\vec{J} = -d(x)\nabla u + \vec{v}(x)u$ and the diffusion equation for u is (1.45), repeated here for convenience: $\partial u/\partial t = -\nabla \cdot \vec{J} = \nabla \cdot d(x)\nabla u - \nabla \cdot (\vec{v}(x)u)$. The standard boundary conditions for (1.45) relate the flux of individuals across a boundary to the density at the boundary. Specifically, let Ω be a bounded region with smooth boundary $\partial\Omega$, and let $\vec{n}$ denote the outward pointing unit normal. Then the flux across the boundary $\partial\Omega$ at any given point is proportional to the density with constant of proportionality $\beta(x)$ if $\vec{J}\cdot\vec{n}=\beta u$, i.e. $[-d(x)\nabla u+\vec{v}(x)u]\cdot\vec{n}=\beta(x)u$. The quantity $\nabla u\cdot\vec{n}$ is the directional derivative of u in the direction of the outward normal vector to $\partial\Omega$; it is often denoted by $\partial u/\partial\vec{n}$, and we will use that notation. The boundary condition for the diffusion model (1.45) would typically be expressed as

$$d(x)\frac{\partial u}{\partial\vec{n}}+[\beta(x)-\vec{v}(x)\cdot\vec{n}(x)]u=0. \tag{1.55}$$

To understand what the boundary condition means, it is useful to return to the form $\vec{J}\cdot\vec{n}=\beta u$. If $\beta=0$ then the condition says that there is no flux across $\partial\Omega$, so that $\partial\Omega$ acts as a perfect barrier to dispersal. As β increases, a larger proportion of individuals who encounter the boundary will cross it in the outward direction. Finally, if we write the condition as $u=(1/\beta)\vec{J}\cdot\vec{n}$, then as $\beta\to\infty$ the boundary condition becomes $u=0$ on $\partial\Omega$, indicating that individuals who encounter the boundary cross it immediately and thereby maintain the density on the boundary at zero. It we want to compare boundary conditions within a single model it may be convenient to write the boundary condition as

$$\alpha[d(x)\partial u/\partial\vec{n}-\vec{v}\cdot\vec{n}u]+(1-\alpha)u=0 \tag{1.56}$$

with $0\le\alpha\le 1$. In that formulation, α measures the fraction of individuals which do not cross the boundary when they encounter it. Thus, $\alpha=1$ corresponds to a situation where no individual crosses $\partial\Omega$, $\alpha=0$ corresponds to one where all individuals who encounter $\partial\Omega$ cross it, and $0<\alpha<1$ corresponds to intermediate situations. The boundary conditions (1.55) or (1.56), along with other sorts of boundary conditions, can be obtained from the theory of stochastic processes (Gardiner, 1985). The interpretation of conditions (1.55), (1.56) is the same as in the derivation from Fick's law, but cast in terms of the probability of an individual crossing $\partial\Omega$ as opposed to the rate at which individuals cross. There is some standard terminology that is used to specify boundary conditions. The condition $u=0$ on $\partial\Omega$ is sometimes called "absorbing" because under that condition $\partial\Omega$ effectively absorbs all individuals encountering it. That boundary condition is typically called a Dirichlet condition in the mathematical literature. In population dynamics, the boundary condition $u=0$ is sometimes said to correspond to a lethal boundary, because it can be interpreted as meaning that all individuals who encounter $\partial\Omega$ die. (For purposes of analyzing what happens inside Ω it doesn't usually matter whether individuals encountering $\partial\Omega$ die or simply leave and don't return.) The boundary condition $d(x)\partial n/\partial\vec{n}-\vec{v}\cdot\vec{n}u=0$ is called a no-flux or reflecting boundary condition, since it means that individuals encountering $\partial\Omega$ are always "reflected" back into Ω so they do not leave. In the mathematical literature the boundary condition $\partial u/\partial\vec{n}=0$ is called a Neumann condition. It corresponds to a reflecting or no-flux boundary condition *if* $\vec{v}=0$, but not otherwise. Boundary conditions of the form $\alpha(x)\partial u/\partial\vec{n}+\beta(x)u=0$ with both α and β positive are called Robin conditions in the mathematical literature.

In some cases individuals may have a preference for crossing a habitat edge in a particular direction. Models for movement which incorporate a preferred direction of motion at an interface have been studied in the context of stochastic processes under the name "skew Brownian motion." The basic process of skew Brownian motion is formulated as a random walk in one dimension, analogous to those described in (1.27)–(1.30), but with $\alpha=1/2$ *except* at a single point where in general $\alpha\ne 1/2$; see Walsh (1978), and Harrison and Shepp (1981). (Recall that α is the probability that an individual moves to the right at any given time step.) According to Walsh (1978) and Harrison and Shepp (1981), the scaling used in (1.27)–(1.30) where $\Delta x, \Delta t \to 0$ with $(\Delta x)^2/2\Delta t = d$ can be applied to this type of motion. If $\alpha = 1/2$ except at $x = 0$, the process is described by a diffusion equation of the form

$$\begin{aligned} &\frac{\partial u}{\partial t} = d\frac{\partial^2 u}{\partial x^2} \quad \text{on} \quad (-\infty, 0) \cup (0, \infty) \\ &\lim_{x\to 0+} \alpha\frac{\partial u}{\partial x}(x,t) = \lim_{x\to 0-} (1-\alpha)\frac{\partial u}{\partial x}(x,t) \\ &\lim_{x\to 0+} \frac{\partial^2 u}{\partial x^2}(x,t) = \lim_{x\to 0-} \frac{\partial^2 u}{\partial x^2}(x,t); \end{aligned} \tag{1.57}$$

see Walsh (1978). The first equation in (1.57) is the same diffusion equation we derived earlier in this section. The next equation implies a discontinuity in the flux at the point zero if $\alpha \neq 1/2$ and $\lim_{x\to 0+} \partial u/\partial x$ or $\lim_{x\to 0-} \partial u/\partial x$ is nonzero. The third equation requires that $\partial^2 u/\partial x^2$ can be extended continuously across the interface at $x = 0$. In Cantrell and Cosner (1998, 1999) we considered models based on skew Brownian motion. In Chapter 2 we discuss them in some detail. The models are set on an interval containing 0; they have the form

$$\begin{aligned} &\frac{\partial u}{\partial t} = D_1\frac{\partial^2 u}{\partial x^2} + su \quad \text{for} \quad x < 0 \\ &\frac{\partial u}{\partial t} = D_2\frac{\partial^2 u}{\partial x^2} + ru \quad \text{for} \quad x > 0 \\ &\lim_{x\to 0+} \alpha D_2\frac{\partial u}{\partial x} = \lim_{x\to 0-} (1-\alpha)D_1\frac{\partial u}{\partial x} \\ &\lim_{x\to 0+} D_2\frac{\partial^2 u}{\partial x^2} + ru = \lim_{x\to 0-} D_1\frac{\partial^2 u}{\partial x^2} + su. \end{aligned}$$

The equations above extend those of Walsh (1978) in a fairly natural way, but the model has some peculiar features. In Cantrell and Cosner (1998) we show that to conserve total population in Walsh's model, which has no explicit birth or death rate, the population distribution must sometimes include a multiple of the Dirac delta, i.e. a point mass, at $x = 0$. In other words, the pure dispersal model for skew Brownian motion derived by Walsh (1978) implies that there may be a nonzero number of individuals *on* the interface at $x = 0$. There are other possible formulations for models with a preferred direction of movement at an interface. One that involves a discontinuity in density but keeps flux continuous has been derived by Ovaskainen (preprint). The general problem of correctly formulating and analyzing models where individuals have a directional preference at an interface deserves further study.

Once we have specified a patch Ω, the dispersal properties and local population dynamics of a species inhabiting Ω, and the behavior (or fate) of individuals encountering the boundary of Ω, we can assemble a complete reaction-diffusion model. A typical example would be a model for the density u of a population whose members disperse throughout Ω by diffusion at a rate $d(x)$ which may vary in space, reproduce (or die) logistically with a net birth or death rate $a(x) - b(x)u$, and leave Ω when they encounter $\partial\Omega$. The model for the population density would then be

$$\begin{aligned} \frac{\partial u}{\partial t} &= \nabla \cdot d(x)\nabla u + [a(x) - b(x)u]u \text{ in } && \Omega \times (0, \infty) \\ u &= 0 && \text{on } \quad \partial\Omega \times (0, \infty). \end{aligned}$$

To obtain a specific solution, we would also have to know the initial state of the system as given by the initial density $u(x, 0)$. In his pioneering work on diffusion models in ecology, Skellam (1951) said of models such as the preceding that "orthodox analytical methods appear inadequate." Much of the remainder of this book is devoted to the presentation of "analytical methods" which have been developed in the half century since Skellam made that remark, and to discussions of how to apply those methods to such models. Although many questions remain open, the new mathematical methods are somewhat closer to being adequate for the treatment of such models than those available in 1951.

1.6 Mathematical Background

1.6.1 Dynamical Systems

The basic units of information about a group or collection of species that we will keep track of in this volume are population counts or population densities for the species in question. The very simplest example imaginable would be to keep track of either the total population of some species over some spatial region or its average population density over the region as time varies. In this case a single nonnegative number $y(t)$ is adequate to designate the total population or the average population density at time t. However, if we want to subdivide the population in some way (e.g., by life stages) and track the total population or average density for each of the divisions over time, or if we want to track the total population or average population density for more than one species simultaneously, a single number for each value of time is inadequate to convey this information. Obviously, a nonnegative number is needed for each distinct subdivision we want to consider in the first instance and for each individual species in question in the second. The most suitable way to encode such information mathematically is to use an *ordered n-tuple* or *vector* of numbers $(y_1(t), y_2(t), \ldots, y_n(t))$ for each value of time.

There are three basic properties of n-vectors that pertain to the analyses of this volume. First of all, two n-vectors $(y_1, y_2, \ldots, y_n)$ and $(y_1^*, y_2^*, \ldots, y_n^*)$ may be added by adding components and an n-vector $(y_1, \ldots, y_n)$ may be multiplied by a scalar c componentwise; i.e.,

$$(y_1, y_2, \ldots, y_n) + (y_1^*, y_2^*, \ldots, y_n^*) = (y_1 + y_1^*, y_2 + y_2^*, \ldots, y_n + y_n^*) \tag{1.58}$$

$$c(y_1, \ldots, y_n) = (cy_1, \ldots, cy_n). \tag{1.59}$$

We say that (1.58) and (1.59) endow the collection of all such n-tuples of real numbers with a *linear structure*, and this structure permits us to do arithmetic on the collection. Of course, since for our purposes the components are always nonnegative, we usually want c in (1.59) to be nonnegative. The second feature of n-vectors that we employ is that we can measure the distance between $(y_1, \ldots, y_n)$ and $(y_1^*, \ldots, y_n^*)$. To this end, we have the formula

$$d((y_1, \ldots, y_n), (y_1^*, \ldots, y_n^*)) = \left(\sum_{i=1}^{n} (y_i - y_i^*)^2 \right)^{1/2}, \tag{1.60}$$

where d in (1.60) stands for distance and the formula is the natural extension of the Pythagorean theorem to n-vectors. It is immediate from (1.60) that

$$d((y_1, \ldots, y_n), (y_1^*, \ldots, y_n^*)) \geq 0 \quad \text{with} \quad d((y_1, \ldots, y_n), (y_1^*, \ldots, y_n^*)) \\ = 0 \quad \text{only if} \quad (y_1, \ldots, y_n) = (y_1^*, \ldots, y_n^*), \tag{1.61}$$

and

$$d((y_1, \ldots, y_n), (y_1^*, \ldots, y_n^*)) = d((y_1^*, \ldots, y_n^*), (y_1, \ldots, y_n)). \tag{1.62}$$

It may not be so immediate but it is nevertheless also the case that if $(y_1^{**}, \ldots, y_n^{**})$ represents some other n-vector, then

$$d((y_1, \ldots, y_n), (y_1^*, \ldots, y_n^*)) \leq d((y_1, \ldots, y_n), (y_1^{**}, \ldots, y_n^{**})) \\ + d((y_1^{**}, \ldots, y_n^{**}), (y_1^*, \ldots, y_n^*)). \tag{1.63}$$

Note that (1.63) can be interpreted as saying that the length of one side of a triangle is less than or equal to the sum of the lengths of the other two sides and hence is known as the *triangle inequality*. Being able to equip the collection of n-vectors with (1.58)–(1.60) enables us to employ methods of real analysis such as calculus on the collection. The third feature of significance in the collection of n-vectors is that there is sometimes a natural way to order two vectors. The most usual choice is to say that

$$(y_1, \ldots, y_n) \leq (y_1^*, \ldots, y_n^*) \Leftrightarrow y_i \leq y_i^* \quad \text{for} \quad i = 1, \ldots, n, \tag{1.64}$$

i.e., the ordering between components of the n-vectors is consistent from one component of the n-tuple to the next. Of course, not all n-vectors are so ordered once $n > 1$ (e.g., (1,2) and (2,1)). Consequently, (1.64) is called a *partial ordering* on the collection of n-vectors.

Of course, our principal aim in this volume is to employ reaction-diffusion models to model and analyze the interaction of biological species in isolated bounded spatial habitats. To do so, we must consider pointwise population densities in place of average densities. So our state spaces are now n-tuples of nonnegative functions $(y_1(x), \ldots, y_n(x))$, where $x \in \overline{\Omega}$, the habitat patch in question, instead of n-tuples of nonnegative numbers $(y_1, \ldots, y_n)$. Of course, to track changes over time in an n-tuple of population densities requires us to consider n-tuples of the form $(y_1(x,t), \ldots, y_n(x,t))$.

Now two functions $y(x)$ and $y^*(x)$ on $\overline{\Omega}$ may be added together in a natural way to form a function $(y + y^*)(x)$ via the formula

$$(y + y^*)(x) = y(x) + y^*(x) \tag{1.65}$$

and a function $y(x)$ may be multiplied by a scalar c to form a function $(c \cdot y)(x)$ in a natural way via the formula

$$(c \cdot y)(x) = c \cdot y(x). \tag{1.66}$$

It follows from (1.65) and (1.66) that just as in the case of the collection of n-tuples of average population densities we can impose a linear structure on the collection of n-tuples of pointwise population densities on the habitat $\overline{\Omega}$ so that we can do arithmetic on the collection. So the first of the three aforementioned features of the collection of n-tuples of average population densities carries over to the collection of n-tuples of pointwise population densities on $\overline{\Omega}$ in a straightforward manner.

The second feature of the collection of n-tuples of average population densities, namely the ability to measure the distance between two such n-tuples, also carries over to the collection of n-tuples $(y_1(x), \ldots, y_n(x))$ of pointwise population densities. However, the inclusion of explicit spatial structure complicates the issue. Should having $(y_1(x), \ldots, y_n(x))$ be close to $(y_1^*(x), \ldots, y_n^*(x))$ mean that the two n-tuples are close in some sense at every point $x \in \overline{\Omega}$ or need they only be close in some way on the average over $\overline{\Omega}$? Should only the values of the population densities matter or should we take into account their spatial derivatives to some order and, if so, to what order? All these possibilities give rather different but nevertheless valid notions of "closeness" in the context of n-tuples of pointwise population densities on the habitat $\overline{\Omega}$. Consequently, we should define distance in enough generality so as to accommodate any of them. We do so via the notion of *metric space* from topology. By a *metric space*, we mean a collection or set Y of objects or points y and a *distance function* or *metric* d to measure distance between points of Y. Mathematically, d is a map that assigns a nonnegative number $d(y, y^*)$ to every ordered pair (y, y^*) of points y, y^* in Y. We require that d satisfy (1.61)–(1.63); i.e., two points at zero distance from each other must coincide, the distance from y to $y^*(d(y, y^*))$ must be the same as the distance from y^* to $y(d(y^*, y))$, and the triangle inequality must hold.

For our purpose, Y will usually be *some* collection of n-tuples of pointwise population densities on $\overline{\Omega}$. The most appropriate collection of such n-tuples in any given situation depends on the choice of the metric d. We will be tracking configurations of population densities as they evolve over time and necessarily will be interested in their limits. We will frequently want a limiting configuration of population densities to inherit the properties of the configurations that approach it. For instance, we may want to insist that all the densities that we consider be continuous on $\overline{\Omega}$. To see why there is an issue as regards the choice of metric, let us look at a specific example. For simplicity, we will consider just a single species distributed across the one-dimensional habitat $[0, 2]$. A possible family of continuous population densities for the species is given by the sequence of functions $f_k(x),\ x \in [0, 2]$, where

$$f_k(x) = \begin{cases} x^k & 0 \le x \le 1 \\ 1 & 1 \le x \le 2. \end{cases}$$

Clearly the pointwise limit of this family of densities is the function $f_0(x)$ with

$$f_0(x) = \begin{cases} 0 & 0 \le x < 1 \\ 1 & 1 \le x \le 2 \end{cases}$$

which is clearly not continuous on $[0, 2]$. If we define $d_1(f, g)$ for real valued functions f, g defined on $[0, 2]$ by

$$d_1(f, g) = \int_0^2 |f(x) - g(x)| dx,$$

then it is easy to establish that d_1 satisfies (1.61)–(1.63). So d_1 is a perfectly acceptable metric. Moreover, the sequence $\{f_k\}$ converges to f_0 under this metric since $d_1(f_k, f_0) = \int_0^1 x^k = 1/(k+1) \to 0$ as $k \to \infty$. So under the metric d_1 the limit of continuous densities on $[0, 2]$ need not be continuous. If preserving continuity in the limiting density is important to us, we need to change the metric. If d_1 is the metric we want, we need to expand the class of density functions we consider beyond continuous ones. The mathematical method of dealing with this issue is to refine the notion of metric space by defining the term *complete metric space*. A metric space Y is complete if all sequences in Y whose elements become arbitrarily close together converge to an element in Y. The technical word mathematicians use for a sequence whose elements become arbitrarily close together is *Cauchy*. A Cauchy sequence $\{y_n\} \subseteq Y$ is one so that given any positive number ε, there will correspond an index N so that for indices $n, m \geq N$ the distance $d(y_n, y_m) < \varepsilon$. Completeness of Y is the property that all Cauchy sequences have limits in Y. In our example, if we switched from d_1 to d_∞ where $d_\infty(f, g)$ is defined by

$$d_\infty(f, g) = \max_{x \in [0,2]} |f(x) - g(x)|$$

for continuous real valued functions f, g defined on $[0, 2]$, a basic result in real analysis (see, e.g., Rudin 1976) is that a Cauchy sequence of such functions converges to a unique continuous real value function defined on $[0, 2]$. So under d_∞, the continuous real-valued functions on $[0, 2]$ are complete. The sequence $\{f_k\}$ simply fails to be Cauchy under the metric d_∞.

A metric space Y does not require a linear structure, only a means of measuring distances between points that satisfies (1.61)–(1.63). However, when there is a linear structure on Y, this feature can be employed in the construction of the metric. To this end, if y and y^* are vectors in Y, $y - y^*$ can be viewed either as a point or vector in its own right or as a line segment joining y and y^*. Consequently, it is natural to suppose that $d(y, y^*) = d(y - y^*, 0)$. In such a case, one need only measure distance to the zero vector of Y. The notation $||y||$ is usually used for this purpose. If $||y||$ has the properties that

$$\begin{aligned} &||y|| \geq 0 \quad \text{and} \quad ||y|| = 0 \Leftrightarrow y = 0, \\ &||cy|| = |c|\,||y|| \quad \text{for all scalars} \quad c, \\ &||y + y^*|| \leq ||y|| + ||y^*|| \quad \text{for all} \quad y, y^* \in Y, \end{aligned} \tag{1.67}$$

then $d(y, y^*) = ||y - y^*||$ defines a metric on Y. The number $||y||$ is referred to as the *norm* of the vector. A linear space Y which is complete as a metric space under such a metric is called a *Banach space*.

The n-tuples of pointwise population densities on the habitat patch $\overline{\Omega}$ that are the primary concern of this volume all belong to Banach spaces. Later in this section we discuss the various Banach spaces that arise in our discussion, including the definition of the norm in each case. For now, we want to make two additional observations. First, Banach spaces are primary objects of study in the branch of mathematics known as *functional analysis*, which provides a collection of very powerful mathematical results for use in analyzing models for several interacting biological species in an isolated bounded habitat. Secondly as noted previously, since our interest is in population densities, we will usually confine ourselves to n-tuples the components of which are nonnegative functions. As a result, we will usually be examining closed subsets of a Banach space rather than the entire space. It is a basic result of functional analysis that closed subsets of Banach spaces are complete metric spaces.

As noted, our primary interest in this volume is to track the predictions of reaction-diffusion models about the long term behavior of the densities for several interacting biological species on an isolated bounded habitat $\overline{\Omega}$. If u_i denotes the density of species i (say i runs from 1 to n), then in most of the models we consider, the time rate of change $\frac{\partial u_i}{\partial t}$ does not explicitly depend upon time; i.e. the growth law is *autonomous*. (Of course, $\frac{\partial u_i}{\partial t}$ does depend upon u_i and its spatial derivatives and may also depend explicitly upon the spatial coordinate x and some or all of the remaining densities.) In the next subsection, we review the basic theory of systems of reaction-diffusion equations. We note there that in all the examples of such systems that we consider, the time evolution of species densities so governed is uniquely determined for all times $t \geq 0$ by the initial configuration of the densities; i.e. $(u_1(x,t), \ldots, u_n(x,t)) = \varphi(u_1(x,0), \ldots, u_n(x,0), t)$ for $t \geq 0$. Moreover, φ is continuous when viewed as a function from $Y \times [0, \infty)$ into Y, where the complete metric space Y is some suitable prespecified collection of n-tuples of possible species densities on an underlying habitat $\overline{\Omega}$. When the reaction-diffusion system is autonomous, the basic theory then guarantees that the configuration to which $(u_1(x,0), \ldots, u_n(x,0))$ evolves after $t + t'$ units of time where $t, t' > 0$ is the same configuration as arises when $\varphi(u_1(x,0), \ldots, u_n(x,0), t)$ is viewed as an initial configuration and then evolves for t' units of time; i.e.,

$$\varphi(u_1(x,0), \ldots, u_n(x,0), t + t') = \varphi(\varphi(u_1(x,0), \ldots, u_n(x,0), t), t'). \tag{1.68}$$

These features mean that the solution trajectories to such autonomous reaction-diffusion models are examples of *continuous time semi-dynamical systems*.

A *(continuous time) dynamical system* or *flow* is a continuous function π defined on $Y \times \mathbb{R}$ and taking values in Y, where for our purposes Y is a metric space, so that

$$\pi(u, 0) = u \tag{1.69}$$

for all $u \in Y$ and

$$\pi(u, t + t') = \pi(\pi(u, t), t') \tag{1.70}$$

for all t, t' in $\mathbb{R}$ so that $\pi(u, t + t')$ and $\pi(u, t)$ are defined. (We refer to (1.70) as the semi-group property.) When the set of real numbers $\mathbb{R}$ in the preceding is replaced by the set $[0, \infty)$, π is called a *semi-dynamical system* or *semiflow*.

If π is a dynamical system and $u \in Y$, then the solution $\pi(u, t)$ exists for all t in some maximal open interval $(t_-(y), t_+(y))$, with $-\infty \leq t_-(y) < 0 < t_+(y) \leq \infty$. The set of points

$$\gamma(u) = \{\pi(u, t) : t_-(y) < t < t_+(y)\} \tag{1.71}$$

is called the *orbit* of π through u. The set of points

$$\gamma^+(u) = \{\pi(u, t) : 0 \leq t < t_+(y)\} \tag{1.72}$$

is called the *positive semi-orbit* of π through u. Of course, in case π is a semi-dynamical system that does not extend to be a dynamical system, only positive semi-orbits are guaranteed to exist for all $u \in Y$.

In order to consider long term behavior of a dynamical or semi-dynamical system, it is legitimate to require at a minimum that the system be such that

$$t_+(u) = +\infty \quad \text{in (1.71) or (1.72) for all } u \in Y. \tag{1.73}$$

Of course, not all dynamical or semi-dynamical systems have this property. For example, in the ordinary differential equation

$$\frac{dy}{dt} = (y - K)^2$$

with $K > 0$, $t_+(m) = \dfrac{1}{m - K}$ for all $m > K$, and the corresponding orbit "blows up" in finite time.

Certainly most ecological models for interacting biological species do not envision orbits that "blow up" in finite time. Moreover, when ecological models are used to describe the long-term interactions for biological species (we are specifically excluding linear models at this point), they usually include mechanisms which preclude unbounded growth in species densities in the long term. Consequently, the hypotheses that we impose on a dynamical or semi-dynamical system to obtain (1.73) will actually restrict the systems much more stringently. Namely, we require that the systems be *dissipative*, by which we mean that there is a bounded subset U of Y so that for any $u \in Y$, $\pi(u, t) \in U$ for all sufficiently large t. (Recall that Y will usually be a closed subset of a Banach space with norm $\|\cdot\|$. Having U be bounded means that there is a positive number M so that if $v \in U$, $\|v\| \leq M$. For an individual element $u \in Y$, how large t must be in order for $\pi(u, t) \in U$ is allowed to depend on u.)

Let us now note a consequence of the dissipativity assumption in the context of the most basic example of a continuous time dynamical system, namely the collection of all solution trajectories (based at $t = 0$) of a system or ordinary differential equations of the form

$$y_i' = f_i(y_1, \ldots, y_n), \tag{1.74}$$

$i = 1, 2, \ldots, n$. Let $(y_1^0, \ldots, y_n^0)$ represent an arbitrary initial configuration for, say, the average population densities of species 1 to n. Assuming (1.74) is dissipative, then the sequence of configurations $\{\pi((y_1^0, \ldots, y_n^0), t_j)\}$ for $j = 1, 2, 3, \ldots$ is contained in the bounded set $U_R = \{(y_1, \ldots, y_n) : \|(y_1, \ldots, y_n)\| \leq R\}$ for some $R > 0$ for any collection of distinct times t_j such that $t_j \to +\infty$ as $j \to +\infty$. If U_R is expressed as the union of some *finite* collection of subsets of itself, at least one of these subsets must contain $\pi((y_1^0, \ldots, y_n^0), t_j)$ for *infinitely* many distinct values of t_j. It follows from this observation that there must be at least one point $(\overline{y}_1, \ldots, \overline{y}_n) \in U_R$ so that some subsequence $\pi((y_1^0, \ldots, y_n^0), t_{j_k})$ of $\pi((y_1^0, \ldots, y_n^0), t_j)$ converges to $(\overline{y}_1, \ldots, \overline{y}_n)$ as the index $k \to \infty$. (This result is one of the main results of advanced calculus and is known as the Bolzano-Weierstrass Theorem (see, e.g., Apostol (1974)). A set P with the property that any sequence of points in P has a convergent subsequence is called *precompact*. If the limit of the subsequence is always actually in P, then P is said to be *compact*. In particular $\{\pi((y_1^0, \ldots, y_n^0), t_j)\}$ is precompact for any distinct collection of times t_j such that $t_j \to \infty$.) Consequently, dissipativity in the context of (1.74) means there will be a bounded subset that attracts the orbits of (1.74) as time tends toward $+\infty$. In the semi-dynamical system context of reaction-diffusion models for interacting biological species in an isolated bounded habitat, the underlying state space Y is infinite dimensional. Consequently, dissipativity by itself does not guarantee precompactness of positive semi-orbits. Rather, dissipativity must be used in conjunction with the smoothing action associated to the elliptic operators in the system to draw this conclusion. This issue is discussed in detail in the next subsection.

In the preceding example (1.74) dissipativity allowed us to conclude that orbits tend toward a bounded attracting set. Of course, if we want to interpret (1.74) as a model describing the temporal evolution of the average population densities of species 1 to n, it must be the case for all $t \neq 0$ that if $(y_1(t), \ldots, y_n(t)) = \pi((y_1^0, \ldots, y_n^0), t)$, then $y_i(t)$ is nonnegative. A condition on (1.74) that guarantees such is the case is for f_i to satisfy

$$f_i(y_1, \ldots, y_n) = y_i \tilde{f}_i(y_1, \ldots, y_n) \tag{1.75}$$

for $i = 1, \ldots, n$. If (1.75) holds, then the principle of the uniqueness of solutions to initial value problems guarantees that $y_i(t) \geq 0$ for all $t \neq 0$ and $i = 1, \ldots, n$ so long as $y_i^0 \geq 0$ for $i = 1, \ldots, n$. Moreover, it also guarantees that $y_i(t)$ can equal 0 for some $i \in \{1, \ldots, n\}$ and some $t \neq 0$ only when $y_i^0 = 0$ and that in such case $y_i(t) = 0$ for all $t \in I\!R$. We then say that the sets $\{(y_1, \ldots, y_n) : y_i \geq 0 \quad \text{for} \quad i = 1, \ldots n\}$, $\{(y_1, \ldots, y_n) : y_i > 0 \quad \text{for} \quad i = 1, \ldots, n\}$ and $\{(y_1, \ldots, y_n) : y_i \geq 0 \quad \text{for} \quad i = 1, \ldots, n \quad \text{and} \quad y_i = 0 \quad \text{for at least one} \quad i \in \{1, \ldots, n\}\}$ are invariant under π.

Precompactness of orbits or positive semi-orbits and invariance of a set under π are essential notions in the analysis of the asymptotic behavior of dynamical and semi-dynamical systems. In general, if we assume that (Y, π) is a dissipative dynamical or semi-dynamical system, we say that a subset U of Y is *forward invariant* under π if $\gamma^+(u) \subseteq U$ for all $u \in U$. If $t_-(u) = -\infty$ in (1.71) and $\gamma(u) \subseteq U$ for all $u \in U$, we say that U is *invariant* under π. (Notice that in the case that (Y, π) is a semi-dynamical system but not a dynamical system Y may nevertheless contain invariant subsets. In general, when backward continuation of an orbit is possible, the continuation may not be uniquely determined. However, it is in the case of the semi-dynamical systems associated with reaction-diffusion models for several interacting biological species on an isolated bounded habitat, so we shall not delve further into this topic.) Now if for $u \in Y$, $\gamma^+(u)$ is precompact, then there is a bounded set in Y which attracts points on the positive semi-orbit of u. This set is referred to as the *omega limit set of* u, denoted by $\omega(u)$ and is defined by

$$\omega(u) = \bigcap_{t \geq 0} \overline{\bigcup_{s \geq t} \{\pi(u, r) : r \geq s\}} \tag{1.76}$$

where $\overline{V}$ denotes the closure of set V. The set $\omega(u)$ consists of all limits of all sequences $\{\pi(u, t_n)\}$ where $t_n \to +\infty$ as $n \to \infty$. It is well-known (e.g. Saperstone, 1981) that $\omega(u)$ is nonempty, compact, connected and invariant under π. (The term *alpha limit set of* u, denoted $\alpha(u)$, consists of all limits of all sequences $\{\pi(u, t_n)\}$ where $t_n \to -\infty$ as $n \to \infty$.) If now for a set U, the set $\{\pi(u, r) : u \in U, \ r \geq s \quad \text{for some} \quad s \geq 0\}$ is precompact, there is a companion notion of the *omega limit set of* U, denoted $\omega(U)$, with $\omega(U)$ defined by

$$\omega(U) = \bigcap_{t \geq 0} \overline{\bigcup_{s \geq t} \{\pi(u, r) : u \in U, r \geq s\}}.$$

The set $\omega(U)$ is the set of all limits of all sequences $\{\pi(u_n, t_n)\}$ where $u_n \in U$ and $t_n \to \infty$ as $n \to \infty$. Notice that $\bigcup_{u \in U} \omega(u) \subseteq \omega(U)$, but that in general $\omega(U)$ is much larger. However, we shall need to use $\bigcup_{u\in U}\omega(u)$ at several places in our discussion. At those places, we denote this set as $\omega(U)$ but make clear that the usage is nonstandard.

A *global attractor* for a dynamical or semi-dynamical system π on the metric space (Y, d) is a set $\mathcal{A}$ which is compact, invariant under π and such that for all bounded subsets V of Y

$$\lim_{t\to\infty} \sup_{v\in V} \inf_{u\in\mathcal{A}} d(\pi(v,t), u) = 0. \tag{1.77}$$

A fundamental result of Bilotti and La Salle (1971) (see Theorem 4.1) says that if (Y, d) is complete, π is dissipative and for all $t > t_0 \geq 0$, the set $\pi(U,t) = \{\pi(u,t) : u \in U\}$ is precompact if U is bounded, then π has a nonempty global attractor. (The function $\pi(\cdot,t) : Y \to Y$ is said to be *compact* if $\pi(U,t)$ is precompact whenever U is bounded. In particular, the image $\{\pi(u_n,t)\}$ of a bounded sequence $\{u_n\}$ is precompact if $\pi(\cdot,t) : Y \to Y$ is compact.) Notice that if $\mathcal{A}$ is a global attractor, $\varepsilon > 0$ is given and $\mathcal{B}(\mathcal{A},\varepsilon)$ is defined by $\mathcal{B}(\mathcal{A},\varepsilon) = \{y \in Y : d(y,u) < \varepsilon \quad \text{for some} \quad u \in \mathcal{A}\}$, then (1.77) implies that if $V \subseteq Y$ is bounded,

$$\pi(V,t) \subseteq \mathcal{B}(\mathcal{A},\varepsilon)$$

for all $t \geq t_0$, where t_0 depends upon V.

A point $u \in Y$ such that $\pi(u,t) = u$ for $t \in I\!R$ is called an *equilibrium* or *equilibrium point* for π. Equilibria play a large role in our discussion of the asymptotic predictions of reaction-diffusion models for interacting biological species in an isolated bounded habitat. Indeed, the most basic example of a global attractor is a *globally attracting equilibrium.* There are two special classes of dynamical or semi-dynamical systems in which the omega limit set $\omega(u)$ for all or almost all $u \in Y$ is an equilibrium. These are *gradient systems* and *monotone dynamical systems.* (We pause to note that a system can be both of gradient type and monotone.) Suppose that π is dissipative and $\gamma^+(u)$ is precompact for every u in Y, where Y is a closed subset of a Banach space. Then π is a gradient system (see, e.g., Henry 1981, Hale (1988)) if there is a *Lyapunov function* for π. By a Lyapunov function we mean a continuous function $V : Y \to I\!R$ so that

$$\begin{aligned}
&\text{(i)} \quad V \text{ is bounded below;}\\
&\text{(ii)} \quad V(u) \to +\infty \quad \text{as} \quad ||u|| \to +\infty;\\
&\text{(iii)} \quad V(\pi(u,t)) \quad \text{is nonincreasing in} \quad t \quad \text{for each} \quad u \in Y;\\
&\text{(iv) If} \quad V(\pi(u,t)) = V(u) \quad \text{for all} \quad t \in I\!R, \quad \text{then} \quad u \quad \text{is an equilibrium for} \quad \pi.
\end{aligned} \tag{1.78}$$

We show in Section 3.1 that many of the reaction-diffusion models for a single biological species an isolated bounded habitat that we consider in this volume may be formulated as gradient systems. In a gradient system, $\omega(u)$ is an equilibrium for all $u \in Y$. (See, e.g., Hale (1988).)

The third feature of the collection of n-tuples of average population densities for species 1 to n that we noted earlier was the partial ordering given by (1.64). The notion of a monotone dynamical or semi-dynamical system is a possibility when Y admits a partial ordering. In the examples we consider Y is a closed subset of a Banach space, and the partial ordering in Y is induced by a closed subset K of the Banach space called a *positive cone* via the equivalence

$$u_1 \leq u_2 \quad \text{in} \quad Y \Leftrightarrow u_2 - u_1 \in K. \tag{1.79}$$

To be a positive cone K must satisfy

$$\begin{aligned} &\text{(i) If} \quad c \quad \text{is a nonnegative scalar and} \\ &\quad u \in K, \quad \text{then} \quad cu \in K; \\ &\text{(ii) If} \quad u_1, u_2 \in K, \quad \text{then} \quad u_1 + u_2 \in K; \\ &\text{(iii) If} \quad u, -u \in K, \quad \text{then} \quad u = 0. \end{aligned} \tag{1.80}$$

In the case where Y is the collection of all n-tuples of nonnegative numbers partially ordered via (1.64), K may be taken equal to Y, and it is then clear that (1.80) holds and that (1.64) is equivalent to (1.79). However, if we consider the collection of all ordered pairs of nonnegative numbers (y_1, y_2) we may define a partial ordering via

$$(y_1, y_2) \leq (y_1^*, y_2^*) \Leftrightarrow y_1 \leq y_1^* \quad \text{but} \quad y_2 \geq y_2^*.$$

In this case (1.79) holds provided we define the positive cone K by

$$K = \{(u_1, u_2) : y_1 \geq 0, y_2 \leq 0\}.$$

So Y need not equal K. This last example will be relevant when we consider two species competition in Sections 4.4 and 5.2.

Once a partial ordering has been established in Y, π will be *monotone* provided

$$u_1 \leq u_2 \quad \text{in} \quad Y \Rightarrow \pi(u_1, t) \leq \pi(u_2, t) \quad \text{in} \quad Y \tag{1.81}$$

for all $t \neq 0$ when π is a dynamical system and for all $t > 0$ when π is a semi-dynamical system. A positive semi-orbit $\gamma^+(u)$ of π is said to be *increasing* provided

$$\pi(u, t_1) \leq \pi(u, t_2) \quad \text{for} \quad 0 \leq t_1 < t_2 \tag{1.82}$$

with an analogous definition of *decreasing* positive semi-orbit. Increasing and decreasing positive semi-orbits are at the heart of the examinations we make of monotone dynamical systems. To that end, suppose π admits an increasing positive semi-orbit $\gamma^+(u_1)$ and a decreasing positive semi-orbit $\gamma^+(u_2)$ and that $u_1 \leq u_2$. Then any positive semi-orbit $\gamma^+(u)$ with $u_1 \leq u \leq u_2$ must lie above the increasing positive semi-orbit $\gamma^+(u_1)$ but below the decreasing positive semi-orbit $\gamma^+(u_2)$. Moreover, if positive semi-orbits for π are precompact, an important result in the theory of monotone dynamical systems (see, e.g., Smith (1995), Theorem 2.1) asserts that $\omega(u_1)$ and $\omega(u_2)$ are equilibria for π. Since $\pi(u_1, t) \leq \pi(u_2, t)$ for all $t > 0$, $\omega(u_1) \leq \omega(u_2)$. So the omega limit set $\omega(u)$ for any u with $u_1 \leq u \leq u_2$ is contained in the set

$$\{v : \omega(u_1) \leq v \leq \omega(u_2)\}$$

which is called the *order interval* from $\omega(u_1)$ to $\omega(u_2)$ and denoted $[\omega(u_1), \omega(u_2)]$.

In the preceding scenario, the positive semi-orbit $\gamma^+(u_2)$ can be viewed as providing a "ceiling" for both the transient and asymptotic phases of the positive semi-orbit $\gamma^+(u)$, while the positive semi-orbit $\gamma^+(u_1)$ can be viewed as providing a "floor". We shall see that in many instances such semi-orbits can be interpreted as representing best and worst case scenarios for the dynamical or semi-dynamical system π. In the context of reaction-diffusion models for interacting biological species in isolated bounded habitats, positive semi-orbits that originate at configurations which are so-called *lower solutions* to the corresponding steady-state elliptic system are increasing, while positive semi-orbits that originate at configurations which are *upper solutions* to the corresponding elliptic systems are decreasing. We discuss upper and lower solutions in the context of ODEs, elliptic *and* parabolic PDEs and systems thereof in the next subsection.

The utility of the notions of monotonicity and comparison extends beyond systems which preserve a partial ordering. For example, in many systems of reaction-diffusion equations for interacting species in an isolated bounded habitat, the growth law for some species density may be such that for $t \geq t_0 \geq 0$ the density is a parabolic upper or lower solution to some single equation reaction-diffusion model. For specificity, assume that the density of the system component is given by u and that v denotes the density in the comparison single equation model. Then if u is an upper solution for the v equation and the density v is chosen so that $v(t_0) = u(t_0)$, $u(t) \geq v(t)$ for all $t \geq t_0$. Likewise, if u is a lower solution, $u(t) \leq v(t)$. Consequently, $v(t)$ provides a "floor" for the density $u(t)$ in the first case and a "ceiling" in the second. Such information can be very useful in obtaining best and worst case scenarios for the density u. We explore this idea in Sections 5.3 and 5.4 and throughout Chapter 7.

Since reaction-diffusion models for interacting biological species in an isolated bounded habitat are continuous time models, the dynamical and semi-dynamical systems we consider in this volume are primarily continuous time dynamical or semi-dynamical systems. However, we do have occasion to employ discrete time systems and consequently we include a few remarks regarding discrete time dynamical and semi-dynamical systems. As in the case of continuous time dynamical or semi-dynamical systems, one would want the underlying state space Y to be a metric space. (For our purposes, Y is a closed subset of a Banach space and hence a complete metric space.) A *discrete time dynamical system* π on Y is a continuous function $\pi : Y \times \mathbb{Z} \to Y$, where $\mathbb{Z}$ denotes the integers, so that discrete time analogues to (1.69) and (1.70) hold. Namely, we require

$$\pi(u, 0) = u \quad \text{for all} \quad u \in Y \tag{1.83}$$

and

$$\pi(u, k + k') = \pi(\pi(u, k), k') \tag{1.84}$$

for all $k, k' \in \mathbb{Z}$ so that $\pi(u, k+k')$ and $\pi(u, k)$ are defined. If π is to be a semi-dynamical system, the integers $\mathbb{Z}$ in the preceding are replaced with the nonnegative integers $\mathbb{Z}^+ \cup \{0\}$.

Suppose now that $S : Y \to Y$ is given by

$$S(u) = \pi(u, 1).$$

Then S is continuous and for $k > 0$ it follows from (1.84) that

$$\pi(u, k) = S^k u$$

where S^k is the composition $S \circ S \circ \cdots \circ S$ (k times). Likewise, if $\pi(u, -1)$ is defined, $\pi(u, -1) = S^{-1}$ and

$$\pi(u, -k) = S^{-k}u.$$

Of course, the restriction of a continuous time dynamical or semi-dynamical system to the integers or nonnegative integers, respectively, is a discrete time dynamical or semi-dynamical system. In the case of an autonomous system of ordinary differential equations of the form (1.74) or reaction-diffusion equations, the associated discrete-time systems track the value along solution trajectories after every 1 unit of time.

An alternative way of viewing such trajectories in the ODE case (1.74), say, is to think of $\pi((y_1^0, \ldots, y_n^0), k+1)$ as the value at time 1 of the solution to the system (1.74) with initial configuration $\pi((y_1^0, \ldots, y_n^0), k)$. Such is always possible since the system is autonomous. Now suppose that (1.74) is replaced with

$$y_i' = f_i(t, y_1, \ldots, y_n) \tag{1.85}$$

$i = 1, \ldots, n$ where $f_i(t+1, y_1, \ldots, y_n) = f_i(t, y_1, \ldots, y_n)$ for all $t \in \mathbb{R}$, $i = 1, \ldots, n$ and all $(y_1, \ldots, y_n) \in \mathbb{R}^n$. Let $\phi((y_1^0, \ldots, y_n^0), t)$ denote the unique solution to (1.85) so that

$$\phi((y_1^0, \ldots, y_n^0), 0) = (y_1^0, \ldots, y_n^0).$$

Now define $\pi : \mathbb{R}^n \times \mathbb{Z} \to \mathbb{R}^n$ by

$$\pi((y_1^0, \ldots, y_n^0), k) = \phi((y_1^0, \ldots, y_n^0), k). \tag{1.86}$$

Since f_i is 1-periodic for all $i \in \{1, \ldots, n\}$, $\phi((y_1^0, \ldots, y_n^0), k+1)$ is the value after 1 time unit of the solution to (1.85) which has the value $\phi((y_1^0, \ldots, y_n^0), k)$. This fact makes π as given in (1.86) into a dynamical system, although $\phi : \mathbb{R}^n \times \mathbb{R} \to \mathbb{R}^n$ is not. To convert (1.85) into a continuous time dynamical system, we would need to employ a construction called a *skew product flow* which tracks the state of the system (1.85) as well as ϕ. We discuss this construction in detail in the periodic reaction-diffusion setting in Section 5.5.

Understanding the asymptotic behavior of discrete time dynamical and semi-dynamical systems involves the same concepts as in the continuous time case. One still has the notions of omega limit set, invariance and so forth. However, there are some subtleties that we should point out. First, if a discrete time dynamical or semi-dynamical system is associated to a system of ODEs or reaction-diffusion equations in the manner we have described, an equilibrium of the discrete dynamical system corresponds to a point on a periodic orbit of the system of differential equations, but need not be an equilibrium to the continuous time system. This observation will be employed when we analyze single species reaction-diffusion models with time-periodic coefficients in Section 3.6. Of course, there are discrete dynamical or semi-dynamical systems which arise in ways other than sampling the solution to a system of differential equations successively at times $0, \Delta t, 2\Delta t, 3\Delta t, \ldots$ for some prespecified length of time Δt. As an example, consider the so-called discrete logistic equation, which arises from discretizing a potential solution y to the logistic equation

$$y' = ry\left(1 - \frac{y}{K}\right) \tag{1.87}$$

by $y'_j = y(j\Delta t)$ and approximating $y'(j\Delta t)$ by the forward difference $y'(j\Delta t) = \dfrac{y_{j+1} - y_j}{\Delta t}$. So doing one arrives at the equation

$$y_{j+1} = \tilde{r} y_j \left(1 - \frac{y_j}{\tilde{K}}\right) \tag{1.88}$$

$j = 0, 1, 2, 3, \ldots$ where $\tilde{r} = 1 + r\Delta t$ and $\tilde{K} = K\left(\dfrac{1 + r\Delta t}{r\Delta t}\right)$. It is immediate that (1.88) defines a discrete semi-dynamical system. However, if for some value of j, it is the case that y_j exceeds $\tilde{K}$, then $y_{j+1} < 0$. Consequently, the invariance of positivity that solutions of (1.87) enjoy does not carry over to solutions of (1.88). The reason for this rather profound difference between the two systems is simple. Positive solutions to the differential equation would have to pass through 0 before they can take on negative values. However, 0 is an equilibrium solution to (1.87) and solutions to initial value problems for (1.87) are unique. Consequently, no solution to (1.87) which is positive at 0 can be negative at some later time. On the other hand, in the discrete model (1.88) solutions may jump across 0, and the only possibility for an invariant set with a meaningful ecological interpretation is $[0, \tilde{K}]$. Note that $[0, K]$ is invariant for (1.87) regardless of the value of r. In (1.88), however, it is not hard to verify that $[0, \tilde{K}]$ is invariant only when $\tilde{r} \leq 4$ (which is equivalent to $r < \dfrac{3}{\Delta t}$ in (1.87)). In (1.87), any solution with $y(0) \in (0, K)$ increases monotonically over time toward the equilibrium K, regardless of the value of r, whereas in (1.88), orbits can be very complicated, even chaotic, depending on the value of $\tilde{r}$. For more detail, see Murray (1993, Section 2.3.).

There are discrete analogues to (1.87) other than (1.88) whose solutions track those to (1.87) more closely. One possibility is just to sample solutions to (1.87) at discrete time intervals. The resulting model is known as the Beverton-Holt model (Beverton and Holt, 1957). Another is the model

$$y_{j+1} = y_j \exp\left(r\left(1 - \frac{y_j}{K}\right)\right) \tag{1.89}$$

which results from solving the initial value problem

$$\frac{dy}{dt} = r\left(1 - \frac{y_j}{K}\right) y$$
$$y(0) = y_j$$

and evaluating the solution at time 1. The model (1.89) is sometimes referred to as a Ricker model (Ricker, 1954, 1975), and is widely used in the study of fisheries (Quinn and Deriso, 1999). Both alternates to (1.88) leave $(0, \infty)$ invariant and are such that

$$y_j < y_{j+1} \quad \text{if} \quad 0 < y_j < K$$

while

$$y_j > y_{j+1} \quad \text{if} \quad y_j > K,$$

just as is the case with solutions to (1.87). However, only the Beverton-Holt model is such that $(0, K)$ and (K, ∞) are invariant, as is the case with solutions to (1.87). Indeed, solutions to (1.89) can be very complicated, even chaotic. For more details, see May (1975).

1.6.2 Basic Concepts in Partial Differential Equations: An Example

The reaction-diffusion models that are the subject of this book are partial differential equations which describe how population densities in space change over time. Since they describe the way that things change over time, it is natural to think of them as dynamical systems; however, as noted, the state space for a reaction-diffusion model will be a set of functions representing the possible spatial densities of a spatially distributed population. Thus, to formulate reaction-diffusion models as dynamical systems we need to define appropriate state spaces of functions and determine how the models act on them. In general we will not be able to solve reaction-diffusion models explicitly, but that is also the case with many nonlinear systems of ordinary differential equations. What we can do in many cases is determine when a model predicts persistence and when it predicts extinction, and perhaps describe some features of its dynamics, by using methods from the theory of dynamical systems. However, there are some new technical issues that arise in formulating reaction-diffusion models as dynamical systems. Many of those are related to the fact that the state spaces for reaction-diffusion models are infinite dimensional. Others have to do with problems such as verifying that the set of nonnegative densities is invariant. (Since negative population densities don't make sense, good models should predict that densities which are initially nonnegative remain so.) To illustrate some of the issues related to formulating reaction-diffusion models as dynamical systems and to introduce some ideas that will be important in their analysis, we briefly review the explicit solutions of some simple linear models and derive a few of their basic properties.

The discussion that follows is a quick review of some standard ideas in the theory of partial differential equations. A complete treatment of this material is given in many standard texts, for example, Strauss (1992). Consider the model

$$\begin{aligned} \frac{\partial u}{\partial t} &= d\frac{\partial^2 u}{\partial x^2} + ru \quad \text{on} \quad (0, 1) \times (0, \infty) \\ u(0, t) &= u(1, t) = 0 \text{ on} \quad (0, \infty), \end{aligned} \tag{1.90}$$

with initial condition $u(x, 0) = f(x)$. Equation (1.90) can be interpreted as a linear growth model for a population that diffuses at rate d, increases in numbers locally at rate r, and inhabits a one-dimensional patch of unit length with an absorbing boundary (i.e. the boundary is lethal to the population). To solve (1.90) one typically looks for solutions in the form $u(x, t) = X(x)T(t)$ and then constructs a general solution as a series of such special solutions. (This is the method of separation of variables; it is discussed in many standard references, again, for example, Strauss (1992).) Substituting $u = X(x)T(t)$ into (1.90) and dividing through by $X(x)T(t)$ yields

$$\frac{T'(t)}{T(t)} = \frac{dX''(x) + rX(x)}{X(x)}. \tag{1.91}$$

The right side of (1.91) depends only on X; the left side only on t. Thus, the only way that (1.91) can hold is if both sides are constant. Suppose that both sides of (1.91) are equal to the constant σ. Then we have

$$dX'' + rX = \sigma X \tag{1.92}$$

and

$$T'(t) = \sigma T(t). \tag{1.93}$$

There are no further restrictions on T, but the boundary conditions $u(0,t) = u(1,t) = 0$ require that

$$X(1) = X(0) = 0. \tag{1.94}$$

It is possible to solve (1.92) explicitly for any given choice of σ, but it turns out that there are nonzero solutions satisfying (1.94) only if $\sigma = r - dn^2\pi^2$ for some positive integer n. In that case we have $X_n(x) = k_1 \sin(n\pi x)$ and $T_n(t) = k_2 e^{(r-dn^2\pi^2)t}$, where k_1 and k_2 are arbitrary constants, so $u_n(x,t) = e^{(r-n^2\pi^2)t}\sin(n\pi x)$ is a solution of (1.90).

Notice that (1.92) can be written as

$$\left[d\left(\frac{\partial^2}{\partial x^2}\right) + r\right]X = \sigma X. \tag{1.95}$$

The differential operator $L = d(\partial^2/\partial x^2) + r$ is linear, so that $L(c_1X_1 + x_2X_2) = c_1L(X_1) + c_2L(x_2)$ for any constants c_1 and c_2. Thus, (1.95) is analogous to an eigenvalue problem for a matrix, with σ as the eigenvalue and L playing the same role as the matrix, but with the eigenvector being replaced by the eigenfunction $X(x)$. The notion of eigenvalues for differential operators will play a central role in our analysis of reaction-diffusion models. To continue the process of solving (1.90), we write $u(x,t)$ as a Fourier series

$$u(x,t) = \sum_{n=1}^{\infty} b_n e^{(r-dn^2\pi^2)t}\sin(n\pi x). \tag{1.96}$$

To satisfy the initial condition $u(x,0) = f(x)$ we need to choose the constants b_n so that

$$f(x) = \sum_{n=1}^{\infty} b_n \sin(n\pi x). \tag{1.97}$$

It turns out that if it is possible to express $f(x)$ in the form (1.97) then we must have $b_n = 2\int_0^1 f(x)\sin(n\pi x)dx$ (see Strauss (1992)). At this point a new problem arises: (1.96) and (1.97) involve infinite series of functions which may fail to converge for some (perhaps all) values of x and t, depending on the coefficients b_n. In the case of (1.96) there is another problem: the series must converge to a function that can be differentiated once in t and twice in x if it is to satisfy (1.90). We can impose conditions on $f(x)$ (or equivalently on $\{b_n\}$) so that (1.97) converges, but there are a number of possible choices for what is meant by convergence. For example, if we require that $\int_0^1 [f(x)]^2 dx$ is finite, then it turns out that $(1/2)\sum_{n=1}^{\infty} b_n^2 = \int_0^1 [f(x)]^2 dx$ (this is a version of Parseval's equality), and the series in (1.97) converges in the sense that

$$\lim_{m\to\infty}\int_0^1\left[f(x) - \sum_{n=1}^{m} b_n \sin(n\pi x)\right]^2 dx = 0. \tag{1.98}$$

However, we could have $\int_0^1 f(x)^2 dx < \infty$ even if $f(x)$ is not continuous, and (1.98) does not guarantee that $\lim_{m \to \infty} \sum_{n=1}^{m} b_n \sin(n\pi x) = f(x)$ for all x. On the other hand, suppose that $f(x)$ is continuous, $f(0) = f(1) = 0$, and $\int_0^1 f'(x)^2 dx < \infty$. Let $a_n = 2\int_0^1 f'(x)\cos(n\pi x)dx$. Parseval's relation applies to cosine series as well as sine series, so $(1/2)\sum_{n=1}^{\infty} a_n^2 = \int_0^1 f'(x)^2 dx < \infty$. On the other hand, integration by parts shows that $nb_n = a_n$ so $b_n = (1/n)a_n$. In that case, we have $|b_n| \leq (1/2)[(1/n)^2 + a_n^2]$ so that $\sum_{n=1}^{\infty} |b_n| < \infty$ (because $\sum_{n=1}^{\infty} 1/n^2 < \infty$). If $\sum_{n=1}^{\infty} |b_n| < \infty$ then since $|b_n \sin(n\pi x)| \leq |b_n|$, the series in (1.97) is majorized by the convergent series $\sum_{n=1}^{\infty} |b_n|$ so it converges uniformly, that is,

$$\lim_{m \to \infty} \sup_{x \in [0,1]} \left| f(x) - \sum_{m=1}^{m} b_n \sin(n\pi x) \right| = 0. \tag{1.99}$$

In general the condition $\sum_{n=1}^{\infty} |b_n| < \infty$ is stronger than the condition $\sum_{n=1}^{\infty} b_n^2 < \infty$; for example $\sum_{n=1}^{\infty} 1/n$ diverges even though $\sum_{n=1}^{\infty} 1/n^2$ converges. Similarly, the conditions $\int_0^1 f'(x)^2 dx < \infty$, $f(x)$ is continuous, $f(0) = f(1) = 0$ are stronger than the condition $\int_0^1 [f(x)]^2 dx < \infty$.

The point here is that to make sense out of (1.90) we need to define a state space for the initial data $f(x)$, and the way we choose a metric on the state space actually determines what the space is. We could interpret (1.97) as an expansion of $f(x)$ in terms of "basis vectors" $\sin(n\pi x)$, in analogy with expressing $\vec{v} \in I\!R^3$ as $\vec{v} = v_1\vec{i} + v_2\vec{j} + v_3\vec{k}$ where $\vec{i}$, $\vec{j}$, and $\vec{k}$ are the standard unit vectors in the x, y, and z directions. However, in $I\!R^3$ the only restriction on v_1, v_2 and v_3 is that they are finite, and $v_1^2 + v_2^2 + v_3^2 < \infty$ if and only if $|v_1| + |v_2| + |v_3| < \infty$, which is true if and only if v_1, v_2, and v_3 are finite. On the other hand, requiring that each b_n is finite, that $\sum_{n=1}^{\infty} b_n^2 < \infty$, and that $\sum_{n=1}^{\infty} |b_n| < \infty$ in (1.97) impose different conditions on $\{b_n\}$, and thus define different subsets of the set of infinite sequences. Thus, in formulating models such as (1.90) we need to choose a metric that allows us to measure the "distance" between functions (or alternatively between sequences $\{b_n\}$), and the way we measure distance actually determines what the space is. The standard metrics corresponding to the notions of convergence shown in (1.98) and (1.99), respectively, are

$d(f,g)=\left[\int_0^1 |f(x)-g(x)|^2dx\right]^{1/2}$ and $d(f,g)=\sup_{x\in[0,1]}|f(x)-g(x)|$, respectively. The standard spaces of functions corresponding to those metrics are denoted by $L^2([0,1])$ and $C([0,1])$, respectively. The space $L^2([0,1])$ consists of all functions $f(x)$ on $[0,1]$ such that the Lebesgue integral of $[f(x)]^2$ over $[0,1]$ exists and is finite. The space $C([0,1])$ consists of all continuous functions on $[0,1]$. (The subspace of $C([0,1])$, consisting of continuous functions with $f(0)=f(1)=0$ is denoted $C_0([0,1])$.) All of these are Banach spaces. In our analysis of reaction-diffusion models we will need to use those spaces (and various others) as state spaces. In many cases the choice of a state space will depend on technical aspects of the mathematics involved in dealing with whatever features are built into a particular model. For example, if a model has coefficients that depend on x, we might need to use different state spaces depending on whether the coefficients were continuous or not.

To verify that $L^2([0,1])$ and $C_0([0,1])$ might be legitimate state spaces for (1.90) we must consider (1.96). We could rewrite (1.96) as $u(x,t)=\sum_{n=1}^{\infty}u_n(t)\sin(n\pi x)$ where $u_n(t)=b_n e^{(r-dn^2\pi^2)t}$. Notice that for $n>\sqrt{r/d}/\pi$ we have $|u_n(t)|\le|b_n|$ for all $t\ge 0$, so that if $\sum_{n=1}^{\infty}b_n^2<\infty$ then $\sum_{n=1}^{\infty}|u_n^2(t)|<\infty$ for all t. Thus, it is reasonable to use either $L^2([0,1])$ or the space of all functions $f(x)$ on $[0,1]$ whose Fourier coefficients $\{b_n\}$ satisfy $\sum_{n=1}^{\infty}|b_n|<\infty$ as state spaces for (1.90). Note also that for $n>\sqrt{r/d}/\pi$ we have $u_n(t)\to 0$ as $t\to\infty$. Thus, for large t, only the coefficients b_n with $n\le\sqrt{r/d}\pi$ are important. The model (1.90) in effect squeezes state space down toward the finite dimensional subspace spanned by $\{\sin(n\pi x): n\le\sqrt{r/d}/\pi\}$.

If we want $u(x,t)$ to satisfy (1.90) we need to verify that $\partial u/\partial t$ and $\partial^2u/\partial x^2$ make sense. Suppose that the coefficients b_n satisfy $|b_n|\le B_0$ for some finite B_0. (This will be true if $\sum_{n=1}^{\infty}b_n^2<\infty$ or $\sum_{n=1}^{\infty}|b_n|<\infty$.) If we differentiate the series in (1.96) by x term by term we obtain the series $\sum_{n=1}^{\infty}b_n n\pi e^{(r-dn^2\pi^2)t}\cos(n\pi x)$. The series is majorized by the series $\sum_{n=1}^{\infty}B_n(t)$ where $B_n(t)=B_0 n\pi e^{(r-dn^2\pi^2)t}$. We have $\lim_{n\to\infty}|B_{n+1}/B_n|=\lim_{n\to\infty}\left([(n+1)/n]\,e^{(n^2-(n+1)^2)d\pi^2 t}\right)=0$ for any $t>0$, since $n^2-(n+1)^2<0$, so $\sum_{n=1}^{\infty}B_n(t)$ converges for all $t>0$ by the ratio test. Hence, the series $\sum_{n=1}^{\infty}b_n n\pi e^{(r-dn^2\pi^2)t}\cos(n\pi x)$ converges uniformly for any given $t>0$. Similar analysis show that the series obtained by differentiating (1.96) term by term any number of times in x or t still converge uniformly. Thus, all of those series converge to continuous functions which are the derivatives of $u(x,t)$.

The key point here is that the model (1.90) has the effect of smoothing out solutions to (1.90). Even if $f(x)$ is discontinuous, $u(x,t)$ will be differentiable infinitely many times in both x and t for $t > 0$. This smoothing property is a key feature of reaction-diffusion models. It often implies that the model maps bounded sets in state space into compact sets. To see how that works, let B_0 be a fixed constant and let F_0 be the subset of $L^2([0,1])$ consisting of functions satisfying $\int_0^1 [f(x)]^2 dx \le B_0^2$. Let F_t be the set of solutions $w(x,t)$ to (1.90), evaluated at time t, such that $w(x,0) \in F_0$. Fix a value of $t > 0$ and suppose that $\{w_k\}$ is a sequence in F_t, so that $w_k(x,0) = f_k(x) \in B_0$ for each k. If $f_k(x) \in B_0$ then we may write $f_k(x) = \sum_{k=1}^{\infty} b_{nk} \sin(n\pi x)$, with coefficients $\{b_{nk}\}$ satisfying $\sum_{k=1}^{\infty} b_{nk}^2 \le B_0^2$. It follows that $|b_{nk}| \le B_0$ for each coefficient b_{nk}. Furthermore, we can write

$$w_k(x,t) = \sum_{n=1}^{\infty} b_{nk} e^{(r-dn^2\pi^2)t} \sin(n\pi x) \text{ and } \partial w_k/\partial x = \sum_{n=1}^{\infty} b_{nk} n\pi e^{(r-dn^2\pi^2)t} \cos(n\pi x),$$

provided those series converge. The series are majorized by the series $\sum_{n=1}^{\infty} B_0 e^{(r-dn^2\pi^2)t}$ and $\sum_{n=1}^{\infty} B_0 n\pi e^{(r-dn^2\pi^2)t}$, which are both convergent for any given $t > 0$; thus, $\{w_k\}$ and $\{\partial w_k/\partial x\}$ are bounded uniformly in k and x. Since $\{\partial w_k/\partial x\}$ is bounded uniformly in x and k, the set of functions $\{w_k\}$ is also equicontinuous by the mean value theorem. Ascoli's theorem (see Rudin (1976)) then implies that the sequence $\{w_k(x,t)\}$ has a subsequence $\{w_{k_i}(x,t)\}$ which converges uniformly on $[0,1]$ to some continuous function $w(x)$. Uniform convergence implies $\int_0^1 (w_{k_i}(x,t) - w(x))^2 dx \to 0$ as $i \to \infty$ so that $w_{k_i}(x,t)$ converges to $w(x)$ in $L^2([0,1])$. This feature of always being able to extract a convergent subsequence is the essence of compactness. Thus, the set F_t of solutions to (1.90) at time t whose initial data belong to the bounded set F_0 is compact. Note that F_0 is not compact. The functions $f_k(x) = \sqrt{B_0} \sin(k\pi x)$ belong to F_0 (since $\int_0^1 [f_k(x)]^2 dx = B_0/2$), but no subsequence of $\{f_k\}$ converges in $L^2([0,1])$ because $\int_0^1 |f_k(x) - f_\ell(x)|^2 dx = B_0$ for any $k \ne \ell$. The compactness of F_t arises from the smoothing property of (1.90).

The solution (1.96) to (1.90) predicts that all solutions to (1.90) will decay to zero if $r/d < \pi^2$, while at least some solutions will grow exponentially if $r/d > \pi^2$. Thus, the model gives a criterion for the growth rate r needed to balance the loss of individuals across the boundary if they disperse by diffusion at rate d. However, before drawing too many conclusions, we should check that solutions to (1.90) which correspond to positive initial densities at least remain nonnegative. Suppose that $w(x,t)$ satisfies (1.90) on $(0,1) \times (0,T]$ and is continuous on $[0,1] \times [0,T]$ with $w(x,0) \ge 0$. Let $v(x,t) = e^{-at} w(x,t)$; then by (1.90) we have

$$\frac{\partial v}{\partial t} = -av + e^{-at}\frac{\partial w}{\partial t} = d\frac{\partial^2 v}{\partial x^2} + (r-a)v. \tag{1.100}$$

Choose a large enough so that $r - a < 0$. Suppose $w(x,t) < 0$ for some (x,t). Then $v(x,t) < 0$ at the same point, so the minimum of v is negative. (Since v is continuous on $[0,1] \times [0,T]$, it must have a minimum.) Since $v \ge 0$ for $x = 0$ or 1 and for $t = 0$, the minimum must occur at a point (x_0, t_0) in $(0, 1)\times(0, T]$ where (1.90) and hence (1.100) are satisfied. If $(x_0, t_0) \in (0, 1) \times (0, T)$ then $\partial v/\partial t = 0$ and $\partial^2 v/\partial x^2 \geq 0$ at $(x, t) = (x_0, t_0)$, so at this point (1.100) implies $0 \geq (r - a)v(x_0, t_0)$. Since $r - a < 0$ and $v(x_0, t_0) < 0$, this is a contradiction. If the minimum occurs on $(0, 1) \times \{T\}$, we have $\partial v/\partial t \leq 0$ at (x_0, t_0), but otherwise the analysis is the same. Thus, if $w(x, t) < 0$ somewhere, we obtain a contradiction, so we must have $w(x, t) \geq 0$ for all $(x, t) \in [0, 1] \times [0, T]$. This type of result is called a maximum principle, although in this case we applied it to a minimum. Suppose now w_1 and w_2 are solutions to (1.90) with $w_1(x, 0) \geq w_2(x, 0)$. Then $w = w_1 - w_2$ is also a solution to (1.90), and $w(x, 0) \geq 0$, so $w(x, t) \geq 0$ for all $t > 0$, and hence $w_1(x, t) \geq w_2(x, t)$ for all t. Thus, the model (1.90) is order preserving.

In the simple case described above we could solve the model (1.90) explicitly by using some calculus, and we could then analyze the solution by using some basic results from real analysis. In general it is not possible to give explicit solutions to reaction-diffusion equations. It *is* possible to show that they have solutions which depend continuously on initial conditions, and that those solutions are smooth (so that reaction-diffusion models send bounded sets of initial data into sets with compact closures). In models where the reaction term for the density u_i of the ith species has the form $f_i(x, t, u_1, \ldots, u_N)u_i$, it is possible to show that the set of $\vec{u} = (u_1, \ldots, u_N)$ with $u_i \geq 0$ for all i is positively invariant; that is, if $u_i(x, t_1) \geq 0$ for all i then $u_i(x, t) \geq 0$ for all i and all $t > t_1$. Thus, we will be able to formulate reaction-diffusion models as dynamical systems on state spaces analogous to $L^2([0, 1])$ or $C_0([0, 1])$. Once we have formulated the models properly, we can obtain criteria for persistence analogous to the criterion $r/d > \pi^2$ for persistence in (1.90). To prove that reaction-diffusion models have the properties described above requires many technicalities. In general we simply state the results we need, and refer the reader to other sources for the details. However, we will try to be precise and accurate in our statements of results about reaction-diffusion models, and in some cases we must use technical terms in definitions and the statements of theorems. One context where this is an issue is in the construction of appropriate state spaces. We have already seen that (1.90) could be formulated in $L^2([0, 1])$ or in $C_0([0, 1])$, and there are many more spaces like those that occur in reaction-diffusion theory. For the reader who is more interested in biological applications than mathematical precision, the good news is that it is often unnecessary to know all the details about the state space to understand the predictions of the model.

1.6.3 Modern Approaches to Partial Differential Equations: Analogies with Linear Algebra and Matrix Theory

The classical approach to partial differential equations is to attempt to find explicit solutions such as (1.96) and then to observe the properties of those solutions. This approach can be very useful but it is limited by the fact that many partial differential equations cannot be solved explicitly, at least in terms of standard functions such as sines and cosines or exponentials. The modern approach to partial differential equations is to prove that solutions exist without computing them explicitly and then to deduce their properties, often by using methods quite different than those used to show their existence. For reaction-diffusion models the basic existence theory can be understood by analogy with systems of ordinary differential equations. For the equilibria of reaction-diffusion equations the basic existence theory is somewhat analogous to the theory of algebraic equations; in particular, in the linear case the theory closely resembles matrix theory. More specifically, recall that for any $n \times n$ matrix the dimension of the nullspace plus the rank of the matrix must equal n. The rank of the matrix is the dimension of the image of $\mathbb{R}^n$ under multiplication by the matrix, so if the dimension of the nullspace is zero (i.e. only the zero vector is mapped to zero by the matrix), then the matrix maps $\mathbb{R}^n$ onto $\mathbb{R}^n$. Also, if the dimension of the nullspace is zero, the mapping of $\mathbb{R}^n$ onto $\mathbb{R}^n$ is one-to-one. Thus, if M is an $n \times n$ matrix, I the identity matrix, and λ a real number, then the matrix $M - \lambda I$ is invertible unless $(M - \lambda I)\vec{v} = 0$ for some nonzero vector $\vec{v}$, i.e. $M - \lambda I$ is invertible unless λ is an eigenvalue of M. If it happens that M is symmetric and positive definite (that is, $M\vec{v} \cdot \vec{v} \geq m_0|\vec{v}|^2$ for all $\vec{v} \in \mathbb{R}^n$, where $m_0 > 0$ is a constant), then $M\vec{v} = \vec{0}$ implies $0 = M\vec{v} \cdot \vec{v} \geq m_0|\vec{v}|^2$, which is possible only if $\vec{v} = \vec{0}$. It follows that if M is symmetric and positive definite then M is invertible, and we can draw this conclusion without computing M^{-1}. Now consider the system of differential equations $d\vec{y}/dt = M\vec{y}$. We have $d|\vec{y}|^2/dt = 2\vec{y} \cdot d\vec{y}/dt = 2(\vec{y} \cdot M\vec{y}) \geq 2m_0|\vec{y}|^2$, so $|\vec{y}|^2 \geq e^{2m_0 t}|\vec{y}(0)|^2$. Hence $|\vec{y}|$ grows exponentially with rate m_0 unless $\vec{y}(0) = \vec{0}$. Again, we did not actually compute $\vec{y}(t)$, but we still could draw conclusions about how $\vec{y}(t)$ behaves. We could represent $\vec{y}(t)$ as $\vec{y}(t) = e^{tM}\vec{y}(0)$, where $e^{tM} = \sum_{j=0}^{\infty}(t^j/j!)M^j$ is the matrix exponential (see Brauer and Nohel (1969), for example.) To show that the related nonlinear system

$$d\vec{y}/dt = M\vec{y} + \vec{f}(\vec{y}) \tag{1.101}$$

has solutions, we might write it as

$$\vec{y}(t) = \vec{y}(0) + \int_0^t [M\vec{y}(s) + \vec{f}(\vec{y}(s))]ds \tag{1.102}$$

and use Picard iteration or a fixed point theorem (again, see Brauer and Nohel (1969)) to conclude that solutions exist. An alternate formulation would be to write the equation as

$$\begin{aligned} d(e^{-tM}\vec{y})/dt &= e^{-tM}(d\vec{y}/dt - M\vec{y}) \\ &= e^{-tM}\vec{f}(\vec{y}(t)) \end{aligned}$$

so that

$$e^{-tM}\vec{y} - \vec{y}(0) = \int_0^t e^{-sM}\vec{f}(\vec{y}(s))ds$$

and hence

$$\vec{y}(t) = e^{tM}\vec{y}(0) + \int_0^t e^{(t-s)M}\vec{f}(\vec{y}(s))ds. \tag{1.103}$$

It turns out that in the context of reaction-diffusion models the formulation analogous to (1.103) is preferable to (1.102). That is because the operator analogous to M will be a differential operator, while the operator analogous to e^{tM} will be the solution operator for a linear diffusion equation. For example, in (1.90), the operator analogous to M is $d\partial^2/\partial x^2 + r$, while the operator analogous to e^{tM} is the solution operator that maps

$u(x,0) = \sum_{n=1}^{\infty} b_n \sin(n\pi x)$ to $u(x,t) = \sum_{n=1}^{\infty} b_n e^{(r-dn^2\pi^2)t} \sin(n\pi x)$. As we noted in the discussion of (1.90), the solution operator makes sense if $\int_0^1 [u(x,0)]^2 dx < \infty$, and because of its smoothing properties it maps bounded sets into compact sets. The operator $d\partial^2/\partial x^2 + r$, however, may not even make sense when applied to an arbitrary function $u(x,0)$ with $\int_0^1 [u(x,0)]^2 dx < \infty$, and when it does make sense it can map bounded sets of functions into unbounded sets. (For example, the operator $d\partial^2/\partial x^2 + r$ maps the bounded sequence $\{\sin(n\pi x)\}$ to the unbounded sequence $\{(-dn^2\pi^2 + r)\sin(n\pi x)\}$.)

Reaction-diffusion models are a subclass of a more general class of equations known as quasilinear parabolic systems. All the models we consider will be second order. The most general type of single species models we study are of the form

$$\frac{\partial u}{\partial t} = \nabla \cdot [d(x,t,u)\nabla u - \vec{b}(x,t)u] + f(x,t,u), \tag{1.104}$$

the most general type of multispecies models we will study are of the form

$$\frac{\partial u_i}{\partial t} = \nabla \cdot [d_i(x,t)\nabla u_i - \vec{b}_i(x,t)u_i] + f_i(x,t,\vec{u}), \tag{1.105}$$

$i = 1 \ldots n$. In most cases of (1.104) we will have $d = d(x,t)$, so that the nonlinear terms in (1.104) will only involve u but not its derivatives, as in (1.105). Such models are called semilinear. In (1.105) the equations are coupled only in the undifferentiated terms; such systems are called weakly coupled. Most of the models we study will have coefficients that do not depend on t, and in those that do have time-dependent coefficients we usually assume that the coefficients are periodic in t. The analysis of systems such as (1.105) will often require rather detailed information about single equations involving the operators that occur in the system. In the subsections that follow we sketch out the modern theory for single equations, and then state some of the corresponding results for systems. The relevant theory for single equations is usually formulated in a setting which is slightly more general than that of diffusion models. Specifically, the theory is usually formulated for equations of the form

$$\frac{\partial u}{\partial t} = \sum_{i,j=1}^{n} a_{ij}(x)\frac{\partial^2 u}{\partial x_i \partial x_j} + \sum_{i=1}^{n} b_i(x)\frac{\partial u}{\partial x_i} + c(x)u \tag{1.106}$$

where the operator

$$L = \sum_{i,j=1}^{n} a_{ij}(x)\frac{\partial^2}{\partial x_i \partial x_j} + \sum_{i=1}^{n} b_i(x)\frac{\partial}{\partial x_i} + c(x) \tag{1.107}$$

is elliptic. We generally assume that the operator L is uniformly strongly elliptic, that is, there are constants $A_1 \geq A_0 > 0$ such that

$$A_0|\vec{\xi}|^2 \leq \sum_{i,j=1}^{n} a_{ij}(x)\xi_i\xi_j \leq A_1|\vec{\xi}|^2 \tag{1.108}$$

for any $\vec{\xi} \in I\!R^n$ and any x for which L is defined. In the case of a diffusion operator $Lu = \nabla \cdot [d(x)\nabla u - \vec{b}(x)u] + r(x)u$, the ellipticity condition (1.108) reduces to $0 < d_0 \leq d(x) \leq d_1$ for some constants d_0 and d_1. We also assume that $a_{ij} = a_{ji}$. This is reasonable because $\partial^2 u/\partial x_i \partial x_j = \partial^2 u/\partial x_j \partial x_i$ for any smooth function u. (In the case of diffusion operators we have $a_{ij} = 0$ for $i \neq j$.)

1.6.4 Elliptic Operators: Weak Solutions, State Spaces, and Mapping Properties

To understand the issues involved in analyzing elliptic operators it is instructive to consider the problem

$$\begin{aligned} \frac{d^2u}{dx^2} &= g(x) \quad 0 < x < 1 \\ u(0) &= u(1) = 0. \end{aligned} \tag{1.109}$$

If $g(x)$ is continuous, then taking

$$u(x) = ax + \int_0^x \int_0^y g(z)dzdy \tag{1.110}$$

with $a = -\int_0^1 \int_0^y g(z)dzdy$ yields a function $u(x)$ which is twice continuously differentiable on (0, 1), continuous on [0, 1], and which satisfies (1.109) on (0, 1). On the other hand, if $g(x)$ is discontinuous, the situation becomes a bit more delicate. (Problems involving discontinuous terms arise in ecological models if the environment changes drastically across some interface and we will want to consider scenarios where that happens.) Suppose for example that $g(x) = 0$ for $0 \leq x < 1/2$ while $g(x) = 1$ for $1/2 \leq x \leq 1$. In that case (1.110) yields $u(x) = (-x/8)$ for $0 \leq x < 1/2$ and $u(x) = (x^2/2) - (5x/8) + 1/8$ for $1/2 \leq x \leq 1$. The derivative $u'(x)$ is given by $u'(x) = -1/8$ for $0 < x < 1/2$, $u'(x) = x - 5/8$ for $1/2 \leq x < 1$, so $u'(x)$ is well defined and continuous. However, if we try to compute $u''(x)$ for $x = 1/2$, we find that at $x = 1/2$ the limit from the right for the difference quotient $[u'(x+h) - u'(x)]/h$ is $\lim_{h \to 0+} [(1/2) + h) - 5/8) - (1/2 - 5/8)]/h = 1$, while the limit from the left is $\lim_{h \to 0-} [(-1/8) - (-1/8)]/h = 0$. Thus, $u''(1/2)$ does not exist so $u(x)$ cannot satisfy (1.109) at $x = 1/2$. However, $u''(x)$ exists and satisfies (1.109) everywhere else. The point is that $u(x)$ satisfies (1.110), which is the integrated version of (1.109), but fails to satisfy (1.109) everywhere because $u(x)$ is not smooth enough. However, $u''(x)$ exists and equals $g(x)$ everywhere except at a single point, so we have

$$\int_0^1 |u''(x) - g(x)|^2 dx = 0. \tag{1.111}$$

Thus, $u''(x) = g(x)$ in terms of the norm on $L^2([0, 1])$.

The preceding example brings us to the general notion of weak solutions. Roughly speaking, a weak solution to a differential equation is a solution to some integrated version of the equation which has the property that smooth solutions to the integrated equation are actual solutions to the original equation. A standard way of defining weak solutions is based on the fact that if $f(x)$ and $g(x)$ are continuous functions such that $\int_0^1 h(x)f(x)dx =$

$\int_0^1 h(x)g(x)dx$ for all smooth functions $h(x)$ on $[0, 1]$, then $f(x) = g(x)$ for all x. (To see that $f(x_0) = g(x_0)$ for any fixed x_0, one would choose a sequence of smooth functions $h_n(x)$ with $h_n(x) = 0$ outside the interval $x_0 - 1/n < x < x_0 + 1/n$, $h_n(x) \geq 0$, and $\int_0^1 h_n(x)dx = 1$. If $f(x)$ is continuous at x_0 then for any $\epsilon > 0$ there is a number N such that $f(x_0) - \epsilon < f(x) < f(x_0) + \epsilon$ for $x \in (x_0 - 1/N, x_0 + 1/N)$; then for $n > N$ we have $\int_0^1 h_n(x)f(x)dx = \int_{x_0-1/n}^{x_0+1/n} h_n(x)f(x)dx \geq \int_{x_0-1/n}^{x_0+1/n} h_n(x)(f(x_0) - \epsilon)dx = (f(x_0) - \epsilon)\int_{x_0-1/n}^{x_0+1/n} h_n(x)dx = f(x_0) - \epsilon$. Similarly, $\int_0^1 h_n(x)f(x)dx \leq f(x_0) + \epsilon$. Since $\epsilon > 0$ was arbitrary, we have $\int_0^1 h_n(x)f(x)dx \to f(x_0)$ as $n \to \infty$. Likewise, $\int_0^1 h_n(x)g(x)dx \to g(x_0)$ as $n \to \infty$, so $f(x_0) = g(x_0)$.) It is possible for the relation $\int_0^1 h(x)f(x)dx = \int_0^1 h(x)g(x)dx$ to hold for all smooth $h(x)$ even if $f(x)$ and $g(x)$ are discontinuous. However, if $f(x)$ or $g(x)$ is discontinuous, it is no longer necessarily the case that $f(x) = g(x)$ for all x. Indeed, the relation may still hold even if there are points where $f(x) \neq g(x)$, provided there are not too many such points. If $f(x) = 1$ for all x and $g(x) = 1$ for all x except, say, $x = 1/4, 1/2$, and $3/4$, where $g(x) = 0$, then $\int_0^1 h(x)f(x)dx = \int_0^1 h(x)g(x)dx$ for all $h(x)$.

We may now define weak derivatives, which provide one way of formulating weak solutions. If $f(x)$ is differentiable and $h(x)$ is a smooth function with $h(0) = h(1) = 0$, then

$$\begin{aligned} \int_0^1 f'(x)h(x)dx &= f(x)h(x)|_0^1 - \int_0^1 f(x)h'(x)dx \\ &= -\int_0^1 f(x)h'(x)dx \end{aligned} \tag{1.112}$$

via integration by parts. We can then define a weak derivative of $f(x)$ to be a function $g(x)$ with the property that for any $h(x)$

$$\int_0^1 g(x)h(x)dx = -\int_0^1 f(x)h'(x)dx. \tag{1.113}$$

If $f'(x)$ exists as a continuous function then (1.112) and (1.113) imply $g(x) = f'(x)$, but (1.113) makes sense in some cases where $f'(x)$ is not continuous. In the case of (1.109) with $g(x) = 0$ for $x < 1/2$ and $g(x) = 1$ for $x \geq 1/2$, a direct calculation shows that the function $u(x)$ defined in (1.110) has a weak second derivative $v(x)$ which is equal to $g(x)$ in the sense that for any smooth $h(x)$, $\int_0^1 h(x)v(x)dx = \int_0^1 h(x)g(x)dx$. The formulation (1.113) is the basis for the theory of distributions; see Strauss (1992). We present it here to illustrate how the notion of solving a differential equation can be extended to cases where some of the terms in the equation are not necessarily smooth. Models with discontinuous coefficients arise naturally in some spatial contexts; moreover, there are some aspects of the mathematical theory of reaction-diffusion models that are facilitated by the use of state spaces such as $L^2([0, 1])$ that contain discontinuous functions.

The example (1.109), (1.110) illustrates some of the mapping properties of the operator d^2/dx^2 and its inverse subject to the given boundary conditions. We can now interpret d^2/dx^2 as an operator on suitable state spaces, that is, as a function between Banach spaces. To define an operator we need to specify its domain and tell how it acts on the elements of its domain. In the case of (1.109) there are a couple of natural choices. The simplest choice for the domain of d^2/dx^2 subject to the boundary conditions of (1.109) is the space of functions on $[0, 1]$ whose first and second derivatives are continuous and which satisfy the boundary conditions. The space $C^2([0, 1])$ is defined to be the set of all functions on $[0, 1]$ which are twice continuously differentiable. The norm on $C^2([0, 1])$ is

$$||u||_{C^2([0,1])} = \sup_{x\in[0,1]} (|u''(x)|) + \sup_{x\in[0,1]} (|u'(x)|) + \sup_{x\in[0,1]} (|u(x)|), \tag{1.114}$$

and under that norm $C^2([0, 1])$ is a Banach space. (We discuss that point in more detail in the next paragraph.) If we let $X = \{u \in C^2([0, 1]) : u(0) = u(1) = 0\}$, retain (1.114) as the norm on X, and let $A = d^2/dx^2$ with the domain of A taken to be X, then A maps X into $C([0, 1])$. Furthermore, if $u, v \in X$ then we have

$$\begin{aligned} ||Au - Av||_{C([0,1])} = ||A(u - v)||_{C([0,1])} &= \sup_{x\in[0,1]} |u''(x) - v''(x)| \\ &\leq ||u - v||_{C^2([0,1])} \end{aligned}$$

so that A is continuous. The inverse operator A^{-1} for A is defined by (1.110), and it is easy to see that if $g \in C([0, 1])$ then $||A^{-1}g||_{C^2([0,1])} \leq C||g||_{C([0,1])}$ for some constant C. Thus, A^{-1} is also continuous as a function from $C([0, 1])$ into X. Finally, A^{-1} maps bounded sets in $C([0, 1])$ into bounded sets in X, but $X \subseteq C([0, 1])$ and by Ascoli's theorem any set that is bounded in X has compact closure in $C([0, 1])$. Thus, if we define E as the operator from X into $C([0, 1])$ which acts as the identity operator on X, that is, $Eu = u$, and combine E with A^{-1} then the operator EA^{-1} maps $C([0, 1])$ into $C([0, 1])$ continuously, and does so in a way that maps bounded sets into sets whose closures are compact. This is exactly the type of mapping property that is required for the application of modern mathematical methods to equilibrium problems for reaction-diffusion models.

The choice of spaces $C([0, 1])$ and $X \subseteq C^2([0, 1])$ above is not the only possibility. We could also choose to work in the space Y of functions which are square integrable and which have weak first and second derivatives in the sense of (1.113) which are also square integrable, with norm

$$||u||_Y = \left[\int_0^1 (|u''(x)|^2 + |u'(x)|^2 + |u(x)|^2)dx\right]^{1/2}. \tag{1.115}$$

If we use Lebesgue integrals then Y turns out to be a Banach space. A problem now arises with the boundary conditions, because (1.115) might still make sense even if $u(0), u(1)$ were undefined or if $u(x)$ were discontinuous at 0 or 1 (or elsewhere). However, by using the Cauchy-Schwartz inequality for integrals (e.g., see Royden (1968)), we can see that for $u \in Y$,

$$\begin{aligned}|u(x_1) - u(x_2)| &= \int_{x_1}^{x_2} u'(x)dx \\ &\le \left(\int_{x_1}^{x_2} dx\right)^{1/2} \left(\int_{x_1}^{x_2} |u'(x)|^2 dx\right) \\ &\le |x_2 - x_1|^{1/2} ||u||_Y.\end{aligned} \tag{1.116}$$

Thus, if $u \in Y$, then u is continuous, so it makes sense to talk about $u(0)$ and $u(1)$, and we may define the space $Y_0 = \{u \in Y : u(0) = u(1) = 0\}$. We can now define an operator B as $B = d^2/dx^2$ with domain Y_0. The operator B is different from A because it acts on a different space, although Au and Bu represent the same function for those choices of u where both make sense. It is clear from (1.115) that B maps Y_0 into $L^2([0, 1])$ continuously, and it is not too hard to show that (1.110) defines an inverse operator B^{-1} that maps $L^2([0, 1])$ continuously into Y_0. It is also true (but harder to show) that the embedding operator E_0 that maps Y_0 to $L^2([0, 1])$ by taking $E_0 u = u$ maps bounded sets in Y_0 to sets with compact closures in $L^2([0, 1])$. Thus, the mapping from $L^2([0, 1])$ into itself defined by $E_0 B^{-1}$ has the same type of mapping properties as the operator EA^{-1} on $C[0, 1])$. This means that we could use many of the same analytic methods while working in either $C([0, 1])$ or $L^2([0, 1])$. The reason for wanting to have both spaces available for use is that each has some advantages and disadvantages from the viewpoint of analysis. The functions in $C([0, 1])$ are continuous, so they will always attain maximum and minimum values, which is an observation that is sometimes useful. On the other hand, the norm in $L^2([0, 1])$ arises from an inner product (analogous to the dot product for vectors) given by $< u, v >= \int_0^1 u(x)v(x)dx$, and having an inner product is also sometimes useful. In what follows we use some state spaces which are generalizations of $C([0, 1])$ and $C^2([0, 1])$, and others that are generalizations of $L^2([0, 1])$ and the space Y whose norm is defined in (1.115).

To verify that $C([0, 1])$ is a Banach space requires only basic ideas from real analysis, as discussed by Rudin (1976), for example. The key idea is that any sequence which converges in the norm of $C([0, 1])$ is uniformly convergent, so taking the limit of a sequence of continuous functions with respect to the norm on $C([0, 1])$ yields another continuous function. To correctly define $L^2([0, 1])$ and verify that it is a Banach space we must interpret integrals in the sense of Lebesgue (see Royden (1968), for example.) The reason is that there are sequences of functions $f_n(x)$ which are integrable in sense of the usual Riemann integral defined in most calculus texts, and for which the sequence $\{f_n(x)\}$ is a Cauchy sequence with respect to the metric defined by $d(f_n, f_m) = ||f_n - f_m|| = \left(\int_0^1 |f_n(x) - f_m(x)|^2 dx\right)^{1/2}$, which do not converge to a Riemann integrable function. The problem is essentially that $\{f_n\}$ might converge pointwise to some function $f(x)$ which is so badly discontinuous that $f(x)$ and $[f(x)]^2$ are not integrable in the sense of Riemann. Lebesgue integrals can handle functions which are too badly discontinuous to be Riemann integrable; for example the function on [0, 1] which is zero for all rational numbers but one for all irrational numbers is not Riemann integrable, but it is Lebesgue integrable. The two notions of integral agree when both make sense. In what follows we will always interpret integrals as Lebesgue integrals. If we use Lebesgue integrals to define $L^2([0, 1])$ then the completeness of $L^2([0, 1])$ (and hence the fact that $L^2([0, 1])$ is a Banach space) follows from the standard convergence theorems for Lebesgue integrals (see Royden (1968)).

When we try to formulate appropriate state spaces for reaction-diffusion models in more than one space dimension we encounter a few more technical issues. The boundary of an interval always consists of two points, but the boundary of a region in two or three dimensions could be very complicated, even fractal. We usually impose some conditions on the boundaries occurring in our models, typically requiring them to be smooth curves or surfaces. Another problem is that the elliptic operators which occur in our models do not have all the mapping properties we might want if we try to define them on $C^2(\overline{\Omega})$. (Here $\Omega \subseteq \mathbb{R}^n$ is a bounded domain and $\overline{\Omega}$ its closure). Finally, the formula (1.116) which implies that functions in Y must be continuous is not valid in exactly the same form in higher space dimensions. Since continuity is a desirable property we will sometimes want to work in spaces whose norms are defined by integrals but whose elements are continuous functions. To do that we must sometimes use spaces constructed from the space $L^p(\Omega)$, whose norm is $||u|| = \left(\int_\Omega |u(x)|^p dx\right)^{1/p}$, where $p \geq 1$ but p is not necessarily equal to 2.

The problem with treating elliptic operators of the form L shown in (1.107) in $C^2(\overline{\Omega})$ is that although L will be continuous as a map from $C^2(\Omega)$ into $C(\overline{\Omega})$, the inverse operator L^{-1} (i.e. the solution operator for the equation $Lu = f$ under suitable boundary conditions) will not in general be continuous as a map from $C(\overline{\Omega})$ to $C^2(\overline{\Omega})$. To define spaces on which L will have a continuous inverse we need a notion of continuity for functions which is stronger than simple continuity. The type of continuity we need is called *Hölder continuity*. Let x_0 be a point in $\mathbb{R}^n$ and let α be a real number with $0 < \alpha \leq 1$. The function $f(x)$ is said to be Hölder continuous with exponent α at x_0 if there exist a neighborhood of x_0 and a constant C such that $|f(x) - f(x_0)| < C|x - x_0|^\alpha$ for all x in the neighborhood of x_0. In the case $\alpha = 1$ the function $f(x)$ is said to be *Lipschitz continuous*. As an example, (1.116) shows that the function u is Hölder continuous with exponent 1/2. Let $\Omega \subseteq \mathbb{R}^n$ be a bounded domain, and denote the closure of Ω by $\overline{\Omega}$. A function $f(x)$ is Hölder continuous with exponent α on $\overline{\Omega}$ if the quantity

$$[f]_\alpha = \sup_{\substack{x,y\in\overline{\Omega} \\ x\neq y}} \frac{|f(x) - f(y)|}{|x - y|^\alpha} \tag{1.117}$$

is finite. The space of functions that are Hölder continuous on $\overline{\Omega}$ with exponent α is denoted by $C^{0,\alpha}(\overline{\Omega})$ or simply $C^\alpha(\overline{\Omega})$. The norm on $C^\alpha(\overline{\Omega})$ is

$$\begin{aligned} ||u||_\alpha = ||u||_{0,\alpha} &= \sup_{x\in\overline{\Omega}}|u(x)| + \sup_{\substack{x,y\in\overline{\Omega} \\ x\neq y}} \frac{|u(x) - u(y)|}{|x - y|^\alpha} \\ &= ||u||_0 + [u]_\alpha, \end{aligned} \tag{1.118}$$

where $||u||_0 = \sup_{\overline{\Omega}}|u(x)|$ is the norm for $C(\overline{\Omega})$. To describe the partial derivatives of a function u defined on a subset of $\mathbb{R}^n$, it is common to use the multi-index notation $\beta = (\beta_1, \ldots, \beta_n)$ where $\beta_1, \ldots, \beta_n$ are integers, to define the order of β, denoted $|\beta|$, by $|\beta| = \beta_1 + \cdots + \beta_n$ and write $\partial^{|\beta|}u/\partial x_1^{\beta_1}\partial x_2^{\beta_2}\ldots\partial x_n^{\beta_n}$ as $\partial^\beta u$. The space of functions on $\overline{\Omega}$ which have spatial derivatives up to order k that are continuous on $\overline{\Omega}$, with the derivatives of order k being Hölder continuous with exponent α, is denoted $C^{k,\alpha}(\overline{\Omega})$ or $C^{k+\alpha}(\overline{\Omega})$. The norm on $C^{k+\alpha}(\overline{\Omega})$ can be written as

$$||u||_{k+\alpha} = \sup_{x\in\overline{\Omega}}|u(x)| + \sum_{|\beta|\leq k}\sup_{x\in\overline{\Omega}}|\partial^{\beta}u(x)| + \sum_{|\beta|=k}[\partial^{\beta}u]_{\alpha} \tag{1.119}$$

where the notation $[\ \ ]_{\alpha}$ is defined in (1.117). The spaces $C^{k+\alpha}(\overline{\Omega})$ are Banach spaces. See Gilbarg and Trudinger (1977) and Friedman (1976) for additional discussion. As noted at the beginning of this discussion, the reason for considering these spaces is that if an elliptic operator L has the form shown in (1.107) and there is a solution operator or inverse operator L^{-1} for L, the inverse operator will typically map $C^{\alpha}(\overline{\Omega})$ into $C^{2+\alpha}(\overline{\Omega})$, but will not necessarily map $C(\overline{\Omega})$ into $C^{2}(\overline{\Omega})$ as was the case in one space dimension. Thus, although L might make sense as a mapping form $C^{2}(\overline{\Omega})$ into $C(\overline{\Omega})$, if we want L to have a continuous inverse we will usually need to think of L as an operator from $C^{2+\alpha}(\overline{\Omega})$ into $C^{\alpha}(\overline{\Omega})$ instead. This point is discussed in some detail by Gilbarg and Trudinger (1977). To correctly state the mapping properties for elliptic operators on spaces of Hölder continuous functions we must put some restrictions on the boundary of the underlying domain Ω. To derive some of the mapping properties of elliptic operators it is necessary to do calculations that involve changes of coordinates which "flatten" parts of the boundary in the sense of mapping them to hyperplanes. When a differential operator is written in terms of the new coordinates derivatives of the functions defining the coordinate changes occur in the coefficients because of the chain rule. Thus, the theory requires that the changes of coordinates are smooth enough that the coefficients of the operator under consideration are well defined and sufficiently smooth in the new coordinate system. Recall that a change of coordinates is a function Φ which maps an open subset of $I\!R^{n}$ into another open subset of $I\!R^{n}$ in such a way that Φ has an inverse function Φ^{-1} and Φ and Φ^{-1} are at least continuous. We say that the boundary $\partial\Omega$ of a domain $\Omega \subseteq I\!R^{n}$ is of class $C^{k+\alpha}$ if for each point $x_0 \in \partial\Omega$ there is a neighborhood U containing x_0 and a change of coordinates Φ on U such that

(i) if $y = (y_1, \ldots, y_n) = \Phi(x)$, the image under Φ of $\partial\Omega \cap U$ lies in the hyperplane $\{y \in I\!R^{n} : y_n = 0\}$ while the image of $\Omega \cap U$ lies in the half-space $\{y \in I\!R^{n} : y_n > 0\}$, and

(ii) each of the coordinates functions $y_1(x), \ldots, y_n(x)$ defining Φ belongs to $C^{k+\alpha}(\overline{U})$ and each of the functions $x_1(y), \ldots, x_n(y)$ defining Φ^{-1} belongs to $C^{k+\alpha}(\Phi(\overline{U}))$, where $\Phi(\overline{U})$ is the image of $\overline{U}$ under Φ.

We can now describe the mapping properties of operators of the form L shown in (1.107), subject to the ellipticity condition (1.108), on spaces of Hölder continuous functions.

Theorem 1.1. Suppose that $\Omega \subseteq I\!R^{n}$ is a bounded domain with $\partial\Omega$ of class $C^{2+\alpha}$ for some $\alpha \in (0, 1)$. Suppose that the coefficients of the operator L shown in (1.107) belong to $C^{\alpha}(\overline{\Omega})$, satisfy (1.108), and also satisfy $c(x) \leq 0$. If $f \in C^{\alpha}(\overline{\Omega})$ and $g \in C^{2+\alpha}(\overline{\Omega})$, the problem

$$\begin{aligned} Lu &= f(x) \text{ in } \Omega \\ u &= g(x) \text{ on } \partial\Omega \end{aligned} \tag{1.120}$$

has a unique solution $u \in C^{2+\alpha}(\overline{\Omega})$ which satisfies the inequality

$$||u||_{2+\alpha} \leq C(||f||_\alpha + ||g||_{2+\alpha}) \tag{1.121}$$

for a constant C independent of f and g where $||\cdot||_\alpha$ and $||\ \ ||_{2+\alpha}$ are the norms of $C^\alpha(\overline{\Omega})$ and $C^{2+\alpha}(\overline{\Omega})$ respectively, as defined in (1.119).

Remark: This result is a version of Theorem 6.14 of Gilbarg and Trudinger (1977) which incorporates some of the remarks given in the discussion of that theorem. A number of related results are derived in Chapter 6 of Gilbarg and Trudinger (1977). Note that the boundary condition is formulated in terms of the restriction of a function g that is defined on all of $\overline{\Omega}$ to the boundary $\partial\Omega$. If we started with a function $g_0 \in C^{2+\alpha}(\partial\Omega)$, then since we are assuming $\partial\Omega$ is of class $C^{2+\alpha}$ we could extend g_0 to a function $g \in C^{2+\alpha}(\overline{\Omega})$ in such a way that $||g||_{C^{2+\alpha}(\overline{\Omega})} \leq C||g_0||_{C^{2+\alpha}(\partial\Omega)}$. The following in a version of Theorem 6.31 of Gilbarg and Trudinger (1977) which treats another type of boundary condition:

Theorem 1.2. Suppose that Ω, L and f are as in Theorem 1.1. Suppose that $g(x), \gamma(x), \beta(x) \in C^{1+\alpha}(\partial\Omega)$ with $\beta(x) > 0$ and $\gamma(x) \geq 0$. Suppose that $c(x) \leq 0$ and either $c(x) \not\equiv 0$ or $\gamma(x) \not\equiv 0$. Let $\partial/\partial\vec{n}$ denote the outward normal derivative on $\partial\Omega$. The problem

$$\begin{aligned} Lu &= f(x) \quad \text{on} \quad \Omega \\ \gamma(x)u + \beta(x)\frac{\partial u}{\partial\vec{n}} &= g(x) \quad \text{on} \quad \partial\Omega \end{aligned} \tag{1.122}$$

has a unique solution $u \in C^{2+\alpha}(\overline{\Omega})$, and there is a constant C independent of f and g so that

$$||u||_{2+\alpha} \leq C(||f||_\alpha + ||g||_{1+\alpha}) \tag{1.123}$$

where $||g||_{1+\alpha}$ is the norm of $C^{1+\alpha}(\partial\Omega)$.

Remark: If $c = \gamma = 0$ then (1.22) may not be solvable for all choices of f. Something of this sort already occurs in the problem $d^2u/dx^2 = f(x),\ u'(0) = u'(1) = 0$. If we integrate the equation from 0 to 1 we obtain $\int_0^1 f(x)dx = \int_0^1 u''(x)dx = u'(1) - u'(0) = 0$ so the equation is not solvable unless $\int_0^1 f(x)dx = 0$. It is also the case that if $c = \gamma = 0$ in (1.122) then the homogeneous problem with $f = g = 0$ has the nontrivial solutions $u = k$ where k is any constant. This is no accident.

Theorem 1.3. Suppose that L, Ω, γ, and β are as in Theorems 1.1. and 1.2., but without the restrictions $c \leq 0$ or $\gamma \geq 0$. Then each of the problems (1.120), (1.122) satisfies one or the other of the following alternatives. Either

(i) the homogeneous problem (i.e. the problem with $f = 0$ and $g = 0$) has only the trivial solution $u = 0$ and the nonhomogeneous problem has a unique solution $u \in C^{2+\alpha}(\overline{\Omega})$ for each $f \in C^\alpha(\overline{\Omega})$ and (for 1.120) $g \in C^{2+\alpha}(\overline{\Omega})$ or (for 1.122) $g \in C^{1+\alpha}(\partial\Omega)$;

or

(ii) the homogeneous problem has nontrivial solutions; in that case the nullspace of L, i.e. the set of solutions to the homogeneous problem, is a finite dimensional subspace of $C^{2+\alpha}(\overline{\Omega})$.

Remark: Theorem 1.3 is a version of Theorem 6.15 of Gilbarg and Trudinger (1977). It is a special case of a more general result about operators known as the Fredholm Alternative which we discuss later in this section. A key point in establishing Theorem 1.3, which will also arise in the derivation of other results, is that an operator satisfying the hypotheses of Theorems 1.1 or 1.2 with homogeneous boundary conditions (i.e. $g = 0$) will have an inverse L^{-1} which is compact as a linear operator on $C^\alpha(\overline{\Omega})$; that is, L^{-1} maps bounded sets in $C^\alpha(\overline{\Omega})$ into sets with compact closures. More precisely, L^{-1} maps $C^\alpha(\overline{\Omega})$ continuously into $C^{2+\alpha}(\overline{\Omega})$, but $C^{2+\alpha}(\overline{\Omega})$ embeds compactly in $C^2(\overline{\Omega})$ by Arzela's theorem, and $C^2(\overline{\Omega})$ then embeds in $C^\alpha(\overline{\Omega})$ by the mean value theorem, so if $E : C^{2+\alpha}(\overline{\Omega}) \to C^\alpha(\overline{\Omega})$ is given by $Eu = u$ then the mapping $EL^{-1} : C^\alpha(\overline{\Omega}) \to C^\alpha(\overline{\Omega})$ is compact since L^{-1} is continuous and E is a compact operator.

The reason for the conditions $c \le 0$ in Theorem 1.1 and $c \le 0$, $\gamma \ge 0$, with $c \not\equiv 0$ or $\gamma \not\equiv 0$ in Theorem 1.2 is that those conditions imply uniqueness for the trivial solutions to (1.120) and (1.122), respectively, when $f = 0$, $g = 0$. Any other condition which guarantees uniqueness of the trivial solutions to those problems could be combined with Theorem 1.3 to yield an existence theorem for solution of the nonhomogeneous problem. The condition $c \le 0$ implies that L satisfies a maximum principle, which means (among other things) that solutions to $Lu = 0$ cannot have a positive maximum or negative minimum inside Ω unless they are constant throughout Ω. The maximum principle can be used to show uniqueness for the homogeneous problem. We discuss maximum principles later in this section.

Theorems 1.1–1.3 provide a description of the mapping properties of elliptic operators on spaces of Hölder continuous functions which will be adequate for our purposes, but they do not address problems along the lines of (1.120) or (1.222) when f is continuous but not Hölder continuous or when f is discontinuous. Some of the situations we want to study are best described by models which have discontinuous terms. Moreover, even in models with no discontinuities built into them having the ability to work with functions that are merely continuous can facilitate our mathematical analysis. Thus, we consider problems such as (1.120) and (1.122) in cases where f is only assumed to belong to some space of integrable functions such as the space $L^2([0, 1])$ we considered in our simple one-dimensional example. To do that requires an appropriate notion of weak derivatives, which in turn is based on the notion of integration by parts in $\mathbb{R}^n$, so we shall digress briefly to discuss those topics and then return to the mapping properties of elliptic operators.

The result for functions on $\mathbb{R}^n$ which gives the closest analogue to the integration by parts formula is the divergence theorem. Recall that if $\Omega \subseteq \mathbb{R}^n$ is a bounded domain with a C^1 boundary and $\vec{F}$ is a C^1 vector field on Ω which is continuous on $\overline{\Omega}$ (i.e. $\vec{F} \in [C^1(\Omega)]^n \cap [C(\overline{\Omega})]^n$), then the divergence theorem states that

$$\int_\Omega \nabla \cdot \vec{F} dx = \int_{\partial\Omega} \vec{F} \cdot \vec{n} dS, \tag{1.124}$$

where $dx = dx_1 \ldots dx_n$ is the volume element on Ω, dS is the surface element on $\partial\Omega$, and $\vec{n}$ is the outward unit normal on $\partial\Omega$. Suppose that u and v are C^1 functions on Ω with either $u = 0$ or $v = 0$ on $\partial\Omega$, and let $\vec{e}_k$ be the unit vector in the x_k direction. We have

$$0 = \int_{\partial\Omega} uv\vec{e}_k \cdot \vec{n} dS = \int_{\Omega} \nabla \cdot (uv\vec{e}_k) dx = \int_{\Omega} \left(\frac{\partial u}{\partial x_k} v + u \frac{\partial v}{\partial x_k} \right) dx$$

so that

$$\int_{\Omega} \frac{\partial u}{\partial x_k} v = - \int_{\Omega} u \frac{\partial v}{\partial x_k} dx. \tag{1.125}$$

The formula (1.125) is the basis for the definition of weak derivatives in $I\!R^n$. If the function $w(x)$ satisfies the relation

$$\int_{\Omega} wv dx = - \int_{\Omega} u \frac{\partial v}{\partial x_k} dx \tag{1.126}$$

for all functions v that are smooth and are equal to zero outside a compact subset of Ω, then w is said to be the *weak derivative of* u *with respect to* x_k *on* Ω. Higher weak derivatives are defined analogously; so $w = \partial^2 u / \partial x_k \partial x_j$ in the weak sense if

$$\int_{\Omega} wv dx = (-1)^2 \int_{\Omega} u \frac{\partial^2 v}{\partial x_k \partial x_j} dx = \int_{\Omega} u \frac{\partial^2 v}{\partial x_k \partial x_j} dx$$

for all smooth functions v that are zero outside some compact subset of Ω, and so on. Certain calculations based on the divergence theorem are widely used in partial differential equations. Some of these are known as Green's formulas. An example is the following. Suppose $u, v, \in C^2(\overline{\Omega})$. Then

$$\begin{aligned} \int_{\Omega} v(\nabla \cdot d(x) \nabla u) dx &= \int_{\Omega} (\nabla \cdot [v d(x) \nabla u] - d(x) \nabla u \cdot \nabla v) dx \\ &= \int_{\partial\Omega} v d(x) \nabla u \cdot \vec{n} dS - \int_{\Omega} d(x) \nabla u \cdot \nabla v dx \\ &= \int_{\partial\Omega} v d(x) \frac{\partial u}{\partial \vec{v}} dS - \int_{\Omega} d(x) \nabla u \cdot \nabla v dx. \end{aligned}$$

Because weak derivatives are defined by (1.126) to satisfy the formula (1.125), they will by definition also satisfy Green's formulas and other formulas based on the divergence theorem when all the derivatives exist and all the integrals make sense and are finite. Weak derivatives are discussed in some detail by Adams (1975), Friedman (1976), and Gilbarg and Trudinger (1977).

The state spaces we need to use to treat models with discontinuous terms can be constructed by using the idea of weak derivatives in the setting of the spaces $L^p(\Omega)$ of functions on Ω whose pth power is integrable in the sense of Lebesgue. The L^p spaces will be familiar to most of those readers who have had a course on real analysis which treats Lebesgue integration; they are discussed in detail, for example, by Royden (1968). For those readers unfamiliar with Lebesgue integrals, one of their essential features is that they agree with the ordinary Riemann integrals of elementary calculus when both types of integrals make sense, but Lebesgue integrals make sense for some functions for which Riemann integrals do not. As a consequence, it is possible to interchange integration with taking the limit of a sequence of functions more readily in the context of Lebesgue integrals than in the context of Riemann integrals. As an example, if the functions $f_n(x)$ are Lebesgue integrable on a bounded domain Ω, are bounded by a Lebesgue integrable function $f_0(x)$ in the sense that $|f_n(x)| \le f_0(x)$ for all n, and $f_n(x) \to f(x)$ for each $x \in \Omega$, then $f(x)$ is Lebesgue integrable on Ω and $\int_\Omega f_n dx \to \int_\Omega f dx$ as $n \to \infty$. (See Royden (1968)). This is not true in general for Riemann integrals, where convergence of the integrals $\int_\Omega f_n dx$ normally requires uniform convergence of the functions.

Lebesgue integrals are defined in terms of what is known as Lebesgue measure. Lebesgue measure is a generalization of the notion of length (in $\mathbb{R}$), area (in $\mathbb{R}^2$), volume (in $\mathbb{R}^3$), etc. The Lebesgue measure of an interval is simply its length, but Lebesgue measure makes sense for sets more complicated than intervals. For example, in one dimension, the set of all rational numbers in $[0, 1]$ has measure 0 while the set of all irrational numbers in $[0, 1]$ has measure 1. This reflects the fact that relatively few points (only countably many) in $[0, 1]$ are rational. Any surface or hypersurface of dimension $n-1$ or lower in $\mathbb{R}^n$ will have Lebesgue measure zero, and any finite or even countable union of sets with Lebesgue measure zero will also have Lebesgue measure zero. Lebesgue integrals are actually defined on equivalence classes of functions which are equal outside of a set of measure zero, and if a function is Lebesgue integrable at all then its integral over any set of measure zero is equal to zero. The appropriate notion of convergence for a sequence of functions relative to Lebesgue integration is convergence almost everywhere, that is, pointwise convergence outside of a set of measure zero. If $\{f_n\}$ is a sequence of Lebesgue integrable functions with $|f_n|$ bounded by a Lebesgue integrable function $f_0(x)$ for all n and $f_n \to f$ almost everywhere in Ω, then f is Lebesgue integrable and $\int_\Omega f_n dx \to \int_\Omega f dx$ as $n \to \infty$. When $f_n \to f$ almost everywhere it is standard to write "$f_n \to f$ a.e.". The significance of the convergence properties of the Lebesgue integral is that they guarantee that spaces of Lebesgue integrable functions are complete as metric spaces, i.e. they are Banach spaces. To be specific, for $1 \le p < \infty$ the space $L^p(\Omega)$ is the set of all functions f on Ω for which $|f|^p$ is integrable, with the norm

$$||f||_p = \left(\int_\Omega |f(x)|^p dx\right)^{1/p}. \tag{1.127}$$

(Once again, to be precise, the elements in $L^p(\Omega)$ are actually equivalence classes of functions, with each equivalence class consisting of functions which are equal outside a set of measure zero.) To be Lebesgue integrable a function must have the property of measurability, which means that if $S \subseteq \mathbb{R}$ is a set such that the Lebesgue measure of S is well defined in $\mathbb{R}$ then the Lebesgue measure of the set $\{x \in \Omega : f(x) \in S\}$ is well defined in $\mathbb{R}^n$. In addition to the spaces $L^p(\Omega)$, we may sometimes want to consider the space $L^\infty(\Omega)$ consisting of Lebesgue measurable functions which are bounded except perhaps on a set of measure zero. The norm on $L^\infty(\Omega)$ is given by

$$\begin{aligned} ||f||_\infty = \inf\{\sup_\Omega |g(x)| : g \quad &\text{is measurable on} \quad \Omega \quad \text{and} \\ g = f \quad &\text{almost everywhere}\}. \end{aligned} \tag{1.128}$$

If $f \in C(\overline{\Omega})$, then $||f||_\infty = \sup_{\overline{\Omega}} |f|$ so that $C(\overline{\Omega})$ is a closed subspace of $L^\infty(\Omega)$. (The space $C(\overline{\Omega})$ is a subspace of $L^p(\Omega)$ for any $p \ge 1$, but in general is not closed.)

We can now define state spaces based on L^p spaces. Denote by $W^{k,p}(\Omega)$ the space of functions whose weak derivatives of order up to k belong to $L^p(\Omega)$. The norm of f in $W^{k,p}(\Omega)$ is denoted by $|| \ \ ||_{k,p}$. It is defined in terms of $L^p(\Omega)$ norms of the weak derivatives of f in much the same way that the norm of f in $C^{k+\alpha}(\overline{\Omega})$ is defined in terms of the norms in $C(\overline{\Omega})$ and $C^\alpha(\overline{\Omega})$ of derivatives of f in (1.119), namely

$$||f||_{k,p} = \sum_{|\beta|\leq k} ||\partial_\beta f||_p, \tag{1.129}$$

where once again $\beta = (\beta_1, \ldots, \beta_n)$, where $\beta_1, \ldots, \beta_n$ are nonnegative integers, $|\beta| = \beta_1 + \cdots + \beta_n$, and $\partial_\beta f$ denotes the derivative $\partial^{|\beta|} f/\partial x_1^{\beta_1} \ldots \partial x_n^{\beta_n}$, now taken in the weak sense as in (1.129). The spaces $W^{k,p}(\Omega)$ are called Sobolev spaces, and they are Banach spaces; see Gilbarg and Trudinger (1997) or Adams (1975). An alternative way to define them is to take the completion of $C^k(\overline{\Omega})$ (or $C^{k+\alpha}(\overline{\Omega})$, for that matter) with respect to the norm (1.129). A particular Sobolev space which we will use in our discussion of eigenvalues for elliptic operators is the space $W^{1,2}(\Omega)$. That space is of interest partly because its norm can be viewed as arising form the inner product

$$< u, v >= \int_\Omega [\nabla u \cdot \nabla v + uv] dx \tag{1.130}$$

(where again ∇u and ∇v are interpreted in the weak sense).

In some cases we will also want to work in the subspace $W_0^{1,2}(\Omega) \subseteq W^{1,2}(\Omega)$, which is obtained by taking the completion in Ω relative to (1.130) of the space of continuously differentiable functions that are zero outside a compact subset of Ω (which means that they are zero on $\partial\Omega$). The space $W_0^{1,2}(\Omega)$ can be viewed as being the subspace of $W^{1,2}(\Omega)$ consisting of functions which are zero on $\partial\Omega$, but some care is required in that interpretation since functions in $W^{1,2}(\Omega)$ are not necessarily continuous on $\overline{\Omega}$.

The fact that functions in Sobolev spaces need not be continuous on $\overline{\Omega}$ means that some care must be taken in formulating boundary conditions on them. For functions $f(x) \in C(\overline{\Omega})$ the mapping from f to the restriction of f to $\partial\Omega$ maps $C(\overline{\Omega})$ continuously to $C(\partial\Omega)$. In the case of Sobolev spaces we need to use the notion of trace. If Y is a space of functions on $\partial\Omega$, the function $u \in W^{m,p}(\Omega)$ has trace $y \in Y$ if for any sequence $\{u_n\}$ of smooth (i.e. infinitely differentiable) functions u_n on $\overline{\Omega}$ which converge to u in the norm of $W^{m,p}(\Omega)$, the functions obtained by restricting u_n to $\partial\Omega$ converge to y in the norm of Y, with $||y||_Y \leq C||u||_{m,p}$ for some constant C independent of u. We have the following (see Adams (1975, Theorems 5.22, 7.53, and 7.57)).

Lemma 1.4. Suppose $\Omega \subseteq I\!R^n$ is a bounded domain with $\partial\Omega$ of class $C^{2+\alpha}$. For $p \in (1, \infty)$ functions in $W^{2,p}(\Omega)$ have traces in $W^{1,p}(\partial\Omega)$. For $m = 1, 2$ and $p < n/m$, functions in $W^{m,p}(\Omega)$ have traces in $L^q(\partial\Omega)$ for $q \in [p, (n-1)p/(n-mp)]$. For $p \geq n/m$, functions in $W^{2,p}(\Omega)$ have traces in $L^q(\partial\Omega)$ for $q \in [p, \infty)$.

Remarks: The spaces $W^{1,p}(\Omega)$ actually have traces which belong to the fractional order Sobolev space $W^{1-1/p,p}(\partial\Omega)$, as defined in Agmon et al. (1959) or Adams (1975); see for example Adams (1975, Theorem 7.53). This is a relatively fine technical point but it is relevant to the formulation of boundary conditions in Sobolev spaces.

We can now state results in Sobolev spaces which are analogous to Theorems 1.1 and 1.2.

Theorem 1.5. Suppose that $\Omega \subseteq I\!R^n$ is a bounded domain with $\partial\Omega$ of class $C^{2+\alpha}$, and that L has the form (1.107), (1.108) holds, and the coefficients of L all belong to $C^{\alpha}(\overline{\Omega})$. Suppose further that $c \leq 0$. Let $p \in (1, \infty)$. If $f \in L^p(\Omega)$ and $g \in W^{2,p}(\Omega)$ then the problem (1.120) (with the condition $u = g$ on $\partial\Omega$ interpreted as meaning that the traces of u and g on $\partial\Omega$ are equal almost everywhere) has a unique solution $u \in W^{2,p}(\Omega)$ satisfying $||u||_{2,p} \leq C(||f||_{0,p} + ||g||_{2,p})$ for some constant C independent of f and g.

Remarks: Results of this type are discussed by Agmon et al. (1959). The key idea is that for any $u \in W^{2,p}(\Omega)$ with $u = 0$ on $\partial\Omega$ (i.e. trace of u on $\partial\Omega$ is 0), we have the estimate $||u||_{2,p} \leq C_1(||u||_{0,p} + ||Lu||_{0,p})$ for a constant C_1 independent of u; see (Agmon et al. 1959). Once this estimate is available we may take $v = u - g$ so that $v = 0$ on $\partial\Omega$, $Lv = f - Lg \in L^p(\Omega)$ in Ω, and then analyze the problem for v in much the same way as in the case considered in Theorem 1.1, which is discussed in detail in Gilbarg and Trudinger (1977, Chapter 6); see Agmon et al. (1959). A similar analysis can be applied in the case of mixed boundary conditions:

Theorem 1.6. Suppose that L and Ω satisfy the hypotheses of Theorem 1.5. Suppose that c, β and γ satisfy the hypotheses of Theorem 1.2, that $f \in L^p(\Omega)$ and $g \in W^{1,p}(\Omega)$. Then the problem (1.122) has a unique solution $u \in W^{2,p}(\Omega)$, which satisfies $||u||_{2,p} \leq C(||f||_{o,p} + ||g||_{1,p})$ for some constant C independent of g or f.

Remarks: The "g" in the boundary condition in (1.122) should be interpreted as the trace of $g \in W^{1,p}(\Omega)$ on $\partial\Omega$. The key estimates by Agmon et al. (1959) are cast in terms of a norm on g which is essentially the norm of the trace of $g \in W^{1,p}(\Omega)$ on $\partial\Omega$ in the space $W^{1-1/p,1/p}(\partial\Omega)$; see Agmon et al. (1959, Formula 14.2) and compare with the discussion by Adams (1975, Chapter 7). If $g_0 \in W^{1,p}(\partial\Omega)$ then g_0 can be extended to $g \in W^{1,p}(\Omega)$ in such a way that $||g||_{W^{1,p}(\Omega)} \leq C||g_0||_{W^{1,p}(\partial\Omega)}$. If we omit the condition $c \leq 0$ from Theorem 1.5 or the conditions on c, γ, and β from Theorem 1.6 we would still be able to obtain a result analogous to Theorem 1.3.

An important feature of the Sobolev spaces $W^{m,p}(\Omega)$ is that they embed in the spaces $C^{k,\alpha}(\overline{\Omega})$ and $W^{j,q}(\Omega)$ under suitable conditions on $\partial\Omega, m, p, k, \alpha, j$, and q. Some of the embeddings are compact. An appropriate condition on $\partial\Omega$ for bounded domains is the local Lipschitz property. That property holds if $\partial\Omega$ is of class $C^{2+\alpha}$ (or even just C^1), but it will also be true if in a neighborhood of each point $x_0 \in \partial\Omega$ the boundary of Ω can be expressed (possibly after rotating coordinates or interchanging the i and j coordinates for some choices of i and j) as the graph of a function $x_n = \Psi(x_1, \ldots, x_{n-1})$ where Ψ is Lipschitz; that is, there is a constant such that $|\Psi(x_1, \ldots, x_{n-1}) - \Psi(y_1, \ldots, y_{n-1})| < C|(x_1 - y_1, \ldots, x_{n-1} - y_{n-1})|$. These domains include domains whose boundaries consist of finitely many smooth surfaces which meet at corners, such as polygons and polyhedra. We have the following:

Theorem 1.7. Suppose that $\Omega \subseteq I\!R^n$ is a bounded domain with $\partial\Omega$ locally Lipschitz. Let m and k denote integers with $k \geq 0$ and $m \geq 1$, and let $p \in [0, \infty)$.

(i) If $mp < n$, then $W^{m+k,p}(\Omega)$ embeds compactly in $W^{k,q}(\Omega)$ for $q \in [1, np/(n - mp))$. (If $q = np/(n - mp)$, the embedding is continuous but will not be compact.)

(ii) If $mp = n$, then $W^{m+k,p}(\Omega)$ embeds compactly in $W^{k,q}(\Omega)$ for $q \in [1, \infty)$.

(iii) If $mp > n$, then $W^{m+k,p}(\Omega)$ embeds compactly in $C^k(\overline{\Omega})$. If $(m - 1)p \leq n < mp$ and $\alpha \in (0, m - (n/p))$ then $W^{m+k,p}(\Omega)$ embeds compactly in $C^{k+\alpha}(\overline{\Omega})$.

Remarks: This theorem is essentially Theorem 6.2 of Adams (1975); it is known as the Rellich-Kondrachov Theorem. In particular, it implies that $W^{2,p}(\Omega)$ embeds compactly in $C^{1+\alpha}(\overline{\Omega})$ for $p > n$ and $\alpha \in (0, 1-(n/p))$ by taking $m = k = 1$. Note that $C^k(\overline{\Omega})$ embeds in $W^{k,q}(\Omega)$ for any $q \geq 1$, so in case (iii) $W^{m+k,p}(\Omega)$ also embeds compactly in $W^{k,q}(\Omega)$ for $q \in [1, \infty)$. There are a number of possible refinements and generalizations of the embedding theorem. In particular, parts (i) and (ii) of Theorem 1.7 remain valid if we merely require $\partial\Omega$ to satisfy a *cone condition*. The boundary $\partial\Omega$ of a bounded domain $\Omega \subseteq \mathbb{R}^n$ satisfies a *cone condition* if there is a fixed cone in $\mathbb{R}^n$ which can be moved by rigid motions (rotations and translations) to a position where its vertex is at $x_o \in \partial\Omega$ and the cone itself lies inside Ω, where $x_0 \in \partial\Omega$ is an arbitrary point on $\partial\Omega$. The point of using this type of condition instead of requiring $\partial\Omega$ to be $C^{2+\alpha}$ is that we can treat domains with corners, etc.

The way that we will typically use Theorem 1.7 is to formulate the solution operator for L as a compact operator from $C(\overline{\Omega})$ into $C(\overline{\Omega})$, or from $C^{1+\alpha}(\overline{\Omega})$ into $C^{1+\alpha}(\overline{\Omega})$.

Theorem 1.8. Suppose L has the form (1.107), with (1.108) satisfied, with all the coefficients of L belonging to $C^{\alpha}(\overline{\Omega})$ and $c \leq 0$. Suppose that $\Omega \subseteq \mathbb{R}^n$ is bounded and that $\partial\Omega$ is of class $C^{2+\alpha}$. The solution operator L^{-1} for the problem $Lu = f$ in Ω, $u = 0$ on $\partial\Omega$ can be extended to $C(\overline{\Omega})$ from $C^{\alpha}(\overline{\Omega})$ (or restricted to $C(\overline{\Omega})$ from $L^p(\Omega)$) to define a compact operator from $C(\overline{\Omega})$ into $C(\overline{\Omega})$; similarly, L^{-1} can be restricted from $C^{\alpha}(\overline{\Omega})$ to $C^{1+\alpha}(\overline{\Omega})$.

Proof: Suppose $f \in C(\overline{\Omega})$. Then $f \in L^p(\Omega)$ for any p. Choose $p > n$. Then by Theorem 1.5, there is a solution operator L^{-1} for L such that $L^{-1} : L^p(\Omega) \to W^{2,p}(\Omega)$ is continuous, which is equivalent to the existence of a constant C such that $||L^{-1}f||_{2,p} \leq C||f||_{0,p}$ for all $f \in L^p(\Omega)$. Since $||f||_{0,p} \leq C_1||f||_0$ for some constant C_1, where $||\cdot||_0$ denotes the norm of $C(\overline{\Omega})$, L^{-1} restricted to $C(\overline{\Omega})$ maps $C(\overline{\Omega})$ continuously into $W^{2,p}(\Omega)$. However, the embedding E of $W^{2,p}(\Omega)$ into $C(\overline{\Omega})$ is compact, so the operator EL^{-1} is continuous and maps bounded subsets of $C(\overline{\Omega})$ into subsets of $C(\overline{\Omega})$ with compact closure. Thus EL^{-1} defines a solution operator which is compact as an operator from $C(\overline{\Omega})$ to $C(\overline{\Omega})$. The case of $C^{1+\alpha}(\overline{\Omega})$ is similar.

Remark: Analogous results hold under the boundary conditions of (1.122). In our treatment of the equilibria of reaction-diffusion models the compactness of solution operators will play an important (if technical) role in the analysis. An important feature of compact operators is that they have eigenvalues. We have the following.

Lemma 1.9. A compact linear operator A from Banach space into itself has a countable set of eigenvalues. The eigenvalues have no limit point other than zero. The multiplicity of each nonzero eigenvalue is finite; that is, the dimension of the nullspace of $A - \sigma I$ is finite for each eigenvalue σ.

Remarks: Lemma 1.9 is essentially Theorem 5.5 of Gilbarg and Trudinger (1977). It is a standard result in functional analysis. So far we have generally considered Banach spaces of real valued functions, but if we consider the analogous spaces of functions which may be complex valued and allow multiplication by complex numbers then Theorem 1.8 still applies. Thus, even if the operator A is originally defined only as a mapping from a space X of real valued functions into itself, A can be extended to the space $Z : \{x+iy : x, y \in X\}$. If viewed as an operator on Z, A may have complex eigenvalues. Such is precisely analogous to the case of matrices with real entries, which may also have complex eigenvalues. If a linear operator is compact it must be continuous (another basic fact from functional analysis), and thus bounded in the sense that $||A|| = \sup\{||Ax||/||x|| : x \in X, x \neq 0\} < \infty$. If σ is an eigenvalue of A there must be $x \in X$, $x \neq 0$, such that $Ax = \sigma x$, so $|\sigma| = ||Ax||/||x|| \leq ||A|| < \infty$. Thus, the set of eigenvalues of a compact operator is bounded. If L is an elliptic operator such that L^{-1} exists as a compact operator on a space X, then for each nonzero eigenvalue μ of L^{-1} we have $L^{-1}x = \mu x$ for some $x \in X$, $x \neq 0$, so that $Lx = (1/\mu)x$. Thus, the elliptic operator L will have eigenvalues corresponding to nonzero eigenvalues of EL^{-1}. If $L^{-1} : X \to Y$ is combined with an embedding $E : Y \to X$ so that EL^{-1} is compact then it turns out that it is still generally true that L will have eigenvalues corresponding to nonzero eigenvalues of EL^{-1}, but some care must be taken in the rigorous mathematical analysis because even if $EL^{-1} : X \to X$ the original operator L may only make sense on the subspace $Y \subseteq X$. This is not usually a problem because if $x \in X$ is an eigenvector for EL^{-1} corresponding to an eigenvalue $\mu \neq 0$ then $x = (1/\mu)EL^{-1}x = (1/\mu)L^{-1}x \in Y$ so Lx makes sense and $(1/\mu)x = Lx$.

In some of the applications we study we encounter the problem of solving equations of the form $Lu - \sigma u = f$ where σ is an eigenvalue of L. Since σ is an eigenvalue, the operator $L - \sigma I$ is not invertible, so we cannot expect to solve $Lu - \sigma u = f$ for arbitrary choices of f. On the other hand, the nullspace of $L - \sigma I$ is typically finite dimensional, so there will be many functions f for which $Lu - \sigma u = f$ has a solution. To address this problem we will need to use the notion of adjoint operators. Recall that if M is an $n \times n$ matrix of real numbers and $\vec{v}, \vec{w} \in I\!R^n$ we have $M\vec{v} \cdot \vec{w} = \vec{v} \cdot M^T\vec{w}$ where M^T is the transpose of M. In general, if H is a Hilbert space (that is, a Banach space whose norm $|| \cdot ||$ arises from the inner product $< \cdot, \cdot >$ as $||u|| = \sqrt{< u, u >}$) and $A : H \to H$ is a continuous (i.e. bounded) linear operator, then the adjoint of A is the operator A^* such that $< Au, v >=< u, A^*v >$ for any $u, v \in H$. In the case of differential operators there is an additional complication, which is that they generally cannot be formulated as continuous linear operators from a given Hilbert or Banach space into itself. For example, if we define the operator L as a differential operator of the form (1.107) acting on $C^{2+\alpha}(\overline{\Omega})$ then L does not map $C^{2+\alpha}(\overline{\Omega})$ into itself but rather into $C^{\alpha}(\overline{\Omega})$. If we attempt to define L on $C^{\alpha}(\overline{\Omega})$ we find that not all Hölder continuous functions have Hölder continuous partial derivatives, so we cannot define L as a map of all of $C^{\alpha}(\overline{\Omega})$ into $C^{\alpha}(\overline{\Omega})$, either. The problem of correctly formulating the adjoint operators of differential operators in a rigorous way as maps between Hilbert or Banach spaces can be resolved (see Kato (1966)), but we do not discuss it further here. In what follows we will sometimes need to use facts about adjoint operators in a technical way, but for many purposes it will suffice to consider the "formal" adjoints operators corresponding to differential operators. Roughly speaking, the formal adjoint of a differential operator is the representation that the actual adjoint operator relative to the Hilbert space $L^2(\Omega)$ would have on its domain of definition. We now introduce the notion of formal adjoint operators and discuss the relationship between solvability of $Lu - \sigma u = f$ when σ is an eigenvalue of L and the eigenfunctions of the formal adjoint operator. To do so we must also interpret the Sobolev space $W^{1,2}(\Omega)$ as a Hilbert space and formulate the problem $Lu = f$ in $W^{1,2}(\Omega)$.

The space $L^2(\Omega)$ is a Hilbert space with respect to the inner product

$$< u, v >_{0,2}= \int_{\Omega} uv dx \tag{1.131}$$

in the case where we require functions in $L^2(\Omega)$ to be real valued and use only real scalars, or as

$$< u, v >_{0,2} = \int_\Omega u\overline{v}dx$$

where $\overline{v}$ denotes the complex conjugate of v in the case where functions in $L^2(\Omega)$ may be complex valued. The space $W^{1,2}(\Omega)$ is also a Hilbert space with respect to the inner product shown in (1.131) (or in the corresponding complex case, with v and ∇v replaced by $\overline{v}$ and $\nabla\overline{v}$) where ∇ is understood in the weak sense. The inner product in $W^{1,2}(\Omega)$ is sometimes denoted $< , >_{1,2}$. All Sobolev spaces built upon $L^2(\Omega)$, i.e. $W^{2,2}(\Omega)$, $W^{3,2}(\Omega)$, etc., can be viewed as Hilbert spaces in an analogous way. To compute the adjoint for an elliptic operator and to formulate an appropriate notion of a weak solution in $W^{1,2}(\Omega)$ we rewrite $< v, Lu >_{0,2}$ via the divergence theorem. For those purposes it is convenient to write the elliptic operator in the form

$$\begin{aligned} Lu = &\sum_{i,j=1}^{n} \frac{\partial}{\partial x_i}\left(a_{ij}(x)\frac{\partial u}{\partial x_j}\right) + \sum_{i=1}^{n}\frac{\partial}{\partial x_i}(B_i(x)u) \\ &+\sum_{i=1}^{n} b_i(x)\frac{\partial u}{\partial x_i} + c(x)u. \end{aligned} \tag{1.132}$$

We still assume $a_{ij} = a_{ji}$ and that (1.108) holds. If the coefficients a_{ij} and B_i are differentiable this form is equivalent to an operator of the form (1.107):

$$Lu = \sum_{i,j=1}^{n} a_{ij}\frac{\partial^2 u}{\partial x_i \partial x_j} + \sum_{i=1}^{n}\left(b_i + B_i + \sum_{j=1}^{n}\frac{\partial a_{ij}}{\partial x_j}\right)\frac{\partial u}{\partial x_i} + \left(c + \sum_{i=1}^{n}\frac{\partial B_i}{\partial x_i}\right)u. \tag{1.133}$$

Assuming for the moment that u and v are smooth, we have

$$\begin{aligned} &\frac{\partial}{\partial x_i}\left(v\left[\sum_{j=1}^{n} a_{ij}\frac{\partial u}{\partial x_j} + B_i u\right]\right) = \\ &\qquad v\frac{\partial}{\partial x_i}\left(\sum_{j=1}^{n} a_{ij}\frac{\partial u}{\partial x_j} + B_i u\right) + \frac{\partial v}{\partial x_i}\left(\sum_{j=1}^{n} a_{ij}\frac{\partial u}{\partial x_j} + B_i u\right) \end{aligned}$$

so that by the divergence theorem

$$\begin{aligned} &\int_\Omega v\sum_{i=1}^{n}\frac{\partial}{\partial x_i}\left(\sum_{j=1}^{n} a_{ij}\frac{\partial u}{\partial x_j} + B_i u\right)dx \\ &= \int_{\partial\Omega} v\sum_{i=1}^{n}\left[\left(\sum_{j=1}^{n} a_{ij}\frac{\partial u}{\partial x_j}\right)\cdot n_i + B_i n_i u\right]dS \\ &\quad - \int_\Omega\left[\sum_{i,j=1}^{n} a_{ij}\frac{\partial u}{\partial x_j}\frac{\partial v}{\partial x_i} + \sum_{i=1}^{n} B_i\frac{\partial v}{\partial x_i}u\right]dx \end{aligned} \tag{1.134}$$

where $\vec{n} = (n_1, \ldots, n_n)$ is the outward unit normal to $\partial\Omega$. If $(a_{ij}(x)) = d(x)I$ as in most of the problems we consider, then (1.134) becomes

$$\int_\Omega v\nabla \cdot \left[d(x)\nabla u + \vec{B}u\right]dx = \int_{\partial\Omega} v\left[d(x)\frac{\partial u}{\partial \vec{n}} + (\vec{B}\cdot\vec{n})u\right]dS \\ - \int_\Omega [d(x)\nabla u \cdot \nabla v + (\vec{B}\cdot\nabla v)u]dx. \tag{1.135}$$

If we multiply Lu times v, integrate over Ω, and apply (1.134) to the first two terms we obtain the relation

$$\int_\Omega vLudx = \int_\Omega \left(-\sum_{i,j=1}^n a_{ij}\frac{\partial u}{\partial x_j}\frac{\partial v}{\partial x_i} - \sum_{i=1}^n B_i\frac{\partial v}{\partial x_i}u + \sum_{i=1}^n vb_i\frac{\partial u}{\partial x_i} + cuv\right)dx \\ + \int_{\partial\Omega} v\sum_{i=1}^n\left[\left(\sum_{j=1}^n a_{ij}\frac{\partial u}{\partial x_j}\right)n_i + B_in_iu\right]dS. \tag{1.136}$$

We can now define weak solutions of $Lu = f$ in $W^{1,2}(\Omega)$. We will need to use the space $W_0^{1,2}(\Omega)$. (Recall that $W_0^{1,2}(\Omega)$ is the closure in $W^{1,2}(\Omega)$ of the set $C_0^1(\Omega)$ of continuously differentiable functions on Ω which are zero except on some compact subset of Ω. Thus, functions in $W_0^{1,2}(\Omega)$ are zero on $\partial\Omega$ in a weak sense.) A function $u \in W^{1,2}(\Omega)$ is a weak solution of the problem (1.120), i.e. $Lu = f \in L^2(\Omega)$ and $u = g$ on $\partial\Omega$ in the sense that $g \in W^{1,2}(\Omega)$ and $u - g \in W_0^{1,2}(\Omega)$, provided

$$\int_\Omega\left[-\sum_{i,j=1}^n a_{ij}\frac{\partial u}{\partial x_j}\frac{\partial v}{\partial x_i} - \sum_{i=1}^n B_i\frac{\partial v}{\partial x_i}u + \sum_{i=1}^n b_i\frac{\partial u}{\partial x_i}v + cuv\right]dx \\ = \int_\Omega vfdx \tag{1.137}$$

holds for all $v \in C_0^1(\Omega)$. The absence of a boundary term in (1.137) reflects the fact that functions in $C_0^1(\Omega)$ are zero in some neighborhood of $\partial\Omega$. Since $C_0^1(\Omega)$ is dense in $W_0^{1,2}(\Omega)$, the formula (1.137) extends to $v \in W_0^{1,2}(\Omega)$. It turns out that if $c + \sum_{i=1}^n \partial B_i/\partial x_i$ (the coefficient of the undifferentiated term in L as in (1.133)) is nonpositive in the weak sense then the problem $Lu = f$ in Ω, $u = g$ on $\partial\Omega$, will have a unique weak solution in $W^{1,2}(\Omega)$ with no additional conditions on $\partial\Omega$, but before stating a result to that effect we will give a weak formulation of another boundary value problem and define the formal adjoint L^* for L. We need to define L^* to state some of the results on the existence of solutions to $Lu = f$ and related problems.

The other boundary condition which is natural for an operator of the form L is

$$\sum_{i=1}^n\left(\sum_{j=1}^n a_{ij}\frac{\partial u}{\partial x_j}\right)n_i + (\vec{B}\cdot\vec{n})u + \gamma(x)u = g \quad \text{on} \quad \partial\Omega. \tag{1.138}$$

This condition may seem complicated but in the case $Lu = \nabla \cdot [d(x)\nabla u - \vec{V}(x)u]$ (which is most relevant to our models) the condition simply states that the flux $[-d(x)\nabla u + \vec{V}u]\cdot\vec{n}$ out of Ω across $\partial\Omega$ is given by a term γu proportional to the density together with an external source term g. Thus, if $\gamma = g = 0$ the boundary condition is a no-flux boundary condition while if $\gamma > 0$ but $g = 0$ the boundary condition says that the flux out of Ω across $\partial\Omega$ is proportional to the local density. Thus, (1.138) is a boundary condition which is relevant for our modeling purposes. The weak formulation for $Lu = f$ under the boundary condition (1.138) is

$$\int_{-\Omega}\left[-\sum_{i,j=1}^{n} a_{ij}\frac{\partial u}{\partial x_j}\frac{\partial v}{\partial x_i} - \sum_{i=1}^{n} B_i\frac{\partial v}{\partial x_i}u + \sum b_i\frac{\partial u}{\partial x_i}v + cuv\right]dx$$
$$= \int_{\Omega} vf\,dx + \int_{\partial\Omega} v[g - \gamma u]dS \tag{1.139}$$

for any $v \in W^{1,2}(\Omega)$. (For the boundary condition (1.138) we require $\partial\Omega$ to be piecewise C^1 so that $\vec{n}$ is well defined almost everywhere on $\partial\Omega$. It is also possible to formulate boundary value problems where u is specified on part of $\partial\Omega$ and (1.138) holds on the remainder of $\partial\Omega$; see Gilbarg and Trudinger (1977).)

It is fairly easy to verify the form L^* should have, but it is less obvious what boundary conditions on v should be associated with L^* to ensure that $\int_\Omega uL^*v\,dx = \int vLu\,dx$ under a given set of boundary conditions on u. If we take

$$L^*v = \sum_{i,j=1}^{n}\frac{\partial}{\partial x_i}\left(a_{ij}\frac{\partial v}{\partial x_j}\right) - \sum_{i=1}^{n}\frac{\partial}{\partial x_i}(b_i v)$$
$$- \sum_{i=1}^{n} B_i\frac{\partial v}{\partial x_i} + cv \tag{1.140}$$

we can calculate as in the derivation of (1.136) to obtain

$$\int_\Omega uL^*v\,dx = \int_\Omega\left(-\sum_{i,j=1}^{n} a_{ij}\frac{\partial u}{\partial x_i}\frac{\partial v}{\partial x_j} + \sum_{i=1}^{n} b_i\frac{\partial u}{\partial x_i}v - \sum_{i=1}^{n} uB_i\frac{\partial v}{\partial x_i} + cuv\right)dx$$
$$+ \int_{\partial\Omega} u\sum_{i=1}^{n}\left[\left(\sum_{j=1}^{n} a_{ij}\frac{\partial v}{\partial x_i}\right)n_i - b_i n_i v\right]dS. \tag{1.141}$$

Since $a_{ij} = a_{ji}$, the integrals over Ω on the right sides of (1.141) and (1.136) are the same. Thus we shall have $\int_\Omega uL^*v\,dx = \int_\Omega vLu\,dx$ if we can impose boundary conditions on v that correspond to those on u in such a way that the boundary integrals in (1.136) and (1.141) are the same. The simplest case is when Lu is augmented with the boundary condition $u = 0$. In that case we would simply restrict our attention to the subspace $W_0^{1,2}(\Omega)$ of $W^{1,2}(\Omega)$. Since functions in $W_0^{1,2}(\Omega)$ are zero on $\partial\Omega$ as far as integral formulas such as (1.136) and (1.141) are concerned, the boundary integrals drop out so that $\int_\Omega vLudx = \int_\Omega uL^*vdx$ if L^* has the form (1.140). For boundary conditions of the form (1.138) (but in the homogeneous case where $g = 0$) the boundary integrals in (1.136) and (1.141) both reduce to $-\int_{\partial\Omega} \gamma uvdS$ if we impose the boundary condition

$$\sum_{i=1}^{n}\left[\left(\sum_{j=1}^{n} a_{ij}\frac{\partial v}{\partial x_j}\right)n_i\right] - \sum_{i=1}^{n} b_i n_i v + \gamma(x)v = 0. \tag{1.142}$$

In the case $Lu = \nabla\cdot[d(x)\nabla u - \vec{V}(x)u]$ with boundary conditions $[d(x)\nabla u - \vec{V}(x)u]\cdot\vec{n} + \gamma(x)u = 0$, the adjoint operator L^* will have boundary conditions $d(x)\partial v/\partial\vec{n} + \gamma(x)v = 0$ on $\partial\Omega$. In particular, no-flux boundary conditions on L lead to Neumann conditions on L^*. We can now state a result on solvability which embodies the eigenvalues of L. An eigenvalue of L is a number σ such that $L\sigma = \sigma\phi$ for some nonzero ϕ satisfying the boundary conditions of L. We will state the result in the case of homogeneous Dirichlet boundary conditions and then remark on how it should be modified for other boundary conditions.

Theorem 1.10. Suppose that $\Omega \subseteq I\!R^n$ is a bounded domain, that L has the form (1.132) with all coefficients bounded and measurable, and that (1.108) holds. The operator L has a countable discrete set $\Sigma \subseteq I\!R$ of real eigenvalues such that if $\sigma \notin \Sigma$, then the problems

$$Lu = \sigma u + f \quad \text{in} \quad \Omega, \quad u = 0 \quad \text{on} \quad \partial\Omega \tag{1.143}$$

and

$$L^*v = \sigma v + f \quad \text{in} \quad \Omega, \quad v = 0 \quad \text{on} \quad \partial\Omega \tag{1.144}$$

have unique solutions in $W^{1,2}(\Omega)$ for any $f \in L^2(\Omega)$. If $\sigma \in \Sigma$ then the operators $L - \sigma I$ and $L^* - \sigma I$ with homogeneous Dirichlet boundary conditions have finite dimensional nullspaces which contain nonzero elements. If $\sigma \in \Sigma$ the problem (1.143) is solvable if and only if

$$\int_\Omega f\phi^* dx = 0 \tag{1.145}$$

for all $\phi^* \in W_0^{1,2}(\Omega)$ with $L^*\phi^* = \sigma\phi^*$.

Remarks: Theorem 1.10 is a version of the Fredholm alternative. A similar version is given as Theorem 8.6 of Gilbarg and Trudinger (1977), but the notation used there is different. Note that Theorem 1.10 does not exclude the possibility of complex eigenvalues. In fact, operators satisfying the theorem may indeed have complex eigenvalues. Usually the condition of formal self-adjointness, i.e. $L = L^*$ in (1.136) and (1.141) and in their boundary conditions, implies that the eigenvalues are real.

As in the case of Theorems 1.2, 1.3, 1.5 and 1.6, if the coefficient of the undifferentiated term in Lu is nonpositive, then $\Sigma \subseteq (-\infty, 0)$ and in particular L is invertible. For operators of the form (1.132) that coefficient is $c + \sum_{i=1}^{n} \partial B_i/\partial x_i$ (see (1.133)). If c or $\partial B_i/\partial x_i$ are not continuous on $\overline{\Omega}$ we may need to interpret nonpositivity in the weak sense; i.e. $\int_\Omega \left[cw - \sum_{i=1}^{n} B_i(\partial w/\partial x_i) \right] dx \leq 0$ for all $w \in C_0^1(\Omega)$ with $w \geq 0$.

If we want to determine the solvability of $Lu = \sigma u + f$ in Ω, $u = g$ on $\partial\Omega$, we would write $\tilde{u} = u - g$ and apply the Fredholm alternative to the problem $L\tilde{u} = \sigma\tilde{u} + f - Lg$ (formulated in the appropriate weak sense). A version of the Fredholm alternative including nonhomogeneous boundary conditions is given as part of Theorem 8.6 of Gilbarg and Trudinger (1977).

The homogeneous Dirichlet conditions on L and L^* may be replaced by homogeneous boundary conditions (1.138) for L and (1.142) for L^* and Theorem 1.10 will remain valid. In that case we would require $\partial\Omega$ to be piecewise C^1, and we would need to impose some additional conditions to conclude that $\Sigma \subseteq (-\infty, 0)$. If $c + \sum_{i=1}^{n} \partial B_i/\partial x_i \leq 0$ and either this last inequality is strict on an open subset of Ω or $\gamma(x) > 0$ on a relatively open subset of $\partial\Omega$ then we may conclude $\Sigma \subseteq (-\infty, 0)$.

The solutions whose existence is asserted in Theorem 1.10 may only be solutions in the weak sense if f and the coefficients of L are not continuous or if $\partial\Omega$ is not smooth. However, if L, f, and $\partial\Omega$ are smooth enough to apply the theory described in Theorems 1.2, 1.3, 1.5, and 1.6, then weak solutions are also classical solutions. Thus, if everything is smooth, the problem $Lu = \sigma u + f$ has a solution $u \in C^{2+\alpha}(\overline{\Omega})$ if and only if it has a solution in $W^{1,2}(\Omega)$. To see why this is so, note that if $u \in W^{1,2}(\Omega)$, then $u \in L^p(\Omega)$ for all $p < \infty$ if $n = 2$ or for $p < 2n/(n-2)$ if $n \geq 3$. (If $n = 1$, $u \in W^{1,2}([a,b])$ implies that $u \in C^{1/2}([a,b])$ by the calculation in (1.116).) If $u \in L^{p_0}(\Omega)$ then Theorem 1.5 (or 1.6) implies $u \in W^{2,p_0}(\Omega)$. Then by Theorem 1.7 we may conclude either $u \in L^{p_1}(\Omega)$ for some $p_1 > p_0$ or $u \in C^\alpha(\overline{\Omega})$ depending on n. If we can only conclude that $u \in L^{p_1}(\Omega)$ we still have $u \in W^{2,p_1}(\Omega)$ by Theorem 1.5 (or 1.6) so we have $u \in L^{p_2}(\Omega)$ for some $p_2 > p_1$ or $u \in C^\alpha(\overline{\Omega})$ by Theorem 1.7, again depending on n. If we repeat this argument enough times we eventually will get $u \in W^{2,p_k}(\Omega)$ where p_k is large enough that $W^{2,p_k}(\Omega)$ embeds in $C^\alpha(\overline{\Omega})$. Then by Theorem 1.1 (or 1.2) we have $u \in C^{2+\alpha}(\overline{\Omega})$.

Even if the coefficients of L are merely bounded and measurable and Ω merely satisfies a cone condition, the solution operator L^{-1} is a continuous map of $L^2(\Omega)$ into $W^{1,2}(\Omega)$, and $W^{1,2}(\Omega)$ embeds compactly in $L^2(\Omega)$ by Theorem 1.7, so we can view the solution operator as a compact operator on the Hilbert space $L^2(\Omega)$. Generally if L is formally self-adjoint we will have $< L^{-1}u, v > = < u, L^{-1}v >$ where $<\ ,\ >$ is the inner product in $L^2(\Omega)$. In other words, L^{-1} is a symmetric compact operator on $L^2(\Omega)$. Such operators have many nice properties which we will sometimes exploit in our treatment of eigenvalues in Chapter 2.

For linear equations such as those defining eigenvalues for elliptic operators, the spaces $W^{1,2}(\Omega)$ and $L^2(\Omega)$ are adequate for most types of analysis. For nonlinear equations of the form $Lu = f(x,u)$ we generally want to work in either the Hölder spaces $C^\alpha(\overline{\Omega})$ or in $W^{2,p}(\Omega)$ for p large enough that $W^{2,p}(\Omega)$ embeds in $C^\alpha(\overline{\Omega})$. The reason is that nonlinear functions of $u \in L^2(\Omega)$ are not necessarily in $L^2(\Omega)$. (As an example consider $u(x) = x^{-1/3}$ on $(0,1)$. We have $u \in L^2(\Omega)$ but $u^2 = x^{-2/3} \notin L^2(\Omega)$.) However, if $u \in C^\alpha(\overline{\Omega})$ then $f(x,u) \in C^\alpha(\overline{\Omega})$ if f is C^α in x and Lipschitz in u, and it follows that $f(x,u) \in L^p(\Omega)$ for all $p \in [1, \infty]$.

1.6.5 Reaction-Diffusion Models as Dynamical Systems

The models we want to study will usually have the form

$$\begin{aligned} \frac{\partial u_i}{\partial t} &= L_i u_i + f_i(x, u_1, \ldots, u_m) \text{ in } && \Omega \times (0, \infty) \\ B_i u_i &= 0 && \text{on } \partial\Omega \times (0, \infty) \\ u_i(x, 0) &= w_i(x) && \text{on } \Omega, \end{aligned} \tag{1.146}$$

$i = 1, \ldots, m$ where for each i, L_i is elliptic and has the form (1.107), the operator B_i defines boundary conditions of the forms shown in (1.120) or (1.122), and L and Ω satisfy the hypotheses of Theorem 1.3. In some cases we will allow L_i and/or f_i to depend on t as well as x. Often we have $L_i u = \nabla \cdot [d_i(x)\nabla u - \vec{b}_i(x)u]$ or even simply $L_i u = d_i \nabla^2 u = d_i \Delta u$. Systems such as (1.146) are called parabolic in the terminology of partial differential equations. There are a number of ways to formulate the existence theory for systems such as (1.146). A classical approach based on *a priori* estimates analogous to those described in Theorems 1.1–1.3 is given by Friedman (1964). Treatments based on the theory of analytic semigroups of operators are given by Friedman (1976), Pazy (1983), and Henry (1981). A rather general version of existence theory has been developed by Amann (1988, 1989, 1990, and the references therein). For our purposes the key issue will be to verify that the system (1.146) generates a semiflow on an appropriate state space. Results to that effect are derived by Mora (1983). Here we just give a sketch of how such results are obtained, omitting the technical details, and state some of the most relevant results. A similar discussion is given in Cantrell et al. (1993a). In the context of systems such as (1.146) there is an important distinction between local and global solutions. A local solution is one that exists for some time interval $[0, T)$ where T is finite. A global solution exists for all $t > 0$. Generally, the way that a local solution can fail to be a global solution is by becoming infinite in a finite time. This occurs, for example, in the ordinary differential equation $du/dt = u^2$ if $u(0) > 0$. For the models we treat densities will generally remain bounded or at most grow exponentially, so the solutions will be global.

Before stating the results we will need about the system (1.146) and related models, we will briefly sketch the ideas behind the results. The usual setting for an abstract treatment of (1.146) is a situation in which we have a pair of Banach spaces X and Y with $X \subseteq Y$ and a linear operator $A : X \to Y$ with certain properties, including the continuity of A and of $(A - \lambda I)^{-1}$ for complex λ satisfying $Re\lambda \leq \Lambda_0$ for some Λ_0. For models such as (1.146) we would normally choose function spaces such as $Y = [L^p(\Omega)]^m$ and $X = \{u \in [W^{2,p}(\Omega)]^m : B_i u_i = 0 \text{ on } \partial\Omega \text{ for } i = 1, \ldots, m\}$, but allow the functions in X and Y to be complex. By subtracting a constant multiple cu_i of u_i from $L_i u_i$ and adding it to $f_i(x, \vec{u})$ we can rearrange (1.146) (if it is necessary) so that each of the operators L_i satisfies the hypotheses of Theorem 1.5 or 1.6. If we let A denote the $m \times m$ matrix of operators with the operators $-L_i$ on the diagonal and off diagonal entries all zero, it follows from Theorems 1.5 and/or 1.6 that A^{-1} exists as a bounded (i.e. continuous) operator from Y into X. Furthermore, the *a priori* estimates underlying Theorem 1.5 and 1.6, i.e. the inequalities $||u||_{2,p} \leq C(||u||_{0,p} + ||L_i y||_{0,p})$, imply that for $\lambda \in \mathbb{C}$ such that $|\lambda| \geq R_0$ for some fixed R_0 and for $-(\pi/2) - \delta < \arg\lambda < (\pi/2) + \delta$ for some $\delta > 0$, we have the estimate

$$||(A + \lambda I)^{-1} u||_Y \leq C||u||_Y / |\lambda|, \tag{1.147}$$

for some constant C independent of u (see Friedman (1976)). It follows from (1.147) that $-A$ generates an analytic semigroup of operators on Y, which is usually denoted e^{-tA}. (If we had a single equation with $L = \nabla^2 = \Delta$ we might write the semigroup as $e^{t\Delta}$.) By a semigroup of operators we mean a family $S(t)$ of operators defined for $t \geq 0$ and depending continuously on t such that for each $t \geq 0$, $S(t)$ is a continuous (i.e. bounded) linear operator from Y to Y, $S(0) = I$, and for $s, t \geq 0$, $S(s)S(t) = S(s+t)$. The operator $-A$ is called the infinitesimal generator of e^{-tA}. (The defining property of the infinitesimal generator B of a semigroup $S(t)$ is that $\lim_{h \to 0+} ([S(h)u - u]/h) = Bu$ for all u in the domain of B, where the limit is taken in the metric induced by the norm on Y.) Results asserting that elliptic operators generate analytic semigroups are discussed in detail by Friedman (1976) and Pazy (1983); see also Henry (1981). The key properties of e^{-tA} are that e^{-tA} defines a semigroup of operators which are continuous (i.e. bounded) on Y, and depend analytically on t for $t \in \mathbb{C}$ with $|\arg t| < \delta$, and for any such t the operators $d(e^{-tA})/dt$ and Ae^{-tA} are also continuous on Y with $d(e^{-tA}u)/dt = Ae^{-etA}u$ for any $u \in Y$. Furthermore, for $t \in \mathbb{C}$ with $|\arg t| < \delta$, we have

$$||Ae^{-tA}u||_Y \leq (C/|t|)||u||_Y \tag{1.148}$$

for $u \in Y$, for some constant C independent of u. (Again, see Friedman (1976), Pazy (1983), or Henry (1981).)

For the nonlinear case (1.146) it is natural to express the problem in an abstract form

$$\frac{du}{dt} = -Au + F(u), \quad u(0) = u_0 \in Y, \tag{1.149}$$

(where $F(u)$ is the matrix with terms $f_i(x, \vec{u})$ on the diagonal and with zeros everywhere else) and express the solution as

$$u(t) = e^{-tA}u_0 + \int_0^t e^{-(t-s)A}F(u(s))ds. \tag{1.150}$$

The problem with (1.150) is that it is not clear that solutions to (1.150) actually make sense in the context of (1.149). The difficulty is that if we apply A to both sides of (1.150) we obtain

$$Au = Ae^{-tA}u_0 + \int_0^t Ae^{-(t-s)A}F(u(s))ds, \tag{1.151}$$

and all we can say about the size of the term in the integral is that $||Ae^{-(t-s)A}F(u(s))|| \leq (c/|t-s|)||F(u(s))||_Y$. That creates a problem because $\int_0^t (1/|t-s|)ds$ does not converge. (If $u_0 \in X$ then $Ae^{-tA}u_0 = e^{-tA}(Au_0)$ is continuous as $t \to 0$, but the integral term remains a problem.) Fortunately, operators which generate analytic semigroups can be analyzed by methods taken from complex variable theory and the theory of Laplace transforms, and thus shown to have fractional powers which have some nice properties. Specifically, we have $||A^\gamma e^{-tA}u||_Y \leq (C/|t|^\gamma)||u||_Y$. Thus, if we apply A^γ to (1.150) we obtain

$$A^\gamma u = A^\gamma e^{-tA}u_0 + \int_0^t A^\gamma e^{-(t-s)A}F(u(s))ds, \tag{1.152}$$

and for $0 < \gamma < 1$ the integral term is well behaved if $F(u)$ is bounded in Y, because $\int_0^t [1/(t-s)^\gamma]ds$ converges if $0 < \gamma < 1$. Furthermore, we can define Banach spaces $X^\gamma \subseteq Y$ by taking X^γ to be the completion of X in Y with respect to the norm $||u||_\gamma = ||A^\gamma u||_Y$, and since A^{-1} is a compact operator on Y it follows that the embedding of X^{γ_1} into X^{γ_2} is compact if $\gamma_1 > \gamma_2$. Thus, we can use (1.150) to define a semiflow on X^γ for $\gamma \in (0, 1)$ provided $F(u(s))$ is smooth by using standard fixed point theorems, just as we might for a system of ordinary differential equations. Also, if we know that orbits which are bounded in X^{γ_2} are also bounded in X^{γ_1} for some $\gamma_1 > \gamma_2$ then we can conclude that bounded orbits in X^{γ_2} are precompact because of the compact embedding of X^{γ_1} into X^{γ_2}. (See Friedman (1976), Pazy (1983), and Henry (1981).) In fact, we can arrange (1.146) so that the operator e^{-tA} satisfies an estimate of the form $||e^{-tA}u||_y \leq Ce^{-\beta t}||u||_Y$ and $||A^\gamma e^{-tA}u||_Y \leq C(e^{-\beta t}/|t|^\gamma)||u||_Y$, so that if $F(u)$ is bounded in Y we have u bounded in X^γ via (1.151). In practice we will usually try to obtain bounds on $\vec{u}(t)$ in $[C(\overline{\Omega})]^m$, but such bounds imply bounds for $f_i(x, \vec{u})$ in $L^p(\Omega)$ for any $p \in [1, \infty]$ if the functions f_i are smooth, so it is generally sufficient to obtain bounds on $\vec{u}$ in $[C(\overline{\Omega})]^m$ to conclude that orbits for the system defined in (1.150) are bounded in X^γ for any $\gamma \in (0, 1)$ and hence are precompact in each space X^γ. Again, the details of this approach to parabolic equations and systems are discussed by Friedman (1976), Pazy (1983), and Henry (1981).

There are two remaining problems with this approach. First, it is not obvious what it means to have $\vec{u} \in X^\gamma$ in terms of the smoothness of $\vec{u}$; secondly, it is not obvious that solutions to (1.150) are classical solutions to (1.146). The first problem is relatively easy to resolve. If we start with $Y = [L^p(\Omega)]^m$ and $X \subseteq [W^{2,p}(\Omega)]^m$ then X^γ embeds in $[C^{1+\alpha}(\overline{\Omega})]^m$ if $0 < \alpha < 2\gamma - (n/p) - 1$. This fact and more precise results about the embeddings of X^γ into Hölder and Sobolev spaces follow from (Pazy, 1983, Theorem 4.3). To be sure that solutions to (1.150) are classical solutions to (1.146) we would need to assume that the functions $f_i(x, u)$ are Hölder continuous in x and at least Lipschitz in u, but in that case there are *à priori* estimates for parabolic equations analogous to the estimates underlying Theorems 1.1, 1.2, 1.5 and 1.6 for elliptic equations which show that solutions as defined by (1.150) are indeed classical solutions. Such estimates are discussed by Friedman (1964). The issue of smoothness is also discussed in Friedman (1976) and Pazy (1983). Thus, while the semiflows defined by (1.150) may need to be defined in X^γ, $C^{1+\alpha}(\overline{\Omega})$, or other subspaces of X^γ such as $C(\overline{\Omega})$ or $C^1(\overline{\Omega})$, the orbits $\vec{u}(t)$ will represent classical solutions of (1.146) if the functions f_i are smooth. All of the theoretical considerations discussed above can be extended in a natural way to the case where the operators L_i and the functions f_i depend on t, as long as all coefficients depending on t are uniformly Hölder continuous; see Friedman (1964, 1976). In that case the models will no longer generate semi-dynamical systems, because of the time dependence in the coefficients, but as we shall see in later chapters, there are ways to circumvent that problem in some cases.

We may now state some results on systems such as (1.146).

Theorem 1.11. Suppose that the domain Ω, operators L_i, and boundary conditions B_i in (1.146) satisfy the hypotheses of Theorem 1.3 for $i = 1, \ldots, m$. Suppose that for each $i = 1, \ldots, m$ the function $f_i(x, \vec{u})$ is measurable and is bounded uniformly in x if $\vec{u}$ is restricted to any bounded subset of $\mathbb{R}^m$, and $f_i(x, \vec{u})$ is Lipschitz continuous in $\vec{u}$, uniformly for $x \in \Omega$ and for $\vec{u}$ restricted to any bounded subset of $\mathbb{R}^m$. Let $Y = [L^p(\Omega)]^m$ for some $p > n$ and let $X = \{\vec{u} \in [W^{2,p}(\Omega)]^m : B_i u_i = 0 \text{ on } \partial\Omega \text{ for } i = 1, \ldots, m\}$. Let A be the matrix of operators with $-L_i$ as the ith diagonal element and with all off-diagonal terms equal to zero. Let $X^\gamma \subseteq Y$ denote the space generated by A^γ for $\gamma \in (0, 1)$, so that $||u||_{X_\gamma} = ||A^\gamma u||_Y$. Then there exists a $\gamma_0 \in (0, 1)$ such that for $\gamma \in (\gamma_0, 1)$ the system (1.146) generates a local semiflow on X^γ. Bounded orbits in X^γ are precompact.

Remarks: This sort of result is discussed by Henry (1981, Ch. 3). The key points in the proof are as follows: the *a priori* estimates of Agmon et al. (1959), as used in Theorem 1.5 and 1.6, imply that $-A$ generates an analytic semigroup on $[W^{2,p}(\Omega)]^m$. The general theory of analytic semigroups allows us to formulate (1.146) as (1.150), and to define X^γ. Since $p > n$ we can choose γ_0 so that for $\gamma \in (\gamma_0, 1)$ we have $0 < 2\gamma - (n/p) - 1$ and hence X^γ embeds in $[C^{1+\alpha}(\overline{\Omega})]^m$ for some $\alpha > 0$ by Theorem 4.3 of Pazy (1983). The hypotheses on f_i imply that if $\vec{u} \in [C^{1+\alpha}(\overline{\Omega})]^m$ then $f_i \in L^p(\Omega)$ for all $p \geq 1$, so in particular $(f_1, \ldots, f_m) \in Y$. Thus, (1.152) makes sense and shows that (1.150) defines a semi-dynamical system on X^γ. Finally, since $(\lambda I - A)^{-1}$ exists as a linear compact operator on $[L^p(\Omega)]^n$ for λ real and sufficiently large (by Theorem 1.5 or 1.6), the embeddings $X^{\gamma_1} \hookrightarrow X^{\gamma_2}$ for $\gamma_> \gamma_2 > \gamma_0$ are compact, and the compactness of bounded orbits follows.

We may sometimes prefer to work in the more concrete spaces $C(\overline{\Omega})$ or $C^1(\overline{\Omega})$. We have the following.

Theorem 1.12. (Mora, 1983) Suppose that the hypotheses of Theorem 1.11 hold and that in addition the functions f_i are all C^2 in both x and $\vec{u}$, i.e. $f_i \in C^2(\overline{\Omega} \times I\!R^m)$ for each i. Then (1.146) generates a local semiflow on $[C(\overline{\Omega})]^m$ under Neumann boundary conditions. Under Dirichlet boundary conditions (1.146) generates a local semiflow on the subspaces of $[C(\overline{\Omega})]^m$ and of $[C^1(\overline{\Omega})]^m$ consisting of functions that are zero on $\partial\Omega$. Under Neumann or Robin conditions (1.146) generates a local semiflow on the subspace of $[C^1(\overline{\Omega})]^m$ consisting of functions satisfying the boundary conditions. In each case bounded orbits are precompact.

We usually work with semilinear models of the form (1.146), but we also consider logistic models with nonlinear diffusion of the form

$$\begin{aligned} \frac{\partial u}{\partial t} &= \nabla \cdot d(x, u)\nabla u + f(x, u) \text{ in } \quad \Omega \\ u &= 0 \qquad\qquad\qquad\qquad\quad \text{on} \quad \partial\Omega. \end{aligned} \tag{1.153}$$

It turns out that these models also generate local semiflows on fractional order Sobolev spaces $W^{s,p}(\Omega)$ where s is not necessarily an integer (see Adams, 1975), which have the property of embedding in $C^{1+\alpha}(\overline{\Omega})$, provided that $d(x, u)$ and $f(x, u)$ are C^2 in both x and u and that $d_0 \leq d(x, u)$ for some $d_0 > 0$. Such a result is derived by Amann (1986). Much more general results on quasilinear systems are given in more recent work by Amann (1988, 1989, 1990). Some other sources for discussions of reaction-diffusion models as dynamical systems are Hirsch (1988) and Smith (1995).

The results we have discussed so far on reaction-diffusion models as dynamical systems are local in time. In general, some additional information is required to conclude that a given orbit exists globally in time. For a semi-dynamical system on a Banach space Z in which bounded orbits are precompact it suffices to show that for each $T > 0$ there is a $B(T) < \infty$ so that $\sup\{||\vec{u}(t)||_Z : 0 \leq t < T\} \leq B(T)$ to conclude that the orbit $u(t)$ exists for all $t > 0$. In the models we consider it will suffice to show that $\sup\{|\vec{u}(x, t)| : x \in \overline{\Omega}, 0 \leq t < T\} < B(T)$ because the bound on $|\vec{u}|$ implies a bound in $[L^p(\Omega)]^m$ for any $p \in [1, \infty]$ and then (1.152) or analogous formulas imply boundedness in spaces X^γ which either play the role of state spaces for the semi-dynamical system or which embed in $[C^{1+\alpha}(\overline{\Omega})]^m$, which in turn embeds in state spaces constructed from $[C(\overline{\Omega})]^m$ or $[C^1(\overline{\Omega})]^m$. For most of the models we consider it is relatively easy to get bounds on $|\vec{u}|$ via arguments based on the maximum principle or related comparison theorems, but in a few cases some work is required. Issues related to global existence are discussed by Henry (1981). See also Amann (1985, 1989), Fitzgibbon et al. (1992), Hollis and Morgan (1991), or Morgan (1990).

1.6.6 Classical Regularity Theory for Parabolic Equations

In some cases we may want to treat solutions to systems such as (1.146) from the viewpoint of classical partial differential equations. To that end we now discuss some results analogous to Theorems 1.1 and 1.2 for parabolic equations. Recall the definitions (1.117)–(1.119) for $C^\alpha(\overline{\Omega})$ and $C^{k+\alpha}(\overline{\Omega})$. For functions on $\overline{\Omega} \times [0, T]$ define

$$[f]_{\alpha,\alpha/2} = \sup_{\substack{(x,t),(y,s)\in\overline{\Omega}\times[0,T] \\ (x,t)\neq(y,s)}} \left(\frac{|f(x,t) - f(y,s)|}{|x-y|^\alpha + |t-s|^{\alpha/2}} \right). \tag{1.154}$$

Functions with $[\ ]_{\alpha,\alpha/2}$ finite form a Banach space $C^{\alpha,\alpha/2}(\overline{\Omega} \times [0, T])$ under the norm

$$||u||_{\alpha,\alpha/2} = \sup_{\overline{\Omega}\times[0,T]} |u(x,t)| + [u]_{\alpha,\alpha/2}. \tag{1.155}$$

Let ∂_x^β denote the derivative with respect to x corresponding to the multi-index $\beta = (\beta_1, \ldots, \beta_n)$, as in (1.119), and let ∂_t denote the derivative with respect to t. Define $C^{2+\alpha,1+\alpha/2}(\overline{\Omega} \times [0, T])$ to be the space of functions on $\overline{\Omega} \times [0, T]$ whose derivatives up to order two in x and order one in t are Hölder continuous, with norm

$$\begin{aligned} ||u||_{2+\alpha,1+\alpha/2} = \sup\nolimits_{(x,t)\in\overline{\Omega}\times[0,T]} |u(x,t)| &+ \sum_{|\beta|\leq 2} \sup_{(x,t)\in\overline{\Omega}\times[0,1]} |\partial_x^\beta u(x,t)| \\ + \sup_{(x,t)\in\overline{\Omega}\times[0,T]} |\partial_t u(x,t)| &+ \sum_{|\beta|=2} [\partial_x^\beta u(x,t)]_{\alpha,\alpha/2} + [\partial_t u(x,t)]_{\alpha,\alpha/2}. \end{aligned} \tag{1.156}$$

The spaces $C^{\alpha,\alpha/2}(\overline{\Omega}\times[0, T])$ and $C^{2+\alpha,1+\alpha/2}(\overline{\Omega}\times[0, T])$ are Banach spaces; see Friedman (1964). We also will need to use the space $C^{1+\delta,\delta/2}(\overline{\Omega} \times [0, T])$ whose norm is given by

$$||u||_{1+\delta,\delta/2} = ||u||_{\delta,\delta/2} + \sum_{|\beta|=1} ||\partial_x^\beta u||_{\delta,\delta/2}. \tag{1.157}$$

We have:

Theorem 1.13. (Friedman, 1964) Suppose that $\partial\Omega$ is a bounded domain with $\partial\Omega$ of class $C^{2+\alpha}$ and that L is an elliptic operator of the form (1.107), satisfying (1.108), but with coefficients a_{ij}, b_i, and c which may depend on t as well as x. Suppose that a_{ij}, b_i, and $c \in C^{\alpha,\alpha/2}(\overline{\Omega}\times[0, T])$. If $f(x,t) \in C^{\alpha,\alpha/2}(\overline{\Omega}\times[0, T])$, $g(x,t) \in C^{2+\alpha,1+\alpha/2}(\overline{\Omega}\times[0, T])$, and $Lg = f$ for $t = 0$ and $x \in \partial\Omega$, then the problem

$$\begin{aligned} \frac{\partial u}{\partial t} - Lu &= f(x,t) \text{ in } \quad \Omega \times (0, T] \\ u(x,t) = g(x,t) &\quad \text{on} \quad \partial\Omega \times [0, T] \cup (\Omega \times \{0\}) \end{aligned} \tag{1.158}$$

has a unique solution $u \in C^{2+\alpha,1+\alpha/2}(\overline{\Omega} \times [0,T])$ which satisfies

$$||u||_{2+\alpha,1+\alpha/2} \leq C(||f||_{\alpha,\alpha/2} + ||g||_{2+\alpha,1+\alpha/2}). \tag{1.159}$$

where C is a constant independent of f and g.

Remarks: Theorem 1.13 result is a version of Theorems 6 and 7 of Friedman (1964, Ch. 3.) See also Ladyzhenskaya et al. (1968). If the compatibility condition $Lg = f$ for $t = 0$ on $\partial\Omega$ is omitted the problem will still have a unique solution but it may not be smooth for $t = 0$. Ladyzhenskaya et al. (1968) also treat the case of boundary conditions analogous to (1.122). In that case we have the following result.

Theorem 1.14. Suppose that Ω, L, and f satisfy the hypotheses of Theorem 1.13. Suppose that $g(x) \in C^{2+\alpha}(\overline{\Omega})$ and $\gamma(x)$, $\beta(x)$, and $h(x) \in C^{1+\alpha}(\partial\Omega)$, with $\gamma \geq 0$ and $\beta > 0$ on $\partial\Omega$. Suppose also that $\gamma(x)g(x) + \beta(x)\partial g/\partial\vec{n} = h(x)$ on $\partial\Omega$. Then the problem

$$\begin{aligned} &\frac{\partial u}{\partial t} - Lu = f(x,t) && \text{in} \quad \Omega \times (0,T] \\ &u(x,0) = g(x) && \text{in} \quad \overline{\Omega} \\ &\gamma(x)u(x,t) + \beta(x)\frac{\partial u}{\partial \vec{n}}(x,t) = h(x) && \text{on} \quad \partial\Omega \times [0,T] \end{aligned} \tag{1.160}$$

has a unique solution in $C^{2+\alpha,1+\alpha/2}(\overline{\Omega} \times [0,T])$ which satisfies

$$||u||_{2+\alpha,1+\alpha/2} \leq C(||f||_{\alpha,\alpha/2} + ||g||_{2+\alpha} + ||h||_{1+\alpha}) \tag{1.161}$$

where the norm on h is taken $C^{1+\alpha}(\partial\Omega)$ and the norm of g is taken in $C^{2+\alpha}(\overline{\Omega})$, and the constant C independent of f, g, and h.

For certain convergence arguments an additional *a priori* estimate called a $1+\delta$ estimate (Friedman, 1964, Theorem 4 of Ch. 7) is also needed. A version of this result is:

Lemma 1.15. (Friedman, 1964) Suppose that Ω and L satisfy the hypotheses of Theorem 1.13 with the additional condition that the coefficients a_{ij} are uniformly Lipschitz in $\overline{\Omega} \times [0,T]$. Suppose that $f(x,t)$ is continuous on $\overline{\Omega} \times [0,T]$ with $f(x,0) = 0$ on $\partial\Omega$. Then for any $\delta < 1$ there is a constant C independent of $f(x,t)$ such that any solution of

$$\begin{aligned} &\frac{\partial u}{\partial t} - Lu = f(x,t) \text{ in } && \Omega \times (0,T] \\ &u(x,t) = 0 && \text{on} \quad (\partial\Omega \times (0,T]) \cup (\overline{\Omega} \times \{0\}) \end{aligned} \tag{1.162}$$

satisfies

$$||u||_{1+\delta,\delta/2} \leq C \sup_{(x,t)\in\overline{\Omega}\times[0,T]} |f(x,t)|. \tag{1.163}$$

Remarks: The analogous result is valid if the boundary condition $u = 0$ on $\partial\Omega \times (0,T]$ is replaced with the boundary condition $\gamma(x)u + \beta(x)\partial u/\partial\vec{n} = 0$ on $\partial\Omega \times (0,T]$ with γ and β as in Theorem 1.14. See Friedman (1964, Ch. 7).

Theorem 1.13 and Lemma 1.15 and the corresponding results for Robin boundary conditions imply that the trajectories of the semi-dynamical systems whose existence is asserted in Theorem 1.11 actually correspond to classical solutions of (1.146) if $\partial\Omega$, f_i, and the coefficients of L_i and B_i are sufficiently smooth. (Theorem 1.13 and Lemma 1.15 are stated for a single parabolic equation but can be applied to (1.146) componentwise because in the system (1.146) there is no coupling between equations in the terms involving derivatives.)

1.6.7 Maximum Principles and Monotonicity

Methods based on comparisons between different solutions of a reaction-diffusion model or between solutions of different models play an important role in the mathematical theory of reaction-diffusion models. Comparisons between solutions are made possible by maximum principles, which typically place restrictions on the nature and location of maximum and minimum points of solutions to partial differential equations. Maximum principles for classical solutions are treated in detail by Protter and Weinberger (1967) and Walter (1970). Extensions to weak solutions are discussed by Gilbarg and Trudinger (1977). Maximum principles are the basis of theoretical approaches to reaction-diffusion models based on monotone iteration (Leung, 1989, Pao, 1992) or monotone semi-dynamical systems (Hirsch, 1988b; Hess, 1991; Smith, 1995).

A version of the maximum principle for elliptic equations is as follows (see Protter and Weinberger (1967)).

Theorem 1.16. Suppose that the operator L has the form (1.107) and satisfies (1.108), with $c(x) \leq 0$. Suppose that $\Omega \subseteq I\!R^n$ is a bounded domain and the coefficients of L are uniformly bounded on Ω.

(i) Suppose that $u \in C^2(\Omega)$ and $Lu \geq 0$ in Ω. If u attains a maximum $M \geq 0$ at any point in the interior of Ω then $u(x) \equiv M$ in Ω.

(ii) Suppose further that $u \in C^2(\Omega) \cap C(\overline{\Omega})$ and that each point on $\partial\Omega$ lies on the boundary of some ball contained in Ω. If $u(x) = M$ at some point $x_0 \in \partial\Omega$ for which $\partial u/\partial\vec{n}$ exists, then either $\partial u/\partial\vec{n} > 0$ at x_0 or $u(x) \equiv M$ in Ω.

Remark: The geometric condition on $\partial\Omega$ in (ii) will hold if $\partial\Omega$ is of class $C^{2+\alpha}$. If $u \in C^{2+\alpha}(\overline{\Omega})$ then $u \in C^2(\Omega) \cap C(\overline{\Omega})$. The corresponding results hold in the case of a minimum $M \leq 0$. In that case if $u(x) = M \leq 0$ at a point $x_0 \in \partial\Omega$ then either $\partial u/\partial\vec{n} < 0$ at x_0 or $u(x) \equiv M$ in Ω. In some cases the condition $c \leq 0$ can be weakened but counter-examples show that some restrictions on c are necessary; see Protter and Weinberger (1967).

The maximum principle has various implications. Suppose $u_1, u_2 \in C^2(\Omega) \cap C(\overline{\Omega})$ are solutions of $Lu = f(x)$ in Ω and $u_1 - u_2 = 0$ on $\partial\Omega$. If L satisfies the hypotheses of Theorem 1.16 then $u_1 - u_2$ cannot have a positive maximum M inside Ω, because then we would have $u_1 - u_2 \equiv M > 0$ in Ω so that $u_1 - u_2 \equiv M > 0$ on $\partial\Omega$. Hence $u_1 - u_2 \leq 0$ in Ω. Similarly, $u_1 - u_2$ cannot have a negative minimum inside Ω, so that $u_1 - u_2 \geq 0$ in Ω. It follows that $u_1 = u_2$, so that the solution to $Lu = f$ is unique. A similar analysis using part (ii) of Theorem 1.16 applies to the case of boundary conditions of the form $\gamma(x)u + \beta(x)\partial u/\partial\vec{n} = g$ on $\partial\Omega$ if $\gamma \geq 0$ and $\beta > 0$ on $\partial\Omega$. These observations explain the condition $c(x) \leq 0$ in Theorems 1.1 and 1.3. If $c(x) \leq 0$ then the maximum principle implies that the solution $u \equiv 0$ to the homogeneous problem $Lu = 0$ in Ω, $u = 0$ (or $\gamma u + \beta\partial u/\partial\vec{n} = 0$) on $\partial\Omega$ is unique, so in that case Theorems 1.1 and 1.2 follow from Theorem 1.3. If $-Lu \geq 0$ in Ω then since $Lu \leq 0$ in Ω the maximum principle implies that u cannot have a nonpositive minimum M inside Ω unless $u \equiv M$. If $u \geq 0$ on $\partial\Omega$

then we must have $u \geq 0$ on Ω, with $u > 0$ on Ω unless $u \equiv 0$. If $Lu_1 \leq Lu_2$ in Ω and $u_1 \geq u_2$ on $\partial\Omega$, then $-L(u_1 - u_2) \geq 0$ in Ω and $u_1 - u_2 \geq 0$ on $\partial\Omega$, so $u_1 \geq u_2$ on Ω. These applications of the maximum principle are representative but far from exhaustive. In some cases it is convenient to use the ideas behind Theorem 1.16 rather than the theorem itself. For example, suppose that u is an equilibrium of a diffusive logistic equation, so that $0 = \nabla \cdot d(x)\nabla u + r[1 - (u/K)]u$ in Ω, and that $u = 0$ on $\partial\Omega$. If u is continuous on $\overline{\Omega}$ then u must attain a maximum somewhere in $\overline{\Omega}$. If u has a maximum $M > K$ at some point $x^* \in \Omega$ then for $x = x^*$ we must have $\partial u/\partial x_i = 0$ and $\partial^2 u/\partial x_i^2 \leq 0$ for $i = 1, \ldots, n$; but then for $x = x^*$ we have $0 \leq -\nabla \cdot d(x)\nabla u = r[1 - (M/K)]M < 0$, which is a contradiction. Thus, since $u = 0 < K$ on $\partial\Omega$, we must have $u \leq K$ for $x \in \Omega$.

The maximum principle extends to parabolic equations. In fact, some results based on the maximum principle for parabolic equations are valid without the hypothesis $c \leq 0$ required for the elliptic case.

Theorem 1.17. Suppose that L has the form (1.107) but with coefficients that may depend on t. Suppose that (1.108) holds and $c(x,t) \leq 0$. Let $\Omega \subseteq I\!R$ be a bounded domain and suppose that the coefficients of L are uniformly bounded on $\Omega \times (0, T]$.

(i) Suppose that $u \in C^{2,1}(\Omega \times (0, T])$, that is, the first derivative of u with respect to t and the derivatives of u of order 2 or less in the space variables are continuous on $\Omega \times (0, T]$. Suppose that $\partial u/\partial t - Lu \leq 0$ in $\Omega \times (0, T]$. If u attains a maximum $M \geq 0$ at a point $(x_0, t_0) \in \Omega \times (0, T]$, then $u(x,t) \equiv M$ on $\Omega \times (0, t_0]$.

(ii) Suppose further that $u \in C^{2,1}(\Omega \times (0, T]) \cap C(\overline{\Omega} \times (0, T])$, and that each point of $\partial\Omega$ lies on the boundary of some ball lying inside Ω. If $u(x_0, t_0) = M$ at some point of $\partial\Omega \times (0, T]$ for which $\partial u/\partial \vec{n}$ exists, then either $\partial u/\partial \vec{n} > 0$ at (x_0, t_0) or $u \equiv M$ on $\overline{\Omega} \times (0, t_0]$.

Remarks: Clearly if $u \in C(\Omega \times [0, T])$ then the conclusion $u \equiv M$ extends to $\Omega \times [0, t_0]$ in (i) and (ii). Results of this type are discussed by Protter and Weinberger (1967); see also Friedman (1964) and Leung (1989). As in the elliptic case, if $u_t - Lu \geq 0$ then u cannot attain a minimum $M \leq 0$ at a point $(x_0, t_0) \in \Omega \times (0, T]$ unless $u \equiv M$ on $\Omega \times (0, t_0]$, and similarly if $u(x_0, t_0) = M$ at a point $(x_0, t_0) \in \partial\Omega \times (0, T]$ then either $\partial u/\partial \vec{n} < 0$ at (x_0, t_0) or $u(x,t) \equiv M$ in $\Omega \times (0, t_0]$.

Corollary 1.18. Suppose that L and $\Omega \times (0, T]$ satisfy the hypotheses of Theorem 1.17 except for the requirement $c(x,t) \leq 0$. Suppose that $\gamma(x)$ and $\beta(x)$ are bounded functions on $\partial\Omega$ with $\gamma(x) \geq 0$ and $\beta(x) > 0$. If $u(x,t) \in C^{2,1}(\Omega \times (0, T]) \cap C(\overline{\Omega} \times [0, T])$ with $u_t - Lu \geq 0$ on $\Omega \times (0, T]$ and $\gamma(x)u(x,t) + \beta(x)\partial u(x,t)/\partial \vec{n} \geq 0$ on $\partial\Omega \times (0, T]$ or $u \geq 0$ on $\partial\Omega \times (0, T]$, then either $u(x,t) > 0$ on $\Omega \times (0, T]$ or $u(x,t) \equiv 0$ on $\overline{\Omega} \times [0, t_0]$ for some $t_0 > 0$. If $u(x, 0) > 0$ for some $x \in \Omega$, or if there is a $t_1 > 0$ such that for each $t \in (0, t_1)$ either $u(x,t) > 0$ or $\gamma u(x,t) + \beta \partial u(x,t)/\partial \vec{n} > 0$ for some $x \in \partial\Omega$, then $u(x,t) > 0$ in $\Omega \times (0, T]$.

Remarks: To eliminate the requirement $c \leq 0$ introduce the new variable $w = e^{-kt}u$. Then w satisfies $w_t - (Lw - kw) = e^{-kt}(u_t - Lu) \geq 0$. If we choose k large enough then $c - k \leq 0$ so that Theorem 1.17 applies to w. If $w < 0$ then w must have a negative minimum M on $\overline{\Omega} \times (0, T]$, but then $w \equiv M < 0$ on $[0, t_0] \times \overline{\Omega}$ for some $t_0 > 0$, contradicting the hypothesis $w(x, 0) = u(x, 0) \geq 0$. Hence $w \geq 0$, and thus $u \geq 0$. If $w = 0$ at any point $(x,t) \in \Omega \times (0, T]$ then Theorem 1.17 implies $w \equiv 0$ on $\overline{\Omega} \times [0, t_0]$, which implies $u \equiv 0$

on $\overline{\Omega} \times [0, t_0]$. If $u(x, 0) > 0$ for some $x \in \Omega$ or either $u(x, t) > 0$ or $\gamma u + \beta \partial u/\partial \vec{n} > 0$ for some $x \in \partial\Omega$ for $t \in [0, t_1]$ then $u \not\equiv 0$ on $\overline{\Omega} \times [0, t_0]$ so we must have $w > 0$ and hence $u > 0$ on $\Omega \times (0, T]$.

An important feature of maximum principles is that they permit comparisons between solutions.

Theorem 1.19. Suppose that L and Ω satisfy the hypotheses of Theorem 1.17 with $c(x, t) \equiv 0$. Suppose that $f(x, t, u)$, $\partial f(x, t, u)/\partial u \in C(\overline{\Omega} \times [0, T] \times I\!R)$. If $\overline{u}, \underline{u} \in C^{2,1}(\overline{\Omega} \times (0, T]) \cap C(\overline{\Omega} \times [0, T]$ with

$$\frac{\partial \overline{u}}{\partial t} - L\overline{u} \geq f(x, t, \overline{u}) \quad \text{in} \quad \Omega \times (0, T], \tag{1.164}$$

$$\frac{\partial \underline{u}}{\partial t} - L\underline{u} \leq f(x, t, \underline{u}) \quad \text{in} \quad \Omega \times (0, T], \tag{1.165}$$

$\overline{u}(x, 0) \geq \underline{u}(x, 0)$ on Ω, and either $\overline{u}(x, t) \geq \underline{u}(x, t)$ or

$$\gamma(x)\overline{u} + \beta(x)\frac{\partial \overline{u}}{\partial \vec{n}} \geq \gamma(x)\underline{u} + \beta(x)\frac{\partial \underline{u}}{\partial \vec{n}} \quad (\gamma \geq 0, \beta > 0)$$

on $\partial\Omega \times (0, T]$, then either $\overline{u} \equiv \underline{u}$ or $\overline{u} > \underline{u}$ on $\Omega \times (0, T]$.

Remarks: Theorem 1.19 follows from Corollary 1.18 by setting $u = \overline{u} - \underline{u}$ so that $\partial u/\partial t - Lu - c(x, t)u \geq 0$, where $c(x, t) = [f(x, t, \overline{u}) - f(x, t, \underline{u})]/(\overline{u} - \underline{u})$ is bounded because $\partial f/\partial u$ is continuous. Results of this type are discussed by Protter and Weinberger (1967), Walter (1970), Fife (1979), Smoller (1982), Leung (1989) and Pao (1992). Comparison theorems such as Theorem 1.19 provide the technical basis for applying the theory of monotone dynamical systems to reaction-diffusion models, as in the work of Hirsch (1988), Smith (1995) and others. Similar results can be obtained for elliptic equations but in that case the condition $c(x) \leq 0$ cannot be completely eliminated, so some additional condition on f is needed. It suffices to assume that $\partial f/\partial u \leq 0$ since $f(x, \overline{u}) - f(x, \underline{u}) = [\partial f(x, \xi)/\partial u](\overline{u} - \underline{u})$ by the mean value theorem. (In general ξ depends on x.) Analogous results hold for systems of reaction-diffusion equations.

Theorem 1.20. Suppose that Ω and the operators L_i, $i = 1, \ldots, m$ satisfy the hypotheses of Theorem 1.17, with the coefficient c_i of the undifferentiated term in L_i equal to zero for each i. Suppose that for each i the functions $f_i(x, t, \vec{u})$ and $\partial f_i(x, t, \vec{u})/\partial u_j$, $j = 1, \ldots, m$, belong to $C(\overline{\Omega} \times [0, T] \times I\!R)$, and that

$$\partial f_i/\partial u_j \geq 0 \quad \text{for} \quad i \neq j. \tag{1.166}$$

If $\vec{w} = (w_1, \ldots, w_m)$ and $\vec{v} = (v_1, \ldots v_m)$ satisfy

$$\frac{\partial w_i}{\partial t} - L_i w_i \geq f_i(x, t, \vec{w}) \quad \text{in} \quad \Omega \times (0, T], \tag{1.167}$$

$$\frac{\partial v_i}{\partial t} - L_i v_i \leq f_i(x, t, \vec{v}) \quad \text{in} \quad \Omega \times (0, T], \tag{1.168}$$

with

$$w_i(x, 0) \geq v_i(x, 0) \quad \text{on} \quad \Omega,$$

and either

$$w_i \geq v_i \quad \text{on} \quad \partial\Omega \times (0, T]$$

or

$$\gamma_i(x)w_i + \beta_i(x)\frac{\partial w_i}{\partial \vec{n}} \geq \gamma_i(x)v_i + \beta_i(x)\frac{\partial v_i}{\partial \vec{n}} \;(\gamma_i \geq 0, \beta_i > 0) \quad \text{on} \quad \partial\Omega \times (0, T]$$

for $i = 1, \ldots, m$, then $w_i \geq v_i$ in $\Omega \times (0, T]$ for $i = 1, \ldots, m$.

Remarks: Condition (1.166) is sometimes called a Kamke condition or quasimonotone condition. Systems satisfying (1.166) are called cooperative. Condition (1.166) is used in the theory of systems of ordinary differential equations in results analogous to Theorem 1.20; see Hirsch (1982) or Smith (1995). In Theorem 1.20 it is possible to have $w_i > v_i$ for some components but $w_j \equiv v_j$ for others. Results analogous to Theorem 1.20 can be derived for elliptic systems but require some additional hypotheses on the functions f_i; see Protter and Weinberger (1967). Other references include Walter (1970), Fife (1979), Smoller (1982), and Smith (1995).

Certain systems which do not satisfy (1.166) still admit comparison principles. Those include many models for two competing species. Suppose that u_1 and u_2 satisfy

$$\partial u_i/\partial t - L_i u_i = f_i(x, t, u_1, u_2) \quad \text{for} \quad i = 1, 2, \tag{1.169}$$

with $\partial f_1/\partial u_2 \leq 0$ and $\partial f_2/\partial u_1 \leq 0$ and with the operators L_i as in Theorem 1.20. If we let $\tilde{u}_2 = k - u_2$ for some constant k then we have

$$\begin{aligned} \partial u_1/\partial t - L_1 u_1 &= \tilde{f}_1(x, t, u_1, \tilde{u}_2) = f_1(x, t, u_1, k - \tilde{u}_2) \\ \partial \tilde{u}_2/\partial t - L_2 \tilde{u}_2 &= \tilde{f}_2(x, t, u_1, \tilde{u}_2) = -f_2(x, t, u_1, k - \tilde{u}_2) \end{aligned} \tag{1.170}$$

so that $\partial \tilde{f}_1/\partial \tilde{u}_2 = -\partial f_1/\partial u_2 \geq 0$ and $\partial \tilde{f}_2/\partial u_1 = -\partial f_2/\partial u_1 \geq 0$. Thus the system (1.169) can be converted to a system (1.170) which satisfies Theorem 1.20. Thus, if (w_1, w_2) and (v_1, v_2) are solutions of (1.169) satisfying homogeneous boundary conditions and $w_1 \geq v_1$, $w_2 \leq v_2$ for $t = 0$ then $w_1 \geq v_1$ and $w_2 \leq v_2$ for all $t > 0$. Models for more than two competitors and predator-prey models usually cannot be converted into forms satisfying Theorem 1.20.

If the functions $\underline{u}, \overline{u}$ satisfy the hypotheses of Theorem 1.19, or if the vector valued functions $\vec{v}$, $\vec{w}$ satisfy the hypotheses of Theorem 1.20, we say that $\underline{u}$ and $\vec{v}$ are subsolutions (or lower solutions) and $\overline{u}$ and $\vec{w}$ are supersolutions (or upper solutions) for their respective models. Sub- and supersolutions can be used in various ways to show the existence of solutions to reaction-diffusion equations and systems and the elliptic equations and systems describing their equilibria. Some versions of the existence theory for systems do not require the condition (1.166). Existence theory via sub- and supersolutions is treated in detail by Leung (1989) and Pao (1992). In the case of equations or systems satisfying comparison principles such as Theorems 1.19 and 1.20, sub- and supersolutions can be used to construct iteration schemes consisting of sequences of linear problems whose solutions converge monotonically to solutions of the original nonlinear problem; see Leung (1989) and Pao (1992). If each of the functions f_i is either monotone increasing or monotone decreasing in u_j for $j \neq i$ then it may still be possible to construct monotone iteration schemes from sub- and supersolutions but such schemes may be more complicated. (Systems in which each function f_i is monotone increasing or monotone decreasing in u_j for $j \neq i$ can be embedded in larger systems for which (1.166) is satisfied; see for example Cosner (1997).) Finally, existence of solutions can be deduced from the existence of sub- and supersolutions without any monotonicity assumptions on f_i by methods based on the Schauder fixed point theorem. (Applications of the Schauder fixed point theorem typically require that the solution operators $(\partial/\partial t - L)^{-1}$ or L^{-1} be compact, but that follows from the regularity properties of elliptic and parabolic operators.) All of these ideas and more are discussed by Leung (1989) and Pao (1992). A simple result on existence via sub- and supersolutions is the following.

Proposition 1.21. Suppose that L, Ω, and f satisfy the hypotheses of Theorem 1.19. Suppose that L and Ω also satisfy the hypotheses of Theorem 1.13 and that $f(x,t,u)$ is Hölder continuous with exponent α with respect to x and exponent $\alpha/2$ with respect to t. Suppose that $h(x) \in C^{2+\alpha}(\overline{\Omega})$ with $Lh = f(x,0,0)$ on $\partial\Omega$, and that $\overline{u}$ and $\underline{u}$ satisfy (1.164), (1.165) respectively with

$$\underline{u}(x,0) \leq h(x) \leq \overline{u}(x,0) \quad \text{on} \quad \overline{\Omega} \times \{0\}$$

and

$$\underline{u}(x,t) \leq 0 \leq \overline{u}(x,t) \quad \text{on} \quad \partial\Omega \times [0,T].$$

Then the problem

$$\begin{aligned} &\frac{\partial u}{\partial t} - Lu = f(x,t,u) \text{ in } && \Omega \times (0,T] \\ &u = 0 && \partial\Omega \times (0,T] \\ &u(x,0) = h(x) && \text{on } \overline{\Omega} \times \{0\} \end{aligned} \tag{1.171}$$

has a solution $u^* \in C^{2+\alpha,1+\alpha/2}(\overline{\Omega} \times [0,T])$ with $\underline{u}(x,t) \leq u^*(x,t) \leq \overline{u}(x,t)$.

Remarks: If the compatibility condition $Lh = f(x,0,0)$ on $\partial\Omega$ is omitted then u^* still will exist but may not be smooth for $t = 0$. Analogous results hold for boundary conditions $\gamma u + \beta\partial u/\vec{n} = 0$. Theorem 1.19 implies that $\underline{u} \leq \overline{u}$ on $\overline{\Omega} \times [0,T]$ and that u^* is unique. Results similar to Proposition 1.21 hold for the corresponding elliptic problems $-Lu = f(x,u)$ in Ω, $u = g(x)$ on $\partial\Omega$ (or $\gamma(x)u + \beta(x)\partial u/\partial\vec{n} = g(x)$ on $\partial\Omega$) but in the elliptic case we must require $\overline{u}(x) \leq \overline{u}(x)$ as an additional hypothesis, and u^* is not necessarily unique.

Sketch of Proof: Choose a constant C large enough that $\partial f/\partial u + C \geq 0$ on $\overline{\Omega} \times [0,T]$ for $\underline{u} \leq u \leq \overline{u}$. To construct the iteration let $\underline{u}^0 = \underline{u}$, $\overline{u}^0 = \overline{u}$, and define $\underline{u}^k$ and $\overline{u}^k$ recursively as the solutions of

$$\begin{aligned} &\frac{\partial u}{\partial t} - Lu + Cu = f(x,t,\underline{u}^{k-1}) + C\underline{u}^{k-1} \text{ in } && \Omega \times (0,T] \\ &u = 0 && \text{on } \partial\Omega \times (0,T] \\ &u(x,0) = h(x) && \text{in } \overline{\Omega} \times \{0\} \end{aligned}$$

and

$$\begin{aligned}
&\frac{\partial u}{\partial t} - Lu + Cu = f(x,t,\overline{u}^{k-1}) + C\overline{u}^{k-1} && \text{in} \quad \Omega \times (0,T] \\
&u = 0 && \text{on} \quad \partial\Omega \times (0,T] \\
&u(x,0) = h(x) && \text{in} \quad \overline{\Omega} \times \{0\}
\end{aligned}$$

respectively. If we let $v^k = \underline{u}^k - \underline{u}^{k-1}$ then we have

$$\begin{aligned}
&\frac{\partial v^1}{\partial t} - Lv^1 + Cv^1 \geq 0 && \text{in} \quad \Omega \times (0,T] \\
&v^1 \geq 0 && \text{on} \quad (\partial\Omega \times (0,T]) \cup (\overline{\Omega} \times \{0\})
\end{aligned}$$

so $v^1 \geq 0$ in $\Omega \times (0,T]$ by Theorem 1.19. If $v^k \geq 0$ then $\underline{u}^k \geq \underline{u}^{k-1}$ so

$$\begin{aligned}
&\frac{\partial v^{k+1}}{\partial t} - Lv^{k+1} + Cv^{k+1} = f(x,t,\underline{u}^k) - f(x,t,\underline{u}^{k-1}) + C(\underline{u}^k - \underline{u}^{k-1}) \\
&\geq 0 \quad \text{in} \quad \Omega \times (0,T]
\end{aligned}$$

and $v^{k+1} = 0$ on $(\partial\Omega \times (0,T]) \cup (\overline{\Omega} \times \{0\})$ so $v^{k+1} \geq 0$ by Theorem 1.19. Hence by induction $v^k \geq 0$ for all $k \geq 1$, so that $\underline{u}^k \geq \underline{u}^{k-1}$. Similarly, since $\bar{u}^k$ is the solution of

$$\begin{aligned}
&\frac{\partial u}{\partial t} - Lu + Cu = f(x,t,\overline{u}^{k-1}) + C\overline{u}^{k-1} && \text{in} \quad \Omega \times (0,T] \\
&u = 0 && \text{on} \quad \partial\Omega \times (0,T] \\
&u(x,0) = h(x) && \text{on} \quad \overline{\Omega} \times \{0\},
\end{aligned}$$

arguments similar to those shown above imply via induction that $\overline{u}^{k+1} \leq \overline{u}^k$ and that for each k $\underline{u}^k \leq \overline{u}^k$. Since $\{\underline{u}^k\}$ is increasing and bounded above, $\{\underline{u}^k\}$ converges pointwise. The convergence of $\{\underline{u}^k\}$ in $C^{2+\alpha,1+\alpha/2}(\overline{\Omega} \times [0,T])$ follows via parabolic regularity, i.e. by Lemma 1.17 and Theorem 1.13. Hence $\{\underline{u}^k\}$ converges to a solution of (1.171). (The sequence $\{\overline{u}^k\}$ also converges to a solution of (1.171).)

Remarks: In the elliptic case we would require $\underline{u} \leq \overline{u}$ on Ω, and the inequalities (1.164) and (1.165) would be replaced by $-L\overline{u} \geq f(x,\overline{u})$ on Ω, $-L\underline{u} \leq f(x,\underline{u})$ on Ω. The iterations will still be monotone because even in the elliptic case the operator $-L + CI$ will satisfy a comparison principle if $C > 0$. Analogous results hold for systems satisfying (1.166), with sub- and supersolutions defined by (1.168) and (1.167) respectively. Again, the case of elliptic systems is quite similar to the parabolic case, except that we must require $w_i \geq v_i$ on Ω as a hypothesis and solutions to elliptic systems of the form $-Lu_i = f(x,\vec{u})$ are not necessarily unique, even under the condition (1.166). Another approach to obtaining solutions of elliptic problems via monotonicity can be developed by explicitly considering the solutions of elliptic problems as equilibria of parabolic models. Such an approach was introduced by Aronson and Weinberger (1975, 1978) and has been applied and extended by many others.

Theorem 1.22. Suppose that L has the form (1.107) and satisfies (1.108), and that L and Ω satisfy the hypotheses of Theorem 1.1, but without the restriction $c \leq 0$. Suppose that the coefficients a_{ij} of L are uniformly Lipschitz in $\overline{\Omega}$, and that $f(x,u)$ and $\partial f(x,u)/\partial u$ are Hölder continuous in x and continuous in u on $\overline{\Omega} \times I\!R$. Finally, suppose that $\overline{u}$ and $\underline{u}$ are super- and subsolutions to the problem

$$\begin{aligned} Lu + f(x,u) &= 0 \text{ in } \quad \Omega \\ u &= 0 \qquad\qquad \text{on} \quad \partial\Omega; \end{aligned} \tag{1.172}$$

that is, $-L\overline{u} \geq f(x,\overline{u})$ and $-L\underline{u} \leq f(x,\underline{u})$ on Ω with $\underline{u} \leq 0 \leq \overline{u}$ on $\partial\Omega$ and $\underline{u} \leq \overline{u}$ on Ω. Let v be the solution to

$$\begin{aligned} \frac{\partial v}{\partial t} &= Lv + f(x,v) \text{ on } \quad \Omega \times (0,\infty) \\ v &= 0 \qquad\qquad \text{on} \quad \partial\Omega \times (0,\infty) \\ v(x,0) &= \underline{u}(x). \end{aligned} \tag{1.173}$$

Then $v(x,t)$ is monotonically increasing with respect to t, and as $t \to \infty$ $v(x,t)$ converges to an equilibrium u^* of (1.173) which is the minimal equilibrium that satisfies $\underline{u}(x) \leq u^*(x) \leq \overline{u}(x)$. If w satisfies (1.173) but with $w(x,0) = \overline{u}(x)$ then $w(x)$ is monotonically decreasing with respect to t and converges to $u^{**}(x)$, the maximal equilibrium for (1.173) which satisfies $\underline{u}^*(x) \leq u^{**}(x) \leq \overline{u}(x)$.

Proof: If $\underline{u}(x)$ is a solution to (1.172) then $v(x,t) = \underline{u}(x)$ so we have $u^*(x) = \underline{u}(x)$. If $\underline{u}(x)$ is not a solution to (1.172) then we cannot have $v = \underline{u}$ we so must have $v(x,t) > \underline{u}(x) = v(x,0)$ for any $t > 0$ by Theorem 1.19. Thus, $v(x,\epsilon) > v(x,0)$ for any $\epsilon > 0$. However, $v_\epsilon(x,t) = v(x,t+\epsilon)$ satisfies (1.173) but with $v_\epsilon(x,0) = v(x,\epsilon) > v(x,0)$. Thus, $v_\epsilon(x,t) > v(x,t)$ for all $t > 0$ by Theorem 1.19, so $v(x,t+\epsilon) > v(x,t)$, i.e. $v(x,t)$ is increasing in t. (Since $v(x,0) \leq \overline{u}(x)$ it follows by Theorem 1.19 that $v(x,t)$ is bounded, so that $v(x,t)$ exists for all $t > 0$.) Since $v(x,t)$ is bounded above and increasing with t, we must have $v(x,t) \to u^*(x)$ for each $x \in \Omega$, where $u^*(x) \leq \overline{u}(x)$. It remains to verify that $u^*(x)$ is indeed a solution of (1.172). Let $u_n(x,t) = v(x,t+n-1)$. Then $u_n(x,1) = v(x,n)$. Also, $\partial u_n/\partial t = Lu_n + f(x,u_n)$ on $\Omega \times [0,1]$. Let $\rho_1(t)$ be a smooth function with $\rho_1(t) = 0$ for $t \leq 1/4$ and $\rho_1(t) = 1$ for $t \geq 1/2$. (Such functions are used in the construction of partitions of unity and mollifiers; see Friedman (1976) and Gilbarg and Trudinger (1977).) We have $\partial(\rho_1 u_n)/\partial t - L(\rho_1 u_n) = \rho_1 f(x,u_n) - u_n \partial\rho_1/\partial t$ in $\overline{\Omega} \times [0,1]$. Moreover, $\rho_1 f - u_n \partial\rho_1/\partial t$ is bounded uniformly in $C(\overline{\Omega} \times [0,1])$ since $\underline{u} \leq u_n \leq \overline{u}$ for all n and is zero for $t < 1/4$. So Lemma 1.15 applies to $\rho_1 u_n$, and thus the sequence $\{\rho_1 u_n\}$ is uniformly bounded in $C^{1+\delta,\delta/2}(\overline{\Omega} \times [0,1])$ for some $\delta > 0$. It follows that the sequence $\{u_n\}$ is uniformly bounded in $C^{1+\delta,\delta/2}(\overline{\Omega} \times [1/2,1])$. Let $\rho_2(t)$ be a smooth function with $\rho_2(t) = 0$ for $t \leq 1/2$ and $\rho_2(t) = 1$ for $t \geq 3/4$. Then $\partial(\rho_2 u_n)/\partial t - L(\rho_2 u_n) = \rho_2 f(x,u_n) - u_n \partial\rho_2/\partial t$ in $\overline{\Omega} \times [1/2,1]$ and $\rho_2 u_n = 0$ for $t = 1/2$. Since $\{u_n\}$ is uniformly bounded in $C^{1+\delta,\delta/2}(\overline{\Omega} \times [1/2,1])$, so is the expression $\rho_2 f(x,u_n) - u_n \partial\rho_2/\partial t$. Thus it follows from Theorem 1.13 that $\{\rho_2 u_n\}$ is uniformly bounded in $C^{2+\delta,1+\delta/2}(\overline{\Omega} \times [1/2,1])$ so that $\{u_n\}$ is uniformly bounded in $C^{2+\delta,1+\delta/2}(\overline{\Omega} \times [3/4,1])$. Hence, u_n, $\partial u_n/\partial t$, and the derivatives of u_n of order two or less in the space variables are uniformly bounded and equicontinuous on $\overline{\Omega} \times [3/4,1]$. The Arzela-Ascoli Theorem implies that $\{u_n\}$ must have a subsequence which converges uniformly along with its first derivative in t and derivatives of order up to two in the space variables on $\overline{\Omega} \times [3/4,1]$. Passing to the subsequence, we have $u_n \to \tilde{u}$ in

$\overline{\Omega} \times [3/4, 1]$ where $\tilde{u}$ satisfies $\partial\tilde{u}/\partial t - L\tilde{u} = f(x, \tilde{u})$ in $\overline{\Omega} \times [3/4, 1]$ because of the convergence of the derivatives of u_n and the fact that $\partial u_n/\partial t - Lu_n = f(x, u_n)$ for each n. However, $u_n(x, t) = v(x, t + n - 1) \to u^*(x)$ as $n \to \infty$ for any fixed $(x, t) \in \overline{\Omega} \times [3/4, 1]$ so we must have $\tilde{u} = u^*$. Thus, $u^*(x)$ satisfies the equation $\partial u/\partial t - Lu = f(x, u)$, but since $\partial u^*/\partial t = 0$ we have that u^* satisfies (1.172). If $z(x)$ is any equilibrium for (1.173) with $z(x) \geq \underline{u}(x)$ then $z(x) \geq v(x, t)$ by Theorem 1.19, but since $v(x, t) \to u^*(x)$ as $t \to \infty$ we must have $z(x) \geq u^*(x)$. The proof that $w(x, t) \to u^{**}(x)$ is analogous.

Remarks: Theorem 1.22 can be extended to systems which have the appropriate monotonicity properties. Such systems include those that satisfy the cooperativity condition (Kamke condition) (1.166), and those which can be converted into systems satisfying (1.166). An important class of examples are models for two competing species such as (1.169). Suppose that $\partial f_1/\partial u_2 < 0$ and $\partial f_2/\partial u_1 < 0$ in (1.169) and that f_1, f_2 do not depend on t. It follows from (1.170) and the related discussion that if (v_1, v_2) and (w_1, w_2) satisfy

$$\frac{\partial v_1}{\partial t} - L_1 v_1 \leq f_1(x, v_1, v_2)$$
$$\frac{\partial v_2}{\partial t} - L_2 v_2 \geq f_2(x, v_1, v_2)$$
$$\frac{\partial w_1}{\partial t} - L_1 w_1 \geq f_1(x, w_1, w_2)$$
$$\frac{\partial w_2}{\partial t} - L_2 w_2 \leq f_2(x, w_1, w_2)$$

in $\Omega \times (0, T]$ with $v_1 \leq w_1$, $v_2 \geq w_2$ on $(\partial\Omega \times (0, T]) \cup (\overline{\Omega} \times \{0\})$ then $v_1 \leq w_1$ and $v_2 \geq w_2$ on $\overline{\Omega} \times [0, T]$. Suppose that $(\underline{u}_1, \overline{u}_2)$ and $(\overline{u}_1, \underline{u}_2)$ depend only on x, with $\underline{u}_i \leq \overline{u}_i$ for $i = 1, 2$, and

$$-L_1\underline{u}_1 \leq f_1(x, \underline{u}_1, \overline{u}_2)$$
$$-L_2\overline{u}_2 \geq f_2(x, \underline{u}_1, \overline{u}_2)$$
$$-L_1\overline{u}_1 \geq f_1(x, \overline{u}_1, \underline{u}_2)$$
$$-L_2\underline{u}_2 \leq f_2(x, \overline{u}_1, \underline{u}_2) \text{ on } \Omega$$

with $\underline{u}_i = 0$ on $\partial\Omega$, $\overline{u}_i \geq 0$ on $\partial\Omega$. Let (v_1, v_2) and (w_1, w_2) be solutions to (1.169) which are zero on $\partial\Omega \times (0, \infty)$ with $v_1(x, 0) = \underline{u}_1(x)$, $v_2(x, 0) = \overline{u}_2(x)$, and $w_1(x, 0) = \overline{u}_1(x)$, $w_2(x, 0) = \underline{u}_2(x)$. Then as $t \to \infty$, we have $v_1(x, t) \uparrow v_1^*(x)$, $v_2(x, t) \downarrow v_2^*(x)$; $w_1(x, t) \downarrow w_1^*(x)$, and $w_2(x, t) \uparrow w_2^*(x)$ where (v_1^*, v_2^*) and (w_1^*, w_2^*) are equilibria for (1.169) with $v_1^* \leq w_1^*$ and $v_2^* \geq w_2^*$. The same result would hold under Robin or Neumann boundary conditions. We return to this point later in our discussion of competition models.

Sometimes we consider models where the reaction term is not continuous with respect to x, so that we must work with weak solutions in $W^{2,p}(\Omega)$ or even $W^{1,2}(\Omega)$. There are versions of the maximum principle that apply in that situation. The following is a version of Theorems 8.1 and 8.19 of Gilbarg and Trudinger (1977):

Theorem 1.23. Suppose that L has the form (1.132) and that L and Ω satisfy the hypotheses of Theorem 1.10. Suppose further that the coefficient $c+\sum_{i=1}^{n} \partial B_i/\partial x_i$ of the undifferentiated term in L is nonpositive in the sense that

$$\int_{\Omega}\left[cw - \sum_{i=1}^{n} B_i \frac{\partial w}{\partial x_i}\right] dx \leq 0$$

for all $w \in C_0^1(\Omega)$ with $w \geq 0$. If $u \in W^{1,2}(\Omega)$ and $Lu \geq 0$ on Ω (in the weak sense) then $\sup_{\Omega} u \leq \sup_{\partial\Omega} u^+$, where u^+ denotes the positive part of u on $\partial\Omega$, i.e. $u^+ = (1/2)(|u| + u)$. If there is an open ball B with $\overline{B} \subseteq \Omega$ such that $\sup_{B} u = \sup_{\Omega} u \geq 0$ then u is constant on Ω.

Remarks: Analogous results can be obtained in the parabolic case. Evidently this has been done by Liang et al. (1983), but we have not read that paper. This type of maximum principle can be used to extend the method of super-and subsolutions to models where the reaction terms are not smooth. The following result can be obtained by the methods used by Berestycki and Lions (1980).

Theorem 1.24. Suppose that L and Ω satisfy the hypotheses of Theorem 1.5, and that $f(x, u)$ is measurable in x for all u, continuous in u for almost all $x \in \Omega$, and bounded on bounded subsets of $\Omega \times I\!R$. Suppose that for any finite interval $I \subseteq I\!R$ there is a constant C such that $f(x, u) + Cu$ is increasing in u on $\Omega \times I$. If $\underline{u}, \overline{u} \in W^{2,p}(\Omega)$ for some $p > n$ and $\underline{u}, \overline{u}$ satisfy

$$-L\underline{u} \leq f(x, \underline{u}) \quad \text{on} \quad \Omega, \quad \underline{u} \leq 0 \quad \text{on} \quad \partial\Omega$$

and

$$-L\overline{u} \geq f(x, \overline{u}) \quad \text{on} \quad \Omega, \quad \overline{u} \geq 0 \quad \text{on} \quad \partial\Omega$$

in the weak sense, then there exists a solution $u^* \in W^{2,p}(\Omega)$ to the problem $Lu = f(x, u)$ in $\Omega, u = 0$ on $\partial\Omega$, with $\underline{u} \leq u^* \leq \overline{u}$.

Remarks: This result and various related results can be derived by using the methods of Berestycki and Lions (1980). Several examples are treated in that paper. An advantage of working in Sobolev space is that neither f nor $\underline{u}$ and $\overline{u}$ need to be as smooth as would be necessary if we were working in spaces of Hölder continuous functions. In particular, this fact makes it possible to use sub- and supersolutions which are constructed by piecing together sub- and supersolutions defined on subdomains of Ω. We have the following result.

Theorem 1.25. (Berestycki and Lions, 1980) Suppose that L and Ω satisfy the hypotheses of Theorem 1.5 and that Ω_1 is a subdomain of Ω with $\overline{\Omega_1} \subseteq \Omega$ and $\partial\Omega_1$ of class $C^{2+\alpha}$. Let $\Omega_2 = \Omega\backslash\Omega_1$ and let $\vec{n} = (\nu_1, \ldots, \nu_n)$ be the outward normal to Ω_1. If $\underline{u}_i \in W^{2,2}(\Omega_i)$ and $f_i(x) \in L^1(\Omega_i)$ satisfy $-L\underline{u}_i \leq f_i$ on Ω_i for $i = 1, 2$, with $\underline{u}_1 = \underline{u}_2$ and $\sum_{i,j=1}^{n} a_{ij}\nu_j(\partial\underline{u}_1/\partial x_i) \leq \sum_{i,j=1}^{n} a_{ij}\nu_j(\partial\underline{u}_2/\partial x_i)$ on $\partial\Omega_1$ then the function u defined by $\underline{u}(x) \equiv \underline{u}_i(x)$ for $x \in \Omega_i, i = 1, 2$, belongs to $W^{1,2}(\Omega)$ and satisfies $-L\underline{u} \leq f$ in Ω, where $f(x) = f_i(x)$ on Ω_i. Similarly, if $\overline{u}_i \in W^{2,2}(\Omega_i)$ and $-L\overline{u}_i \geq f_i$ on Ω_i for $i = 1, 2$ with $\overline{u}_1 = \overline{u}_2$ and $\sum_{i,j=1}^{n} a_{ij}\nu_j(\partial\overline{u}_1/\partial x_i) \geq \sum_{i,j=1}^{n} a_{ij}\nu_j(\partial\overline{u}_2/\partial x_i)$ on $\partial\Omega_1$ then the function $\overline{u}(x) \equiv \overline{u}_i(x)$ on Ω_i, $i = 1, 2$, belongs to $W^{1,2}(\Omega)$ and satisfies $-L\overline{u} \geq f$ on Ω.

Remarks: The inequalities are to be interpreted in the weak sense. The point of the result is that it provides a way to link together local sub- and supersolutions to form global ones. In solving problems like $-Lu = f(x, u)$ we would choose $f_i(x) = f(x, \underline{u}_i)$ when defining $\underline{u}$ and $f_i(x) = f(x, \overline{u}_i)$ when defining $\overline{u}$.

There are many possible variations and extensions of the maximum principle, comparison theorems, and the method of sub- and supersolutions, but those we have discussed are adequate for the purposes of this book. There are also a number of alternative ways to obtain results on positivity of solutions, comparison principles, etc. An elliptic or parabolic operator will satisfy a maximum principle if and only if it has a positive Green's function. Green's functions sometimes can be formulated in terms of integral kernels, which may make sense even when applied to functions that are not continuous. Green's functions for parabolic problems are treated in detail by Friedman (1964). Alternatively, if L is an elliptic operator of the form (1.107) then for Ω sufficiently large the resolvent operator $(\lambda I - L)^{-1}$ will exist and be positive, in the sense of mapping positive functions to positive functions. The semigroup e^{tL} corresponding to L, which gives the solution to $\partial u/\partial t = Lu$, can be expressed as a limit of a series of powers of the resolvent of L which all have positive coefficients. See Friedman (1976, Part 2, Section 1). Thus, the semigroup inherits some of the positivity properties of the resolvent $(\lambda I - L)^{-1}$. Finally, there are some types of models other than reaction-diffusion systems which satisfy results analogous to maximum principles. In later chapters, we discuss how some of them can be treated by methods similar to those used to study reaction-diffusion models.

2

Linear Growth Models for a Single Species: Averaging Spatial Effects Via Eigenvalues

2.1 Eigenvalues, Persistence, and Scaling in Simple Models

To make simple predictions about the persistence, extinction, or stability of a population or community inhabiting a spatially heterogenous environment it is usually necessary somehow to "average" spatially varying demographic parameters, for example birth and death rates, over the environment. In the context of spatially explicit population models, such averages can often be defined in a natural way in terms of the principal eigenvalues of differential operators, matrices, or other operators describing the dispersal and demographics of populations. This is analogous to the case of matrix models for age or stage structured populations with a finite number of discrete ages or stages, as described by Caswell (1989). For matrix models the intrinsic growth rate of the population is given by the principal eigenvalue of the Leslie matrix, which in effect averages fecundity and survivorship over age or stage classes; see Caswell (1989). In this chapter we shall see how the principal eigenvalues of differential operators can give the intrinsic growth rates for spatially structured populations described by reaction-diffusion models. The same approach can be applied to other sorts of models for spatially structured populations; specifically metapopulation models as in the work of Hanski (1997). We discuss that point later in this chapter. In this section we shall illustrate how principal eigenvalues can be used to assess the effects of spatial variability and scale in diffusion models by reviewing the pioneering work of Skellam (1951) and Kierstead and Slobodkin (1953) on the minimal patch size needed to sustain a population in an environment with a hostile exterior. In the course of that review we shall give informal descriptions of a number of properties and features of principal eigenvalues. In later sections of this chapter we give more general and precise descriptions of such properties, along with references and some proofs. Most of the facts about eigenvalues that are stated in this section are discussed in detail in standard references on partial differential equations, e.g. Strauss (1992).

The simplest models of the sort introduced by Skellam (1951) and Kierstead and Slobodkin (1953) are basic population growth models with diffusion where the population is assumed to inhabit a finite region with a lethal exterior. Specifically, if $u(x, t)$ is the population density on a region Ω where the intrinsic growth rate of the population is r and

Spatial Ecology via Reaction-Diffusion Equations R.S. Cantrell and C. Cosner
 ISBN: 0-471-49301-5

the diffusion coefficient is d, then the simplest sort of model takes the form

$$\begin{aligned} u_t &= d\Delta u + ru \text{ in } && \Omega \times (0,\infty) \\ u &= 0 && \text{on } \partial\Omega \times (0,\infty). \end{aligned} \tag{2.1}$$

(Models of this type, possibly with more general density dependent growth terms, are sometimes called KISS models, where KISS is an acronym for Kierstead and Slobodkin and Skellam.) Note that the model (2.1) is spatially heterogenous because of the boundary condition. Solutions to (2.1) may be found by separation of variables in terms of the solutions to the related eigenvalue problem

$$\begin{aligned} d\Delta\psi + r\psi &= \sigma\psi \text{ in } && \Omega \\ \psi &= 0 && \text{on } \partial\Omega; \end{aligned} \tag{2.2}$$

see for example Strauss (1992). It turns out that (2.2) admits a nonzero solution ψ only for certain values of the parameter σ. A solution to (2.2) consists of a number σ called an *eigenvalue* and a nonzero function $\psi(x)$ called an *eigenfunction*. If the boundary $\partial\Omega$ of Ω is reasonably smooth (corners are permissible but features such as fractal boundaries require more careful treatment) then (2.2) will have an infinite sequence of eigenvalues

$$\sigma_1 > \sigma_2 \geq \sigma_3 \geq \cdots \geq \sigma_k \geq \cdots \quad \text{with} \quad \sigma_k \to -\infty \tag{2.3}$$

as $k \to \infty$; e.g. see Strauss (1992). Notice that if ψ is a solution to (2.2) then so is $c\psi$ for any constant c, so usually the eigenfunctions are normalized by requiring $\int_\Omega \psi^2 dx = 1$. Solutions to (2.1) can then be written as

$$u(x,t) = \sum_{k=1}^{\infty} u_k e^{\sigma_k t} \psi_k(x), \tag{2.4}$$

where the coefficients u_k depend on the initial data $u(x,0)$. The *principal eigenvalue* of (2.2) is the largest eigenvalue, σ_1. We shall see that the associated eigenfunction ψ_1 is always positive inside Ω, which is a feature that characterizes principal eigenvalues in general. In view of (2.3), the solution shown in (2.4) will grow exponentially if $\sigma_1 > 0$ but decay exponentially if $\sigma_1 < 0$. Thus, the model (2.1) predicts persistence or extinction depending on the sign of the principal eigenvalue σ_1 of (2.2). In this case, as in many others, a prediction of persistence is equivalent to the instability of the equilibrium $u \equiv 0$ in (2.1).

Example 2.1. If Ω is the interval $(0,\ell)$ in one space dimension then (2.2) becomes $d\psi'' + r\psi = \sigma\psi$, which can be solved explicitly in terms of exponentials, trigonometric functions, or polynomials depending on the sign of $r - \sigma$. There are nonzero solutions satisfying the boundary conditions only if $\sigma = \sigma_k = r - d\pi^2k^2/\ell^2$ for some integer k, and in that case $\psi_k = (2/\ell)\sin(\pi kx/\ell)$. Thus, $\sigma_1 = r - d\pi^2/\ell^2$ so $\sigma_1 > 0$ only if $\ell > (\sqrt{d/r})\pi$. The number $(\sqrt{d/r})\pi$ is the minimum patch size needed to support a population in this model. For smaller patches, the population is close enough to the boundary $\partial\Omega$ that the loss rate $d\pi^2/\ell^2$ of individuals from dispersal out of Ω is greater than the local population growth rate, r. In the case where Ω is a square with area A and hence side $\sqrt{A}$ the principal eigenvalue σ_1 of (2.2) is $\sigma_1 = r - 2d\pi^2/A$, with $\psi_1 = (4/A)\sin(\pi x/\sqrt{A})\sin(\pi y/\sqrt{A})$, so $\sigma_1 > 0$ only if $A > 2d\pi^2/r$.

The eigenvalues in Example 2.1 depend on both the biological parameters d and r and the size and geometry of Ω. It is sometimes possible to separate the biology from the geometry by considering a related eigenvalue problem, namely

$$\begin{aligned} \Delta\phi + \lambda\phi = 0 \text{ in } \quad & \Omega \\ \phi = 0 \qquad\qquad \text{on} \quad & \partial\Omega. \end{aligned} \tag{2.5}$$

If ϕ is an eigenfunction for (2.5) then setting $\psi = \phi$ yields $d\Delta\psi + r\psi = (r - d\lambda)\psi$, so the eigenvalues of (2.2) are related to those of (2.5) by $\sigma = r - d\lambda$. The condition $\sigma_1 > 0$ can be expressed as $r/d > \lambda_1$, where λ_1 is the principal eigenvalue of (2.5). (In general, (2.5) has eigenvalues $\lambda_1 < \lambda_2 \leq \lambda_3 \leq \cdots$ where λ_1 is the only eigenvalue with a positive eigenfunction.) In the case of the interval $(0, \ell)$, $\lambda_1 = \pi^2/\ell^2$. In the case of the square of area A, $\lambda_1 = 2\pi^2/A$. Both of these scale as $1/\ell^2$, which is no accident. Suppose that Ω is rescaled to $\tilde{\Omega} = \{\ell x : x \in \Omega\}$. Let $\tilde{\phi}(x) = \phi((1/\ell)x)$ where ϕ is a solution of (2.5), so that $\tilde{\phi}$ is defined on $\tilde{\Omega}$ and by the chain rule $\Delta\tilde{\phi}(x) = (1/\ell)^2\Delta\phi((1/\ell)x) = (-\lambda/\ell^2)\phi((1/\ell)x) = (-\lambda/\ell^2)\tilde{\phi}(x)$. Hence $\tilde{\phi}$ satisfies

$$\begin{aligned} \Delta\tilde{\phi} + (\lambda/\ell^2)\tilde{\phi} = 0 \text{ in } \quad & \tilde{\Omega} \\ \tilde{\phi} = 0 \qquad\qquad\qquad \text{on} \quad & \partial\tilde{\Omega}. \end{aligned} \tag{2.6}$$

It follows that the eigenvalues of (2.6) are given by $\tilde{\lambda}_k = \lambda_k/\ell^2$ where λ_k is an eigenvalue of (2.5). The condition for persistence of a population described by $u_t = d\Delta u + ru$ in $\tilde{\Omega} \times (0, \infty)$ with $u = 0$ on $\partial\tilde{\Omega} \times (0, \infty)$ is thus

$$r/d > \tilde{\lambda}_1 = \lambda_1/\ell^2. \tag{2.7}$$

This behavior under scaling is independent of space dimension and is one of the reasons why the eigenvalues of problems analogous to (2.5) are useful.

We have seen that the principal eigenvalue σ_1 of (2.2) determines the stability or instability of the solution $u \equiv 0$ to (2.1), and that the principal eigenvalue λ_1 of (2.5) can be used to describe the dependence of σ_1 on the geometry and size of the underlying habitat region Ω via the relationship $\sigma_1 = r - d\lambda_1$. These are some of the essential features of how the principal eigenvalues of problems analogous to (2.2) and (2.5) can be used to understand the behavior of a reaction-diffusion model in terms of the parameters and geometric properties of the underlying spatial domain of the model. In the remainder of this chapter we shall examine how the notion of a principal eigenvalue can be extended to more general models with spatial heterogeneity. We shall also describe how principal eigenvalues can be estimated and how their dependence on parameters can be determined even in cases where they cannot be computed analytically. Some of the methods for estimating eigenvalues can be used to provide deeper insight into the way that $\lambda_1(\Omega)$ depends on the shape of Ω, and we shall address that topic in the next section.

2.1.1 An Application: Species-Area Relations

We now turn to an application of the scaling (2.7) to a model for the species-area relationship. It is well known that larger islands typically support more species than smaller islands if other factors such as the types of habitat on the islands and their distance from a continent are similar; see Williamson (1981), MacArthur and Wilson (1967), and Cantrell and Cosner (1994). The standard explanation for this phenomenon is the theory of island biogeography introduced by MacArthur and Wilson (1967). That theory is based on the hypothesis that island communities are structured by a dynamic equilibrium between colonizations and extinctions. (This is conceptually similar to the hypotheses underlying metapopulation models.) Larger islands should be easier to find and colonize, and they should support larger populations which are less susceptible to extinction, so they can be expected to sustain more species at equilibrium. This mechanism is plausible and supported by empirical evidence in the case of true islands, but the species-area relationship also seems to be present in relictual communities (see Brown (1971) and Patterson and Atmar (1986)) which arise when a fragment of habitat becomes isolated due to climatic or geological changes or human activity which renders the surrounding areas utterly inhospitable to the populations inhabiting the fragment. Many nature reserves are in effect relictual communities. Island biogeography theory has been invoked in discussions of such communities, but in many cases the hostility of the surrounding environment renders colonization effectively impossible and thus calls into question the dynamic equilibrium hypothesis; see Brown (1971). An alternative approach can be formulated from the scaling (2.7). Suppose we consider a collection of species whose population dynamics are described (at least at low densities) by models of the form (2.1). If the parameter combination r/d appearing in (2.7) is distributed among the species according to the distribution $s(p)$, then the number of species with $r/d > \lambda$ will be given by $\int_\lambda^\infty s(p)dp$. Now consider a collection of regions which correspond to rescalings of a fixed region Ω into $\tilde{\Omega} = \ell\Omega$. Then the area of $\tilde{\Omega}$ is $A = |\Omega|\ell^2$, where $|\Omega|$ = area of Ω. The principal eigenvalue of (2.5) on $\tilde{\Omega}$ is $\lambda_1(\tilde{\Omega}) = \lambda_1(\Omega)/\ell^2 = \lambda_1(\Omega)|\Omega|/A$. A species can be expected to persist in $\tilde{\Omega}$ if $r/d > \lambda_1(\tilde{\Omega})$, as in (2.7). Thus, the number of species expected to persist in $\tilde{\Omega}$ is given by

$$S(A) = \int_{\lambda_1(\tilde{\Omega})}^\infty s(p)dp = \int_{\lambda_1(\Omega)|\Omega|/A}^\infty s(p)dp. \tag{2.8}$$

To fit data to (2.8) one would assume that the distribution $s(p)$ takes some particular form (e.g. lognormal) and then fit the parameters of the distribution (e.g. by least squares). This was done for a number of data sets in Cantrell and Cosner (1994); the formulation (2.8) fit the data about as well as the species-area relation $S = CA^z$ deduced from the MacArthur-Wilson (1967) theory. Other sorts of spatially explicit models for biogeography include those of Holt (1992), which is based on a metapopulation model, Durrett and Levin (1996) which is based on an interacting particle system model, and Olson and Hood (1994).

The mechanism that leads to the phenomenon of critical patch size and the scaling (2.7) in KISS models is passive diffusion across a patch boundary into a hostile exterior region. There are other mechanisms, in particular certain types of environmental heterogeneity in a patch with a closed (i.e. reflecting) boundary, which lead to eigenvalue problems where the eigenvalues have similar scaling properties and thus yield a minimal patch size. We shall return to that point in the next section.

2.2 Variational Formulations of Eigenvalues: Accounting for Heterogeneity

The simple KISS models discussed in the previous section can account only for the type of spatial heterogeneity imposed on an otherwise homogeneous patch by a deadly boundary.

There are various other forms of spatial heterogeneity, and many of them lead to models where the coefficients vary in space. It is usually impossible to solve the eigenvalue problems associated with such models explicitly, even in simple geometries. Hence, we need methods of approximating and comparing eigenvalues even if we cannot compute them analytically. Many important features of eigenvalues can be deduced from their variational formulation as the critical values of what is known as a functional, that is, a function whose range is the real or complex numbers, but whose domain is a space of functions.

The classical variational theory of eigenvalues was treated by Courant and Hilbert (1953), and is described in many standard texts, e.g. Strauss (1992). The theory can be used to characterize *all* the eigenvalues of problems with the appropriate form, but we will be concerned almost exclusively with principal eigenvalues so we shall limit our attention to those. The eigenvalue problems which can be treated by variational methods are those which involve operators arising as the derivatives of a functional. Such operators can involve spatial heterogeneities of various sorts, but their form is somewhat restricted; in particular models involving advection may be difficult to treat variationally. The variational formulation of eigenvalues is useful in large measure because it provides a means of comparing eigenvalues arising from different models, or from the same model on different spatial domains. We shall use such comparisons to describe how the geometry of the underlying spatial domain Ω affects the principal eigenvalue in (2.5).

Eigenvalue problems that can be treated by variational methods typically arise from models of the form

$$\begin{aligned} u_t &= \nabla \cdot d(x)\nabla u + m(x)u \text{ in } \quad \Omega \times (0,\infty) \\ d(x)\frac{\partial u}{\partial \vec{n}} &+ \beta(x)u = 0 \qquad \text{on} \quad \partial\Omega \times (0,\infty), \end{aligned} \tag{2.9}$$

or the corresponding model with the homogeneous Dirichlet boundary condition $u = 0$ on $\partial\Omega \times (0,\infty)$. (In general x represents a point in 1, 2, or 3 space dimensions.) The eigenvalue problem associated with (2.9) which is analogous to (2.2) is

$$\begin{aligned} \nabla \cdot d(x)\nabla\psi + m(x)\psi &= \sigma\psi \text{ in } \quad \Omega \\ d(x)\frac{\partial \psi}{\partial \vec{n}} + \beta(x)\psi &= 0 \qquad \text{on} \quad \partial\Omega \end{aligned} \tag{2.10}$$

or with $\psi = 0$ on $\partial\Omega$ if $u = 0$ on $\partial\Omega \times (0,\infty)$ in (2.9). As in the case of (2.1), the solution to (2.9) can be expressed by separation of variables in the form (2.4) where $\{\sigma_k\}$ is now the set of eigenvalues for (2.10), and the eigenvalues of (2.10) satisfy (2.3) (see Strauss (1992) and Courant and Hilbert (1953)). Thus we again may determine whether (2.9) predicts growth or extinction in terms of the sign of the principal eigenvalue σ_1.

Note that (2.9) and (2.10) do not include advection (taxis) terms. Generally, the analysis of models with advection via variational methods is more complicated than when advection is absent. We discuss models with advection at the end of this section, and describe why the forms (2.9) and (2.10) are natural for variational problems in the Appendix to this chapter.

To make sense out of (2.10) in classical terms we would need to assume $d(x)$ is differentiable. For technical reasons we shall assume slightly more. The following result is classical (see Courant and Hilbert (1953)), so we shall not repeat the proof, but we have stated it in a fairly general form. The fact that σ_1 depends continuously on d, m, and β with respect to $L^\infty(\Omega)$ for d and m and $L^\infty(\partial\Omega)$ for β is a classical result. The fact that σ_1 depends continuously on m with respect to $L^p(\Omega)$ for $p < \infty$ follows as in de Figueiredo

(1982). (See Weinberger (1974) for a modern treatment of eigenvalue problems from the variational viewpoint. Some of the hypotheses are technical assumptions about the smoothness of $\partial\Omega$ or the coefficients of (2.10); we explicate those in the mathematical discussion following the theorem.)

Theorem 2.1. Suppose that $\partial\Omega$ is piecewise of class $C^{2+\alpha}$ and that Ω satisfies the interior cone condition. Suppose that $d(x) \in C^{1+\alpha}(\overline{\Omega})$ with $d(x) \geq d_0 > 0$; that $m(x) \in L^\infty(\Omega)$, and $\beta(x) \in L^\infty(\partial\Omega)$ with $\beta(x) \geq 0$. The principal eigenvalue of (2.10) is given by

$$(a)\ \sigma_1 = \max_{\substack{\psi\in W^{1,2}(\Omega)\\ \psi\neq 0}} \left[\frac{-\int_\Omega d(x)|\nabla\psi|^2 dx + \int_\Omega m(x)\psi^2 dx - \int_{\partial\Omega} \beta(x)\psi^2 dS}{\int_\Omega \psi^2 dx} \right]$$

or alternatively

$$(b)\ \sigma_1 = \max\left\{ -\int_\Omega d(x)|\nabla\psi|^2 dx + \int_\Omega m(x)\psi^2 dx - \int_{\partial\Omega} \beta(x)\psi^2 dS : \psi \in W^{1,2}(\Omega),\ \int_\Omega \psi^2 dx = 1 \right\}. \tag{2.11}$$

In the case of Dirichlet boundary conditions on a subset $\Gamma \subseteq \partial\Omega$, the space $W^{1,2}(\Omega)$ is replaced by the subspace of $W^{1,2}(\Omega)$ consisting of functions which are zero (in the appropriate sense) on Γ. (This space is $W_0^{1,2}(\Omega)$ if Dirichlet conditions hold on all of $\partial\Omega$.) The eigenfunction ψ_1 corresponding to σ_1 can be chosen so that $\psi_1 > 0$ on Ω (in fact, $\psi_1 > 0$ on $\overline{\Omega}$ except for Dirichlet boundary conditions) and σ_1 is the only eigenvalue admitting a positive eigenfunction. The eigenvalue σ_1 depends continuously on $m(x)$ with respect to $L^p(\Omega)$ for any $p \in (0, \infty]$ in the case of one or two space dimensions and for $p > N/2$ in the case of $N > 2$ space dimensions; also, σ_1 depends continuously on $d(x)$ and $\beta(x)$ with respect to $L^\infty(\Omega)$ and $L^\infty(\partial\Omega)$. Finally, σ_1 is simple, i.e. the eigenspace for σ_1 is one dimensional.

Remarks:

1. *What do the hypotheses and notation mean?*
   The conditions on Ω essentially require that the boundary $\partial\Omega$ consists of a finite number of smooth curves or surfaces, which may form corners where they meet but may not form sharp cusps. (The interior cone condition means that there is a cone of fixed size and shape that can be oriented to fit inside Ω if its vertex is at any point of $\partial\Omega$.) The space $W^{1,2}$ consists of functions ψ with $|\nabla\psi|^2$ integrable over Ω but not necessarily continuous.
2. *Why are the hypotheses needed? (A few technicalities.)*
   The hypotheses on Ω imply that any function $\psi \in W^{1,2}(\Omega)$ has a trace on $\partial\Omega$ which is square integrable, so the boundary terms in (2.11a,b) make sense. (See Adams (1975, Theorem 5.22).) The condition $d(x) \in C^{1+\alpha}(\overline{\Omega})$ together with the conditions on $\partial\Omega$ allow the application of standard elliptic regularity results as discussed by Gilbarg and Trudinger (1977) or Friedman (1976) to (2.10). Those in turn imply that if $\psi \in W^{1,2}(\Omega)$ then in fact $\psi \in W^{2,p}(\Omega)$ for any p and hence $\psi \in C^{1+\alpha}(\overline{\Omega})$ so that the boundary condition in (2.10) makes sense in classical terms. It may be necessary to apply the

regularity results repeatedly to obtain the desired degree of smoothness. This and related issues involving function spaces and the regularity of solutions are discussed in Chapter 1.

Part of the significance of Theorem 2.1 is that it permits comparisons between the principal eigenvalues of different problems.

Corollary 2.2. If we denote the principal eigenvalue of (2.10) as $\sigma_1(d, m, \beta)$ then $\sigma_1(d, m, \beta)$ is increasing with respect to m in the sense that if $m_1 \geq m_2$ then $\sigma_1(d, m_1, \beta) \geq \sigma_1(d, m_2, \beta)$, and if $m_1 > m_2$ on a subset of positive measure then $\sigma_1(d, m_1, \beta) > \sigma_1(d, m_2, \beta)$. Similarly, $\sigma_1(d, m, \beta)$ is decreasing with respect to β in the same sense, and is decreasing with respect to d in the same sense unless $\beta \equiv 0$ and $m(x)$ is a constant. (In this last case the eigenfunction ψ_1 turns out to be a constant so the terms involving d in (2.11a,b) will have no effect.)

Proof: Let ψ_1 be the eigenfunction associated with $\sigma_1(d, m_2, \beta)$. Then ψ_1 is the maximizer of the quotient (2.11a) for $\sigma_1(d, m_2, \beta)$ and we have

$$\begin{aligned}
\sigma_1(d, m_2, \beta) &= \frac{-\int_\Omega d|\nabla\psi_1|^2 dx + \int_\Omega m_2\psi_1^2 dx - \int_{\partial\Omega} \beta\psi_1^2 dS}{\int_\Omega \psi_1^2 dx} \\
&\leq \frac{-\int_\Omega d|\nabla\psi_1|^2 dx + \int_\Omega m_1\psi_1^2 dx - \int_{\partial\Omega} \beta\psi_1^2 dS}{\int_\Omega \psi_1^2 dx} \\
&\leq \max_{\substack{\psi\in W^{1,2}(\Omega)\\ \psi\neq 0}} \frac{-\int_\Omega d|\nabla\psi|^2 dx + \int_\Omega m_1\psi^2 dx - \int_{\partial\Omega} \beta\psi^2 dS}{\int_\Omega \psi^2 dx} \\
&= \sigma_1(d, m_1, \beta).
\end{aligned}$$

Since $\psi_1 > 0$ in Ω, the first inequality is strict if $m_1(x) > m_2(x)$ on a set of positive measure. The arguments for β and d are similar.

Remarks: The parameter $m(x)$ describes the local growth rate of the population in (2.9), so increasing $m(x)$ should increase the average growth rate, as measured by σ_1, and in fact it does. The coefficient β describes how likely it is for an individual reaching $\partial\Omega$ to leave Ω. In the case $d(x) = 1$, $\beta(x) = \beta_0$ we can write the boundary condition in (2.9) as $(1 - p)\partial u/\partial\vec{n} + pu = 0$ where $p = \beta_0/(\beta_0 + 1)$. In that case $\beta_0/(\beta_0 + 1) = p$ is precisely the fraction of individuals reaching $\partial\Omega$ which leave Ω. Thus, increasing β has the effect of increasing the rate at which individuals are lost by dispersal out of Ω. Increasing d increases the movement rate within Ω, which increases the rate of contact with $\partial\Omega$, and hence the rate of loss through $\partial\Omega$ if $\beta \neq 0$. Increasing the loss rate of individuals across $\partial\Omega$ reduces the average population growth rate σ_1.

Theorem 2.1 also allows comparisons between models with the same coefficients but on different domains, and thus can help us understand how σ_1 depends on Ω. Recall that for the model $u_t = d\Delta u + ru$ with $u = 0$ on $\partial\Omega$, the principal eigenvalue is $\sigma_1 = r - d\lambda_1$, where λ_1 is the principal eigenvalue of (2.5). We can write (2.5) as $\Delta\phi = -\lambda\phi$ in $\Omega, \phi = 0$ on $\partial\Omega$, so that $\lambda_1 = -\sigma_1$ for the problem $\Delta\psi = \sigma\psi$ in Ω, $\psi = 0$ on $\partial\Omega$. Hence

$$\lambda_1 = -\max_{\substack{\psi\in W_0^{1,2}(\Omega)\\ \psi\neq 0}} \left(-\int_\Omega |\nabla\psi|^2 dx \Big/ \int_\Omega \psi^2 dx\right) = \min_{\substack{\psi\in W_0^{1,2}(\Omega)\\ \psi\neq 0}} \left(\int_\Omega |\nabla\psi|^2 dx \Big/ \int_\Omega \psi^2 dx\right) \quad (2.12)$$

Formula (2.12) allows us to compare the values of λ_1 on different regions. Let $\lambda_1(\Omega)$ denote the principal eigenvalue for (2.5) on Ω. We have:

Corollary 2.3. If $\Omega_1 \subseteq \Omega_2$ then $\lambda_1(\Omega_1) \geq \lambda_1(\Omega_2)$, with strict inequality if $\Omega_2\backslash\Omega_1$ is an open set.

Proof: If $\psi \in W_0^{1,2}(\Omega_1)$ then ψ can be extended to be zero on $\Omega_2\backslash\Omega_1$, and the resulting function will belong to $W_0^{1,2}(\Omega_2)$. Take ψ to be the eigenfunction for $\lambda_1(\Omega_1)$ and let $\tilde{\psi}_1$ be the extension of ψ_1 to Ω_2 which is zero on $\Omega_2\backslash\Omega_1$. We have $\tilde{\psi} \in W_0^{1,2}(\Omega_2)$ and

$$\begin{aligned}\lambda_1(\Omega_2) &= \min_{\substack{\psi\in W_0^{1,2}(\Omega_2)\\ \psi\neq 0}} \int_{\Omega_2} |\nabla\psi|^2 dx \int_{\Omega_2} \psi^2 dx \leq \int_{\Omega_2} |\nabla\tilde{\psi}_1|^2 dx \Big/ \int_{\Omega_2} \tilde{\psi}_1^2 dx \\ &= \int_{\Omega_1} |\nabla\psi_1|^2 dx \Big/ \int_{\Omega_1} \psi_1^2 dx = \lambda_1(\Omega_1).\end{aligned}$$

To obtain the strict inequality we note that the eigenfunction for $\lambda_1(\Omega_2)$ is positive on Ω_2 but $\tilde{\psi}_1$ is not, so $\tilde{\psi}_1$ cannot be the minimizer of the quotient for $\lambda_1(\Omega_2)$.

We can now determine more precisely what aspects of Ω the eigenvalue $\lambda_1(\Omega)$ measures. Suppose that $\Omega_1 \subseteq \Omega$. Then $\lambda_1(\Omega) \leq \lambda_1(\Omega_1)$ no matter what the perimeter/area ratio is for Ω_1. If $\Omega_1 \subseteq \Omega \subseteq \Omega_2$ and $\Omega_1 \subseteq \tilde{\Omega} \subseteq \Omega_2$ where Ω_2 is similar in shape to Ω_1 but slightly larger, then $\lambda_1(\Omega_1)$ and $\lambda_1(\Omega_2)$ will have similar values, and $\lambda_1(\Omega), \lambda_1(\tilde{\Omega})$ will lie in the interval $(\lambda_1(\Omega_2), \lambda_1(\Omega_1))$ and thus will have similar values. See Figure 2.1. On the other hand, if the maximum width of Ω in any direction is less than a, then (perhaps after rotating coordinates) Ω can be seen to fit inside a rectangle of width a and of some height, say b. The principal eigenvalue for such a rectangle is $\pi^2[1/a^2 + 1/b^2] \geq \pi^2/a^2$, so as the width of Ω (and hence maximum distance from a point inside Ω to $\partial\Omega$) decreases, $\lambda_1(\Omega)$ tends to infinity. This will be true even if the length of Ω increases so that the area of Ω is held fixed. The point of these comments is that $\lambda_1(\Omega)$ essentially measures the *core area* of Ω, taking into account the size of that area and its average distance from $\partial\Omega$. On the other hand, $\lambda_1(\Omega)$ is relatively insensitive to changes that increase the length of $\partial\Omega$, but do not affect the core area; for example, adding many small "wiggles" to $\partial\Omega$ can change the perimeter/area ratio dramatically, but will have little effect on $\lambda_1(\Omega)$. Some of these features are indicated in Figure 2.1. The notion of core area for a habitat fragment has been used to some extent in the literature on conservation; see Noss et al. (1989) and Lovejoy et al. (1986). It may be a more accurate index of fragmentation than perimeter/area ratios or other measures; see Groom and Schumaker (1990). Additionally, the results of some simulations suggest that the amount core area may influence emigration, persistence, and other features of the simulation; see Stamps et al. (1987) and McKelvey et al. (1986).

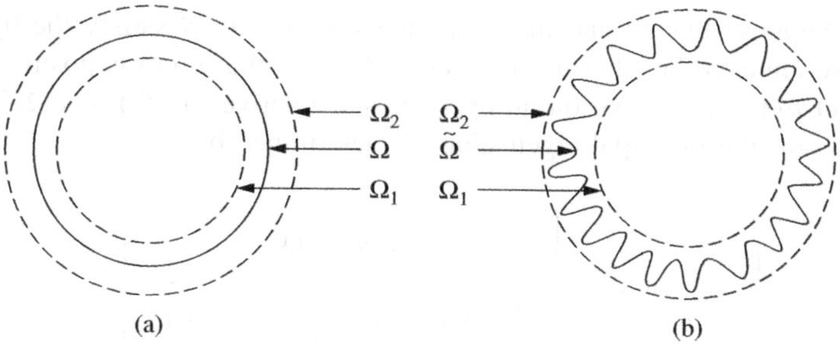

Figure 2.1 The dotted lines show Ω_1 and Ω_2 in both (a) and (b). The solid lines show Ω in (a) and $\tilde{\Omega}$ in (b). The perimeter to area ratio is larger for $\tilde{\Omega}$ than for Ω. However, since Ω and $\tilde{\Omega}$ both contain Ω_1 and are contained in Ω_2, the principal eigenvalues ($\lambda_1(\Omega_2)$ and $\lambda_1(\Omega_2)$) must both have values in the interval $(\lambda_1(\Omega_2), \lambda_1(\Omega_1))$ no matter how much the perimeter to area ratios of Ω and $\tilde{\Omega}$ might differ. The region Ω_1 could be viewed as the common core area of Ω and $\tilde{\Omega}$.

The proofs of Corollaries 2.2 and 2.3 illustrate an important method for comparing eigenvalues associated with different problems, namely using carefully selected test functions in variational formulations such as (2.11a,b) or (2.12) to obtain estimates. We shall employ that method in various contexts. There are many other methods for estimating eigenvalues. See Weinberger (1974), Bandle (1980), Schaefer (1988), Sperb (1981) and Belgacem (1997), among others. One interesting result which sheds some light on what λ_1 measures is the isoperimetric inequality (Bandle, 1980), which says that among all planar regions Ω of a given area, $\lambda_1(\Omega)$ is smallest when Ω is a circular disc.

The eigenvalue σ_1 is nicely characterized by (2.11a,b). In the case of the Laplace operator the eigenvalue λ_1 can be recast as $-\sigma_1$ and the characterization (2.11a,b) can be used to yield (2.12). However, the corresponding problem in the case of variable coefficients requires more sophisticated treatment. The eigenvalue problem which has the same relation to (2.10) that (2.5) has to (2.2) is

$$\begin{aligned} \nabla \cdot d(x)\nabla\phi + \lambda m(x)\phi = 0 \text{ in } \quad \Omega \\ d(x)\frac{\partial\phi}{\partial\vec{n}} + \beta(x)\phi = 0 \qquad \text{on} \quad \partial\Omega. \end{aligned} \tag{2.13}$$

If $m(x)$ is strictly positive then we can characterize the principal eigenvalue λ_1 for (2.13) as in (2.11a) but corresponding to $-\sigma_1$ as in (2.12):

$$\lambda_1 = \min_{\substack{\phi \in W^{1,2}(\Omega) \\ \phi \neq 0}} \left[\left(\int_\Omega d(x)|\nabla\phi|^2 dx + \int_{\partial\Omega} \beta(x)\phi^2 dx \right) \Big/ \int_\Omega m(x)\phi^2 dx \right].$$

The problem with this formulation is that it may not make sense if $m(x)$ changes sign, since the denominator might be zero or negative. This problem can be addressed by looking at $1/\lambda_1$ as the maximum of the reciprocal of the form in (2.13). This approach to determining the eigenvalues for (2.13) when $m(x)$ changes sign was introduced by Manes and Micheletti (1973) for the case of Dirichlet boundary conditions. (They obtained characterizations of all the eigenvalues, not just the first one.) The formulation given here allows for more general boundary conditions, but the case of the pure Neumann conditions is more subtle and must be treated separately.

Theorem 2.4. Suppose that Ω and the coefficients $d, m,$ and β satisfy the hypotheses of Theorem 2.1. Assume further that $\partial\Omega$ is of class C^1, and that $m(x)$ is positive on an open subset of Ω and $\beta(x)$ is positive on an open subset (relative to $\partial\Omega$) of $\partial\Omega$. The problem (2.13) admits a positive principal eigenvalue λ_1^+ determined by

$$\frac{1}{\lambda_1^+} = \max_{\substack{\phi \in W^{1,2}(\Omega) \\ \phi \neq 0}} \left[\frac{\int_\Omega m\phi^2 dx}{\int_\Omega d|\nabla\phi|^2 dx + \int_{\partial\Omega} \beta(x)\phi^2 dS} \right]. \tag{2.14}$$

In the case of Dirichlet boundary conditions

$$\frac{1}{\lambda_1^+} = \max_{\substack{\phi \in W_0^{1,2}(\Omega) \\ \phi \neq 0}} \left[\frac{\int_\Omega m\phi^2 dx}{\int d|\nabla\phi|^2 dx} \right]. \tag{2.15}$$

The principal eigenvalue is the only positive eigenvalue admitting a positive eigenfunction, and it is a simple eigenvalue. The principal eigenvalue depends continuously on $m(x)$ with respect to $L^p(\Omega)$ for any $p \in (1, \infty]$ in the case of one or two space dimensions, and for $p > N/2$ in the case of $N > 2$ space dimensions.

Remarks: The problem (2.13) also admits a negative principal eigenvalue if $m(x)$ is negative on an open subset of Ω. Theorem 2.4 is essentially due to Manes and Micheletti (1973), along with various related results. See also de Figueiredo (1982). A sketch of the proof of the existence of λ_1^+ is given in the Appendix to this chapter.

In the case of pure Neumann boundary conditions $\beta(x) \equiv 0$ so the denominator of the right side of (2.14) would vanish if ϕ where constant, and hence the maximum might not exist. It turns out that the existence of a positive (or negative) principal eigenvalue for (2.13) in the case of Neumann boundary conditions depends on the integral of $m(x)$ over Ω.

Theorem 2.5. (Brown and Lin, 1980) In the case of Neumann boundary conditions ($\beta \equiv 0$) the problem (2.13) admits a positive principal eigenvalue if and only if

$$\int_\Omega m(x)dx < 0. \tag{2.16}$$

In that case the positive principal eigenvalue is characterized by (2.14) with $\beta \equiv 0$. The positive principal eigenvalue is the only positive eigenvalue which admits a positive eigenfunction, and it is a simple eigenvalue.

Remarks: A sketch of the proof of the existence of λ_1^+ will be given in the Appendix to this chapter. The result is due to Brown and Lin (1980). It is clear from (2.14) and (2.15) that λ_1^+ is increasing in d and β and decreasing in m.

The eigenvalues σ_1 and λ_1^+ are related as follows.

Theorem 2.6. Suppose that λ is a positive parameter and that either $\beta(x) > 0$ on an open subset of $\partial\Omega$, or the boundary condition is a Dirichlet condition on part of $\partial\Omega$, or (2.16) holds. The principal eigenvalue σ_1 of

$$\begin{aligned} \nabla \cdot d(x)\nabla\psi + \lambda m(x)\psi &= \sigma\psi \text{ in } \quad \Omega \\ \frac{\partial\psi}{\partial\vec{n}} + \beta(x)\psi &= 0 \qquad\qquad \text{on} \quad \partial\Omega \end{aligned} \tag{2.17}$$

is positive if and only if $0 < \lambda_1^+ < \lambda$ where λ_1^+ is the positive principal eigenvalue of (2.13).

If $\beta(x) \equiv 0$ and the inequality (2.16) is reversed then $\sigma_1 > 0$ for all $\lambda > 0$.

Proof: Suppose $\beta(x) > 0$ on an open subset of $\partial\Omega$. The cases of Dirichlet boundary conditions on part of $\partial\Omega$ or of Neumann conditions ($\beta \equiv 0$) with (2.16) holding are similar. Let ψ_1 be the eigenfunction for σ_1. Multiplying by ψ_1 in (2.17), integrating over Ω, and applying Green's formula and the boundary conditions yields

$$-\int_\Omega d|\nabla\psi_1|^2 dx - \int_{\partial\Omega} \beta\psi_1^2 dS + \lambda \int_\Omega m\psi_1^2 dx = \sigma_1 \int_\Omega \psi_1^2 dx.$$

By (2.14)

$$\int_\Omega m\psi_1^2 dx \le \frac{1}{\lambda_1^+}\left[\int_\Omega d|\nabla\psi_1|^2 dx + \int_{\partial\Omega} \beta\psi_1^2 dS\right]$$

so

$$\left(\frac{\lambda}{\lambda_1^+} - 1\right)\left[\int_\Omega d|\nabla\psi_1|^2 dx + \int_{\partial\Omega} \beta\psi_1^2 dS\right] \ge \sigma_1 \int_\Omega \psi_1^2 dx,$$

so $\sigma_1 < 0$ if $\lambda < \lambda_1^+$. On the other hand, if ϕ_1 is the eigenfunction for λ_1^+ we can multiply (2.13) by ϕ_1 and integrate to obtain

$$-\int_\Omega d|\nabla\phi_1|^2 dx - \int_{\partial\Omega} \beta\phi_1^2 dS = -\lambda_1^+ \int_\Omega m\phi_1^2 dx$$

so that by (2.11) as it applies to (2.17) we have

$$\begin{aligned} \sigma_1 &\ge \left(-\int_\Omega d|\nabla\phi_1|^2 dx + \lambda \int_\Omega m\phi_1^2 dx - \int_{\partial\Omega} \beta\phi_1^2 dS\right) \Big/ \int_\Omega \phi_1^2 dx \\ &= (\lambda - \lambda_1^+)\left(\int_\Omega m\phi_1^2 dx \Big/ \int_\Omega \phi_1^2 dx\right). \end{aligned}$$

By (2.14) we have $\displaystyle\int m\phi_1^2 dx > 0$ since $\lambda_1^+ > 0$, so $\sigma_1 > 0$ for $\lambda > \lambda_1^+$.

If $\beta(x) \equiv 0$ and (2.16) is reversed then we can use the test function $\psi \equiv 1$ in (2.11a,b) and obtain $\sigma_1 > 0$ immediately if $\lambda > 0$.

Remarks: By writing (2.17) as $\nabla \cdot (d(x)/\lambda)\nabla\psi + m(x)\psi = (\sigma_1/\lambda)\psi$ in Ω we can see that the condition $\lambda > \lambda_1^+$, implying $\sigma_1 > 0$, can be interpreted as requiring that the diffusion coefficient be sufficiently small. The mechanism underlying the condition is that if $\beta(x) > 0$ on part of Ω or if Dirichlet conditions are imposed on part of Ω then a large diffusion coefficient might lead to a rapid loss of individuals across $\partial\Omega$ and hence to a prediction of extinction (i.e. $\sigma_1 < 0$.) A small diffusion coefficient would induce a slower rate of loss by dispersal out of Ω, and hence would allow persistence ($\sigma_1 > 0$.) The Neumann case ($\beta \equiv 0$) is more subtle. In that case $\partial\Omega$ is perfectly reflecting, i.e. any individual that reaches $\partial\Omega$ simply turns around, so dispersal out of Ω plays no role. The only mechanism which might cause loss of population is dispersal into regions where the local population growth rate is negative. If condition (2.16) holds then the average local growth rate is negative. If individuals disperse rapidly they effectively average the local growth rate and as a result the population will decline ($\sigma_1 < 0$) if (2.16) holds and λ is too small. On the other hand, if the inequality in (2.16) is reversed then the average of $m(x)$ is positive so that even if individuals disperse throughout Ω rapidly enough to effectively experience the average of $m(x)$ the population will still grow ($\sigma_1 > 0$).

In the previous section we showed how the scaling properties of the principal eigenvalue of the Laplacian can be used to determine the minimal patch size needed to sustain a population in terms of the dispersal and local growth rates of the population. Hypotheses about how those rates are distributed among a collection of species then yield species-area relations as in (2.8). In that case the mechanism leading to a critical patch size was dispersal out of a homogeneous patch into a lethal exterior. The same sort of area effects can also arise if the patch boundary is strictly reflecting (Neumann condition), and the local growth rate in the patch is positive in some places but negative on the average. In such a scenario the mechanism leading to a critical patch size is the proximity of bad habitat to good. In a small patch where bad regions are always near good ones, the chances of dispersal into an unfavorable region are greater than when good and bad regions can be farther apart. Suppose that a population has dispersal rate d and local growth rate $rm(x)$ where m is positive on part of Ω but (2.16) holds. Consider the model

$$\begin{aligned} u_t &= d\Delta u + rm(x)u \quad \text{in} \quad \Omega \times (0,\infty) \\ \frac{\partial u}{\partial \vec{n}} &= 0 \qquad\qquad \text{on} \quad \partial\Omega \times (0,\infty). \end{aligned} \tag{2.18}$$

This model predicts population growth if $\sigma_1 > 0$ in the problem $d\Delta\psi + rm(x)\psi = \sigma\psi$ in Ω, $\partial\psi/\partial\vec{n} = 0$ on $\partial\Omega$. By Theorem 2.6 that will be the case if $r/d > \lambda_1^+$ where λ_1^+ is the principal eigenvalue of

$$\begin{aligned} \Delta\phi + \lambda m(x)\phi &= 0 \quad \text{in} \quad \Omega, \\ \frac{\partial \phi}{\partial \vec{n}} &= 0 \qquad \text{on} \quad \partial\Omega. \end{aligned} \tag{2.19}$$

If we rescale Ω to $\tilde{\Omega} = \ell\Omega$ by "stretching" the coordinates (so that the size of the region where the growth rate is positive is multiplied by ℓ) then we can rescale (2.17) to $\tilde{\Omega}$ as

$$\begin{aligned} u_t &= d\Delta u + rm((1/\ell)x)u \quad \text{on} \quad \tilde{\Omega} \times (0,\infty), \\ \frac{\partial u}{\partial \vec{n}} &= 0 \qquad\qquad \text{on} \quad \partial\tilde{\Omega} \times (0,\infty). \end{aligned} \tag{2.20}$$

The scaling in m expands the region where $m > 0$ by a factor of ℓ. The associated eigenvalue problem for $\tilde{\lambda}_1^+$ can be solved in terms of (2.19) by letting $\tilde{\phi}(x) = \phi_1((1/\ell)x)$ where $\phi_1 > 0$ is the eigenfunction for λ_1^+. We have $\Delta\tilde{\phi} = (1/\ell^2)\Delta\phi_1((1/\ell)x) = -(\lambda_1^+/\ell^2)m((1/\ell)x)\phi_1((1/\ell)x) = (-\lambda_1^+/\ell^2)m((1/\ell)x)\tilde{\phi}$ so that $\tilde{\phi} > 0$ is the eigenfunction for $\tilde{\lambda}_1^+ = (\lambda_1^+/\ell^2)$. If $|\Omega|$ = area of Ω and A = area of $\tilde{\Omega}$ then $A = \ell^2|\Omega|$ so $\tilde{\lambda}_1^+ = |\Omega|\lambda_1^+/A$ and the model (2.20) predicts population growth when $r/d > \tilde{\lambda}_1^+ = |\Omega|\lambda_1^+/A$, that is, when $A > (d/r)|\Omega|\lambda_1^+$. The formulas for $\tilde{\lambda}_1^+$ and hence for the critical patch size in terms of λ_1^+ are exactly the same as in the case we discussed in the previous section, and hence would yield the same species-area relations via (2.8) under the same assumptions about the distribution of r/d among species. This formulation assumes that the environmental factors affecting local growth rates are the same for all species, so that "good habitat" has the same meaning for all species, so it may be more appropriate for modeling the expected number of species within a given guild rather than across an entire community. Some additional discussion of this type of scaling (among others) is given in Cantrell and Cosner (2001). In the case of Robin boundary conditions ($\beta > 0$) the scaling becomes slightly more complicated because the boundary condition changes, but the general trend is similar. Any boundary conditions with $\beta > 0$ will lead to a critical patch size.

For models with advection the simple variational formulas (2.11a,b), (2.14), and (2.15) do not apply. It is possible to characterize the principal eigenvalues of operators with advective terms via methods based on positive operator theory, and we will do that in a later section. It is also possible to characterize the principal eigenvalues of operators with general advective terms variationally, but the characterization is complicated; see Holland (1977, 1978) and Belgacem (1997). However, there is a class of advective models which can be treated variationally via simple changes of coordinates. The key feature is that the advection term must be related to the diffusion coefficient via a gradient. Such models arise naturally as descriptions of populations that disperse along environmental gradients. Consider a model for dispersal based on a flux $\vec{J} = -d(x)\nabla u + \vec{b}(x)u$. The linear population growth model with the dispersal term is

$$u_t = -\nabla \cdot \vec{J} + m(x)u = \nabla \cdot [d(x)\nabla u - \vec{b}(x)u] + m(x)u \quad \text{in} \quad \Omega \times (0, \infty). \tag{2.21}$$

The boundary conditions we have used in models without advection are based on the hypothesis that some fraction of the individuals reaching the boundary of a patch will cross it. In other words, the flux across the boundary at each point is proportional to the density. In the advective case this can be formulated as $\vec{J} \cdot \vec{n} = \beta(x)u$, or equivalently

$$d(x)\partial u/\partial \vec{n} + [\beta(x) - \vec{b}(x) \cdot \vec{n}]u = 0, \quad \text{on} \quad \partial\Omega \times (0, \infty). \tag{2.22}$$

Suppose that there is a function $B(x)$ such that

$$\vec{b}(x) = -d(x)\nabla B(x). \tag{2.23}$$

The first step in making (2.21) and (2.22) more tractable is the change of variables $v = e^B u$, which (after some computation) converts (2.21) to

$$v_t = \nabla \cdot d(x)\nabla v + \vec{b}(x) \cdot \nabla v + m(x)v \quad \text{in} \quad \Omega \times (0, \infty) \tag{2.24}$$

and (2.22) to

$$d(x)\frac{\partial v}{\partial \vec{n}} + \beta(x)v = 0 \quad \text{on} \quad \partial\Omega \times (0, \infty). \tag{2.25}$$

The eigenvalue problems of interest relative to (2.24) and (2.25) are

$$\nabla \cdot d(x)\nabla\psi + \vec{b} \cdot \nabla\psi + m(x)\psi = \sigma\psi \quad \text{in} \quad \Omega \tag{2.26}$$

$$\nabla \cdot d(x)\nabla\phi + \vec{b} \cdot \nabla\phi + \lambda m(x)\phi = 0 \quad \text{in} \quad \Omega \tag{2.27}$$

with boundary condition (2.25) in either case. The second step in obtaining variational formulas for eigenvalues is to multiply by e^{-B} and note that $\nabla \cdot de^{-B}\nabla\psi = e^{-B}\nabla \cdot d\nabla\psi - de^{-B}\nabla B \cdot \nabla\psi = e^{-B}\nabla \cdot d\nabla\psi + e^{-B}\vec{b} \cdot \nabla\psi$, so that (2.26) and (2.27) become

$$\nabla \cdot d(x)e^{-B(x)}\nabla\psi + m(x)e^{-B(x)}\psi = \sigma e^{-B(x)}\psi \tag{2.28}$$

and

$$\nabla \cdot d(x)e^{-B(x)}\nabla\phi + \lambda m(x)e^{-B(x)}\phi = 0. \tag{2.29}$$

To fit the forms of (2.28) and (2.29) correctly we rewrite the boundary conditions as $de^{-B}\partial\psi/\partial\vec{n} + \beta e^{-B}\psi = 0$ on $\partial\Omega$. Then (2.29) has precisely the form of (2.13) and so if $\beta \not\equiv 0$ the principal eigenvalue of (2.29) and hence (2.27) is determined by (2.14):

$$\frac{1}{\lambda_1^+} = \max_{\substack{\phi\in W^{1,2} \\ \phi\neq 0}} \left[\frac{\int_\Omega me^{-B}\phi^2 dx}{\int_\Omega de^{-B}|\nabla\phi|^2 dx + \int_{\partial\Omega} \beta e^{-B}\phi^2 dS} \right]. \tag{2.30}$$

In the case $\beta \equiv 0$ condition (2.16) is required for the existence of λ_1^+. In the present setting, (2.16) becomes $\int_\Omega me^{-B}dx < 0$. Equation (2.28) deviates slightly from the form (2.10) but the factor e^{-B} multiplying σ is strictly positive, so we can still apply the classical theory of Courant and Hilbert (1953) to obtain

$$\sigma_1 = \max_{\substack{\psi\in W^{1,2}(\Omega) \\ \psi\neq 0}} \left[\frac{-\int_\Omega de^{-B}|\nabla\psi|^2 dx + \int_\Omega me^{-B}\psi^2 dx - \int_{\partial\Omega} \beta e^{-B}\psi^2 dS}{\int_\Omega e^{-B}\psi^2 dx} \right]. \tag{2.31}$$

(In the Dirichlet case the terms involving β are absent in (2.30) and (2.31) and the maximum is taken on $W_0^{1,2}(\Omega)$ rather than $W^{1,2}(\Omega)$.)

2.3 Effects of Fragmentation and Advection/Taxis in Simple Linear Models

The variational characterizations of eigenvalues in the previous section make it possible to analyze how spatial heterogeneity and advection (or taxis) affect the predictions of simple linear models for population growth and dispersal. (We shall see in later chapters that predictions from linear models typically have important implications for the analysis of nonlinear models.) We first consider the effects of habitat fragmentation.

2.3.1 Fragmentation

One of the most obvious ways that human activities affect natural systems is by dividing large regions of contiguous habitat into smaller fragments. The effects of fragmentation are complex and varied, but often are detrimental to at least some populations. See, for example, Lovejoy et al. (1986), McKelvey et al. (1986), Quinn and Karr (1986) and Fagan et al. (1999). To understand the effects of fragmentation, we consider a situation where the local population growth rate $m(x)$ is positive in favorable regions but negative in unfavorable ones, and examine how the arrangement of those regions affects the eigenvalues λ_1^+ and σ_1, and hence the predictions of models such as (2.9):

$$\begin{aligned} u_t &= \nabla \cdot d(x)\nabla u + m(x)u \text{ in } \quad \Omega \times (0, \infty) \\ d(x)\frac{\partial u}{\partial \vec{n}} &+ \beta(x)u = 0 \qquad \text{on} \quad \partial\Omega \times (0, \infty). \end{aligned}$$

It is reasonable to expect that an environment where favorable and unfavorable habitat regions are closely intermingled will be less suitable than one where there are large regions of favorable habitat, even if the total amounts of favorable and unfavorable habitat are the same in the two environments. The reason why this should be the case is that the chances of an individual dispersing into an unfavorable region are greater if the favorable regions are small and are close to unfavorable regions. However, since rearranging the location of favorable regions will typically increase $m(x)$ in some places and decrease it in others, a more sophisticated form of analysis is required than the simple monotonicity result of Corollary 2.2. Recall that the model (2.9) (or the corresponding model with boundary condition $u = 0$ on part or all of $\partial\Omega$) predicts persistence if $\sigma_1 > 0$ in (2.10) and extinction if $\sigma_1 < 0$. Furthermore, by Theorem 2.6, we have $\sigma_1 > 0$ in (2.10) precisely when the positive principal eigenvalue $\lambda_1^+(m)$ for (2.13) satisfies $\lambda_1^+(m) < 1$. We can use the following result to understand how $\lambda_1^+(m(x))$ depends on $m(x)$:

Theorem 2.7. Suppose that the domain Ω and the functions $d(x)$, $\beta(x)$, and $m_n(x)$, $n = 1, 2, \ldots$ satisfy the hypotheses of Theorem 2.1, and that $\beta(x) > 0$ on a subset of $\partial\Omega$ or that the boundary condition $d(x)\partial u/\partial\vec{n} + \beta(x)u = 0$ is replaced with the condition $u = 0$ on all or part of $\partial\Omega$. Suppose further that there is a constant M_1 so that $||m_n||_\infty \le M_1$ for all n. Denote the positive principal eigenvalue of (2.13) with $m = m_n(x)$ by $\lambda_1^+(m_n)$. We have $\lambda_1^+(m_n) \to \infty$ as $n \to \infty$ if and only if

$$\limsup_{n\to\infty} \int_\Omega m_n \psi \, dx \le 0 \tag{2.32}$$

for all $\psi \in L^1(\Omega)$ with $\psi \ge 0$ a.e. If $\beta(x) \equiv 0$ on $\partial\Omega$ then the conclusion remains valid under the additional hypothesis that there exists $M_0 > 0$ such that $\int_\Omega m_n dx \le -M_0$ for all n.

Remarks: This result is proved in Cantrell and Cosner (1989) in the case of a Dirichlet boundary condition $u = 0$ on $\partial\Omega$. The proof is given in the Appendix to this chapter. We discuss the interpretation of condition (2.32) below, but first we describe the implications of Theorem 2.7 for persistence in the model (2.9).

Corollary 2.8. If (2.32) holds then for $m = m_n(x)$ with n sufficiently large the model (2.9) predicts extinction, i.e. $u \to 0$ as $t \to \infty$.

Proof: By Theorem 2.6, we have $\sigma_1 > 0$ if and only if $\lambda_1^+(m) < 1$. Since $\lambda_1^+(m_n) \to \infty$ as $n \to \infty$, we must have $\lambda_1^+(m_n) > 1$ and hence $\sigma_1 < 0$ for n sufficiently large, so that u decays exponentially as $t \to \infty$.

Remarks: The condition (2.32) requires that for any nonnegative integrable function ψ on Ω the sequence of numbers $\left\{\int_\Omega m_n \psi \, dx\right\}$ must have the property that for any $\epsilon > 0$,

$\int_\Omega m_n \psi dx < \epsilon$ for all sufficiently large n. (If we take ψ = constant then (2.32) implies that the average of $m_n(x)$ must become small or negative as $n \to \infty$.) Condition (2.32) will hold if $m_n(x)$ is small on all of Ω, e.g. $m_n(x) = 1/n$, but it may also hold in cases where $m_n(x)$ is large on parts of Ω, but negative on others provided that the regions where $m_n > 0$ and $m_n < 0$ become smaller and closer together as $n \to \infty$.

Suppose, for example, Ω is the interval $(0, \ell)$ and that $m_n(x) = \sin(nx)$. It follows from the Riemann-Lebesgue lemma (e.g. see Royden (1968) and Rudin (1966)) that (2.32) holds. Thus, the model $u_t = du_{xx} + \sin(nx)u$ on $(0, \ell) \times (0, \infty)$, $u(0, t) = u(\ell, t) = 0$ predicts extinction if n is sufficiently large. The functions $\sin(nx)$ will have maximum value 1 and will be positive on approximately half of $(0, \ell)$ for large n. However, as $n \to \infty$ the regions where $\sin(nx) > 0$ become smaller and they alternate with regions where $\sin(nx) < 0$. Thus, subdividing the interval $(0, \ell)$ more and more finely into favorable and unfavorable regions can result in a prediction of extinction, even if the total amount of favorable habitat remains roughly fixed. This analysis supports the general viewpoint that a highly fragmented landscape will often be less suitable for a given population than a landscape with large regions of favorable habitat. In fact, any distribution of favorable and unfavorable habitat regions that preserves a sufficiently large region of sufficiently good habitat will support a population, independent of what happens outside the favorable region. That idea can be formalized as follows.

Proposition 2.9. Suppose that σ_1 is the principal eigenvalue of (2.10). Suppose that there is a subdomain $\Omega' \subseteq \Omega$ such that the principal eigenvalue σ_1' for

$$\begin{aligned} \nabla \cdot d(x)\nabla\psi + m(x)\psi &= \sigma\psi \text{ in } \Omega \\ \psi &= 0 \quad \text{on } \partial\Omega \end{aligned} \tag{2.33}$$

is positive. Then $\sigma_1 > 0$, so (2.9) predicts persistence.

Proof: Let ψ' be the eigenfunction for the principal eigenvalue σ_1' of (2.33) on Ω'. Extend ψ' to be identically equal to zero outside Ω'. The resulting function ψ is then in $W^{1,2}(\Omega)$, and thus may be used as a test function in the quotient in (2.11a) or the expression on the right in (2.11b). Thus, the maximum over all test functions in $W^{1,2}$ must be at least as large as the value achieved by ψ, which is σ_1'. Since σ_1 is that maximum, $\sigma_1 \geq \sigma_1' > 0$.

2.3.2 Advection/Taxis

We now consider how advection can affect eigenvalues and hence predictions of extinction and persistence, at least in cases where the advection term can be expressed in terms of a gradient as in (2.23). Specifically, suppose a population grows and disperses according to the model

$$\begin{aligned} u_t &= \nabla \cdot [d(x)\nabla u - \vec{b}(x)u] + m(x)u \text{ in } \Omega \times (0,\infty) \\ u &= 0 \quad \text{on } \partial\Omega \times (0,\infty) \end{aligned} \tag{2.34}$$

and that (2.23) holds, so that $\vec{b} = -d\nabla B$. Assume that $\vec{b}$ is continuously differentiable, so that $\nabla\cdot\vec{b}$ makes sense. Then (2.34) be rewritten as in (2.24), so that the principal eigenvalue

σ_1 is determined by (2.28) and can be expressed as in (2.31) as

$$\sigma_1 = \sup_{\substack{\psi \in W_0^{1,2}(\Omega) \\ \psi \neq 0}} \left[\frac{\int_\Omega -de^{-B}|\nabla\psi|^2 dx + \int_\Omega me^{-B}\psi^2 dx}{\int_\Omega e^{-B}\psi^2 dx} \right]. \tag{2.35}$$

If $\psi \in W_0^{1,2}(\Omega)$ (i.e. ψ and $|\nabla\psi|$ are square integrable on Ω with $\psi = 0$ on $\partial\Omega$ in a suitable sense), then $\phi = e^{-B/2}\psi \in W_0^{1,2}(\Omega)$ also, and the mapping $\psi \mapsto e^{-B/2}\psi$ is bijective, so that we can rewrite (2.35) by taking $\psi = e^{B/2}\phi$ and taking the supremum over $\phi \in W_0^{1,2}(\Omega)$. We have $\nabla\psi = (1/2)e^{B/2}\phi\nabla B + e^{B/2}\nabla\phi$, so (2.35) becomes

$$\sigma_1 = \sup_{\substack{\phi \in W_0^{1,2}(\Omega) \\ \phi \neq 0}} \left[\frac{\int_\Omega -d[|\nabla\phi|^2 + \nabla B \cdot (\phi\nabla\phi) + (1/4)|\nabla B|^2\phi^2]dx + \int_\Omega m\phi^2 dx}{\int_\Omega \phi^2 dx} \right]. \tag{2.36}$$

Now, $\phi = 0$ on $\partial\Omega$ so by Green's formula

$$\int_\Omega d\nabla B \cdot (\phi\nabla\phi)dx = \int_\Omega d\nabla B \cdot \nabla(\phi^2/2)dx = \int_\Omega (-\nabla \cdot (d\nabla B))(\phi^2/2)dx. \tag{2.37}$$

(To be precise, $\phi \in W_0^{1,2}$ is zero on $\partial\Omega$ only in the weak or distributional sense, but that means ϕ behaves as if it were zero on $\partial\Omega$ as far as integration via Green's formula is concerned.) Substituting (2.37) into (2.36) yields

$$\sigma_1 = \sup_{\substack{\phi \in W_0^{1,2}(\Omega) \\ \phi \neq 0}} \frac{\int_\Omega -d|\nabla\phi|^2 + \int_\Omega \{(1/4)[2\nabla \cdot (d\nabla B) - d|\nabla B|^2] + m\}\phi^2 dx}{\int_\Omega \phi^2 dx}. \tag{2.38}$$

Equation (2.38) allows us to assess the effects of drift. The quotient in (2.38) is the same as the quotient in (2.35) with $B = 0$, which corresponds to the case of no drift, except that there is the extra term $(1/4)[2\nabla \cdot (d\nabla B) - d|\nabla B|^2]$ multiplying ϕ^2 in the second integral in the numerator. Another way to interpret (2.38) is that (2.38) asserts $\sigma_1 = \tilde{\sigma}_1$, where $\tilde{\sigma}_1$ is the principal eigenvalue for the problem

$$\begin{aligned} &\nabla \cdot d(x)\nabla\psi + \{(1/4)[2\nabla \cdot (d\nabla B) - d|\nabla B|^2] + m\}\psi = \sigma\psi \quad \text{in} \quad \Omega \\ &\psi = 0 \quad \text{on} \quad \partial\Omega. \end{aligned} \tag{2.39}$$

Since taking $B = 0$ in (2.35) gives the principal eigenvalue for (2.34) with no drift, we can compare (2.35) with (2.39) to obtain the following result.

Proposition 2.10. Let σ_0 denote the principal eigenvalue corresponding to the case of no drift in (2.34) (i.e. $\vec{b} = 0$ in (2.34) and hence $B = 0$ in (2.35)). Let

$$\overline{\gamma} = \sup_\Omega\{(1/4)[2\nabla \cdot (d\nabla B) - d|\nabla B|^2]\}$$

$$\underline{\gamma} = \inf_\Omega\{(1/4)[2\nabla \cdot (d\nabla B) - d|\nabla B|^2]\}.$$

If σ_1 is the principal eigenvalue corresponding to the case of (2.34) with drift $\vec{b}$ which can be written as $\vec{b} = -d\nabla B$, i.e. if σ_1 is as defined by (2.35), then

$$\sigma_0 + \underline{\gamma} \le \sigma_1 \le \sigma_0 + \overline{\gamma}. \tag{2.40}$$

Remark: Proposition 2.10 is very close to results obtained in Murray and Sperb (1983). Related results are obtained in Cantrell and Cosner (1991a). Murray and Sperb (1983) considered the case where d and $\vec{b}$ are constant. In that case, if $\vec{b} = (b_1, \ldots, b_n)$ we can take $B = (-1/d)[b_1x_1 + b_2x_2 + \cdots + b_nx_n]$ so that $|\nabla B| = |\vec{b}|/d$ and $\nabla^2 B = 0$. Thus, $\underline{\gamma} = \overline{\gamma} = -|\vec{b}|^2/4d < 0$. It follows that in the case of a completely hostile exterior environment and constant diffusion, any constant drift as might arise from prevailing winds is detrimental to a population inhabiting a finite patch Ω.

Proof: We have by 2.38 and the definition of $\underline{\gamma}$ that

$$\sigma_1 \ge \sup_{\substack{\phi \in W_0^{1,2} \\ \phi \ne 0}} \left[\frac{\int_\Omega -d|\nabla\phi|^2 dx + \int_\Omega m\phi^2 dx + \underline{\gamma}\int_\Omega \phi^2 dx}{\int_\Omega \phi^2 dx} \right]$$

$$= \sup_{\substack{\phi \in W_0^{1,2} \\ \phi \ne 0}} \left[\frac{\int_\Omega -d|\nabla\phi|^2 dx + \int_\Omega m\phi^2}{\int_\Omega \phi^2 dx} + \underline{\gamma} \right] = \sigma_0 + \underline{\gamma}.$$

The inequality $\sigma_1 \le \sigma_0 + \overline{\gamma}$ is proved in a similar way.

Proposition 2.10 shows that in general advection may be helpful, harmful, or neither because the term $(1/4)(2\nabla \cdot (d\nabla B) - d|\nabla B|^2)$ could be strictly negative or positive or could change sign. Proposition 2.10 and the calculations in (2.35)–(2.38) give a fairly general way of assessing the effects of advection or drift due to a gradient. A special case occurs when individuals can sense and respond to gradients in the local environmental quality $m(x)$. If individuals preferentially move in the direction of most rapid increase in $m(x)$, then $\vec{b} = \alpha\nabla m$ for some constant α. In the case of constant diffusion, we may set $B = -(\alpha/d)m$ and recover $\vec{b} = -d\nabla B$. Suppose we consider the case of (2.34) with $\vec{b} = \alpha\nabla m$ and no-flux boundary conditions. We can again compute the principal eigenvalue σ_1 of the differential operator on the right side of (2.34) via (2.28). In that case

$$\sigma_1 = \sup_{\substack{\psi \in W^{1,2}(\Omega) \\ \psi \ne 0}} \left[\frac{-\int_\Omega d e^{(\alpha/d)m} |\nabla\psi|^2 dx + \int_\Omega m e^{(\alpha/d)m} \psi^2 dx}{\int_\Omega e^{(\alpha/d)m} \psi^2 dx} \right]. \tag{2.41}$$

In (2.41) the sup is taken over all of $W^{1,2}(\Omega)$, and constant functions belong to $W^{1,2}(\Omega)$, so by taking $\psi = 1$ in the quotient in (2.41) we obtain

$$\sigma_1 \ge \int_\Omega e^{(\alpha/d)m} m\,dx \Big/ \int_\Omega e^{(\alpha/d)m} dx. \tag{2.42}$$

In (2.42) the integral in the denominator is always positive. If $m(x_0) < 0$ then $\lim_{\alpha\to\infty} e^{(\alpha/d)m(x_0)} m(x_0) = 0$, while if $m(x_0) > 0$ then $\lim_{\alpha\to\infty} e^{(\alpha/d)m(x_0)} m(x_0) = \infty$. Thus, the integral in the numerator of (2.42) is positive for α large, so $\sigma_1 \geq 0$ in (2.42) for α large provided $m(x_0) > 0$ for some $x_0 \subset \Omega$. Roughly, α measures the rate of advection up the gradient of m.

Remarks: In the case of constant diffusion and no-flux boundary conditions, a population can survive (i.e. $\sigma_1 > 0$) in any environment where $m(x) > 0$ somewhere and $m(x)$ depends differentially on x (so that the gradient of m exists) by moving rapidly enough up the gradient of $m(x)$, i.e. if α is large. Results of this type are obtained in Belgacem and Cosner (1995).

The question of how advection up the habitat gradient affects population dynamics under general boundary conditions and variable diffusion is much more delicate. In that case advection upward along the gradient of $m(x)$ may be either helpful or harmful. (It can be harmful, for example, if regions where $m(x)$ is large lie adjacent to a lethal boundary, since then individuals moving toward regions where $m(x)$ is large are more likely to encounter the boundary.) The methods described above can address the more general situation, but the specific predictions will depend on details of the boundary conditions, diffusion rate, and spatial distribution of habitat types as described by $m(x)$. For additional discussion, see Belgacem and Cosner (1995), Belgacem (1997) and Cosner and Lou (2001).

In cases where there is advective transport but the vector $\vec{b}$ describing the advection is not of the form $-d\nabla B$ for some B, it becomes much more complicated to treat eigenvalue problems via the variational approach developed in Section 2.2. (Variational approaches are still possible in principle, see Holland (1977, 1978) and Belgacem (1997), just difficult to use in practice.) To extend the notion of eigenvalues naturally to such cases, and to the case of time periodic models with diffusion, it is natural to use an alternative approach based on the theory of positive operators. That approach is the topic of a later section.

2.4 Graphical Analysis in One Space Dimension

In the case of models in one space dimension where the local growth rate $m(x)$ is piecewise constant, it is possible to characterize the eigenfunction corresponding to a principal eigenvalue explicitly in terms of trigonometric and hyperbolic functions. By matching the eigenfunction and its derivative across the jumps in $m(x)$ we can obtain a transcendental equation which determines the principal eigenvalue. By studying that equation we can determine how the principal eigenvalue depends on various features of the growth rate. Although this approach is limited to simple cases, it can provide insights that are more detailed than those that can be obtained by more general methods. We illustrate the method in some detail in a particular case, then give a somewhat less detailed treatment of several other cases. The analysis is taken from Cantrell and Cosner (1991b, 1999).

2.4.1 The Best Location for a Favorable Habitat Patch

Consider the model

$$\begin{aligned} \frac{\partial u}{\partial t} &= D\frac{\partial^2 u}{\partial x^2} + m(x)u \quad \text{on} \quad (0,1)\times(0,\infty) \\ -\delta\frac{\partial u}{\partial x}(0,t) + \beta u(0,t) &= \delta\frac{\partial u}{\partial x}(1,t) + \beta u(1,t) = 0 \quad \text{for} \quad t \in (0,\infty). \end{aligned} \tag{2.43}$$

This model predicts growth for the population if the principal eigenvalue σ_1 of the problem

$$\begin{aligned} &D\frac{d^2\psi}{dx^2} + m(x)\psi = \sigma\psi \quad \text{in} \quad (0,1) \\ &-\delta\frac{d\psi}{dx}(0) + \beta\psi(0) = \delta\frac{d\psi}{dx}(1) + \beta\psi(1) = 0 \end{aligned} \tag{2.44}$$

is positive. We may write (2.44) as $\frac{d^2\psi}{dx^2}+(m/D)\psi = (\sigma/D)\psi$. It follows from Theorem 2.6 that $\sigma_1/D > 0$, and hence $\sigma_1 > 0$ if and only if $(1/D) > \lambda_1^+(m(x))$, where $\lambda_1^+(m(x))$ is the principal eigenvalue of

$$\begin{aligned} &\frac{d^2\phi}{dx^2} + \lambda m(x)\phi = 0 \quad \text{in} \quad (0,1) \\ &-\delta\frac{d\phi(0)}{dx} + \beta\phi(0) = \delta\frac{d\phi(1)}{dx} + \beta\phi(1) = 0. \end{aligned} \tag{2.45}$$

A smaller value of $\lambda_1^+(m(x))$ reflects a more favorable environment than a larger one, because the condition $1/D > \lambda_1^+(m(x))$ is more easily satisfied. Thus, the way that changes in $m(x)$ affect the quality of the environment can be measured by how they affect $\lambda_1^+(m(x))$.

The first question we shall address is the following: if we can select only a single subinterval of length $T < 1$ which is favorable for population growth (i.e. $m(x) > 0$), and the rest of the interval $(0, 1)$ is unfavorable, what is the optimal location for the interval representing the favorable patch? Since we are interested only in comparing different configurations we may normalize $m(x)$, so we shall assume that for some $k > 0$ and $a \in [0, 1 - T]$,

$$m_a(x) = \begin{cases} -1 & \text{for } x \in [0,a) \\ k & \text{for } x \in [a, a+T) \\ -1 & \text{for } x \in [a+T, 1]. \end{cases}$$

The parameter a determines the location of the favorable patch, so we need to determine how $\lambda_1^+(m(x))$ depends on a. We start with the case of Dirichlet boundary conditions, i.e. a lethal or absorbing boundary, so that $\delta = 0$ and $\beta = 1$. On each of the subintervals $[0, a)$, $[a, a + T)$ and $[a + T, 1]$ we can solve explicitly for the eigenfunction ϕ in (2.45). If we let $\alpha = \sqrt{\lambda_1^+(m(x))}$ we may take

$$\phi(x) = \begin{cases} \sinh(\alpha x) & \text{for } x \in [0,a) \\ A\cos(\alpha\sqrt{k}(x-c)) & \text{for } x \in [a, a+T) \\ B\sinh(\alpha(1-x)) & \text{for } x \in [a+T, 1]. \end{cases}$$

Here we have used the boundary conditions $\phi(0) = \phi(1) = 0$ to determine the form of ϕ on $[0, a]$ and $[a + T, 1]$. The constant c is a phase shift that enables us to write the general solution to $\phi'' + \alpha^2 k\phi = 0$ in terms of a single function. By computing $d\phi/dx$ and matching the values of ϕ and $d\phi/dx$ across the interfaces at $x = a,\ a+T$ we obtain the relations

$$\begin{aligned}
&\sinh \alpha a = A \cos \alpha\sqrt{k}(a-c)\\
&\alpha \cosh \alpha a = -\alpha A\sqrt{k} \sin \alpha\sqrt{k}(a-c)\\
&A \cos \alpha\sqrt{k}(a+T-c) = B \sinh \alpha(1-(a+T))\\
&-\alpha A\sqrt{k} \sin \alpha\sqrt{k}(a+T-c) = -\alpha B \cosh \alpha(1-(a+T))
\end{aligned}$$

Dividing the first equation by $\cosh \alpha a$ and using the second equation to rewrite the right side of the result yields

$$\tanh \alpha a = (-1/\sqrt{k}) \cot(\alpha\sqrt{k}(a-c)). \tag{2.46}$$

Treating the third and fourth equations similarly yields

$$\tanh(\alpha(1-(a+T))) = (1/\sqrt{k}) \cot(\alpha\sqrt{k}(a+T-c)). \tag{2.47}$$

Using (2.46) and (2.47) we have

$$\begin{aligned}
\cot(\alpha\sqrt{k}T) &= \cot(\alpha\sqrt{k}(a+T-c) - \alpha\sqrt{k}(a-c))\\
&= \frac{\cot(\alpha\sqrt{k}(a+T-c))\cot(\alpha\sqrt{k}(a-c)+1)}{\cot(\alpha\sqrt{k}(a-c)) - \cot(\alpha\sqrt{k}(a+T-c))}\\
&= \frac{k \tanh(\alpha(1-a-T))\tanh(\alpha a) - 1}{\sqrt{k}[\tanh(\alpha a) + \tanh \alpha(1-a-T)]}\\
&= F(a, \alpha).
\end{aligned} \tag{2.48}$$

We know that for $\alpha^2 = \lambda = \lambda_1^+(m)$ we may assume $\phi > 0$ in $(0, 1)$, so it must be the case that $\alpha \in (0, \pi/\sqrt{k}T)$ since otherwise ϕ changes sign in the interval $(a, a+T)$. The function $\cot(\alpha\sqrt{k}T)$ is strictly decreasing in α, with $\cot(\alpha\sqrt{k}T) \to \infty$ as $\alpha \to 0+$ and $\cot(\alpha\sqrt{k}/T) \to -\infty$ as $\alpha \to \pi/\sqrt{k}T-$. Also, for any fixed a, $F(\alpha, a) \to -\infty$ as $\alpha \to 0+$ since the denominator in $F(\alpha, a)$ approaches 0 while the numerator approaches -1. Moreover, a somewhat messy calculation shows that $\partial F/\partial\alpha > 0$ for $\alpha > 0$. It follows that the graphs of $F(\alpha, a)$ and $\cot \alpha\sqrt{k}T$ intersect exactly once in the interval $(0, \pi/\sqrt{k}T)$, thus determining α and hence $\lambda_1^+(m)$. Furthermore, changes in the parameter a which increase $F(\alpha, a)$ move the intersection to the left, because $\cot(\alpha\sqrt{k}T)$ is decreasing in α and is independent of a (see Figure 2.2).

Hence, to determine the optimal value of a, we want to find the value that maximizes $F(\alpha, a)$. If we compute the derivative of $F(\alpha, a)$ with respect to a, then the quotient rule produces a fraction in which the denominator is a square, so the sign of $\partial F(\alpha, a)/\partial a$ is determined by the numerator. The numerator is $R(\alpha, a) = \text{sech}^2(\alpha a)[1 + k\tanh^2(\alpha(1-a-T))] - \text{sech}^2(\alpha(1-a-T)[1 + k\tanh^2 \alpha a]$. Setting $R(\alpha, a) = 0$ yields the relation

$$\frac{1 + k\tanh^2(\alpha(1-a-T))}{\text{sech}^2(\alpha(1-a-T))} = \frac{1 + k\tanh^2(\alpha a)}{\text{sech}^2(\alpha a)}.$$

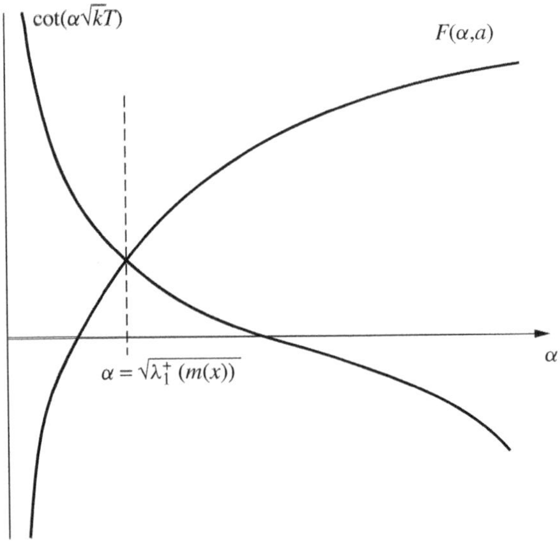

Figure 2.2 Graphical analysis for determining $\alpha = \sqrt{\lambda_1^+(m(x))}$ in (2.48).

It is easy to see that the function $(1 + k\tanh^2(x))/\operatorname{sech}^2(x)$ is increasing for $x > 0$, so $R(\alpha, a) = 0$ implies $\alpha(1 - a - T) = \alpha a$, so that $a = (1 - T)/2$. Furthermore, $(1 + k\tanh^2(x))/\operatorname{sech}^2(x) = 1$ when $x = 0$, so since it is increasing in x we have $1 + k\tanh^2(x) - \operatorname{sech}^2(x) > 0$ for $x > 0$. We may conclude that

$$\lim_{a\to 0^+} R(\alpha, a) = 1 + k\tanh^2(\alpha(1 - T)) - \operatorname{sech}^2\alpha(1 - T) > 0$$

and

$$\lim_{a\to(1-T)-} R(\alpha, a) = \operatorname{sech}^2(\alpha(1 - T)) - 1 - k\tanh^2(\alpha(1 - T)) < 0.$$

Thus, we see that for any $\alpha \in (0, \ \pi/\sqrt{kT})$, the value of $F(\alpha, a)$ is largest when $a = (1 - T)/2$. Increasing $F(\alpha, a)$ decreases the value of α which is determined by (2.48), which in turn decreases $\lambda_1^+(m(x))$, so the smallest possible value for $\lambda_1^+(m(x))$ for any given choice of k or T occurs when $a = (1 - T)/2$. Geometrically, that corresponds to the favorable region being centered in the middle of the interval (0, 1).

Remark: In the case of Dirichlet boundary conditions something similar to this turns out to be true in more general situations. For the model

$$\begin{aligned} \frac{\partial u}{\partial t} &= D\nabla^2 u + m(x)u \quad &&\text{in} \quad \Omega \times (0, \infty) \\ u &= 0 &&\text{on} \quad \partial\Omega \times (0, \infty), \end{aligned} \tag{2.49}$$

the eigenvalue problem corresponding to (2.45) is

$$\begin{aligned} \nabla^2\phi + \lambda m(x)\phi &= 0 \quad &&\text{in} \quad \Omega \\ \phi &= 0 &&\text{on} \quad \partial\Omega. \end{aligned}$$

If we constrain $m(x)$ by requiring that $\int_\Omega m(x)dx = m_0, \quad -m_1 \leq m(x) \leq m_2$, where m_0, m_1, and m_2 are positive constants, it turns out that the minimum value for $\lambda_1^+(m(x))$ occurs when $m(x)$ has the form $m(x) = m_2\chi_{\Omega_1}(x) - m_1\chi_{\Omega\setminus\Omega_1}(x)$ where Ω_1 is a domain inside Ω such that $\overline{\Omega}_1 \subseteq \Omega$ (that is, Ω_1 is subdomain of Ω lying in the interior of Ω) and $\chi_{\Omega_1}, \chi_{\Omega\setminus\Omega_1}$ are the characteristic functions of Ω_1 and its complement, respectively. In other words, the optimal arrangement for $m(x)$ is obtained when $m(x) = m_2$ on some region in the interior of Ω, and $m(x) = -m_1$ elsewhere. (The constraint on the integral of $m(x)$ means that the average of $m(x)$ is fixed.) This result is obtained in Cantrell and Cosner (1989) by means of an iteration process. In general it is difficult to give a simple description of Ω_1, except to the extent that Ω_1 lies inside Ω, away from $\partial\Omega$.

We now consider the case of no-flux, i.e. Neumann, boundary conditions. In this case we must have $kT - (1 - T) < 0$, so that the integral of $m(x)$ will be negative and hence $\lambda_1^+(m(x))$ will exist. The model is still given by (2.45) but with $\beta = 0$ and $\delta = 1$. The analysis proceeds in the same general way as in the case of Dirichlet boundary conditions, but now to satisfy the boundary conditions we must construct the eigenfunction $\phi(x)$ as

$$\phi(x) = \begin{cases} \cosh(\alpha x) & \text{for} \quad x \in [0, a) \\ A\cos(\alpha\sqrt{k}(x - c)) & \text{for} \quad x \in [a, a + T) \\ B\cosh\alpha(1 - x) & \text{for} \quad x \in [a + T, 1]. \end{cases}$$

Matching $\phi(x)$ and $\phi'(x)$ across the interfaces and solving as before yields (after some calculation)

$$\begin{aligned} \cot(\alpha\sqrt{k}T) &= \frac{k\cosh(\alpha a)\cosh(\alpha(1 - a - T)) - \sinh(\alpha a)\sinh(\alpha(1 - a - T))}{\sqrt{k}\sinh\alpha(1 - T)} \\ &= \frac{(k - 1)\cosh(\alpha(1 - T)) + (k + 1)\cosh(\alpha(1 - 2a - T))}{\sqrt{k}\sinh\alpha(1 - T)} \\ &= G(a, \alpha). \end{aligned}$$

It turns out that $G(a, \alpha)$ is decreasing in α, with $G(a, \alpha) \to (k - 1)\sqrt{k}$ as $\alpha \to \infty$, and $G(a, \alpha) \to \infty$ as $\alpha \to 0+$ with order $\sqrt{k}/\alpha(1 - T)$. Since $\cot(\alpha\sqrt{k}T) \to \infty$ with order $1/\sqrt{k}T\alpha$ as $\alpha \to 0+$, it follows from our assumption $kT < 1 - T$ that $\cot(\alpha\sqrt{k}T)$ grows more rapidly as $\alpha \to 0+$ than does $G(a, \alpha)$. Thus, the graphs of $\cot(\alpha\sqrt{k}T)$ and $G(a, \alpha)$ must intersect. Since $\cot(\alpha\sqrt{k}T)$ is decreasing in α, increasing $G(a, \alpha)$ moves the graph of $G(a, \alpha)$ (with respect to α) upward and hence causes the curves to intersect at a smaller value of α. Computing $\partial G(a, \alpha)/\partial a$ yields a fraction with positive denominator and numerator equal to $-2\alpha(k+1)\sinh\alpha(1-2a-T)$. Thus, $\partial G(a, \alpha)/\partial a$ is negative for all α if $0 < a < (1 - T)/2$ and positive for $(1 - T)/2 < a < 1 - T$. Thus, the graph of $G(a, \alpha)$ is *lowest*, leading to the largest value of α at the intersection point, if $a = (1 - T)/2$. It follows that the best location for the favorable patch under Dirichlet boundary conditions is the worst under Neumann conditions. The cases $a = 0$ and $a = 1 - T$ give the same value for α at the intersection point. That value corresponds to the optimal location of the favorable patch, which in this case is at either end of the interval $[0, 1]$.

The same sort of analysis can be carried out under Robin, i.e. third kind, boundary conditions (2.45). Such an analysis was performed in Cantrell and Cosner (1991b) for the case where $k = 1$, with $\delta = 1$ and $\beta > 0$ but otherwise arbitrary. The conclusion of the analysis was that for $\beta < \pi/2T$ and $\delta = 1$, the best location for the favorable patch was at either end of $[0, 1]$, just as in the case of Neumann conditions. For $\beta = \pi/2T$ the intersection point of the curves determining α was at $\alpha = \beta = \pi/2T$ independent of a, so that the location of the favorable patch was irrelevant. For $\beta > \pi/2T$, the optimal location for the favorable patch was in the center of the interval, just as in the case of Dirichlet boundary conditions. Note that as the boundary conditions change from being close to Neumann (reflecting) to being close to Dirichlet (absorbing), there is a sudden shift in the optimal location for a single favorable patch. The explanation is that both the regions where $m(x) < 0$ and the loss of individuals across the boundary of the overall environment $[0, 1]$ have a detrimental effect on the population. For the case considered in Cantrell and Cosner (1991b), with $\delta = 1$ and $m(x)$ taking the value 1 on the favorable patch and -1 elsewhere, it turns out that for the value $\beta = \pi/2T$ the loss rate across the boundary has the same effect as the loss rate due to the presence of regions where $m(x)$ is negative. Hence, in that case, the two effects are equal in strength and the location of the favorable patch is irrelevant. When $\beta > \pi/2T$ the loss of individuals across the boundary of $[0, 1]$ is the stronger effect, so the best location of the favorable patch is at the center of the interval where it is the most insulated from the effects of the boundary. If $\beta < \pi/2T$ then the effect of the regions where $m(x) = -1$ is stronger than the effect of loss across the boundary, so the optimal location of the favorable patch is at one of the ends of the interval, where it is only exposed to the region where $m(x) = -1$ at one of its ends.

2.4.2 Effects of Buffer Zones and Boundary Behavior

In most of the models we consider, individuals are assumed to be equally likely to cross interfaces between habitat types in either direction. However, in some cases, individuals may have a preferred direction at an interface; see Goszczynski (1979a,b), Wegner and Merriam (1979), Yahner (1983), Bach (1984) and Kareiva (1985). A model for diffusion with an interface which particles or individuals cross with a preferred direction was proposed and developed by Walsh (1978) and Harrison and Shepp (1981), known as skew Brownian motion. A description of how to formulate models based on skew Brownian motion is given in Chapter 1. In the present discussion our goal is to gain insight into the ways that the quality of the habitat in regions surrounding a favorable patch interacts with the behavior of individuals at the patch boundary to influence population dynamics in the overall environment. The question is motivated by issues in refuge design. Generally, it might be expected that having buffer zones of moderate quality around a refuge would be better than having buffer zones of low quality. However, if individuals respond to the quality of the buffer zones in their behavior at the refuge boundary, the situation is less clear. In particular, if individuals have an aversion to crossing the patch boundary which increases as the quality of the buffer zone decreases, it is plausible that a moderately hostile buffer zone might be less effective than a very hostile one, because individuals might be much more averse to crossing the refuge boundary into a very hostile region than a moderately hostile one. A moderately hostile region that individuals enter freely thus might cause a higher loss rate than a very hostile region that individuals usually avoid. In one space dimension, it is sometimes possible to bring the approach of the preceding subsection to bear. To address this question and related issues we consider a one-dimensional environment consisting of a favorable region of length $2L$ bounded on each side by less favorable regions of length ℓ, with the exterior of the overall environment assumed to be lethal. Let the growth rate in the more favorable region be given by $m(x) = r > 0$ and in the less favorable regions by $m(x) = s < r$. Similarly, suppose that the diffusion rate in the less favorable region is D_1, while that in more favorable region is D_2. Finally, assume that at the interfaces the probability of moving from the less favorable region into the more favorable region is γ, so that the probability of moving from the more favorable region into the less favorable region is $1 - \gamma$. We may choose coordinates so that one of the interfaces is at $x = 0$ and the other at $x = 2L$, so that the overall environment consists of the interval $[-\ell, 2L + \ell]$ and is symmetric about $x = L$. This symmetry will be exploited in our analysis of the model. In this setting we are thinking of the favorable region $(0, 2L)$ as a refuge and the less favorable regions $(-\ell, 0)$ and $(2L, 2L + \ell)$ as buffer zones. The model now becomes

$$\frac{\partial u}{\partial t} = \begin{cases} D_1 \dfrac{\partial^2 u}{\partial x^2} + su & \text{for} \quad -\ell < x < 0, \ 2L < x < 2L + \ell, \ t \in (0, \infty) \\ D_2 \dfrac{\partial^2 u}{\partial x^2} + ru & \text{for} \quad 0 < x < 2L, \ t \in (0, \infty) \end{cases} \tag{2.50}$$

with boundary conditions

$$u(-\ell, t) = u(2L + \ell, t) = 0. \tag{2.51}$$

To describe the matching conditions at interfaces we will use the notation $f|_{x=a+} = \lim_{x \to a+} f(x)$, and similarly $f|_{x=a-} = \lim_{x \to a-} f(x)$. The matching conditions are

$$\begin{aligned} &(1-\gamma) D_1 \frac{\partial u}{\partial x}\bigg|_{x=0-} = \gamma D_2 \frac{\partial u}{\partial x}\bigg|_{x=0+} \\ &\gamma D_2 \frac{\partial u}{\partial x}\bigg|_{x=2L-} = (1-\gamma) D_1 \frac{\partial u}{\partial x}\bigg|_{x=2L+} \\ &\left(D_1 \frac{\partial^2 u}{\partial x^2} + su\right)\bigg|_{x=0-} = \left(D_2 \frac{\partial^2 u}{\partial x^2} + ru\right)\bigg|_{x=0+} \\ &\left(D_2 \frac{\partial^2 u}{\partial x^2} + ru\right)\bigg|_{x=2L-} = \left(D_1 \frac{\partial^2 u}{\partial x^2} + su\right)\bigg|_{x=2L+} . \end{aligned} \tag{2.52}$$

The eigenvalue problem we shall study for this model is then given by

$$\begin{aligned} &D_1 \frac{d^2\psi}{dx^2} + s\psi = \sigma\psi \quad \text{for} \quad x \in (-\ell, 0) \cup (2L, 2L + \ell) \\ &D_2 \frac{d^2\psi}{dx^2} + r\psi = \sigma\psi \quad \text{for} \quad x \in (0, 2L). \end{aligned} \tag{2.53}$$

The interface conditions (2.52) now reduce to

$$\begin{aligned} &(1-\gamma) D_1 \frac{d\psi}{dx}\bigg|_{x=0-} = \gamma D_2 \frac{d\psi}{dx}\bigg|_{x=0+} \\ &\gamma D_2 \frac{d\psi}{dx}\bigg|_{x=2L-} = (1-\gamma) \frac{d\psi}{dx}\bigg|_{2L+} \\ &\psi|_{x=0-} = \psi|_{x=0+} \\ &\psi|_{x=2L-} = \psi|_{x=2L+} . \end{aligned} \tag{2.54}$$

We can reduce the eigenvalue problem further by using the symmetry of the problem. If we construct a function $\tilde{\psi}(x)$ that satisfies (2.53) on $(-\ell, L)$ and the interface conditions (2.54) at $x = 0$, with $\tilde{\psi} = 0$ at $x = -\ell$ and $d\tilde{\psi}/dx = 0$ at $x = L$, then we can extend $\tilde{\psi}$ by reflection about $x = L$ to obtain an eigenfunction $\psi(x)$ for the original problem by taking $\psi = \tilde{\psi}(x)$ for $x \in [-\ell, L]$ and $\psi = \tilde{\psi}(2L - x)$ for $x \in [L, 2L + \ell]$. Making this reduction and dropping the tilde on $\tilde{\psi}$ leads to the problem

$$\begin{aligned}
&D_1 \frac{d^2\psi}{dx^2} + s\psi = \sigma\psi \quad \text{for} \quad x \in (-\ell, 0) \\
&D_2 \frac{d^2\psi}{dx^2} + r\psi = \sigma\psi \quad \text{for} \quad x \in (0, L) \\
&\psi|_{x=0-} = \psi|_{x=0+} \\
&(1-\gamma)D_1 \frac{d\psi}{dx}\bigg|_{x=0-} = \gamma D_2 \frac{d\psi}{dx}\bigg|_{x=0+} \\
&\psi(-\ell) = 0, \quad \frac{d\psi}{dx}(L) = 0.
\end{aligned} \tag{2.55}$$

We may now analyze (2.55) in the same manner as in our examination of (2.45); i.e. by using calculus to examine the intersection of two transcendental curves. The analysis is more intricate in the present case, however, and the reader may find it useful to consult Figure 2.3 upon occasion.

To begin our analysis of (2.55), note that for $0 < \gamma < 1$ we will have $\sigma_1 < r$. That follows since for $\gamma = 1$ the favorable subinterval $(0, L)$ will have purely reflecting boundary conditions, so we would have $\psi \equiv 1$ and $\sigma_1 = r$ in that case; but the loss into the interval $(-\ell, 0)$ via the matching condition at $x = 0$ and the absorbing boundary condition at $x = -\ell$ means that ψ must have $-d\psi/dx < 0$ at $x = 0$, so that ψ in effect satisfies a third-kind boundary condition at $x = 0$, and hence must correspond to a value of σ_1 which is less than the value $\sigma_1 = r$ which would occur for a reflecting boundary. Thus, we may assume that on $(0, L)$ $\psi(x)$ has the form

$$\psi(x) = A\cos(\sqrt{(r-\sigma)/D_2}(x - L)). \tag{2.56}$$

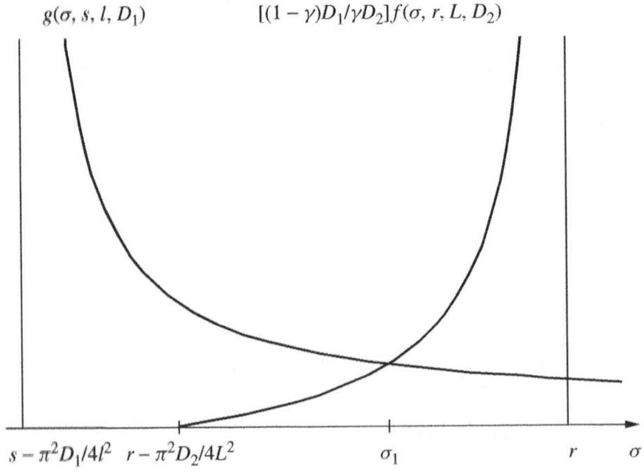

Figure 2.3 Graphical analysis for determining σ_1 in (2.58).

To prevent $\psi(x)$ from changing sign in the interval $(0, L)$ we must have $(r-\sigma)/D_2 < \pi^2/4L^2$ or $\sigma > r - \pi^2 D_2/4L^2$. The value $r - \pi^2 D_2/4L^2$ is the principal eigenvalue that would be obtained if $\ell = 0$ so that $\psi(0) = 0$, so any buffer at all will be of some benefit. The form $\psi(x)$ will take on $(-\ell, 0)$ depends on how σ is related to s. We have for $x \in (-\ell, 0)$

$$\psi(x) = \begin{cases} B\sinh(\sqrt{(\sigma-s)/D_1}(x+\ell) & \text{if } \quad \sigma > s \\ B(x+\ell) & \text{if } \quad \sigma = s \\ B\sin(\sqrt{(s-\sigma)/D_1}(x+\ell)) & \text{if } \quad \sigma < s. \end{cases} \tag{2.57}$$

It turns out that each of the cases $\sigma_1 < s$, $\sigma_1 = s$, $\sigma_1 > s$ can occur for some values of the remaining parameters. If $\sigma_1 < s$ we must have $\sigma_1 \geq s - \pi^2 D_1/4\ell^2$ so that $\psi(x)$ does not change sign on $(-\ell, 0)$. By substituting the forms (2.56), (2.57) into the matching conditions at $x = 0$ in (2.55) and doing some algebra to eliminate A and B we obtain the relation

$$\left(\frac{(1-\gamma)D_1}{\gamma D_2}\right) f(\sigma, r, L, D_2) = g(\sigma, s, \ell, D_1) \tag{2.58}$$

where

$$f(\sigma, r, L, D_2) = \frac{\cot(\sqrt{(r-\sigma)/D_2}L)}{\sqrt{(r-\sigma)/D_2}} \tag{2.59}$$

and

$$g(\sigma, s, \ell, D_1) = \begin{cases} \dfrac{\tan(\sqrt{(s-\sigma)/D_1}\ell)}{\sqrt{(s-\sigma)/D_1}} & \text{if } \quad \sigma < s \\ \ell & \text{if } \quad \sigma = s \\ \dfrac{\tanh(\sqrt{(\sigma-s)/D_1}\ell)}{\sqrt{(\sigma-s)/D_1}} & \text{if } \quad \sigma > s. \end{cases} \tag{2.60}$$

It turns out that for $\sigma > s - \pi^2 D_1/4\ell^2$ the function $g(\sigma, s, \ell, D_1)$ is continuous and decreasing in σ, with a vertical asymptote at $\sigma = s - \pi^2 D_1/4\ell^2$. For $r - \pi^2 D_2^2/4L^2 < \sigma < r$ the function $f(\sigma, r, L, D_2)$ is increasing in σ, with $f = 0$ when $\sigma = r - \pi^2 D_2^2/4L^2$ and with a vertical asymptote at $\sigma = r$. Since $r > s > s - \pi^2 D_1/4\ell^2$, the graphs of f and g as functions of σ must intersect at some value of σ between $\min(r - \pi^2 D_2/4L^2,\ s - \pi^2 D_1/4\ell^2)$ and r. Furthermore, since the term $[(1-\gamma)D_1/\gamma D_2]f(\sigma, r, L, D_2)$ is increasing in σ, while $g(\sigma, s, \ell, D_1)$ is decreasing, any change in the other parameters that raises the graph of f or lowers the graph of g will decrease the value σ_1 determined by the intersection of the graphs. Conversely, raising the graph of g or lowering the graph of $[(1-\gamma)D_1/\gamma D_2]f$ will increase σ_1.

We can use the behavior of the graphs of f and g to study how various parameters affect the value of σ_1 and hence the overall quality of the environment. First, notice that for $\sigma > s - \pi^2 D_1/4\ell^2$ the function $g(\sigma, s, \ell, D_1)$ is increasing in ℓ. Similarly, $[(1-\gamma)D_1/\gamma D_2]f(\sigma, r, L, D_2)$ is decreasing in L. *Thus, increasing the size of ℓ or L will increase the average population growth rate σ_1 on the overall environment $[-\ell, 2L+\ell]$.* However, there are limits to the amount σ_1 can increase. We always have $r - \pi^2 D_2/4L_2 < \sigma_1 < r$, so as $L \to \infty$ we have $\sigma_1 \to r$. Thus, by increasing L we can bring σ_1 as close as we like to r; that is, we can effectively eliminate the effects of the lethal boundaries $x = -\ell, 2L + \ell$ by making L sufficiently large. To understand what happens as $\ell \to \infty$ requires a bit more analysis. First, notice that for $\sigma = s$ we have $g(s, s, \ell, D_1) = \ell$ while $[(1 - \gamma)D_1/\gamma D_2]f(s, r, L, D_2)$ does not depend on ℓ. Thus, for ℓ large we have $g(s, s, \ell, D_1) > [(1 - \gamma)D_1/\gamma D_2]f(s, s, L, D_2)$, so that the intersection point σ_1 must lie to the right of $\sigma = s$ since g is decreasing in σ and f is increasing in σ. Thus, for large ℓ, the formula for g near the intersection point $\sigma = \sigma_1$ is $g(\sigma, s, \ell, D_1) = \tanh \sqrt{(\sigma - s)/D_1}\ell)/\sqrt{(\sigma - s)/D_1}$ so that $g(\sigma, s, \ell, D_1) \uparrow \sqrt{D_1/(\sigma - s)}$ as $\ell \to \infty$. Thus, as $\ell \to \infty$, $\sigma_1 \uparrow \sigma_1^*$ where σ_1^* is the value of σ for which $[(1-\gamma)D_1/\gamma D_2]f(\sigma, r, L, D_2) = \sqrt{D_1/(\sigma - s)}$. Since $f(\sigma, r, L, D_2) \to \infty$ as $\sigma \to r$ we have $\sigma_1^* < r$. *Thus, there is ultimately a diminishing benefit to increasing the size of the buffer zones, because the size of the buffer zones cannot completely compensate for their lower quality.*

We can also examine the role of the parameter γ which describes how individuals behave at the interface between more favorable and less favorable regions. Recall that $\gamma < 1/2$ corresponds to a preference to move from the more favorable region into the less favorable region, while $\gamma > 1/2$ corresponds to a preference for the more favorable region. As $\gamma \to 1$, the graph of $[(1 - \gamma)D_1/\gamma D_2]f$ decreases toward zero for each value of $\sigma < r$, but $f(\sigma, r, L, D_2) \to \infty$ as $\sigma \to r$, so as $\gamma \to 1$ the intersection point σ_1 for the graphs of $[(1 - \gamma)D_1/\gamma D_2]f$ and g moves to the right toward r; that is, $\sigma_1 \to r$ *as* $\gamma \to 1$. As $\gamma \to 0$, the function $[(1 - \gamma)D_1/\gamma D_2]f \uparrow \infty$ for all $\sigma \in (r - \pi^2 D_2/4L^2, r)$ but $f(\sigma, r, L, D_2) = 0$ for $\sigma = r - \pi^2 D_2/4L^2$. Thus, as $\gamma \to 0$, σ_1 moves to the left. The intersection of $[(1 - \gamma)D_1/\gamma D_2]f$ and g must occur for $\sigma > \max\{r - \pi^2 D_2/4L^2,\ s - \pi^2 D_1/4\ell^2\}$, because $f(\sigma, r, L, D_2) = 0$ at $\sigma = r - \pi^2 D_2/4L^2$ and $g(\sigma, s, \ell, D_1) \uparrow \infty$ as $\sigma \downarrow s - \pi^2 D_1/4\ell^2$. For any value of $\sigma_0 > \max\{r - \pi^2 D_2/4L^2,\ s - \pi^2 D_1/4\ell^2\}$, the value of $[(1 - \gamma)D_1/\gamma D_2]f(\sigma_0, r, L, D_2)$ will be greater than that of $g(\sigma_0, s, \ell, D_1)$ for γ close enough to zero, so for γ sufficiently small we will have $\sigma_1 < \sigma_0$. Since $\sigma_0 > \max\{r - \pi^2 D_2/4L^2,\ s - \pi^2 D_1/4\ell^2\}$ is arbitrary, $\sigma_1 \downarrow \max\{r - \pi^2 D_2/4L^2,\ s - \pi^2 D_1/4\ell^2\}$ *as* $\gamma \downarrow 0$. This makes sense, because $r - \pi^2 D_2/4L^2$ is the principal eigenvalue for $D_2 d^2\psi/dx^2 + r\psi = \sigma\psi$ on $(0, 2L)$ with absorbing boundary conditions while $s - \pi^2 D_1/4\ell^2$ is the principal eigenvalue for $D_1 d^2\psi/dx^2 + s\psi = \sigma\psi$ on $(-\ell, 0)$ (or on $(2L, 2L + \ell)$) with an absorbing boundary condition at one end and a reflecting boundary condition at the other. If we set $\gamma = 0$, that would correspond to a situation where any individual reaching 0 or $2L$ from the interior of $(0, 2L)$ would move out into $(-\ell, 0)$ or $(2L, 2L + \ell)$, which means that $(0, 2L)$ would in effect have absorbing boundaries. An individual in $(-\ell, 0)$ which reached 0 would turn back into $(-\ell, 0)$ while an individual reaching $-\ell$ would perish because of the assumption that the exterior of the overall environment $(-\ell, 2L + \ell)$ is lethal. That situation corresponds to a reflecting boundary condition at $x = 0$ and an absorbing boundary condition at $x = -\ell$. (The same situation would occur on $(2L, 2L + \ell)$ with reflection at $x = 2L$ and absorption at $x = 2L + \ell$.) The larger of the growth rates on $(0, 2L)$ and $(-\ell, 0) \cup (2L, 2L + \ell)$ would give the effective growth rate on the overall interval, since the faster growing population would eventually become much larger than more slowly growing population. Either the population in $(0, 2L)$ or in $(-\ell, 0) \cup (2L, 2L + \ell)$ might grow faster if $\gamma = 0$, even though $r > s$, because of effects of the boundary and interface conditions. Various other aspects of the dependence of σ_1 on L and ℓ can be analyzed as above; see Cantrell and Cosner (1999).

If individuals can assess how the quality of their habitat changes across an interface and respond to that assessment by crossing the interface with greater or lesser probability, then γ will depend on r and s. Suppose that $\gamma = \gamma(r - s)$, so that the probability of an individual crossing the interfaces at $x = 0, 2L$ depends on the difference in habitat quality across the interface. Suppose further that $\gamma(0) = 1/2$, so that if the regions on the two sides of the interface have the same habitat quality there is no preferred direction at the interface. Suppose further that $\gamma(z) \to 1$ as $z \to \infty$. Some possible forms are $\gamma(z) = e^{\delta z}/(1 + e^{\delta z})$ or $\gamma(z) = (1 + z^p)/(2 + z^p)$, among many others. If we take $\gamma(z) = e^{\delta z}/(1 + e^{\delta z})$ then δ measures how strongly $\gamma(z)$ responds to changes in z. We have $(1 - \gamma)/\gamma = e^{\delta(r-s)}$ in that case, so (2.58) may be written as

$$f(\sigma, r, L, D_2) = e^{-\delta(r-s)}(D_2/D_1)g(\sigma, s, \ell, D_1). \tag{2.61}$$

We know that we must always have $r - \pi^2 D_2/4L^2 < \sigma_1 < r$, so if $s < r - \pi^2 D_2/4L^2$ then $s < \sigma_1$ so the relevant form for g in (2.61) is $g(\sigma, s, \ell, D_1) = \tanh(\sqrt{(\sigma - s)/D_1}\ell)/\sqrt{(\sigma - s)/D_1}$. It follows that for $s \leq s_0 < r - \pi^2 D_2/4L^2$ we must have $\sigma - s \geq r - \pi^2 D_2/4L^2 - s_0 > 0$ so that $\partial g/\partial s$ is bounded for $(-\infty, s_0)$. We also have $g > 0$ on $(-\infty, s_0)$. Our goal is to see how changing s affects σ_1. We can determine that if we know how changing s affects $e^{\delta(r-s)}g(\sigma, s, \ell, D_1)$. We have

$$\frac{\partial}{\partial s}(e^{\delta(r-s)}g(\sigma, s, \ell, D_1)) = \left(-\delta g + \frac{\partial g}{\partial s}\right) e^{\delta(r-s)}g. \tag{2.62}$$

It follows from (2.62) that for δ sufficiently large $e^{\delta(r-s)}g(\sigma, s, \ell, D_1)$ is decreasing in s, so that making s smaller makes $e^{\delta(r-s)}g(\sigma, s, \ell, D)$ larger and hence increases σ_1. If δ is small, the reverse is true and making s smaller decreases σ_1. *The implication is that if the strength of aversion to unfavorable habitat, as measured by δ, is sufficiently great then it is beneficial to the population to decrease the quality of habitat in the buffer zone.* This may seem counter-intuitive, but the mechanism is that if the degree of aversion to entering the buffer zone increases more rapidly than the quality of the buffer zone decreases, a buffer zone of low quality may lead to a lower loss rate across the interface with the refuge than a buffer zone of moderate but still relatively low quality. This phenomenon seems to be robust with respect to the details of how γ depends on $r - s$, provided $\gamma(z) \to 1$ rapidly enough as $z \to \infty$. For example, it turns out that if $\gamma(z) = (1 + z^p)/(2 + z^p)$ then for $\gamma = \gamma(r - s)$ we have $\sigma_1 \to r$ as $s \to -\infty$ if $p > 1/2$; see Cantrell and Cosner (1999).

Remarks: In the examples discussed above we have examined a few specific issues in some detail. The methods used to understand how eigenvalues depend on parameters could be used to address other questions about the effects of boundary conditions, the arrangement of favorable and unfavorable habitats, etc. The main limitation of the method is that it can only be used in one space dimension, with piecewise constant coefficients. Some questions related to those discussed here are treated in Cantrell and Cosner (1991b, 1999).

2.5 Eigenvalues and Positivity: Advective and Time-Periodic Models

We have seen that if a differential operator describes dispersal as a combination of diffusion and drift (i.e. advection or taxis), and the advection term arises as a gradient, then the principal eigenvalue of the operator can be characterized variationally as a critical value (typically a maximum or minimum) of a certain functional (see for example (2.30), (2.31). However, if the advection term is not in the form of a gradient, there may be no simple variational form for the eigenvalues. There may be some type of variational formulation, as discussed by Holland (1977, 1978) and Belgacem (1997), but in general such formulations are complicated or hard to analyze. The same issues already arise in the context of matrices. If M is a symmetric $n \times n$ matrix then the eigenvalues of M can be described as critical values of the expression $M\vec{x} \cdot \vec{x}$ subject to the constraint $|\vec{x}| = 1$, but this is no longer true if M is not symmetric. It turns out that principal eigenvalues can be shown to exist for nonsymmetric positive matrices and for diffusion-advection operators where the advection term is not a gradient, and/or the coefficients are time dependent but periodic in time, by methods based on the theory of positive matrices and operators. The issues related to extending the notion of principal eigenvalues to such operators are primarily technical and mathematical; the interpretation of the eigenvalues is the same as in the cases which can be treated variationally. For that reason, the material in this section is presented in the form of statements of mathematical results and some proofs or sketches of proofs. Specific applications of the results in this section to ecological models will be given in later chapters.

As noted above, there are large classes of nonsymmetric matrices and differential operators which do not admit simple variational formulas for their eigenvalues but which still possess a real eigenvalue with a positive eigenvector or eigenfunction, i.e. a principal eigenvalue. Those are, roughly speaking, the matrices or operators which are positive, or which can be expressed in terms of inverses of matrices or operators which are positive. In this context, a positive matrix is defined to be a matrix with positive entries, so that it maps the set of componentwise positive vectors into itself. A positive operator maps positive functions to positive functions. The maximum principle implies that many elliptic differential operators have positive inverses, and something similar is true for parabolic operators with periodic coefficients. In this section we shall see how ideas based on positivity allow us to extend the idea of principal eigenvalues to such operators, even if they are not symmetric. The derivations of many of the results about operators are quite technical, so the reader who is primarily interested in applications may want to focus mostly on the statements of results rather than the analysis. To illustrate the essential ideas in a less technical setting we begin with a discussion of results for nonnegative matrices.

A matrix A with nonnegative entries is said to be *primitive* if A^k has all of its entries positive for some integer k. Such matrices satisfy the following well-known theorem:

Theorem 2.11. (Perron-Frobenius Theorem) Suppose that A is a primitive matrix. Then the problem $A\vec{v} = \sigma\vec{v}$ has a principal eigenvalue $\sigma_1 > 0$ which admits right and left eigenvectors with all entries positive. If σ is any other eigenvalue of A then $|\sigma| < \sigma_1$.

Discussion: The Perron-Frobenius theorem is a basic result on positive matrices; see Caswell (1989) and Berman and Plemmons (1979). One major application of the theory of nonnegative matrices in ecology and population dynamics is to matrix models for age- or stage-structured populations. In those models populations are structured by age or developmental stage, rather than by spatial location, and the matrix describes transition rates between stages. The principal eigenvalue of the matrix in such models can typically be interpreted as an intrinsic growth rate for the overall population, just as the principal eigenvalue σ_1 in (2.2) can be interpreted as the intrinsic growth rate in (2.1). For a detailed discussion of matrix population models see Caswell (1989). Another way in which nonnegative matrices arise in population dynamics is in spatially discrete models, including some types of metapopulation models such as those discussed in Hanski (1997). In such models the nonnegative matrix typically describes dispersal or colonization, and the interpretation of the principal eigenvalue is typically similar to the interpretation of the principal eigenvalue of the operator describing dispersal in a reaction-diffusion model. We discuss connections between different types of spatial models in more detail in a later section.

There are results analogous to the Perron-Frobenius theorem which can be applied to differential operators, but there are a number of technical issues that arise in applying those results. The first issue is determining what positivity should mean for operators acting on functions. In the case of $n \times n$ matrices, a matrix A has all its elements positive if and only if the vector $A\vec{v}$ has all its elements positive whenever $\vec{v}$ is a vector whose components are nonnegative and are not all zero. This formulation makes sense for operators, too. Roughly, an operator L is positive if Lf is positive for any function f which is nonnegative and nonzero. At this point the second issue arises. Differential operators generally are not positive. However, by the maximum principle, many differential operators have inverses which are positive. If the operator L has an inverse L^{-1} and $L^{-1}\phi = \mu\phi$ for some function $\phi(x) > 0$ and $\mu \neq 0$ then $L\phi = (1/\mu)\phi$, so if μ is a principal eigenvalue for L^{-1}, then $(1/\mu)$ is a principal eigenvalue for L. Hence we can proceed by considering the inverse operators of the differential operators we want to analyze. A third issue is that for some of the operators arising in applications it is not clear whether the inverse operator exists or is positive. Sometimes that issue can be addressed by considering the adjoint operator. The adjoint of an operator is analogous to the transpose of a matrix. If A is an $n \times n$ matrix then any right eigenvector of A^T is a left eigenvector for A. The Perron-Frobenius theorem could be restated as asserting that if A is primitive, then both A and A^T have the same principal eigenvalue λ_1, and this formulation extends to operators. In general the set of functions on which a differential operator acts is determined partly by the boundary conditions. In that sense the boundary conditions are part of the definition of the operator. Usually, the adjoint L^* of a differential operator L can be computed formally by writing the relation

$$\int_\Omega vLu\,dx = \int_\Omega uL^*v\,dx \tag{2.63}$$

and determining L^* and the necessary boundary conditions on v by integration by parts via the divergence theorem. The operator L^* computed in this way typically will coincide with true adjoint of L defined in terms of duality as long as v is a smooth function. The actual adjoint operator will often have its domain of definition expanded or restricted in some way. For our present purposes it is enough to be able to identify L^* as a differential operator. For more complete treatments of adjoint and inverse operators, unbounded operators, duality, and other topics in operator theory, see, for example, Kato (1966) and Schaefer (1971).

2.5.1 Advective Models

The operators we want to study will typically have the form shown in (2.21), with boundary conditions as in (2.22), but without the assumption that $\vec{b}/d$ is a gradient:

$$Lu = \nabla \cdot [d(x)\nabla u - \vec{b}(x)u] + m(\vec{x})u \quad \text{in} \quad \Omega, \tag{2.64}$$

$$d(x)\frac{\partial u}{\partial \vec{n}} + [\beta(x) - \vec{n} \cdot \vec{b}(x)]u = 0 \quad \text{on} \quad \partial\Omega. \tag{2.65}$$

Normally we assume $\beta(x) \geq 0$. Recall that $\beta(x) = 0$ is the no-flux case. In general, it is unclear whether or not L will satisfy a maximum principle. In particular, we cannot tell the sign of $-\vec{b}(x) \cdot \vec{n} + \beta(x)$, but that term usually must be nonnegative for a maximum principle to hold. However, if we calculate the adjoint L^* via (2.63) we obtain

$$L^*u = \nabla \cdot d(x)\nabla u + \vec{b} \cdot \nabla u + m(x)u \quad \text{in} \quad \Omega \tag{2.66}$$

with boundary condition

$$d(x)\frac{\partial u}{\partial \vec{n}} + \beta(x)u = 0 \quad \text{in} \quad \partial\Omega. \tag{2.67}$$

If $m(x) < 0$ then the maximum principle applies to the operator L^* in (2.66). Thus, if L^*u is continuous with $L^*u \geq 0$ on Ω and $L^*u \not\equiv 0$ then $u < 0$ on $\overline{\Omega}$. (If (2.67) is replaced with $u \equiv 0$ on $\partial\Omega$ then $u < 0$ on Ω.) If we choose C to be a constant large enough that $m(x)-C < 0$ then the solution to $-L^*u+Cu = f$ in Ω with boundary condition (2.67) will be strictly positive in Ω if $f \geq 0$, $f \not\equiv 0$ in Ω. Thus, if the inverse operator $(-L^* + C)^{-1}$ exists it must map positive functions into other positive functions. To show that L has a principal eigenvalue we will apply an abstract result analogous to the Perron-Frobenius theorem to the operator $B = (-L^* + C)^{-1}$ since that is the operator with the appropriate properties. The abstract theorem we shall use is known as the *Krein-Rutman Theorem.* It implies that both B and B^* have principal eigenvalues. Now $B^* = (-L + C)^{-1}$ and so if $B^*u = \mu u$ and u is smooth then $(-L + C)B^*u = \mu(-L + C)u$ or $u = -\mu Lu + \mu Cu$ so $Lu = [-(1/\mu) + C]u$. Thus, eigenvalues of B^* correspond to eigenvalues of L. (Since B^* depends on C, so does μ. However, the eigenvalues $-(1/\mu) + C$ of L are actually independent of C, as we shall soon see.)

We now state a version of the Krein-Rutman theorem. The statement requires a few technical definitions. Some of those are given in the discussion of mathematical background material in Chapter 1, but we recall them here for convenience. A *Banach space* is a vector space E with a norm $||\ \ ||$ such that the metric $d(u, v) = ||u - v||$ makes E a complete metric space. A linear operator on E is a function whose domain and range are subsets of E and which has the properties $B(u + v) = Bu + Bv$ and $B(cu) = cBu$ for $u, v \in$ domain B and c a constant. Generally we will consider cases where $c \in I\!R$ but we may allow $c \in \mathbb{C}$ in some cases. Recall that an operator B is *bounded* if domain $B = E$ and there is a constant b_0 such that $||Bu|| \leq b_0||u||$ for all $u \in E$. An operator B is *compact* if the image under B of any bounded set has compact closure, i.e. $\{Bu : u \in U\}$ is precompact in E if $U \subseteq E$ is bounded. It is a standard result that compact linear operators are bounded, but not all bounded operators are compact. A linear operator is bounded if and only if it is continuous relative to the metric defined by $||\ \ ||$. To define positivity for operators we need a definition of positivity on the underlying space E. Positivity in E is defined by a *positive cone* $P \subseteq E$, i.e. a set with the properties that $u, v \in P$ implies $u + v \in P$ and $cu \in P$ for any positive constant c, and that if $u \in P$ and $-u \in P$ then $u = 0$. The prototype would be the positive orthant $P = \{\vec{v} \in I\!R^n : v_i \geq 0, i = 1, \ldots, n\} \subseteq I\!R^n$. The cone P defines an ordering on E if we define $u \geq v$ to mean $u - v \in P$. An *ordered Banach space* is a Banach space with a positive cone P which is closed in the metric induced by the norm $||\ \ ||$. If the positive cone in E has nonempty interior, then the operator B is said to be *strongly positive* if $Bu \in$ the interior of P whenever $u \in P - \{0\}$. We can now state the Krein-Rutman theorem.

Theorem 2.12. (Krein-Rutman) Suppose that B is a compact linear operator on the ordered Banach space E with positive cone P. Suppose further that P has nonempty interior and that B is strongly positive. The eigenvalue problem $Bu = \mu u$ admits a unique eigenvalue μ_1 which has a positive eigenvector u_1. The eigenvalue μ_1 is positive and is simple, that is, the only nonzero solutions to $(B - \mu_1 I)^k v = 0$ for any $k = 1, 2, 3, \ldots$ are multiples of the eigenvector u_1 associated with μ_1. Any other eigenvalue μ of B must satisfy $|\mu| \leq \mu_1$. Furthermore, μ_1 is also a simple eigenvalue for the adjoint operator B^* and the associated eigenvector is positive with respect to the dual cone P^*. Finally, if $v \in P, v \neq 0$, the problem

$$\mu w - Bw = v \tag{2.68}$$

has a unique positive solution if $\mu > \mu_1$ and no positive solution if $\mu \leq \mu_1$.

Remarks: The version of the Krein-Rutman theorem presented here is similar to the one presented in Amann (1976). Proofs of the Krein-Rutman theorem can be found in Krasnosel'skii (1964) and Schaefer (1971). The simplicity of the eigenvalue μ_1 is not directly of interest in linear models but is relevant to the analysis of nonlinear models via bifurcation theory, as we shall see in the next chapter. For our current purposes the existence of principal eigenvalues is the key point. There are a number of technical issues related to choosing the space E and cone P and verifying that L and/or L^* have the appropriate mapping properties on E to allow the construction of $(-L + C)^{-1}$ and/or $(-L^* + C)^{-1}$. Some of those discussed in Section 1.6. For now we merely note some cases in which suitable spaces can be chosen so that the operators $(-L^* + C)^{-1}$ and perhaps also $(-L + C)^{-1}$ are well defined, compact, and strongly positive.

Corollary 2.13. Suppose that Ω is a bounded domain with $\partial\Omega$ of class $C^{2+\alpha}$. Suppose that $d(x) \in C^{1+\alpha}(\overline{\Omega})$, $\vec{b}(x) \in [C^{1+\alpha}(\overline{\Omega})]^N$ (where N is the dimension of Ω), $m(x) \in C^{\alpha}(\overline{\Omega})$ and $\beta(\vec{x}) \in C^{1+\alpha}(\partial\Omega)$. Suppose also that $d(x) \geq d_0 > 0$ and $\beta(x) \geq 0$. If L has the form shown in (2.64) then the eigenvalue problem

$$L\psi = \sigma\psi \quad \text{in} \quad \Omega$$

subject to the boundary condition (2.65) has a principal eigenvalue σ_1, which is also the principal eigenvalue for

$$L^*\psi = \sigma\psi \quad \text{in} \quad \Omega$$

subject to boundary condition (2.67).

Remark: In the case where $\vec{b}(x)/d(x)$ is a gradient, σ_1 is the same eigenvalue as is determined by (2.31).

Sketch of Proof: We begin with the problem $L^*\psi = \sigma\psi$ in Ω under boundary condition (2.67). Choose C large enough that $C - m(x) \geq 0$ in $\overline{\Omega}$. The maximum principle then applies to the problem $(-L^* + C)u = f(x)$ under boundary condition (2.67). Thus, if a solution exists it is unique, and if, in addition, $f(x)$ is continuous on $\overline{\Omega}$ with $f(x) \geq 0$ on $\overline{\Omega}$ and $f(x_0) > 0$ for some $x_0 \in \overline{\Omega}$ then $u > 0$ on $\overline{\Omega}$. Standard results in the theory of elliptic partial differential equations then imply that the operator $(-L^* + C)$ has an inverse on $C(\overline{\Omega})$ and that $(-L^* + C)^{-1} : C(\overline{\Omega}) \to C(\overline{\Omega})$ is compact; see Gilbarg and Trudinger (1977) and Amann (1976). If we take the E to be the Banach space $C(\overline{\Omega})$ with positive cone $P = \{u \in C(\overline{\Omega}) : u \geq 0 \quad \text{on} \quad \overline{\Omega}\}$ then (E, P) is an ordered Banach space and the Krein-Rutman theorem applies to $(-L^* + C)^{-1}$. (If $f \in P - \{0\}$ then $f(x) \geq 0$ in Ω with $f(x) > 0$ somewhere, so a version of the strong maximum principle implies that $(-L^* + C)^{-1}f > 0$ on $\overline{\Omega}$; see the discussion of mathematical background in Chapter 1.) The Krein-Rutman theorem implies the existence of a principal eigenvalue $\mu_1 > 0$ for $(-L^* + C)^{-1}$, so that there is a function $\psi^* \in C(\overline{\Omega})$ with $\psi^* > 0$ on $\overline{\Omega}$ such that

$(-L^*+C)^{-1}\psi^* = \mu_1\psi^*$, so ψ^* belongs to the domain of the operator $(-L^*+C)$, and hence $(-L^*+C)\psi^* = (1/\mu_1)\psi^*$, or $L^*\psi^* = \sigma_1\psi^*$ with $\sigma_1 = C-(1/\mu_1)$. Furthermore, the Krein-Rutman theorem implies that $[(-L^*+C)^{-1}]^*$ also has μ_1 as a principal eigenvalue. When the domains of the operators are suitably defined, calculations using the definitions of adjoint operators show that $[(-L^*+C)^{-1}]^*p = q$ if and only if $p = (-L+C)q$. (See, for example, Kato (1966).) Thus, there is a function $\psi > 0$ such that $[(-L^*+C)^{-1}]^*\psi = \mu_1\psi$, so $(1/\mu_1)\psi = (-L+C)\psi$ or $L\psi = \sigma_1\psi$.

Remarks: The interpretation of σ_1 is the same as it is in the cases which can be treated variationally. If $\sigma_1 > 0$ then the linear growth and dispersal model $u_t = Lu$ predicts instability of the equilibrium $u = 0$, so that populations are expected to increase exponentially. If $\sigma_1 < 0$, the model predicts exponential decay toward extinction.

In the sketch of the proof of Corollary 2.13, μ_1 is uniquely determined for any given value of C, but in general μ_1 will depend on C. However, the principal eigenvalue σ_1 for L^* is unique and independent of the choice of C. If $L^*\psi_1 = \sigma_1\psi_1$ and $L^*\psi_2 = \sigma_2\psi_2$ with $\psi_1, \psi_2 > 0$ then we can choose some fixed value for C such that $C - \sigma_i > 0$ for $i = 1, 2$; then $(-L^*+C)\psi_i = (-\sigma_i + C)\psi_i$ so $(-L^*+C)^{-1}\psi_i = [1/(C-\sigma_i)]\psi_i > 0$. Since the principal eigenvalue μ_1 of $(-L^*+C)^{-1}$ is unique for any fixed C, we must have $\mu_1 = 1/(C-\sigma_i)$ for $i = 1, 2$ so $\sigma_1 = \sigma_2$. Also, the eigenvalue σ_1 is simple, because of the simplicity of μ_1, which follows from the Krein-Rutman theorem.

In the case of Dirichlet boundary conditions Corollary 2.13 is still valid, but since functions satisfying Dirichlet boundary conditions are zero on $\partial\Omega$, the operator $(-L^*+C)^{-1}$ does not map nonnegative nonzero functions into functions which are strictly positive on $\overline{\Omega}$. The point is that the operator $(-L^*+C)^{-1}$ is not strongly positive with respect to the cone P used in the derivation of Corollary 2.13, although it does map $C(\overline{\Omega})$ into itself. That problem can be fixed by using $E = C_0^1(\overline{\Omega})$ and $P = \{u \in C_0^1(\overline{\Omega}) : u \geq 0 \text{ on } \Omega, \ \partial u/\partial\vec{n} \leq 0 \text{ on } \partial\Omega\}$. Functions in the interior of P have $u > 0$ on Ω *and* $\partial u/\partial\vec{n} < 0$ on $\partial\Omega$, so the interior of P is an open subset of $C_0^1(\Omega)$. The strong maximum principle implies that if $f \in C_0^1(\overline{\Omega})$ with $f \geq 0$ and $f(x) > 0$ for some $x \in \overline{\Omega}$ then $(-L^*+C)^{-1}f$ lies in the interior of P.

In the case of Dirichlet boundary conditions or boundary conditions of the form (2.65) with $\vec{b}(x)\cdot\vec{n} \leq 0$ on $\partial\Omega$, the operator $(-L+C)$ satisfies a maximum principle if $C - m(x) + \nabla\cdot\vec{b}(x) > 0$, so under such boundary conditions we could apply the proof of Corollary 2.13 directly to $-L+C$ without having to consider L^*. In that case we can take $B = (-L+C)^{-1}$ in (2.68). As a result, we obtain the following:

Corollary 2.14. Suppose that L satisfies the hypotheses of Corollary 2.13 and that either the boundary condition is $u = 0$ on $\partial\Omega$ or is given by (2.65) with $\vec{b}(x)\cdot\vec{n} \leq 0$ on $\partial\Omega$. If $f(x) \in C(\overline{\Omega})$ is nonnegative on $\overline{\Omega}$ and positive for some $x \in \overline{\Omega}$, the problem $(-L-\tau)u = f$ has a positive solution if and only if $\tau < -\sigma_1$.

Proof: Choose C large enough that $C + \tau > 0$ and $C > \max(m(\vec{x}) - \nabla\cdot\vec{b}(x))$. The last condition implies that $-L+C$ satisfies a maximum principle, so that $B = (-L+C)^{-1}$ exists and satisfies the hypotheses of the Krein-Rutman theorem by standard elliptic theory (e.g. see Gilbarg and Trudinger (1977) and Amann (1976)). Consider the equation $(-L-\tau)u = f$. We may write it as $(-L+C)u - (\tau+C)u = f$, so that

$$[1/(C+\tau)]u - (-L+C)^{-1}u = [1/(C+\tau)](-L+C)^{-1}f \in P. \qquad (2.69)$$

Equation (2.69) is equivalent to (2.68) with $\mu = 1/(C + \tau)$ and $B = (-L + C)^{-1}$. Recall from the proof of Corollary 2.13 that σ_1 is related to the principal eigenvalue μ_1 of $(-L + C)^{-1}$ via $\sigma_1 = C - (1/\mu_1)$, i.e. $\mu_1 = 1/(C - \sigma_1)$. Thus, by Theorem 2.12, (2.69) has a unique positive solution if $\mu > \mu_1$, which is equivalent to $1/(C + \tau) > 1/(C - \sigma_1)$ or $\tau < -\sigma_1$, and (2.69) has no positive solution when $\mu \leq \mu_1$, i.e. $\tau \geq -\sigma_1$, so the same is true for the equation $(-L - \tau)u = f$.

2.5.2 Time Periodicity

Corollary 2.13 extends to the case of diffusion operators with time-periodic coefficients. Because the operators explicitly involve the time variable, the formulation of the eigenvalue problem is slightly different than in the elliptic case. Also, the existing form of the theory does not allow for time dependence in the boundary conditions, which means that some types of physically natural boundary conditions cannot be treated in the time-periodic context. The sorts of eigenvalue problems which can be treated have the form

$$\begin{aligned} L\psi &= \frac{\partial \psi}{\partial t} - \nabla \cdot [d(x,t)\nabla u - \nabla \vec{b}(x,t)\psi] - m(x,t)\psi = \mu\psi, \\ &\text{for} \quad (x,t) \in \Omega \times I\!R, \\ &\delta(x)\frac{\partial \psi}{\partial \vec{n}} + \beta(x)\psi = 0 \quad \text{on} \quad \partial\Omega \times I\!R, \\ &\psi(x, t+T) = \psi(x,t) \quad \text{for some fixed} \quad T, \\ &\text{for all} \quad (x,t) \in \overline{\Omega} \times I\!R, \end{aligned} \tag{2.70}$$

where we assume $\delta(x), \beta(x) \geq 0$ with $\delta(x) + \beta(x) > 0$ on $\partial\Omega$. We shall also want to treat the corresponding problem for the formal adjoint operator:

$$L^*\psi = -\frac{\partial \psi}{\partial t} - \vec{\nabla} \cdot d(x,t)\nabla\psi + \vec{b}(x,t) \cdot \nabla\psi - m(x,t)\psi = \mu\psi \tag{2.71}$$

under the same boundary and periodicity conditions as in (2.70). Recall that a function $u(x,t)$ belongs to class $C^{i+\alpha, j+\alpha/2}(\overline{\Omega}, I\!R)$ if u has continuous derivatives up to order i in x with the ith order derivatives Hölder continuous with exponent α, and continuous and uniformly bounded derivatives up to order j in t, with $\partial^j u/\partial t^j$ uniformly Hölder continuous with exponent $\alpha/2$. We have the following lemma, which is a version of a result due to A.C. Lazer (Lazer, 1982; Castro and Lazer, 1982). (See also Hess (1991).)

Lemma 2.15. Suppose that L has the form shown in (2.70). Suppose that the coefficients of L satisfy $d(x,t) \in C^{1+\alpha,\alpha/2}(\overline{\Omega}, I\!R), \vec{b}(x,t) \in [C^{1+\alpha,\alpha/2}(\overline{\Omega} \times I\!R)]^n, m(x,t) \in C^{\alpha,\alpha/2}(\overline{\Omega}, I\!R)$, with $d(x,t) \geq d_0 > 0$ and with all coefficients T-periodic in t. Suppose also that $\delta(x), \beta(x) \in C^{1+\alpha}(\overline{\Omega})$. The problem (2.70) admits a unique principal eigenvalue μ_1 with associated T-periodic eigenfunction $\psi_1(x,t) > 0$ on $\Omega \times I\!R$ (in fact on $\overline{\Omega} \times I\!R$ except in the case of Dirichlet boundary conditions). The adjoint problem (2.71) also has μ_1 as a principal eigenvalue with corresponding T-periodic eigenfunction $\psi_1^*(\vec{x}, t) > 0$. The eigenvalue μ_1 depends continuously on $m(x,t)$ with respect to $C(\overline{\Omega} \times [0,T])$ and is decreasing with respect to m, that is, if $m_1 \geq m_2$ and the principal eigenvalues corresponding to m_1 and m_2 are denoted $\mu_1(m_1)$, $\mu_1(m_2)$ then $\mu_1(m_1) \leq \mu_1(m_2)$, with strict inequality if $m_1 > m_2$.

Remarks: This result is based on positivity arguments, but they are more complicated than those sketched in the discussion of Corollary 2.13. See Lazer (1982), Castro and Lazer (1982) and Hess (1991) for details. In the case where the coefficients of L do not depend on t, so that they can be viewed as T-periodic for all periods T, the eigenfunctions must also be periodic in t for all periods, so they must be constant in t. In that case we have $L = \partial/\partial t - E$ where E is an elliptic operator of the form shown in (2.64), so we have $\mu_1 = -\sigma_1$ where σ_1 is the principal eigenvalue of E. More generally, note that $u = e^{-\mu_1 t}\psi_1(x,t) > 0$ is a solution to the linear growth/dispersal model

$$\begin{aligned} \frac{\partial u}{\partial t} &= \nabla \cdot [d(x,t)\nabla u - \vec{b}(x,t)u] + m(x,t)u \quad \text{in} \quad \Omega \times I\!R, \\ \delta(x)\frac{\partial u}{\partial \vec{n}} &+ \beta(x)u = 0 \qquad\qquad \text{on} \quad \partial\Omega \times I\!R, \end{aligned} \tag{2.72}$$

so the prediction is growth if $\mu_1 < 0$ but decline if $\mu_1 > 0$. (Since u is a solution to (2.72), so is Cu for any constant C, and Cu can be used as either a sub- or supersolution for purposes of comparison with other solutions. Thus, μ_1 plays the same role as $-\sigma_1$.)

For the purposes of analyzing nonlinear (i.e. density dependent) population models, sometimes it is necessary to consider eigenvalues analogous to the eigenvalue λ_1 (or λ_1^+) defined in (2.13). The results that are currently available are stated in terms of operators which have the form of L^* in (2.66) and (2.67). Specifically, suppose that we consider

$$Lu = \nabla \cdot d(x)\nabla u + \vec{b}(x) \cdot \nabla u + c(x)u \tag{2.73}$$

with boundary condition

$$\delta(x)\frac{\partial u}{\partial \vec{n}} + \beta(x)u = 0 \quad \text{on} \quad \partial\Omega$$

where $c(x) \leq 0$, $c(x) \in C^\alpha(\overline{\Omega})$, $\vec{b}(x)$ and $d(x)$ are as in the statement of Corollary 2.13, and $\delta(x)$ and $\beta(x)$ are as in Lemma 2.15. We have:

Lemma 2.16. Suppose that $m(x)$ is continuous in x on $\overline{\Omega}$ with $m(x_0) > 0$ at some point of $\overline{\Omega}$. If $\beta(x) \equiv 0$ and $c(x) \equiv 0$ then suppose also that $\int_\Omega m(x)dx < 0$. Then the eigenvalue problem

$$\begin{aligned} L\phi + \lambda m(x)\phi &= 0 \qquad \text{in} \quad \Omega \\ \delta(x)\frac{\partial \phi}{\partial \vec{n}} + \beta(x)\phi &= 0 \qquad \text{on} \quad \partial\Omega \end{aligned} \tag{2.74}$$

has a unique necessarily positive principal eigenvalue $\lambda_1^+(m(x))$. Furthermore, if we consider the operator $Ku = L^{-1}m(x)u$ on $C(\overline{\Omega})$ (or $C^1(\overline{\Omega})$) then $1/\lambda_1^+(m)$ is an algebraically simple eigenvalue of K. If λ is any other eigenvalue of (2.74) (possibly complex) then $Re\lambda > \lambda_1^+(m(x))$.

Remarks: A version of this result for the time periodic case is given in Hess (1991, Theorem 16.1). If $m(x) > 0$, this result can be obtained from the Krein-Rutman Theorem via arguments similar to the proof sketched for Theorem 2.12 and a clever use of the maximum principle; see Protter and Weinberger (1966) and Amann (1976). The essential ideas for the case where $m(\vec{x})$ changes sign in Ω, that is, the case of an indefinite weight function, were developed in the paper (Hess and Kato, 1980). The result extends to the case of operators with T-periodic coefficients as in Lemma 2.15, but the requirement $m(x_0) > 0$ for some $x_0 \in \overline{\Omega}$ must be replaced by the requirement

$$\int_0^T \left[\max_{x\in\overline{\Omega}} m(x,t)\right] dt > 0; \tag{2.75}$$

see Hess (1991), §16.

2.5.3 Additional Results on Eigenvalues and Positivity

There are many other results on the behavior of eigenvalues which are related to positivity. Some of them are stated here for reference.

Proposition 2.17. Suppose the hypotheses of Lemma 2.16 hold. The problem

$$\begin{aligned}
&[-L - \lambda m(x)]u = f(x) \quad \text{in} \quad \Omega \\
&\delta(x)\frac{\partial u}{\partial \vec{n}} + \beta(x)u = 0
\end{aligned}$$

with $f(x) \geq 0$, $f(x_0) > 0$ for some $x_0 \in \overline{\Omega}$, and $f(x)$ continuous, has a unique positive solution if $0 < \lambda < \lambda_1^+(m(x))$ and no positive solution if $\lambda \geq \lambda_1^+(m(x))$.

Remarks: This is essentially Theorem 16.6 of Hess (1991). The theorem in Hess (1991) is stated for the periodic-parabolic case, where the operator L has the form shown in (2.70), so the notation differs from that of Hess (1991) by the presence of a minus sign on L. By Corollary 2.14 and Proposition 2.17 we have the following:

Corollary 2.18. Suppose L satisfies the hypotheses of Corollary 2.14 and Lemma 2.16. For $\lambda > 0$ let $\sigma_1(\lambda)$ be the principal eigenvalue and $\psi_1(x,\lambda) > 0$ the corresponding eigenfunction for

$$\begin{aligned}
&L\psi + \lambda m(x)\psi = \sigma\psi \quad \text{in} \quad \Omega \\
&\delta(x)\frac{\partial u}{\partial \vec{n}} + \beta(x)u = 0 \quad \text{on} \quad \partial\Omega.
\end{aligned} \tag{2.76}$$

Then $\sigma_1(\lambda) < 0$ if and only if $\lambda < \lambda_1^+(m)$.

Sketch of proof: If $\sigma_1(\lambda) < 0$ then the problem $[-L - \lambda m(x)]u = -\sigma_1(\lambda)\psi_1(x,\lambda) > 0$ has the positive solution $u = \psi_1(x,\lambda)$ so by Proposition 2.17 we must have $\lambda < \lambda_1^+(m)$. If $\lambda < \lambda_1^+(m)$ then by Proposition 2.17 the problem $[-L - \lambda m(x) - \tau]u = f > 0$ must have a positive solution for $\tau = 0$, but by Corollary 2.14 that is possible only if $0 = \tau < -\sigma_1(\lambda)$ so $\sigma_1(\lambda) < 0$.

Remarks: Positivity can also be used to obtain comparisons between eigenvalues corresponding to different growth rates (i.e. weight functions) $m_1(x)$, $m_2(x)$. As with other results based on positivity, the comparison results extend to the periodic-parabolic case treated in Hess (1991). We have:

Corollary 2.19. Suppose the hypotheses of Lemma 2.16 hold. If $m_2(x) \geq m_1(x)$ and $m_2(x_0) > m_1(x_0)$ for some $x_0 \in \overline{\Omega}$ then $\lambda_1^+(m_2) < \lambda_1^+(m_1)$. Similarly, if σ_i is the principal eigenvalue of the problem $L\psi + \lambda m_i \psi = \sigma\psi$ in Ω, then $\sigma_2 > \sigma_1$.

Proof: Let ϕ_2 be the eigenfunction for $L\phi_2 + \lambda_1^+(m_2)m_2(x)\phi_2 = 0$. Then $L\phi_2 + \lambda_1^+(m_2)m_1(x)\phi_2 = \lambda_1^+(m_2)[m_1(x) - m_2(x)]\phi_2$, so $[-L-\lambda_1^+(m_2)m_1]\phi_2 = \lambda_1^+(m_2)[m_2(x) - m_1(x)]\phi_2$ where the expression on the right is nonnegative and is positive at $x = x_0$. Since $\phi_2 > 0$, by Proposition 2.17 we must have $\lambda_1^+(m_2) < \lambda_1^+(m_1)$. A similar argument based on Corollary 2.14 establishes the relation $\sigma_2 > \sigma_1$.

There are a number of related results on how the eigenvalues σ_1 and λ_1^+ depend on the weight function. Those include:

Proposition 2.20. If the hypotheses of Corollary 2.14 and Lemma 2.16 are satisfied, then the principal eigenvalue σ_1 for (2.76) is analytic as a function of λ, and the corresponding eigenfunction can be chosen to be analytic in λ. Furthermore, for fixed λ, σ_1 depends continuously on $m(x)$ with respect to the norm $\sup_{\overline{\Omega}}|m(x)|$; that is, if $\lim_{n\to\infty} \sup_{\overline{\Omega}}|m(x) - m_n(x)| = 0$ then $\sigma_1(m_n(x)) \to \sigma_1(m(x))$ as $n \to \infty$.

Discussion: This result extends to the periodic-parabolic case. In that setting, the analytic dependence of σ_1 on λ is shown in Hess (1991, Lemma 15.1) and the continuous dependence on m in Hess (1991, Lemma 15.7).

Another result involving the dependence of eigenvalues on the weight function (i.e. local growth rate) $m(x)$ which extends to the non-selfadjoint case is Theorem 2.7. We have:

Theorem 2.21. Suppose that the operator L has the form (2.73) and that the hypotheses of Corollary 2.14 and Lemma 2.16 are satisfied. Suppose further that for some constant $d_1 > 0$

$$\int_\Omega [d(x)|\nabla\phi|^2 - \phi\vec{b}(x)\cdot\nabla\phi - c(x)\phi^2]dx \geq d_1 \int_\Omega |\nabla\phi|^2 dx \tag{2.77}$$

for all $\phi \in W^{1,2}(\Omega)$, that $\delta(x) > 0$ on $\partial\Omega$ and $\beta(x) > 0$ on a subset of $\partial\Omega$ which is open relative to $\partial\Omega$. If $\{m_n(x)\}$ is a sequence of continuous functions with $m_n(x_0) > 0$ for some $x_0 \in \overline{\Omega}$, such that $\max|m_n(x)| \leq M_0$ for some constant M_0, and satisfying hypotheses (2.32) then $\lambda_1^+(m_n(x)) \to \infty$ as $n \to \infty$.

Remarks: The proof is similar to that of Theorems 2.7 and 2.8; see the Appendix to this chapter. The result remains valid under Dirichlet boundary conditions. It and related results are discussed in some detail for the case of Dirichlet boundary conditions ($\delta(x) = 0$) in Cantrell and Cosner (1991a).

A number of related results on elliptic eigenvalue problems which are not self-adjoint are given in Cantrell and Cosner (1991a). Other aspects of how eigenvalue problems can be treated via positivity are given in Hess (1991) and Nussbaum (1984).

2.6 Connections with Other Topics and Models

2.6.1 Eigenvalues, Solvability, and Multiplicity

One of the implications of the Krein-Rutman theorem is the solvability result stated in Proposition 2.17. More generally, let Ω be a bounded domain and suppose that the operator

L has the form

$$Lu = \nabla \cdot d(x)\nabla u + \vec{b}(x) \cdot \nabla u + m(x)u$$

with boundary conditions

$$\delta(x)\frac{\partial u}{\partial \vec{n}} + \beta(x)u = 0 \quad \text{on} \quad \partial\Omega,$$

where $\partial\Omega$ and the coefficients of L satisfy the regularity hypotheses of Corollary 2.13. If σ_0 is not an eigenvalue of $L\psi = \sigma\psi$, then the problem $Lu - \sigma_0 u = f(x)$ with the given boundary conditions will have a unique solution $u \in C^{2+\alpha}(\overline{\Omega})$ for any $f(x) \in C^{\alpha}(\overline{\Omega})$. In general, the sign of u is not related to the sign of f. If σ_0 is a simple eigenvalue for L and the adjoint operator L^*, then the situation is more delicate but something can still be said. Suppose that ψ_0, ψ_0^* are eigenfunctions of $L\psi = \sigma_0\psi$ and $L^*\psi = \sigma_0\psi$ respectively. Then the problem $Lu - \sigma_0 u = f$ is solvable if and only if $\int_\Omega \psi_0^* f\,dx = 0$, and in that case if u is a solution then so is $u + s\psi_0$ for any $s \in I\!R$. These facts are consequences of the Fredholm alternative in Hilbert space and the fact that an operator such as L will have a compact resolvent when L is viewed as an unbounded operator on $L^2(\Omega)$. See Friedman (1976) and Gilbarg and Trudinger (1977) for a discussion of the Fredholm alternative and related issues. We shall use the relation between eigenvalues and solvability in a number of discussions in later chapters.

We have discussed the notion of simplicity for an eigenvalue. There is a more general notion of the multiplicity of an eigenvalue which appears in Theorem 2.12 and which is relevant for some of the results on bifurcation of solutions which we shall discuss later. If σ is an eigenvalue of a linear operator L, then the solutions to the problem $L\psi - \sigma\psi = 0$, i.e. the eigenfunctions, form a vector space. The dimension of that space is the *geometric multiplicity* of σ. If L is a continuous, i.e. bounded, linear operator from a Banach space Y into itself, we can also define the *algebraic multiplicity* of an eigenvalue σ. Differential operators generally cannot be expressed as bounded operators, but they may have inverse operators which are bounded, so to apply the notion of algebraic multiplicity in the case of differential operators we would typically work with their inverses. Suppose L^{-1} exists as a bounded operator, so that 0 is not an eigenvalue of L. We may write $L\psi = \sigma\psi$ as $T\psi = \mu\psi$ where $T = L^{-1}$ and $\mu = 1/\sigma$. In that setting we define the algebraic multiplicity of μ as the dimension of the vector space $\cup_{n=0}^{\infty} N([T - \mu I]^n)$, where I is the identity operator, and $N([T - \mu I]^n)$ denotes the nullspace of $(T - \mu I)^n$. An eigenvalue with algebraic multiplicity equal to one is said to be algebraically simple. In the case where L and hence T can be interpreted as self-adjoint operators on a Hilbert space, then the algebraic multiplicity of μ will be the same as its geometric multiplicity, and hence the same as the geometric multiplicity of σ. (See, for example, Kato (1966).) Also, the principal eigenvalues for the elliptic operators discussed in this chapter correspond to algebraically simple eigenvalues of their inverse operators. This includes eigenvalues for problems of the form $L\phi + \lambda m(x)\phi = 0$ if those problems are written as $(1/\lambda)\phi + L^{-1}m(x)\phi = 0$. (See Hess and Kato (1980) and DeFigueiredo (1982).)

2.6.2 Other Model Types: Discrete Space and Time

The reaction-diffusion models which are the main subject of this book treat space and time as continuous variables. There are various other sorts of spatial models which treat space and/or time as discrete variables. Several of those types of models are discussed in Chapter 1, including "stepping stone" or "island chain" models, spatially explicit metapopulation models, and models for dispersal based on integral kernels. (See Chapter 1 for a detailed discussion of these types of models and for references.) In some cases, the discrete aspects of such models can have profound effects on their predictions; see for example Durrett and Levin (1994). However, such models can often be treated by methods analogous to those used to study reaction-diffusion models.

Recall that "island chain" or "stepping stone" models typically view space as a collection of discrete patches, treat each patch as a point, and view the overall population as a vector whose components are the populations of the patches. In such models dispersal is often treated as a discrete analogue of diffusion. The models thus take the form of a system of ordinary differential equations, with each equation describing the population on some specific patch, and with coupling between equations via the dispersal terms. A typical linear model of this sort is

$$\frac{du_i}{dt} = \sum_{j \neq i} d_{ij} u_j - d_{ii} u_i + r_i u_i, \quad i = 1, \ldots, n, \tag{2.78}$$

where u_i represents the population on the ith patch, r_i the local growth rate on the ith patch, d_{ii} the rate at which individuals disperse out of the ith patch, and d_{ji} the rate at which individuals dispersing from patch i enter patch j. For the model to make sense, we must have $d_{ij} \geq 0$ for all i, j. The system (2.78) may be written in matrix notation as $d\vec{u}/dt = M\vec{u}$. If, for example, $d_{i+1 \cdot i} > 0$ for $i = 1, \ldots, n-1$ and $d_{i-1 \cdot i} > 0$ for $i = 2, \ldots, n$ (so that some individuals leaving patch i enter patches $i-1$ and $i+1$) then $M + cI$ will be primitive for c sufficiently large, so $M + cI$ will have a principal eigenvalue $\sigma_0(c)$ with positive eigenvector $\vec{v}$ by the Perron-Frobenius theorem. The original matrix M will thus have principal eigenvalue $\sigma_1 = \sigma_0(c) - c$ with the same eigenvector. The solution to $d\vec{u}/dt = M\vec{u}$ is $\vec{u} = e^{tM}\vec{u}(0)$, where e^{tM} is the matrix exponential of M. Standard results on matrix exponentials (e.g. see Brauer and Nohel (1969)) then imply that positive solutions to (2.78) will grow if $\sigma_1 > 0$ and decline if $\sigma_1 < 0$. If the fraction of individuals leaving patch i which go to patch j depends only on the distance between those patches, and if the growth rate r_i is the same on each patch, the system (2.78) takes the form

$$\frac{du_i}{dt} = d\left[\sum_{j \neq i} \delta_{ij} u_j - u_i\right] + r u_i, \tag{2.79}$$

where δ_{ij} are nonnegative constants. The system (2.79) may be written as $d\vec{u}/dt = M\vec{u} = dM_0\vec{u} + r\vec{u}$ where the ij entry of M_0 is δ_{ij} for $i \neq j$ and is -1 if $i = j$. Assuming that enough of the entries δ_{ij} are positive, $M + cI$ and $M_0 + cI$ will be primitive matrices for c sufficiently large. Suppose that M and M_0 are primitive. If the principal eigenvalue for M_0 is μ_0, with corresponding eigenvector $\vec{v}$, then $M\vec{v} = dM_0\vec{v} + r\vec{v} = (d\mu_0 + r)\vec{v}$ so the principal eigenvalue for M is $\sigma_1 = d\mu_0 + r$. It follows that the model (2.79) predicts population growth when

$$r/d > -\mu_0. \tag{2.80}$$

This is precisely analogous to the condition (2.7) for simple reaction-diffusion models. Recall that for reaction-diffusion models the eigenvalue problem (2.5) is written as $\Delta\phi + \lambda\phi = 0$, i.e. $-\Delta\phi = \lambda\phi$, so that the analogous problem for M_0 would be

$M_0\vec{v} + \lambda\vec{v} = 0$, leading to the relation $-\mu_0 = \lambda_1$. (Here λ_1 is the principal eigenvalue for $M_0\vec{v} + \lambda\vec{v} = 0$ which is analogous to (2.5).) Both (2.7) and (2.80) allow us to separate the parameters r and d, which describe biological properties of the population, from the eigenvalues λ_1 (or $-\mu_0$) which describe the underlying spatial environment. Many of the same scenarios that we have described in terms of reaction-diffusion models can be treated from the viewpoint of "island chain" or "stepping stone" models such as (2.78). For example, Seno (1988) used models of the form (2.78) to study a situation where all the patches are the same except for one singular patch which is more (or less) favorable than the others. This is quite analogous to the situation analyzed in Cantrell and Cosner (1991b) which is discussed in detail in Section 2.4, and Seno (1988) obtained similar results from his modeling approach.

The philosophical basis of metapopulation models is quite different from that of reaction-diffusion models or "island chain" models, but some of their mathematical features are quite similar. The underlying spatial structure for metapopulation models, as for "island chain" models is a discrete set of patches. Instead of describing the population of each patch in terms of growth, immigration, or emigration, metapopulation models describe the probability that each patch is occupied in terms of colonization and extinction rates. Recently, Hanski (1997, 1999) has formulated metapopulation models that capture some of the details of the patch network. If A_i represents the area of each patch, d_{ij} the distance between patches i and j, and $p_i(t)$ the probability that the ith patch is occupied at time t, the models take the form

$$\frac{dp_i}{dt} = C\sum_{i \neq j} A_j e^{-\alpha d_{ij}} p_j - (E/A_i)p_i + \text{quadratic terms}, \tag{2.81}$$

where C represents a species-specific colonization rate, E an extinction rate, and α can be viewed as describing the extent to which individuals are lost while dispersing between patches. Let M be the matrix with entries $A_i A_j e^{-\alpha d_{ij}}$ for $i \neq j$ and with diagonal entries all zero. Again, $M + cI$ is a primitive matrix for c large, so M has a principal eigenvalue μ_1. Hanski and Ovaskainen (2000) show that the model (2.81) predicts persistence if $\mu_1 > E/C$. Since the matrix M describes the patch network but not the biological characteristics of the population inhabiting it, this criterion again separates the geometric information carried by μ_1 from the population characteristics described by E and C. Hanski and Ovaskainen (2000) refer to μ_1 as the metapopulation capacity of a fragmented landscape.

If a population has discrete generations and disperses but does not reproduce between episodes of recruitment, it may be appropriate to use a discrete time model to describe population dynamics. Suppose that a population inhabits a network of discrete patches, recruits at times $t = 1, 2, 3$, etc. and disperses and experiences mortality during the interval between recruitment events according to

$$\frac{du_i}{dt} = M\vec{u} - Du_i, \tag{2.82}$$

where D is the mortality rate and M has entries $d_{ij} \geq 0$ for $i \neq j$ and diagonal entries $-d_{ii}$. If $M+cI$ is primitive for some constant c, then the matrix exponential e^{Mt} will have positive entries. The solution to (2.82) is given by $\vec{u}(t) = e^{Mt}e^{-Dt}\vec{u}(0)$, so $\vec{u}(1) = e^{-D}e^{M}\vec{u}(0)$. Suppose that if there are u^* individuals prior to an episode of recruitment there will be Ru^* individuals afterward. Combining this with the effects of dispersal, we obtain the model

$$\vec{u}(T+1) = (Re^{-D})e^{M}\vec{u}(T). \tag{2.83}$$

If μ_2 is the principal eigenvalue for the matrix e^M, then (2.83) predicts population growth if $Re^{-D}\mu_2 > 1$ but decline if $Re^{-D}\mu_2 < 1$. The condition for growth can be written as $\mu_2 > e^D/R$, which again separates the spatial structure encoded by M from the demographic parameters R and D. Something analogous can be done in the context of continuous space. If a population diffuses throughout a region Ω according to the equation

$$\begin{aligned} u_t &= d\nabla^2 u - Du \quad \text{in} \quad \Omega \\ d\frac{\partial u}{\partial \vec{n}} &+ \beta u = 0 \quad \text{on} \quad \partial\Omega \end{aligned} \tag{2.84}$$

then $u(x,t)$ can be expressed as

$$u(x,t) = \int_\Omega G(x,y,t)u(y,0)dy$$

where G is the Green's function for (2.84). Setting $t = 1$ gives a formula for $u(1)$ in terms of $u(0)$. More generally, dispersal and mortality could be described by some positive kernel $K(x,y)$ other than the Green's function of a diffusion operator. This approach to dispersal has been used extensively by Lewis and his co-workers (Lewis, 1997; Van Kirk and Lewis, 1997), especially in the context of studying invasions. If we combine reproduction with dispersal as in the spatially discrete case we obtain models of the form

$$u(x, T+1) = R(x)\int_\Omega K(x,y)u(y,T)dy. \tag{2.85}$$

If the kernel $K(x,y)$ is positive, bounded, and continuous (or even if K is not too singular) then the operator mapping $v(x)$ to $R(x)\int_\Omega K(x,y)v(y)dy$ will be positive and compact on suitable spaces of functions, so that it will have a positive principal eigenvalue μ_3 by virtue of the Krein-Rutman theorem. In that case (2.84) predicts population growth if $\mu_3 > 1$ but decay if $\mu_3 < 1$. A detailed discussion of this approach and a derivation of this type of result for a more general class of models is given in Van Kirk and Lewis (1997).

The preceding examples show that in a variety of different types of spatial models, eigenvalues can be used to "average" the effects of dispersal, habitat geometry, and local rates of recruitment and/or mortality to yield criteria for persistence or extinction of populations on a larger spatial scale. The listing of examples is not exhaustive, but we hope that it is adequate to convince the reader that by using eigenvalues it is possible to treat a variety of spatial models from a single unified viewpoint.

Appendix

In this appendix we shall present some proofs and mathematical discussions which are somewhat technical. The first of these will address the variational formulation of eigenvalues. Theorem 2.1 is not proved here, because it is a well-known classical result. See Courant and Hilbert (1953). However, the methods used to prove Theorem 2.4 could also be readily adapted to prove Theorem 2.1.

To obtain the existence of a principal eigenvalue in Theorem 2.4 we work in the Hilbert space $W^{1,2}(\Omega)$ (or $W_0^{1,2}(\Omega)$ in the case of Dirichlet boundary conditions.) The essential abstract result we shall use is the following. (See de Figueiredo (1982).)

Lemma 2.A.1 Let H be a Hilbert space. Denote the inner product on H by $\langle\cdot,\cdot\rangle$ and the corresponding norm by $\|\ \ \|$. Let T be a compact symmetric operator on H. If

$$\mu_1 = \sup\{\langle Tu, u\rangle : \|u\| = 1\} > 0 \tag{2.A.1}$$

then μ_1 is an eigenvalue of T, i.e. there exists $\phi_1 \in H,\ \ \phi_1 \neq 0$ so that $T\phi_1 = \mu_1\phi_1$.

Proof: If T is compact it must be bounded, so $\mu_1 \leq \|T\| < \infty$. Choose a sequence $\{u_n\}$ with $\|u_n\| = 1$ and $\lim\limits_{n\to\infty}\langle Tu_n, u_n\rangle = \mu_1$. A bounded sequence in a Hilbert space must have a weakly convergent subsequence, and a compact operator on a Hilbert space maps weakly convergent sequences into strongly convergent sequences, so we may pass to a subsequence and obtain $u_n \to \phi_1$ weakly and $Tu_n \to T\phi_1$ strongly for some $\phi_1 \in H$. Then $\mu_1 = \lim\limits_{n\to\infty}\langle Tu_n, u_n\rangle = \langle T\phi_1, \phi_1\rangle$. Since $\mu_1 \neq 0$ we must have $\phi_1 \neq 0$ so we may assume that $\|\phi_1\| = 1$.) If $w \in H,\ w \neq 0$ then

$$\begin{aligned}\langle (T - \mu_1)w, w\rangle &= \langle Tw, w\rangle - \mu_1\|w\|^2\\ &= \|w\|^2[\langle T(w/\|w\|), (w/\|w\|)\rangle - \mu_1]\\ &\leq 0.\end{aligned}$$

(The inequality clearly holds for $w = 0$ as well.) For any fixed $v \in H$, let $w = \phi_1 - tv$. We have for all $t \in I\!R$

$$\begin{aligned}0 \leq\ &\langle (\mu_1 - T)(\phi_1 - tv), (\phi_1 - tv)\rangle = \langle (\mu_1 - T)\phi_1, \phi_1\rangle - 2t\langle (\mu_1 - T)\phi_1, v\rangle\\ &+t^2\langle (\mu_1 - T)v, v\rangle,\end{aligned}$$

but $\langle (T - \mu_1)\phi_1, \phi_1\rangle = 0$ so we have $2t(\langle (\mu_1 - T)\phi_1, v\rangle) \leq t^2(\langle (\mu_1 - T)v, v\rangle)$. The last inequality cannot be valid for both positive and negative values of t near $t = 0$ unless $\langle (T - \mu_1)\phi_1, v\rangle = 0$. Since $v \in H$ was arbitrary, we must have $\langle (T - \mu_1)\phi_1, v\rangle = 0$ for all $v \in H$, so $(T - \mu_1)\phi_1 = 0$, i.e. $T\phi_1 = \mu_1\phi_1$.

We want to work in the Hilbert space $W^{1,2}(\Omega)$, but we want to replace the standard inner product

$$\langle u, v\rangle_{1,2} = \int_\Omega (\nabla u \cdot \nabla v + uv)dx$$

with the inner product

$$\langle u, v\rangle = \int_\Omega d(x)\nabla u \cdot \nabla v dx + \int_{\partial\Omega} \beta(x)uvdS. \tag{2.A.2}$$

The following is adapted from Mikhailov (1978, Ch. III, §5).

Lemma 2.A.2. Suppose that Ω is a bounded domain with $\partial\Omega$ of class C^1, $d(x)$ is a bounded measurable function with $d(x) \geq d_0$ almost everywhere on Ω, and $\beta(x)$ is a bounded measurable function on $\partial\Omega$ with $\beta(x) \geq 0$ almost everywhere and $\beta(x) > 0$ almost everywhere on a subset of $\partial\Omega$ that is open relative to $\partial\Omega$. Then the inner products $\langle \cdot, \cdot \rangle_{1,2}$ and $\langle \cdot, \cdot \rangle$ generate equivalent norms.

Proof: Since $d(x)$ is bounded it is clear that for some constant C,

$$\int_\Omega d(x)|\nabla u|^2 dx \leq C \int_\Omega (|\nabla u|^2 + u^2) dx.$$

The regularity of $\partial\Omega$ implies that any function $u \in W^{1,2}(\Omega)$ has a trace $tr(u)$ on $\partial\Omega$ which belongs to $L^2(\partial\Omega)$ with $||tr(u)||_{L^2(\partial\Omega)} \leq C||u||_{1,2}$ for some constant C, and the trace operator $tr : W^{1,2}(\Omega) \to L^2(\partial\Omega)$ is compact; see Adams (1975, Theorem 5.22) and Mikhailov (1978, Ch. III). It follows from these relations and the boundedness of $\beta(x)$ that $\langle u, u \rangle \leq C\langle u, u \rangle_{1,2}$.

Suppose that there does not exist any constant C_1 with $\langle u, u \rangle_{1,2} \leq C_1 \langle u, u \rangle$. We may then choose a sequence $\{u_n\}$ with $\langle u_n, u_n \rangle = 1$ and $\langle u_n, u_n \rangle_{1,2} = n$; letting $v_n = (1/\sqrt{n})u_n$ we have $\langle v_n, v_n \rangle_{1,2} = 1$ but $\langle v_n, v_n \rangle = 1/n$. Since $\{v_n\}$ is bounded in $W^{1,2}(\Omega)$, the compact embedding of $W^{1,2}(\Omega) \hookrightarrow L^2(\Omega)$ implies that there is a subsequence which converges in $L^2(\Omega)$. Passing to the subsequence we have

$$\begin{aligned}
||v_n - v_m||_{1,2}^2 &= ||v_n - v_m||_{L^2(\Omega)}^2 + ||\nabla(v_n - v_m)||_{L^2(\Omega)}^2 \\
&\leq ||v_n - v_m||_{L^2(\Omega)}^2 + \int_\Omega |\nabla(v_n - v_m)|^2 dx \\
&\leq ||v_n - v_m||_{L^2(\Omega)}^2 + (2/d_0) \int_\Omega d(x)(|\nabla v_n|^2 + |\nabla v_m|^2) dx \\
&\leq ||v_n - v_m||_{L^2(\Omega)}^2 + (2/d_0)(\langle v_n, v_n \rangle + \langle v_m, v_m \rangle) \\
&\leq ||v_n - v_m||_{L^2(\Omega)}^2 + (2/d_0)[(1/n) + (1/m)],
\end{aligned}$$

so that the subsequence is Cauchy and hence convergent in $W^{1,2}(\Omega)$. It follows that $v_n \to v$ in $W^{1,2}(\Omega)$ with $\langle v, v \rangle_{1,2} = 1$. We also have

$$0 \leq \int_\Omega d(x)|\nabla v_n|^2 dx \leq \langle v_n, v_n \rangle = 1/n,$$

$$0 \leq \int_{\partial\Omega} \beta(x) v_n^2 dS \leq \langle v_n, v_n \rangle = 1/n,$$

and $tr(v_n) \to tr(v)$ on $\partial\Omega$; thus

$$\int_\Omega d(x)|\nabla v|^2 dx = 0$$

and

$$\int_{\partial\Omega} \beta(x) v^2 dS = 0.$$

Since $d(x) \geq d_0$ almost everywhere, we must have $\nabla v = 0$ almost everywhere, so v must be a constant. Since $\beta(x) > 0$ on an open subset of $\partial\Omega$, the constant must be zero. However, $v \equiv 0$ contradicts $\langle v, v \rangle_{1,2} = 1$. Thus, our assumption that $\langle u, u \rangle_{1,2}$ is not bounded by $C_1\langle u, u\rangle$ for any C_1 leads to a contradiction, so there must be a C_1 such that $\langle u, u\rangle_{1,2} \leq C_1\langle u, u\rangle$.

Proof of Theorem 2.4: For any $u \in W^{1,2}(\Omega)$ we can define a functional $f(u) : v \mapsto \int_\Omega m(x)uv dx$. Since $m(x) \in L^\infty(\Omega)$ the functional $f(u)$ is bounded on $W^{1,2}(\Omega)$, because the Schwartz inequality implies $||f(u)v|| \leq ||m||_\infty ||u||_2 ||v||_2$ (where $||\cdot||_\infty$ and $||\cdot||_2$ denote the norms of L^∞ and L^2, respectively), and the norm on $W^{1,2}(\Omega)$ dominates the norm on $L^2(\Omega)$. Thus, by the Riesz representation theorem, there is an element Tu such that $\langle Tu, v\rangle = f(u)v$. Since $f(u)v = f(v)u$ the operator $T : W^{1,2}(\Omega) \to W^{1,2}(\Omega)$ is symmetric, and since $f(u)$ is bounded as a functional with norm $||f(u)|| \leq ||m||_\infty ||u||_2 \leq c||m||_\infty ||u||$ (where $||\ \ ||$ is the norm defined by the inner product in (2.A.2)) the operator T is also bounded. Suppose that $\{u_n\}$ is a bounded sequence in $W^{1,2}(\Omega)$. Since $W^{1,2}(\Omega)$ is a Hilbert space, $\{u_n\}$ has a weakly convergent subsequence. Passing to the subsequence and reindexing, we have $u_n \rightharpoonup u$ (i.e. $\{u_n\}$ converges weakly to u) for some $u \in W^{1,2}(\Omega)$. The space $W^{1,2}(\Omega)$ embeds compactly in $L^2(\Omega)$ (see Adams (1975, Theorem 6.2); see also Gilbarg and Trudinger (1977)) so we have $u_n \to u$ in $L^2(\Omega)$. However,

$$\begin{aligned}
||Tu_n - Tu||^2 &= \langle Tu_n - Tu, Tu_n - Tu\rangle \\
&= \int_\Omega m(u_n - u)(Tu_n - Tu)dx \\
&\leq ||m||_\infty ||u_n - u||_2 ||Tu_n - Tu||_2 \\
&\leq c||m||_\infty ||u_n - u||_2 ||Tu_n - Tu||.
\end{aligned}$$

Thus, $||Tu_n - Tu|| \leq c||m||_\infty ||u_n - u||_2$ so $Tu_n \to Tu$ in $W^{1,2}(\Omega)$. This establishes that T is compact. Since $m(x) > 0$ on an open subset of Ω, we can take u to be a function which is nonzero on a set of positive measure inside that subset and zero elsewhere, so that $\langle Tu, u\rangle = \int_\Omega mu^2 dx > 0$. It then follows from Lemma 2.A.1 that there exist a principal eigenvalue $\mu_1 > 0$ and corresponding eigenfunction ϕ_1 for T such that $T\phi_1 = \mu_1\phi_1$. Thus, for any $v \in W^{1,2}(\Omega)$, $\langle T\phi_1, v\rangle = \mu_1\langle\phi_1, v\rangle$ or alternatively

$$(1/\mu_1)\int_\Omega m\phi_1 v dx = \int_\Omega d(x)\nabla\phi_1 \cdot \nabla v dx + \int_{\partial\Omega} \beta(x)\phi_1 v dS. \tag{2.A.3}$$

Since relation (2.A.3) holds for any $v \in W^{1,2}(\Omega)$, ϕ_1 is a weak solution to

$$\begin{aligned}
\nabla \cdot d(x)\nabla\phi_1 &= (1/\mu_1)m(x)\phi_1 \quad \text{in} \quad \Omega \\
d(x)\frac{\partial\phi_1}{\partial\vec{n}} + \beta(x)\phi_1 &= 0 \qquad\qquad \text{on} \quad \partial\Omega.
\end{aligned}$$

By standard elliptic regularity theory, ϕ_1 will belong to $W^{2,p}(\Omega)$ for any p. (If $m(x) \in C^\alpha(\overline{\Omega})$ then $\phi_1 \in C^{2+\alpha}(\overline{\Omega})$.) (See Gilbarg and Trudinger (1977) or Section 1.6.) The positivity of the eigenfunction ϕ follows as in the derivation of Theorem 2.1 in Courant and Hilbert (1953).

Remarks: The space $W^{1,2}(\Omega)$ embeds compactly in L^{p^*} for any $p^* < \infty$ if the dimension of the underlying domain Ω is 1 or 2 and for $p^* < 2N/(N-2)$ if $N \geq 3$. Thus, for such

p we have by repeated applications of Hölder's inequality that

$$|\langle Tu, v\rangle| = |\int_\Omega muv\,dx| \le ||m||_{L^r(\Omega)}||u||_{L^p(\Omega)}||v||_{L^q(\Omega)}$$

if $\frac{1}{p}+\frac{1}{q}+\frac{1}{r} = 1$. Taking $p = q = p^* > 2$ we have $\langle Tu, v\rangle \le C||m||_{L^r(\Omega)}||u||||v||$ with $r = p^*/(p^*-2)$. The point is that the quadratic form $\langle Tu, v\rangle$ is bounded on $W^{1,2}(\Omega)\times W^{1,2}(\Omega)$ not only in terms of $||m||_\infty$ but in terms of $||m||_{L^r(\Omega)}$ for $r > p^*/(p^*-2)$. In the case of $N = 1, 2$ any $r > 1$ is admissible. If $N = 3$ we must have $r > N/2$ because of the restriction on p^*. Thus, we could replace the condition $m \in L^\infty(\Omega)$ with $m \in L^r(\Omega)$, and also we can obtain the continuity of $\langle Tu, v\rangle$ and hence μ_1 with respect to m relative to the norm in L^r. This is done in detail in deFigueiredo (1982).

The formula (2.A.1) as applied in the proof of Theorem 2.4 gives

$$\mu_1 = \frac{1}{\lambda_1^+(m)} = \sup\left\{\int_\Omega mu^2dx : \int_\Omega d|\nabla u|^2dx + \int_{\partial\Omega}\beta u^2dS = 1\right\}. \tag{2.A.4}$$

If $w \in W^{1,2}(\Omega)$ with $w \ne 0$, use the norm defined by (2.A.2) and let $\tilde{w} = w/||w||$. We have

$$1 = ||\tilde{w}||^2 = \int_\Omega d|\nabla\tilde{w}|^2dx + \int_{\partial\Omega}\beta(x)\tilde{w}^2dS.$$

Also,

$$\int_\Omega m\tilde{w}^2dx = \frac{1}{||w||^2}\int_\Omega mw^2dx = \frac{\int_\Omega mw^2dx}{\int_\Omega d|\nabla w|^2dx + \int_{\partial\Omega}\beta w^2dS}.$$

Thus, taking

$$\sup\left\{\frac{\int_\Omega mw^2dx}{\int_\Omega d|\nabla w|^2dx + \int_{\partial\Omega}\beta w^2dS} : w \in W^{1,2}(\Omega),\ w \ne 0\right\} \tag{2.A.5}$$

gives the same value $\mu_1 = 1/\lambda_1^+(m)$ as in (2.A.4).

The case of Neumann boundary conditions, as in Theorem 2.5, requires special treatment because $\left(\int d|\nabla u|^2dx\right)^{1/2}$ is not equivalent to the standard norm on $W^{1,2}(\Omega)$ since it is equal to zero for u constant but nonzero. To prove Theorem 2.5 we shall need an auxiliary result which is roughly equivalent to a lemma introduced by Fleming (1975), and is discussed in the context of deriving Theorem 2.5 by Brown and Lin (1980).

Lemma 2.A.3. Suppose that $\int_\Omega m(x)dx < 0$. Let $\{\phi_n\} \in W^{1,2}(\Omega)$ be a sequence with $\int_\Omega \phi_n^2dx = 1$ and $\int_\Omega m\phi_n^2dx > 0$. There is a constant $c_0 > 0$ such that $\int_\Omega d(x)|\nabla\phi_n|^2dx \ge c_0$ for all n.

Proof: Suppose not. Then by passing to a subsequence we may choose ϕ_n so that $\int_\Omega d(x)|\nabla\phi_n|^2 dx \leq 1/n$. Since $\int_\Omega \phi_n^2 dx = 1$, the subsequence is bounded in $W^{1,2}(\Omega)$. Since $W^{1,2}(\Omega)$ embeds compactly in $L^2(\Omega)$, there is a subsequence which converges in $L^2(\Omega)$. Passing to that subsequence we have $\phi_n \to \phi$ in $L^2(\Omega)$ so that $\int_\Omega \phi^2 dx = \lim_{n\to\infty}\int_\Omega \phi_n^2 dx = 1$. Also,

$$
\begin{aligned}
||\phi_n - \phi_m||_{1,2}^2 &= \int_\Omega |\phi_n - \phi_m|^2 dx + \int_\Omega |\nabla\phi_u - \nabla\phi_m|^2 dx \\
&\leq \int_\Omega |\phi_n - \phi_m|^2 dx + (2/d_0)\int_\Omega (d|\nabla\phi_n|^2 + d|\nabla\phi_m|^2) dx \\
&\leq \int_\Omega |\phi_n - \phi_m|^2 dx + \frac{2}{d_0}\left(\frac{1}{n} + \frac{1}{m}\right),
\end{aligned}
$$

so the sequence is Cauchy and hence convergent in $W^{1,2}(\Omega)$. We have

$$\int_\Omega d|\nabla\phi|^2 dx = \lim_{n\to\infty}\int_\Omega d|\nabla\phi_n|^2 dx = 0.$$

Thus, ϕ must be constant and since $\int_\Omega \phi^2 dx = 1$ we must have $\phi = 1/|\Omega|^{1/2} \neq 0$ so

$$\int_\Omega m\phi^2 dx = 1/|\Omega| \int_\Omega m dx < 0.$$

However, since $m(x)$ is a bounded measurable function, the mapping $\phi_n \mapsto \int_\Omega m\phi_n^2 dx$ is continuous as a mapping from $L^2(\Omega)$ into $I\!R$, so $\int_\Omega m\phi^2 dx = \lim_{n\to\infty}\int_\Omega m\phi_n^2 dx \geq 0$. This yields a contradiction. Thus, there must exist a constant $c_0 > 0$ as asserted by the lemma.

Proof of Theorem 2.5: Our approach follows that of Brown and Lin (1980). Observe that λ_1^+ is a positive principal eigenvalue for

$$
\begin{aligned}
&\nabla \cdot d(x)\nabla\phi + \lambda m(x)\phi = 0 \text{ on } \quad \Omega \\
&\frac{\partial\phi}{\partial\vec{n}} = 0 \qquad\qquad\qquad \text{on } \quad \partial\Omega
\end{aligned}
\tag{2.A.6}
$$

if and only if $\sigma_1 = 0$ is a principal eigenvalue for

$$
\begin{aligned}
&\nabla \cdot d(x)\nabla\psi + \lambda_1^+ m(x)\psi = \sigma\psi \text{ in } \quad \Omega, \\
&\frac{\partial\psi}{\partial\vec{n}} = 0 \qquad\qquad\qquad \text{on } \quad \partial\Omega.
\end{aligned}
\tag{2.A.7}
$$

If $\int_\Omega m(x)dx > 0$ then by substituting $\psi = 1$ into the formula (2.11a) we see that $\sigma_1 \geq \int_\Omega m dx/|\Omega| > 0$. (Recall that $\beta(x) = 0$ in the case of Neumann conditions.) Suppose

$\int_\Omega m(x)dx = 0$. If $\sigma_1 = 0$ in (2.A.7) let $\psi_1 > 0$ be the associated eigenfunction, multiply (2.A.7) by $1/\psi_1$ and integrate by the divergence theorem to obtain

$$-\int_\Omega \nabla\left(\frac{1}{\psi_1}\right) \cdot d\nabla\psi_1 = 0. \tag{2.A.8}$$

(Here we have used the assumption that $\int_\Omega m(x)dx = 0$.) Since $\nabla(1/\psi_1) = -\nabla\psi_1/\psi_1^2$, we have

$$\nabla(1/\psi_1)\cdot\nabla\psi_1 = -|\nabla_1\psi_1|^2/\psi_1^2 = -|\nabla(\ln\psi_1)|^2$$

so that (2.A.8) implies

$$\int_\Omega d|\nabla(\ln\psi_1)|^2 dx = 0$$

so $\nabla \ln\psi_1 = 0$ and ψ_1 is constant. We then have from (2.A.7) that $m(x) = 0$ on Ω, but we are assuming that $m(x) > 0$ on an open subset of Ω, so that is impossible. Thus, we must have $\int_\Omega m dx < 0$ if (2.A.6) is to have a positive principal eigenvalue.

Suppose that $\int_\Omega m dx < 0$. Let

$$\mu_1 = \sup_{\substack{\phi\in W^{1,2}(\Omega)\\ |\nabla\phi|\neq 0}} \left\{\frac{\int_\Omega m\phi^2 dx}{\int_\Omega d|\nabla\phi|^2 dx}\right\}.$$

Since $m(x)$ is positive on an open set we have $\mu_1 > 0$, and we may choose a sequence $\{\phi_n\}$ such that

$$0 < \frac{\int_\Omega m\phi_n^2}{\int_\Omega d|\nabla\phi_n|^2 dx} \to \mu_1 \quad \text{as} \quad n\to\infty.$$

We may normalize the sequence by requiring $\int_\Omega \phi_n^2 dx = 1$. Since $\int_\Omega m dx < 0$, Lemma 2.A.3 implies that $\int_\Omega d|\nabla\phi_n|^2 dx \geq c_0 > 0$ for some $c_0 > 0$, so that $\mu_1 \leq ||m||_\infty/c_0 < \infty$. By passing to a subsequence if necessary we may assume that

$$\frac{\int_\Omega m\phi_n^2 dx}{\int_\Omega d|\nabla\phi_n|^2 dx} \geq \mu_1/(1+1/n),$$

so that

$$-\mu_1\int_\Omega d|\nabla\phi_n|^2 dx + \int_\Omega m\phi_n^2 dx \geq (-1/n)\int_\Omega m\phi_n^2 dx \geq (-1/n)||m||_\infty.$$

It follows that by taking $\lambda_1^+ = 1/\mu_1$ we get

$$\sigma_1 = \sup\left\{(1/\mu_1)\left[-\mu_1\int_\Omega d|\nabla\phi|^2dx + \int_\Omega m\phi^2dx\right] : \int_\Omega \phi^2 = 1\right\} \geq 0.$$

Also, $\mu_1\int_\Omega d|\nabla\phi|^2 \geq \int_\Omega m\phi^2dx$ for any ϕ, so $-\mu_1\int_\Omega d|\nabla\phi|^2dx + \int_\Omega m\phi^2dx \leq 0$, for any ϕ, so that $\sigma_1 \leq 0$. It follows that for $\lambda_1^+ = 1/\mu_1$, the principal eigenvalue σ_1 in (2.A.7) is zero, so that $\lambda_1^+ > 0$ is indeed a principal eigenvalue for (2.A.6).

Proof of Theorem 2.7: We consider first the case where $\beta(x) > 0$ on an open subset of $\partial\Omega$ or where Dirichlet boundary conditions are imposed on all or part of $\partial\Omega$. Let $\{m_n(x)\}$ be a sequence of bounded measurable functions with $||m_n||_\infty < M_1$ and such that (2.32) holds for $\psi \in L^1(\Omega)$ provided $\psi \geq 0$ almost everywhere. Suppose that contrary to the assertion of Theorem 2.7 we have $\lambda_1^+(m_n) \not\to \infty$ as $n \to \infty$. By passing to a subsequence we may assume that $\lambda_1^+(m_n) \leq \Lambda_0$ for some $\Lambda_0 < \infty$. Let ϕ_n be the eigenfunction corresponding to $\lambda_1^+(m_n)$, normalized by $\int_\Omega d|\nabla\phi_n|^2dx + \int_{\partial\Omega}\beta\phi_n^2dS = 1$ (or, in the case of Dirichlet boundary conditions, $\int_\Omega d|\nabla\phi_n|^2dx = 1$.) By Lemma 2.A.2, or in the case of Dirichlet boundary conditions, by Poincaré's Inequality, the sequence $\{\phi_n\}$ is uniformly bounded in $W^{1,2}(\Omega)$ (or $W_0^{1,2}(\Omega)$ in the Dirichlet case) so that by the compact embedding of $W^{1,2}(\Omega)$ into $L^2(\Omega)$ there is a subsequence of $\{\phi_n\}$ which converges in $L^2(\Omega)$. Passing to the subsequence we may assume $\phi_n \to \phi$ in $L^2(\Omega)$. By multiplying (2.13) by ϕ_n and by using the definition of a $W^{1,2}$ solution to (2.13) (which means, in effect, multiplying (2.13) by ϕ_n and using Green's formula) we have

$$1 = \int_\Omega d|\nabla\phi_n|^2dx + \int_{\partial\Omega}\beta\phi_n^2dS = \lambda_1^+(m_n)\int_\Omega m_n\phi_n^2dx \tag{2.A.9}$$

(omit the integral involving β in the Dirichlet case). Recall that we supposed $\lambda_1^+(m_n) \leq \Lambda_0$ for some Λ_0. By (2.A.9) we have

$$1 = \lambda_1^+(m_n)\left[\int_\Omega m_n(\phi_n^2 - \phi^2)dx + \int_\Omega m_n\phi^2dx\right]. \tag{2.A.10}$$

Since $\phi_n \to \phi$ in $L^2(\Omega)$, the first integral on the right in (2.A.10) goes to zero as $n \to \infty$. By hypothesis (2.32), we may assume the second integral also goes to zero as $n \to \infty$. Letting $n \to \infty$ in (2.A.10) thus yields the contradiction $1 = 0$, so the sequence $\{\lambda_1^+(m_n)\}$ cannot have a bounded subsequence and we must therefore have $\lim_{n\to\infty}\lambda_1^+(m_n) = \infty$.

In the case of Neumann boundary conditions, we must proceed slightly differently. Again, suppose that $\lambda_1^+(m_n) \not\to \infty$ as $n \to \infty$, so that by passing to a subsequence we may assume $\lambda_1^+(m_n) \leq \Lambda_0$. Normalize the sequence of eigenfunctions so that $\int_\Omega \phi_n^2dx = 1$. We must then have $\int_\Omega d|\nabla\phi_n|^2dx \geq c_0 > 0$. If not we may proceed as in the proof of Lemma 2.A.3 and by passing to a subsequence obtain a sequence $\{\phi_n\}$ with $\int d|\nabla\phi_n|^2dx \leq 1/n$. As in the proof of Lemma 2.A.3 that sequence must have a subsequence converging in $W^{1,2}(\Omega)$.

If the subsequence converges to ϕ, we must have $\int_\Omega d|\nabla\phi|^2 dx = 0$ but $\int_\Omega \phi^2 dx = 1$ so $\phi = 1/|\Omega|^{1/2}$. We then have

$$0 \leq 1/\lambda_1^+(m_n)\int_\Omega d|\nabla\phi_n|^2 dx = \int_\Omega m_n\phi_n^2 dx$$
$$= \left[\int_\Omega m_n(\phi_n^2 - \phi^2)dx + \int_\Omega m_n\phi^2\right].$$

By hypothesis the second integral on the right is bounded by $-M_0/|\Omega| < 0$ for all n. The first integral on the right goes to zero as $n \to \infty$. Thus, by letting $n \to \infty$ we obtain $0 \leq -M_0/|\Omega|$, a contradiction. Thus, we must have $\int d|\nabla\phi_n|^2 dx \geq c_0 > 0$ for some c_0. We can renormalize the sequence $\{\phi_n\}$ so that $\int_\Omega d|\nabla\phi_n|^2 dx = 1$ and $\int_\Omega \phi_n^2 dx \leq 1/c_0$ by dividing each element ϕ_n by $\left(\int_\Omega d|\nabla\phi_n|^2 dx\right)^{1/2}$, so that the sequence is uniformly bounded in $W^{1,2}(\Omega)$. From this point on we may proceed as in the case of $\beta > 0$ on part of $\partial\Omega$ or the Dirichlet case.

Conversely, let us assume that $\lambda_1^+(m_n) \to \infty$ as $n \to \infty$. If there exists $\psi \in L^1(\Omega)$ with $\psi \geq 0$ and $\limsup_{n\to\infty}\int_\Omega m_n\psi dx = \epsilon > 0$ then we have $\int_\Omega m_n\psi dx \geq \epsilon/2$ for some subsequence of $\{m_n\}$. We can approximate $\sqrt{\psi}$ as closely as we wish in $L^2(\Omega)$ with a smooth function ϕ_0, so that $\int_\Omega m_n\phi_0^2 dx \geq \epsilon/4$. We have

$$\frac{1}{\lambda_1^+(m_n)} = \sup_{\substack{\phi\in W^{1,2}(\Omega)\\ \phi\neq 0}}\left(\frac{\int_\Omega m_n\phi^2 dx}{\int_\Omega d|\nabla\phi|^2 dx + \int_{\partial\Omega}\beta\phi^2 dS}\right)$$
$$\geq \frac{\int_\Omega m_n\phi_0^2 dx}{\int_\Omega d|\nabla\phi_0|^2 dx + \int_{\partial\Omega}\beta\phi_0^2 dS} \tag{2.A.11}$$
$$\geq \frac{\epsilon/4}{\int_\Omega d|\nabla\phi_0|^2 dx + \int_{\partial\Omega}\beta\phi_0^2 dS}.$$

Inequality (2.A.11) implies that $1/\lambda_1^+(m_n)$ is bounded below by a positive quantity independent of n, so we cannot have $\lambda_1^+(m_n) \to \infty$. Thus, if $\lambda_1^+(m_n) \to \infty$ as $n \to \infty$ we must have $\limsup_{n\to\infty}\int_\Omega m_n\psi dx \leq 0$ for any $\psi \in L^2(\Omega)$ with $\psi \geq 0$ almost everywhere. This part of the proof is essentially the same in all cases. In the Neumann case, if $\int_\Omega m_n\psi dx \geq 0$ then ψ and hence ϕ_0 must be nonconstant since $\int_\Omega m_n dx \leq -M_0$, so we will have $\int_\Omega d|\nabla\phi_0|^2 dx > 0$.

Remarks on the proof of Theorem 2.21:

If we start with the eigenvalue problem

$$\begin{aligned} &\nabla \cdot d(x)\nabla\phi + \vec{b}(x)\cdot\nabla\phi + c(x)\phi + \lambda_1 m_n(x)\phi = 0 \text{ in } \quad \Omega, \\ &\delta(x)\frac{\partial\phi}{\partial\vec{n}} + \beta(x)\phi = 0 \qquad\qquad\qquad\qquad \text{on} \quad \partial\Omega \end{aligned}$$

then multiplying by ϕ and using Green's formula and the boundary conditions gives

$$\lambda_1\int_\Omega m_n\phi^2 dx = \int_\Omega [d|\nabla\phi|^2 - \phi\vec{b}\cdot\nabla\phi - c\phi^2]dx + \int_{\partial\Omega}[d(x)\beta(x)/\delta(x)]\phi^2 dS.$$

so by (2.77)

$$\lambda_1\int m_n\phi^2 dx \geq d_1\int_\Omega |\nabla\phi|^2 dx + \int_{\partial\Omega}[d(x)\beta(x)/\delta(x)]\phi^2 dS.$$

This last inequality may be used in a manner analogous to (2.A.9) to arrive at an inequality which can be exploited in the same way as (2.A.10) to prove Theorem 2.21.

3

Density Dependent Single-Species Models

3.1 The Importance of Equilibria in Single Species Models

Nonlinear reaction-diffusion equations arise as models for population dynamics when linear growth and dispersal models such as those discussed in the preceding chapter are augmented with density-dependent terms describing the effects of resource depletion, crowding, social interactions among the members of a species, interactions between species, or various other processes. If the density-dependent mechanisms are intraspecific, or if interactions with other species can be scaled out of the model (e.g. because they occur on much faster or slower time scales than do the population dynamics of the focal species), then the resulting models typically take the form of a single reaction-diffusion equation. The prototype for such equations is the logistic equation with diffusion.

The key point of this section is easy to state, but somewhat technical to prove: the dynamics of a single reaction-diffusion equation are essentially determined by its equilibria and their stability properties. That is not the case with systems of reaction-diffusion equations or even ordinary differential equations, or discrete time models, since those sorts of models may have stable periodic solutions or more complex dynamics. The rest of this section will be devoted to a discussion of why the dynamics of a single reaction-diffusion model are determined by its equilibria. The next few sections are then devoted to the analysis of equilibria of single reaction-diffusion equations and the biological interpretation of that analysis. The reader who is primarily interested in biological applications may wish to skip the remainder of this section and go directly to the analysis of equilibria.

To understand why the dynamics of a single reaction-diffusion equation are determined by its equilibria, it is useful to take the viewpoint of dynamical systems theory. Reaction-diffusion equations with Lipschitz nonlinearities can be interpreted as semi-dynamical systems, typically on function spaces such as $W^{1,p}(\Omega), C(\overline{\Omega})$, or $C^1(\overline{\Omega})$, where Ω is the underlying spatial domain; see Henry (1981) and Smith (1995), or the mathematical preliminaries in Chapter 1. (The systems are "semi-dynamical" instead of being true dynamical systems, because they typically are not continuous backward in time.) The semi-dynamical system generated by a reaction-diffusion equation will be infinite dimensional, because its state space will be some space of functions describing the spatial distribution of the population. The fact that the state space is infinite dimensional

Spatial Ecology via Reaction-Diffusion Equations R.S. Cantrell and C. Cosner
 ISBN: 0-471-49301-5

might suggest that the dynamics could be complicated, since chaos can already occur in some three dimensional systems, but some special features of single reaction-diffusion equations insure that is not the case. It will turn out that generally the dynamics can be characterized completely or almost completely in terms of the equilibria and their stability properties.

Many of the models we shall consider will have the form

$$\begin{aligned} \frac{\partial u}{\partial t} &= \nabla \cdot d(x)\nabla u + \vec{b}(x) \cdot \nabla u + f(x, u) \text{ in } && \Omega \times (0, \infty) \\ d(x)\frac{\partial u}{\partial \vec{n}} &+ \beta(x)u = 0 && \text{on } \quad \partial\Omega \times (0, \infty). \end{aligned} \tag{3.1}$$

(We discuss cases where the diffusion rate d can depend on the density u or where the coefficients are time periodic in later sections.) Typically, in (3.1) we assume that the coefficients and the nonlinearity are such that (3.1) generates a semiflow on the fractional power spaces X^γ as in Theorem 1.11 of Section 1.6. That will be the case if $d(x) \in C^{1+\alpha}(\overline{\Omega})$, $\vec{b}(x) \in C^\alpha(\overline{\Omega})$, $f(x, u)$ is Lipschitz in u and is a measurable function in x which is bounded if u is restricted to a bounded set, Ω is bounded, $\partial\Omega$ is of class $C^{2+\alpha}$, $\beta(x) \in C^{1+\alpha}(\partial\Omega)$ and $d(x) \geq d_0 > 0$, $\beta(x) \geq 0$. For some results we require more smoothness on the coefficients. Generally, we will be interested only in positive solutions, so we will assume that $f(x, 0) = 0$ so that $u \equiv 0$ is a subsolution and hence the subset of the state space with $u \geq 0$ is positively invariant. A common choice for $f(x, u)$ is the logistic form $f(x, u) = a(x)u - c(x)u^2$, but we shall consider some other forms as well. In general, solutions to equations of the form (3.1) need not be bounded nor exist for all time; however, we generally assume that there is a positive number K analogous to the carrying capacity for the logistic equation such that $f(x, u) < 0$ for $u > K$. In that case all positive solutions of (3.1) will be bounded as $t \to \infty$, because any constant larger than K will be a supersolution. We shall discuss the construction and implications of sub- and supersolutions in more detail in later sections. The key point in the present discussion is that in general the relevant solutions to models such as (3.1) are nonnegative and ultimately bounded, i.e. the set of nonnegative densities is positively invariant and the semiflow restricted to the set is dissipative. It follows from the regularity theory of parabolic partial differential equations that in the standard state spaces bounded orbits are precompact and hence the semiflow will have a compact attractor, and the ω-limit set of any initial state $u_0(x)$ will be a compact invariant set. (See Bilotti and LaSalle (1971) and Henry (1981), or the discussion in Chapter 1.)

Suppose that model (3.1) has the properties described above, and that $\vec{b} \equiv 0$. Let $F(x, u) = \int_0^u f(x, r)dr$, and consider the functional

$$V(u) = \int_\Omega \left[\frac{1}{2}d(x)|\nabla u|^2 - F(x, u)\right] dx + \int_{\partial\Omega} \frac{1}{2}\beta u^2 dS. \tag{3.2}$$

We can choose to work in a state space that embeds in $W^{1,2}(\Omega) \cap C(\overline{\Omega})$ (for example, $W^{1,p}(\Omega)$ for p large) so that V is bounded on bounded subsets of the state space. If $u(x, t)$ is an orbit of (3.1) starting at $u(x, 0) = u_0$,

$$\begin{aligned}\frac{dV}{dt} &= \int_{\Omega} [d(x)\nabla u \cdot \nabla u_t - f(x,u)u_t]dx + \int_{\partial\Omega} \beta(x)uu_t dS \\ &= \int_{\Omega} [-\nabla \cdot d(x)\nabla u - f(x,u)]u_t dx + \int_{\partial\Omega} [d(x)\frac{\partial u}{\partial \vec{n}} + \beta(x)u]u_t dS \\ &= -\int_{\Omega} u_t^2 dx \leq 0.\end{aligned} \tag{3.3}$$

Thus, V decreases monotonically along the orbit. The boundedness and precompactness of orbits imply that $V(u(x,t))$ is bounded below, so there exists $V_0 = \lim_{t\to 0} V(u(x,t))$. If w is in the ω-limit set of $u_0(x)$ then $V(w) = V_0$. Let $w^*(x,t)$ be the orbit starting at some point $w_0(x)$ in the ω-limit set of u_0. Since that set is invariant, the entire orbit $w^*(x,t)$ must belong to it, so $V(w^*(x,t)) = V_0$. In view of (3.3), that is possible only if $w_t^* \equiv 0$. Thus, $w_0(x)$ must be an equilibrium of (3.1). Since w_0 was an arbitrary element of the ω-limit set of an arbitrary point u_0, the ω-limit set for the semiflow generated by (3.1) must consist entirely of equilibria.

Remark: The same sort of analysis applies to the case of Dirichlet boundary conditions. The boundary integral in (3.2) no longer appears in that case, and the state space must be restricted to functions vanishing on $\partial\Omega$, but the rest of the analysis is the same. Under the assumption $\vec{b} \equiv 0$, the equation (3.1) can be interpreted as a gradient system. In the finite dimensional case, gradient systems have the form $d\vec{y}/dt = \nabla G(\vec{y})$. In the setting of (3.1) we would take $G = -V$ and replace the gradient with a Fréchet derivative. The important thing about gradient systems is that the analysis given above applies so that their ω-limit sets consist only of equilibria; see Hofbauer and Sigmund (1988). A class of such systems analogous to (3.1) but with discrete diffusion (where the states are finite dimensional vectors and diffusion is replaced by something like a second difference) is discussed in the context of population models by Glendinning (1994).

If we allow $\vec{b}(x) \neq 0$ in (3.1) or allow density dependence in d so that $d = d(x,u)$, then the analysis based on V fails because the computation in (3.3) cannot be carried out in the same way. However, the maximum principle implies that the semiflow generated by (3.1), is order preserving. In that case the set of elements in the state space whose ω-limit sets consist of equilibria is open and dense; see Hirsch (1988) or Smith (1995) for detailed and precise discussions of these points. What is important here is that even if we allow drift or density-dependent diffusion in (3.1), we can still understand the asymptotic behavior of almost all solutions by understanding the set of equilibria. Most of this chapter will be devoted to doing the latter.

In the case of models based on a single reaction-diffusion equation with time-periodic coefficients, it is no longer sufficient to consider only equilibria for the original model, but even in that case it is often possible to gain a good understanding of the model from an analysis of its periodic steady states. These states can be viewed as equilibria of the Poincaré map (i.e. the period map), and the Poincaré map is typically order preserving. (See Hess (1991).) For models with nonperiodic time dependent coefficients less can be said, although some results are available in the almost periodic case (Shen and Yi, 1998). In the case of general time dependent coefficients, it is sometimes possible to get information by making comparisons with simpler models, but usually the results are much less precise than in the autonomous or time-periodic cases.

Much of our analysis of models with coefficients which are constant in time can be carried out in terms of eigenvalues of related linear problems. In many cases the effects of patch size, boundary conditions, and spatial heterogeneity on the predictions of (3.1) can be understood by using the results in Chapter 2 which describe how those factors affect eigenvalues. We take that approach whenever possible.

3.2 Equilibria and Stability: Sub- and Supersolutions

3.2.1 Persistence and Extinction

Many population models of the form (3.1) share two key features of the logistic model: $f(x,0) = 0$ and $f(x,u) < 0$ for u sufficiently large. If $f(x,u)$ is continuous and is differentiable with respect to u at $u = 0$, these features imply that $f(x,u) = g(x,u)u$ for some function $g(x,u)$ which is bounded above for all u by a function of x, for example $\sup\{f(x,u)/u : u > 0\}$. We have the following criterion for extinction in such cases:

Proposition 3.1. Suppose that

$$f(x,u) \leq g_0(x)u \quad \text{for} \quad x \in \Omega, \tag{3.4}$$

where $g_0(x)$ is a bounded measurable function if $\vec{b}/d$ is a gradient and $g_0(x) \in C^{\alpha}(\overline{\Omega})$ if $\vec{b}/d$ is not a gradient. If the principal eigenvalue σ_1 of

$$\begin{aligned} \nabla \cdot d(x)\nabla\psi + \vec{b}(x)\cdot\nabla\psi + g_0(x)\psi &= \sigma\psi \text{ in } \quad \Omega \\ d(x)\frac{\partial\psi}{\partial\vec{n}} + \beta(x)\psi &= 0 \qquad\qquad \text{on} \quad \partial\Omega \end{aligned} \tag{3.5}$$

is negative, then (3.1) has no positive equilibria and all nonnegative solutions decay exponentially to zero as $t \to \infty$.

Remarks: The result remains true for Dirichlet boundary conditions. The different regularity requirements on g_0 in the cases where $\vec{b}/d$ is or is not a gradient arise from the conditions needed to obtain the existence of σ_1 from variational methods or methods based on positivity, respectively. The meaning of Proposition 3.1 is essentially that the *per capita* growth rate $f(x,u)/u$ must be large enough at *some* point x and *some* density u so that the corresponding spatially averaged growth rate given by the eigenvalue σ_1 for $g_0(x) = \sup\{f(x,u)/u : u > 0\}$ is positive.

Proof: Let $\overline{u} = ce^{\sigma_1 t}\psi_1$, where $\psi_1 > 0$ is the eigenfunction corresponding to σ_1 in (3.5) and $c > 0$ is a constant. We have

$$\begin{aligned} &\frac{\partial\overline{u}}{\partial t} - \nabla\cdot d(x)\nabla\overline{u} - \vec{b}\cdot\nabla\overline{u} - f(x,\overline{u}) \\ &= \sigma_1\overline{u} - [\nabla\cdot d(x)\nabla\overline{u} + \vec{b}(x)\cdot\nabla\overline{u} + g_0(x)\overline{u}] + g_0(x)\overline{u} - f(x,\overline{u}) \\ &= g_0(x)\overline{u} - f(x,\overline{u}) \geq 0, \end{aligned}$$

so that $\overline{u}$ is a supersolution to (3.1). If $u(x,t)$ is any nonnegative solution to (3.1) we may choose c so large that $\overline{u}(x,0) > u(x,0)$. Then $\overline{u}(x,t) > u(x,t)$ for all $t > 0$, and since $\sigma_1 < 0$ we have $\overline{u}(x,t) \to 0$ exponentially as $t \to \infty$, so $u(x,t)$ must decay toward zero exponentially as well. This rules out any positive equilibria for (3.1).

Proposition 3.1 gives a criterion for extinction in models such as (3.1). In the density-independent (i.e. linear) case, we have a prediction of extinction if $\sigma_1 < 0$ and growth at any density if $\sigma_1 > 0$. The density-dependent case is more complicated, because density-dependent models may predict extinction for some initial densities and persistence for others. Such behavior is already present in nonspatial models for populations with an Allee effect, e.g. $du/dt = f(u)$ where $f(0) = 0$, $f(u) < 0$ for $0 < u < u_0$, $f(u) > 0$ for $u_0 < u < K$, and $f(u) < 0$ for $u > K$. However, the sharp switch between predictions of extinction and predictions of persistence which occurs in linear models also occurs in models of logistic type, among others. The reason is that logistic models attain their highest population growth rates at low densities, so that if they allow population growth at any density at all then they predict invasibility at low densities. The next result shows that invasibility implies persistence in models such as (3.1).

Proposition 3.2. Suppose that $f(x,u) = g(x,u)u$ with $g(x,u)$ of class C^2 in u and C^α in x, and there exists a $K > 0$ such that $g(x,u) < 0$ for $u > K$. If the principal eigenvalue σ_1 is positive in the problem

$$\begin{aligned} &\nabla \cdot d(x)\nabla\psi + \vec{b}\cdot\nabla\psi + g(x,0)\psi = \sigma\psi \quad \text{in} \quad \Omega \\ &d(x)\frac{\partial\psi}{\partial\vec{n}} + \beta(x)\psi = 0 \qquad\qquad \text{on} \quad \partial\Omega \end{aligned} \tag{3.6}$$

then (3.1) has a minimal positive equilibrium u^*, and all solutions to (3.1) which are initially positive on an open subset of Ω are eventually bounded below by orbits which increase toward u^* as $t \to \infty$.

Remark: Proposition 3.2 still holds under Dirichlet boundary conditions.

Proof: The assumptions on $f(x,u)$ imply that we can write $f(x,u) = [g(x,0) + g_1(x,u)u]u$ where $g_1(x,u)$ is C^1 in u. Let ψ_1 be an eigenfunction for (3.6) with $\psi_1 > 0$ on Ω. For $\epsilon > 0$ sufficiently small,

$$\begin{aligned} &\nabla \cdot d(x)\nabla(\epsilon\psi_1) + \vec{b}(x)\cdot\nabla(\epsilon\psi_1) + f(x,\epsilon\psi_1) \\ &= \epsilon[\nabla\cdot d(x)\nabla\psi_1 + \vec{b}(x)\nabla\psi_1 + g(x,0)\psi_1] + g_1(x,\epsilon\psi_1)\epsilon^2\psi_1^2 \\ &= \sigma_1\epsilon\psi_1 + g_1(x,\epsilon\psi_1)\epsilon^2\psi_1^2 \\ &= \epsilon\psi_1[\sigma_1 + \epsilon g_1(x,\epsilon\psi_1)\psi_1] > 0. \end{aligned}$$

It follows that for $\epsilon > 0$ small, $\epsilon\psi_1$ is a subsolution for the elliptic problem

$$\begin{aligned} &\nabla \cdot d(x)\nabla u + \vec{b}(x)\cdot\nabla u + f(x,u) = 0 \text{ in } \quad \Omega \\ &d(x)\frac{\partial u}{\partial\vec{n}} + \beta(x)u = 0 \qquad\qquad \text{on} \quad \partial\Omega \end{aligned}$$

corresponding to (3.1). If $\underline{u}(x,t)$ is a solution to (3.1) with $\underline{u}(x,0) = \epsilon\psi_1$, then $\partial\underline{u}/\partial t|_{t=0} > 0$ on Ω and general properties of sub- and supersolutions imply that $\underline{u}(x,t)$ is increasing in t. Since $K > \underline{u}$ is a supersolution to (3.1) we must have $\underline{u}(x,t) \uparrow u^*(x)$ as $t \to \infty$, where u^* is the minimal positive solution of (3.1). (We can be sure that $u^*(x)$ is minimal because $\epsilon\psi_1$ will be a strict subsolution for *all* $\epsilon > 0$ sufficiently small.) If $u(x,t)$ is a solution to (3.1) which is initially nonnegative and is positive on an open subset of Ω, then the strong maximum principle implies $u(x,t) > 0$ on $\overline{\Omega}$ for $t > 0$. Choosing any $t_0 > 0$ we can take $\epsilon > 0$ so small that $\epsilon\psi_1 < u(x,t_0)$ on $\overline{\Omega}$; then $\underline{u}(x,t-t_0) < u(x,t)$ for $t = t_0$ and thus by the maximum principle for $t > t_0$. Hence, $u(x,t)$ is bounded below by $\underline{u}(x,t-t_0)$ and $\underline{u}(x,t-t_0) \uparrow u^*$ as $t \to \infty$, as desired.

Remarks: The regularity hypotheses with respect to x can be weakened if we work in Sobolev spaces such as $W^{1,p}(\Omega)$ instead of seeking classical solutions. If our model has Dirichlet boundary conditions on part or all of $\partial\Omega$ then nonnegative nonzero solutions to (3.1) cannot satisfy $u(x,t) > 0$ on $\overline{\Omega}$ for $t > 0$; but in that case we have $u(x,t) > 0$ on Ω and those parts of $\partial\Omega$ where the boundary conditions are not Dirichlet. On the parts of the boundary where $u = 0$, the strong maximum principle implies $\partial u/\partial\vec{n} < 0$, so for $\epsilon > 0$ sufficiently small we have $\partial u/\partial\vec{n}|_{t=t_o} < \epsilon\partial\psi_1/\partial\vec{n} = \partial\underline{u}/\partial\vec{n}|_{t=0}$. Hence, we may still conclude that if ϵ is sufficiently small $u(x,t_0) > \underline{u}(x,0)$ so that $u(x,t) > \underline{u}(x,t-t_0)$ as before.

Suppose that $f(x,u) = m(x)u + h(x,u)u^2$ where $m(x)$ is a bounded measurable function, $h(x,u)$ is bounded and measurable on bounded subsets of $\Omega \times I\!R$, $h(x,u)$ is continuous in u for almost all x, and that $m(x) + h(x,u)u < 0$ for $u > K_0$. In that case the existence of a positive equilibrium when $\sigma_1 > 0$ in (3.6) follows from results such as Theorem 1.24. (The sub- and supersolutions would be the same as in the proof of Proposition 3.2.) Models where $f(x,u)$ is piecewise C^α in x can be treated by dividing Ω into subsets on which $f(x,u)$ is C^α in x and then imposing matching conditions on the density and flux across the interfaces between subsets. This approach is somewhat analogous to the methods used in Section 2.4 to study eigenvalues. Treatments of piecewise smooth models are given by Freedman et al. (1989), Freedman and Kriztin (1992) and Cui and Freedman (1999). Some related results are derived by Kriztin and Freedman (1993).

3.2.2 Minimal Patch Sizes

Propositions 3.1 and 3.2 show that the effects of spatial heterogeneity, drift, patch size, and boundary conditions on the predictions of models of the form (3.1) can be understood in terms of their effects on the principal eigenvalues in (3.5) and (3.6). The ways that spatial effects influence eigenvalues are discussed at length in Chapter 2, but in the context of (3.1) a new issue arises: the eigenvalue problem (3.5) used to predict extinction may be different from the problem (3.6) used to determine persistence. The two eigenvalue problems will coincide when

$$g_0(x) = \max_{u\geq 0}[f(x,u)/u] = \lim_{u\to 0+}[f(x,u)/u] = \frac{\partial f}{\partial u}|_{u=0} = g(x,0) \tag{3.7}$$

where g_0 and g are as in (3.5) and Proposition 3.2, respectively. The relation (3.7) will hold in the logistic case ($f(x,u)/u) = a(x) - c(x)u$ with $c > 0$ but may fail if $(f(x,u)/u)$ is increasing in u for some values of u. (The property of having $f(x,u)/u$ increasing at some densities is sometimes called depensation.) If the coefficients in (3.1) are constant in x, then as in Chapter 2 we can account for the size of Ω by writing $\Omega = \ell\Omega_0$ where $|\Omega_0| = 1$ and then rescaling the model on Ω back to Ω_0, as in Section 2.1. Suppose for example that our model is

$$\begin{aligned} u_t &= d\Delta u + f(u) \text{ in } && \Omega\times(0,\infty) = \ell\Omega_0\times(0,\infty) \\ u &= 0 && \text{on } \partial\Omega\times(0,\infty), \end{aligned}$$

and let λ_0 be the principal eigenvalue for $\Delta\phi + \lambda\phi = 0$ in Ω_0, $\phi = 0$ on $\partial\Omega_0$. Let $g_0 = \max_{u \geq 0}(f(u)/u)$ and $g = \lim_{u \to 0} f(u)/u = f'(0)$. The principal eigenvalue in (3.5) is then given by $g_0 - d\lambda_0/\ell^2$ while the principal eigenvalue in (3.6) is $g - d\lambda_0/\ell^2$. If we take $f(u) = u(1-u)$ then these coincide, with $g_0 = g = 1$. We have extinction in (3.1) from any initial density if $\ell < \sqrt{d\lambda_0}$ and persistence starting at any initial density that is positive somewhere if $\ell > \sqrt{d\lambda_0}$. However, if we take $f(u) = u(1 + u - u^2)$ then $g_0 = \max_{u \geq 0}(1 + u - u^2) = 5/4$ while $g = 1$, so we get a prediction of extinction from any initial density for $\ell < \dfrac{2\sqrt{5}}{5}\sqrt{d\lambda_0}$, but a prediction of persistence starting at any positive density only if $\ell > \sqrt{d\lambda_0}$. For $(2\sqrt{5}/5)\sqrt{d\lambda_0} < \ell < \sqrt{d\lambda_0}$ our analysis so far does not give any information. We shall see that $\sigma_1 < 0$ in (3.6) means that $u \equiv 0$ is locally stable, so that populations whose initial densities are too small will go extinct, but that does not rule out the possibility that larger populations may persist. To summarize, in the density dependent case we have two notions of minimal patch size: the size below which no population can survive, and the size above which the patch can be invaded by a population introduced at low density. These notions coincide in the case of logistic models with passive diffusion, or more generally in the case of models with $f(x, u)/u$ decreasing in u, but they may not in cases where $f(x, u)/u$ is increasing for some values of u_0. Some biological mechanisms that could cause $f(x, u)/u$ to increase with u are group defense or cooperative hunting, among others. We shall see that even for logistic equations, the two notions of minimal patch size need not coincide if the diffusion rate d is density dependent.

Proposition 3.2 gives a condition implying the existence of a minimal positive equilibrium for (3.1) in terms of the instability of the equilibrium $u \equiv 0$. The same sort of stability analysis can be applied to any equilibrium. Suppose that the hypotheses of Proposition 3.2 are satisfied and that $\tilde{u}$ is an equilibrium of (3.1) with $\tilde{u} > 0$ on $\bar{\Omega}$, which is isolated in the sense that for δ small enough there is no other equilibrium u^* with $\tilde{u} - \delta < u^* < \tilde{u} + \delta$. If $v(x)$ is small we may write

$$f(x, \tilde{u} + v) = f(x, \tilde{u}) + \frac{\partial f}{\partial u}(x, \tilde{u})v + f_1(x, \tilde{u}, v)v^2 \tag{3.8}$$

for some continuous function f_1. Let σ_1 and $\psi_1 > 0$ be the principal eigenvalue and eigenfunction for the linearized problem

$$\begin{aligned} \nabla \cdot d(x)\nabla\psi + \vec{b} \cdot \nabla\psi + \frac{\partial f}{\partial u}(x, \tilde{u})\psi &= \sigma\psi \quad \text{in} \quad \Omega \\ d(x)\frac{\partial\psi}{\partial\vec{n}} + \beta(x)\psi &= 0 \quad \text{on} \quad \partial\Omega. \end{aligned} \tag{3.9}$$

If we let $u = \tilde{u} + \epsilon\psi_1$ and calculate using (3.8), (3.9), and the fact that $\tilde{u}$ is an equilibrium, we obtain

$$\begin{aligned} &\nabla \cdot d(x)\nabla u + \vec{b} \cdot \nabla u + f(x, u) = \\ &\nabla \cdot d(x)\nabla\tilde{u} + \vec{b} \cdot \nabla\tilde{u} + f(x, \tilde{u}) + \epsilon(\nabla \cdot d(x)\nabla\psi_1 + \vec{b} \cdot \nabla\psi_1 + \frac{\partial f}{\partial u}(x, \tilde{u})\psi_1) \\ &+ f(x, \tilde{u} + \epsilon\psi_1) - f(x, \tilde{u}) - \frac{\partial f}{\partial u}(x, \tilde{u})\epsilon\psi_1 \\ &= \epsilon\sigma_1\psi_1 + f_1(x, u, \epsilon\psi_1)\epsilon^2\psi_1^2 \\ &= \epsilon\psi_1(\sigma_1 + \epsilon f_1(x, \tilde{u}, \epsilon\psi_1)\psi_1). \end{aligned} \tag{3.10}$$

The last expression in (3.10) will have the same sign as σ_1 for $\epsilon > 0$ small. Thus, if $\sigma_1 < 0$, $u = \tilde{u} + \epsilon\psi_1$ will be a supersolution for the equilibrium problem corresponding to (3.1), and $\tilde{u} - \epsilon\psi_1$ will be a subsolution. If $\sigma_1 > 0$ then $\tilde{u} + \epsilon\psi_1$ is a subsolution and $\tilde{u} - \epsilon\psi_1$ is a supersolution. In the case $\sigma_1 < 0$, let $\underline{u}$ and $\overline{u}$ be solutions of (3.1) which start out at time $t = t_0$ with $\underline{u} = \tilde{u} - \epsilon\psi_1$ and $\overline{u} = \tilde{u} + \epsilon\psi_1$. Since $\tilde{u} - \epsilon\psi_1$ and $\tilde{u} + \epsilon\psi_1$ are sub and supersolutions, respectively, for the equilibrium problem for (3.1), we must have $\underline{u} \uparrow$ and $\overline{u} \downarrow$ as t increases. Since $\tilde{u}$ is isolated, if $\epsilon > 0$ is small then $\underline{u} \uparrow \tilde{u}$ and $\overline{u} \downarrow \tilde{u}$, so that if $\tilde{u} - \epsilon\psi_1 \leq w \leq \tilde{u} + \epsilon\psi_1$ at $t = t_0$ then $w \to \tilde{u}$ as $t \to \infty$. Hence $\tilde{u}$ is stable if $\sigma_1 < 0$ in (3.9). If $\sigma_1 > 0$, then for $\epsilon > 0$ small enough the solution $v(x,t)$ to (3.1) with $v(x,0) = \tilde{u} + \epsilon\psi_1$ is increasing in t and starts out larger than $\tilde{u}$, so $v(x,t)$ moves away from $\tilde{u}$ and cannot return. By making ϵ small we can have $v(x,0)$ as close to $\tilde{u}$ as we want. Hence, for $\sigma_1 > 0$, $\tilde{u}$ is unstable. This sort of connection between the stability of equilibria for a nonlinear problem such as (3.1) and the stability of the zero solutions (as determined by the sign of σ_1) of corresponding linearized problems is sometimes called the *principle of linearized stability*.

In the analysis leading to Proposition 3.2 we used the fact that solutions to (3.1) which start at densities that are subsolutions to the equilibrium problem must increase, and if bounded must approach another equilibrium. We also used the instability of the equilibrium $u \equiv 0$, as expressed by the condition $\sigma_1 > 0$ in (3.6), to construct subsolutions to the equilibrium problem lying arbitrarily close to $u \equiv 0$. This construction can be interpreted as a special case of a general property of order preserving dynamical systems with bounded orbits, namely that unstable equilibria will have orbits which connect from them to other equilibria; see Hess (1991) or Smith (1995).

3.2.3 Uniqueness of Equilibria

If the hypotheses of Proposition 3.2 are satisfied then there will be orbits of (3.1) starting arbitrarily close to zero (namely at $\epsilon\psi_1$, where ψ_1 is determined by (3.6)) which increase toward a minimal equilibrium u^* of (3.1). If $f(x,u) < 0$ for $u > K$ then any constant $C > K$ will be a supersolution to the equilibrium problem for (3.1). Solutions to (3.1) with initial data $u(x,0) = C$ for some $C > K$ will decrease toward a maximal equilibrium $u^{**} \geq u^*$ for (3.1). If (3.1) has a unique equilibrium, so that $u^{**} = u^*$, then all orbits of (3.1) will be trapped and "squeezed" between orbits increasing toward u^* from below and decreasing toward u^* from above, so that u^* will be globally asymptotically stable among positive solutions to (3.1). In general, models such as (3.1) may have multiple equilibria, and we shall discuss some models where that is the case. However, we first give a condition under which (3.1) can have at most one positive equilibrium, so that if the equilibrium exists it must be unique. The key idea used in the next result was introduced by Hess (1977) in a slightly different context.

Proposition 3.3. Suppose that the hypotheses of Proposition 3.2 are satisfied and that $f(x,u) = g(x,u)u$ with $g(x,u)$ strictly decreasing in u for $u \geq 0$. Then the minimal positive equilibrium u^* is the only positive equilibrium for (3.1). If $g_1(x,u)$ and $g_2(x,u)$ are both strictly decreasing in u and if u_1^* and u_2^* are the unique positive equilibria corresponding to g_1 and g_2 respectively then $g_1(x,u) \leq g_2(x,u)$ for $u > 0$ implies that $u_1^* \leq u_2^*$. If $g_1(x,u) < g_2(x,u)$ for $u > 0$ then $u_1^* < u_2^*$.

Proof: If u^{**} is a positive equilibrium of (3.1) with $u^{**} \neq u^*$ then since u^* is minimal we must have $u^{**} > u^*$ somewhere on Ω. Since $u^* > 0$ is an equilibrium of (3.1) it is a positive solution to

$$\nabla \cdot d(x)\nabla\psi + \vec{b}(x)\cdot\nabla\psi + g(x,u^*)\psi = \sigma\psi \text{ in } \Omega$$
$$d(x)\frac{\partial\psi}{\partial\vec{n}} + \beta(x)\psi = 0 \quad \text{on } \partial\Omega \tag{3.11}$$

with $\sigma = 0$, so that $\sigma_1 = 0$ must be the principal eigenvalue for (3.11). Similarly, $u^{**} > 0$ satisfies

$$\nabla \cdot d(x)\nabla\psi + \vec{b}(x)\cdot\nabla\psi + g(x,u^{**})\psi = \sigma\psi \text{ in } \Omega$$
$$d(x)\frac{\partial\psi}{\partial\vec{n}} + \beta(x)\psi = 0 \quad \text{on } \partial\Omega, \tag{3.12}$$

with $\sigma = 0$, so $\sigma_1 = 0$ in (3.12) also. However, since $g(x,u)$ is strictly decreasing in u and $u^{**} > u^*$ on at least part of Ω, the principal eigenvalue in (3.12) must be less than the principal eigenvalue in (3.11). (This follows from Corollary 2.2 if $\vec{b} = 0$ or from Corollary 2.19 in the general case.) Thus, we cannot have $\sigma_1 = 0$ in both (3.11) and (3.12), so (3.1) cannot have any equilibrium other than the minimal equilibrium u^*. If $g_1(x,u) \le g_2(x,u)$ then u_1^* is a subsolution of the equilibrium problem for (3.1) with $f = g_2(x,u)u$, while any constant larger than the constant K of Proposition 3.2 is a supersolution. Thus, the model (3.1) with $f = g_2(x,u)u$ has an equilibrium u^{**} with $u^{**} \ge u_1^*$. Since u_2^* is the unique positive equilibrium for (3.1) with $f = g_2(x,u)u$, we must have $u_2^* = u^{**} \ge u_1^*$. If $g_1 < g_2$ then u_1^* is a strict subsolution so $u_2^* = u^{**} > u_1^*$.

Remarks: As usual, the same argument applies in the case of Dirichlet boundary conditions. The result also holds if we have $g(x,u)$ strictly increasing in u; that was the situation treated by Hess (1977). That situation does not usually occur in ecological models but may arise in other contexts. If $g(x,u)$ is increasing in u for some positive densities and decreasing for others then (3.1) may have multiple equilibria. A general discussion of the existence and multiplicity of equilibria for models similar to (3.1) is given in Lions (1982). Other sorts of conditions implying uniqueness can be given; we shall discuss some of them in a later section.

If, for example, we consider the logistic equation

$$\frac{\partial u}{\partial t} = \nabla \cdot d(x)\nabla u + \vec{b}(x)\cdot\nabla u + a(x)u - c(x)u^2 \text{ in } \Omega\times(0,\infty)$$
$$d(x)\frac{\partial u}{\partial\vec{n}} + \beta(x)u = 0 \quad \text{on } \partial\Omega\times(0,\infty) \tag{3.13}$$

we have $g(x,u) = a(x) - c(x)u$, which is strictly decreasing in u if $c(x) > 0$. Let σ_1 be the principal eigenvalue for

$$\nabla \cdot d(x)\nabla\psi + \vec{b}(x)\cdot\nabla\psi + a(x)\psi = \sigma\psi \text{ in } \Omega,$$
$$d(x)\frac{\partial\psi}{\partial\vec{n}} + \beta(x)\psi = 0 \quad \text{on } \partial\Omega. \tag{3.14}$$

If $\sigma_1 < 0$ then by Proposition 3.1 all positive solutions of (3.13) decline toward zero as $t \to \infty$. If $\sigma_1 > 0$ then (3.13) has a unique positive equilibrium u^*, and all positive solutions approach u^* as $t \to \infty$. Suppose $\sigma_1 = 0$. Then let $\psi^* > 0$ be the eigenfunction corresponding to the principal eigenvalue for the adjoint problem for (3.14). By the arguments used to prove Corollary 2.13 that eigenvalue will also be zero if $\sigma_1 = 0$. Multiplying (3.13) by ψ_1^*, integrating by parts using the divergence theorem, and using $\sigma_1 = 0$ gives

$$\frac{d}{dt}\int_\Omega \psi_1^* u dx = -\int_\Omega c(x)\psi^* u^2 dx. \tag{3.15}$$

It follows from (3.15) that if $c(x) > 0$ then $\int_\Omega \psi_1^* u dx$ must be strictly decreasing if $u > 0$, so all positive solutions to (3.13) must decline toward zero in that case. In summary, if $\sigma_1 \leq 0$ in (3.14) then all positive solutions of (3.13) decline toward zero as $t \to \infty$, while if $\sigma_1 > 0$ then (3.13) has a unique positive equilibrium which is globally attracting among positive solutions. If the problem

$$\begin{aligned} \nabla \cdot d(x)\nabla\phi + \vec{b}(x)\cdot\nabla\phi + \lambda a(x)\phi &= 0 \text{ in } \quad \Omega \\ d(x)\frac{\partial \phi}{\partial \vec{n}} + \beta(x)\phi &= 0 \qquad \text{on} \quad \partial\Omega \end{aligned} \tag{3.16}$$

admits a positive principal eigenvalue $\lambda_1^+(a(x))$ (which it will except in the case of no-flux boundary conditions with $\int_\Omega a(x)dx \geq 0$), then by Theorem 2.6 or Corollary 2.18 (depending on the absence or presence of the advection term $\vec{b}(x)$), we have $\sigma_1 > 0$ if and only if $\lambda_1^+(a(x)) < 1$.

In the case of spatially constant coefficients we can use rescalings of the sort discussed in detail in Section 2.1 to observe that for a patch of fixed shape but variable size (i.e. $\Omega = \ell\Omega_0$ for some fixed domain Ω_0) there will be a critical size ℓ^* such that for $\ell \leq \ell^*$ all positive solutions to (3.13) decline toward zero, but for $\ell > \ell^*$ all positive solutions approach a unique positive equilibrium. This sort of analysis would apply to other models where $f(x,u) = ug(x,u)$ where $g(x,u)$ is decreasing in u, for example $f(x,u) = \dfrac{a(x)u}{1+c(x)u} - e(x)u$. We shall see in the next sections that if $g(x,u)$ is not decreasing in u then other things can happen. Assembling the scaling arguments from Section 2.1 with the preceding analysis, we have the following:

Corollary 3.4. If the underlying spatial domain for the reaction-diffusion model

$$\begin{aligned} u_t &= d\Delta u + g(u)u \text{ on} \quad \Omega \times (0,\infty) \\ u &= 0 \qquad\qquad\quad \text{on} \quad \partial\Omega \times (0,\infty) \end{aligned} \tag{3.17}$$

is $\Omega = \ell\Omega_0 = \{\ell x : x \in \Omega_0\}$, and if $g(u)$ is Lipschitz and decreasing in u, with $g(0) > 0$ and $g(u) < 0$ for $u > K > 0$, then there is a number $\ell^* > 0$ such that for $\ell \leq \ell^*$ all positive solutions of (3.17) approach 0 as $t \to \infty$, while if $\ell > \ell^*$ there is a unique positive equilibrium $u^*(x)$ (depending on ℓ) such that all positive solutions approach u^* as $t \to \infty$.

If we linearize (3.17) about $u = 0$ we obtain the eigenvalue problem

$$\begin{aligned} d\Delta\psi + g(0)\psi &= \sigma\psi \text{ on} \quad \Omega \\ \psi &= 0 \qquad\quad \text{on} \quad \partial\Omega. \end{aligned} \tag{3.18}$$

If λ_0 is the principal eigenvalue for the problem

$$\begin{aligned} \Delta\phi + \lambda\phi &= 0 \text{ on } \quad \Omega_0 \\ \phi &= 0 \qquad \text{on} \quad \partial\Omega_0, \end{aligned}$$

then the scaling process of Section 2.1 implies $\sigma_1 = g(0) - d\lambda_0/\ell^2$ is the principal eigenvalue for (3.18). We have $\sigma_1 \leq 0$ if $\ell \leq \sqrt{d\lambda_0/g(0)}$ and $\sigma_1 > 0$ if $\ell > \sqrt{d\lambda_0/g(0)}$ so the result follows by Lemmas 3.1–3.3 and the associated discussion, with $\ell^* = \sqrt{d\lambda_0/g(0)}$.

3.3 Equilibria and Scaling: One Space Dimension

3.3.1 Minimum Patch Size Revisited

For simple reaction-diffusion models in one space dimension it is possible to give a somewhat more explicit analysis of how the equilibria behave in terms of the size of the underlying spatial interval. The discussion in this section will follow the approach of Aronson and Weinberger (1975, 1978) and Ludwig et al. (1979). Related ideas are discussed in Fife (1979). Consider the model

$$\begin{aligned} \frac{\partial u}{\partial t} &= \frac{\partial^2 u}{\partial x^2} + f(u) \text{ on } \quad (-\ell/2, \ell/2) \times (0, \infty) \\ u(x,t) &= 0 \qquad \text{for} \quad x = \pm\ell/2. \end{aligned} \tag{3.19}$$

(If we start with a diffusion coefficient $d \neq 1$ we can rescale x to in effect absorb d into ℓ.) Taking the equilibrium equation for (3.19) and multiplying by du/dx leads to

$$\begin{aligned} 0 &= \frac{d^2u}{dx^2}\frac{du}{dx} + f(u)\frac{du}{dx} \\ &= \frac{d}{dx}\left[\frac{1}{2}\left(\frac{du}{dx}\right)^2 + F(u)\right], \end{aligned}$$

where

$$F(u) = \int_0^u f(s)ds, \tag{3.20}$$

so that

$$\frac{1}{2}\left(\frac{du}{dx}\right)^2 + F(u) = \text{constant}. \tag{3.21}$$

If u is a positive equilibrium of (3.19) then u must attain a maximum on $(-\ell/2, \ell/2)$, which we denote as $||u||$. (Since $u > 0$, the maximum of u is also the norm of u in $C(\overline{\Omega})$.) Since $du/dx = 0$ at the maximum of u, we must have

$$\frac{1}{2}\left(\frac{du}{dx}\right)^2 + F(u) = F(||u||). \tag{3.22}$$

Since $u(x)$ takes on all values between 0 and $||u||$, (3.22) implies that we must have $F(||u||) \geq F(u)$ for $0 \leq u \leq ||u||$. In many cases of interest, for example if $f(u) > 0$ for

$0 < u < K$ but $f(u) < 0$ for $u > K$, we have $F(u) < F(||u||)$, except when $u = ||u||$. In such cases, if we have $u(x_0) = ||u||$, we must have

$$\frac{du}{dx} = \begin{cases} \sqrt{2}(\sqrt{F(||u||) - F(u)} & \text{for} \quad -\ell/2 < x < x_0 \\ -\sqrt{2}(\sqrt{F(||u||) - F(u)} & \text{for} \quad x_0 < x < \ell/2. \end{cases} \tag{3.23}$$

We may rewrite (3.23) and integrate to obtain for $x < x_0$

$$\int_x^{x_0} \left[\frac{du/dx}{\sqrt{F(||u||) - F(u)}} \right] dx = \sqrt{2} \int_x^{x_0} dx;$$

by using the substitution rule with $v = u(x)$ and computing the integrals we obtain

$$\int_{u(x)}^{||u||} \frac{dv}{\sqrt{F(||u||) - F(v)}} = \sqrt{2}(x_0 - x). \tag{3.24}$$

Similarly, for $x > x_0$, we have

$$\int_{u(x)}^{||u||} \frac{dv}{\sqrt{F(||u||) - F(v)}} = \sqrt{2}(x - x_0). \tag{3.25}$$

Taking $x = -\ell/2$ in (3.24) and $x = \ell/2$ in (3.25) and using the boundary conditions of (3.19) we have

$$\int_0^{||u||} \frac{dv}{\sqrt{F(||u||_- F(v)}} = \sqrt{2}[x_0 + (\ell/2)] = \sqrt{2}[(\ell/2) - x_0]$$

so that we must have $x_0 = 0$, and hence

$$\sqrt{2} \int_0^{||u||} \frac{dv}{\sqrt{F(||u||) - F(v)}} = \ell. \tag{3.26}$$

Taking $x_0 = 0$ in (3.24) and (3.25) gives an implicit formula for $u(x)$ in terms of $||u||$. For specific choices of $f(u)$ we can use (3.26) to see how equilibria of (3.19), as parameterized by $||u||$, depend on the length of ℓ of the underlying spatial domain, i.e. on patch size.

Example 3.1. (Logistic growth). (Ludwig et al., 1979) In this case $f(u) = r(1-(u/K))u = ru - (r/K)u^2$; so $F(u) = (ru^2/2) - (r/3K)u^3$. The condition $F(u) \le F(||u||)$ restricts the range of values $||u||$ may take to the interval $[0, K]$. Note that for $u = K$, $F(K) - F(u) = 0$, $(F(K) - F(u))' = -f(u) = 0$, so that we must have $F(K) - F(u) = (K - u)^2 h(u)$ for some nonnegative function $h(u)$. It follows that $1/\sqrt{F(K) - F(u)} = (K - u)\sqrt{h(u)}$, so that for $||u|| = K$ the integral in (3.26) diverges to infinity. (The fact that $F(K) - F(u)$ is of order $(u - K)^2$ can be seen from Figure 3.1 by observing that the horizontal line $F = F(K)$ is tangent to the graph of $F = F(u)$ where they intersect.) On the other hand, we may substitute $w = v/||u||$ in (3.26) to obtain

$$\begin{aligned} \ell &= \sqrt{2} \int_0^1 \frac{||u|| dw}{\sqrt{F(||u||) - F(||u||w)}} \\ &= \sqrt{2} \int_0^1 \frac{dw}{\sqrt{[F(||u||) - F(||u||w)]/||u||^2}}. \end{aligned} \tag{3.27}$$

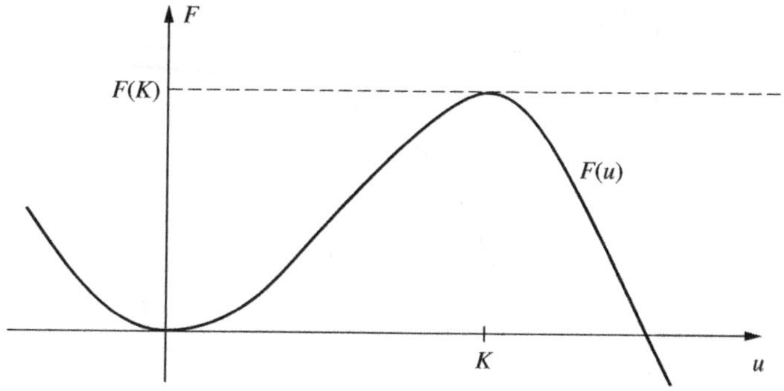

Figure 3.1 Sketch of the graph of $F(u)$ as in (3.20)–(3.26) for $f(u) = ru(1 - [u/K])$.

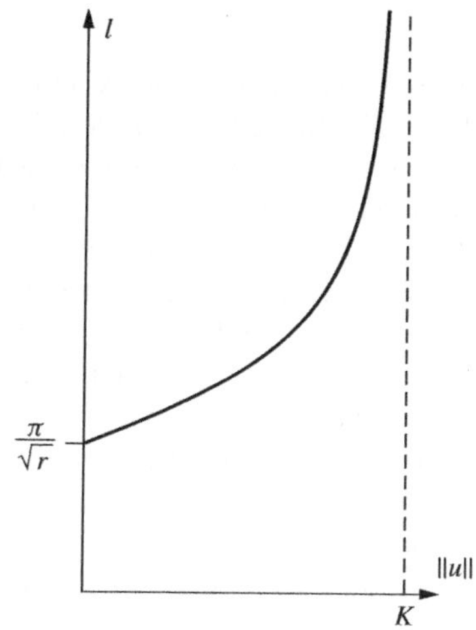

Figure 3.2 Graph of 1 (habitat length) versus $||u||$ (maximum density) obtained from (3.27) for the logistic model with $f(u) = ru[1 - (u/K)]$. In this case $F(u)$ has the form shown in Figure 3.1.

We have $[F(||u||) - F(||u||w)]/||u||^2 = (r/2)(1 - w^2) - (r||u||/3K)(1 - w^3)$. Thus, the integral in (3.27) is increasing with $||u||$, and as $||u|| \to 0$ the integrand approaches $\sqrt{2/r}/\sqrt{1 - w^2}$, so $\ell \to \pi/\sqrt{r}$. (Recall that the principal eigenvalue σ_1 for the linearized problem $\psi'' + r\psi = \sigma\psi$ on $(-\ell/2, \ell/2)$, $\psi(-\ell/2) = \psi(\ell/2) = 0$, is given by $r - \pi^2/\ell^2$ so $\sigma_1 = 0$ when $\ell = \pi/\sqrt{r}$.) Hence, as $||u||$ goes from 0 to K, ℓ goes from $\pi/\sqrt{r}$ to infinity (see Figure 3.2). We can see how the equilibria of (3.19) depend on ℓ by interchanging the axes of ℓ and $||u||$ in Figure 3.2. This leads to Figure 3.3, which is a representation of equilibria of (3.19) in terms of ℓ. Figures such as 3.3 are a standard way of expressing how solutions to an equation depend on a parameter; they are known as bifurcation diagrams, and will be discussed further in the next section.

Example 3.2. (Allee effect). If organisms must maintain some minimal density to engage in social behavior such as group hunting or group defense, or to find mates, their population may be subject to decline at low densities but be able to grow at higher densities. This phenomenon is known as an Allee effect (Allee, 1931). In simple population models an Allee effect can be captured phenomenologically by using a population growth function

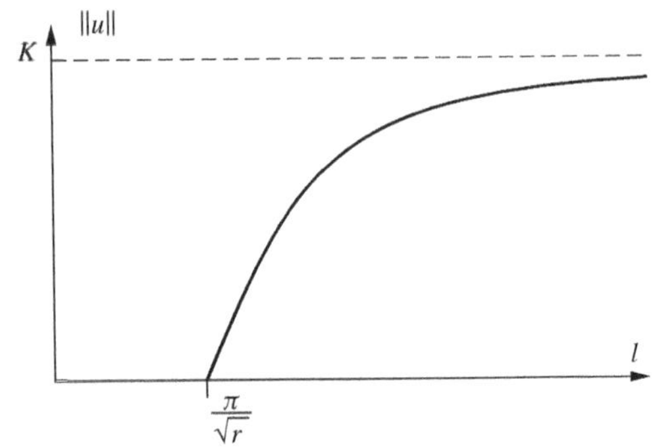

Figure 3.3 Bifurcation diagram for $||u||$ (maximum density) versus 1 for the logistic model with $f(u) = ru[1 - (u/K)]$. The plot is obtained by switching the axes in Figure 3.2.

$f(u)$ which is negative for small positive values of u. A simple example of such a choice of $f(u)$ is the cubic $f(u) = r(1-u/K)u(u-a) = -(r/K)u^3+r(1+a/K)u^2-rau$, where $a \in (0, K)$ is the threshold density needed for population growth. Since $F(0) = 0$ we must have $F(||u||) \geq 0$ for any attainable value of $||u||$. The maximum of $F(u)$ occurs at $u = K$; we have $F(u) = (-r/4K)u^4+[r(1+a/K)/3]u^3-(ra/2)u^2$ so $F(K) = rK^3/12-rK^2a/6$ and thus $F(K) > 0$ only if $a < K/2$. If $a < K/2$ then there is a number $u_0 > a$ such that $F(u) < 0$ for $0 < u < u_0$, $F(u_0) = 0$, and $F(u) > 0$ on some interval $u_0 < u < u_1$ with $u_1 > K$ (see Figure 3.4A). For values of $||u||$ with $u_0 < ||u|| < K$ the horizontal line $F = F(||u||)$ intersects $F(u)$ transversely, so that $F(||u||) - F(u) = (||u|| - u)h(u)$ with $h(u)$ positive and the integral in (3.26) converges to a finite value for ℓ. As $||u|| \uparrow K$, the line $F = F(||u||)$ moves upward to the line $F = F(K)$, which is tangent to $F(u)$ at $u = K$. Thus, for $||u|| = K$, we have $F(K) - F(u) = (u - K)^2h_1(u)$ for some $h_1(u)$ so that the integral in (3.26) diverges to infinity. Similarly, as $||u|| \downarrow u_0$ the line $F = F(||u||)$ moves downward toward $F = 0$, which intersects $F = F(u)$ tangentially at 0, so that $F(u_0) - F(u) = (u)^2h_2(u)$ for some $h_2(u)$ and again (3.26) diverges to infinity. Since ℓ depends continuously on $||u||$, there is a minimal value ℓ^* on the graph of ℓ as a function of $||u||$ which occurs at some value of $||u|| \in (u_0, K)$. See Figure 3.4B. Interchanging the ℓ and $||u||$ axes yields the bifurcation diagram 3.4C. The interpretation of those diagrams is that (3.19) has no positive equilibria for $\ell < \ell^*$ and two positive equilibria for $\ell > \ell^*$. In this case $f'(0) = -ra$ so that for the linearized problem $\psi'' + f'(0)\psi = \sigma\psi$ with $\psi = 0$ for $x = \pm\ell/2$ we have $\sigma_1 = -ra - \pi^2/\ell^2 < 0$ for all ℓ. Thus, the equilibrium $u = 0$ is always stable under small perturbations, so the patch is never invasible at low densities. An invading or colonizing population must achieve some threshold density to be successful.

Example 3.3. (The spruce budworm). (Ludwig et al., 1979) The paper by Ludwig et al. (1979) presents an analysis of the effects of patch size on the population of the spruce budworm. The spruce budworm may be present in forests at low densities in an endemic state, but may also have outbreaks in which the densities become much higher, possibly to the point of causing economically significant damage to the forest. The model used by Ludwig et al. (1979) is

$$\frac{\partial u}{\partial t} = d\frac{\partial^2 u}{\partial x^2} + r\left(1 - \left(\frac{u}{K'S}\right)\right)u - \frac{\beta u^2}{(\alpha' S)^2 + u^2}. \tag{3.28}$$

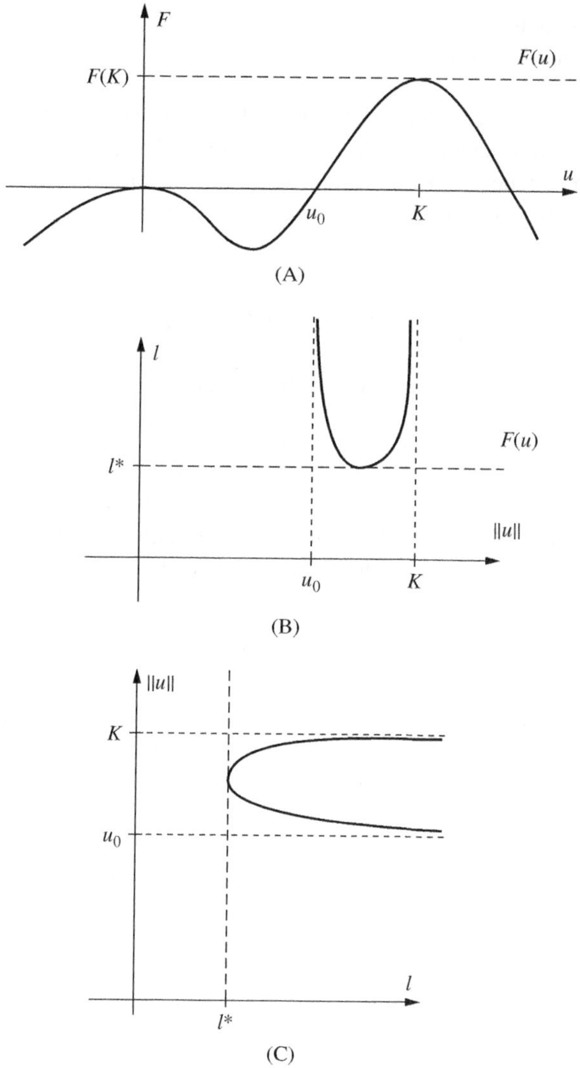

Figure 3.4 Figures 3.4A, 3.4B, and 3.4C correspond to Figures 3.1, 3.2, and 3.3, respectively, but for the case of a function $f(u) = ru[1 - (u/K)][u - a]$ which builds an Allee effect into the population dynamics; see the text.

This is a diffusive logistic equation modified by the addition of an extra loss term to account for the effects of predation by birds. The term assumes a constant predator density, and uses a Holling type 3 functional response to account for the predators switching to alternate prey if the budworm density is too low. (See Ludwig et al. (1978, 1979) for further discussion.) The parameters in the model (3.28) have units as follows:

$d \sim (\text{length})^2/\text{time}$
$r \sim 1/\text{time}$
$K' \sim \text{budworms/branch}$
$S \sim \text{branches/area} = \text{branches}/(\text{length})^2$

$\alpha' \sim$ budworms/branch
$\beta \sim$ budworms/(length)^2 time.

To understand the qualitative features of a model it is a standard (and wise) practice to reduce the number of parameters to as few as possible, and to put the remaining parameters into forms that do not depend on the system of units, i.e. are dimensionless. To that end, we introduce new independent variables $\tilde{x} = \sqrt{r/d}x$, $\tilde{t} = rt$, so that (3.28) is transformed into

$$\frac{\partial u}{\partial \tilde{t}} = \frac{\partial^2 u}{\partial \tilde{x}^2} + \left(1 - \frac{u}{K'S}\right)u + \frac{(\beta/r)u^2}{(\alpha' S)^2 + u^2}. \tag{3.29}$$

We then introduce the new independent variable $\tilde{u} = u/\alpha' S$ so that (3.29) becomes

$$\frac{\partial \tilde{u}}{\partial \tilde{t}} = \frac{\partial^2 \tilde{u}}{\partial \tilde{x}^2} + \left(1 - \frac{\tilde{u}}{Q}\right)\tilde{u} - \frac{1}{R}\frac{\tilde{u}^2}{1 + \tilde{u}^2} \tag{3.30}$$

with $Q = K'/\alpha'$ and $R = r\alpha' S/\beta$. Note that $\tilde{x}, \tilde{t}, \tilde{u}, Q$, and R are dimensionless. Dropping the tildes in (3.30) yields a model of the form (3.19) with

$$f(u) = u - \frac{u^2}{Q} - \frac{1}{R}\left(\frac{u^2}{1 + u^2}\right).$$

The behavior of $f(u)$ can be understood via graphical analysis (see Ludwig et al. 1979). It turns out that if $Q < 3\sqrt{3}$ then $f(u)$ has only one positive root, which in the nonspatial case would correspond to a small equilibrium density; see Figure 3.5A. If $Q > 3\sqrt{3}$ then the behavior of $f(u)$ depends on R. For R small, $f(u)$ still has only a single small positive root as in Figure 3.5A. For intermediate values of R, $f(u)$ has three positive roots; for large values of R, $f(u)$ has only a single large positive root. (See Figures 3.5B and 3.5C.) Recall that R is proportional to S, which measures the density of branches in the forest, so in the nonspatial case the interpretation of the dependence of $f(u)$ on R for $Q > 3\sqrt{3}$ is that for sparse forests the budworms can only persist at a low endemic equilibrium; for moderately dense forests there are multiple equilibria corresponding to both endemic and outbreak states, and for very dense forests only the outbreak state can occur. In the spatial model, the cases with Q small or with Q large and R either small or large yield bifurcation diagrams which are qualitatively similar to those for the logistic model shown in Figures 3.2 and 3.3. That is because in those cases the relevant qualitative features of $f(u)$ and hence $F(u)$ are similar to the logistic case. In the case $Q > 3\sqrt{3}$ with intermediate values of R such that $f(u)$ is as shown in Figure 3.5B, more complicated relations between $||u||$ and ℓ are possible. In particular, $F(u)$ may have the form shown in Figure 3.6.

In that case we can satisfy the relation $F(u) \leq F(||u||)$ for $0 \leq u \leq ||u||$ if $0 \leq ||u|| \leq u_1$ or $u_4 \leq ||u|| \leq u_3$. For $||u|| < u_1$ the line $F = f(||u||)$ intersects $F = F(u)$ transversely, so that as in the logistic case the integral in (3.26) converges to a finite value of ℓ. At $||u|| = u_1$ the line $F = F(||u||)$ intersects $F = F(u)$ tangentially, so that $F(u_1) - f(u) = (u_1 - u)^2 h(u)$ for some $h(u)$ and the integral in (3.26) diverges to infinity. For $u_4 < ||u|| < u_3$, the line $F = F(||u||)$ again intersects $F = F(u)$ transversely, so that the integral in (3.26) is finite. For $||u|| = u_4$ the line $F = F(u_4)$ intersects $F = F(u)$ tangentially at $u = u_1$; thus we have $F(u_4) - F(u) = F(u_1) - F(u) = (u_1 - u)^2 h(u)$ and again the integral in (3.26) diverges. Finally, at $||u|| = u_3$, $F = F(u_3)$ intersects $F = F(u)$ tangentially and the integral in (3.26) diverges. These features are shown in the graph of

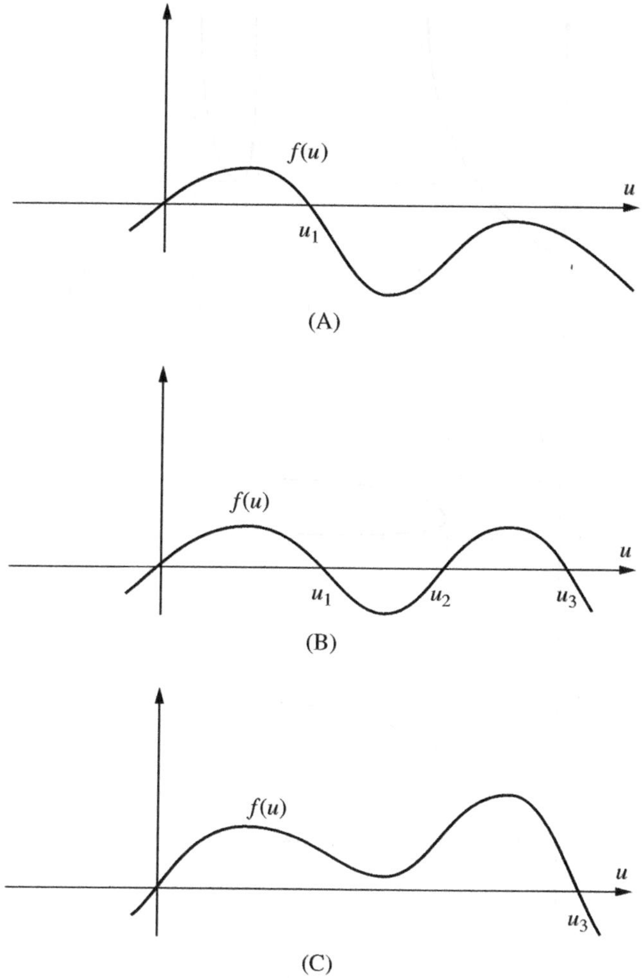

Figure 3.5 Figures 3.5A, B, and C show the form of $f(u)$ in the model (3.30) for the spruce budworm in cases where the parameter R is small, intermediate, and large.

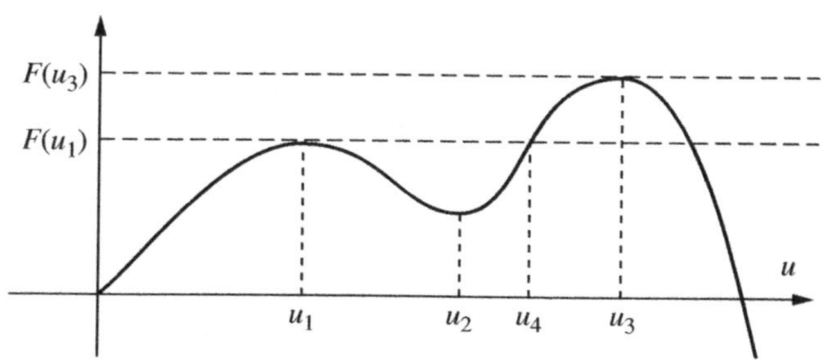

Figure 3.6 This graph shows a possible form for $F(u)$ in the model for the spruce budworm when $f(u)$ has the form shown in Figure 3.5B. It is analogous to the graphs shown in Figures 3.1 and 3.4A.

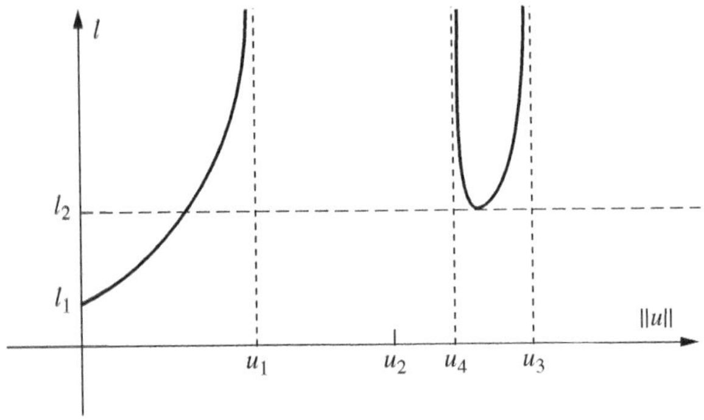

(A)

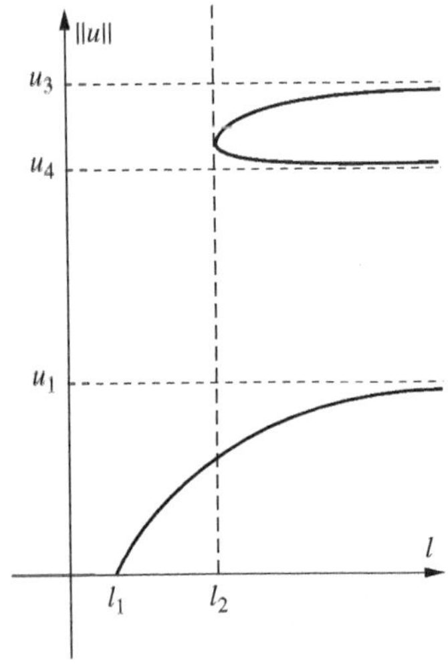

(B)

Figure 3.7 These graphs show the relationship between the maximum values of positive equilibria and the length of the intervals where such positive equilibria exist in the case of the spruce budworm model where $F(u)$ has the form shown in Figure 3.7. They are analogous to the graphs shown in Figures 3.2 and 3.3 or 3.4B and 3.4C. It turns out that $l_1 = \pi$ and $l_2 > l_2$; see Ludwig et al. (1979).

the dependence of ℓ on $||u||$ in Figure 3.7A and the corresponding bifurcation diagram (Figure 3.7B), obtained by interchanging the axes of Figure 3.7A. The interpretation of Figures 3.7A and 3.7B is that there are two critical patch sizes ℓ_1 and ℓ_2 with $\ell_1 < \ell_2$ for the budworm model. For $\ell < \ell_1$ no population can persist. For $\ell_1 < \ell < \ell_2$ a population can persist at low density. For $\ell > \ell_2$ a population can persist either at a low (endemic) density or a larger (outbreak) density.

The methods of this section can be applied to other models of the form shown in (3.19), but they extend only to certain special cases of models in two or more space dimensions (Aronson and Weinberger, 1978) and generally do not extend to models involving spatial heterogeneity, i.e. models with coefficients depending on x. However, by using powerful abstract results from nonlinear functional analysis, it is possible to construct bifurcation diagrams analogous to those discussed in this section for much more general models. That will be the topic of the next section.

3.4 Continuation and Bifurcation of Equilibria

3.4.1 Continuation

In the preceding section we showed how the relationship between the set of equilibria of certain reaction-diffusion models and a parameter occurring in the models (specifically, the size of the underlying spatial domain) can be described by bifurcation diagrams such as Figures 3.3, 3.4C, and 3.7B. The methods we used there only apply in the case of one space dimension, but similar results hold in much more general situations. However, more general situations require more technicalities, which we will discuss in this section. The reader who is primarily interested in biological applications may want to focus on the results stated in this section rather than the mathematical details of the discussion. To proceed with our analysis we must state and apply some abstract results which are cast in terms of mappings (i.e. functions) between Banach spaces. Recall that a Banach space is simply a vector space with a norm that defines a metric with the topological property of completeness. The Euclidean space $I\!R^n$ with the standard norm is a Banach space, as are many standard spaces of functions, including L^P spaces, Sobolev spaces, and various spaces of continuous and differentiable functions.

To illustrate the essential mathematical issues related to continuation and bifurcation it is instructive to consider some simple examples before treating the general theory. Suppose that x and y are real numbers satisfying the relation $x^2 - y^2 = 0$, and we want to think of y as depending on x. If we start at a solution (x_0, x_0) with $x_0 \neq 0$, we see that the set of solutions near (x_0, x_0) is given by the smooth curve $(x, y) = (x, y(x))$ with $y(x) = x$. However, at $(0, 0)$ that set of solutions intersects another set of solutions given by $(x, y) = (x, y(x))$ with $y(x) = -x$, so that near the point $(0, 0)$ the full set of solutions cannot be described as a curve$(x, y(x))$ for any single function $y(x)$. In the terminology of continuation/bifurcation theory, one would say that the solution to the relation $x^2 - y^2 = 0$ passing through (x_0, x_0) can be continued in the variable x as the curve (x, x) until it reaches the point $(0, 0)$, where another curve of solutions $(x, -x)$ bifurcates from it. In this case the curve (x, x) continues on through the point $(0, 0)$ and is valid for all x. What is special about $(0, 0)$? The relation $f(x, y) = 0$ with $f(x, y) = x^2 - y^2$ has the property that $\partial f/\partial y = 0$ at $(0, 0)$ but not at any other solution (x_0, x_0) or $(x_0, -x_0)$. In general, if $f(x, y)$ is smooth, $f(x_0, y_0) = 0$, and $\partial f/\partial y \neq 0$ at (x_0, y_0) then the Implicit Function Theorem (see below for a formal statement of the theorem) implies that near (x_0, y_0) the relation $f(x, y) = 0$ has a unique solution $y = y(x)$. What happens when $\partial f/\partial y = 0$ depends more delicately on $f(x, y)$. Consider the relation $x - y^2 = 0$. For $x_0 > 0$, solutions to this relation near $(x_0, \sqrt{x_0})$ are given by $(x, y(x))$ with $y(x) = \sqrt{x}$. At $(0, 0)$, $\partial f/\partial y = 0$, but instead of having another curve of solutions branch off from $(x, y(x))$ while the original solution continues for $x < 0$, in this case the curve $(x, \sqrt{x})$ cannot be extended for $x < 0$. It does however connect smoothly to another branch of solutions, $(x, -\sqrt{x})$, at the point $(0, 0)$. Bifurcation theory is the abstract theory which makes it possible to analyze the set of solutions to relations like $f(x, y) = 0$ at points where the implicit function theorem fails. In the present context our main goal will be to obtain bifurcation diagrams of the sort derived in Section 3.3 for models even more general than those considered in Section 3.1. A secondary goal will be to develop some mathematical tools which can also be used to study other questions related to parameter dependence. The first of those tools is a version of the Implicit Function Theorem.

Theorem 3.5. (Implicit Function Theorem) Let X, Y, and Z be Banach spaces and let F be a function from an open subset $U \subseteq X \times Y$ into Z. Suppose that $F(x, y)$ and the partial derivative $F_y(x, y)$ are continuous on U, and that there is some $(x_0, y_0) \in U$ where $F(x_0, y_0) = 0$. If the linear map $F_y(x_0, y_0)$ from Y onto Z has a continuous inverse, then for each point $x \in X$ which is sufficiently close to x_0, there is a unique $y(x) \in Y$ such that $F(x, y(x)) = 0$, and the function $y(x)$ is differentiable with respect to x.

Remark: Versions of this theorem are stated and proved in Deimling (1985) and Zeidler (1985). If the original function $F(x, y)$ has continuous derivatives of order up to m then so does $y(x)$. In the case $X = Y = Z = I\!R$, the mapping $F_y(x_0, y_0)$ is simply multiplication by the number $\partial F/\partial y|_{(x_0, y_0)}$ so that the mapping is invertible if $\partial F/\partial y|_{(x_0, y_0)} \neq 0$; hence the relation $F(x, y) = 0$ will uniquely define y as a function of x for x close to x_0 if $\partial F/\partial y|_{(x_0, y_0)} \neq 0$.

Example 3.4. Suppose that Ω is a bounded region in $I\!R^n$ (in applications $n = 1, 2$, or 3) and that $\partial\Omega$ is smooth. Let $\tilde{\Omega}$ be the region obtained by rescaling the coordinates Ω by a factor of ℓ, so that $\tilde{\Omega} = \{(\ell x_1, \ldots, \ell x_n) : x = (x_1, \ldots, x_n) \in \Omega\}$, and let $\tilde{x} = \ell x = (\ell x_1, \ldots, \ell x_n)$. (This is the same type of rescaling as in Example 2.1.) Consider the equilibrium equation for a model of the form (3.1) but with constant coefficients and with Dirichlet boundary conditions on $\tilde{\Omega}$:

$$\begin{aligned} 0 &= d\nabla^2 u + f(u) \text{ on } \tilde{\Omega} \\ u &= 0 \qquad\qquad\quad \text{on } \partial\tilde{\Omega}. \end{aligned} \tag{3.31}$$

Since $x = \tilde{x}/\ell$ we have $\partial u/\partial \tilde{x}_i = (1/\ell)\partial u/\partial x_i$, so that we may rewrite (3.31) as

$$\begin{aligned} 0 &= (d/\ell^2)\nabla^2 u + f(u) \text{ on } \Omega \\ u &= 0 \qquad\qquad\qquad\quad \text{on } \partial\Omega. \end{aligned}$$

Using $\lambda = \ell^2$, this can be rewritten as

$$\begin{aligned} 0 &= d\nabla^2 u + \lambda f(u) \text{ on } \Omega \\ u &= 0 \qquad\qquad\quad\;\; \text{on } \partial\Omega \end{aligned} \tag{3.32}$$

where $\lambda = \ell^2$. (Note that the dependence of λ on ℓ does not depend on the number of space dimensions.) In (3.32) the parameter λ represents the size of $\tilde{\Omega}$. An analogous parameter can be introduced in the case of variable coefficients, but the interpretation is less clear:

$$\begin{aligned} 0 &= \nabla \cdot d(x)\nabla u + \lambda f(x, u) \text{ in } \Omega \\ u &= 0 \qquad\qquad\qquad\qquad\quad \text{on } \partial\Omega. \end{aligned} \tag{3.33}$$

We now analyze (3.33). Suppose that $d(x) \in C^{1+\alpha}(\overline{\Omega})$, and that $f(x, u)$ and $\dfrac{\partial f}{\partial u}(x, u)$ are Hölder continuous (i.e. of class C^α) in x and u. Suppose that for $\lambda = \lambda_0$ the problem (3.33) admits a positive solution u_0. To apply Theorem 3.5 we need to cast (3.33) in terms of appropriate spaces. Let $X = I\!R$, $Y = C_0^{2+\alpha}(\overline{\Omega})$, and $Z = C^\alpha(\overline{\Omega})$. (Recall that $C_0^{2+\alpha}(\overline{\Omega})$ is the space of functions which are zero on $\partial\Omega$ and whose derivatives up to second order are Hölder continuous on $\overline{\Omega}$.) Let

$$F = F(\lambda, u) = \nabla \cdot d(x)\nabla u + \lambda f(x, u). \tag{3.34}$$

Solutions to (3.33) are then characterized by $F(\lambda, u) = 0$. The smoothness assumptions on $d(x)$ and $f(x, u)$ ensure that F is a continuous map from $X \times Y$ into Z. To compute $F_u(\lambda, u)$, we can calculate $F_u(\lambda, u)v$ for any $v \in Y$ as follows: let ϵ be a real parameter and compute

$$F_u(\lambda, u)v = \frac{d}{d\epsilon} F(\lambda, u + \epsilon v)|_{\epsilon=0} = \nabla \cdot d(x)\nabla v + \lambda \frac{\partial f}{\partial u}(x, u)v. \tag{3.35}$$

Suppose now that $\lambda_0, u_0(x)$ are such that $F(\lambda_0, u_0) = 0$, i.e. $\lambda = \lambda_0$ and $u = u_0$ satisfy (3.33). Theorem 3.5 will imply that for (λ, u) close to (λ_0, u_0) the relation $F(\lambda, u) = 0$ will uniquely determine u as a function of λ, provided $F_u(\lambda_0, u_0)$ is invertible. To invert $F_u(\lambda_0, u_0)$ we must be able to solve

$$\begin{aligned} &\nabla \cdot d(x)\nabla v + \lambda_0 \frac{\partial f}{\partial u}(x, u_0(x))v = h(x) \text{ in } \quad \Omega \\ &v = 0 \qquad\qquad\qquad\qquad\qquad\qquad\qquad \text{on } \quad \partial\Omega \end{aligned} \tag{3.36}$$

uniquely for any $h(x) \in C^\alpha(\overline{\Omega})$. (The boundary condition $v = 0$ arises because the inverse of $F_u(\lambda_0, u_0)$ must map Z into Y, and since $Y = C_0^{2+\alpha}(\overline{\Omega})$ and $v \in Y$, v must be zero on $\partial\Omega$.) Standard *a priori* estimates for elliptic equations imply that if (3.36) has a unique solution $v(x)$ for each $h(x) \in C^\alpha(\overline{\Omega})$ then the mapping from $h(x)$ to $v(x)$ is continuous as a mapping from Z into Y).

To guarantee the solvability of (3.36) we must put some conditions on $f(x, u)$. Suppose that $f(x, u) = g(x, u)u$, so that $\partial f/\partial u = g(x, u) + u\partial g/\partial u$, and that u_0 is a solution of (3.33). In that case, for $\lambda = \lambda_0$ and $u = u_0$, we have

$$\begin{aligned} &\nabla \cdot d(x)\nabla u_0 + \lambda_0 g(x, u_0)u_0 = 0 \text{ on } \quad \Omega \\ &u_0 = 0 \qquad\qquad\qquad\qquad\qquad\quad \text{on } \quad \partial\Omega. \end{aligned} \tag{3.37}$$

It follows from (3.37) that $\psi = u_0$ is a solution to the eigenvalue problem

$$\begin{aligned} &\nabla \cdot d(x)\nabla \psi + \lambda_0 g(x, u_0(x))\psi = \sigma\psi \text{ in } \quad \Omega \\ &\psi = 0 \qquad\qquad\qquad\qquad\qquad\qquad\quad \text{on } \quad \partial\Omega \end{aligned} \tag{3.38}$$

with $\sigma = 0$, so that $\sigma = 0$ is an eigenvalue of (3.38), and since $u_0 > 0$ in Ω, it must be the principal eigenvalue $\sigma_1(d(x), \lambda_0 g(x, u_0(x))$ for (3.38). (See Theorem 2.1 and the related discussion in Chapter 2.) In this case (3.36) becomes

$$\begin{aligned} &\nabla \cdot d(x)\nabla v + \lambda_0[g(x, u_0) + u_0 \frac{\partial g}{\partial u}(x, u_0)]v = h(x) \text{ in } \quad \Omega \\ &v = 0 \qquad\qquad\qquad\qquad\qquad\qquad\qquad\qquad\qquad \text{on } \quad \partial\Omega. \end{aligned} \tag{3.39}$$

If $\partial g/\partial u < 0$, since $u_0 > 0$, it follows from Corollary 2.2 and (3.39) that the eigenvalue problem

$$\begin{aligned} &\nabla \cdot d(x)\nabla \psi + \lambda_0[g(x, u_0) + u_0 \frac{\partial g}{\partial u}(x, u_0)]\psi = \sigma\psi \text{ in } \quad \Omega \\ &\psi = 0 \qquad\qquad\qquad\qquad\qquad\qquad\qquad\qquad\qquad \text{on } \quad \partial\Omega \end{aligned} \tag{3.40}$$

has a principal eigenvalue σ_1^*, with

$$\begin{aligned}\sigma_1^* &= \sigma_1(d(x), \lambda_0[g(x, u_0(x)) + u_0(x)\frac{\partial g}{\partial u}(x, u_0(x))]) \\ &< \sigma_1(d(x), \lambda_0 g(x, u_0(x))) = 0.\end{aligned} \tag{3.41}$$

Since σ_1^* is the largest eigenvalue of (3.40), zero cannot be an eigenvalue for (3.40), so by standard results on elliptic differential operators (see Section 2.6), the problem (3.39) is uniquely solvable in $C^{2+\alpha}(\overline{\Omega})$ for any $h(x) \in C^\alpha(\overline{\Omega})$, and the solution mapping $h(x) \mapsto v(x)$ is continuous. Thus, for $f(x, u) = ug(x, u)$ with $\partial g/\partial u < 0$ the mapping $F(\lambda, u)$ defined by (3.34) satisfies the hypotheses of Theorem 3.5 for any solution $(\lambda_0, u_0(x))$ of (3.33) with $\lambda_0 > 0$, $u_0(x) > 0$ on Ω. Theorem 3.5 then implies:

Proposition 3.6. If $f(x, u) = ug(x, u)$ with $\partial g/\partial u < 0$ for $u > 0$ and $(\lambda_0, u_0(x))$ is a solution of (3.33) with $\lambda_0 > 0$ and $u_0(x) > 0$ on Ω, then all solutions to (3.33) which are sufficiently close to $(\lambda_0, u_0(x))$ in $I\!R \times C_0^{2+\alpha}(\overline{\Omega})$ must lie on a smooth curve $(\lambda, u(\lambda, x))$ with $u(\lambda_0, x) = u_0(x)$. Furthermore, $u(\lambda, x)$ is continuously differentiable in λ as a mapping from $I\!R$ into $C_0^{2+\alpha}(\overline{\Omega})$.

Remarks: This result extends to models with advection and with other boundary conditions, for example

$$\begin{aligned}&\nabla \cdot d(x)\nabla u + \vec{b}(x) \cdot \nabla u + \lambda f(x, u) = 0 \text{ in } \quad \Omega \\ &d(x)\frac{\partial u}{\partial \vec{n}} + \beta(x)u = 0 \qquad \text{on} \quad \partial\Omega\end{aligned} \tag{3.42}$$

with $\beta(x) \in C^{2+\alpha}(\partial\Omega)$, $\beta(x) \geq 0$, and $\vec{b}(x) \in [C^{1+\alpha}(\overline{\Omega})]^n$. To treat this case we would replace the space $Y = C_0^{2+\alpha}(\overline{\Omega})$ with $Y = \{y(x) \in C^{2+\alpha}(\overline{\Omega}) : d(x)\frac{\partial y}{\partial \vec{n}} + \beta(x)y = 0 \text{ on } \partial\Omega\}$ and use Corollary 2.19 instead of Corollary 2.2 for the eigenvalue comparison (3.41). We could also weaken the regularity conditions on $f(x, u)$ with respect to x; for example, if we assume that $f(x, u) = m(x)u - cu^2$ with $m(x)$ a bounded measurable function then Proposition 3.6 is still valid if we replace $Y = C_0^{2+\alpha}(\overline{\Omega})$ with $W_0^{2,p}(\Omega)$ and $Z = C^\alpha(\overline{\Omega})$ with $L^p(\Omega)$.

Example 3.5. The Implicit Function Theorem can also be used to show that eigenvalues of the sort discussed in Chapter 2 depend differentiably on the parameters in the differential operators which describe dispersal or recruitment rates, spatial scale, boundary conditions, etc. For example, let $\sigma_1(\lambda)$ be the principal eigenvalue and $\psi_1(\lambda, x)$ the corresponding eigenfunction for the problem

$$\begin{aligned}&\nabla \cdot d(x)\nabla\psi + \lambda m(x)\psi = \sigma\psi \text{ in } \quad \Omega \\ &\psi = 0 \qquad \text{on} \quad \partial\Omega\end{aligned} \tag{3.43}$$

with

$$\int_\Omega \psi_1^2 dx = 1. \tag{3.44}$$

Equations (3.43) and (3.44) can be rewritten as $F(\lambda, \psi_1, \sigma_1) = 0$ where $F(\lambda, \psi_1, \sigma_1) = (\nabla \cdot d(x)\nabla\psi_1 + \lambda m(x)\psi_1 - \sigma_1\psi_1, \int_\Omega \psi_1^2 dx - 1)$. In this case we take $X = I\!R$, $Y = C_0^{2+\alpha}(\overline{\Omega}) \times I\!R$, and $Z = C^\alpha(\overline{\Omega}) \times I\!R$. The function F is a smooth mapping from $X \times Y$ into Z. Also, we can compute $F_y(\lambda, \psi_1, \sigma_1)(\rho, \tau)$ as in (3.35):

$$\frac{d}{d\epsilon}F(\lambda, \psi_1 + \epsilon\rho, \sigma_1 + \epsilon\tau)|_{\epsilon=0} = (\nabla \cdot d(x)\nabla\rho + \lambda m(x)\rho - \sigma_1\rho - \tau\psi_1,\ 2\int_\Omega \psi_1\rho dx)$$

In this case inverting $F_y(\lambda_0, \psi_1(\lambda_0), \sigma_0(\lambda_0)$ amounts to solving

$$\begin{aligned} &\nabla \cdot d(x)\nabla\rho + \lambda_0 m(x)\rho - \sigma_1(\lambda_0)\rho - \tau\psi_1(\lambda_0) = h(x), \\ &2\textstyle\int \psi_1(\lambda_0)\rho dx = r \end{aligned} \tag{3.45}$$

uniquely for arbitrary $h \in C^\alpha(\Omega)$ and $r \in I\!R$. Since $\sigma_1(\lambda_0)$ is an eigenvalue of (3.43) with eigenfunction $\psi_1(\lambda_0)$, the problem

$$\nabla \cdot d(x)\nabla\rho + \lambda_0 m(x)p - \sigma_1(\lambda_0)\rho = h(x) + \tau\psi_1(\lambda_0, x) \tag{3.46}$$

will have a solution ρ^* only if

$$\int_\Omega [h(x) + \tau\psi_1(\lambda_0, x)]\psi_1(\lambda_0, x)dx = 0.$$

Since $\int_\Omega \psi_1^2(\lambda_0)dx = 1$, this requirement is equivalent to

$$\tau = -\int_\Omega h\psi_1(\lambda_0)dx \tag{3.47}$$

(see Theorem 1.10). However, for any specific solution ρ^*, the problem (3.46) will have infinitely many solutions of the form $\rho^* + s\psi_1(\lambda_0)$ for any $s \in I\!R$. The second equation in (3.45) requires $2\int_\Omega \psi_1(\lambda_0)(\rho^* + s\psi_1(\lambda_0))dx = r$ so that $s = (r/2) - \int_\Omega \rho^*\psi_1(\lambda_0)dx$. Fixing ρ^* and then using $\rho = \rho^* + s\psi_1$ and τ as defined by (3.47) yields a unique solution to (3.45). The mapping from (h, r) to (ρ, τ) will be continuous by standard elliptic *a priori* estimates. Thus, for $(\lambda, \psi_1, \sigma_1)$ near $(\lambda_0, \psi_1(\lambda_0), \sigma_1(\lambda_0))$ the pair (ψ_1, σ_1) depends differentiably on λ. If we differentiate the relations (3.43) and (3.44) with respect to λ and denote derivatives with respect to λ by the subscript λ, we obtain

$$\begin{aligned} &\nabla \cdot d(x)\nabla\psi_{1\lambda} + \lambda m(x)\psi_{1\lambda} + m(x)\psi_1 = \sigma_{1\lambda}\psi_1 + \sigma_1\psi_{1\lambda} \\ &2\int_\Omega \psi_1\psi_{1\lambda}dx = 0. \end{aligned} \tag{3.48}$$

Multiplying the first equation by ψ_1, integrating by parts via the divergence theorem, and using the second equation in (3.48) and equations (3.43) and (3.44) yields

$$\sigma_{1\lambda} = \int_\Omega m(x)\psi_1^2 dx. \tag{3.49}$$

Equation (3.49) measures the sensitivity of $\sigma_1(\lambda)$ with respect to changes in λ. Relations such as (3.49) can be used to examine the sensitivity of eigenvalues to various parameters; this is done in Cantrell and Cosner (1987) and Belgacem and Cosner (1995), for example.

3.4.2 Bifurcation Results

Returning to the problem of extending the one-dimensional results of the previous section to more general settings, let us consider the branch of solutions $(\lambda, 0)$ to (3.33). Assuming that $f(x, u) = g(x, u)u$, (3.35) becomes $F_u(\lambda, 0)v = \nabla \cdot d(x)\nabla v + \lambda g(x, 0)v$. Thus, as in Example 3.4, $F_u(\lambda, 0)$ will be invertible if zero is not an eigenvalue of

$$\begin{aligned} &\nabla \cdot d(x)\nabla \psi + \lambda g(x, 0)\psi = \sigma \psi \text{ in } \quad \Omega \\ &\psi = 0 \qquad\qquad\qquad\qquad\qquad\quad \text{on } \quad \partial\Omega; \end{aligned} \tag{3.50}$$

in other words, if λ is not an eigenvalue of

$$\begin{aligned} &\nabla \cdot d(x)\nabla \phi + \lambda g(x, 0)\phi = 0 \text{ in } \quad \Omega \\ &\phi = 0 \qquad\qquad\qquad\qquad\quad \text{on } \quad \partial\Omega. \end{aligned} \tag{3.51}$$

If $g(x, 0) > 0$ on an open subset of Ω then (3.51) will have an infinite sequence of positive eigenvalues $0 < \lambda_1 < \lambda_2 \leq \lambda_3 \ldots$. As long as $\lambda_0 \neq \lambda_j$ for any j, then we may conclude as in Example 3.4 that there will be a neighborhood of $(\lambda_0, 0)$ where $F(\lambda, u) = 0$ uniquely determines u as a function of λ. Since a branch of solutions near $(\lambda_0, 0)$ is given by $(\lambda, 0)$, it follows that there can be no solutions to (3.33) in a neighborhood of $(\lambda_0, 0)$, so that in particular there is an interval $(\lambda_0 - \epsilon, \lambda_0 + \epsilon)$ on which solutions to (3.33) with $u \not\equiv 0$ are bounded away from 0. At the values $\lambda = \lambda_j$, $j = 1, \ldots, \infty$, the operator $F_u(\lambda, 0)$ will have a nontrivial nullspace containing the eigenfunction corresponding to λ_j, so it will not be invertible, and the Implicit Function Theorem will not apply. As shown by the analysis of the one-dimensional case in the preceding section, it is possible that a curve of solutions with $u > 0$ branches off of the set $(\lambda, 0)$ at the value $\lambda = \lambda_1$. This phenomenon turns out to be quite general, but to understand it in any generality requires the use of bifurcation theory. In the present setting, a point $(\lambda_0, 0)$ is a bifurcation point if every neighborhood of $(\lambda_0, 0)$ contains solutions to $F(\lambda, u) = 0$ with $u \neq 0$. The only candidates for bifurcation points are $(\lambda_j, 0)$ where λ_j is an eigenvalue of (3.51), but it is not clear that they are indeed bifurcation points. To determine that we shall use some ideas and results formulated by Rabinowitz (1971, 1973); see also Deimling (1985) and Zeidler (1985). The Rabinowitz theory is formulated in terms of mappings from $I\!R \times Y$ into Y, where Y is a Banach space. Like the Implicit Function Theorem, many results in bifurcation theory have hypotheses involving the linearization of a nonlinear operator about some point. The result we will want to apply is stated for problems of the form

$$u = \lambda L u + H(\lambda, u) \tag{3.52}$$

where L is a linear operator and $H(\lambda, u)$ has the property that $||H(\lambda, u)||/||u|| \to 0$ as $||u|| \to 0$, uniformly in λ for λ restricted to any bounded interval. We must require that L and H are compact in the sense that they map bounded sets into sets whose closures are compact. The result we need is formulated in terms of characteristic values of the operator L. The number μ is a *characteristic value* of L if there is a nonzero $\phi \in Y$ with $\phi = \mu L\phi$; that is, $1/\mu$ is a nonzero eigenvalue of L. The algebraic multiplicity of a characteristic value μ is the dimension of $\cup_{n=0}^{\infty} N(I - \mu L)^n$, where $N(I - \mu L)^n$ is the nullspace of the operator $(I - \mu L)^n$. Each nonzero eigenvalue of a compact linear operator, and thus every characteristic value, has finite algebraic multiplicity; see, for example, Schechter (2002).

If the algebraic multiplicity of a characteristic value is one, it is said to be algebraically simple. We can now state a basic result for (3.52):

Theorem 3.7. (Rabinowitz 1971, 1973) Suppose that L and H are operators as described above. If μ_0 is a simple characteristic value of L then $\lambda = \mu_0$ is a bifurcation point for (3.52). Let $\mathcal{S} \subset I\!R \times Y$ be the set of nontrivial solutions to (3.52) (i.e. solutions with $u \neq 0$). Let $\mathcal{C}$ be the connected component of the set $\mathcal{S} \cup \{(\mu_0, 0)\}$ containing $(\mu_0, 0)$. Then in a neighborhood of $(\mu_0, 0)$, $\mathcal{C} = \mathcal{C}^+ \cup \mathcal{C}^-$ where $\mathcal{C}^+ \cap \mathcal{C}^- = \{(\mu_0, 0)\}$ and each of $\mathcal{C}^+, \mathcal{C}^-$ satisfies one of the alternatives

(i) $\mathcal{C}^+$ (resp. $\mathcal{C}^-$) is unbounded in $I\!R \times Y$, or

(ii) $\overline{\mathcal{C}}^+$ (resp. $\overline{\mathcal{C}}^-$) contains $(\mu_1, 0)$ where μ_1 is a characteristic value of L and $\mu_1 \neq \mu_0$.

Remark: If μ_0 is merely a characteristic value of L with odd multiplicity, we may still conclude that $(\mu_0, 0)$ is a bifurcation point and that the continuum $\mathcal{C}$ must satisfy either (i) or (ii), but we are no longer guaranteed $\mathcal{C} = \mathcal{C}^+ \cup \mathcal{C}^-$ as in the case where μ_0 is simple. See Rabinowitz (1971, 1973).

To apply Theorem 3.7 to (3.33) we must write (3.33) in the form (3.52). If $f(x, u) = g(x, u)u$ we may write (3.33) as

$$-\nabla \cdot d(x)\nabla u = \lambda g(x, 0)u + \lambda[g(x, u) - g(x, 0)]u. \tag{3.53}$$

If we define A as the operator $-\nabla \cdot d(x)\nabla$ on the space $C_0^{2+\alpha}(\overline{\Omega})$ then A maps $C_0^{2+\alpha}(\overline{\Omega})$ into $C^{\alpha}(\overline{\Omega})$, and by standard results about elliptic operators A has a continuous inverse $A^{-1} : C^{\alpha}(\overline{\Omega}) \to C_0^{2+\alpha}(\overline{\Omega})$. The operator A^{-1} is not compact, but the space $C_0^{2+\alpha}(\overline{\Omega})$ embeds compactly in $C_0^{\alpha}(\Omega)$ (also in $C_0^{1+\alpha}(\overline{\Omega})$) so the inclusion operator $E : C_0^{2+\alpha}(\overline{\Omega}) \to C_0^{\alpha}(\overline{\Omega})$ defined by $Eu = u$ is compact. (These are standard results about elliptic operators; see the discussion in Chapter 1, or see Gilbarg and Trudinger (1977) and Friedman (1976) for more detailed discussions.) We may now choose the space $Y = C_0^{\alpha}(\overline{\Omega})$ and write (3.53) as

$$u = \lambda E A^{-1}(g(x, 0)u) + \lambda E A^{-1}([g(x, u) - g(x, 0)]u). \tag{3.54}$$

This equation now has the form of (3.52), with $Y = C_0^{\alpha}(\overline{\Omega})$, $Lu = EA^{-1}(g(x, 0)u)$ and $H(\lambda, u) = \lambda E A^{-1}([g(x, u) - g(x, 0)]u)$. The operators L and H are compact because of the compactness of the embedding E. The operator H has the property $||H(\lambda, u)||/||u|| \to 0$ as $||u|| \to 0$ because $g(x, u) - g(x, 0) \to 0$ as $||u|| \to 0$ and EA^{-1} is continuous. (Here $||\cdot||$ denotes the norm of $C_0^{\alpha}(\overline{\Omega})$.) The value μ is a characteristic value of L if $\mu L\phi = \phi$, which can be expressed as $\mu E A^{-1}[g(x, 0)\phi] = \phi$. Since $A^{-1} : C^{\alpha}(\overline{\Omega}) \to C_0^{2+\alpha}(\overline{\Omega})$ and $Eu = u$ on $C_0^{2+\alpha}(\overline{\Omega})$, we have $EA^{-1}[g(x, 0)\phi] = A^{-1}g(x, 0)\phi$ so $\phi \in C_0^{2+\alpha}(\overline{\Omega})$ and thus we have $\mu g(x, 0)\phi = A\phi$. This is equivalent to (3.51), so the characteristic values of L are just the eigenvalues of (3.51). In particular, the principal eigenvalue $\lambda_1^+(d(x), g(x, 0))$ of (3.51) is a characteristic value of L. Since the principal eigenvalue is simple as an eigenvalue of (3.51), it is also simple as a characteristic value of L. Thus we may apply Theorem 3.7 to (3.33) via (3.54) to obtain:

Proposition 3.8. The point $(\lambda_1^+(d(x), g(x, 0)), 0)$ is a bifurcation point for nonzero solutions of (3.33) in the sense of Theorem 3.7.

Remarks: We need some additional information about the branch of nonzero solutions which bifurcates from the point $(\lambda_1^+(d(x), g(x, 0)), 0)$. To be relevant in the context of ecological models, the bifurcating solutions must be positive. Also, we may want to determine which of alternatives (i) and (ii) of Theorem 3.7 applies to the branch of solutions bifurcating from $(\lambda_1^+(d(x), g(x, 0)), 0)$. To resolve these questions requires a more detailed examination of what happens near the bifurcation point. We shall turn to that issue shortly.

The analysis leading to Proposition 3.8 can be applied to more general models of the type shown in (3.42) with only minor changes. It is also possible to handle cases where the diffusion rate is density dependent, that is $d = d(x, u)$. In such cases the formulation (3.54) becomes more complicated, but otherwise the analysis is similar; see Cantrell and Cosner (1991a) for details. To handle the boundary conditions of (3.42) we would choose $Y = \{u \in C^{1+\alpha}(\overline{\Omega}) : \alpha(x)\partial u/\partial \vec{n} + \beta(x)u = 0 \text{ on } \partial\Omega\}$ instead of using $Y = C_0^\alpha(\overline{\Omega})$. Also, if $g(x, 0)$ is merely assumed to be bounded and measurable, the analysis would be quite similar, except for the function spaces involved. In that case we would define A as the operator $\nabla \cdot d(x)\nabla$ acting on a Sobolev space of some sort. Recall that for p sufficiently large the space $W^{2,p}(\Omega)$ embeds compactly in $C^{1+\alpha}(\overline{\Omega})$ (see Gilbarg and Trudinger (1977), Adams (1975)) so by choosing p to be large we can define the space $X = \{u \in W^{2,p}(\Omega) : u = 0 \text{ on } \partial\Omega\}$. Then A is bounded (i.e. continuous) linear operator from X into $L^p(\Omega)$ with a continuous inverse. For a situation where $g(x, 0)$ is merely bounded and measurable, i.e. $g(x, 0) \in L^\infty(\Omega)$, the bifurcation theory still could be done in the space Y of functions in $C^{1+\alpha}(\overline{\Omega})$ satisfying the boundary conditions. The difference in that case is that the operator taking u to $g(x, 0)u$ maps functions $u \in C^{1+\alpha}(\overline{\Omega})$ into $L^\infty(\Omega)$ and hence into $L^p(\Omega)$ for any p, but it does not map $C^\alpha(\overline{\Omega})$ into itself. Thus, A^{-1} would have to be taken as an operator from $L^p(\Omega)$ into X rather than from $C^\alpha(\overline{\Omega})$ into $C^{2+\alpha}(\overline{\Omega})$. Otherwise, the analysis would be unchanged.

In the case of no-flux boundary conditions the operator $-\nabla \cdot d(x)\nabla$ is not invertible, but if λ_0 is not an eigenvalue for the problem $\nabla \cdot d(x)\nabla\phi + \lambda g(x, 0)\phi = 0$ with no-flux boundary conditions then the operator defined by $Au = -\nabla \cdot d(x)\nabla u - \lambda_0 g(x, 0)u$ is invertible. We may write (3.33) as $-\nabla \cdot d(x)\nabla u - \lambda_0 g(x, 0)u = (\lambda - \lambda_0)g(x, 0)u + \lambda[g(x, u) - g(x, 0)]u$, let $\mu = \lambda - \lambda_0$, then obtain $u = \mu A^{-1}g(x, 0)u + (\mu + \lambda_0)A^{-1}([g(x, u) - g(x, 0)]u)$ and proceed as before, using μ as the bifurcation parameter. If the Dirichlet boundary condition in (3.51) is replaced with a no-flux (i.e. Neumann) condition then a positive principal eigenvalue for (3.51) exists only in the case $\int_\Omega g(x, 0)dx < 0$, so the present discussion only applies to that case.

Putting together the various extensions of Proposition 3.8 allows us to treat models of the form

$$\begin{aligned} \nabla \cdot d(x, u)\nabla u + \vec{b}(x) \cdot \nabla u + \lambda g(x, u)u &= 0 \quad \text{in} \quad \Omega \\ \alpha(x)\frac{du}{\partial \vec{n}} + \beta(x)u &= 0 \qquad\qquad \text{on} \quad \partial\Omega \end{aligned} \tag{3.55}$$

(or $u = 0$ on $\partial\Omega$) provided $d(x, u) \geq d_0 > 0$, $d(x, u)$ is of class $C^{1+\alpha}$ in x and u, $\vec{b}(x)$ is C^α in x, and $g(x, u)$ is continuous in x and C^1 in u. If $\vec{b}(x) = \vec{0}$ we can allow $g(x, u)$ to be merely bounded and measurable in x. (Continuity is needed in the case $\vec{b} \neq 0$ so that we can apply the methods of Hess and Kato (1980) to conclude that the linearized problem

$$\begin{aligned} \nabla \cdot d(x,0)\nabla\phi + \vec{b}(x)\cdot\nabla\phi + \lambda g(x,0)\phi &= 0 \text{ in } \quad \Omega \\ \alpha(x)\frac{\partial u}{\partial \vec{n}} + \beta(x)u &= 0 \qquad \text{on} \quad \partial\Omega \end{aligned} \tag{3.56}$$

(or $u = 0$ on $\partial\Omega$) actually has a principal eigenvalue. In summary, we have:

Proposition 3.9. Suppose the coefficients of (3.55) have the regularity properties described above, with $d(x,u) \geq d_0 > 0$, $\beta(x) \neq 0$, and $g(x,0)$ positive on an open subset of Ω, and let λ_1^+ denote the positive principal eigenvalue of (3.56). Then $(\lambda_1^+, 0)$ is a bifurcation point for nonzero solutions of (3.56) in the sense of Theorem 3.7.

Remarks: The case of (3.55) with Dirichlet boundary conditions is treated in (Cantrell and Cosner 1991b); the general case is similar. Proposition 3.9 extends Proposition 3.8 to the case of (3.55). However, Proposition 3.6 does not extend to (3.55) without additional assumptions on $d(x,u)$.

We now describe the solution set of (3.55) near the bifurcation point $(\lambda_1^+, 0)$. The key point here is that λ_1^+ is a simple eigenvalue. In the case of bifurcation from a simple eigenvalue, the solution set near the bifurcation point can often be described in detail. However, the theory behind such a description is somewhat abstract and complicated, so we will discuss it in the Appendix to this chapter. The relevant results are due to Crandall and Rabinowitz (1971, 1973). To state the conclusion of the abstract theory we must recall the notion of a complement to a subspace of a vector space. If X is a vector space with subspace $Y \subseteq X$, a *complement* of Y is any other subspace $Z \subseteq X$ with $Y \cap Z = \{0\}$ and $Y + Z = X$.

Proposition 3.10. Suppose that $d(x,u) \geq d_0 > 0$, $d(x,u)$ is of class $C^{1+\alpha}$ in x and u for $x \in \overline{\Omega}$, that $\vec{b}(x)$ is C^α in x and that $g(x,u)$ is C^α in x and C^1 in u. Let λ_1^+ be the positive principal eigenvalue of (3.56) and let $\phi_1 > 0$ be an eigenfunction corresponding to λ_1^+. Let $Z \subseteq C^{2+\alpha}(\overline{\Omega})$ be any closed complement of the span of $\{\phi_1\}$. Then there is a neighborhood of $(\lambda_1^+, 0)$ in $I\!R \times C^{2+\alpha}(\overline{\Omega})$ within which the set of pairs $(\lambda, u) \in I\!R \times C^{2+\alpha}(\overline{\Omega})$ satisfying (3.55) with $u \neq 0$ consists of the parametric curve $(\lambda, u) = (\lambda(s), s\phi_1 + s\rho(s))$ where s is a real parameter which takes values in an open interval I_0 containing $s = 0$, $\lambda(s)$ is a real valued function with $\lambda(0) = \lambda_1^+$, and $\rho(s)$ is a function from I_0 into Z with $\rho(0) = 0$. Furthermore, $\lambda(s)$ and $\rho(s)$ are differentiable as functions of s.

Remarks: A statement of the abstract results used to obtain Proposition 3.10 and a discussion of how they apply in this context are given in the Appendix to this chapter. Proposition 3.10 implies that the set of solutions to (3.55) near $(\lambda_1^+, 0)$ consists of the line of trivial solutions $(\lambda, 0)$ and a curve of solutions (λ, u) with $u \neq 0$ bifurcating at $\lambda = \lambda_1^+$. The key element required for this result is that the eigenvalue λ_1^+ is simple. If $\vec{b}(x) = 0$ we may allow $g(x,u)$ to be merely bounded and measurable in x; in that case we would have to replace $C^{2+\alpha}(\overline{\Omega})$ with the Sobolev space $W^{2,p}(\Omega)$. A common choice for the complement Z of the span of ϕ_1 is the set of all functions ρ so that $\int_\Omega \phi_1 \rho\, dx = 0$. Since ϕ_1 does not change sign in Ω, we may choose $\phi_1 > 0$ on $\overline{\Omega}$ except in the case of Dirichlet boundary conditions. In the Dirichlet case $\phi_1 > 0$ on Ω and $\partial\phi_1/\partial\vec{n} < 0$ on $\partial\Omega$. Either of these conditions on ϕ_1 implies that for the solutions $(\lambda(s), s\phi_1 + s\rho(s))$ bifurcating from $(\lambda_1^+, 0)$ we have $u = s\phi_1 + s\rho(s) > 0$ on Ω for s small and positive. Thus, the bifurcating solutions are biologically relevant.

Proposition 3.10 gives us some local information about solutions to (3.55) near the bifurcation point $(\lambda_1^+, 0)$ immediately, and we can use it to get more. However, it also allows us to answer the global question left open by Propositions 3.8 and 3.9.

Proposition 3.11. Suppose that the coefficients of (3.55) satisfy the hypotheses of Proposition 3.9 and that the boundary conditions have $\beta(x) > 0$. Suppose further that there is a constant $K > 0$ such that $g(x, u) < 0$ for $u > K$. Then the branch of positive solutions to (3.55) bifurcating from $(\lambda_1^+, 0)$ extends to positive infinity in λ.

Remarks: If $d = d(x)$ and $g(x, u)$ is decreasing in u then Proposition 3.11 can be deduced from the continuation result given in Proposition 3.6. However, the conclusions of Proposition 3.6 may not hold if these conditions are violated. The case of $\beta(x) = 0$ (Neumann boundary conditions) requires a slightly different approach so we will treat it separately.

Proof: (Sketch) To prove Proposition 3.11 we must rule out alternative (ii) in Theorem 3.7 and obtain some bounds on λ and u which prevent the branch of positive solutions bifurcating at $(\lambda_1^+, 0)$ from reaching infinity in u or in the negative λ direction. We first note that if u is a solution to (3.55) and $\lambda = 0$ then $u = 0$ by the strong maximum principle. Similarly, zero cannot be an eigenvalue for (3.56), so $(0, 0)$ is not a bifurcation point for (3.55) and there is a neighborhood of $(0, 0)$ where the only solutions to (3.55) are of the form $(\lambda, 0)$. Thus, the continuum $\mathcal{C}^+$ cannot intersect $\lambda = 0$ and hence must lie in the region where $\lambda > 0$. If $\lambda > 0$ then the maximum principle implies that $u \leq K$. (If u has a maximum larger than K at a point $x_0 \in \Omega$, then $\nabla u = 0$ and $\nabla^2 u \leq 0$ at x_0, while $\lambda g(x, u(x_0)) < 0$, contradicting (3.55). If u has a positive maximum at $x_0 \in \partial\Omega$ then by the strong maximum principle either $u \equiv$ constant or $\partial u/\partial\vec{n} > 0$ at x_0. In the first case the maximum would also occur at each $x \in \Omega$, so by the previous argument it could not be larger than K. The second case contradicts the boundary conditions of (3.55).) By Proposition 3.10 we know that one of the continua $\mathcal{C}^+$, $\mathcal{C}^-$ of nontrivial solutions bifurcating from $(\lambda, 0)$ at $(\lambda_1^+, 0)$ must initially consist of positive solutions. Let us identify that component at $\mathcal{C}^+$. If (λ, u) is allowed to move continuously around $\mathcal{C}^+$, then the only way for u to stop being positive is for (λ, u) to pass through a point (λ_0, u_0) where $u_0 \geq 0$ but either $u_0(x_0) = 0$ for some $x_0 \in \Omega$ or $\partial u_0/\partial\vec{n} = 0$ at some point on $\partial\Omega$. In either case, the strong maximum principle implies $u_0 \equiv 0$. Thus, the only way that solutions on $\mathcal{C}^+$ can become negative is if the branch passes through $(\mu_1, 0)$, for some $\mu_1 \neq \lambda_1^+$, which means that alternative (ii) would hold in Theorem 3.7. We will now show that alternative (ii) does not occur. The argument is based on the compactness of the operators L and H; similar arguments are common in the applications of bifurcation and continuation methods to differential equations. We have seen that for solutions on the continuum $\mathcal{C}^+$ to change sign they must do so by passing through a bifurcation point $(\mu_1, 0)$ with $\mu_1 \neq \lambda_1^+$, so if alternative (ii) occurs there must be at least one characteristic value $\mu_1 \neq \lambda_1^+$ which is a bifurcation point for positive solutions of (3.55). Thus, there must be a sequence $\{(\lambda_j, u_j)\} \subseteq \mathcal{C}^+$ with $\lambda_j \to \mu_1$, $u_j \to 0$ as $j \to \infty$, with $u_j > 0$ in Ω. Let $w_j = u_j/||u_j||$. Then

$$w_j = \lambda_j L w_j + H(\lambda_j, u_j)/||u_j|| \tag{3.57}$$

with L and H as in (3.52). Since $u_j \to 0$, the set $\{u_j\}$ is bounded, so by (3.57) and the compactness of L and H, $\{w_j\}$ must have a convergent subsequence. Passing to the subsequence we have $w_j \to w$ as $j \to \infty$. The general formulation of (3.52) implies that $H(\lambda_j, u_j)/||u_j|| \to 0$ as $j \to \infty$, so we have in the limit

$$w = \mu_1 Lw. \tag{3.58}$$

We have $||w_j|| = 1$ for every j, and $w_j > 0$ on Ω, so $||w|| = 1$ and $w \geq 0$ on Ω. Also, by (3.58), μ_1 is an eigenvalue for (3.56). However, the principal eigenvalue λ_1^+ is the only positive eigenvalue for (3.56) admitting a nonnegative eigenfunction, so we must have $\mu_1 = \lambda_1^+$, a contradiction. It follows that alternative (ii) does not hold in Theorem 3.7. We have already established that $\lambda > 0$ on C^+ and that $0 \leq u \leq K$. The bounds on u and standard results on elliptic regularity imply that u is bounded in $C^{1+\alpha}(\overline{\Omega})$ for λ restricted to any finite interval, so the only remaining possibility in Theorem 3.7 is that C^+ extends to infinity in the positive λ direction.

In the case of pure Neumann conditions the situation is somewhat different. In general there may or may not be a positive principal eigenvalue λ_1^+, but $\lambda = 0$ is always an eigenvalue. Also, the simple problem $\nabla^2 u + \lambda u[1 - (u/K)] = 0$ with Neumann conditions has a branch of nonzero solutions (λ, K) passing through $\lambda = 0$. However, something similar to Proposition 3.11 holds under slightly more restrictive conditions on (3.55).

Proposition 3.12. Suppose that $\beta(x) = 0$ and in addition to the hypotheses of Proposition 3.11 we have $\int_\Omega g(x,0)dx < 0$ with $g(x,u) \leq g_0(x)$ where $\int_\Omega g_0(x)dx < 0$ and $g_0(x) > 0$ on an open subset of Ω. Suppose further that $\vec{b}(x) = 0$. Then (3.55) with Neumann boundary conditions has a branch C^+ of positive solutions bifurcating from $(\lambda, 0)$ at $(\lambda_1^+, 0)$ which extends to positive infinity in λ.

Proof: (Sketch) The condition $\int_\Omega g(x,0)dx < 0$ is needed for λ_1^+ to exist under Neumann boundary conditions. The only way that the proof of Proposition 3.11 must be changed is that a different argument is needed to show that C^+ does not connect to $\lambda = 0$. If (λ, u) satisfies (3.55) with $u > 0$ then λ is the positive principal eigenvalue of

$$\begin{aligned} &\nabla \cdot d(x,u)\nabla\phi + \lambda g(x,u)\phi = 0 \text{ in } && \Omega \\ &\frac{\partial \phi}{\partial \vec{n}} = 0 && \text{on } \partial\Omega. \end{aligned}$$

However, since $d(x,u) \geq d_0 > 0$ and $g(x,u) \leq g_0(x)$, we must have $\lambda \geq \lambda_1^+(d_0, g_0(x))$ by the eigenvalue comparisons following from the variational formulation in Theorem 2.5. Thus, C^+ is bounded away from $\lambda = 0$ in that case.

Remarks: We could also treat the case of Neumann conditions if $\vec{b}(x) \neq 0$ provided $d(x,u) = d(x)$. In that case we would need to have $\int_\Omega g(x,0)\psi^*(x)dx < 0$ and similarly $\int_\Omega g_0(x)\psi^*(x)dx < 0$ where ψ^* is the eigenfunction corresponding to the principal eigenvalue $\sigma_1^* = 0$ of the adjoint problem

$$\begin{aligned} \nabla \cdot d(x)\nabla\psi - \nabla \cdot \vec{b}(x)\psi &= \sigma\psi \text{ in } \Omega \\ \frac{\partial\psi}{\partial\vec{n}} &= 0 \qquad \text{on } \partial\Omega. \end{aligned}$$

(Recall that $\sigma_1^* = \sigma_1$ where σ_1 is the eigenvalue of the original operator $\nabla \cdot d(x)\nabla + \vec{b} \cdot \nabla$ with Neumann conditions; in this case, $\sigma_1 = 0$.) Again, the integral conditions on g and g_0 are needed so that the eigenvalue λ_1^+ exists. In this situation the eigenvalue comparison would be based on positivity arguments along the lines of those used in Corollary 2.19. See Senn and Hess (1982). In the case $\int_\Omega g(x,0)dx > 0$ there will typically be a branch of positive solutions (λ, u) that extends to $(0, \overline{u})$ for some positive constant $\overline{u}$. See for example Cantrell et al. (1996). These solutions are analogous to the solutions (λ, K) of $\nabla^2 u + \lambda[1 - (u/K)]u = 0$ under Neumann conditions.

We have now shown that for the diffusive logistic equation and its generalizations as given in (3.55) there is a continuum of positive solutions bifurcating from the line of trivial solution $(\lambda, 0)$ at $\lambda = \lambda_1^+$ and extending to positive infinity in λ. In the case of models with passive diffusion, i.e. $d = d(x)$, and with $g(x,u)$ decreasing in u, the results of Section 3.2 imply that (3.55) has a positive solution only if $\sigma_1 > 0$ in

$$\begin{aligned} \nabla \cdot d(x)\nabla\psi + \vec{b}(x) \cdot \nabla\psi + \lambda g(x,0)\psi &= \sigma\psi \text{ in } \Omega \\ \alpha(x)\frac{\partial\psi}{\partial\vec{n}} + \beta(x)\psi &= 0 \qquad \text{on } \partial\Omega, \end{aligned}$$

and if there is a positive solution it is unique and stable. (By the results of Chapter 2, $\sigma_1 > 0$ if and only if $\lambda > \lambda_1^+$.) In the proof of the uniqueness result Proposition 3.3, the hypothesis $d = d(x)$ could be replaced with the condition $\vec{b}(x) \equiv 0$, $d = d(x,u)$ with $d(x,u)$ nondecreasing in u as in Proposition 3.12. (See also Cantrell and Cosner (1991a).) If we merely have $g(x,u) \le g(x,0)$ and $d = d(x)$ or $\vec{b} \equiv 0$ and $d(x,u) \ge d(x,0)$ then the arguments used in Proposition 3.12 would imply $\lambda > \lambda_1^+$ for any (λ, u), where u is a positive solution of (3.55), but the proof of uniqueness fails. In general, it need not be the case that $\lambda > \lambda_1^+$ nor that the positive solution to (3.55) is unique. The global structure and properties of the set of solutions to (3.55) with λ and u positive can be complicated and hard to analyze; see Lions (1982). However, we may obtain some local information near $(\lambda_1^+, 0)$ by exploiting Proposition 3.10 and related ideas.

Suppose that $(\lambda(s), u(s))$ is the branch of positive solutions to (3.55) bifurcating from $(\lambda_1^+, 0)$. By Proposition 3.10, $u(s) = s\phi_1 + s\rho(s)$ where ϕ_1 is a positive eigenfunction for (3.56). If we use subscripts u and s to indicate derivatives with respect to those variables, we can differentiate (3.55) with respect to s and obtain

$$\begin{aligned} \nabla \cdot d_u u_s \nabla u + \nabla \cdot d\nabla u_s + \vec{b} \cdot \nabla u_s + \lambda[g_u u_s u + g u_s] + \lambda_s g u &= 0 \text{ in } \Omega \\ \alpha(x)\frac{\partial u_s}{\partial\vec{n}} + \beta(x)u_s &= 0 \qquad \text{on } \partial\Omega. \end{aligned} \tag{3.59}$$

Setting $s = 0$ in (3.59) simply recovers (3.56) since $\lambda = \lambda_1^+$, $u = 0$, and $u_s = \phi_1$ at $s = 0$. Differentiating again in (3.59) yields

$$
\begin{aligned}
&\nabla \cdot (d_{uu}u_s^2 + d_u u_{ss})\nabla u + 2\nabla \cdot d_u u_s \nabla u_s + \nabla \cdot d\nabla u_{ss} + \vec{b} \cdot \nabla u_{ss} \\
&\quad + \lambda[g_{uu}u_s^2 u + g_u u_{ss} u + 2g_u u_s^2 + g u_{ss}] \\
&\quad + 2\lambda_s[g u_s + g_u u_s u] + \lambda_{ss} g u = 0 \qquad \text{in} \quad \Omega \\
&\alpha(x)\frac{\partial u_{ss}}{\partial \vec{n}} + \beta(x) u_{ss} = 0 \qquad \text{on} \quad \partial\Omega.
\end{aligned} \tag{3.60}
$$

Setting $s = 0$ in (3.60) and denoting u_{ss} and λ_s evaluated at $s = 0$ by $u_{ss}(0)$, $\lambda_s(0)$ respectively yields

$$
\begin{aligned}
&2\nabla \cdot d_u(x,0)\phi_1 \nabla\phi_1 + \nabla \cdot d\nabla u_{ss}(0) + \vec{b} \cdot \nabla u_{ss}(0) + \\
&\lambda_1^+ g(x,0) u_{ss}(0) + 2\lambda_1^+ g_u(x,0)\phi_1^2 + 2\lambda_s(0) g(x,0)\phi_1 = 0 \text{ in } \quad \Omega, \\
&\alpha(x)\frac{\partial u_{ss}}{\partial \vec{n}}(0) + \beta(x) u_{ss}(0) = 0 \qquad \text{on} \quad \partial\Omega.
\end{aligned} \tag{3.61}
$$

Let ϕ_1^* be a positive eigenfunction corresponding to the eigenvalue λ_1^+ for the adjoint problem to (3.56):

$$
\begin{aligned}
&\nabla \cdot d(x,0)\nabla\phi_1^* - \nabla \cdot (\vec{b}(x)\phi_1^*) + \lambda g(x,0)\phi_1^* = 0 \text{ in } \quad \Omega \\
&\alpha(x)\frac{\partial \phi_1^*}{\partial \vec{n}} + \beta(x)\phi_1^* = 0 \qquad \text{on} \quad \partial\Omega
\end{aligned} \tag{3.62}
$$

If we multiply (3.61) by ϕ_1^* and apply divergence theorem, we obtain

$$
\begin{aligned}
&-2\int_\Omega d_u(x,0)\phi_1 \nabla\phi_1 \cdot \nabla\phi_1^* + \int_\Omega [\nabla \cdot d(x,0)\nabla\phi_1^* - \nabla \cdot (\vec{b}\phi_1^*) \\
&+ \lambda_1^+ g(x,0)\phi_1^*] u_{ss}(0) dx + 2\lambda_1^+ \int_\Omega g_u(x,0)\phi_1^2\phi_1^* dx + 2\lambda_s(0)\int_\Omega g(x,0)\phi_1\phi_1^* dx = 0.
\end{aligned} \tag{3.63}
$$

By (3.62), the terms involving $u_{ss}(0)$ in (3.63) drop out, so we have

$$
\lambda_s(0)\int_\Omega g(x,0)\phi_1\phi_1^* dx = \int_\Omega d_u(x,0)\phi_1 \nabla\phi_1 \cdot \nabla\phi_1^* dx - \lambda_1^+ \int_\Omega g_u(x,0)\phi_1^2\phi_1^* dx. \tag{3.64}
$$

In the general case it is not clear what the signs of the integral terms in (3.64) will be, but those can be determined in various special cases. If $d = d(x)$ and $g(x,0) > 0$ then we have $d_u = 0$, and since $\phi,\ \phi^* > 0$,

$$
\lambda_s(0) = \frac{-\lambda_1^* \displaystyle\int_\Omega g_u(x,0)\phi_1^2\phi_1^* dx}{\displaystyle\int_\Omega g(x,0)\phi_1\phi_1^* dx}. \tag{3.65}
$$

If $g_u(x,0) < 0$ then $\lambda_s(0) > 0$ so the continuum $\mathcal{C}^+$ bends in the positive λ direction as it leaves $(\lambda_1^+, 0)$. Alternatively, if $g_u(x,0) > 0$ then $\mathcal{C}^+$ bends in the negative λ direction. (In the terminology of bifurcation theory, these two cases are sometimes called subcritical bifurcation and supercritical bifurcation, respectively.) In the case $g_u(x,0) > 0$, $g(x,0) > 0$, the continuum $\mathcal{C}^+$ exits from $(\lambda_1^+, 0)$ in the negative λ direction but must still extend to infinity in the positive λ direction. Thus, there must exist a solution (λ_1^+, u_1) with $u_1 > 0$

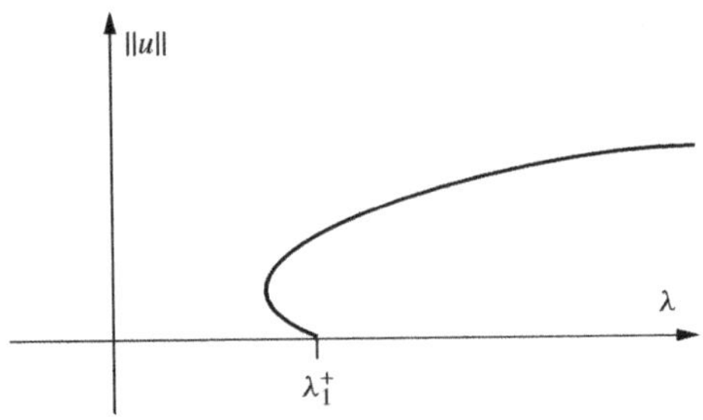

Figure 3.8 The simplest possible bifurcation diagram for (3.55) when $\lambda_s(0) < 0$. This graph is analogous to figures 3.3, 3.4C, and 3.7B.

which is part of the continuum C^+. The continuum C^+ also contains solutions with $\lambda < \lambda_1^+$. By Proposition 3.10 there is a neighborhood of $(\lambda_1^+, 0)$ in which all nontrivial solutions of (3.55) are given by $\lambda = \lambda(s)$, $u = s\phi_1 + s\rho(s)$. If we take a small enough neighborhoods U around $(\lambda_1^+, 0)$ then all solutions of (3.55) in the neighborhood U will have $u < u_1$. Thus, there must be other solutions in C^+ with $\lambda < \lambda_1^+$ which lie outside of the neighborhood U but are part of the continuum C^+ connecting $(\lambda_1^+, 0)$ to (λ_1^+, u_1). It follows that for some values of $\lambda < \lambda_1^+$ there are at least two positive solutions of (3.55). See Figure 3.8. Also, for $\lambda < \lambda_1^+$, the principal eigenvalue of

$$\begin{aligned} \nabla \cdot d(x,0)\nabla\psi + \vec{b}(x)\cdot\nabla\psi + \lambda g(x,0)\psi &= \sigma\psi \text{ in } \Omega \\ \alpha(x)\frac{\partial\psi}{\partial\vec{n}} + \beta(x)\psi &= 0 \quad \text{on } \partial\Omega \end{aligned} \tag{3.66}$$

is negative (see Theorem 2.6 and Corollary 2.18), so the solution $u = 0$ is locally stable.

Another case where it is possible to tell more about $\lambda_s(0)$ is when $\vec{b} \equiv 0$. In that case the eigenvalue problem (3.55) is formally self-adjoint, so that $\phi_1 = \phi_1^*$ and λ_1^+ has the variational formulation (2.14). From the formula (2.14) we have $\int_\Omega g(x,0)\phi_1^2 dx > 0$, so that in the case $\vec{b} \equiv 0$ (3.64) yields

$$\lambda_s(0) = \frac{\int_\Omega d_u(x,0)\phi_1|\nabla\phi_1|^2 dx - \lambda_1^+ \int_\Omega g_u(x,0)\phi_1^3 dx}{\int_\Omega g(x,0)\phi_1^2 dx} \tag{3.67}$$

with the sign of $\lambda_s(0)$ determined by the sign of the numerator. Note that λ_1^+ and ϕ_1 depend on $d(x,0)$ and $g(x,0)$ but not on $d_u(x,0)$ or $g_u(x,0)$, so the factors influencing the location of the bifurcation point $(\lambda_1^+, 0)$ are distinct from those influencing the direction of bifurcation. We will return to (3.67) and study its implications in detail in the next section, but it is clear that the sign of $\lambda_s(0)$ could be either positive or negative, depending on $d_u(x,0)$ and $g_u(x,0)$.

It turns out that when the relation $F(\lambda, u) = 0$ holds for all λ if $u = 0$ and a branch of solutions $(\lambda(s), u(s))$ with $u(s) \neq 0$ bifurcates from the line $(\lambda, 0)$ at a point $(\lambda_0, 0)$ where zero is a simple eigenvalue of the linearized equation $F_u(\lambda_0, 0)\psi = \sigma\psi$, then the stability of the bifurcating solution can often be related to the sign of $\lambda_s(0)$. There are some additional technical conditions which F must satisfy, but they are essentially the same as those needed to justify Proposition 3.10 and they are satisfied in the problems we are studying. The abstract theory relating the sign of $\lambda_s(0)$ and the stability of bifurcating solutions is due to Crandall and Rabinowitz (1973). A discussion of the theory is given in the Appendix to this chapter. In the present context, we take $F(\lambda, u)$ to be equal to the terms on the left side of (3.55), so that $\lambda_0 = \lambda_1^+$ and the linearized problem $F_u(\lambda(s), u(s))\psi = \sigma\psi$ is given by

$$\begin{aligned}
&\nabla \cdot d(x,u)\nabla\psi + \nabla \cdot d_u(x,u))\psi\nabla u + \vec{b}(x)\cdot\nabla\psi \\
&+\lambda[g(x,u) + g_u(x,u)u]\psi = \sigma\psi \qquad \text{in} \quad \Omega \qquad (3.68)\\
&\alpha(x)\frac{\partial\psi}{\partial\vec{n}} + \beta(x)\psi = 0 \qquad \text{on} \quad \partial\Omega.
\end{aligned}$$

On the line of solutions $(\lambda, 0)$, (3.68) reduces to (3.66). The principal eigenvalue of $F_u(\lambda, 0)$ depends on λ, and that of $F_u(\lambda(s), u(s))$ on s, so we can write $F_u(\lambda, 0)\psi = \sigma(\lambda)\psi$ and $F_u(\lambda(s), u(s))\psi = \mu(s)\psi$, where $\sigma(\lambda)$ and $\mu(s)$ would represent the principal eigenvalues for their respective problems. In our context $\sigma(\lambda)$ would be the principal eigenvalue for (3.66) and $\mu(s)$ the principal eigenvalue of (3.68). The results of Crandall and Rabinowitz (1973) imply that for solutions $(\lambda(s), u(s))$ sufficiently near $(\lambda_1^+, 0)$,

$$\operatorname{sign}(\mu(s)) = \operatorname{sign}(-s\lambda_s(s)\sigma_\lambda(\lambda_1^+)). \qquad (3.69)$$

(Here σ_λ denotes the derivative of σ with respect to λ.) We have arranged things so that $s > 0$ for $u(s)$ positive. The general results of Crandall and Rabinowitz (1973) imply that $\sigma_\lambda(\lambda_1^+) \neq 0$, and since $\sigma(\lambda) > 0$ for $\lambda > \lambda_1^+$ but $\sigma(\lambda) < 0$ if $\lambda < \lambda_1^+$ (by Theorem 2.6 or Corollary 2.18) we must have $\sigma_\lambda(\lambda_1^+) > 0$. (This can also be seen from direct calculation as in (3.49) in many cases.) Thus, (3.69) implies that $\mu(s) < 0$ when $\lambda_s(0) > 0$ but $\mu(s) > 0$ when $\lambda_s(0) < 0$. Since the sign of $\mu(s)$ determines the linearized stability of $u(s)$, we have:

Proposition 3.13. Suppose that the hypotheses of Proposition 3.10 are satisfied. If the branch $(\lambda(s), u(s))$ of positive solutions has $\lambda_s(0) < 0$ (i.e. the bifurcation is subcritical) then there is a neighborhood U of $(\lambda_1^+, 0)$ such that for bifurcating solutions $(\lambda(s), u(s))$ in U, $u(s)$ is unstable in the sense of linearized stability. If $\lambda_s(0) > 0$ (supercritical bifurcation) then there is a neighborhood U of $(\lambda_1^+, 0)$ such that as long as $(\lambda(s), u(s))$ remains in U, $u(s)$ is stable in the sense of linearized stability.

Remarks: Because of the order preserving properties of models consisting of a single reaction-diffusion equation, linearized stability can generally be shown to imply stability in the sense of Lyapunov by using sub- and supersolutions constructed from eigenfunctions, for example, as in (3.10).

3.4.3 Discussion and Conclusions

By combining the results of this section we can give a fairly good description of the set of equilibria for models such as (3.55) in terms of the parameter λ. We have limited our discussion to models which permit population growth at low densities, at least in some locations. (This is the meaning of the hypothesis that $g(x, 0) > 0$ on some subset of Ω.) In that situation, models such as (3.55) will have a continuum C^+ of positive solutions bifurcating at $\lambda = \lambda_1^+$ from the line of trivial solutions $(\lambda, 0)$.

If we assume that $d(x,u) \geq d_0 > 0$ (so the models are nondegenerate as parabolic or elliptic partial differential equations), $g(x,0) > 0$ on some open subset of Ω, (so that λ_1^+ exists in (3.56)) and the coefficients of (3.55) are smooth (so that the technical conditions for the existence and compactness of inverses to elliptic operators are satisfied) we can summarize the results describing $\mathcal{C}^+$ as follows:

Corollary 3.14. Suppose that $\beta(x) \neq 0$ in (3.55) (i.e., suppose that the boundary conditions are not pure Neumann conditions). Then (3.55) has a continuum $\mathcal{C}^+$ of positive solutions bifurcating from the line of trivial solutions $(\lambda, 0)$ at $\lambda = \lambda_1^+$, where λ_1^+ is the positive principal eigenvalue of (3.56). Furthermore:

(i) If there is a constant K such that $g(x,u) < 0$ for $u > K$ then $\mathcal{C}^+$ is bounded from above in u, bounded away from zero in λ, and extends to positive infinity in λ.

(ii) If $g(x,u) \leq g(x,0)$ for $u > 0$ and either $d = d(x)$, or $\vec{b} = 0$ and $d(x,u) \geq d(x,0)$ for $u > 0$ then $\mathcal{C}^+$ is contained in the set of (λ, u) with $\lambda \geq \lambda_1^+$. (In fact, (3.55) can have a positive solution only if $\lambda \geq \lambda_1^+$. If $g(x,u) < g(x,0)$ for $u > 0$ then $\lambda > \lambda_1^+$.)

(iii) If $g_u(x,u) < 0$ for all $u \geq 0$ and either $d = d(x)$, or $\vec{b}(x) = 0$ and $d_u(x,0) \geq 0$ then for each $\lambda > \lambda_1^+$ the positive solution with $(\lambda, u) \in \mathcal{C}^+$ is the unique positive solution to (3.55) and $\mathcal{C}^+$ is a smooth arc; i.e., $u(\lambda)$ depends differentiably on λ for $\lambda > \lambda_1^+$.

Remarks: Part (iii) of Corollary 3.14 implies that the bifurcation diagram in Figure 3.2 extends to logistic equations (and other models with $g_u \leq 0$) in any number of space variables; see Figure 3.9. Part (iii) is based on the continuation result Proposition 3.6. The remaining parts are based on bifurcation theory as in Propositions 3.9-3.12. (The case $\vec{b}(x) = 0$, $d_u(x,0) \geq 0$ in Part (iii) is proved in the same way as Proposition 3.6. The specific eigenvalue comparisons used to treat that case are shown in Cantrell and Cosner (1991a).)

In the case of Neumann conditions ($\beta(x) \equiv 0$), we must require $\int_\Omega g(x,0)dx < 0$ or λ_1^+ does not exist and in general there may be nonzero solutions which connect to the solution branch $(\lambda, u) = (0, c)$ (where c is any constant.) To obtain Part (i) in that case we need to assume that $g(x,u) \leq g_0(x)$ with $\int_\Omega g_0(x)dx < 0$ and that either $d = d(x)$ or $\vec{b} \equiv 0$. The remaining parts of Corollary 3.14 extend to the Neumann case. Unless we are in the

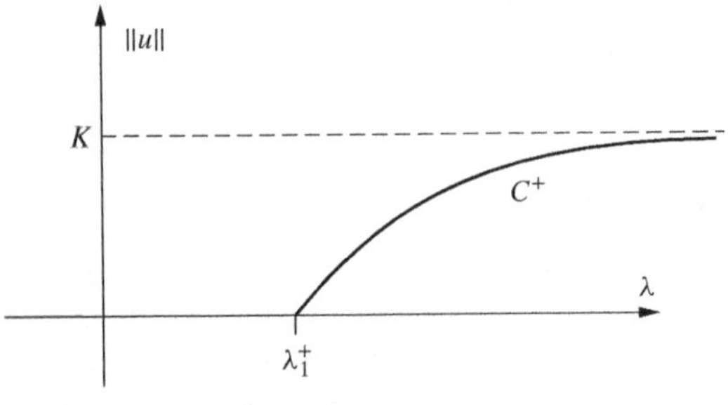

Figure 3.9 Bifurcation diagram corresponding to case 3 of Corollary 3.14. Note the similarity to Figure 3.3.

situation of Part (iii), the continuum $\mathcal{C}^+$ does not necessarily include all positive solutions of (3.55). (Such can already be seen in the case of one space dimension, as in Figure 3.7.) Under various specific hypotheses on d and/or g we can obtain more information about the behavior of $\mathcal{C}^+$ near λ_1^+ as in Proposition 3.13.

The results summarized in Corollary 3.14 do not address situations where $g(x, 0) \leq 0$ or where the population dynamical term in the model has a form other than $g(x, u)u$. The case $g(x, 0) \leq 0$ occurs, for example, in the model

$$d\nabla^2 u + \lambda u(u - a)(1 - u) = 0$$

for population growth with an Allee effect built into the population dynamics. In that case a bifurcation diagram similar to Figure 3.4C has been derived for models in more than one space dimension by Ouyang and Shi (1998).

3.5 Applications and Properties of Single Species Models

3.5.1 How Predator Incursions Affect Critical Patch Size

Empirical studies of predation on bird nests in forest fragments suggest that incursions into the fragments by generalist predators can be a significant source of mortality (Gates and Gysel, 1972; Wilcove, 1985; Angelstam, 1986; Brand and George, 2000); see Cantrell et al. (2001, 2002) for additional references. In some situations the predators mainly live in the matrix surrounding the forest and may gain only minor or incidental benefits from predation within the fragments, so that the predator population is not significantly influenced by the size or density of the prey population there. Thus, the matrix acts as a source of predators which may then penetrate to a greater or lesser extent into the fragments (Soderstrom et al. 1998).

This type of situation is analyzed via diffusion models in Cantrell et al. (2001, 2002). The starting point for the models is a simple diffusive logistic model for a single species inhabiting a bounded patch Ω in two space dimensions. If u represents the density of the species, the basic model is

$$\begin{aligned} \frac{\partial u}{\partial t} &= d\Delta u r[1 - (u/K)]u \text{ in } \quad \Omega \times (0, \infty) \\ (1 - \gamma)\frac{\partial u}{\partial x} &+ \gamma u = 0 \qquad \text{on} \quad \partial\Omega \times (0, \infty) \end{aligned} \tag{3.70}$$

so that $\gamma = 0$ corresponds to no-flux (Neumann) boundary conditions and $\gamma = 1$ corresponds to lethal (Dirichlet) boundary conditions. We assume that the predators penetrate some distance δ across the patch boundary $\partial\Omega$ into the patch Ω, with density $p(x)$. (The assumption is that the predators maintain this density by exploiting resources outside of the patch Ω.) Depending on details of predator behavior, $p(x)$ might be constant for a distance up to δ from $\partial\Omega$, or might decline linearly or exponentially with distance from $\partial\Omega$. Let $\Omega_\delta = \{x \in \Omega :$ the distance from x to $\partial\Omega$ is $< \delta\}$. For the moment we will assume only that $p(x)$ is smooth and positive on Ω_δ and that $p(x) = 0$ on $\Omega \backslash \Omega_\delta$. Suppose that the predator's functional response, that is, the rate of prey consumption by a single predator, is given by $h(u)$. Typical examples are $h(u) = Au$ (Lotka-Volterra), $h(u) = Au/(1 + Bu)$ (Holling type II), and $h(u) = Au^2/(1 + Bu^2)$ (Holling type III). (See Chapter 1 for a discussion of predator-prey models and functional responses in general.) If we account for predation in (3.70) the model becomes

$$\begin{aligned} \frac{\partial u}{\partial t} &= d\Delta u + r[1-(u/K)]u - h(u)p(x) \text{ in } \quad \Omega \times (0,\infty) \\ (1-\gamma)\frac{\partial u}{\partial \vec{n}} + \gamma u &= 0 \qquad \text{on} \quad \partial\Omega \times (0,\infty). \end{aligned} \tag{3.71}$$

The model (3.71) is a typical single-species model of the type analyzed in the earlier sections of this chapter. We can write the reaction term in (3.71) as $ug(x,u)$ where $g(x,u) = r[1-(u/K)] - (h(u)/u)p(x)$. Assuming that $h'(0)$ exists $g(x,u)$ will be well defined for $u \geq 0$ with $g(x,0) = r - h'(0)p(x)$. Thus, (3.71) will predict persistence for $\sigma_1 > 0$ but will predict that the extinction equilibrium $u \equiv 0$ is locally stable (so that (3.71) predicts extinction for small populations) if $\sigma_1 < 0$, where σ_1 is the principal eigenvalue for

$$\begin{aligned} d\Delta\psi + [r - h'(0)p(x)]\psi &= \sigma\psi \text{ in } \quad \Omega, \\ (1-\gamma)\frac{\partial \psi}{\partial \vec{n}} + \gamma\psi &= 0 \qquad \text{on} \quad \partial\Omega. \end{aligned} \tag{3.72}$$

In the Lotka-Volterra case or in the Holling type II case under the additional condition $(r/K) > ABp(x)$ we have $g(x,u) \leq g(x,0)$ for $u > 0$ so that $\sigma_1 < 0$ implies extinction for any population.

In the original model (3.70) without predation we have $p(x) = 0$, so in that case $\sigma_1 = -d\lambda_\gamma + r$ where λ_γ is the principal eigenvalue of

$$\begin{aligned} \Delta\phi + \lambda\phi &= 0 \qquad \text{in} \quad \Omega \\ (1-\gamma)\frac{\partial \phi}{\partial \vec{n}} + \gamma\phi &= 0 \text{ on } \quad \partial\Omega. \end{aligned} \tag{3.73}$$

As discussed in Chapter 2, λ_γ depends on the size of Ω. In the case $\gamma = 1$ we have $\lambda_1 = \lambda^*/|\Omega|$ if λ^* is the principal eigenvalue for the equation (3.73) on a domain with the same shape as Ω but with unit area. Thus, the requirement $\sigma_1 > 0$ typically imposes a minimum size on the patch Ω if it is to sustain a population. We now come to our first observation about the role of the functional response $h(u)$ on the model (3.71): If the functional response $h(u)$ has $h'(0) = 0$ then $\sigma_1 = -d\lambda_1 + r$ for any value of $p(x)$, so that predation has no effect on the minimal size of a patch needed to sustain the prey species, nor on the model's predictions regarding persistence versus extinction. This is not to say that predation has no effect in this case. Suppose that the patch Ω is small enough that the predator density is a constant p throughout Ω. The analysis in Section 3.2 shows that if $r(1-u/K) - p(h(u)/u) < 0$ for $u > k$ then all equilibria for (3.71) are bounded above by k. If $p(x) = 0$ then we have $k = K$; if $p > 0$ and $(h(u)/u) > 0$ then $k < K$. Depending on how $h(u)$ behaves for $u > 0$ and how large p is, k in principle could be arbitrarily small. The point is that even though predation would not affect minimal patch size or predictions of invasibility and hence persistence in a given patch in the case $h'(0) = 0$, it could still effectively reduce the carrying capacity and thus make stochastic extinction more likely in practice. Among standard functional responses, the only one with $h'(0) = 0$ is the Holling type III.

In the case where $h'(0) > 0$, equation (3.72) and the comparison principles for eigenvalues discussed in Chapter 2 imply that $\sigma_1 < -d\lambda_\gamma + r$, so that the presence of predators reduces the effective population growth rate in that case and could shift the prediction of (3.71) from persistence to extinction. We have $h'(0) > 0$ for Lotka-Volterra and Holling type II functional responses. The importance of the distinction between the Holling type III response and the Holling type II or Lotka-Volterra response can be understood by moving to a larger spatial scale and thinking of our patch as one of many patches in a metapopulation. Metapopulations models assume that local populations on patches may become extinct but then empty patches can be recolonized. The effects of incursions by predators with Holling II or Lotka-Volterra functional response can include making a patch noninvasible ($\sigma_1 < 0$) even if it could sustain a prey population in the absence of predators. That would prevent colonization of the patch and would thereby effectively remove it from the metapopulation. The loss of one or more patches might be significant for the persistence of the metapopulation, especially if removing those patches had the effect of disconnecting parts of the metapopulation from each other by eliminating the pathways for colonizations from one part to another. On the other hand, a predator with a Holling type III functional response might reduce local populations and thus make stochastic local extinction more likely, but would not prevent recolonization and thus would not remove patches from the metapopulation.

In cases where $h'(0) > 0$ it is natural to ask how σ_1 in (3.72) depends on $h'(0)$, $p(x)$, and δ. By the variational formulation for σ_1 given in Theorem 2.1, we have

$$\sigma_1 = \sup_{\substack{\psi \in W^{1,2}(\Omega) \\ \psi \neq 0}} \left[\frac{-d\int_\Omega |\nabla\psi|^2 dx + \int_\Omega [r - (h'(0)p(x)]\psi^2 dx - d[\gamma/(1-\gamma)]\int_{\partial\Omega} \psi^2 dS}{\int_\Omega \psi^2 dx} \right] \tag{3.74}$$

If $\gamma = 1$, the boundary integral no longer appears in the formula and $W^{1,2}(\Omega)$ is replaced by $W_0^{1,2}(\Omega)$, so that we have $\psi = 0$ on $\partial\Omega$. In any case we may construct a test function from the eigenfunction ψ_δ corresponding to the principal eigenvalue σ_δ of

$$\begin{aligned} d\Delta\psi + r\psi &= \sigma_\delta\psi \text{ in } && \Omega\backslash\Omega_\delta \\ \psi &= 0 && \text{on } \partial(\Omega\backslash\Omega_\delta). \end{aligned} \tag{3.75}$$

(Recall that Ω_δ is the strip near $\partial\Omega$ where $p(x) > 0$ and $\Omega\backslash\Omega_\delta$ is the core area of Ω where $p(x) = 0$.) Let ψ be the test function obtained by extending ψ_δ to be zero outside $\Omega\backslash\Omega_\delta$. It then follows from (3.74), (3.75), and the fact that $\psi = 0$ on Ω_δ that

$$\sigma_1 > \frac{-d\int_{\Omega\backslash\Omega_\delta} |\nabla\psi_\delta|^2 dx + r\int_{\Omega\backslash\Omega_\delta} \psi_\delta^2 dx}{\int_{\Omega\backslash\Omega_\delta} \psi_\delta^2 dx} = \sigma_\delta. \tag{3.76}$$

It follows from (3.76) that if $\sigma_\delta > 0$ then $\sigma_1 > 0$ so that (3.71) predicts persistence. In other words, whenever the prey species could persist in the part of the patch where predators do not penetrate (i.e. $\Omega\backslash\Omega_\delta$) if it had a lethal boundary, then the prey can persist in the full patch Ω, independent of $h'(0)$ or $p(x)$.

The fact that σ_1 is characterized as the supremum of the expression on the right in (3.74) makes it fairly easy to obtain lower bounds on σ_1 by substituting test functions into (3.74).

Using $\psi \equiv 1$ yields the estimate

$$\sigma_1 \geq r - h'(0)\left(\frac{1}{|\Omega|}\int_\Omega p(x)dx\right) - [d\gamma/(1-\gamma)]\frac{|\partial\Omega|}{|\Omega|}. \tag{3.77}$$

The term $(1/|\Omega|)\int p(x)dx$ is just the average of $p(x)$. Since $p(x) = 0$ outside of Ω_δ, we have $(1/|\Omega|)\int p(x)dx \leq p_0|\Omega_\delta|/|\Omega|$, where $p_0 = \sup p(x)$, so that (3.77) becomes

$$\sigma_1 \geq r - h'(0)p_0(|\Omega_\delta|/|\Omega|) - [d\gamma/(1-\gamma)]|\partial\Omega|/|\Omega|. \tag{3.78}$$

The term $|\partial\Omega|/|\Omega|$ is the perimeter-to-area ratio of Ω; the term $|\Omega_\delta|/|\Omega|$ is the fraction of Ω which is subject to predator incursions. Thus, although the estimate (3.78) may be somewhat rough, it gives an explicit bound on σ_1 which depends only on simple aspects of the geometry of Ω, the size of δ, and the parameters of (3.72). In general it is much more difficult to find good upper bounds on σ_1. In the case of one space dimension, it is sometimes possible to get more information about σ_1 by using the methods of section 2.4. This type of analysis is done in Cantrell et al. (2002). Additional discussion and references are given in Cantrell et al. (2001, 2002).

3.5.2 Diffusion and Allee Effects

In some situations the local growth rate of a population may increase with population density at low densities. This phenomenon is known as an Allee effect (Allee, 1931). Allee effects can occur for various reasons. At very low densities individuals may have trouble finding mates. Some species engage in group defense, cooperative hunting, or other beneficial social behaviors which are not possible if the population density is too low. Some authors distinguish between weak Allee effects, where the local growth rate is always positive but increases at low densities, and strong Allee effects, where the local growth rate is actually negative at low densities. If an Allee effect occurs in the local population growth rate it can be described in terms of models such as (3.55) by taking $g(x, u)$ to have $g_u(x, 0) = \partial g/\partial u|_{u=0} > 0$ and, for strong Allee effects, $g(x, 0) < 0$. It turns out that diffusion can amplify Allee effects and in some cases can create Allee effects at the level of the overall population even when such effects are not present locally.

To understand the way that Allee effects can be created or modified by diffusion, it is useful to think of the strong Allee effect in terms of the stability of equilibria. The statement "small populations will decrease toward zero but larger populations will persist" is a qualitative description of a strong Allee effect. Translated into mathematical terminology, the statement becomes "the equilibrium where the population is zero is locally stable but there is a threshold such that populations starting above the threshold are bounded away from zero." For simple nonspatial models of the form $du/dt = g(u)u$, this is equivalent to having $g(0) < 0$ but $g(a) = 0$ for some $a > 0$, so that if $u(0) > a$ then $u(t) > a$ for all t. In the context of spatial models the local stability of the zero solution will depend on various factors which are averaged by the principal eigenvalue of the linearization of the model at zero, and showing existence or nonexistence of a positive equilibrium is more difficult, but the same general principles apply.

Our analysis follows from relations (3.64) and (3.65) describing the way that a nonzero equilibrium bifurcates from the zero solution of a model such as (3.55), together with the relation (3.69) which describes the stability or instability of the bifurcating solution. The models we shall consider have the form (at equilibrium)

$$\begin{aligned} \nabla \cdot d(u)\nabla u + \lambda g(u)u &= 0 \text{ in } \quad \Omega \\ u &= 0 \qquad\qquad\quad \text{on} \quad \partial\Omega. \end{aligned} \tag{3.79}$$

(A similar analysis would apply with any boundary conditions with some loss of individuals across the boundary, i.e. any boundary conditions except no-flux conditions. We could also allow d and g to depend on x.) If $g \leq 0$ for all u then (3.79) cannot have any positive solutions, and if $g(u) > 0$ for some $u > 0$ but $g(0) < 0$ then the nonspatial model $du/dt = g(u)u$ will already have a strong Allee effect. Solutions to $du/dt = g(u)u$ with $u > 0$ will be supersolutions to the time-dependent problem corresponding to (3.79), so if solutions with small initial data tend toward zero in the nonspatial model then they will do so in the spatial model as well. Thus, we shall assume that $g(0) > 0$. This includes the case of logistic growth, among others. Recall that rescaling Ω by a factor of ℓ in its linear dimensions modifies (3.79) by dividing d by ℓ^2, or alternatively multiplying λ by ℓ^2, so that we can think of λ as a scale factor describing the size of Ω. By absorbing the appropriate constant factor into $g(u)$ we can arrange the scaling so that if $\Omega = \ell\Omega_1$ with $|\Omega_1| = 1$ (that is, $\Omega = \{\ell x : x \in \Omega_1\}$ where $|\Omega_1| = 1$) then $\lambda = \ell^2$.

As shown in the preceding section, (3.79) will have a continuum of solutions $(\lambda(s), u(s))$ bifurcating from the line of solutions $(\lambda, 0)$ at $\lambda = \lambda_1^+$ where λ_1^+ is the principal eigenvalue of the problem linearized at $u = 0$, that is,

$$\begin{aligned} \nabla \cdot d(0)\nabla\phi + \lambda g(0)\phi &= 0 \text{ in } \quad \Omega \\ \phi &= 0 \qquad\qquad\quad \text{on} \quad \partial\Omega. \end{aligned} \tag{3.80}$$

The solution $u = 0$ to (3.79) is locally stable if $\sigma_1 < 0$ and locally unstable if $\sigma_1 > 0$, where σ_1 is the principal eigenvalue of

$$\begin{aligned} \nabla \cdot d(0)\nabla\psi + \lambda g(0)\psi &= \sigma\psi \text{ in } \quad \Omega \\ \psi &= 0 \qquad\qquad\quad \text{on} \quad \partial\Omega. \end{aligned} \tag{3.81}$$

We have $\sigma_1 < 0$ if and only if $\lambda < \lambda_1^+$ (Theorem 2.6). We can determine the direction relative to λ_1^+ taken by $\lambda(s)$ for the bifurcating branch of solutions $(\lambda(s), u(s))$ using formula (3.64) or (3.65) to determine the sign of $d\lambda/ds|_{s=0}$. If $d\lambda/ds|_{s=0} > 0$ then the diagram of the bifurcating solution branch will look like Figure 3.9, at least locally near $(\lambda_1^+, 0)$. If $d\lambda/ds|_{s=0} < 0$ then the diagram will look like Figure 3.8. In the latter case the model (3.79) will have solutions $(\lambda(s), u(s))$ with $u(s) > 0$ but $\lambda(s) < \lambda_1^+$. However, if $\lambda < \lambda_1^+$, then $\sigma_1 < 0$ in (3.81) so the solution $u \equiv 0$ is locally stable. Thus, if $d\lambda/ds|_{s=0} < 0$ then for λ slightly less than λ_1^+ the solution $u = 0$ to (3.79) is locally stable in the corresponding time-dependent model, but there is a positive equilibrium $\underline{u}$ (i.e., a positive solution to (3.79)). Comparison principles based on the maximum principle imply that any solution to the time dependent model that is initially larger than $\underline{u}$ will remain larger than $\underline{u}$. Thus, a time dependent model corresponding to (3.79) with $d\lambda/ds|_{s=0} < 0$ will have a strong Allee effect for some values of λ, independent of any other properties of $d(u)$ and $g(u)$. Furthermore, if we use the scaling $\ell^2 = \lambda$, there will be two values of ℓ related to minimum patch size. The smallest value of ℓ for which the patch will be

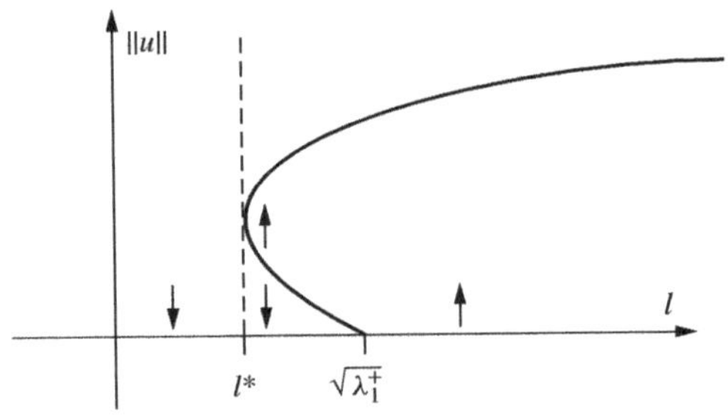

Figure 3.10 For $l > \sqrt{\lambda_1^+}$ invasibility implies persistence. For $l^* < l < \sqrt{\lambda_1^+}$ there is still conditional persistence but invasibility fails, so there is an Allee effect. For $l < l^*$ no population can persist.

invasible (i.e. $\sigma_1 > 0$) is still $\ell = \sqrt{\lambda_1^+}$. However, if the minimum value for ℓ such that a sufficiently large population will persist is given by ℓ^*, then $\ell^* < \sqrt{\lambda_1^+}$. (See Figure 3.10.)

We now consider the implications of the formulas (3.64) and (3.65). If $d(u)$ is actually constant in u we may use (3.65). Since (3.80) is formally self-adjoint, the eigenfunction ϕ^* corresponding to the principal eigenvalue of the adjoint problem to (3.80) is the same as ϕ. Thus, in that case, (3.65) yields

$$\frac{d\lambda}{ds}|_{s=0} = \lambda_s(0) = -\lambda_1^+(g_u(0)/g(0))\left[\int_\Omega \phi_1^3 dx \Big/ \int_\Omega \phi_1^2 dx\right]. \tag{3.82}$$

We can see from (3.82) that $d\lambda/ds|_{s=0} < 0$ if $g_u(0) > 0$. In that case, the time-dependent model corresponding to (3.79) will have a strong Allee effect for $\ell^* < \ell < \sqrt{\lambda_1^+}$. The case $g_u(0) > 0$ corresponds to a weak Allee effect. Thus we have our first biological conclusion: passive diffusion can convert a weak Allee effect at the local level to a strong Allee effect on the population level for some patch sizes near the critical size for invasibility.

Suppose now that $d = d(u)$. In that case we must start with (3.64), which yields

$$\frac{d\lambda}{ds}|_{s=0} = \lambda_s(0) = \frac{d_u(0)\int_\Omega \phi_1|\nabla\phi_1|^2 dx - \lambda_1^+ g_u(0)\int_\Omega \phi_1^3 dx}{g(0)\int_\Omega \phi_1^2 dx}. \tag{3.83}$$

Notice that the value of λ_1^+ depends on $d(0)$ and $g(0)$ but not on $d_u(0)$ or $g_u(0)$, so that any values of $d_u(0)$ and $g_u(0)$ could occur in (3.83) with any particular value of λ_1^+, depending on $d(u)$ and $g(u)$. If $d_u(0)$ is sufficiently negative, we could have $d\lambda/ds|_{s=0} < 0$ even if $g_u(0) < 0$. The biological interpretation of having $d_u(0) < 0$ is that at low densities individuals reduce their movement rate in response to the presence of conspecifics. That in turn leads to a tendency for the population to aggregate; see Turchin (1989). We have now arrived at our second biological conclusion: density dependent diffusion of a type that reflects a tendency to aggregate can create a strong Allee effect even in cases where there is no Allee effect at all in the corresponding nonspatial model, e.g. in the case of logistic growth. This phenomenon, where the population dynamics of the overall population viewed on a large scale are different from those of local populations viewed on a small scale, occurs in various spatial models. It has been given the name "scale transition" by Chesson (1997).

It is of some interest to focus on the logistic equation. Suppose that $g(u) = r(1 - u/K)$, so $g(0) = r$ and $g_u(0) = -r/K$. If we put $\phi = \phi_1, \lambda = \lambda_1^+$ in (3.80), multiply (3.80) by ϕ_1^2, and integrate via the divergence theorem we obtain

$$\begin{aligned} 0 &= \int_\Omega d(0)\phi_1^2 \nabla^2 \phi_1 dx + \lambda_1^+ \int_\Omega g(0)\phi_1^3 dx \\ &= -2d(0)\int_\Omega \phi_1 |\nabla \phi_1|^2 dx + \lambda_1^+ g(0) \int_\Omega \phi_1^3 dx \end{aligned}$$

so that

$$\int_\Omega \phi_1 |\nabla \phi_1|^2 dx = \frac{\lambda_1^+ g(0)}{2d(0)} \int_\Omega \phi_1^3 dx. \tag{3.84}$$

Substituting (3.84) into (3.83), using $g(0) = r$ and $g_u(0) = -r/K$, and doing some algebra yields

$$\frac{d\lambda}{ds}|_{s=0} = \frac{\lambda_1^+}{2}\left[\frac{d_u(0)}{d(0)} + \frac{2}{K}\right]\left(\int_\Omega \phi_1^3 dx \Big/ \int_\Omega \phi_1^2 dx\right). \tag{3.85}$$

Thus, we get $d\lambda/ds|_{s=0} < 0$ if $d_u(0)/d(0) < (-2/K)$. This last relation implies that the situations most likely to support an Allee effect induced by aggregative density dependent diffusion are those where the tendency to aggregate increases rapidly with density at low densities (that is, $d_u(0)/d(0)$ is negative but large in magnitude) and the carrying capacity K is large. Again, a strong Allee effect induced by diffusion will typically occur on patches that are slightly smaller than the minimal size for invasibility; see Figure 3.10. Aggregation can induce a strong Allee effect in a population with a logistic local growth rate in this scenario because at moderate densities the diffusion rate $d(u)$ will be reduced, so that the population will tend to aggregate where the density is not too small, and those regions will be in the interior of Ω away from the lethal boundary. At low densities $d(u) \approx d(0)$, so if there are not enough individuals present in Ω to reduce $d(u)$ significantly there will be less of a tendency to aggregate inside Ω (because there are not enough conspecifics around to produce the cues for aggregation) and thus there will be less protection from the lethal boundary.

The fact that density-dependent dispersal can induce a strong Allee effect in a population with a logistic local growth rate is interesting from the theoretical viewpoint because it shows how changing spatial scales from local to global can change population dynamics. There is also a more practical lesson about managed populations contained in Figures 3.8 and 3.10, as opposed to Figure 3.9. Recall that Figure 3.9 describes the dependence on patch size of a population with logistic growth and passive diffusion. In that case, as the size of the patch shrinks the equilibrium population declines gradually toward zero. In the case of an aggregating population, the dependence of the possible equilibria on the size of the patch is described by Figures 3.8 and 3.10. In that situation, as patch size decreases there may remain a fairly large population until the size of the patch shrinks below the minimal size needed to support a population (ℓ^* in Figure 3.10), at which point the population will collapse to zero. Furthermore, at that patch size, $u = 0$ is stable, so the population cannot be replaced via colonization at low densities. The lesson for population management is that in aggregating populations a decrease in habitat area may endanger the population even though the size or density of the population remains large, so that there is no warning that the population is in trouble.

3.5.3 Properties of Equilibria

We have seen that many single species models have a continuum of equilibria depending on a parameter λ that can be viewed as measuring the spatial scale of the environment. It is sometimes biologically interesting and/or mathematically useful to examine how the equilibria behave as λ increases toward infinity or decreases toward the minimal value that will support a positive equilibrium. We will begin with an estimate for the total population at equilibrium for the diffusive logistic model

$$\begin{aligned} \nabla \cdot d(x)\nabla u + \lambda[m(x)u - cu^2] &= 0 \text{ in } \Omega \\ d(x)\frac{\partial u}{\partial \vec{n}} + \beta(x)u &= 0 \quad \text{on } \partial\Omega \end{aligned} \tag{3.86}$$

where in general $m(x)$ may change sign in Ω, but where $m(x) > 0$ at least on some subset of Ω with positive measure. The model has the general form given in (3.55), with $g(x, u) = m(x) - cu$, so the results of Section 3.4 imply that (3.86) will have a positive solution if and only if $\lambda > \lambda_1^+$, that solution will be unique, and that solution will approach zero as $\lambda \searrow \lambda_1^+$. By multiplying (3.86) by u and integrating via the divergence theorem (in the classical case where the coefficients are smooth enough that all solutions are C^2) or by using the definition of a weak solution (in the case where $m(x)$ is merely bounded and measurable, for example) we obtain

$$-\int_\Omega d(x)|\nabla u|^2 dx - \int_{\partial\Omega} \beta(x)u^2 dS + \lambda \int_\Omega m(x)u^2 dx - c\int_\Omega u^3 dx = 0. \tag{3.87}$$

If we solve for $\int_\Omega mu^2 dx$ in (3.87) we can see that quantity is positive. By the variational formula (2.14) we have

$$\frac{\int_\Omega mu^2 dx}{\int_\Omega d(x)|\nabla u|^2 dx + \int_{\partial\Omega} \beta(x)u^2 dS} \leq \frac{1}{\lambda_1^+}$$

so that

$$\lambda_1^+ \int mu^2 dx \leq \int_\Omega d(x)|\nabla u|^2 dx + \int_{\partial\Omega} \beta(x)u^2 dS$$

and thus (3.87) yields

$$(\lambda - \lambda_1^+) \int_\Omega m(x)u^2 dx \geq c\int_\Omega u^3 dx. \tag{3.88}$$

If we let $m_+(x)$ be the positive part of $m(x)$, i.e. $m_+(x) = m(x)$ if $m(x) > 0$ and $m_+(x) = 0$ if $m(x) \leq 0$, then $\int_\Omega mu^2 dx \leq \int_\Omega m_+ u^2$, and by Hölder's inequality we have

$$\int_\Omega m_+ u^2 dx \leq \left[\int_\Omega |m_+|^3 dx\right]^{1/3} \left[\int_\Omega (u^2)^{3/2} dx\right]^{2/3} = ||m_+||_3 ||u||_3^2 \tag{3.89}$$

where $||\cdot||_3$ denotes the norm of $L^3(\Omega)$. Using the estimate (3.89) in (3.88) yields the inequality $(\lambda-\lambda_1^+)||m_+||_3||u||_3^2 \geq c||u||_3^3$ so $||u||_3 \leq (\lambda-\lambda_1^+)||m_+||_3/c$. Hölder's inequality also implies $\int_\Omega u dx \leq \left(\int_\Omega u^3 dx\right)^{1/3} \left(\int_\Omega 1^{3/2} dx\right)^{2/3} = ||u||_3|\Omega|^{2/3}$ so we have

$$\int_\Omega u dx \leq (\lambda - \lambda_1^+)||m_+||_3|\Omega|^{2/3}/c. \tag{3.90}$$

Since u represents the population density on Ω, (3.90) gives a bound on the total population.

Note that if $m(x) = r$ and $c = r/K$ (the usual parameterization for a logistic equation with spatially constant coefficients) we have $||m_+||_3 = ||r||_3 = r|\Omega|^{1/3}$ so (3.90) becomes

$$\int_\Omega u dx \leq (\lambda - \lambda_1^+)|\Omega|K. \tag{3.91}$$

If a population had density equal to its carrying capacity K over all of Ω, the total population would be $K|\Omega|$, so (3.91) shows how the loss of population across the boundary of Ω reduces the total population for λ near λ_1^+. A similar type of analysis should be possible for other models where the local growth rate $g(x, u)$ is decreasing in u, but the details would depend on g. An estimate similar to (3.90) is derived in Cantrell and Cosner (1989), but the notation in that paper is different from that used here.

The approach leading to (3.90) provides some information about how equilibria behave as λ approaches the bifurcation point λ_1^+, i.e., as patch size decreases toward the minimal patch size that will support a population. It is also of interest to examine what happens as $\lambda \to \infty$, which in the equilibrium situation is equivalent to having the diffusion rate approach zero. Specifically, the model

$$\begin{aligned} d\Delta u + \lambda u g(u) &= 0 \text{ in } \Omega \\ u &= 0 \text{ on } \partial\Omega, \end{aligned} \tag{3.92}$$

may be rewritten in the form

$$\begin{aligned} -\epsilon^2 \Delta u &= u g(u) \text{ in } \Omega \\ u &= 0 \text{ on } \partial\Omega \end{aligned} \tag{3.93}$$

where $\epsilon^2 = d/\lambda$. The analysis of models such as (3.93) as $\epsilon \to 0$ is part of the branch of applied analysis known as singular perturbation theory. See Chang and Howes (1984) and Kevorkian and Cole (1996).

From the present discussion assume $g(u)$ is twice continuously differentiable, $g(u) > 0$ for $0 < u < K$, where K is a positive constant, but $g(K) = 0$, $g(u) < 0$ for $u > K$, and $g'(u) < 0$. Note that (3.92) will then have a unique positive solution for $\lambda > \lambda_1^+$ by the results of Section 3.4. In this case $\lambda_1^+ = \lambda_0 d/g(0)$ where λ_0 is the principal eigenvalue of $-\Delta\phi = \lambda_0\phi$ in Ω, $\phi = 0$ on $\partial\Omega$, so (3.92) will have a unique positive solution if $\lambda > \lambda_0 d/g(0)$, which will be true if $\epsilon^2 = d/\lambda < g(0)/\lambda_0$. Thus, it makes sense to ask how the solution to (3.93) behaves as $\epsilon \to 0$. Note that $\epsilon \to 0$ if either $d \to 0$ or $\lambda \to \infty$. Under the hypotheses listed above we have the following:

Proposition 3.15. Suppose that Ω is bounded and $\partial\Omega$ is of class C^1 in the sense that $\partial\Omega = \{x : F(x) = 0\}$ for some continuously differentiable function $F(x)$ with $\nabla F \neq 0$ on

$\partial\Omega$. Under the hypotheses stated above, the unique positive solution $u_\epsilon(x)$ of (3.93) has the property that $u_\epsilon(x) \to K$ uniformly on every closed subset of Ω as $\epsilon \to 0$.

The proof of Proposition 3.15 is given in the Appendix to this chapter. Notice that the logistic model has $g(u) = r[1 - (u/K)]$ and hence satisfies the hypotheses of Proposition 3.15, as do various other models. The biological interpretation is that if the diffusion coefficient (i.e. dispersal rate) d for the population is sufficiently small, or the spatial scale of Ω as measured by λ is sufficiently large, then the population density in Ω will be close to its carrying capacity K on Ω except for a relatively narrow strip near the boundary of Ω. (In the terminology of fluid mechanics or classical asymptotic analysis and singular perturbation theory, the solution to (3.93) is said to have a boundary layer.) Proposition 3.15 can be extended to general boundary conditions $\alpha(x)\partial u/\partial\vec{n} + \beta(x)u = 0$ on $\partial\Omega$ by noting that the solution u_ϵ to (3.92) with $u_\epsilon = 0$ on the boundary is a subsolution to (3.92) with boundary conditions changed to $\alpha(x)\partial u/\partial\vec{n} + \beta(x)u = 0$, and $u \equiv K$ is a supersolution to (3.92) with those boundary conditions, so that (3.92) with modified boundary conditions has a solution $\tilde{u}$ lying between u_ϵ and K. Since the solution to (3.92) with such boundary conditions is unique, $\tilde{u}$ is the only positive solution, so because $u_\epsilon < \tilde{u} < K$ and $u_\epsilon \to K$ as $\epsilon \to 0$, we have $\tilde{u} \to K$ as $\epsilon \to 0$. As is the case for u_ϵ, the convergence is uniform on any closed subset of Ω.

In the case where g depends on x the behavior of u as $\epsilon^2 = d/\lambda \to 0$ is similar but somewhat more complicated. Consider the equation

$$\begin{aligned} -\epsilon^2\Delta u &= g(x,u)u \quad \text{in} \quad \Omega \\ \alpha(x)\frac{\partial u}{\partial\vec{n}} + \beta(x)u &= 0 \text{ on } \quad \partial\Omega. \end{aligned} \tag{3.94}$$

Suppose that $g(x,u)$ is decreasing in u and that $g(x,0) > 0$ on some open subset of Ω. Suppose further that there is a function $K(x)$ defined on the subset of Ω where $g(x,0) > 0$ such that $g(x, K(x)) = 0$, and assume that $K(x) \le K_0$ for some positive constant K_0. Then (3.94) will have a unique solution for ε sufficiently small.

Proposition 3.16. Suppose that E_1 and E_2 are open subsets of Ω with $g(x,0) \ge g_1 > 0$ on $\overline{E}_1$ and $g(x,0) \le -g_2 < 0$ on $\overline{E}_2$, and that $g(x,u)$ is of class C^2 in x and u for $x \in E_1 \cup E_2$. Assume that on $\overline{E}_1$ $g(x,u)$ has the property that the functions $\min_{x\in S} g(x,u)$ and $\max_{x\in S} g(x,u)$ are C^2 and decreasing in u for any closed subset $S \subseteq \overline{E}_1$. If u is the positive solution to (3.94) then as $\epsilon \to 0$, $u \to 0$ uniformly on any closed subset of E_2 and $u \to K(x)$ uniformly on any closed subset of E_1.

The proof is given in the Appendix to this chapter. The point of the proposition is that when the diffusion coefficient becomes small the solution to (3.94) approaches zero in those regions where the local growth rate at low densities is negative and approaches the local carrying capacity $K(x)$ in those regions where the local growth rate at low densities is positive. A typical model satisfying the hypotheses of Proposition 3.16 would be the logistic model given by $g(x,u) = m(x) - c(x)u$ with $c(x) \ge c_0 > 0$ and $m(x)$ piecewise smooth. If $m(x) = 0$ only on a set of measure zero (e.g. on a one-dimensional curve in a two-dimensional domain Ω) then $K(x) = m(x)/c(x)$ for x with $m(x) > 0$, and $u \to m_+(x)/c(x)$ almost everywhere in Ω as $\epsilon \to 0$. (Here $m_+(x)$ denotes the positive part of $m(x)$.) The hypotheses of Proposition 3.16 are also satisfied by $g(x,u) = g_1\chi_{\Omega_1}(x) - g_2\chi_{\Omega\setminus\Omega_1}(x) - cu$ if Ω_1 is an open subset of Ω with smooth boundary. ($\chi_S(x)$ denotes the characteristic function of the set S.) In that case, we have $u \to g_1$ uniformly on any closed subset of Ω_1 but $u \to 0$ on any closed subset of $\Omega\backslash\overline{\Omega}_1$ as $\epsilon^2 = d/\lambda \to 0$. So in that case the positive solution to (3.94) will develop a sharp internal transition at $(\partial\Omega_1) \cap \Omega$. There are other classes of models where Proposition 3.16 can be applied; for example, if $g(x, u) = [a(x)/(1+b(x)u)] - d(x)$ where a, b, and d are positive. (This form of model would occur if we considered a population of predators that engaged in mutual interference when feeding, as described by a Beddington-DeAngelis functional response, in a situation where the supply of prey was maintained at a fixed level.)

In general, when d/λ is too large models such as (3.92) will not have any positive equilibria. However, in the case of no-flux boundary conditions there may be a positive equilibrium for all values of d/λ, so it makes sense in such cases to consider what happens as d/λ approaches infinity. An example of that type of situation is considered in Cantrell et al. (1996). The analysis of Section 3.1 implies that the model

$$\begin{aligned} u_t &= d\Delta u + \lambda[m(x)u - cu^2] \text{ in } && \Omega \times (0, \infty) \\ \frac{\partial u}{\partial \vec{n}} &= 0 && \text{on } \partial\Omega \times (0, \infty) \end{aligned} \tag{3.95}$$

will have a unique positive equilibrium for all positive values of d and λ provided $\int_\Omega m(x)dx > 0$, because in that case the principal eigenvalue σ_1 of

$$\begin{aligned} d\Delta\psi + \lambda m(x)\psi &= \sigma\psi \text{ in } && \Omega \\ \frac{\partial \psi}{\partial \vec{n}} &= 0 && \text{on } \partial\Omega \end{aligned}$$

will be positive for all d and λ positive (see Theorem 2.6). We have the following result (Cantrell et al. 1996, Theorem A.2):

Proposition 3.17. Assume that $\int_\Omega m(x)dx > 0$. Let u^* be the unique positive equilibrium for (3.95). As $d/\lambda \to \infty$, $u^* \to \int_\Omega m(x)dx/(c|\Omega|)$ uniformly on $\overline{\Omega}$.

Remarks: This result essentially says that as $d/\lambda \to \infty$, u^* converges to the equilibrium of the nonspatial model $du/dt = \overline{m}u - cu^2$ where $\overline{m} = (1/|\Omega|)\int_\Omega m(x)dx$ is the spatial average of $m(x)$. That is not surprising, because if d is large, individuals move rapidly and randomly in Ω so that each individual effectively experiences the average of $m(x)$ over Ω by moving around Ω.

3.6 More General Single Species Models

The key feature of reaction-diffusion models that allows us to use the analytic approach of Section 3.1 is that a single reaction-diffusion equation without time dependence in its coefficients will typically generate a semi-dynamical system which is order preserving. This property is shared by a number of other types of spatial models, including models that are discrete in space or time or models with periodic time dependence in their coefficients. In this section we will sketch how the analysis of Section 3.1 could be extended to other types of models and we will state a result on time-periodic reaction-diffusion models that will be needed later. The types of models we will discuss in this section will include models based on systems of ordinary differential equations such as island chain models (in which the Laplace operator is replaced by a spatially discrete "diffusion" operator) and the spatially structured metapopulation models of Hanski and his coworkers (Hanski and Ovaskainen, 2000, 2001). We will also discuss discrete time matrix models and models based on integral kernels (Van Kirk and Lewis, 1997, 1999). To give a unified treatment of several types of models we must take a somewhat abstract approach, which involves order-preserving dynamical systems.

To discuss order properties of dynamical systems we need the notion of a positive cone as discussed in Section 2.5. Suppose that X is a Banach space with positive cone P. We say $u > v$ if $u - v \in P$ and $u - v \neq 0$. Let $\pi(u_0, t)$ denote the dynamical or semi-dynamical system corresponding to a model of the form

$$\frac{du}{dt} = f(u), \quad u(0) = u_0. \tag{3.96}$$

The dynamical or semi-dynamical system is said to be strictly monotone if $u > v$ implies $\pi(u, t) > \pi(v, t)$ for all $t > 0$. (See Smith (1995).) Comparison principles based on the maximum principle typically can be used to show that the semi-dynamical system generated by a single reaction-diffusion equation is strictly monotone if it is set in the appropriate Banach space of functions; see the discussion in Chapter 1. For systems of ordinary differential equations we may write (3.96) as $du_i/dt = f_i(u_1, \ldots, u_n)$, $i = 1, \ldots, n$. In that case the corresponding dynamical system will be strictly monotone relative to the standard ordering generated by the cone $P = \{\vec{u} : u_i \geq 0, i = 1, \ldots, n\}$ if f satisfies the Kamke condition: for $i = 1, \ldots, n$, $f_i(u) \geq f_i(v)$ if $u \geq v$ with $u_i = v_i$. See Smith (1995, Ch. 3). In cases where f is smooth this follows from the condition $\partial f_i / \partial u_j \geq 0$ for $i \neq j$. Systems satisfying the Kamke condition are often called cooperative in the mathematical literature. If each component u_i represents the density of a population or subpopulation then the condition $\partial f_i / \partial u_j \geq 0$ for $i \neq j$ reflects some type of mutualism. It turns out that systems of two reaction-diffusion equations or ordinary differential equations which describe two competing populations will often generate strictly monotone semi-dynamical or dynamical systems, but the ordering that must be used is equivalent to defining $(u_1, u_2) \geq (v_1, v_2)$ if $u_1 \geq v_1$ but $u_2 \leq v_2$. This point will be discussed in detail in later chapters. Generally predator-prey models or models for three or more competitors are not monotone or order preserving.

The discrete dynamical system $S^n(u_0)$ generated by the iteration $u_{n+1} = S(u_n), n = 0, 1, \ldots$, is strictly order preserving if $u > v$ implies $S^n(u) > S^n(v)$ for all $n > 0$. (See Hess (1991).) In the discrete case the point u_0 is said to be a strict subequilibrium for the dynamical or semi-dynamical system if it is a strict subequilibrium of the fixed point equation $S(u) = u$, i.e. if $S(u_0) > u_0$. (See Hess (1991, Ch. 1).) This definition and the assumption that $S^n(u)$ is strictly order preserving imply $S^n(u_0) > u_0$ for all n. In the continuous case, u_0 is a strict subequilibrium if $\pi(u_0, t) > u_0$ for all $t > 0$. (See Smith (1995, Ch. 2).) If $\pi(u, t)$ is generated by a model of the form (3.96) then u_0 would be a strict subequilibrium for the fixed point equation $f(u) = 0$ if $f(u_0) > 0$. In the case of systems of ordinary differential equations we would typically assume that $f(u)$ is at least continuous, so if u_0 were a strict subequilibrium to the fixed point equation we would have $f(\pi(u_0, t)) > 0$ for t small (since $\pi(u_0, t) \to u_0$ as $t \to 0$) and thus $d(\pi(u_0, t))/dt > 0$

for t small. Hence, we would have $\pi(u_0, \epsilon) > \pi(u_0, 0) = u_0$ for any $\epsilon > 0$ sufficiently small. Since π is a monotone semiflow we have $\pi(u_0, t+\epsilon) > \pi(u_0, t)$ for all $t \geq 0$, so that $\pi(u_0, t) > \pi(u_0, s)$ for $t > s \geq 0$. In particular, $\pi(u_0, t) > u_0$. Thus, u_0 will also be a strict subequilibrium for $\pi(u, t)$ in the sense of Smith (1991). A similar result is valid for a single reaction diffusion equation but since the operator $\Delta = \nabla^2$ is not a continuous function from X into X the derivation is more complicated. This point is addressed in the discussion of sub- and supersolutions in Chapter 1. The point u_0 is a strict superequilibrium for $S^n(u)$ if $S(u_0) < u_0$; for $\pi(u, t)$ a strict superequilibrium satisfies $\pi(u_0, t) < u_0$ for all $t > 0$. Again, for reaction-diffusion equations or systems of ordinary differential equations a strict superequilibrium of $f(u) = 0$ (i.e. a point u_0 with $f(u_0) < 0$) will be a strict superequilibrium for $\pi(u, t)$.

We can now state the fundamental result on strictly monotone (in the continuous case) or strictly order preserving (in the discrete case) dynamical or semi-dynamical systems. If $a, b \in X$ with $a < b$, the order interval $[a, b]$ is defined as $\{u \in X : a \leq u \leq b\}$.

Theorem 3.18. Suppose that the dynamical or semi-dynamical system $\pi(u, t)$ (respectively $S^n(u)$) is strictly monotone (respectively, strictly order preserving). Suppose that $\pi(u, t)$ (respectively $S^n(u)$) maps the order interval $[a, b] \subseteq X$ into itself, and that trajectories of $\pi(u, t)$ with $u \in [a, b]$ have compact closures relative to X (respectively, $S([a, b])$ is relatively compact in X.) If a is a strict subequilibrium for $\pi(u, t)$ (respectively, $S^n(u)$) then there is an equilibrium u^* with $a < u^* \leq b$ such that $\pi(a, t) \to u^*$ as $t \to \infty$ (respectively $S^n(a) \to u^*$ as $n \to \infty$.) The equilibrium u^* is the minimal equilibrium in $[a, b]$ for the dynamical or semi-dynamical system. If in addition b is a strict superequilibrium then there is also an equilibrium u^{**} with $a < u^{**} < b$ such that $\pi(b, t) \to u^{**}$ as $t \to \infty$ (respectively $S^n(b) \to u^{**}$ as $n \to \infty$) and u^{**} is the maximal equilibrium in $[a, b]$.

Remarks: The discrete case of this theorem is essentially Lemma 1.1 of (Hess 1991). The continuous version is essentially Lemma 5.1 of Smith (1991). The minimal equilibrium u^* is minimal in the sense that there can be no other equilibrium $\tilde{u}$ with $a < \tilde{u} < u^*$; the equilibrium u^{**} is maximal in the analogous sense. In the context of finding conditions which imply a model predicts persistence (and has a positive equilibrium) we would typically assume that the extinction equilibrium $u \equiv 0$ is unstable. In the reaction-diffusion case, and for some other types of models in continuous time, the instability of the equilibrium $u \equiv 0$ is characterized by the positivity of a principal eigenvalue analogous to σ_1. If ψ_1 is the corresponding eigenfunction or eigenvector then in many cases $\sigma_1 > 0$ will imply that for all $\epsilon > 0$ sufficiently small $\epsilon\psi_1$ will be a strict subequilibrium for the fixed point (i.e. equilibrium) equation and hence for the dynamical or semi-dynamical system. In such a situation, where there is an unstable equilibrium u_1 with a line segment consisting of strict subequilibria emanating from it in a positive direction, it is sometimes possible to show that there is a full orbit connecting the unstable u_1 to another (minimal) equilibrium $u_2 > u_1$. A result to that effect is Theorem 5.3 of Smith (1995). (See also Hess (1991, Proposition I.9.1).) For discrete time models, the instability of the equilibrium $u \equiv 0$ is characterized by the spectral radius of the model linearized at zero being larger than 1; see Hess (1991, Ch. I.8). If the linearized operator satisfies the hypotheses of the Krein-Rutman theorem (see Chapter 2) then it will have a principal eigenvalue equal to the spectral radius. If we denote the positive eigenvector or eigenfunction of the model linearized about $u \equiv 0$

by ψ_1 then if $u \equiv 0$ is unstable it is again often the case that $\epsilon\psi_1$ is a strict subequilibrium if $\epsilon > 0$.

We can now describe how the ideas used in Section 3.1 can be extended to other sorts of systems. A typical discrete-diffusion model has the form

$$\frac{du_i}{dt} = \sum_{\substack{j=1 \\ j \neq i}}^{n} d_{ij} u_j + g_i(u_i) u_i \quad i = 1, \ldots, n \tag{3.97}$$

where u_i represents the density of the population on the ith patch in a system of n patches. The coefficients d_{ij} for $j \neq i$ are normally taken to be nonnegative; d_{ij} represents the rate of immigration into the ith patch from the jth patch. If the pattern of dispersal is such that dispersers from each patch will eventually reach all other patches, then the matrix $((d_{ij})) + cI$ will be primitive if c is sufficiently large. (One example is when d_{ij} is positive for $j = i+1, i-1$ unless $i = 1$ or $i = n$ so that dispersers move to neighboring patches.) The system (3.97) will satisfy the Kamke condition if $d_{ij} \geq 0$, and hence in that case it will be strictly monotone relative to the standard ordering where $\vec{u} \geq \vec{v}$ means $u_i \geq v_i$ for all i, that is, P is the positive orthant in $I\!R^n$. The linearized system at $\vec{u} = 0$ is given by

$$\frac{dv_i}{dt} = \sum_{\substack{j=1 \\ j \neq i}}^{n} d_{ij} v_j + g_i(0) v_i, \quad i = 1, \ldots, n. \tag{3.98}$$

The eigenvalue problem corresponding to (3.98) is

$$\sum_{\substack{j=1 \\ j \neq i}}^{n} d_{ij} \psi_j + g_i(0) \psi_i = \sigma \psi_i, \quad i = 1, \ldots, n. \tag{3.99}$$

If $((d_{ij})) + cI$ is primitive for some c, there will be a principal eigenvalue σ_1 for (3.99) with a corresponding eigenvector $\vec{\psi}$ whose components are positive. If $\sigma_1 > 0$ then we may use $\epsilon\vec{\psi}$ as a strict subequilibrium which is positive in all components if $\epsilon > 0$ is small. If there is a constant K_0 such that $g_i(u) \leq -g_0 < 0$ for $i = 1, \ldots, n$ if $u \geq K_0$, then we may use $(K_0, K_0, \ldots, K_0)$ as a strict superequilibrium. In that case Theorem 3.18 implies that there is a positive equilibrium for (3.97). In fact, if $\vec{u}$ is initially positive then $\epsilon\vec{\psi} < \vec{u}(0)$ for ϵ sufficiently small, so $\vec{u}(t)$ is bounded below by the trajectory starting at $\epsilon\vec{\psi}$, which approaches a minimal positive equilibrium as $t \to \infty$. This is all quite analogous to the reaction-diffusion case.

Another class of models with a similar structure but different interpretation are the metapopulation models of Hanski and his co-workers (Hanski and Ovaskainen, 2000, 2001). They envision a network of patches which are subject to local colonizations and extinctions. The models describe the probability p_i that the ith patch is occupied. If A_i denotes the area of the ith patch and d_{ij} the distance between the ith and jth patches then the models typically have the form

$$\begin{aligned}\frac{dp_i}{dt} &= (\text{rate of colonization})(1 - p_i) - (\text{rate of extinction}) p_i \\ &= \left(\sum_{\substack{j=1 \\ j \neq i}}^{n} C e^{-\alpha d_{ij}} A_j p_j \right) (1 - p_i) - (E/A_i) p_i \end{aligned} \tag{3.100}$$

where C, α, and E are positive rate constants describing the propensity of the species in question to send out colonists, to die during transit through the matrix between patches, and to experience local extinctions, respectively. Since p_i represents a probability, the model is meaningful only on the set $Y = \{\vec{p} : 0 \leq p_i \leq 1 \quad \text{for} \quad i = 1, \ldots, n\}$. It follows from the structure of the system that the set Y is invariant with respect to trajectories going forward in time. The model (3.100) satisfies the Kamke condition for $\vec{p} \in Y$. It is clear from the structure of the model that $\vec{p} = (1, 1, \ldots, 1)$ is a strict superequilibrium. The eigenvalue problem for the system linearized about $\vec{p} = 0$ is

$$\sum_{\substack{j=1 \\ j \neq i}}^{n} Ce^{-\alpha d_{ij}} A_j \psi_j - (E/A_i)\psi_i = \sigma \psi_i. \tag{3.101}$$

Let $L = ((\ell_{ij}))$ where $\ell_{ij} = Ce^{-\alpha d_{ij}} A_j$ if $j \neq i$ and $\ell_{ii} = -E/A_i$. Since the numbers $Ce^{-\alpha d_{ij}} A_j$ are positive for $j \neq i$, the matrix $L + cI$ will be primitive for $c > 0$ sufficiently large, so there will be a principal eigenvalue σ_1 for (3.101) with $\psi_i > 0$ for all i. Again, if $\sigma_1 > 0$, then $\epsilon\vec{\psi}$ is a strict subequilibrium when $\epsilon > 0$ is small enough. Thus, as in the analysis of Section 3.1, we find that (3.100) predicts persistence (and has a positive equilibrium) if $\sigma_1 > 0$.

Hanski and Ovaskainen (2001, 2002) consider the system (3.100) and show it has a positive equilibrium if $\lambda_M > (E/C)$, where λ_M is the principal eigenvalue of the matrix $M = ((m_{ij}))$ where $m_{ij} = A_i A_j e^{-\alpha d_{ij}}$ if $j \neq i$ and $m_{ii} = 0$. They call λ_M the *metapopulation capacity* of the patch network. The condition $\lambda_M > (E/C)$ can be seen to be equivalent to $\sigma_1 > 0$ via results related to the Perron-Frobenius Theorem (or its generalization the Krein-Rutman Theorem). Direct calculations show that $\vec{v} = (v_1, \ldots, v_n)$ satisfies $[(E/C)I - M]\vec{v} = \vec{h}$ for $\vec{h} = (h_1, \ldots, h_n)$ if and only if $-L\vec{v} = \vec{k}$ with $\vec{k} = (Ch_1/A_1, \ldots, Ch_n/A_n)$. Thus, $(E/C)I - M$ has a positive inverse, i.e. $[(E/C)I - M]\vec{v} = \vec{h}$ has a unique positive solution for any positive $\vec{h}$, if and only if $-L\vec{v} = \vec{k}$ has a positive solution for any positive $\vec{k}$, i.e. $-L$ has a positive inverse. Let $c_0 > 0$ be arbitrary. Then $M + c_0 I$ is a positive matrix with principal eigenvalue $\lambda_M + c_0$. By the Perron-Frobenius Theorem, and its corollaries, the matrix $(E/C)I - M = [(E/C) + c_0]I - [M + c_0 I]$ has a positive inverse if and only if $(E/C) + c_0 > \lambda_M + c_0$, i.e. $E/C > \lambda_M$. (See, for example, Berman and Plemmons (1979).) If $c_1 > \max_i(E/A_i)$ then $L + c_1 I$ is a positive matrix, with principal eigenvalue $\sigma_1 + c_1$. The Perron-Frobenius Theorem and its corollaries then imply that $-L = c_1 I - (L + c_1 I)$ has a positive inverse if and only if $c_1 > c_1 + \sigma_1$, that is, $0 > \sigma_1$. Since $-L$ has a positive inverse if and only if $(E/C)I - M$ does, it follows that $\sigma_1 < 0$ if and only if $\lambda_M < E/C$. Direct calculation shows that there is a solution to $L\vec{\psi} = 0$ with $\vec{\psi}$ positive, i.e. we have $\sigma_1 = 0$, if and only if $(E/C)\vec{\psi} = M\vec{\psi}$ with $\vec{\psi}$ positive, i.e. we have $\lambda_M = E/C$. Thus, $\sigma_1 > 0$ if and only if $\lambda_M > E/C$. Hence, the formulation of Hanski and Ovaskainen (2000, 2001) for persistence is equivalent to the one we have been using.

The notion of a cooperative system of ordinary differential equations extends naturally to systems of reaction-diffusion equations. In fact, the use of monotone methods for showing the existence of solutions to elliptic and parabolic systems somewhat predates the formulation of the abstract theory of monotone dynamical systems. Comprehensive treatments of this approach to reaction-diffusion systems are given by Leung (1989) and Pao (1992). A treatment of reaction-diffusion systems from the viewpoint of the theory of monotone dynamical systems is given by Smith (1995). The approach of Section 3.1 is most directly applicable to reaction-diffusion systems of the form

$$\begin{aligned} \frac{\partial u_i}{\partial t} &= \nabla \cdot d_i(x)\nabla u_i + \vec{b}(x)\cdot \nabla u + f_i(x,\vec{u}) \text{ in } && \Omega \times (0,\infty) \\ \alpha_i(x)\frac{\partial u}{\partial \vec{n}} &+ \beta_i(x)u = 0 && \text{on } \partial\Omega \times (0,\infty), \\ i &= 1,\dots,n \end{aligned} \tag{3.102}$$

where $\partial f_i/\partial u_j \geq 0$ for $i \neq j$; that is, for each $x \in \Omega$ the reaction terms in (3.102) satisfy the Kamke condition. Such systems are often called quasimonotone in the literature on partial differential equations. As in the case of cooperative systems of ordinary differential equations, such systems typically generate monotone semiflows. The condition $\partial f_i/\partial u_j \geq 0$ for $i \neq j$ implies mutualistic interactions between the species described by the model (3.102). In some cases, nonmutualistic models can be converted to models satisfying that condition or can otherwise be shown to generate monotone semiflows; specifically, that is the case for models of two competing species. Such models will be discussed in some detail in the next chapter. It turns out that similar ideas can also be used to treat models with time delays. (See Smith (1995).) The examples discussed above by no means exhaust the possibilities, but they illustrate to some extent the scope of monotone methods.

We will now turn our attention to models in discrete time. It is well known that simple discrete-time models for a single species can display complex dynamics, including chaos. (See Murray (1993).) However, some discrete-time models have relatively simple dynamics and can be treated in essentially the same way as continuous time models. The simplest examples are nonspatial single-species models of the form $u_{n+1} = G(u_n)u_n$ where u_n represents a population or population density at time n and $G(u)$ is a positive continuous function. If $G(u)u$ is increasing then $u > v$ implies $G(u)u > G(v)v$ so that the model generates a strictly order preserving discrete dynamical system. If $G(0) > 1$ then $G(\epsilon) > 1$ for all ϵ sufficiently small, so that $u = \epsilon$ is a strict subequilibrium. If $G(u) < 1$ for all $u > K_0$ then $u = K > K_0$ will be a strict superequilibrium. In that case, all positive solutions to the model can be bounded below by solutions u_n with $u_1 = \epsilon$ where u_n increases toward the minimal positive equilibrium as $n \to \infty$. An example of such a model is the Beverton-Holt version of the logistic equation in discrete time, where $G(u) = a/(1+bu)$ for a and b positive. In that case $G(u)u$ is increasing, $G(0) > 1$ if $a > 1$, and $G(u) < 1$ if $u > (a-1)/b$, so the theory applies when $a > 1$. (If $a < 1$ then since $u_{n+1} \leq au_n$ we would have all solutions approaching zero as $n \to \infty$.) The same sort of analysis can be applied to matrix models of the form

$$\vec{u}(n+1) = G(\vec{u}(n))\vec{u}(n), \tag{3.103}$$

where $G(\vec{u})$ is a nonnegative matrix. Models for age structured populations with discrete generations often have the form (3.103). (See Caswell (1989).) If G depends continuously on $\vec{u}$ and $G(\vec{u})\vec{u}$ is componentwise increasing in $\vec{u}$, then (3.103) will generate a strictly order preserving discrete dynamical system. (The assumption that $G(\vec{u})\vec{u}$ is componentwise increasing can be weakened, but the point here is simply to give some examples indicating the scope of the methods.) If $G(\vec{u})$ is primitive for $\vec{u}$ near 0 and $G(0)$ has a principal eigenvalue $\lambda_G > 1$ with eigenvector $\vec{\psi}$ then $\epsilon\vec{\psi}$ will be a strict subequilibrium for ϵ small. If there is a positive vector $\vec{v}$ such that $G(k\vec{v})\vec{v} < \vec{v}$ componentwise for $k \geq k_0 > 0$ then $k\vec{v}$ will be a strict superequilibrium for $k \geq k_0$, and by Theorem 3.18 the model (3.103) will have a positive equilibrium. Furthermore, since $\epsilon\vec{\psi} < \vec{u}$ for any positive vector $\vec{u}$ provided

$\epsilon > 0$ is small, the model will predict persistence. A more detailed treatment of this type of model from essentially this viewpoint is given in Cosner (1996).

Models such as (3.103) can also occur in the context of spatial modeling. If we have several patches inhabited by some species and at each time step some fraction f_{ji} of the population of the ith patch migrates to the jth patch, we obtain a dispersal model of the form

$$u_i(n+1) = \sum_{\substack{j=1 \\ j \neq i}}^{n} f_{ij} u_j(n) + \left[1 - \left(\sum_{\substack{j=1 \\ j \neq i}}^{n} f_{ji} \right) \right] u_i(n). \tag{3.104}$$

(We must have $\sum_{\substack{j=1 \\ j \neq i}}^{n} f_{ji} \leq 1$ for all i for this to make sense.) Combining the dispersal model (3.104) with population dynamical models for each patch would yield a model of the form (3.103).

Models for populations in continuous space may involve discrete time. Recently Lewis and his co-workers (Lewis, 1997; Van Kirk and Lewis, 1997, 1999) have made extensive use of models of the form

$$u_{n+1}(x) = \int_{\Omega} K(x, y) f(u_n(y), y) dy \tag{3.105}$$

where $K(x, y)$ is kernel describing the dispersal of the population. Models of the form (3.105) arise if a population alternates discrete recruitment events with periods of continuous dispersal. If the dispersal is diffusive, then $K(x, y)$ will be the Green's function $G(x, y, t)$ for the corresponding diffusion equation evaluated at $t = T$, where T is the time period between recruitment events. Other formulations for $K(x, y)$ will be positive and the mapping from $w(x)$ to $\int_{\Omega} K(x, y) w(y) dy$ will be a compact operator on an appropriate function space if the underlying domain Ω is bounded. A typical choice for the function space is $X = L^2(\Omega)$, because the theory of integral operators on $L^2(\Omega)$ is well developed; see, for example, Van Kirk and Lewis (1997) and Hardin et al. (1988a,b, 1990). In the case where $f(u, y)$ is increasing in u the model (3.105) will typically generate a strictly order preserving semi-dynamical system on $L^2(\Omega)$. If $f(u, y) = g(u, y)u$ then the operator corresponding to the linearization of the model (3.105) around $u = 0$ is $Aw = \int_{\Omega} K(x, y) g(0, y) w(y) dy$. If $g(0, y) > 0$ on Ω then Aw will be a positive operator satisfying the hypotheses of the Krein-Rutman Theorem and hence will have a principal eigenvalue λ_A with positive eigenfunction ϕ. As in other types of discrete time models, if $\lambda_A > 1$ then $\epsilon\phi$ will be a strict subequilibrium for $\epsilon > 0$ small. If there is a constant k_0 such that the operator $B(u)w = \int_{\Omega} K(x, y) g(u, y) w(y) dy$ has spectral radius (and hence principal eigenvalue) less than one for $u(x) > k_0$ then any sufficiently large constant will be a strict superequilibrium and the methods of Section 3.1 can be applied via Theorem 3.18 to conclude that the model predicts persistence and has a positive equilibrium. As usual, these hypotheses can be weakened and a more detailed analysis is possible; see Hardin et al. (1988a,b, 1990) and Van Kirk and Lewis (1997).

The last example leading to a strictly order-preserving semi-dynamical system returns us to the context of reaction-diffusion models. This is the case of time-periodic reaction-diffusion models treated in detail by Hess (1991). Since the treatment by Hess (1991) is comprehensive, we shall just describe the approach in fairly general terms and state some specific results for future reference. The models we consider are T-periodic diffusive logistic equations of the form

$$\begin{aligned} \frac{\partial u}{\partial t} &= \nabla \cdot d(x,t)\nabla u + \vec{b}(x,t) \cdot \nabla u + m(x,t)u - c(x,t)u^2 \text{ in } \quad \Omega \times (0,\infty) \\ \alpha(x)\frac{\partial u}{\partial \vec{n}} &+ \beta(x)u = 0 \qquad \text{on} \quad \partial\Omega \times (0,\infty) \end{aligned} \tag{3.106}$$

where as usual $\Omega \subseteq I\!R^n$ is a bounded domain with $\partial\Omega$ of class $C^{2+\gamma}$, and where the coefficients of (3.106) are T-periodic, with $d(x,t) \geq d_0 > 0$, $c(x,t) \geq c_0 > 0$, and $d(x,t) \in C^{1+\gamma,\gamma/2}(\overline{\Omega} \times I\!R)$, $\vec{b}(x,t) \in [C^{\gamma,\gamma/2}(\overline{\Omega} \times I\!R)]^n$, and $m(x,t), c(x,t) \in C^{\gamma,\gamma/2}(\overline{\Omega} \times I\!R)$ for some $\gamma \in (0,1)$. Associated with (3.106) is the eigenvalue problem

$$\begin{aligned} &\frac{\partial u}{\partial t} - \nabla \cdot d(x,t)\nabla\psi - \vec{b}(x,t) \cdot \nabla\psi - m(x,t)\psi = \mu\psi \text{ in } \quad \Omega \times I\!R \\ &\alpha(x)\frac{\partial \psi}{\partial \vec{n}} + \beta(x)\psi = 0 \qquad \text{on} \quad \partial\Omega \times I\!R \\ &\psi \text{ is } T-\text{periodic in} \quad t. \end{aligned} \tag{3.107}$$

As is stated is Lemma 2.5 and discussed in detail by Hess (1991), the problem (3.107) has a unique principal eigenvalue μ_1 with eigenfunction $\psi_1 > 0$ (Lazer, 1982; Castro and Lazer, 1982). In the case where d, b, and m are independent of t, $\mu_1 = -\sigma_1$ where σ_1 is the principal eigenvalue of

$$\begin{aligned} &\nabla \cdot d(x)\nabla\psi + \vec{b}(x) \cdot \nabla\psi + m(x)\psi = \sigma\psi \text{ in } \quad \Omega \\ &\alpha(x)\frac{\partial \psi}{\partial \vec{n}} + \beta(x)\psi = 0 \qquad \text{on} \quad \partial\Omega. \end{aligned} \tag{3.108}$$

The discrete semi-dynamical system related to (3.106) is the period map or Poincaré map S, where $S(v(x)) = u(x,T)$ where $u(x,t)$ is the solution to (3.106) with $u(x,0) = v(x)$. It follows from the maximum principle that S is strictly order preserving. Parabolic regularity theory implies that if we set the semi-dynamical system in an appropriate Sobolev space then the image under S of a bounded set will have compact closure, so the hypotheses of Theorem 3.18 are satisfied. If $\mu_1 < 0$ in (3.107) then $\epsilon\psi_1$ will be a strict subequilibrium for S if $\epsilon > 0$ is small. Any sufficiently large constant will be a strict superequilibrium. A point worth noting is that if $u(x,t)$ is a solution to (3.106) with $u(x,0) = v(x)$ where $v(x)$ is a strict subequilibrium then $u(x,(n+1)T) = S^{n+1}(v) > S^n(v) = u(x,nT)$ so that $u < u(x,T) < u(x,2T) < \ldots$, but in general it is *not* true that $u(x,t) > u(x,s)$ for all $t > s$. If $v^*(x)$ is a fixed point for $S(u)$ then the solution to (3.106) with $u(x,0) = v^*(x)$ will be T periodic since $u(x,T) = S(v^*) = v^* = u(x,0)$, etc. Thus, by using Theorem 3.18 and some of the other ideas from Section 3.1 we can obtain:

Proposition 3.19. If $\mu_1 \geq 0$ in (3.107) then all positive solutions to (3.106) approach zero as $t \to \infty$. If $\mu_1 < 0$ in (3.107) then (3.106) has a unique positive T-periodic solution $u^*(x,t)$ which is globally attracting among positive solutions.

Remarks: This result and related results are given by Hess (1991, Ch. III.28). Similar sorts of results have been obtained in the almost periodic case but the formulation is more delicate, partly because strong and simple results on principal eigenvalues are not available in that setting. (See Shen and Yi (1998).)

Note that Proposition 3.19 includes a result on the uniqueness of the positive steady state u^*, as do some of the results in previous sections of this chapter. In the setting of more general types of models it is sometimes possible to use ideas related to eigenvalue comparison to show that the minimal equilibrium is the only equilibrium. The key idea is this: many of the continuous time models discussed in this section have equilibrium equations of the form $A(u^*)u^* = 0$ where A is a matrix or a differential or integral operator. If $u^* > 0$, it follows that $\sigma = 0$ *is an eigenvalue of* $A(u^*)$ *with a positive eigenvector or eigenfunction.* For many of the sorts of operators we have considered the principal eigenvalue σ_1 is unique and is characterized by having a positive eigenvector or eigenfunction. Furthermore, many such operators admit eigenvalue comparison theorems. (These can be deduced via the Krein-Rutman or Perron-Frobenius Theorems, or in some cases via variational formulas, as in Chapter 2.) The result used in Section 3.1 to the effect that the principal eigenvalue for the problem $\nabla \cdot d(x)\nabla\psi + \vec{b}(x) \cdot \nabla\psi + m(x)\psi = \sigma\psi$ is increasing with respect to $m(x)$ is a typical example. If $A(u)(v) = \nabla \cdot d(x)\nabla v + \vec{b}(x) \cdot \nabla v + g(x, u)v$ and $g(x, u)$ is decreasing in u, then $\sigma_1(A(u^{**})) < \sigma_1(A(u^*))$ if $u^{**} > u^*$. The argument for uniqueness is then as follows: if $u^* > 0$ is the minimal positive equilibrium, then any other positive equilibrium must satisfy $u^{**} > u^*$, so $\sigma_1(A(u^{**})) < \sigma_1(A(u^*))$. However, since $A(u^{**})u^{**} = A(u^*)u^* = 0$ we must have $\sigma_1(A(u^{**})) = \sigma_1(A(u^*)) = 0$, a contradiction. Under suitable hypotheses on the nonlinear terms this sort of analysis can be applied to models such as (3.97). In the discrete-time case the equilibrium equation typically has the form $A(u^*)u^* = u^*$, so that the principal eigenvalue for $A(u^*)$ is one rather than zero, but otherwise the same ideas apply to that case as well.

Appendix

Bifurcation and Stability

The bifurcation and stability results in Propositions 3.10 and 3.13 are derived from general results based on the Implicit Function Theorem which are due to Crandall and Rabinowitz (1971, 1973). The setting for those results is essentially the same as for the Implicit Function Theorem, Theorem 3.5, except that one of the Banach spaces is taken to be $\mathbb{R}$. Specifically, suppose that Y and Z are Banach spaces, V is an open subset of Y containing zero, $(a, b) \subseteq \mathbb{R}$ is an open interval, and F is a twice differentiable function from $(a, b) \times V \subseteq \mathbb{R} \times Y$ into Z. In what follows, we denote the nullspace and range of a linear operator A by $N(A)$ and $R(A)$ respectively, and denote dimension and codimension by dim and codim. A version of the result underlying Proposition 3.10 is:

Theorem 3.A.1. (Crandall and Rabinowitz, 1971) Suppose that $F(\lambda, 0) = 0$ for $\lambda \in (a, b)$, and that for some $\lambda_0 \in (a, b)$, 0 is a simple eigenvalue of $F_y(\lambda_0, 0)$; that is,

$$\dim\ N(F_y(\lambda_0, 0)) = \operatorname{codim}\ R(F_y(\lambda_0, 0)) = 1. \tag{3.A.1}$$

Let y_0 be an element of Y that spans $N(F_y(\lambda_0, 0))$. Suppose further that

$$F_{\lambda y}(\lambda_0, 0)y_0 \notin R(F_y(\lambda_0, 0)). \tag{3.A.2}$$

Let $W \subseteq Y$ be any closed complement of the span of $\{y_0\}$. Then there exists an open interval I_0 containing 0 and continuously differentiable functions $\lambda : I_0 \to I\!R$ and $\rho : I_0 \to W$ with $\lambda(0) = \lambda_0$, $\rho(0) = 0$, such that if $\lambda = \lambda(s)$ and $y = y(s) = sy_0 + s\rho(s)$ then $F(\lambda(s), sy_0 + s\rho(s)) = 0$ for $s \in I_0$. Furthermore, the entire solution set for $F(\lambda, y) = 0$ in any sufficiently small neighborhood of $(\lambda_0, 0)$ in $I\!R \times Y$ consists of the line $(\lambda, 0)$ and the curve $(\lambda(s), sy_0 + s\rho(s))$.

Remarks: Theorem 3.A.1 is a version of the main theorem, Theorem 1.7, of Crandall and Rabinowitz (1971). A formulation very similar to the one above is given as Lemma 1.1 of Crandall and Rabinowitz (1973). If F has continuous derivatives up to order n in λ and up to order $n+1$ in y, then $\lambda(s)$ and $\rho(s)$ are n times continuously differentiable in s; see Crandall and Rabinowitz (1971, Theorem 1.18). Versions of Theorem 3.A.1 are given in Deimling (1985) and Zeidler (1985).

Proof of Proposition 3.10: As in some of the examples involving the Implicit Function Theorem, we let Y be the subspace of $C^{2+\alpha}(\overline{\Omega})$ consisting of functions satisfying the boundary conditions; that is, $Y = \{u \in C^{2+\alpha}(\overline{\Omega}) : \alpha(x)\partial u/\partial\vec{n} + \beta(x)u = 0 \quad \text{on} \quad \partial\Omega\}$. We then let $Z = C^{\alpha}(\overline{\Omega})$ and define $F : I\!R \times Y \to Z$ by (3.55), so that $F(\lambda, u) = \nabla \cdot d(x, u)\nabla u + \vec{b}(x) \cdot \nabla u + \lambda g(x, u)u$. In this situation, we have $F_y(\lambda, 0)v = \nabla \cdot d(x, 0)\nabla v + \vec{b}(x) \cdot \nabla v + \lambda g(x, 0)v$, so that the relation $F_y(\lambda, 0)v = 0$ defining $N(F_y(\lambda, 0))$ is just (3.56). Recall from Chapter 2 that the principal eigenvalue λ_1^+ for (3.56) is simple, so dim $N(F(\lambda_1^+, 0)) = 1$. It follows from the Fredholm Alternative that codim $R(F_y(\lambda_1^+, 0)) = 1$; see for example Friedman (1976) and Gilbarg and Trudinger (1977). (The Fredholm alternative applies here because if L is a second order elliptic operator and I is the identity operator then for appropriate choices of the constant C the operator $L + CI$ has an inverse which can be interpreted as a compact linear operator on $C^{\alpha}(\overline{\Omega})$, $L^p(\Omega)$, or various other spaces. Some authors prefer the term "completely continuous" to compact.) This point is discussed in Chapter 2 in the context of using the Krein-Rutman theorem to show the existence of principal eigenvalues.

Since the relation $F_y(\lambda, 0)v = 0$ is just (3.56), the element $\phi \in Y$ spanning $N(F_y(\lambda_1^+, 0))$ can be taken to be any eigenfunction for (3.56). Thus, (3.A.2) will hold if there is no solution $v \in V$ to the problem $F_y(\lambda_1^+, 0)v = F_{\lambda y}(\lambda_1^+, 0)\phi_0$. The operator $F_\lambda(\lambda, u) : I\!R \to Z$ acts on real numbers via multiplication by $g(x, u)u$. The derivative $F_{\lambda y}(\lambda, u)$ applied to ϕ_0 can be computed as

$$\begin{aligned} F_{\lambda y}(\lambda, u)\phi_0 &= \partial F_\lambda(\lambda, u + \epsilon\phi_0)/\partial\epsilon|_{\epsilon=0} = \partial[g(x, u + \epsilon\phi_0)(u + \epsilon\phi_0)]/\partial\epsilon|_{\epsilon=0} \\ &= g(x, u)\phi_0 + g_u(x, u)u\phi_0. \end{aligned}$$

(Technically perhaps one should write $F_{\lambda y}(\lambda, u)(1, \phi_0)$ since $F_{\lambda,y}$ acts on $I\!R \times Y$, but this is not usually done. The notation used here is fairly standard in the literature on this type of problem.) Thus, the problem $F_y(\lambda_1^+, 0)v = F_{\lambda y}(\lambda_1^+, 0)\phi_0$ may be written as

$$\begin{aligned} \nabla \cdot d(x, 0)\nabla v + \vec{b}(x) \cdot \nabla v + \lambda_1^+ g(x, 0)v &= g(x, 0)\phi_0 \text{ in } \Omega \\ \alpha(x)\frac{\partial v}{\partial \vec{n}} + \beta(x)v = 0 \qquad &\text{on } \partial\Omega. \end{aligned} \tag{3.A.3}$$

To verify 3.A.2 we must show that 3.A.3 has no solution. In the case where $\vec{b}(x) = 0$ we can multiply (3.A.3) by ϕ_0 and integrate over Ω. Applying the divergence theorem then yields (via (3.56))

$$0 = \int_\Omega [\nabla \cdot d(x,0)\nabla\phi_0 + \lambda_1^+ g(x,0)\phi_0]v = \int_\Omega g(x,0)\phi_0^2 dx.$$

The variational characterization of λ_1^+ is valid in this case, and implies that $\int_\Omega g(x,0)\phi_0^2 dx > 0$, so that (3.A.3) cannot have any solution. If $g(x,0) > 0$ but $\vec{b} \neq 0$ we would multiply by ϕ_0^* where ϕ_0^* is an eigenfunction corresponding to λ_1^+ in the adjoint problem to (3.56) which is chosen to have the same sign as ϕ_0, and then argue in the same way. The case where $\vec{b} \neq 0$ but $g(x,0)$ changes sign is more subtle, and is related to the algebraic simplicity of λ_1^+. If we write $Lu = \nabla \cdot d(x,0)\nabla u + \vec{b}(x) \cdot \nabla u$ and $Mu = g(x,0)u$, then (3.56) and (3.A.3) may be written as $L\phi + \lambda M\phi = 0$ and $Lv + \lambda_1^+ Mv = M\phi$ respectively. In the case of Dirichlet boundary conditions, Hess and Kato (1980) showed that λ_1^+ is algebraically simple as a characteristic value of $L^{-1}M$, which means that $N([I + \lambda_1^+ L^{-1}M]^n)$ is spanned by $\{\phi_0\}$ for all n. (An eigenvalue of this type is sometimes said to be an *M-simple eigenvalue of L*.) This result extends to problems with more general boundary conditions and even to periodic-parabolic eigenvalues; see Hess (1991, Ch. II Section 16). In our case, we could write (3.A.3) as $(I + \lambda_1^+ L^{-1}M)v = L^{-1}M\phi_0$, so that $(I + \lambda_1 L^{-1}M)^2 v = (I + \lambda_1^+ L^{-1}M)(L^{-1}M)\phi_0 = (L^{-1}M)(I + \lambda_1^+ L^{-1}M)\phi_0 = 0$. Hence $v \in N([I + \lambda_1^+ L^{-1}M]^2)$ so $v = c\phi_0$ for some constant c, since λ_1^+ is a simple characteristic value for $L^{-1}M$. But then $(L + \lambda_1^+ M)v = 0$, so that (3.A.3) becomes $0 = M\phi_0 = g(x,0)\phi_0$, which cannot hold since $\phi_0 > 0$ on Ω and $g(x,0)$ must be positive on part of λ. Thus, (3.A.3) has no solutions and hence (3.A.2) is satisfied.

Proposition 3.13 is based on results from Crandall and Rabinowitz (1973). The setting for those results is the same as for Theorem 3.A.1, and we shall assume that the hypotheses of Theorem 3.A.1 are satisfied in what follows.

Theorem 3.A.2 (Crandall and Rabinowitz, 1973). Suppose that $F(\lambda, y)$ satisfies the hypotheses of Theorem 3.A.1, and let $\lambda_0, y_0, I_0, \lambda(s), y(s)$ and W be defined as in that theorem. Then there are a subinterval $I_1 \subseteq (a,b)$ and continuously differentiable functions $\sigma(\lambda): I_1 \to I\!R$, $\psi(\lambda): I_1 \to Y$, $\mu(s): I_0 \to I\!R$, and $\tau(s): I_0 \to Y$ such that

$$F_y(\lambda, 0)\psi(\lambda) = \sigma(\lambda)\psi(\lambda)$$

$$F_y(\lambda(s), y(s))\tau(s) = \mu(s)\tau(s),$$

with $\sigma(\lambda_0) = 0$, $\psi(\lambda_0) = y_0 = \tau(0)$, and $\mu(0) = 0$. Furthermore, $\psi(\lambda) - y_0 \in W$ and $\tau(s) - y_0 \in W$. Additionally, $\sigma'(\lambda_0) \neq 0$, and near $s = 0$ the functions $\mu(s)$ and $-s\lambda'(s)\sigma'(\lambda_0)$ have the same zeros, and if $\mu(s) \neq 0$, they have the same sign. More precisely

$$\lim_{\substack{s\to 0\\ \mu(s)\neq 0}} \frac{-s\lambda'(s)\sigma'(\lambda_0)}{\mu(s)} = 1.$$

Remarks: This theorem is given as Corollary 1.13 and Theorem 1.16 in Crandall and Rabinowitz (1973). Formula (3.69) is an immediate consequence of this result, obtained by taking F to be the same as in the proof of Proposition 3.10. The derivation of Proposition 3.13 from (3.69) is given in the text of Section 3.4.

Singular Perturbation

These results on the behavior of solutions to reaction-diffusion models as the diffusion coefficient goes to zero are based on a result of DeSanti (1986). A version of that result that was used in Cantrell and Cosner (1989) is as follows:

Theorem 3.A.3. (De Santi, 1986) Consider the problem

$$\begin{aligned} -\epsilon^2 \Delta w &= h(x, w) \text{ in } \quad \Omega \\ w &= p(x) \qquad\qquad \text{on} \quad \partial\Omega \end{aligned} \tag{3.A.4}$$

where $\Omega \subseteq I\!R^n$ is a bounded domain with $\partial\Omega$ given by $F(x) = 0$ for some function $F \in C^1(I\!R^n)$ with $\nabla F \neq 0$ on $\partial\Omega$, and where $h(x, w) \in C^2(\Omega \times I\!R)$, $p(x) \in C^2(\partial\Omega)$. Let $H(x, w) = \int_0^w h(x, s)ds$. Suppose there exists a function $q(x) \in C^2(\Omega)$ so that

(i) $h(x, q(x)) = 0$,

(ii) $h_w(x, q(x)) \leq -h_0 < 0$ in Ω for some constant $h_0 > 0$, and

(iii) $[H(x, w) - H(x, q(x))][p(x) - q(x)] < 0$ for all $w \in (q(x), p(x)]$ (if $p(x) > q(x)$) or for all $w \in [p(x), q(x))$ (if $p(x) < q(x)$) and $x \in \partial\Omega$.

Then for ϵ sufficiently small, (3.A.4) has a classical solution $w(x, \epsilon)$ such that $w(x, \epsilon) \to q(x)$ as $\epsilon \to 0$, uniformly on each closed subset of Ω.

Proof of Proposition 3.15: Let $w = K - u$. Then $\Delta w = -\Delta u$, so equation (3.93) becomes

$$\begin{aligned} -\epsilon^2 \Delta w &= -(K - w)g(K - w) \text{ in } \quad \Omega \\ w &= K \qquad\qquad\qquad\qquad \text{on} \quad \partial\Omega. \end{aligned} \tag{3.A.5}$$

Take $h(w) = -(K-w)g(K-w) = (w-K)g(K-w)$. Also, $w \equiv 0$ satisfies the differential equation in (3.A.5) (since $g(K) = 0$) and $w = K$ satisfies the boundary condition. Let $p(x) \equiv K$ and $q(x) \equiv 0$. Since $h(0) = -Kg(K) = 0$, condition (i) holds in Theorem 3.A.3. We have $h_w(w) = g(K-w) - (w-K)g'(K-w)$ so $h_w(0) = g(K) + Kg'(K) = Kg'(K)$, so (ii) holds for $h_0 = -Kg'(K) > 0$. Finally, since $q = 0$, condition (iii) can be written as $\left(\int_0^w (s-K)g(K-s)ds\right) K < 0$ for $0 < w \leq K$; but $(s-K)g(K-s) < 0$ on $(0, K)$ so (iii) holds, and Proposition 3.15 thus follows from Theorem 3.A.3.

Proof of Proposition 3.16: The essential idea is to construct local sub- and supersolutions which satisfy Theorem 3.A.3 and then extend them to global sub- and supersolutions. The global sub- and supersolutions will be taken in the weak sense. Specifically, they will typically have "corners" where they are not classically differentiable. The general notion of sub- and supersolutions extends to this case. (See, for example, Berestycki and Lions (1980).) The pertinent results are stated in Chapter 1. Some of the arguments based on sub- and supersolutions yield results that are local (even though the sub- and supersolutions are global), but the local results can be extended to arbitrary closed subsets of Ω where $g(x, 0)$ is of one sign via a compactness argument.

We first consider the set E_2 on which $g(x,u) \leq -g_2 < 0$. For $x \in E_1$, let $\underline{v}_\epsilon(x)$ be the unique positive solution to

$$\begin{aligned} -\epsilon^2 \Delta v &= g_0(v)v \text{ on } \quad E_1 \\ v &= 0 \qquad\qquad \text{on} \quad \partial E_1, \end{aligned} \tag{3.A.6}$$

where $g_0(v) = \min_{x \in E_1} g(x,v)$. (We have $g(x,0) \geq g_1 > 0$.) For $x \in \Omega \backslash E_1$ let $\underline{v}_\epsilon(x) = 0$. Since $\underline{v}_\epsilon \geq 0$ in E_1 we have $\partial \underline{v}_\epsilon / \partial \vec{n} \leq 0$ on ∂E_1. Since $\underline{v}_\epsilon$ satisfies the differential equation in (3.94) for $x \in \Omega \backslash E_1$ and is a subsolution for $x \in E_1$, with $\underline{v}_\epsilon = 0$ on $\partial\Omega$, it follows that $\underline{v}_\epsilon$ is a weak subsolution to (3.94). Also, $g_0(K_0) < 0$ so we must have $\underline{v}_\epsilon < K_0$ by the maximum principle. For $x \in E_2$ let $\overline{v}_\epsilon$ be the solution to

$$\begin{aligned} -\epsilon^2 \Delta v &= -g_2 v \text{ on } \quad E_2 \\ v &= K_0 \qquad\quad \text{on} \quad \partial E_2, \end{aligned} \tag{3.A.7}$$

and let $\overline{v}_\epsilon = K_0$ on $\Omega \backslash E_2$. Recall that $K(x) \leq K_0$ so $g(x, K_0) \leq 0$ on Ω. To verify that $\overline{v}_\epsilon \geq \underline{v}_\epsilon$ on E_2 and that $\overline{v}_\epsilon$ is a weak supersolution to (3.94) we use the maximum principle. If $\overline{v}_\epsilon < 0$ in E_2 then $\overline{v}_\epsilon$ has negative minimum at some $x_0 \in E_2$. At x_0 we would have $-\epsilon \Delta \overline{v}_\epsilon \leq 0$, but $-g_2 \overline{v}_\epsilon > 0$, a contradiction. Thus $\overline{v}_\epsilon \geq 0$ on E_2. Similarly, $\overline{v}_\epsilon$ cannot have a positive maximum inside E_2, and by (3.A.7) v cannot be constant, so the maximum of $\overline{v}_\epsilon$ will be K_0. This value is attained everywhere on ∂E_2, so by the strong maximum principle $\partial \overline{v}_\epsilon / \partial \vec{n} > 0$ on ∂E_2. Taking the normal derivative $\partial \overline{v}_\epsilon / \partial \vec{n}$ relative to $\Omega \backslash E_2$ yields 0 since $\overline{v}_\epsilon = K_0$ on $\Omega \backslash E_2$. Thus, $\overline{v}_\epsilon \geq 0$ and $\overline{v}_\epsilon = K_0$ on E_1 since $E_1 \subseteq \Omega \backslash E_2$, so $\overline{v}_\epsilon \geq \underline{v}_\epsilon$, and $\overline{v}_\epsilon$ is a weak supersolution to (3.94). It follows that (3.94) has a solution $\tilde{u}$ with $\underline{v}_\epsilon \leq \tilde{u} \leq \overline{v}_\epsilon$, and since $\underline{v}_\epsilon > 0$ on E_1, $\tilde{u}$ is positive. The positive solution to (3.94) is unique so if u is the positive solution to (3.94) then $0 \leq \underline{v}_\epsilon \leq u \leq \overline{v}_\epsilon$. On E_2 the weak supersolution $\overline{v}_\epsilon$ satisfies the hypotheses of Theorem 3.A.3 with $p(x) = K_0$, $q(x) = 0$, and $h(x,w) = -g_2 w$, so as $\epsilon \to 0$ we have $\overline{v}_\epsilon \to 0$ uniformly on any closed subset of E_2. This establishes the assertion $u \to 0$ as $\epsilon \to 0$ uniformly on any closed subset of E_2.

On E_2 the limiting value of u is the constant zero, so we could treat all of E_2 with one pair of sub- and supersolutions. On E_1 we need to track the variation in $K(x)$ so we will need to obtain local estimates in the neighborhood of each point and then extract a global estimate on any given closed subset of E_1 via compactness. Specifically, we will show that if S is a closed subset of E_1 then for any $\delta > 0$ and $x_0 \in S$ there is a ball $B(x_0)$ about x_0 and a number $\epsilon_0(x_0) > 0$, such that for $0 < \epsilon < \epsilon_0(x_0)$ the positive solution to (3.94) satisfies $|u(x) - K(x)| < \delta$ on $B(x_0)$. Since S is compact there will be a finite subcollection of such balls, $\{B(x_1), \ldots, B(x_n)\}$ that covers S, so for $0 < \epsilon < \min\{\epsilon_0(x_1), \ldots, \epsilon_0(x_n)\}$ we have $|u(x) - K(x)| < \delta$ on S. Since $\delta > 0$ is arbitrary, this argument implies that $u(x) \to K(x)$ uniformly on S as $\epsilon \to 0$. It remains to show that the ball $B(x_0)$ and number $\epsilon_0(x_0)$ do indeed exist for any $x_0 \in S$ and $\delta > 0$.

If $x_0 \in S$ then by the continuity of $K(x)$ there is a ball $B_0(x_0) \subseteq E_1$ such that $K(x_0) - \delta/4 < K(x) < K(x_0) + \delta/4$ on $B_0(x_0)$. By hypothesis, $\max_{x \in B_0(x_0)} g(x,u)$ is decreasing in u, and by the choice of B_0 we have $\max_{x \in B_0(x_0)} g(x,u) < 0$ for $u \geq K(x_0) + \delta/4$ and $\max_{x \in B_0(x_0)} g(x,u) > 0$ for $u \leq K(x_0) - \delta/4$. Thus, we must have $\max_{x \in B_0(x_0)} g(x, \overline{K}) = 0$ for some $\overline{K} \in (K(x_0) - \delta/4, K(x_0) + \delta/4)$, with $\max_{x \in B_0(x_0)} g(x,u) > 0$ for $u < \overline{K}$ and

$\max_{x \in B_0(x_0)} g(x,u) < 0$ for $u > \overline{K}$. Let $\overline{g}(u) = \max_{x \in B_0(x_0)} g(x,u)$ and define $\overline{f}(u)$ as $\overline{f}(u) = \overline{g}(u)u$ for $u > \overline{K}/2$. Extend $\overline{f}(u)$ so that $\overline{f}(u) > 0$ for $u \leq \overline{K}/2$. Let K_1 be a constant with $K_1 > \overline{K}$. Consider the problem

$$\begin{aligned} -\epsilon^2 \Delta w &= \overline{f}(w) \text{ in } \quad B_0(x_0) \\ w &= K_1 \qquad\quad \text{on} \quad \partial B_0(x_0). \end{aligned} \tag{3.A.8}$$

Since $\overline{f}(w) = \overline{g}(w)w$ for $w > \overline{K}/2$, $\overline{f}(w)$ satisfies the hypotheses of Theorem 3.A.3 with $p(x) = K_1$ and $q(x) = \overline{K}$, so (3.A.8) has a solution w^* with $w^* \to \overline{K}$ uniformly on closed subsets of $B(x_0)$ as $\epsilon \to 0$. Also, the maximum principle implies that w^* cannot have a maximum larger than $\overline{K}$ or a minimum smaller than $\overline{K}$ inside Ω, so w^* satisfies $-\epsilon^2 \Delta w = \overline{g}(w)w$ in $B_0(x_0)$ and the maximum of w^* is K_1. Since $w^* = K_1$ everywhere on $\partial B_0(x_0)$, the strong maximum principle implies that $\partial w^*/\partial \vec{n} > 0$ on $\partial B_0(x_0)$. Let $\overline{w} = w^*$ in $B_0(x_0)$ and $\overline{w} = K_1$ on $\Omega \backslash B_0(x_0)$. Then $\overline{w}$ is a weak supersolution to (3.94) and $\overline{w} \geq \overline{K}$ on $B_0(x_0)$ with $\overline{w} \to \overline{K}$ as $\epsilon \to 0$ uniformly on closed subsets of $B_0(x_0)$. To construct a corresponding weak subsolution we can let $\underline{g}(u) = \min_{x \in B_0(x_0)} g(x,u)$ and take $\underline{w}$ to be the unique positive solution to

$$\begin{aligned} -\epsilon^2 \Delta w &= \underline{g}(w)w \text{ on } \quad B_0(x_0) \\ w &= 0 \qquad\qquad\quad \text{on} \quad \partial B_0(x_0) \end{aligned} \tag{3.A.9}$$

and $\underline{w} = 0$ outside of $B_0(x_0)$. Since $\underline{g}(u) < 0$ for $u > K(x_0) + \delta/4$ and $\underline{g}(u) > 0$ for $u < K(x_0) - \delta/4$, there is a constant $\underline{K} \in (K(x_0) - \delta/4,\ K(x_0) + \delta/4)$ such that $\underline{g}(u) > 0$ for $u < \underline{K}$ and $\underline{g}(u) < 0$ for $u > \underline{K}$. (Note that $\overline{g}(u) \geq \underline{g}(u)$ so that $\underline{K} \leq \overline{K}$.) Theorem 3.A.3 (in fact, Proposition 3.15) can be applied to (3.A.9), so $\underline{w} \to \underline{K}$ uniformly on closed subsets of $B_0(x_0)$. By the maximum principle $\underline{w} \leq \underline{K}$ on $B_0(x_0)$ so $\underline{w} \leq \overline{w}$ on Ω. Since $\underline{w}$ is a weak subsolution and $\overline{w}$ is a weak supersolution to (3.94), there must be a solution $\tilde{u}$ to (3.94) with $\underline{w} \leq \tilde{u} \leq \overline{w}$, but since (3.94) has a unique positive solution u we must have $\underline{w} \leq u \leq \overline{w}$. If we let $B(x_0)$ be any ball about x_0 with $\overline{B(x_0)} \subseteq B_0(x_0)$ and choose $\epsilon_0(x_0)$ small enough that for $0 < \epsilon < \epsilon_0(x)$ we have $\underline{w} > \underline{K} - \delta/4$ and $\overline{w} < \overline{K} + \delta/4$ on $\overline{B(x_0)}$, then on $B(x_0)$ we have $u(x) > \underline{K} - \delta/4 \geq K(x_0) - \delta/2 \geq K(x) - 3\delta/4$ and $u(x) < \overline{K} + \delta/4 \leq K(x) + 3\delta/4$, so that $|u(x) - K(x)| < \delta$ on $B(x_0)$ as required. This completes the proof of Proposition 3.16.

4

Permanence

4.1 Introduction

4.1.1 Ecological Overview

The use of mathematics to model the interaction of biological species began with the work of Alfred Lotka (1925) and Vito Volterra (1931) on certain systems of ordinary differential equations which now bear both their names. Let us suppose that $u_1(t), u_2(t), \ldots, u_n(t)$ represent the population densities at time t for species $1, 2, \ldots, n$, respectively, in an n-species ecological community under consideration. The salient feature of a Lotka-Volterra model for the interaction of these n species is that for *each* species in the community, the relative or *per capita* rate of population growth (i.e., $\dfrac{du_i(t)}{dt}/u_i(t)$, where i ranges from 1 to n) is a linear expression in terms of the population densities of *all* the species in the community.

As a consequence, a Lotka-Volterra ordinary differential equation (ODE) model for the n-species community with population densities $u_1 = u_1(t), u_2 = u_2(t), \ldots, u_n = u_n(t)$ has the form

$$\frac{du_i}{dt} = u_i \left(a_i + \sum_{j=1}^{n} b_{ij} u_j \right) \tag{4.1}$$

for $i = 1, 2, \ldots, n$. The coefficients in a Lotka-Volterra ODE system (i.e., the a_i's and the b_{ij}'s) may depend on time (t) or space (x) or other parameters, but not on the population densities ($u_i, i = 1, 2, \ldots, n$) themselves. (In the earliest examples, the coefficients were simply constants.) The term $b_{ij}u_iu_j$ indicates the impact of the interaction of species i with species j upon the growth rate of species i. Its form in the work of Lotka and Volterra was no accident. Having the impact of the interaction of species i and j upon the growth rate of species i be proportional to the product of the densities of species i and j is a version of the *principle of mass action*. Thus, model (4.1) illustrates how "our traditions of analysis in theoretical and empirical ecology have been largely inherited from developments in classical physics and its applied variants." (Holling, 1973) It also reflects the "conviction [of Lotka and Volterra, along with other pioneers of the theoretical ecology such as G.F. Gause, A.J. Nicholson and V. Bailey] that biology in general, and ecology in particular, would be more scientific if it could be made more mathematical." (Kareiva, 1989) The branch of mathematics that seeks to analyze the behavior of solutions to systems such as (4.1) over time is called *dynamical systems*. It is a field in which the state of the mathematical art

Spatial Ecology via Reaction-Diffusion Equations R.S. Cantrell and C. Cosner
 ISBN: 0-471-49301-5

has advanced rather dramatically since the time Lotka and Volterra derived their models. Nevertheless, the mathematical theory was then advanced enough so that early analyses of (4.1) "provided insights unanticipated by field ecologists (e.g. predator-prey cycles)" (Kareiva, 1989).

The most natural place to begin analysis of (4.1) is to find a componentwise constant solution $(u_1^*, u_2^*, \ldots, u_n^*)$ for the system (i.e., an equilibrium) and then ask what happens when this solution is perturbed slightly. Does the system have the tendency to restore itself over time to the equilibrium state, or might it exhibit some more complicated behavior, such as a tendency toward a periodic cycle of fluctuation? Moreover, if the system does tend toward a periodic oscillation about the equilibrium, what are the amplitude and period of this oscillation? Such an approach is not only mathematically natural but also is particularly well suited to a field such as physics where quantitative precision is a very high priority. However, the central question one would wish to ask regarding the predictions of a model for interacting biological species is inherently a *qualitative* rather than quantitative one, although its answer necessarily has quantitative overtones. Namely, does the model predict the long term survival of the several species under consideration, i.e., *coexistence*? Knowing that the solutions of the system tend to a componentwise positive equilibrium or periodic cycle is a clear indication that the model predicts coexistence of the ecological community, but strictly speaking, such knowledge is more than one needs to know to answer the underlying question. More importantly, it is not always the case that solutions to (4.1) tend to either an equilibrium or a periodic cycle (Hirsch and Smale, 1974), so that some more general criterion for addressing the question of coexistence is needed even in this particular class of models.

The fact that an equilibrium approach to the analysis of Lotka-Volterra ODE models does not yield a satisfactory criterion for deciding when such models predict coexistence in an ecological community not only calls into question the value of the approach and prompts a search for a more suitable criterion. It also raises the issue of the extent of the utility of models essentially derived from concepts in classical physics in the discipline of theoretical ecology (Holling, 1973). For at the scales at which one wants to observe a system of interacting biological species, such a system is more "profoundly affected by changes external to it, and continually confronted by the unexpected" (Holling, 1973) than are most systems of interest in classical physics. The word *extent* is important here, for as noted by Kareiva (1989), Lotka-Volterra models have from their inception provided useful and unexpected insights to ecologists and indeed continue to do so. Moreover, although the equilibrium theory of Lotka-Volterra systems cannot be used to address fully the question of when such models predict coexistence of interacting biological species, it has led to numerous major results in theoretical ecology (e.g., the principle of competitive exclusion), particularly in the seminal work of Robert MacArthur (Kareiva, 1989). Nevertheless, the principle of mass action cannot capture the demographic effects of all types of interactions between pairs of species, nor can Lotka-Volterra ODE models capture such obvious features of possible species interactions as dispersal or discrete generations, to note but two. Theoretical ecology thus requires a range of demographic models, each able to focus on particular aspects of particular kinds of systems, and any criterion for establishing when a Lotka-Volterra ODE model predicts coexistence should have an analogue for a wide range of other models (Cosner, 1996). Moreover, any such criterion should be robust within a given modeling regime in another sense. The demographics of species interactions can only be approximated in any given model, and can be difficult and perhaps even impossible to detect and formulate explicitly at the mid-to-upperranges of abundance for the species in question (Strong, 1986a,b). Consequently, a criterion for deciding when a model for interacting biological species predicts coexistence should exhibit some degree of continuous dependence on the parameters of the model when one of the species is at low densities and should not rely too heavily on the particulars of the rules for the demographics of species interactions when all species in the community are abundant.

These last considerations refer to the aspect of "vagueness and variability" (Strong, 1986a,b; Cosner, 1996) of models for interacting biological species. They also suggest an even deeper question regarding biological communities that has been a subject for debate among ecologists essentially as long as there have been models for interacting species. Namely, "do the populations at a site consist of all those that happened to arrive there [the view of Gleason (1926)], or of only a special subset – those with properties allowing their coexistence [the view of Elton (1933)]?" (Roughgarden, 1989). In other words, do species interactions play a role in determining the structure of ecological communities? It is evident from the subject matter of this chapter that the authors presuppose the answer to the question is yes, at least some of the time. And while a thorough retelling of the history of the debate over the role of species interactions in determining community structure is not the point of the present discussion (the interested reader is directed to the excellent article of Roughgarden (1989)), it is nevertheless worthwhile to report that at the time of the writing of the aforementioned article, a synthesis of the two views seems to have emerged from 60^+ years of theoretical and empirical work. Namely, in Roughgarden's well-turned phrase "a community reflects both its applicant pool and its admission policies" (Roughgarden, 1989).

Returning to the discussion of a possible criterion for determining when a model for interacting biological species predicts coexistence, we have identified three important features of such a criterion:

1. Robustness over a range of models.
2. Robustness with respect to qualitative features of a model when all species in question are in abundance.
3. Robustness with respect to quantitative changes in the parameters of a model when at least one species is at low densities.

The criterion proposed here is known as *permanence*. There is one other feature that permanence should have that we now note. Clearly, a situation of major theoretical and empirical ecological interest is the outcome of introducing an exotic species into an established ecological community (Law and Morton, 1996). Quite possibly, the introduction of the exotic may cause one or more of the species in the community to decrease in abundance, perhaps dramatically. What will be critical for the coexistence of the community will be whether the abundances of such species can recover over time, if not to their initial abundances, at least to some viable lower density levels. In other words, "the main issue for coexistence" "is whether the densities of rare species tend to increase" (Law and Morton, 1996). These considerations lead us to conclude that permanence should be:

4. Independent of the initial state of the system so long as it is componentwise positive.

This fourth feature of permanence means that a species whose persistence involves exceeding an abundance threshold could not be part of a system that we would describe as *permanent*; for example, any species subject to a form of Allee effect (Allee, 1931, 1938). Put more generally,the term *permanence* is not applicable when accidents of history matter.

When they do, there is a related notion called *conditional persistence* that is sometimes applicable which we will also explore.

Permanence is essentially a *qualitative* criterion for addressing the qualitative issue of whether a model for interacting biological species predicts the coexistence of all the species in question. However, to determine that a certain species is expected to be present in a community in the long term is to determine that its density will be above some positive level in the long term. How positive "positive" needs to be is a quantitative issue that will vary from situation to situation, and which permanence, *per se*, does not address. The notion *practical persistence* refers to criteria for a prediction of existence in a system of interacting species, along with some means of estimating asymptotic lower bounds ("floors" in the parlance of Strong (1986a,b)) on the densities of the species in question. Permanence does assert the existence of asymptotic floors on the densities of the species in the system, but without any information on the "heights" of the floors. The extra information afforded in a practical persistence estimate may come at the cost of more stringent conditions on the parameters of the model. Whether permanence or practical persistence is the more useful tool in a particular investigation will depend on the objectives of the investigation. When predictions concerning the ultimate disposition of a very specific ecological community are desirable, as they frequently are in ecology, practical persistence estimates provide more information. However, if biogeographic inferences are to be drawn from the dependence of a prediction of coexistence upon the parameters of a model and the model is being viewed as describing the interactions of a range of ecological communities, permanence is usually a more flexible tool. We shall revisit this point when we discuss practical persistence in detail in Chapter 5.

4.1.2 ODE Models as Examples

As a means of illustrating some of the issues raised in the preceding subsection, we consider three ODE models. First, let us begin with a detailed analysis of the predator-prey model

$$\begin{aligned} \frac{du}{dt} &= u(1-u) - \frac{Auv}{1+Bu+Cv} \\ \frac{dv}{dt} &= \frac{Euv}{1+Bu+Cv} - Dv. \end{aligned} \tag{4.2}$$

We shall see that a number of the issues and phenomena which can arise in models for interacting populations occur in the model (4.2). Here u is the density of the prey species and v that of the predator, and the parameters A, B, C, D, and E are positive constants. The functional response term in the model, $\frac{Auv}{1+Bu+Cv}$, is commonly referred to in the literature as being of Beddington-DeAngelis type (Beddington, 1975; DeAngelis et al., 1975) and may be derived mechanistically via considerations of time utilization (Beddington, 1975; Ruxton et al., 1992) or of spatial limits on predation (Cosner et al., 1999). The term Bu may be regarded as reflecting the time it takes for a predator to handle an item of prey, while the term Cv reflects mutual interference among predators. When this term is neglected (i.e., when $C = 0$), a Beddington-DeAngelis functional response reduces to a Holling functional response of Type II.

It follows from the fact that solutions to initial value problems for (4.2) are unique that if $u_0 \geq 0$ and $v_0 \geq 0$ are population densities for a prey and a predator subject to (4.2) at some initial time and $u(t)$ and $v(t)$ denote their densities at a time t units later, then $u(t) \geq 0$ and $v(t) \geq 0$ with $u(t) = 0$ or $v(t) = 0$ if and only if $u_0 = 0$ or $v_0 = 0$, respectively. Put in other words, the first quadrant of the so called uv-phase plane, its interior and its boundary are invariant under the flow generated by (4.2). This fact makes it convenient to analyze (4.2) by delineating the phase portraits of (4.2) that occur as the system parameters vary.

Consider the predator's numerical response term $\dfrac{Euv}{1 + Bu + Cv}$. Since $\dfrac{Eu}{1 + Bu + Cv} \leq \dfrac{Eu}{Bu} = \dfrac{E}{B}$, the predator's density is a lower solution of the linear equation

$$\frac{dw}{dt} = \left(\frac{E}{B} - D\right) w. \tag{4.3}$$

Consequently, if $\dfrac{E}{B} - D < 0$, the predator density tends to 0 overtime at an exponential rate, and so we postulate that

$$\frac{E}{D} > B. \tag{4.4}$$

To determine the equilibrium for (4.2) we set $\dfrac{du}{dt} = 0$ and $\dfrac{dv}{dt} = 0$. The intersection of the resulting curves (called the prey and predator *isoclines*, respectively) give the equilibria. When $\dfrac{du}{dt} = 0$, either $u = 0$ (the v-axis) or we have the hyperbola

$$v = \frac{(1 - u)(1 + Bu)}{A - C + Cu}. \tag{4.5}$$

It is easy to see from (4.5) that the hyperbola has a vertical asymptote when $u = \dfrac{C - A}{C} < 1$ and that the u-intercept is 1 with $v < 0$ when $u > 1$. When $\dfrac{dv}{dt} = 0$, either $v = 0$ (the u axis) or we have the line

$$v = \frac{(E - BD)u - D}{CD} \tag{4.6}$$

with slope $E - BD$ (>0 from (4.4)) and u-intercept $\dfrac{D}{E - BD}$. It is now immediate that the isoclines (4.5) and (4.6) intersect in the interior of the first quadrant (so that we have a componentwise positive equilibrium) if and only if

$$\frac{E}{D} > B + 1. \tag{4.7}$$

In this case, it is easy to determine that there is a unique such equilibrium (u_*, v_*) with

$$u_* = \frac{2AD}{[A(E - BD) - CE] + \sqrt{[A(E - BD) - CE]^2 + 4ACDE}}. \tag{4.8}$$

The possible configurations of the isoclines are shown in Figure 4.1. It is clear from Figure 4.1 that we always have the extinction states $(0, 0)$ and $(1, 0)$ as well.

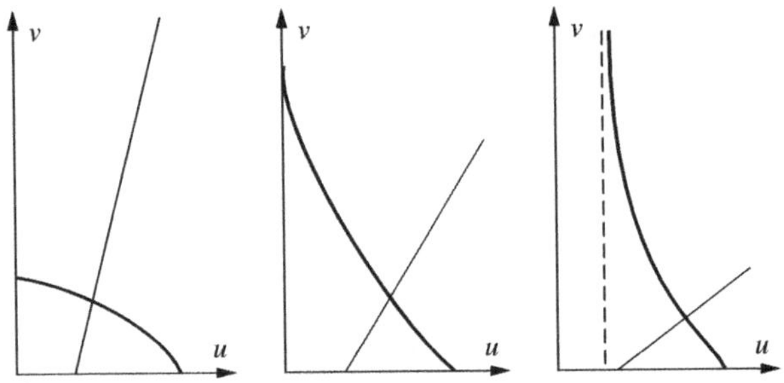

Figure 4.1 These graphs show possible forms of the isoclines for a predator-prey model with Beddington-DeAngelis functional response corresponding to small, medium, and large values of the intraspecific interference parameter C. Note that as C increases the slope of the predator isocline decreases while the prey isocline goes from convex to concave and then develops a vertical asymptote.

It is easy to check that any orbit for (4.2) is bounded with an ultimate bound independent of initial prey and predator densities. The prey density is a lower solution to the logistic equation

$$\frac{dy}{dt} = y(1-y). \tag{4.9}$$

The dynamics of the logistic equation (4.9) are well-known to any student in an introductory course to ordinary differential equations. Namely, the orbit corresponding to any positive initial data y_0 tends to the carrying capacity 1 as time progresses. Consequently, for any $\varepsilon > 0$, there must be a time $t_1 = t_1(u_0)$ (i.e., depending only on the initial prey density) such that $u(t) < 1 + \varepsilon$ for $t \geq t_1$. In particular, there must be a t_1 such that $u(t) < 2$ for $t \geq t_1$. For $t \geq t_1$, the predator density is then a lower solution to the equation

$$\frac{dw}{dt} = \left(\frac{2E}{1+2B+Cw} - D\right) w. \tag{4.10}$$

Clearly $\dfrac{2E}{1+2B+Cw} < \dfrac{2E}{1+2B}$, so that if $\dfrac{2E}{1+2B} - D \leq 0$, w tends to 0 exponentially with time. If $\dfrac{2E}{1+2B} - D > 0$, all solutions to (4.10) with $w(t_1) > 0$ converge to the root of the equation $\dfrac{2E}{1+2B+Cw} - D = 0$; i.e., $w = \dfrac{2E - 2BD - D}{CD}$. In this case, for any $\varepsilon > 0$, there will be a $t_2 = t_2(v(t_1)) = t_2(v_0) > t$, such that $v(t) < \dfrac{2E + 2BD - D}{CD} + \varepsilon$ for $t \geq t_2$. In either case, there will be $t_* > 0$ depending only on u_0 and v_0 so that

$$u(t) \leq 2$$
$$v(t) \leq \max\left\{1, \frac{2E - 2BD - D}{CD} + 1\right\}$$

for all $t \geq t_*$. In the parlance of Strong (1986a,b), the values 2 and $\max\left\{1, \dfrac{2E - 2BD - D}{CD} + 1\right\}$ give asymptotic *ceilings* on the densities of a prey and predator whose interactions are governed by (4.2). Note also that if the condition (4.7) for the existence of a componentwise positive equilibrium to (4.2) is reversed, then for $\varepsilon > 0$

and sufficiently small, $\frac{E(1+\varepsilon)}{1+B(1+\varepsilon)} - D < 0$. Since $u < 1+\varepsilon$ for all large enough t, it follows that $v \to 0$ exponentially as time increases if $\frac{E}{D} < B+1$.

So now let us assume that (4.7) holds. The simplest dynamics for (4.2) would be to have all solution orbits $(u(t), v(t))$ with $u_0 > 0$ and $v_0 > 0$ tend to the equilibrium (u_*, v_*) as $t \to \infty$. One means of establishing that such is the case is to construct a Lyapunov function for (4.2). To this end, we seek a real-valued function $V(u, v)$, defined when $u > 0$ and $v > 0$, so that

(i) $V(u_*, v_*) = 0$.

(ii) $V(u, v) > 0$ if $u \neq u_*$ or $v \neq v_*$.

(iii) $\frac{dV}{dt}(u(t), v(t)) < 0$ for any solution trajectory $(u(t), v(t))$ with $u_0 > 0$, $v_0 > 0$ and $(u_0, v_0) \neq (u_*, v_*)$.

As a general rule, the construction of a Lyapunov function is as much an art as a science. However, there are certain forms for $V(u, v)$ that have been widely used and hence make good initial guesses. One such possibility (see, for example, Goh (1976)) is an appropriate positive linear combination of $V_1(u) = u - u_* - u_* \ln\left(\frac{u}{u_*}\right)$ and $V_2(v) = v - v_* - v_* \ln\left(\frac{v}{v_*}\right)$. It is very easy to check that any $V(u, v)$ of the form

$$V(u, v) = V_1(u) + \alpha V_2(v)$$

satisfies (i) and (ii) so long as $\alpha > 0$. If we experiment calculating $\frac{d}{dt}V(u(t), v(t))$ using (4.2), we find that the choice of $\alpha = \frac{A(1+Bu_*)}{E(1+Cv_*)}$ allows us to cancel all uv terms and we find that

$$\begin{aligned}\frac{d}{dt}V(u(t), v(t)) &= \left[-1 + \frac{ABv_*}{(1+Bu_*+Cv_*)(1+Bu+Cv)}\right](u-u_*)^2 \\ &\quad - \left[\frac{ACE(1+Bu_*)}{E(1+Cv_*)(1+Bu_*+Cv_*)(1+Bu+Cv)}\right](v-v_*)^2.\end{aligned} \tag{4.11}$$

In (4.11) the coefficient of $(v - v_*)^2$ is always negative, so we need only focus on the coefficient of $(u - u_*)^2$.

Notice that

$$\begin{aligned}-1 &+ \frac{ABv_*}{(1+Bu_*+Cv_*)(1+Bu+Cv)} \\ &\leq -1 + \frac{ABv_*}{1+Bu_*+Cv_*} \\ &= -1 + B(1-u_*)\end{aligned}$$

by (4.2). We conclude that V is a Lyapunov function and that (u_*, v_*) is globally asymptotically stable so long as

$$B(1-u_*) < 1. \tag{4.12}$$

One may readily check from (4.8) that $u_* < 1$. Consequently, so long as $B \le 1$ and (4.7) holds, (4.2) admits a globally attracting componentwise positive equilibrium (u_*, v_*) independent of the values of C or A. From the modeling standpoint, if the handling time for a prey item by a predator is short enough, the dynamics for (4.2) are that of a globally attracting positive equilibrium independent of the level of mutual interference among the predators. However, we do not suggest that the level of mutual interference among the predators plays no role in the outcome of the interaction. Indeed, we find from (4.8) that $\lim_{C\to\infty} u_* = 1$ and hence from (4.6) that $\lim_{C\to\infty} v_* = \lim_{C\to\infty} \frac{(E-BD)u_* - D}{CD} = 0$. Hence the equilibrium (u_*, v_*) becomes arbitrarily and uncomfortably close to the extinction state $(1, 0)$ as C becomes large.

Assuming (4.7), by (4.12) we must have $B > 1$ in order to have the possibility of more complicated dynamics. However, since $\lim_{C\to\infty} u^* = 1$, for any given $B > 1$, (4.12) will hold for all large enough C. Increasing C serves to stabilize the interaction given by (4.2), though not necessarily to promote coexistence. Moreover, we should also note that having $B(1-u^*) \ge 1$ does not by itself imply that the dynamics of (4.2) are something other than convergence of every orbit to a componentwise positive equilibrium. It merely opens the possibility of some other outcome. The range of possibilities in this instance are governed by the famous Poincaré-Bendixson Theorem (see, for example, Hale and Kocak (1991)). Indeed, given initial data (u_0, v_0) with $u_0 > 0$ and $v_0 > 0$, the orbit $(u(t), v(t))$, $t \ge 0$, to (4.2) with $(u(0), v(0)) = (u_0, v_0)$ either

(i) converges to one of (u_*, v_*), $(1, 0)$ or $(0, 0)$ as $t \to \infty$;

(ii) spirals with increasing time toward one side of a periodic orbit to (4.2) as $t \to \infty$; or

(iii) converges as $t \to \infty$ toward a compact connected set which consists of some subcollection of $\{(u_*, v_*), (1, 0), (0, 0)\}$ and full orbits to (4.2) (i.e., with time ranging from $-\infty$ to ∞) linking the equilibria in the subcollection together.

We may invoke (4.7) to limit the possibilities further. If $(u(t), v(t))$ were an orbit of (4.2) converging to $(0, 0)$ as $t \to \infty$, we would have $\frac{du}{dt}/u = 1 - u - \frac{Av}{1+Bu+Cv}$ converging to 1 as $u(t)$ and $v(t)$ converge to 0. Consequently, if $(u(t), v(t))$ is near enough to $(0, 0)$, the u component of the trajectory increases with time, moving *away* from 0. So no orbit to (4.2) corresponding to positive initial densities u_0 and v_0 can converge to $(0, 0)$. Likewise, no such orbit can converge to $(1, 0)$, since $\frac{dv}{dt}/v = \frac{Eu}{1+Bu+Cv} - D$ converges to $\frac{E}{1+B} - D > 0$ (by (4.7)) when u converges to 1 and v converges to 0. Clearly then, (4.7) eliminates $(1, 0)$ and $(0, 0)$ from alternative (i). Moreover, it leaves only one way for alternative (iii) to hold; namely, the existence of a full orbit to (4.2) which converges to (u_*, v_*) as $t \to -\infty$ and also $t \to \infty$. (Such an orbit is referred to as a *homoclinic orbit*). Alternative (ii) is left untouched by the assumption that (4.7) holds.

To explore the possibility of destabilizing (u_*, v_*), set $f(u, v) = u(1-u) - \frac{Auv}{1+Bu+Cv}$ and $g(u, v) = \frac{Euv}{1+Bu+Cv} - Dv$, and let $J(u, v) = \begin{pmatrix} \frac{\partial f}{\partial u}(u, v) & \frac{\partial f}{\partial v}(u, v) \\ \frac{\partial g}{\partial u}(u, v) & \frac{\partial g}{\partial v}(u, v) \end{pmatrix}$ be the linearization of the vector field $\begin{pmatrix} f(u,v) \\ g(u,v) \end{pmatrix}$ at the point (u, v). It is a basic fact of dynamical systems that (u_*, v_*) is linearly unstable if $J(u_*, v_*)$ admits an eigenvalue with positive real part. The quantity $\frac{\partial f}{\partial u}(u_*, v_*) + \frac{\partial g}{\partial v}(u_*, v_*)$, called the trace of $J(u_*, v_*)$ and denoted $TrJ(u_*, v_*)$, is the sum of the eigenvalues. If $TrJ(u_*, v_*) > 0$, then at least one of the two eigenvalues of $J(u_*, v_*)$ has positive real part. (Complex valued eigenvalues occur in conjugate pairs, so that their sum is necessarily real.) One easily calculates that

$$TrJ(u_*, v_*) = 1 - 2u_* - \left[\frac{Av_* + ACv_*^2}{(1 + Bu_* + Cv_*)^2}\right] + \left[\frac{Eu_* + BEu_*^2}{(1 + Bu_* + Cv_*)^2}\right] - D \quad (4.13)$$

and with a little bit of care using the facts $\frac{Av_*}{1 + Bu_* + Cv_*} = 1 - u_*$ and $\frac{Eu_*}{1 + Bu_* + Cv_*} = D$, one can reduce (4.13) to

$$TrJ(u_*, v_*) = -u_* + (1 - u_*)\left[\frac{BD}{E} - \frac{CD}{A}\right]. \quad (4.14)$$

In addition to (4.7) (i.e., $E/D > B + 1$), we must already assume $B - 1 > 0$; otherwise (u_*, v_*) is a globally attracting equilibrium regardless of the values of C and A. Moreover, even when $B - 1 > 0$, we can guarantee that (u_*, v_*) is a globally attracting equilibrium by making C large enough. These facts suggest that if we are to destabilize (u_*, v_*) we must do so by lessening the effect of mutual interference among predators. So we let $C \to 0$. From (4.8), it follows that $\lim_{C\to 0} u_* = \frac{D}{E - BD}$, so that from (4.14)

$$\begin{aligned}\lim_{C\to 0} TrJ(u_*, v_*) &= -\frac{D}{E - BD} + \left(1 - \frac{D}{E - BD}\right)\left(\frac{BD}{E}\right) \\ &= D\left[\frac{(B - 1)(E - BD) - 2BD}{E(E - BD)}\right].\end{aligned}$$

It follows that if $(B - 1)(E - BD) - 2BD > 0$, $TrJ(u_*, v_*) > 0$ for $0 < C \ll 1$. In such case, there either is a periodic orbit or a homoclinic orbit for (4.2) as described in the Poincaré-Bendixson Theorem. It is not difficult to calculate that $(B-1)(E-BD)-2BD > 0$ is equivalent to

$$\frac{E}{D} > \left(\frac{B}{B - 1}\right)(B + 1). \quad (4.15)$$

Clearly (4.15) is a more strenuous requirement upon the ratio E/D than is (4.7). By making E/D larger still, we may guarantee, again for $B > 1$ and small values of C, that in fact both eigenvalues of $J(u_*, v_*)$ have positive real part and that (u_*, v_*) is an unstable spiral point for (4.2), so that all orbits corresponding to positive initial densities spiral to a periodic orbit as $t \to \infty$. Such a trajectory spiraling out from (u_*, v_*) eliminates the possibility of a homoclinic orbit.

We now see that the dynamics of (4.2) may well be more complicated than convergence of all trajectories with positive initial densities to (u_*, v_*). A mechanism for producing more complicated dynamics is as follows. First, there must be a "predisposition" in the form of a long enough handling time of prey items by predators as reflected in the requirement that

B exceed 1. In that case, more complicated dynamics will be possible if the comparatively long handling time is compensated for by a diminished level of mutual interference among the predators and a favorable enough ratio of feeding efficiency (E) to death rate (D) for the predators. Such a mechanism is akin to the notion of a "paradox of enrichment" (Arditi and Berryman, 1991; Berryman, 1992; Ginzburg, 1998; Rosenzweig, 1971; Tanner, 1975). However, far more important than the observation that dynamics more complicated than convergence to a componentwise positive equilibrium are possible for the predator-prey system (4.2) is the realization that destabilization of such an equilibrium is not tantamount to a loss of coexistence. Indeed, (4.2) illustrates that coexistence is less about the complexity of the dynamics of a system of interacting species than it is about the location of the dynamics longterm relative to the extinction states of the system. For example, a periodic orbit to (4.2) that attracts all solution trajectories for (4.2) corresponding to positive initial prey and predator densities (except for the equilibrium (u_*, v_*)) and is located some distance from the uv-axes may be a better harbinger of coexistence of the two species than is having (u_*, v_*) globally attracting but extremely close to $(0, 1)$ (which we see is possible if C is large enough). (The model is deterministic; but when densities become too low stochastic factors which might result in extinction become more important, so the model's prediction of persistence becomes less reliable relative to the real system when densities become too small.) The condition $\frac{E}{D} > B + 1$ (i.e. (4.7)) for the existence of (u_*, v_*) also guarantees that the extinction equilibria $(0, 0)$ and $(1, 0)$ repel orbits corresponding to positive initial densities. We shall see that *in this case* this feature is all that is required for us to know that "the densities of rare species tend to increase" (Law and Morton, 1996) and that the long term dynamics are bounded away from the uv-axes in a suitable sense; i.e., the system is permanent. It is true that we must continue to be concerned with the issue of how close a periodic or homoclinic orbit may come to the uv-axes. However, we have that concern even in the case when all solution trajectories corresponding to positive initial densities converge to (u_*, v_*) when (u_*, v_*) is uncomfortably close to $(1, 0)$. Finally, we note in reference that the treatment of this example follows that in Cantrell and Cosner (2001b).

A permanent system may exhibit even more complicated dynamics, as we note with our next example. Hastings and Powell (1991) derived the three component simple food chain model:

$$\begin{aligned}\frac{dR}{dt} &= R(1-R) - x_c y_c \frac{CR}{R+R_0}\\ \frac{dC}{dt} &= x_c C\left[-1 + \frac{y_c R}{R+R_0}\right] - x_p y_p \frac{PC}{C+C_0}\\ \frac{dP}{dt} &= x_p P\left[-1 + y_p \frac{C}{C+C_0}\right].\end{aligned} \tag{4.16}$$

In (4.16), R represents the density of a resource species (i.e. the bottom of the food chain), C the density of a consumer species feeding upon the resource, and P the density of a predator feeding in turn upon the consumer. In the functional response term $x_c y_c \frac{CR}{R+R_0}$, the parameter R_0 is the half-saturation density measured in terms of the units of resource carrying capacity which is scaled to 1, x_c is the mass-specific metabolic rate of the consumer measured relative to the production-to-biomass ratio of the resource species, and y_c is a measure of the ingestion rate of the consumer per unit metabolic rate (McCann and Yodzis, 1994). (The parameters C_0, x_p and y_p are defined analogously.) McCann and Yodzis (1994) note that if $x_c = 0.4$, $y_c = 2.01$, $R_0 = 0.161$, $x_p = 0.071$, $y_p = 5.0$ and $C_0 = 0.5$, the system (4.16) exhibits chaotic solutions for which nevertheless "all three densities remain bounded away from zero by some 'healthy' amount". They refer the this phenomenon as "persistent chaos". It is a form of permanence, as we shall see. For more examples in this vein, see Klebanoff and Hastings (1994 a,b) and Schaffer et al. (1986).

The third example we wish to consider is the celebrated May-Leonard model for three competing species (May and Leonard, 1975):

$$\begin{aligned}\frac{du_1}{dt} &= u_1[1 - u_1 - \alpha u_2 - \beta u_3]\\ \frac{du_2}{dt} &= u_2[1 - \beta u_1 - u_2 - \alpha u_3]\\ \frac{du_3}{dt} &= u_3[1 - \alpha u_1 - \beta u_2 - u_3]\end{aligned} \tag{4.17}$$

where u_i denotes the density of the ith competitor. The (necessarily for a competition model) positive parameters α and β are assumed to satisfy

$$\alpha < 1 < \beta, \quad \alpha + \beta > 2. \tag{4.18}$$

This model is a special case of (4.1); it is of Lotka-Volterra form. It played an important role in the development of the mathematics of permanence, as we describe below and in the next subsection.

It is easy to establish that under assumption (4.18), (4.17) has exactly 5 equilibria with $u_i \geq 0$ for $i = 1, 2, 3$: $(0, 0, 0), (1, 0, 0), (0, 1, 0), (0, 0, 1)$ and $\left(\frac{1}{1+\alpha+\beta}, \frac{1}{1+\alpha+\beta}, \frac{1}{1+\alpha+\beta}\right)$. Notice that near $(0, 0, 0)$ the relative growth rates $\frac{du_i}{dt}/u_i$ are all approximately equal to 1 and that near each of $(1, 0, 0)$, $(0, 1, 0)$ and $(0, 0, 1)$ the relative growth rate for *one* of the three competitors is approximately $1 - \alpha > 0$. Consequently, as was the case with (4.2) (under the assumption (4.7)), no solution trajectory to (4.17) (under the assumption (4.18)) with all initial densities positive can converge to an equilibrium with one or more of the interacting species absent. However, in this case, the system fails to be permanent, so that it cannot be the case that prohibiting solution trajectories from converging to extinction equilibria is a sufficient condition for a prediction of permanence.

To understand why (4.17) should fail to be permanent under assumption (4.18), consider the subsystem which arises when u_3 is set to zero:

$$\begin{aligned}\frac{du_1}{dt} &= u_1[1 - u_1 - \alpha u_2]\\ \frac{du_2}{dt} &= u_2[1 - \beta u_1 - u_2].\end{aligned} \tag{4.19}$$

The system (4.19) has only the three equilibria $(0, 0)$, $(1, 0)$ and $(0, 1)$. Since $\frac{du_i}{dt}/u_i = 1$ at $(0, 0)$ for $i = 1, 2$, $\frac{du_1}{dt}/u_1 = 1 - \alpha > 0$ at $(0, 1)$ and $\frac{du_2}{dt}/u_2 = 1 - \beta < 0$ at $(1, 0)$, all solution trajectories $(u_1(t), u_2(t))$ to (4.19) with $u_1(0) > 0$ and $u_2(0) > 0$ must converge to $(1, 0)$ as $t \to \infty$. Moreover, there is a full orbit Γ_{21} to (4.19) such that along Γ_{21}, $\lim_{t\to-\infty} (u_1(t), u_2(t)) = (0, 1)$ and $\lim_{t\to+\infty} (u_1(t), u_2(t)) = (1, 0)$. (See, for example, Hirsch and Smale (1974) and Schuster et al. (1979a).) Consequently, all solutions to (4.17) with $u_3(t) \equiv 0$ converge to $(1, 0, 0)$ as $t \to \infty$ and there is an orbit $\gamma_{21} (= \Gamma_{21} \times \{0\})$ to (4.17) such that $\lim_{t \to -\infty} \gamma_{21}(t) = (0, 1, 0)$ and $\lim_{t \to \infty} \gamma_{21}(t) = (1, 0, 0)$. Analogously, all solution trajectories to (4.17) with $u_2(t) \equiv 0$ converge to $(0, 0, 1)$ as $t \to \infty$ and there is an orbit γ_{13} to (4.17) such that $\lim_{t \to -\infty} \gamma_{13}(t) = (1, 0, 0)$ and $\lim_{t \to +\infty} \gamma_{13}(t) = (0, 0, 1)$. Likewise, all solution trajectories to (4.17) with $u_1(t) \equiv 0$ converge to $(0, 1, 0)$ as $t \to \infty$ and there is an orbit γ_{32} to (4.17) such that $\lim_{t \to -\infty} \gamma_{32}(t) = (0, 0, 1)$ and $\lim_{t \to \infty} \gamma_{32}(t) = (0, 1, 0)$. Define γ by

$$\gamma = \{(1, 0, 0)\} \cup \gamma_{13} \cup \{(0, 0, 1)\} \cup \gamma_{32} \cup \{(0, 1, 0)\} \cup \gamma_{21} \cup \{(1, 0, 0)\}. \tag{4.20}$$

Notice that γ in (4.20) is a 3-dimensional analogue of the type of compact, connected set referred to in alternative (iii) of the Poincaré-Bendixson Theorem (see Figure 4.2a). We say that γ is a *heteroclinic cycle* for (4.17) in the boundary of the positive octant of $\mathbb{R}^3$. As is suggested by analogy with the Poincaré-Bendixson Theorem, it may well be possible for a solution trajectory to (4.17) with positive initial densities to spiral out toward γ. Indeed, Schuster et al. (1979a) prove that such is the case for any componentwise positive initial data other than the equilibrium $\left(\frac{1}{1+\alpha+\beta}, \frac{1}{1+\alpha+\beta}, \frac{1}{1+\alpha+\beta}\right)$. Consequently, solution trajectories (again, other than the componentwise positive equilibrium) move in an infinitely repeating pattern from being uncomfortably close to extinction for species 2 and 3 to being uncomfortably close to extinction for species 1 and 2 to being uncomfortably close to extinction for species 3 and 1 to once again being uncomfortably close to extinction for species 2 and 3 with a level of discomfort that rises with progressing time. (See Figure 4.2b. For more detail about the nature of such trajectories, see May and Leonard (1975).) Certainly, in no reasonable sense would a trio of interacting species modeled by (4.17) (under the assumption (4.18)) be considered coexistent.

This last example demonstrates that to make a prediction of coexistence in the context of ODE models for n interacting species, we must not only be able to eliminate the possibility that a componentwise positive solution trajectory can converge over time to a state with one or more extinct components but must also be able to exclude the existence of a cycle in the boundary of the positive orthant in $\mathbb{R}^n$ which attracts orbits with all components initially positive. This observation has lead directly to one of the main mathematical tools for detecting permanence in a model for interacting species, as we note below and as we shall present in Section 3.

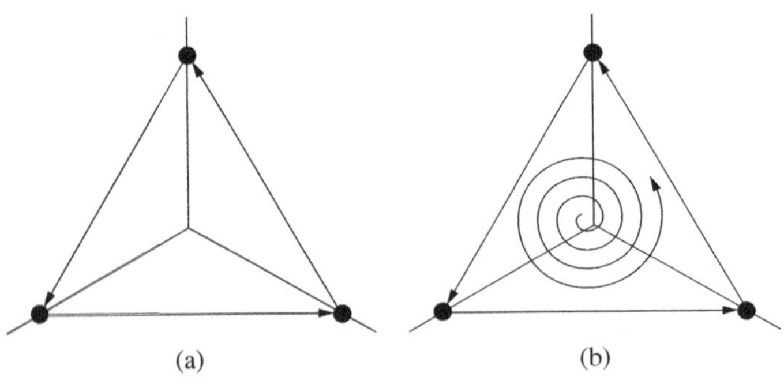

Figure 4.2 (a) The boundary cycle in the May-Leonard model, (b) a trajectory approaching the boundary cycle in the May-Leonard model.

4.1.3 A Little Historical Perspective

Notions of persistence are absolutely indispensable to many of the results we present in this volume. However, our aim is to examine how interacting communities of biological species are influenced by spatial features of their habitats and their interactions, with persistence theory as a principal tool. We do not intend an exploration of persistence theory *per se*, and would recommend the interested reader to the two books of Hofbauer and Sigmund (1989, 1998) or the introductory survey of Hutson and Schmitt (1992). Indeed, our main objective in this chapter and the next is to obtain persistence theory results in the context of reaction-diffusion models for bounded habitats, expressed in a manner that is as ecologically "user friendly" as is feasible, for use in our subsequent analyses.

Neither is it our intent at this juncture to present a comprehensive history of the mathematical development of persistence theory. Other treatments already exist in the literature, all written or co-written by pioneers in the subject. In addition to the work of Hofbauer and Sigmund and Hutson and Schmitt noted above, we would call the reader's attention to the papers of Waltman (1991), Freedman and Moson (1990) and Janzen and Sigmund (1998). However, we do believe it is both interesting and useful to make a few observations regarding the development of the mathematics of persistence. To a large extent, this development is a tale of two schools of thought (permanence and uniform persistence) on two continents (Europe and North America) that arose in the mid-to-late 1970s and that had to a large extent coalesced by the late 1980s to early 1990's. Each school of thought is responsible for a major theoretical tool for the detection of permanence in a model for interacting biological species. The permanence school is primarily associated with the *average Lyapunov function approach*, while the uniform persistence school is primarily associated with the *acyclicity approach*. Once we have given both formal ecological and mathematical definitions of permanence (in Section 2), we shall describe in detail both approaches for detecting permanence (in Section 3). But for now, let us note some of the landmark papers in the development of each school. If we may be allowed an editorial remark at this point, we are of the opinion that the pioneers in both schools are due the gratitude of subsequent workers in mathematical and/or theoretical ecology, ourselves included.

Roughly speaking, permanence in a model system for the densities of a collection of interacting species means the system possesses both an asymptotic "ceiling" and a positive asymptotic "floor" on the densities of all the species in question, the "heights" of which are independent of the initial state of the system so long as each component is positive. The notion of permanence first appeared in the work of Schuster, Sigmund and Wolff (1979 a,b), along with the essence of the "average Lyapunov function" approach. "The essential feature[s] of the technique were [subsequently] extracted and made explicit in the context of ordinary differential equations in Hofbauer (1981)" (Hutson and Schmitt, 1992) and "put into a dynamical systems framework in Hutson (1984)." (Hutson and Schmitt, 1992). The basic ideas are not hard to understand. View the collection of ordered n-tuples of species densities as a set whereupon there is some means of measuring distance between n-tuples. If all the densities are positive, such an n-tuple is "inside"or "interior" to the set, whereas if one or more species is extinct, the n-tuple lies on the "boundary" of the set. An average Lyapunov function associated with the governing system of equations for the densities of the interacting species is a nonnegative (hence necessarily real-valued) function which is zero on the boundary of the set and positive on the interior with the additional property that its time average grows along every interior solution trajectory sufficiently close to the boundary "so that if one waits for a sufficiently long time, one can be sure to be further away from the boundary that one was before" (Janzen and Sigmund, 1998). Asymptotically, such a function prevents any solution trajectory from coming within some positive distance from the boundary, which provides the asymptotic "floor" for the system. The asymptotic "ceiling" arises usually from self-regulation on some component of the system, and the existence of the asymptotic "ceiling" may well play a role in constructing the average Lyapunov function and hence establishing the asymptotic "floor".

Uniform persistence in the system in essence means the system possesses a positive asymptotic "floor" on the densities of all the species involved, again independent of the initial state of the system so long as each component is initially positive. There is no requirement for an asymptotic "ceiling". So on the surface, uniform persistence is a less stringent requirement on a model for interacting species than is permanence. However, in most of the systems that we consider in this volume, there is effectively little difference. For example, in systems of competition, self-regulation is usually imposed on each component species so that an asymptotic "ceiling" on the density of anyone of the interacting species can be determined without reference to the values of the other densities. On the other hand, a system of predation lacking some form of dissipation may well experience "boom-bust" cycles which tend to inhibit the establishment of a positive asymptotic "floor".

The notion of uniform persistence evolved through a series of papers published between 1977 and 1986. In Freedman and Waltman (1977), the concept of *weak persistence* was introduced in the context of three species food chain models. Weak persistence in a model for interacting species requires that there be a positive threshold that is exceeded infinitely often as time progresses by each component of a solution trajectory. Weak persistence does not force all densities to exceed the threshold simultaneously nor does it prevent a density from becoming uncomfortably small infinitely often. As a result, weak persistence is not a strong enough notion of coexistence, as the May-Leonard competition model (4.17) is weakly persistent under assumption (4.18), but should not be regarded as predicting coexistence. These drawbacks to the notion of weak persistence as a predictor of coexistence in models for interacting species can be addressed by requiring that each component of a solution trajectory for the system must eventually exceed some positive threshold once and for all. This concept is referred to as *persistence or strong persistence* and was introduced in Freedman and Waltman (1984). In particular, the May-Leonard example (4.17)–(4.18) is not strongly persistent. However, there is a difficulty with strong persistence as a predictor of coexistence in a model for interacting species. Namely, the positive threshold is allowed to depend on the solution trajectory. In this event, the Lotka-Volterra predator-prey system

$$\begin{aligned}\frac{du_1}{dt} &= (a - bu_2)u_1 \\ \frac{du_2}{dt} &= (-c + du_1)u_2\end{aligned} \tag{4.21}$$

is strongly persistent, since its phase portrait consists of an infinite family of periodic orbits surrounding the equilibrium $(c/d, a/b)$ and filling up the first quadrant of u_1u_2-space (Janzen and Sigmund, 1998). However, since periodic orbits become arbitrarily close to extinction states, we can not assert that the model predicts coexistence of the species in question. Requiring that the threshold be independent of the initial state of the system so long as each density is initially positive alleviates the difficulty and leads us to uniform persistence. All three terms "were defined for the first time for dynamical systems in a locally compact metric space with respect to sets with boundary and nonempty interior" (Freedman and Moson, 1990) in Butler, Freedman and Waltman (1986).

Just as the development of the notion of uniform persistence was strongly influenced by the May-Leonard example (4.17)–(4.18), so was the parallel development of the acyclicity approach to detecting permanence in models for interacting species. In the acyclicity approach, we again view the collection of ordered n-tuples of densities for the species in question as a set, with densities that are componentwise positive viewed as "interior" to the set and densities with one or more zero components viewed as being on the "boundary" of the set. Particular emphasis is placed on the dynamics of the system on the boundary, where we effectively can consider systems which are simpler by virtue of involving no more than $n - 1$ interacting species at a time. Limit sets of these resulting subsystems are then identified. The Acyclicity Theorem allows us to assert that the original system is permanent provided that two requirements are met. First, no solution trajectory from the interior converges to one of the boundary limits with increasing time. Second, there can be no heteroclinic (or homoclinic) cycle in the boundary which links up various of the limits, as was the case in the May-Leonard example.The acyclicity approach was first used to establish strong persistence in Freedman and Waltman (1984). A version sufficient to guarantee permanence in ODE systems was given in Butler, Freedman and Waltman (1986). The extension of the result to a setting appropriate for the consideration of systems of reaction-diffusion equations appeared in Hale and Waltman (1989).

4.2 Definition of Permanence

Our hope is that the preceding discussion and examples have served to give the reader some intuition about what should be meant when one says that a model for interacting biological species *exhibits permanence* or *is permanent*. Indeed, it is relatively straightforward to state a definition in terms of the notion of ecological "ceilings" and "floors" on species densities that is both accessible to a broad scientific audience and applicable to a rather wide class of continuous-time and discrete-time models for interacting species. All that is required is that solution trajectories corresponding to tuples of nonnegative densities exist, be uniquely determined and componentwise nonnegative for all positive time. Consequently, the definition is pertinent to appropriate general models from matrix theory, ordinary differential equations, reaction-diffusion equations and delay differential equations, just to name some of the more obvious "suspects". In particular, the definition allows in principle for explicit spatial dependence in the model and for the model to be free of explicit reference to time (autonomous) or to depend in an explicit manner on a time variable (nonautonomous). The real issue is whether the mathematics which exists to deal with a particular model is developed enough so that tractable conditions for permanence can ultimately be expressed in terms of various parameters of the model in a way that is in principle measurable by experiment or field work. Mathematical treatment of permanence for such a model requires that the model be recast as a dynamical or semi-dynamical system, which is the suitable mathematical context for permanence as an abstract mathematical notion which will in turn imply permanence in the "ceiling-floor" sense. To avoid confusion, we shall refer when necessary to permanence in the "ceiling-floor" sense as *ecological permanence*, and in the dynamical systems context as *abstract permanence*. In the case of models without explicit reference to time, recasting the model as a dynamical or semi-dynamical system is straightforward. Namely, the collection of solution trajectories to the model forms the requisite dynamical or semi-dynamical system. (Recall the definition of dynamical system from Chapter 1.) In this case, as we shall report throughout this volume, there has been substantial progress in analyzing abstract permanence and in applying the results of this analysis to increase theoretical understanding in ecology. Recasting a nonautonomous model for interacting species as a dynamical or semi-dynamical system requires a mathematically intricate construction called a *skew product flow* or *semiflow* (Sell, 1967). In this construction, one must embed the model in question into a family of related models and track the time evolution of each of these models and their solution trajectories simultaneously to have the semigroup property which is the defining characteristic of a dynamical or semi-dynamical system. The theorems of abstract permanence may or may not be available in the resulting system and when they are, the level of understanding of abstract permanence in this context is lower, and there are far fewer ecological applications to report. One case we consider is that of reaction-diffusion models with time periodic coefficients. We treat such problems in Section 5 of Chapter 5, at which point we discuss more formally the notion of a skew-product flow or semiflow. We also observe in Chapter 5 how ecological permanence may sometimes be obtained for nonautonomous models via the previously mentioned alternative to permanence called practical persistence.

4.2.1 Ecological Permanence

Let $u_1, \dots, u_n$ represent the densities of n biological species whose interaction is tracked over time via a model expressed in terms of their densities. We distinguish models on the basis of how time is kept. If time is viewed as a continuous variable, the general form of model is

$$\frac{\partial u_i}{\partial t} = F_i(t, x, u_1, \dots, u_n) \tag{4.22}$$

for $i = 1, \dots, n$. The form (4.22) allows us to consider ODEs, reaction-diffusion equations, delay differential equations and various other modeling regimes. We always assume that

$$F_i(t, x, u_1, \dots, u_n) = 0 \quad \text{if} \quad u_i = 0. \tag{4.23}$$

The continuous and/or differentiable dependence of F_i in (4.22) upon its arguments will vary according to the requirements of a given model as may the notion of what constitutes an initial condition. We assume here that such dependence is suitable to guarantee that solution trajectories are uniquely determined by the initial condition upon densities and components of solution trajectories remain nonnegative with $u_i(x,t) = 0$ possible for some $x \in \Omega$, $t > 0$ only when the ith component of the initial condition vanishes identically and $u_i(x,t) \equiv 0$ for all $x \in \Omega$ and all $t > 0$. If there is a spatial aspect to the model, we assume that $x \in \Omega$, a bounded open set in $\mathbb{R}^1, \mathbb{R}^2$ or $\mathbb{R}^3$, whose geometry is suitable to the model at hand. When a side requirement on the values of u_i along the boundary $\partial\Omega$ of Ω is necessary, we denote it by

$$B_i u = 0 \tag{4.24}$$

for $x \in \partial\Omega$, $t \geq 0$.

We are now able to define ecological permanence for (4.22). If there is no spatial aspect to the model requiring a boundary condition on any of the densities, or if there is such a spatial aspect but for no $x \in \partial\Omega$ and for no $i \in \{1, \dots, n\}$ does (4.24) reduce to the Dirichlet condition

$$u_i(x,t) = 0, \tag{4.25}$$

we shall say that (4.22) is *ecologically permanent* if there are positive numbers m and M with $m < M$ such that if $(u_1(x,t), \ldots, u_n(x,t))$ denotes the solution trajectory to (4.22) with initial condition $(u_1^0, \ldots, u_n^0)$ and $u_i^0 > 0$ appropriately for $i = 1, \ldots, n$, then there is a $t_0 > 0$ depending only on $(u_1^0, \ldots, u_n^0)$ such that

$$m \leq u_i(x,t) \leq M \tag{4.26}$$

for all $i \in \{1, \ldots, n\}$, all $x \in \Omega$ and all $t \geq t_0$. If there is an $i \in \{1, \ldots, n\}$ and an $x \in \partial\Omega$ such that (4.24) reduces to (4.25), no positive constant can serve as a lower bound on the component $u_i(x,t)$ of the solution trajectory and the bound (4.26) on u_i must be modified to

$$me_i(x) \leq u_i(x,t) \leq M \tag{4.27}$$

again for all $x \in \Omega$ and all $t \geq t_0$, where $e_i(x)$ is a smooth function on $\overline{\Omega}$ such that $e_i(x) > 0$ for all $x \in \Omega$, $e_i(x)$ satisfies boundary condition (4.24), and $\dfrac{\partial e_i(x)}{\partial \vec{n}} < 0$ for all $x \in \partial\Omega$ for which $e_i(x) = 0$. A particular choice of e_i is the solution of the boundary value problem

$$\begin{aligned} -\Delta u &= 1 \text{ in } \quad \Omega \\ B_i u &= 0 \text{ on } \quad \partial\Omega \end{aligned} \tag{4.28}$$

It is well-known (e.g., Gilbarg and Trudinger, 1977), that e_i has the required properties. If $\Omega = (0,1)$ and $e_i(0) = 0 = e_i(1)$, then $e_i(x) = \dfrac{x(1-x)}{2}$. Note that the positive lower bound in (4.27) is the strongest that is possible if the model is subject to (4.25) for some density u_i.

If time is viewed as a discrete variable, say k, the general form of the model becomes

$$u_i(k+1) = F_i(k, x, u_1, \ldots, u_n) \tag{4.29}$$

for $i = 1, \ldots, n$ and k a nonnegative integer.

In analogy to (4.23) we assume that

$$F_i(k, x, u_1, \ldots, u_n) = 0 \quad \text{if} \quad u_i(k) = 0. \tag{4.30}$$

In the continuous time case, (4.23), the structure of the model and the relevant underlying theory (such as for ODEs, reaction-diffusion equations or delay differential equations) serve to guarantee that solution trajectories are uniquely determined by initial data and that components of solution trajectories remain nonnegative with $u_i(x,t) = 0$ possible for some $x \in \Omega$ and $t > 0$ only if the u_i^0 component of the initial condition vanishes and $u_i(x,t) \equiv 0$ for all $x \in \Omega$ and $t > 0$. In (4.29), the fact that solution trajectories are uniquely determined by initial data follows from the fact that (4.29) describes an iteration. However, there is nothing in the structure of (4.29) to guarantee solution trajectories emanating from componentwise nonnegative data remain componentwise nonnegative. So we must restrict the class of models we can consider by requiring as well that

$$F_i(k, x, u_1, \ldots, u_n) > 0 \quad \text{if} \quad u_i(k) > 0. \tag{4.31}$$

The definition of ecological permanence for (4.29) is now analogous to that for (4.22).

4.2.2 Abstract Permanence

We now assume that the model (4.22) (or (4.29)) has been recast as a dynamical system (or flow) or semi-dynamical system (or semiflow) π on an underlying metric space (Y, d). Our focus here is on autonomous models, in which case Y corresponds to the collection of n-tuples of densities for (4.22) or (4.29) or to some subset thereof. In the nonautonomous case, Y is a more complicated object that we shall describe when we consider reaction-diffusion models with time periodic coefficients.

Assume in that which follows that time is a continuous variable. (Analogous results obtain when time is discrete.) Consistent with Chapter 1, we let $\pi(u, t)$ denote the value of π in Y at time $t > 0$ corresponding to $\pi(u, 0) = u$. We require for our purposes that $\pi(u, t)$ be defined in Y for every $u \in Y$ and $t > 0$ and denote by $\gamma^+(u)$ the subset of Y given by

$$\gamma^+(u) = \{\pi(u, t) : t \in [0, \infty)\}. \tag{4.32}$$

We refer to $\gamma^+(u)$ as the positive semi-orbit through u.

In the case π is a dynamical system, the positive semi-orbit $\gamma^+(u)$ may be continued backward in time through some interval $(t_-(u), 0)$, and $\{\pi(u, t) : t > t_-(u)\}$ is the orbit through u. If $t_-(u) = -\infty$, $\pi(u, t)$ is globally defined and we set

$$\gamma(u) = \{\pi(u, t) : t \in (-\infty, \infty)\}. \tag{4.33}$$

If π is a semi-dynamical system, backward continuation is not necessarily possible for all $u \in Y$. However, we shall *assume* that for the semi-dynamical system π, the backward continuation $\pi(u, t)$ of u is unique whenever it is possible. Such an assumption is verifiable in the reaction-diffusion models of primary interest to us in this volume (e.g. see, Henry (1981)), but may place substantial constraints upon models which may be considered here, in particular in the case of discrete time models (4.29) when the iteration fails to be invertible.

We call a subset U of Y *forward invariant* if it has the property:

$$\text{For all} \quad u \in U, \quad \text{the positive semi-orbit satisfies } \gamma^+(u) \subseteq U. \tag{4.34}$$

If we let $\gamma^+(U) = \{\gamma^+(u) | u \in U\}$, then U being forward invariant means $\gamma^+(U) \subseteq U$. We say that a subset U of Y is *invariant* if it has the property:

$$\text{For all} \quad u \in U, \quad \text{the global orbit} \quad \gamma(u) \quad \text{exists and} \quad \gamma(u) \subseteq U. \tag{4.35}$$

In this case we set $\gamma(U) = \{\gamma(u) | u \in U\}$, and U invariant means $\gamma(U) \subseteq U$.

We may now define *permanence for* π, which is what we mean by *abstract permanence*. We henceforth assume that the metric space Y is complete with respect to the topology induced by d (see Chapter 1) and that Y can be written

$$Y = Y_0 \cup \partial Y_0 \tag{4.36}$$

where Y_0 is open in Y and both Y_0 and ∂Y_0 are forward invariant. In terms of the original model (4.22), Y_0 refers to n-tuples of densities which are componentwise positive, while ∂Y_0 consists of those n-tuples of densities which are zero in at least one component, meaning that at least one of the species is absent. With this terminology, we say that π is *permanent* if there is a subset U of Y_0 with the following properties:

(i) U is bounded,

(ii) $\inf_{u \in U} d(u, \partial Y_0) > 0$,

(iii) $\lim_{t \to \infty} d(\pi(v, t), U) = 0$ for all $v \in Y_0$,

where for a point $y \in Y$ and a subset $V \subseteq Y$, $d(y, V) = \inf_{v \in V} d(y, v)$ is the usual notion of the distance from a point to a set in a metric space.

Now let $c = \inf_{u \in U} d(u, \partial Y_0)$. By (ii), $c > 0$. Let $0 < \varepsilon < \frac{c}{2}$ and define the ε-neighborhood $\mathcal{B}(U, \varepsilon)$ by

$$\mathcal{B}(U, \varepsilon) = \{y \in Y : \text{there is some} \quad u \in U \quad \text{such that} \quad d(u, y) < \varepsilon\}.$$

Now let $v \in Y_0$ be arbitrary. By (iii), there is a $t_0(v)$ such that $d(\pi(v, t), U) < \varepsilon$ for all $t \geq t_0(v)$. Let $w \in \partial Y_0$ be arbitrary and $t \geq t_0(v)$. Choose $u \in U$ such that $d(\pi(v, t), u) < \varepsilon$. Then by (ii),

$$\begin{aligned} c \leq d(w, u) &\leq d(w, \pi(v, t)) + d(\pi(v, t), u) \\ &< d(w, \pi(v, t)) + \varepsilon \end{aligned}$$

so that $d(\pi(v, t), w) > c - \varepsilon > \frac{c}{2}$ for any $w \in \partial Y_0$. It follows that if $t \geq t_0(v)$,

$$d(\pi(v, t), \partial Y_0) \geq \frac{c}{2}. \tag{4.37}$$

The inequality in (4.37) suggests strongly that permanence of π implies (4.26) or (4.27); i.e. ecological permanence. In Section 6, we shall verify that such is indeed the case for reaction-diffusion models on bounded spatial habitats.

4.3 Techniques for Establishing Permanence

To assert that a flow or semiflow π is permanent is to guarantee that all forward orbits for π starting in Y_0 are ultimately bounded, as well as ultimately bounded away from the extinction states ∂Y_0, with both bounds independent of the orbit. The property of ultimate uniform boundedness of forward orbits of π starting in Y_0 indicates that ultimate uniform boundedness of all forward orbits for π starting in Y is a precondition for the permanence of π. This property is referred to as the dissipativity of π on Y.

When π arises from an ODE model, Y is a subset of $\mathbb{R}^n$. Recall from Chapter 1 that in this case, bounded subsets of Y have compact closure, i.e. Y is locally compact. It follows that for any fixed $t \geq 0$, $\pi(\cdot, t)$ regarded as a map from Y to Y is compact; i.e., if $V \subseteq Y$ is bounded, then $\pi(V, t)$ has compact closure.

When Y is complete, π is dissipative and $\pi(\cdot, t)$ is compact for $t > t_0$ for some $t_0 \geq 0$, a classical result of Bilotti and LaSalle (1971) (see also Hale and Waltman (1989)) asserts that π has a *global attractor*; i.e., there is a *compact invariant subset U of Y* such that

$$\lim_{t \to \infty} \sup_{v \in V} d(\pi(v, t), U) = 0$$

for any bounded subset V of Y. In such a case, given any ε (> 0) neighborhood $\mathcal{B}(U, \varepsilon)$ of U, there will be a positive time $t_0 = t_0(\varepsilon)$ depending only on ε such that by time t_0 all forward orbits beginning in V have reached $\mathcal{B}(U, \varepsilon)$ and remain there for all subsequent times.

For model formulations such as reaction-diffusion equations on bounded spatial domains or delay differential equations, Y is no longer a finite dimensional set and hence is no longer locally compact. In particular, $\pi(\cdot, 0)$ is not a compact map. However, when π is dissipative, the underlying mathematical features of such models allow us to assert that $\pi(\cdot, t)$ is compact for $t > t_0$ for some $t_0 \geq 0$, so that the Bilotti-LaSalle Theorem for the existence of a compact attractor is still applicable. Consequently, the two established techniques for asserting permanence, the *average Lyapunov function approach* and the *acyclicity approach*, which were initially established in the context of ODE models and rely on the existence of a global attractor for π, extend to many cases where Y is infinite dimensional. In particular, both approaches apply in principle to all the models we shall consider in this volume. Notice, however, that in the definition of permanence, there is no explicit mention of compactness *per se*. Exploring persistence results without compactness is hence conceivable, but would require a different set of techniques.

In the results that follow we assume that time is a continuous variable. Companion results hold in case time is measured discretely; see Hutson and Schmitt (1992) for more details.

4.3.1 Average Lyapunov Function Approach

We assume that (4.22) is a model for interacting species satisfying the conditions of Section 2 and that $\pi : Y \times [0, \infty) \to Y$ is a semiflow resulting from (4.22) with

(i) Y a complete metric space;

(ii) π dissipative; (4.38)

(iii) $\pi(\cdot, t) : Y \to Y$ compact for $t \geq t_0$ for some $t_0 > 0$.

Then there is a global attractor for π, which we denote by $\mathcal{A}$. The definition of global attractor guarantees that if $\varepsilon > 0$ is prescribed, then for any bounded subset V of Y, there is a $t_V > 0$ such that $\pi(V, t) \subseteq \mathcal{B}(\mathcal{A}, \varepsilon)$ for all $t \geq t_v$. Consequently, to determine if π is permanent, it suffices to consider its restriction to $\mathcal{B}(\mathcal{A}, \varepsilon)$. To this end, we define $\tilde{X}$ by

$$\tilde{X} = \overline{\pi(\mathcal{B}(\mathcal{A}, \varepsilon), [t_0, \infty))} \tag{4.39}$$

where $\overline{U}$ denotes the closure of a set U. Then define X by

$$X = \pi(\tilde{X}, t') \tag{4.40}$$

for some $t' > 0$. The salient features of $\tilde{X}$ and X are given in the following result, the proof of which appears in the Appendix to the chapter.

Theorem 4.1. Assume that (4.22) is a model for interacting species satisfying the conditions of Section 2, and that $\pi : Y \times [0, \infty) \to Y$ is its realization as a semiflow. Assume that (4.38) holds for π. Let $\mathcal{A}$ denote the global attractor for π whose existence is guaranteed by (4.38) and let $\tilde{X}$ and X be as in (4.39) and (4.40), respectively. Then the following hold:

(i) $\tilde{X}$ and X are compact and forward invariant under π.

(ii) If $S = X \cap \partial Y_0$, then S and $X \backslash S$ are forward invariant under π.

Notice that if $u \in S$ corresponds to a point $(u_1(x, t'), \ldots, u_n(x, t'))$ along a solution trajectory for (4.22), then $u_i(x, t') \equiv 0$ in Ω for at least one i and moreover, $u_i(x, t') \equiv 0$ in Ω for any i for which $u_i(x, t') = 0$ for some $x \in \Omega$. That such is the case follows by (4.40), the definition of S and our assumptions regarding (4.22).

Once Theorem 4.1 is established, we may invoke Corollary 2.3 of Hutson (1984) to establish the Average Lyapunov Function Test for Permanence of π. In that which follows, the ω-limit sets $\omega(u)$ and $\omega(U)$ for a point u and a subset U of Y, respectively, are as given in Chapter 1.

Theorem 4.2. (Average Lyapunov Function Test for Permanence) Assume that the conditions of Theorem 4.1 hold and let X and S be as defined in (4.40) and Theorem 4.1 (ii), respectively. Suppose that $P : X \to [0, \infty)$ is continuous with $P(u) = 0$ if and only if $u \in S$. For $u \in S$, define

$$a(t, u) = \liminf_{\substack{v \to u \\ v \in X \backslash S}} \left(\frac{P(\pi(v, t))}{P(v)} \right).$$

Then π is permanent if

$$\sup_{t>0} \; a(t, u) > \begin{cases} 1, & u \in \omega(S) \\ 0, & u \in S. \end{cases}$$

4.3.2 Acyclicity Approach

The basic set-up remains the same as with the Average Lyapunov Function Approach. We assume that (4.22) is a model for interacting species satisfying the conditions of Section 2 and that $\pi : Y \times [0, \infty) \to Y$ is its realization as a semiflow satisfying (4.38). We let $\mathcal{A}$ denote the resulting global attractor for π and let $\tilde{X}$ and X be as in (4.39) and (4.40), respectively, and S be as in Theorem 4.1. We then consider the semiflow π on the complete metric space $X = (X \backslash S) \cup S$, where X, $X \backslash S$ and S are all forward invariant under π. $X \backslash S$ corresponds to tuples of species densities which are componentwise positive and S corresponds to tuples of species densities where components are either positive or vanish identically with at least one zero component. Since X is compact and S is closed in X, S is necessarily compact as well.

Let M denote a compact invariant subset of X. We define the *stable set of* M, denoted $W^s(M)$, by

$$W^s(M) = \{u : u \in X, \; \omega(u) \neq \phi, \; \omega(u) \subseteq M\} \tag{4.41}$$

and the corresponding *unstable set of* M, denoted $W^u(M)$, by

$$W^u(M) = \{u : u \in X, \; \alpha(u) \neq \phi, \alpha(u) \subseteq M\}, \tag{4.42}$$

where $\alpha(u)$, the alpha-limit set of u, is as defined in Chapter 1. The set M is said to be an *isolated invariant* set if it has a neighborhood U in X, called an *isolating neighborhood* such that M is the maximal invariant subset of U.

Now consider two not necessarily distinct isolated invariant subsets M and N of X. *M is said to be chained to N*, written $M \to N$, provided there exists $u \notin M \cup N$ with $u \in W^u(M) \cap W^s(N)$. (In this case, the global orbit through u exists and links up to M as $t \to -\infty$ and to N as $t \to \infty$. This notion is a generalization of an orbit connecting two equilibria.) A finite sequence of isolated invariant subsets of X, say $M_1, \ldots, M_k$ is called a *chain* if

$$M_1 \to M_2 \to \ldots \to M_k.$$

A chain $M_1, \ldots, M_k$ is called a *cycle* if $M_k = M_1$.

Now consider the set $\omega(S)$. We say that $\omega(S)$ is *isolated* if $\omega(S) = \cup_{n=1}^{k} M_n$, where $M_1, \ldots, M_k$ are pairwise disjoint compact invariant subsets of S which are *isolated with respect to π and with respect to the semiflow π_S which results from restricting π to S*. The collection $\{M_1, \ldots, M_k\}$ is called an *isolated covering* of $\omega(S)$. We say that $\omega(S)$ is *acyclic* provided there is an isolated covering $\{M_1, \ldots, M_k\}$ of $\omega(S)$ such that no subcollection of $\{M_1, \ldots, M_k\}$ forms a cycle.

With these definitions in hand, we may state the Acyclicity Test for Permanence. The proof of the result in the generality we need for this volume is due to Hale and Waltman, and may be found in Hale and Waltman (1989). See also Cantrell et al. (1993b).

Theorem 4.3. (Acyclicity Test for Permanence) Assume that the conditions of Theorem 4.1 hold and let X and S be as defined in (4.40) and Theorem 4.1 (ii), respectively. Then if $w(S)$ is isolated and acyclic, π is permanent provided

$$W^s(M_n) \cap (X \backslash S) = \phi$$

for $n = 1, \ldots, k$.

It is instructive at this point to reconsider briefly the May-Leonard example (4.17)–(4.18). In this case, X is a compact subset of the first octant in $\mathbb{R}^3$ with S a compact subset of its faces containing the four equilibria $\{(0,0,0), (1,0,0), (0,1,0), (0,0,1)\}$ which comprise $\omega(S)$. Our analysis in Section 1 demonstrates that $\omega(S)$ is isolated and that the stable set for each equilibrium in $\omega(S)$ is disjoint from the interior of the first octant. Since the system fails to be permanent, we would deduce from Theorem 4.3 that $\omega(S)$ must fail to be acyclic. Indeed, it follows from the definition of γ as a heteroclinic cycle in (4.20) that

$$\{(1,0,0)\} \to \{(0,0,1)\} \to \{(0,1,0)\} \to \{(1,0,0)\}.$$

Hence the subcollection $\{(1,0,0), (0,1,0), (0,0,1)\}$ of $\omega(S)$ forms a cycle. Note here also that the isolated covering of $\omega(S)$ in this case is unique up to the order in which the equilibria are listed.

4.4 Invasibility Implies Coexistence

Invasibility is an important concept in ecology and other biological fields, prominent among them genetics and evolution. Indeed, the central notion of evolutionary game theory (Maynard Smith, 1982), *evolutionarily stable strategy* or ESS, is defined in terms of invasibility, or more precisely, noninvasibility. In evolutionary game theory, "[a] 'strategy' is a behavioral phenotype; i.e. it is a specification of what an individual will do in any situation in which it may find itself. An ESS is a strategy such that, if all members of a population adopt it, then no mutant strategy could invade the population under the influence of natural selection." (Maynard Smith, 1982). In ecology, one considers a community of species in some habitat with a particular configuration of species densities instead of a strategy, with a species external to the community, usually called an exotic, playing the role of the mutant strategy. Of course, evolutionary time scales are usually much longer than and not commensurate with ecological time scales, and one usually does not consider natural selection explicitly in ecological time. Instead, features such as biotic interactions, among them competition, predation and mutualism, and environmental conditions come into play. We say that the configuration of the preexisting community in the given habitat is invasible by the exotic if the exotic always increases its density when it is introduced into the habitat at a low density.

Elucidating the connection between the concepts of invasibility and permanence is a primary goal of this section. Indeed, there is a dictum in ecological theory that invasibility implies coexistence. We will examine how and to what extent permanence can be regarded as a quantitative realization of this idea. We shall do so in the process of fulfilling another of the primary goals of this section (and the next). Namely, we shall illustrate how to apply the techniques for establishing permanence which were given in Section 3 in particular systems of interacting species, including the reaction-diffusion models of principal interest in this volume. (For specificity, we consider models for two competing species in this section, employing the Acyclicity Approach to establishing permanence. In the next section, we focus on predator-prey models for autonomous reaction-diffusion equations and take an Average Lyapunov Function Approach.) What we see is that the conditions that we must impose on specific models to apply the techniques for establishing permanence given in Section 3 are naturally interpreted as invasibility conditions. Specifically, we find that we must require that in a suitable number of instances (depending on the particular model) the growth rate at low densities for one of the species in question be positive when the remaining species' densities have a prespecified configuration. Mathematically, such a growth rate is expressed as the principal eigenvalue of a suitable operator whose coefficients are determined by the densities of the other species as well as the environmental features of the underlying model. (In the case of ODE models, for example, the operators in question will all be of the form $x \to cx$, where x is a real variable and c is a specified real number, while in the case of autonomous reaction-diffusion models, the operators are second-order elliptic operators on suitable function spaces.) Consequently, in demonstrating the connection between permanence and the notion that invasibility implies coexistence, we also illustrate the fundamental importance of eigenvalue estimation in establishing permanence in models of interacting biological species. Moreover, due to the dependence of the operators in question on the environmental features of the model and particular configuration of species densities, eigenvalue estimation simultaneously becomes a powerful tool for ecological analysis of the general features of the model, as we have already seen in Chapter 2 and shall see again and again throughout this volume.

4.4.1 Acyclicity and an ODE Competition Model

Let us examine the Lotka-Volterra competition model

$$
\begin{aligned}
\frac{du_i}{dt} &= u_1[a_1 - u_1 - b_1 u_2] \\
\frac{du_2}{dt} &= u_2[a_2 - b_2 u_1 - u_2],
\end{aligned}
\tag{4.43}
$$

where u_1 denotes the density of competing species 1 and u_2 that of competing species 2. Here a_i denotes the intrinsic rate of growth for species i and b_i indicates the magnitude of the competitive impact of species j on species i, where $i \neq j$ and a_i, b_i are positive constants. Notice that in (4.43) the self-regulation (or intraspecific competition) coefficient is set equal to 1 for both species. This model can be derived from the more general Lotka-Volterra competition model

$$\begin{aligned}\frac{dU_1}{dt} &= U_1[A_1 - C_1U_1 - B_1U_2]\\ \frac{dU_2}{dt} &= U_2[A_2 - B_2U_1 - C_2U_2]\end{aligned} \tag{4.44}$$

by the rescalings $u_i = C_iU_i$, $a_i = A_i$ and $b_i = \dfrac{B_i}{C_j}$, $i \neq j$.

To approach the question of permanence in (4.43) via Acyclicity (or via the construction of an Average Lyapunov Function for that matter), we must first show that conditions (i)–(iii) of (4.38) hold when we define $\pi((u_1^0, u_2^0), t)$ to be the unique solution $(u_1(t), u_2(t))$ of (4.43) with $u_1(0) = u_1^0$ and $u_2(0) = u_2^0$. The natural choice for Y in (4.43) is the positive cone $[0, \infty) \times [0, \infty)$ in $I\!R^2$, so that Y is complete, satisfying (i) of (4.38). Since Y is also locally compact, as noted in the previous section, $\pi((\cdot, \cdot), t)$ is compact as a map from Y to Y for any $t \geq 0$, so that (iii) of (4.38) also holds. Hence we need only establish (ii) of (4.38), namely, that π is dissipative. To this end, consider the well-known logistic equation

$$\frac{dw_i}{dt} = w_i(a_i - w_i) \tag{4.45}$$

which results from the assumption that $u_j \equiv 0$. The solution to (4.45) is given by

$$w_i(t) = \begin{cases} \dfrac{\alpha a_i}{e^{-a_it} + \alpha} & \text{if} \quad 0 \leq w_i(0) < a_i \text{ or } \quad w_i(0) > a_i \\ a_i & \text{if} \quad w_i(0) = a_i \end{cases} \tag{4.46}$$

with $\alpha > 0$ when $0 < w_i(0) < a_i$ and $\alpha < -1$ when $w_i(0) > a_i$. For any choice of $w_i(0) > 0$, we have

$$\lim_{t\to\infty} w_i(t) = a_i. \tag{4.47}$$

Hence, for any choice of $w_i(0) \geq 0$, there is a $t = t(w_i(0))$ such that $w_i(t) \leq a_i + 1$ for all $t \geq t(w_i(0))$. Recall the method of upper and lower solutions for ODE's from Section 6 of Chapter 1. It follows that if $\pi((u_1^0, u_2^0), t) = (u_1(t), u_2(t))$ then $u_i(t)$ is a lower solution to (4.45) and that if $w_i(t)$ is the unique solution to (4.45) with $w_i(0) = u_i^0$, then $u_i(t) \leq w_i(t)$ for all $t \geq 0$, so that if $t \geq t(w_1(0))$ and $t \geq t(w_2(0))$, we have that $||\pi(u_1^0, u_2^0), t)|| \leq \sqrt{(a_1+1)^2 + (a_2+1)^2}$, establishing that π is dissipative.

Consequently, Theorem 4.1 guarantees that there is a global attractor $\mathcal{A}$ for π contained in $Y = [0, \infty) \times [0, \infty)$. The preceding considerations regarding (4.45) (specifically (4.47)) enable us to conclude that $\omega(S)$ in Theorem 4.3 is given by

$$\omega(S) = \{(0, 0), (a_1, 0), (0, a_2)\}. \tag{4.48}$$

It is immediate from (4.47) that $\omega(S)$ is acyclic, and that $\{(0, 0)\}$, $\{(a_1, 0)\}$, and $\{(0, a_2)\}$ are isolated invariant subsets with respect to the restriction of π to $S \subseteq ([0, \infty) \times \{0\}) \cup$

$(\{0\} \times [0, \infty))$. To assert that π is permanent via Theorem 4.3 we need to establish in addition that $\{(0,0)\}$, $\{(a_1, 0)\}$, and $\{(0, a_2)\}$ are isolated with respect to π on Y and that

$$\begin{aligned} W^s(\{(0,0)\}) \cap [(0,\infty) \times (0,\infty)] &= \emptyset \\ W^s(\{(a_1,0)\}) \cap [(0,\infty) \times (0,\infty)] &= \emptyset \\ W^s(\{(0,a_2)\}) \cap [(0,\infty) \times (0,\infty)] &= \emptyset. \end{aligned} \tag{4.49}$$

Let us first consider $\{(0,0)\}$. Let $\pi((u_1^0, u_2^0), t) = (u_1(t), u_2(t))$ be an orbit in $(0,\infty) \times (0,\infty)$. If the inequalities

$$u_i(t) + b_i u_j(t) < \frac{a_i}{2} \tag{4.50}$$

for $i = 1, 2$ and $j \neq i$, hold for some $t_0 > 0$, then $\dfrac{du_i(t_0)}{dt} > \dfrac{a_i}{2} u_i(t_0)$.

It follows that

$$u_i(t) \geq u_i(t_0) e^{\left(\frac{a_i}{2}\right)(t - t_0)} \tag{4.51}$$

so long as (4.50) continues to hold for $i = 1, 2$. Hence the u_1 and u_2 components of any solution trajectory increase at least exponentially near $\{(0,0)\}$. So either species can invade when the other is absent and any solution trajectory in $(0,\infty) \times (0,\infty)$ is repelled away from $\{(0,0)\}$. It follows that $\{(0,0)\}$ is an isolated invariant set with respect to π on Y and $W^s(\{(0,0)\}) \cap [(0,\infty) \times (0,\infty)] = \emptyset$.

The preceding results hold precisely because $a_i > 0$. Notice that the linearization of the ith equation in (4.43) about $(0,0)$ is

$$\frac{dw_i}{dt} = a_i w_i \tag{4.52}$$

so that a_i is the intrinsic growth rate for species i. Viewing the right-hand side of (4.52) as a linear operator on $I\!R$, the intrinsic growth rate $a_i > 0$ is also the principal eigenvalue of the operator. So having the principal eigenvalue of the right-hand side of (4.52) be positive implies that species i can invade species j when $u_j = 0$. Either of these invasibilities in turn guarantee that $\{(0,0)\}$ is isolated with respect to π on Y with $W^s(\{(0,0)\}) \cap ((0,\infty) \times (0,\infty)) = \emptyset$.

Let us now consider $\{(a_1, 0)\}$. The linearization of the second equation of (4.43) about $(a_1, 0)$ is

$$\frac{dw_2}{dt} = w_2(a_2 - b_2 a_1) \tag{4.53}$$

so that $a_2 - b_2 a_1$ is the *per capita* rate of growth for species 2 when $u_1 = a_1$. Viewing the right hand side of (4.53) as a linear operator on $I\!R$, the *per capita* rate of growth $a_2 - b_2 a_1$ again corresponds to the principal eigenvalue of the operator. If

$$a_2 - b_2 a_1 > 0, \tag{4.54}$$

then

$$\begin{aligned} \frac{du_2}{dt} &= u_2(a_2 - b_2 u_1 - u_2) \\ &= u_2(a_2 - b_2 a_1 - b_2(u_1 - a_1) - u_2) \\ &\geq u_2 \left(\frac{a_2 - b_2 a_1}{2} \right) \end{aligned}$$

whenever

$$b_2(u_1 - a_1) + u_2 < \frac{a_2 - b_2 a_1}{2}. \tag{4.55}$$

Arguing as with $\{(0, 0)\}$, it follows from (4.54) and (4.55) that the second component of any solution trajectory for π increases near $(a_1, 0)$ so that species 1 is invasible at its carrying capacity a_1 by species 2. It follows that $\{(a_1, 0)\}$ is isolated with respect to π on Y and that $W^s(\{(a_1, 0)\}) \cap ((0, \infty) \times (0, \infty)) = \phi$.

Finally, for $\{(0, a_2)\}$, the linearization of the first equation of (4.43) about $(0, a_2)$ is

$$\frac{dw_1}{dt} = w_1(a_1 - b_1 a_2) \tag{4.56}$$

with the *per capita* growth rate $a_1 - b_1 a_2$ again the principal eigenvalue for the right hand side of (4.56) viewed as a linear operator on $I\!R$. If

$$a_1 - b_1 a_2 > 0, \tag{4.57}$$

we have that species 2 is invasible at carrying capacity by species 1, so that $\{(0, a_2)\}$ is an isolated invariant set for π on Y with $W^s(\{(0, a_2)\}) \cap ((0, \infty) \times (0, \infty)) = \phi$. So if the principal eigenvalues for the right-hand sides of (4.52), (4.53) and (4.56) are positive (i.e. if (4.54) and (4.57) hold), Theorem 4.3 (Acyclicity) implies π and hence (4.43) is permanent.

The remark concerning the May-Leonard example following Theorem 4.3 shows that the invasibility of every component of $w(S)$ is not sufficient *per se* for permanence. If any one of $\{(1, 0, 0), (0, 1, 0), (0, 0, 1)\}$ had been invasible by a second competitor, the system would have been acyclic and hence permanent. But of course, such would have been a different system from the May-Leonard example. So one might more precisely say "a sufficient number of invasibilities implies coexistence."

It may seem to some readers that interpreting $a_1, a_2, a_1 - b_1 a_2$ and $a_2 - b_2 a_1$ as eigenvalues in the preceding discussion is somewhat artificial. Our next example should serve to demonstrate that such is not the case.

4.4.2 A Reaction-Diffusion Analogue

Let us alter the preceding example (4.43) only by adding diffusion over a bounded spatial habitat assuming that the boundary of the habitat is lethal to both competitors. So doing yields the system

$$\begin{aligned} \frac{\partial u_1}{\partial t} &= \Delta u_1 + u_1[a_1 - u_1 - b_1 u_2] \\ \frac{\partial u_2}{\partial t} &= \Delta u_2 + u_2[a_2 - b_2 u_1 - b_2] \end{aligned} \tag{4.58}$$

in $\Omega \times (0, \infty)$ with

$$u_1 = 0 = u_2 \tag{4.59}$$

on $\partial\Omega \times (0, \infty)$, where Ω is the bounded habitat in question and a_i and b_i are once again positive constants. The more general system

$$\begin{aligned} \frac{\partial U_1}{\partial t} &= D_1 \Delta U_1 + U_1[A_1 - C_1 U_1 - B_1 U_2] \\ \frac{\partial U_2}{\partial t} &= D_2 \Delta U_2 + U_2[A_2 - B_2 U_1 - C_2 U_2] \end{aligned} \tag{4.60}$$

in $\Omega \times (0, \infty)$ can be reduced to (4.58) in much the same way that (4.44) reduces to (4.43) so long as $D_1 = D_2$ and time is rescaled by $s = D_1 t$. If $D_1 \neq D_2$ in (4.60), (4.58) must be modified to incorporate the Laplace operator in the second equation of (4.60) being multiplied by the ratio D_2/D_1 or the Laplace operator in the first equation of (4.60) being multiplied by its reciprocal. But, again, our purpose is an illustration of how we may apply the Acyclicity Theorem to assert permanence in such a system, and for such it suffices to focus on (4.58)–(4.59).

Following Section 6 of Chapter 1, we may employ the theory of analytic semi-groups and parabolic partial differential equations to convert (4.58)–(4.59) into a semi-dynamical system $\pi((u_1^0, u_2^0), t)$ on an appropriate function space, where $\pi((u_1^0, u_2^0), t)$ denotes the unique solution $(u_1(x,t), u_2(x,t))$ of (4.58)–(4.59) such that $(u_1(x,0), u_2(x,0)) = (u_1^0(x), u_2^0(x))$. In this instance, a suitable function space is the Cartesian product $C_0^1(\overline{\Omega}) \times C_0^1(\overline{\Omega})$ of smooth real-valued functions on $\overline{\Omega}$ which vanish on $\partial\Omega$, equipped with the norm $||(\cdot,\cdot)||_{C^1(\overline{\Omega})\times C^1(\overline{\Omega})}$ where

$$||(f,g)||_{C^1(\overline{\Omega})\times C^1(\overline{\Omega})} = ||f||_{C^1(\overline{\Omega})} + ||g||_{C^1(\overline{\Omega})} \tag{4.61}$$

and in turn

$$||f||_{C^1(\overline{\Omega})} = \sup_{x\in\overline{\Omega}} |f(x)| + \sum_{i=1}^{\dim\Omega} \left[\sup_{x\in\overline{\Omega}} |\partial_{x_i} f(x)| \right]. \tag{4.62}$$

Since the components of $\pi((u_1^0, u_2^0), t)$ represent population densities on $\overline{\Omega}$, we restrict ourselves to the cone K of $C_0^1(\overline{\Omega}) \times C_0^1(\overline{\Omega})$ where both components are nonnegative on $\overline{\Omega}$. K is a complete metric space, as required by Theorem 4.1 (see, for example, Amann (1976) and Cantrell et al. (1993a)).

We may establish that $\pi((u_1^0, u_2^0), t)$ exists for all $t > 0$ for any $(u_1^0, u_2^0) \in K$ and that in fact π is dissipative by means of arguments which are analogous (albeit mathematically more technical) to the arguments we employed to establish the corresponding observations regarding the flow for the ODE system (4.43). To this end, we begin with the diffusive logistic model

$$\begin{aligned} \frac{\partial u_i}{\partial t} &= \Delta u_i + u_i(a_i - u_i) \text{ in } && \Omega \times (0,\infty) \\ u_i &= 0 && \text{on } \partial\Omega \times (0,\infty) \end{aligned} \tag{4.63}$$

which describes the population dynamics of species i in the absence of competition with species j. The asymptotic predictions of this model are well understood (e.g., Cosner and Lazer, 1984; Cantrell and Cosner, 1989a; Hess, 1991) and are described in the following theorem, which follows from Propositions 3.1–3.3; see also Corollary 3.14.

Theorem 4.4. Consider the eigenvalue problem

$$\begin{aligned} \Delta w_i + a_i w_i &= \sigma_i w_i \text{ in } && \Omega \\ w_i &= 0 && \text{on } \partial\Omega \\ w_i(x) &> 0 && \text{in } \Omega \end{aligned} \tag{4.64}$$

corresponding to the linearization of (4.63) about $u_i = 0$.

(i) If $\sigma_i > 0$ in (4.64), there is a unique $\tilde{w}_i \in C_0^1(\overline{\Omega})$ with $\tilde{w}_i(x) > 0$ for $x \in \Omega$ and $\frac{\partial \tilde{w}_i}{\partial \vec{n}}(x) < 0$ on $\partial\Omega$ such that for any $w_i^0 \in C_0^1(\overline{\Omega})$ with $w_i^0 \gneqq 0$ in $\overline{\Omega}$ the unique solution $w_i(x, t)$ of (4.63) with $w_i(x, 0) = w_i^0(x)$ converges in the norm of $C_0^1(\overline{\Omega})$ to $\tilde{w}_i(x)$ as $t \to \infty$. This $\tilde{w}_i(x)$ is an equilibrium solution to (4.63) and satisfies the pointwise bound $\tilde{w}_i(x) \leq a_i$.

(ii) If $\sigma_i \leq 0$ in (4.64), 0 is the only nonnegative equilibrium solution to (4.63), and for any $w_i^0 \in C_0^1(\overline{\Omega})$ with $w_i^0 \geq 0$ in $\overline{\Omega}$ the unique solution $w_i(x, t)$ of (4.63) with $w_i(x, 0) = w_i^0(x)$ converges in the norm of $C_0^1(\overline{\Omega})$ to 0 as $t \to \infty$.

(iii) $\sigma_i > 0$ in (4.64) if and only if $a_i > \lambda_0^1(\Omega)$ where $\lambda_0^1(\Omega)$ denotes the principal eigenvalue of the negative Laplacian on Ω subject to zero Dirichlet boundary conditions.

Note: σ_i in (4.64) should be regarded as the average intrinsic growth rate of species i over Ω.

It is an immediate consequence of Theorem 4.4 that if $\varepsilon_i > 0$ and $w_i^0(x) \geq 0$ in $C_0^1(\overline{\Omega})$ are prescribed and $w_i(x, t)$ is the unique solution of (4.63) with $w_i(x, 0) = w_i^0(x)$, then there is a t_i depending only on ε_i and $w_i^0(x)$ such that $w_i(x, t) \leq a_i + \varepsilon_i$ for all $t \geq t_i$. Suppose now that $(u_1(x, t), u_2(x, t))$ is the unique solution to (4.58)–(4.59) with $(u_1(x, 0), u_2(x, 0)) = (u_1^0(x), u_2^0(x))$, where $u_i^0(x) \geq 0$ in $C_0^1(\overline{\Omega})$. Then

$$\frac{\partial u_i}{\partial t} \leq \Delta u_i + u_i(a_i - u_i) \text{ in } \quad \Omega \times (0, \infty).$$

It follows from the method of upper and lower solutions that $u_i(x, t) \leq w_i(x, t)$ where $w_i(x, t)$ is the unique solution of (4.63) with $w_i(x, 0) = u_i^0(x)$. Hence there is a $\bar{t}$ depending only on (u_1^0, u_2^0) such that

$$u_i(x, t) \leq a_i + 1 \tag{4.65}$$

for all $t \geq \bar{t}$ for $i = 1, 2$. Inequalities (4.65) demonstrate that the solution to (4.58)–(4.59) corresponding to $(u_1^0, u_2^0) \in K$ is eventually bounded in the topology of $C^0(\overline{\Omega}) \times C^0(\overline{\Omega})$, independent of (u_1^0, u_2^0). Dissipativity of π, however, requires the bound be in the topology of $C_0^1(\overline{\Omega}) \times C_0^1(\overline{\Omega})$. Such a result now follows from (4.65) by Theorems 1.11 and 1.12 and the Remarks following, as does the compactness of $\pi(\cdot, \cdot), t) : K \to K$ for any $t > 0$. (See also Cantrell et al. (1993a, Theorem 3.3).)

We may now invoke Theorem 4.1 to assert the existence of a global attractor $\mathcal{A}$ for π contained in K, the positive cone of $C_0^1(\overline{\Omega}) \times C_0^1(\overline{\Omega})$. Theorem 4.4 enables us to conclude that $w(S)$ in Theorem 4.3 is given by

$$w(S) = \{(0, 0), (\tilde{w}_1(x), 0), (0, \tilde{w}_2(x))\} \tag{4.66}$$

in analogy with (4.48), so long as the average intrinsic growth rates σ_1 and σ_2 in (4.64) are positive. (If either σ_1 or σ_2 is nonpositive in (4.64), the relevant species goes extinct in the model independent of the competition, and permanence is not possible. The results of Chapter 2 show that in this case the habitat Ω lacks sufficient core area relative to the dissipative effect of the lethal or absorbing boundary condition.) Moreover, Theorem 4.4 also enables us to conclude, again in analogy with (4.48), that $\omega(S)$ is acyclic and that $\{(0, 0)\}$, $\{(\tilde{w}_1(x), 0)\}$ and $\{(0, \tilde{w}_2(x))\}$ are isolated invariant subsets with respect to the restriction of π to S.

So, as with the model (4.43), to invoke Theorem 4.3 to assert that π and hence (4.58)–(4.59) are permanent, we must show that $\{(0,0)\}$, $\{(\tilde{w}_1(x),0)\}$ and $\{(0,\tilde{w}_2(x))\}$ are isolated invariant subsets for π on K and that

$$\begin{aligned} &W^s(\{(0,0)\}) \cap \text{int}K = \phi \\ &W^s(\{(\tilde{w}_1(x),0)\}) \cap \text{int}K = \phi \\ &W^s(\{0,\tilde{w}_2(x))\}) \cap \text{int}K = \phi \end{aligned} \tag{4.67}$$

where, as in the Remarks preceding Corollary 2.14, $\text{int}K = \{(f_1, f_2) \in K : f_i(x) > 0$ for $x \in \Omega$ and $\frac{\partial f_i}{\partial \vec{n}}(x) < 0$ for $x \in \partial\Omega, i = 1, 2\}$. All the relevant observations will be seen to follow from the following lemma, which provides the exponential growth analogue to (4.51) in this context and whose proof is given in the Appendix of the Chapter. (See also Cantrell et al. (1993b, Lemma 4.2).)

Lemma 4.5. Suppose $f \in C^\alpha(\overline{\Omega})$ and $\mu > 0$. Let σ be the unique eigenvalue of

$$\begin{aligned} &\mu\Delta\phi + f(x)\phi = \sigma\phi \quad \text{in} \quad \Omega \\ &\phi = 0 \qquad\qquad\qquad \text{on} \quad \partial\Omega \end{aligned} \tag{4.68}$$

possessing an eigenfunction $\phi > 0$. Assume $\sigma > 0$. Suppose for some $\varepsilon \in (0, \sigma)$, $t_* \geq 0$ and $\delta > 0$ the function $u(x, t)$ satisfies

$$\frac{\partial u}{\partial t} \geq \mu\Delta u + [f(x) - \varepsilon]u \tag{4.69}$$

for $t \in (t_*, t_* + \delta)$. If there is a $k > 0$ so that

$$u(x, t_*) \geq k\phi(x) \tag{4.70}$$

for $x \in \overline{\Omega}$, then $u(x, t) \geq ke^{(\sigma-\varepsilon)(t-t_*)}\phi(x)$ for $t \in (t_*, t_* + \delta)$. (The existence of such a $k > 0$ is guaranteed by the strong maximum principle if $u(x, 0) \geq 0$, $u(x, 0) \neq 0$, and $t^* > 0$.)

Now consider $\{(0, 0)\}$, assuming σ_1 and σ_2 in (4.64) are positive. If $0 < \varepsilon_i < \sigma_i$ and w_i is as in (4.64), then

$$\Delta w_i + (a_i - \varepsilon_i)w_i = (\sigma_i - \varepsilon_i)w_i \quad \text{in} \quad \Omega$$

with $\sigma_i - \varepsilon_i > 0$. Suppose now that $(u_1(x, t), u_2(x, t))$ is a solution to (4.58)–(4.59) (equivalently a positive orbit for π) with

$$u_i(x, t_*) + b_i u_j(x, t_*) < \varepsilon_i/2 \tag{4.71}$$

for some $t_* > 0$, $i = 1, 2$, $j \neq i$. By continuity of π, there is $\delta_i > 0$ so that $u_i(x,t)+b_iu_j(x,t) < \varepsilon_i$ for $t \in (t_*, t_*+\delta_i)$, $i = 1, 2$, $j \neq i$. For such t, u_i satisfies (4.69) with $\mu = 1$, $f(x) = a_i$ and $\varepsilon = \varepsilon_i$. Since $u_i(x, t_*) > 0$ for $x \in \Omega$ and $\frac{\partial u_i}{\partial \vec{n}}(x, t_*) < 0$ for $x \in \partial\Omega$, there is a $k_i > 0$ so that $u_i(x, t_*) \geq k_i w_i(x)$ for $x \in \overline{\Omega}$. Lemma 4.5 implies that

$$u_i(x, t) \geq k_i e^{(\sigma_i-\varepsilon_i)(t-t_*)} w_i(x) \tag{4.72}$$

for $t \in (t_*, t_*+\delta_i)$. So again each species density increases exponentially or faster whenever both densities are small, which says that species i can invade when $u_j \equiv 0$. It follows that $\{(0,0)\}$ is an isolated invariant set for π on K and that $W^s(\{(0,0)\}) \cap \text{int}K = \phi$.

Next consider $\{(\tilde{w}_1(x), 0)\}$. Notice that the second equation in (4.58) can be expressed

$$\frac{\partial u_2}{\partial t} = \Delta u_2 + u_2(a_2 - b_2\tilde{w}_1 - b_2(u_1 - \tilde{w}_1) - u_2)$$

so that (4.69) holds with $\mu = 1$, $f(x) = a_2 - b_2\tilde{w}_1(x)$ so long as $b_2(u_1(x,t) - \tilde{w}_1(x)) + u_2(x,t) < \varepsilon$. Let σ_3 denote the average growth rate over Ω for species 2 when $u_1 \equiv \tilde{w}_1$. Then σ_3 is the principal eigenvalue of

$$\begin{aligned} &\Delta w_3 + w_3(a_2 - b_2\tilde{w}_1(x)) = \sigma_3 w_3 && \text{in} \quad \Omega \\ &w_3 = 0 && \text{on} \quad \partial\Omega \\ &w_3 > 0 && \text{in} \quad \Omega. \end{aligned} \tag{4.73}$$

If $\sigma_3 > 0$ in (4.73) and $\varepsilon \in (0, \sigma_3)$, Lemma 4.5 implies that if $b_2(u_1(x,t_*) - \tilde{w}_1(x)) + u_2(x,t_*) < \varepsilon$ and $u_2(x,t_*) \geq kw_3(x)$ on $\overline{\Omega}$ for some $t_* \geq 0$, then $u_2(x,t) \geq ke^{(\sigma_3-\varepsilon)(t-t_*)}w_3(x)$ for as long after t_* as $b_2(u_1(x,t) - \tilde{w}_1(x)) + u_2(x,t) < \varepsilon$ in $\overline{\Omega}$ continues to hold. So the density of species 2 increases at least exponentially whenever it is small and the density of species 1 is near $\tilde{w}_1(x)$, its equilibrium in the absence of competition. Thus species 2 can invade species 1 at equilibrium, and $\{(\tilde{w}_1(x), 0)\}$ is an isolated invariant set for π on K with $W^s(\{(\tilde{w}_1(x), 0)\}) \cap \text{int}K = \phi$.

It should now be clear to the reader that species 1 can invade species 2 at equilibrium, $\{(0, \tilde{w}_2(x))\}$ is an isolated invariant set for π on K and $W^s(\{(0, \tilde{w}_2(x))\}) \cap \text{int}K = \phi$ provided that the average growth rate σ_4 for species 1 over Ω when $u_2 \equiv \tilde{w}_2(x)$ is positive. Moreover, σ_4 is given as the principal eigenvalue of

$$\begin{aligned} &\Delta w_4 + w_4(a_1 - b_1\tilde{w}_2(x)) = \sigma_4 w_4 && \text{in} \quad \Omega \\ &w_4 = 0 && \text{on} \quad \partial\Omega. \end{aligned} \tag{4.74}$$

Theorem 4.3 now implies π and hence (4.58)–(4.59) are permanent so long as σ_1 and σ_2 are positive in (4.64), σ_3 is positive in (4.73) and σ_4 is positive in (4.74).

4.4.3 Connection to Eigenvalues

In the preceding examples, we determined that a model exhibited permanence on the basis that in a suitable number of cases the average growth rate of one of the species was positive when the remaining species' densities had a prescribed quantitative configuration. The positivity of the average rate of growth of the distinguished species has the interpretation that the prescribed configuration of the remaining species' densities is invasible by the distinguished species. This way of conceptualizing a condition for permanence has the immediate appeal of giving a condition for permanence that is explicitly ecological. (Recall Law and Morton (1996).) It also has two additional features that may be less immediately apparent, but which are essential to the analyses of this volume. As we noted in dealing with (4.43) and then with (4.58)–(4.59), the average growth rates correspond to principal eigenvalues for certain linear operators which depend on species' densities, interaction parameters and environmental parameters. The positivity of all the eigenvalues in question is the condition we require for a finding of permanence. It is a *sharp* condition in the sense that if one of the eigenvalues in question is negative, it is not only the case that we are unable to predict permanence. Indeed, we know then that the system is not permanent. The reason is that there is a result corresponding to Lemma 4.5 in the case of a negative principal eigenvalue. Lemma 4.5 was used to show that a "small enough" species density exceeds a positive function which grows exponentially in time. Its analogue in the case of a negative principal eigenvalue can be used to show that a small species density lies below a positive function which decays exponentially. Hence the species in question is driven to extinction if its density becomes too low, ruling out a prediction of coexistence for the system. We should be quick to point out, however, that failure of a model for interacting biological species to be permanent does not necessarily mean that the model predicts extinction of at least one species in the community whatever the initial configuration of the species' densities. Such a prediction is sometimes possible, however, and we shall address the issue in Chapter 5.

The second feature of this way of conceptualizing a condition for permanence results from the connection (particularly in the context of reaction-diffusion models) between the eigenvalues measuring average growth (the σ's) and the companion principal eigenvalues for weighted eigenvalue problem (the $\lambda_+^1(m)$'s) established in Theorem 2.6 of Chapter 2. For instance, σ_3 in (4.73) and σ_4 in (4.74) are positive if and only if

$$\lambda_+^1(a_2 - b_2\tilde{w}_1(x)) < 1 \tag{4.75}$$

and

$$\lambda_+^1(a_1 - b_1\tilde{w}_2(x)) < 1, \tag{4.76}$$

respectively. Such a connection makes it possible to examine explicitly how predictions of coexistence in the model relate to or depend on the parameters in the model. Such examinations are at the heart of the ecological analyses we present in this volume. To give a simple illustration, let us consider (4.58)–(4.59). Note that (4.73) and (4.74) can be rewritten as

$$-\Delta w_3 + b_2\tilde{w}_1(x)w_3 = (a_2 - \sigma_3)w_3 \text{ in } \quad \Omega \tag{4.77}$$

and

$$-\Delta w_4 + b_1\tilde{w}_2(x)w_4 = (a_1 - \sigma_4)w_4 \text{ in } \quad \Omega \tag{4.78}$$

where w_3 and w_4 are positive in Ω and vanish on $\partial\Omega$. Comparison of (4.77) with

$$\begin{aligned} -\Delta z &= \lambda_0^1(\Omega)z \text{ in } && \Omega \\ z &= 0 && \text{on } \quad \partial\Omega \\ z &> 0 && \text{in } \quad \Omega \end{aligned} \tag{4.79}$$

via Corollary 2.2 shows that the average growth rate over Ω of species 2 when species 1 has density $\tilde{w}_1(x)$, namely σ_3, satisfies the inequality

$$\sigma_3 < a_2 - \lambda_0^1(\Omega) \tag{4.80}$$

where a_2 is its local growth rate over Ω and $\lambda_0^1(\Omega)$ in (4.79) is the principal eigenvalue of the negative Laplace operator on Ω subject to Dirichlet (i.e. absorbing or lethal) homogeneous boundary conditions. Hence (4.80) says that the average growth rate of species 2 over Ω

in this circumstance is always less than the amount its local growth rate exceeds $\lambda_0^1(\Omega)$, reflecting a loss due to dissipation at the boundary of the habitat. Likewise, if we compare (4.78) and (4.79) we get that

$$\sigma_4 < a_1 - \lambda_0^1(\Omega). \tag{4.81}$$

The inequalities (4.80) and (4.81) are straightforward to obtain and have a not too surprising interpretation. With only a bit more work, we can make a more substantial examination of the range of growth rates a_1 and a_2 for which (4.58)–(4.59) predicts competitive coexistence via permanence when the competition coefficients are held fixed. To do so, note first that the equilibrium density $\tilde{w}_i(x)$ for species i in the absence of species j is completely determined by $a_i > \lambda_0^1(\Omega)$. Consequently, inequalities (4.75) and (4.76) may be regarded for fixed values of competition coefficients b_1 and b_2 as a coupled system of inequalities in a_1 and a_2, the solution of which gives the locus of permanence in those parameters. Since

$$\begin{aligned} 0 = \Delta\tilde{w}_i + (a_i - \tilde{w}_i)\tilde{w}_i &\text{ in } \Omega \\ \tilde{w}_i = 0 \qquad &\text{ on } \partial\Omega \\ \tilde{w}_i > 0 \qquad &\text{ in } \Omega, \end{aligned} \tag{4.82}$$

it follows that $\lambda_+^1(a_i - \tilde{w}_i) = 1$. The monotonicity property of $\lambda_+^1(m)$ (Theorem 2.4 and the discussion that follows it) imply that

$$\lambda_+^1(a_i - k\tilde{w}_i) < 1 \quad \text{if and only if} \quad k < 1. \tag{4.83}$$

It follows from (4.83) that if $a_1 = a_2 > \lambda_0^1(\Omega)$ we get permanence in (4.58)–(4.59) if and only if $b_1 < 1$ and $b_2 < 1$.

Suppose now that $b_1 < 1$ and $b_2 < 1$ are fixed. The monotonicity property of $\lambda_+^1(m)$ implies that if $\lambda_+^1(a_i - b_i\tilde{w}_j) = 1$, then $\lambda_+^1(c - b_i\tilde{w}_j) < 1$ if and only if $c > a_i$. Since $b_1 < 1$ and $b_2 < 1$, we have $\lambda_+^1(a_2 - b_1\tilde{w}_2) < 1$ and $\lambda_+^1(a_1 - b_2\tilde{w}_1) < 1$. So in order for $\lambda_+^1(a_1 - b_1\tilde{w}_2) = 1$, it must be the case that $a_1 < a_2$. In terms of an $a_1 - a_2$ Cartesian plane, the point (a_1, a_2) for which $\lambda_+^1(a_1 - b_1\tilde{w}_2) = 1$ lies above the line $a_1 = a_2$. Similarly, for $\lambda_+^1(a_2 - b_2\tilde{w}_1) = 1$, it must be the case that $a_1 > a_2$, so that (a_1, a_2) lies below the line $a_1 = a_2$. It follows that the region (a_1, a_2) of ordered pairs of local growth rates for which (4.58)–(4.59) is permanent for this fixed $b_1 < 1$ and $b_2 < 1$ is the open region between the two curves $\lambda_+^1(a_1 - b_1\tilde{w}_2) = 1$ and $\lambda_+^1(a_2 - b_2\tilde{w}_1) = 1$. (See Figure 4.3.) By working somewhat harder, one may obtain estimates of the location of the boundary curves of the region of permanence. In Chapter 6, we make use of such estimates to demonstrate how a change in the hostility of the matrix surrounding a refuge habitat can lead to a reversal of competitive advantage between two competitors within the habitat (Cantrell et al., 1998) and there show how to derive the necessary estimates.

On the other hand, if $b_1 > 1$ and $b_2 > 1$ are fixed, our analysis shows that the curve $\lambda_+^1(a_1 - b_1\tilde{w}_2) = 1$ is now below the line $a_1 = a_2$ while the curve $\lambda_+^1(a_2 - b_2\tilde{w}_1) = 1$ is above the line $a_1 = a_2$. As a consequence, there are no pairs (a_1, a_2) for which (4.58)–(4.59) is permanent in this case.

When $a_1 = a_2 > \lambda_0^1(\Omega)$ and b_1 and b_2 lie in (0, 1), Cosner and Lazer (1984) have in fact shown that (4.58)–(4.59) has the unique globally attracting componentwise positive equilibrium $\left(\dfrac{1-b_1}{1-b_1b_2}\tilde{w}_1, \dfrac{1-b_2}{1-b_1b_2}\tilde{w}_1\right)$. Cantrell and Cosner (1989) showed that for

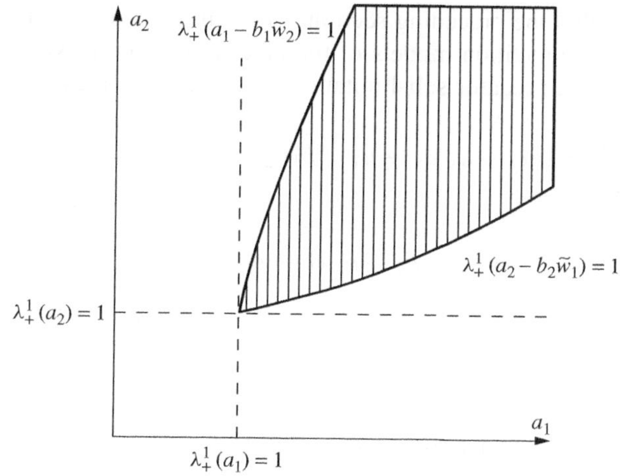

Figure 4.3 The shaded area is the region of the a_1, a_2 plane in parameter space for which (4.58) and (4.59) predict permanence.

fixed b_1 and b_2 in (0, 1), there is a unique globally attracting equilibrium to (4.58)–(4.59) for (a_1, a_2) "near enough" to the line $a_1 = a_2$. However, their result does not quantify how "near" is "near enough". More recent results (Furter and Lopez-Gomez, 1995, 1997) describe somewhat explicit conditions on the parameters a_1, a_2, b_1 and b_2 with $a_1 \neq a_2$ for which (4.58)–(4.59) has a unique globally attracting equilibrium. However, to the best of our knowledge, it remains an open and apparently difficult question as to whether (4.58)–(4.59) has a globally attracting equilibrium whenever it is permanent. It will follow from Section 2, of Chapter 5, however, that if (4.58)–(4.59) is permanent and has a unique componentwise positive equilibrium, then this equilibrium is a global attractor for componentwise positive solutions to (4.58)–(4.59). When (4.58)–(4.59) fails to be permanent, it may have multiple componentwise positive equilibria (Cantrell and Cosner, 1987; Dancer, 1991; Eilbeck et al., 1994; Gui and Lou, 1994) Other notable studies of the componentwise positive equilibria of (4.58)–(4.59) include those of Blat and Brown (1984), Dancer (1984, 1985), Korman and Leung (1986), Leung (1980), Pao (1981) and Schiafino and Tesei (1982).

4.5 Permanence in Reaction-Diffusion Models for Predation

Let us now consider the question of permanence in diffusive models of predation. Our main purpose here is to illustrate the Average Lyapunov Function Approach for determining permanence in systems of interacting biological species. As with the Acyclicity Approach, we shall see that the Average Lyapunov Function Approach leads to a condition for permanence that in particular models is realized as the positivity of the principal eigenvalue for each of a requisite collection of linear operators. Again, the principal eigenvalue corresponds to an average growth rate over the habitat patch so that a positive principal eigenvalue lends itself to the ecological interpretation that one of the species may invade the remainder of the community when the densities of the other species have a prespecified configuration. Moreover, the positivities of the average growth rates correspond, as in Section 4, to conditions among the species' densities, interaction parameters and environmental conditions that serve as a basis for ecological analysis. For the sake of specificity and to parallel the preceding section, we shall focus on predator-prey analogues to the Lotka-Volterra competition models of the preceding section. We shall consider both models with and without predator self-limitation. Specifically, we examine

$$\begin{aligned} \frac{\partial u_1}{\partial t} &= \Delta u_1 + u_1(a_1 - u_1 - b_1 u_2) \\ & \qquad\qquad\qquad\qquad \text{in} \quad \Omega \times (0, \infty) \\ \frac{\partial u_2}{\partial t} &= d\Delta u_2 + u_2(a_2 + b_2 u_1 - u_2) \\ u_1 &= 0 = u_2 \qquad\qquad\qquad \text{on} \quad \partial\Omega \times (0, \infty) \end{aligned} \tag{4.84}$$

and

$$\begin{aligned} \frac{\partial u_1}{\partial t} &= \Delta u_1 + u_1(a_1 - u_1 - b_1 u_2) \\ & \qquad\qquad\qquad\qquad \text{in} \quad \Omega \times (0, \infty) \\ \frac{\partial u_2}{\partial t} &= d\Delta u_2 + u_2(a_2 + b_2 u_1) \\ u_1 &= 0 = u_2 \qquad\qquad\qquad \text{on} \quad \partial\Omega \times (0, \infty). \end{aligned} \tag{4.85}$$

In (4.84) and (4.85), u_1 denotes the density of the prey species in the model and u_2 the predator density. In (4.84) and (4.85), the intrinsic rate of growth for the prey a_1 and the predation interaction coefficients b_1 and b_2 are positive constants. In (4.84), the intrinsic rate of growth for the predator a_2 is constant and may be positive or negative. However, in (4.85) the constant a_2 must be assumed negative in order for the predator density to remain bounded in time.

As with the Acyclicity Approach, the Average Lyapunov Function Approach for determining permanence requires that the system in question be recast as a semi-dynamical system on a metric space Y so that conditions (i)–(iii) of (4.38) hold. For either (4.84) or (4.85), to recast the model as a semi-dynamical system $\pi((u_1^0, u_2^0), t)$, one proceeds exactly as with (4.58)–(4.59). Namely, $\pi((u_1^0, u_2^0), t)$ is the unique solution to (4.84) or (4.85) $(u_1(x,t), u_2(x,t))$ with $(u_1(x,0), u_2(x,0)) = (u_1^0(x), u_2^0(x))$. Once again, π may be regarded as a map from $Y \times [0, \infty)$ into Y, where Y is taken to be the cone K of componentwise nonnegative functions in the space $C_0^1(\overline{\Omega}) \times C_0^1(\overline{\Omega})$, satisfying (i) of (4.38). As with (4.58)–(4.59), for either (4.84) or (4.85), dissipativity in $C_0^1(\overline{\Omega}) \times C_0^1(\overline{\Omega})$ ((ii) of (4.38)) and the compactness of $\pi((\cdot,\cdot), t)$ on K for any $t > 0$ ((iii) of (4.38)) follow from the theory of reaction-diffusion systems once we establish that there are positive constants c_1 and c_2 so that for any $(u_1^0, u_2^0) \in K$, there is a $t_0 = t_0(u_1^0, u_2^0)$ such that

$$\begin{aligned} u_1(x,t) &\le c_1 \\ u_2(x,t) &\le c_2 \end{aligned} \tag{4.86}$$

for all $t \ge t_0$. For either (4.84) or (4.85), c_1 may be taken as $a_1 + 1$, just as with (4.58)–(4.59). Moreover, the argument that such is the case is precisely the same. For (4.84), observe that for any $\varepsilon > 0$, there is a $t^* \ge 0$ depending only on u_1^0 so that $u_1(x,t) \le a_1 + \varepsilon$ for $t \ge t^*$. Then

$$\frac{\partial u_2}{\partial t} \le d\Delta u_2 + u_2(a_2 + b_2(a_1 + \varepsilon) - u_2) \tag{4.87}$$

in $\Omega \times (t^*, \infty)$. The arguments of the preceding section now apply to establish (4.86) with c_2 taken to be any positive number exceeding $a_2 + b_2 a_1$. Consequently, (i)–(iii) of (4.38) hold in the case of (4.84). Notice that if $a_2 + b_2 a_1 < 0$, $a_2 + b_2(a_1 + \varepsilon) < 0$ in (4.87) for $\varepsilon > 0$ and sufficiently small. Hence, if $a_2 + b_2 a_1 < 0$, the predator species goes extinct independent of the choice of (u_1^0, u_2^0) and permanence certainly will not occur.

The issue of the bound c_2 in (4.86) for the model (4.85) is more delicate. Notice first of all that the argument for the case (4.84) will not work, since the absence of a self-regulation term in (4.85) would yield

$$\frac{\partial u_2}{\partial t} \leq d\Delta u_2 + (a_2 + b_2(a_1 + \varepsilon))u_2 \tag{4.88}$$

for $t > t^*$, in place of (4.87). Assuming $a_2 + b_2 a_1 \geq 0$, we may only conclude from (4.88) and Lemma 4.5 that the predator species is bounded above by a positive function which grows exponentially in time. If $a_2 + b_2 a_1 < 0$, the predator species goes extinct and the system fails to be permanent. So we must seek another way to establish (4.86).

It may be tempting to think that the failure of the dissipativity argument for (4.84) to carry over to (4.85) is merely a mathematical artifact. However, predator-prey systems such as (4.85) may well experience "boom-bust" transient cycles, even when they are theoretically permanent (Cantrell and Cosner, 1997). Predator-prey systems with predator self-limitation such as (4.84) are far less subject to such cycles. Indeed, in Cantrell and Cosner (1997) we show how the transient dynamics for such systems may sometimes be bounded. We shall return to this issue in Chapter 5 when we discuss the notion of practical persistence.

If $d = 1$ in (4.85), then $v = b_2 u_1 + b_1 u_2$ satisfies a differential inequality which leads to a bound on v and hence on u_2 via the maximum principle. However, that approach fails if $d \neq 1$ since the differential operators in the two equations in (4.85) are different, so linear combinations of u_1 and u_2 no longer satisfy any simple differential inequality. To obtain an asymptotic upper bound c_2 on the density u_2 in (4.85) with $d \neq 1$, we first obtain an asymptotic upper bound on the total predator population, i.e. $\int_\Omega u_2(x,t)dx$. From this bound, one may obtain the desired asymptotic upper bound on the species density itself. It is not very difficult to obtain an upper bound for $\int_\Omega u_2(x,t)dx$ for large t. We present that argument next. However, obtaining the resulting bound on the density $u_2(x,t)$ for large t is mathematically somewhat challenging. One argues along the lines of Moser's proof (Moser, 1960) of the famous Nash-DeGiorgi estimates for PDE's (Nash, 1958, DeGiorgi, 1957), specialized to the current context, as in Alikakos (1979). For a detailed proof see Cantrell et al. (1993a, Lemma 4.7).

Notice that if α and β are positive constants so that $\beta b_2 - \alpha b_1 < 0$ and γ is a positive constant so that $a_2 + \gamma < 0$, then

$$\begin{aligned} &\alpha u_1(a_1 - u_1 - b_1 u_2 + \gamma) + \beta u_2(a_2 + b_2 u_1 + \gamma) \\ &\leq \alpha u_1(a_1 + \gamma - u_1) \end{aligned} \tag{4.89}$$

so long as $u_1 \geq 0$ and $u_2 \geq 0$. Let $(u_1(x,t), u_2(x,t))$ be a solution to (4.85) with $u_1(x,0) \gneqq$ and $u_2(x,0) \gneqq 0$. Define a function $G : [0,\infty) \to (0,\infty)$ by

$$G(t) = \int_\Omega [\alpha u_1(x,t) + \beta u_2(x,t)]dx \tag{4.90}$$

Differentiating (4.90) and surpressing the argument (x, t) on the right-hand side, we have that

$$
\begin{aligned}
G'(t) &= \int_\Omega \left(\alpha \frac{\partial u_1}{\partial t} + \beta \frac{\partial u_2}{\partial t}\right) dx \\
&= \int_\Omega (\alpha \Delta u_1 + \beta d \Delta u_2) dx \\
&+ \int_\Omega [\alpha u_1(a_1 - u_1 - b_1 u_2 + \gamma) + \beta u_2(a_2 + b_2 + \gamma)] dx \\
&- \int_\Omega \gamma(\alpha u_1 + \beta u_2) dx.
\end{aligned}
\tag{4.91}
$$

Green's First Identity implies that $\int_\Omega \Delta u_i dx = \int_{\partial\Omega} \frac{\partial u_i}{\partial \vec{n}} dS < 0$ for $i = 1, 2$. Consequently, it follows from (4.89) and (4.91) that

$$G'(t) \le \int_\Omega \alpha u_1(a_1 + \gamma - u_1) dx - \gamma G(t). \tag{4.92}$$

Since $\alpha u_1(a_1 + \gamma - u_1) \le \alpha \frac{(a_1 + \gamma)^2}{4}$ for all u_1, (4.92) yields

$$G'(t) \le G_0 - \gamma G(t) \tag{4.93}$$

for all $t \ge 0$, where $G_0 = \alpha \frac{(a_1+\gamma)^2}{4} |\Omega|$. Multiplying (4.93) by $e^{\gamma t}$ and integrating from 0 to t yields

$$G(t) \le G(0) e^{-\gamma t} + \frac{G_0}{\gamma}(1 - e^{-\gamma t}). \tag{4.94}$$

It now follows from (4.94) that there is a $t_* = t_*(G(0))$ so that if $t \ge t_*$, then $G(t) \le \frac{G_0}{\gamma} + 1$. Hence if $t \ge t_*$, then $\int_\Omega u_2(x, t) dx \le \frac{1}{\beta}\left[\frac{G_0}{\gamma} + 1\right]$, establishing the required asymptotic upper bound on $\int_\Omega u_2(x, t) dx$.

So now (4.84) and (4.85) have been recast as semi-dynamical systems on $Y = K$ satisfying (i)–(iii) of (4.38), and we now construct average Lyapunov functions for the systems. We consider two cases, corresponding to $a_2 > 0$ and $a_2 < 0$. (When $a_2 > 0$, we only consider (4.84), as previously noted.) For convenience, we now set $d = 1$, since the value of d does not affect the rest of the analysis.

First let us assume that $a_2 > 0$ and consider (4.84). Theorem 4.4 applies to the population models

$$
\begin{aligned}
\frac{\partial u_i}{\partial t} &= \Delta u_i + u_i(a_i - u_i) \text{ in } \Omega \times (0, \infty) \\
u_i &= 0 \qquad\qquad\qquad \text{on } \partial\Omega \times (0, \infty)
\end{aligned}
\tag{4.95}
$$

$i = 1, 2$, which arise from the assumption that one of the species is absent. We assume as in Theorem 4.4 that the principal eigenvalue σ_i is positive in

$$
\begin{aligned}
\Delta w_i + a_i w_i &= \sigma_i w_i \text{ in } \Omega \\
w_i &= 0 \qquad\quad \text{on } \partial\Omega.
\end{aligned}
\tag{4.96}
$$

Then Theorem 4.4 implies (4.95) has a globally attracting equilibrium $\overline{u}_i$ which is positive in Ω. From the discussion of the previous section, the prey species can invade when the predator has density $\overline{u}_2$ provided that the principal eigenvalue σ_3 of

$$\begin{aligned} &\Delta w_3 + (a_1 - b_1\overline{u}_2)w_3 = \sigma_3 w_3 && \text{in} \quad \Omega \\ &w_3 = 0 && \text{on} \quad \partial\Omega \end{aligned} \tag{4.97}$$

is positive, and the predator species can invade when the prey has density $\overline{u}_1$ provided that the principal eigenvalue σ_4 of

$$\begin{aligned} &\Delta w_4 + (a_2 + b_2\overline{u}_1)w_4 = \sigma_4 w_4 && \text{in} \quad \Omega \\ &w_4 = 0 && \text{on} \quad \partial\Omega. \end{aligned} \tag{4.98}$$

is positive. Under the assumption that $\sigma_1 > 0$ and $\sigma_2 > 0$ in (4.96), $\sigma_3 > 0$ in (4.97) and $\sigma_4 > 0$ in (4.98), we now construct an average Lyapunov function. We restrict our attention to the compact set X given in Theorem 4.1, and define $P : X \to [0, \infty)$ by

$$P((v_1, v_2)) = \left(\int_\Omega w_3 v_1 dx\right)^{\beta_1} \left(\int_\Omega w_4 v_2 dx\right)^{\beta_2} \tag{4.99}$$

where $w_3 > 0$ and $w_4 > 0$ are the eigenfunctions corresponding to $\sigma_3 > 0$ in (4.97) and $\sigma_4 > 0$ in (4.98), respectively, and β_1 and β_2 are positive constants, yet to be determined. Notice that we need $\sigma_1 > 0$ and $\sigma_2 > 0$ to have (4.97) and (4.98) in the first place, so that w_3 and w_4 reflect $\sigma_1 > 0$ and $\sigma_2 > 0$. We must hypothesize $\sigma_3 > 0$ and $\sigma_4 > 0$. Without these invasibility conditions, we shall not be able to establish that P in (4.99) is an average Lyapunov function.

Now let $S \subseteq X$ be as in Theorem 4.1. Recall that if $(u_1, u_2) \in X$, then $(u_1, u_2) \in S$ if and only if $u_1 \equiv 0$ in Ω or $u_2 \equiv 0$ in Ω. For $(u_1, u_2) \in S$, define $a(t, (u_1, u_2))$ by

$$a(t, (u_1, u_2)) = \liminf_{\substack{(v_1,v_2)\to(u_1,u_2) \\ (v_1,v_2)\in X\setminus S}} \left(\frac{P(\pi((v_1, v_2), t))}{P(v)}\right). \tag{4.100}$$

Theorem 4.2 will guarantee that π and hence (4.84) are permanent provided that $a(t, (u_1, u_2))$ in (4.100) satisfies

$$\sup_{t>0} a(t, (u_1, u_2)) > \begin{cases} 0 & \text{for} \quad (u_1, u_2) \in S \\ 1 & \text{for} \quad (u_1, u_2) \in w(S). \end{cases} \tag{4.101}$$

What we shall see is that to show $\sup_{t>0} a(t, (u_1, u_2)) > 0$ for $(u_1, u_2) \in S$ it is not very difficult, but to show $\sup_{t>0} a(t, (u_1, u_2)) > 1$ if $(u_1, u_2) \in w(S)$ is substantially more delicate.

Notice first that (4.99) can be rewritten as

$$P((v_1, v_2)) = \exp\left[\beta_1 \log \int_\Omega w_3 v_1 dx + \beta_2 \log \int_\Omega w_4 v_2 dx\right]. \tag{4.102}$$

For $(v_1, v_2) \in X\backslash S$, let $(z_1(x,t), z_2(x,t)) = \pi((v_1(x), v_2(x)), t)$. Then from (4.102)

$$\begin{aligned}
&\frac{P(\pi((v_1, v_2), t))}{P((v_1, v_2))} \\
&= \exp[\beta_1(\log \int_\Omega w_3(x) z_1(x,t)dx - \log \int_\Omega w_3(x) z_1(x,0)dx) \\
&\quad + \beta_2(\log \int_\Omega w_4(x) z_2(x,t)dx - \log \int_\Omega w_4(x) z_2(x,0)dx)].
\end{aligned} \tag{4.103}$$

Since $\log f(t) - \log f(0) = \int_0^t f'(s)/f(s)ds$ for a positive smooth function, the right-hand side of (4.103) equals

$$\begin{aligned}
&\exp\left[\beta_1 \int_0^t \left[\int_\Omega w_3(x)\frac{\partial z_1}{\partial s}(x,s)dx \Big/ \int_\Omega w_3(x) z_1(x,s)dx\right] ds\right. \\
&\left. +\beta_2 \int_0^t \left[\int_\Omega w_4(x)\frac{\partial z_2}{\partial s}(x,s)dx \Big/ \int_\Omega w_4(x) z_2(x,s)dx\right] ds\right] \\
&= \exp\left[\beta_1 \int_0^t \left[\frac{\int_\Omega w_3(x)\{\Delta z_1(x,s) + z_1(x,s)[a_1 - z_1(x,s) - b_1 z_2(x,s)]\}dx}{\int_\Omega w_3(x) z_1(x,s)dx}\right] ds\right. \\
&\left. +\beta_2 \int_0^t \left[\frac{\int_\Omega w_4(x)\{\Delta z_2(x,s) + z_2(x,s)[a_2 + b_2 z_1(x,s) - z_2(x,s)dx]\}dx}{\int_\Omega w_4(x) z_2(x,s)dx}\right] ds\right].
\end{aligned} \tag{4.104}$$

Using the facts that $\int_\Omega w_3(x)\Delta z_1(x,s)dx = \int_\Omega \Delta w_3(x) z_1(x,s)dx = \int_\Omega (\sigma_3 - a_1 + b_1\overline{u}_2(x)) w_3(x) z_1(x,s)dx$ and $\int_\Omega w_4(x)\Delta z_2(x,s)dx = \int_\Omega \Delta w_4 z_2(x,s)dx = \int_\Omega(\sigma_4 - a_2 - b_2\overline{u}_1(x)) w_4(x) z_2(x,s)dx$, (4.104) simplifies to

$$\begin{aligned}
&\exp\left[\beta_1 \int_0^t \left[\frac{\int_\Omega w_3(x) z_1(x,s)\{\sigma_3 + b_1\overline{u}_2(x) - b_1 z_2(x,s) - z_1(x,s)\}dx}{\int_\Omega w_3(x) z_1(x,s)dx}\right] ds\right. \\
&\left. +\beta_2 \int_0^t \left[\frac{\int_\Omega w_4(x) z_2(x,s)\{\sigma_4 - b_2\overline{u}_1(x) + b_2 z_1(x,s) - z_2(x,s)\}dx}{\int_\Omega w_4(x) z_2(x,s)dx}\right] ds\right].
\end{aligned} \tag{4.105}$$

(Notice that we have exploited (4.97) and (4.98) to bring σ_3 and σ_4 into the expression $\dfrac{P(\pi((v_1, v_2), t))}{P((v_1, v_2))}$; this will be key to establishing that P is an average Lyapunov functional.) We have restricted ourselves to the compact set X which contains the global attractor for the system. The functional expressions $\sigma_3 + b_1\overline{u}_2(x) - b_1\tilde{v}_2(x) - \tilde{v}_1(x)$ and $\sigma_4 - b_2\overline{u}_1(x) + b_2\tilde{v}_1(x) - \tilde{v}_2(x)$ appearing in (4.105) must be bounded below as $(\tilde{v}_1, \tilde{v}_2)$ range over X, since X is compact. In particular, if

$$\sigma_3 + b_1\overline{u}_2(x) - b_1\tilde{v}_2(x) - \tilde{v}_1(x) \geq d_1$$

$$\sigma_4 - b_2\overline{u}_1(x) + b_2\tilde{v}_1(x) - \tilde{v}_2(x) \geq d_2$$

independent of $(\tilde{v}_1, \tilde{v}_2) \in X$, it follows from (4.105) that

$$\frac{P(\pi((v_1, v_2), t)}{P((v_1, v_2))} \geq \exp((\beta_1 d_1 + \beta_2 d_2)t). \tag{4.106}$$

Notice that d_1 and d_2 need not be positive, so the ratio $P(\pi(v_1, v_2), t)/P(v_1, v_2)$ need only exceed a decaying exponential. It is clear from (4.106) that if $(u_1, u_2) \in S$

$$a(t, (u_1, u_2)) \geq \exp((\beta_1 d_1 + \beta_2 d_2)t). \tag{4.107}$$

It is immediate from (4.107) that $\sup_{t>0} a(t, (u_1, u_2)) > 0$ if $(u_1, u_2) \in S$. In fact, we can argue that $\sup_{t>0} a(t, (u_1, u_2)) > c$ for any number $c \in (0, 1)$, even when d_1 and d_2 are negative, simply by choosing t close enough to 0. If d_1 and d_2 were positive, we would be through at this point, for then we get $a(t, (u_1, u_2)) > 1$ for any t. Unfortunately, we have no means of discerning that d_1 and d_2 are positive, nor do we necessarily have any reason to believe that they should be.

We now turn to the question of showing that $\sup_{t>0} a(t, (u_1, u_2)) > 1$ if $(u_1, u_2) \in w(S)$. From Theorem 4.4, it follows that $w(S) = \{(0, 0), (\overline{u}_1(x), 0), (0, \overline{u}_2(x))\}$. Consequently, we need to examine (4.105) when (v_1, v_2) nears $(0, 0)$ or $(\overline{u}_1, 0)$ or $(0, \overline{u}_2)$. The basic problem in the limiting procedure is that we encounter an indeterminate form. How so? Since X is compact, $\pi : X \times [0, 1] \to X$ is uniformly continuous. So if $(v_1, v_2) = (z_1(x, 0), z_2(x, 0))$ is near enough to $(\overline{u}_1, 0) = (\tilde{z}_1(x, 0), \tilde{z}_2(x, 0))$, for example, their images under the semiflow $(z_1(x, t), z_2(x, t))$ and $(\tilde{z}_1(x, t), \tilde{z}_2(x, t))$ will remain uniformly close on $\overline{\Omega}$ for all $t \in [0, 1]$. But since $(\overline{u}_1, 0)$ is an equilibrium, $(\tilde{z}_1(x, t), \tilde{z}_2(x, t)) = (\overline{u}_1(x), 0)$ for all $t \in [0, 1]$. Consequently, in all three cases, at least one of the ratios $\dfrac{\int_\Omega w_3(x) z_1(x, s)\{\sigma_3 + b_1 \overline{u}_2(x) - b_1 z_2(x, s) - z_1(x, s)\}dx}{\int_\Omega w_3 z_1(x, s)}$ and $\dfrac{\int_\Omega w_4(x) z_2(x, s)\{\sigma_4 - b_2 \overline{u}_1(x) + b_2 z_1(x, s) - z_2(x, s)\}dx}{\int_\Omega w_4(x) z_2(x, s)}$ is a $\dfrac{0}{0}$ indeterminate form as (v_1, v_2) approaches $(0, 0)$ or $(\overline{u}_1, 0)$ or $(0, \overline{u}_2)$. Indeed, such was actually the case for any $(u_1, u_2) \in S$. But to get $a(t, (u_1, u_2)) > 0$ for $(u_1, u_2) \in S$, we only needed fairly crude lower bounds on $\sigma_3 + b_1 \overline{u}_2(x) - b_2 z_2(x, s) - z_1(x, s)$ and $\sigma_4 - b_2 \overline{u}_1(x) + b_2 z_1(x, s) - z_2(x, s)$ arising from the compactness and invariance of X. In particular, d_1 and d_2 did not need to be positive. We will need positive lower bounds in the arguments of the exponential in (4.105) to get, say, $a(1, (\overline{u}_1, 0)) > 1$. Such will require a more careful approximation process using the fact that the limiting densities are known. The positivity will require that the average growth rates σ_3 and σ_4 over Ω be positive.

Consider $\{(\overline{u}_1, 0)\}$ and set $t = 1$ in (4.105). For all $s \in [0, 1]$, the ratio

$$\frac{\displaystyle\int_\Omega w_3(x) z_1(x, s)\{\sigma_3 + b_1 \overline{u}_2(x) - b_1 z_2(x, s) - z_1(x, s)\}dx}{\displaystyle\int_\Omega w_3(x) z_1(x, s)dx}$$

converges to

$$\frac{\displaystyle\int_\Omega w_3(x) \overline{u}_1(x)\{\sigma_3 + b_1 \overline{u}_2(x) - \overline{u}_1(x)\}dx}{\displaystyle\int_\Omega w_3(x) \overline{u}_1(x)dx} \tag{4.108}$$

as (v_1, v_2) converges to $(\overline{u}_1, 0)$. Observe that (4.108) may be written as

$$\frac{\int_\Omega [w_3(x)\overline{u}_1(x)\{\sigma_3 - a_1 + b_1\overline{u}_2(x)\} + w_3(x)\overline{u}_1(x)(a_1 - \overline{u}_1(x))]dx}{\int_\Omega w_3(x)\overline{u}_1(x)dx}$$
$$= \frac{\int_\Omega [\overline{u}_1(x)\Delta w_3(x) - \Delta\overline{u}_1(x)w_3(x)]dx}{\int_\Omega w_3(x)\overline{u}_1(x)dx} \qquad (4.109)$$
$$= 0.$$

It follows from (4.109) that the $\exp(\beta_1 \int_0^1 \cdot ds)$ term in (4.105) tends to 1 as (v_1, v_2) tends to $(\overline{u}_1, 0)$.

So now consider the remaining factor of (4.105). We have that $\sigma_4 - b_2\overline{u}_1(x) + b_2 z_1(x, s) - z_2(x, s)$ converges to $\sigma_4 - b_2\overline{u}_1(x) + b_2\overline{u}_1(x) = \sigma_4$ uniformly on $\overline{\Omega}$ for all $s \in [0, 1]$ as (v_1, v_2) converges to $(\overline{u}_1, 0)$. So if (v_1, v_2) is close enough to $(\overline{u}_1, 0)$, $\sigma_4 - b_2\overline{u}_1(x) + b_2 z_1(x, s) - z_2(x, s)$ will exceed $\frac{\sigma_4}{2} > 0$ for all $s \in [0, 1]$ and

$$a(1, (\overline{u}_1, 0)) > \exp\left(\beta_2 \frac{\sigma_4}{2}\right) > 1 \qquad (4.110)$$

for any choice of $\beta_2 > 0$.

If $(u_1, u_2) = (0, \overline{u}_2)$, a completely analogous argument allows us to conclude

$$a(1, (0, \overline{u}_2)) > \exp\left(\beta_1 \frac{\sigma_3}{2}\right) > 1 \qquad (4.111)$$

for any choice of $\beta_1 > 1$. At $(0, 0)$, both integrals in (4.105) involve indeterminate forms and the estimation procedure is more complicated. Nevertheless we obtain

$$a(1, (0, 0)) > \exp\left\{\beta_1 \frac{\sigma_3}{2} + \beta_2\left(\sigma_4 - b_2 \max_{\overline{\Omega}} \overline{u}_1 - 1\right)\right\}. \qquad (4.112)$$

In (4.112), the fact that we may choose β_1 and β_2 in (4.99) independently of one another is crucial. Indeed, we get $\alpha(1, (0, 0)) > 1$ by fixing $\beta_2 > 0$ and then choosing β_1 large enough relative to β_2 and the remaining quantities in the argument of the exponential, since $\sigma_3 > 0$. It follows from (4.110), (4.111) and (4.112) that (4.84) is permanent, assuming σ_1 and σ_2 in (4.96), σ_3 in (4.97) and σ_4 in (4.98) are all positive. The details of the estimation procedure in the case of $\{(0, 0)\}$ are given in the Appendix to the chapter.

We now have permanence in (4.84) when $a_2 > 0$. If $a_2 < 0$ (so that we consider (4.85) as well as (4.84)), it follows from part (iii) of Theorem 4.4 that $\sigma_2 < 0$ in (4.96). As a consequence, the population dynamical model that results when the prey species is absent has 0 as a globally attracting equilibrium; i.e., if the prey species is essentially the sole resource of the predator, the predator tends to extinction when the prey species is absent. So we have no equilibrium $(\overline{u}_2, 0)$ in this case and $\omega(S)$ reduces to $\{(0, 0), (\overline{u}_1, 0)\}$, so long as $\sigma_1 > 0$ in (4.96). Assuming such to be the case and also that $\sigma_4 > 0$ in (4.97), we may define a candidate for average Lyapunov function $P : X \to [0, \infty)$ by

$$P((v_1, v_2)) = \left(\int_\Omega w_1 v_1 dx\right)^{\beta_1} \left(\int_\Omega w_4 v_2 dx\right)^{\beta_2} \qquad (4.113)$$

where w_1 in (4.96) and w_4 in (4.98) are chosen positive in Ω and β_1 and β_2 are positive constants, yet to be determined. Proceeding in a manner analogous to that for the case $a_2 > 0$, we can show that P is indeed an average Lyapunov function, so that (4.84) or (4.85) is permanent when $a_2 < 0$, so long as the average growth rates σ_1 and σ_4 are positive in (4.96) and (4.98), respectively. We leave the details of the argument to the interested reader.

We should note that the assertion of permanence in the models (4.84) and (4.85) represent special cases of Theorem 5.3 and Theorem 5.7 in Cantrell et al. (1993a). We refer the interested reader to Cantrell et al. (1993a) for a more general treatment of permanence in two species reaction-diffusion systems. (In particular, Cantrell et al. (1993a) treat models with different diffusion rates, spatially varying coefficients, and other boundary conditions.)

4.6 Ecological Permanence and Equilibria

4.6.1 Abstract Permanence Implies Ecological Permanence

We claimed in Section 2 that in the context of autonomous reaction-diffusion models for interacting species on a bounded habitat, abstract permanence implies ecological permanence. Our first task in this section will be to show why such is the case. Following that, we show that if an autonomous reaction-diffusion model for interacting species on a bounded habitat is permanent, then it has a (not necessarily stable) componentwise positive equilibrium.

In showing that abstract permanence implies ecological permanence in the context of autonomous reaction-diffusion models for interacting species in a bounded habitat, we will focus on the representative system

$$\frac{\partial u_i}{\partial t} = d_i \Delta u_i + u_i f_i(x, u_1, \ldots, u_n) \text{ in } \quad \Omega \times (0, \infty) \tag{4.114}$$

$$u_i = 0 \text{ on } \quad \partial\Omega \times (0, \infty) \tag{4.115}$$

for $i = 1, \ldots, n$. It is certainly possible to consider more complicated dispersal mechanisms than the purely diffusive terms in (4.114). In particular, we could add spatial variability and/or advection terms to the dispersal terms in (4.114). However, once we recast the problem as a semiflow π, we shall not be drawing on the form of (4.114) in a really explicit way and therefore we choose a form that appears to be a natural extension of the 2-species models we examined in Sections 4 and 5. Our choice of Dirichlet or absorbing boundary conditions is deliberate, since they force us to adopt (4.27) as our definition of ecological permanence, in place of (4.26). We could use (4.26) if, for example, we replaced (4.115) with the reflecting or Neumann condition

$$\nabla u_i \cdot \vec{n} = 0 \text{ on } \quad \partial\Omega \times (0, \infty) \tag{4.116}$$

for $i = 1, \ldots, n$. The system (4.114)–(4.116) could then be recast as a semiflow on n-tuples of continuous functions on the bounded habitat $\overline{\Omega}$. Mathematically, such a choice is simpler than the choice of continuously differentiable functions which vanish on the boundary $\partial\Omega$ of the habitat as the underlying function space for π that (4.115) forces upon us. So by choosing (4.115) for the boundary conditions, we confront rather than avoid a difficulty, so to speak.

So let us suppose that (4.114)–(4.115) has been recast as a semiflow on the cone K of n-tuples of smooth nonnegative functions on $\overline{\Omega}$ which vanish on the boundary $\partial\Omega$ and that the semiflow π is permanent. Let X and S be as in Theorem 4.1. Then we may take $Y = X$, $Y_0 = X \backslash S$ and $\partial Y_0 = S$, where Y, Y_0 and ∂Y_0 are as in (4.36). Since π is permanent there will be a subset U of $X \backslash S$ such that U is bounded, $\inf_{u \in U} d(u, S) > 0$ and $\lim_{t \to \infty} d(\pi(v, t), U) = 0$ for all $v \in X \backslash S$, where we use u and v to represent the n-tuples of densities $(u_1, \ldots, u_n)$ and $(v_1, \ldots, v_n)$, respectively. Since $\inf_{u \in U} d(u, S) > 0$, $\overline{U} \cap S = \emptyset$. Since $\overline{U} \subseteq X \backslash S$, any $u \in \overline{U}$ must have the property that for $i = 1, \ldots, n$, $u_i(x) > 0$ for all $x \in \Omega$ and $\nabla u_i \cdot \vec{n} < 0$ for all $x \in \partial\Omega$. Since X is compact, so is $\overline{U}$. From the compactness of $\overline{U}$, we can establish that (4.114)–(4.115) is ecologically permanent. We present the details of the argument in the Appendix to the chapter.

4.6.2 Permanence Implies the Existence of a Componentwise Positive Equilibrium

A permanent autonomous reaction-diffusion model for interacting species on a bounded habitat also exhibits a componentwise positive equilibrium. Such an equilibrium need not be in and of itself stable. However, it necessarily must be contained within the global attractor for the set U above.

While the observation that permanence in this class of models implies the existence of a componentwise positive equilibrium is primarily of mathematical interest, it does provide a criterion for excluding permanence. *Namely, if the model fails to exhibit a componentwise positive equilibrium, it cannot be permanent.* Just as in the case of the observation that abstract permanence of a reaction-diffusion model for interacting species in a bounded habitat implies ecological permanence, we shall establish that permanence implies the existence of a componentwise positive equilibrium for the representative system (4.114)–(4.115). Just as before, the result remains true if simple Fickian diffusion is replaced by a more general dispersal mechanism and for a range of homogeneous boundary conditions.

Suppose that (4.114)–(4.115) is permanent. Our proof of the fact that abstract permanence implies ecological permanence allows us to conclude that a variation of (4.27) holds. Namely, there are positive constants m and M, with $m < M$ so that for any solution $(u_1, \ldots, u_n)$ of (4.114)–(4.115) with $u_i(x, 0) \underset{\neq}{>} 0$ on $\overline{\Omega}$, there is a $t_0 > 0$ depending only on $(u_1(x, 0), \ldots, u_n(x, 0))$ so that

$$me(x) < u_i(x, t) < Me(x) \text{ in } \Omega \tag{4.117}$$

and

$$M\nabla e(x) \cdot \vec{n} < \nabla u_i(x, t) \cdot \vec{n} < m\nabla e(x) \cdot \vec{n} \text{ on } \partial\Omega \tag{4.118}$$

for $t \geq t_0$ and $i = 1, \ldots, n$, where $e(x)$ is a fixed function in $C_0^1(\Omega)$ having the properties that $e(x) > 0$ in Ω and $\nabla e(x) \cdot \vec{n} < 0$ on $\partial\Omega$. The collection of n-tuples $(w_1(x), \ldots, w_n(x))$ in $[C_0^1(\overline{\Omega})]^n$ that satisfy (4.117) and (4.118) constitute an open and convex subset of $[C_0^1(\overline{\Omega}]^n$. Indeed, the same is true if we "fatten up" the collection of such n-tuples by replacing m with $\frac{m}{2}$ and M with $2M$. Let us now denote by $\tilde{U}$ the set of n-tuples $(w_1(x), \ldots, w_n(x))$ in $[C_0^1(\overline{\Omega})]^n$ which satisfy

$$\frac{m}{2} e(x) < w_i(x) < 2Me(x) \text{ in } \Omega \tag{4.119}$$

and

$$2M\nabla e(x)\cdot\vec{n} < \nabla w_i(x)\cdot\vec{n} < \frac{m}{2}\nabla e(x)\cdot\vec{n} \text{ on } \partial\Omega \tag{4.120}$$

for $i = 1, \ldots, n$. Now let

$$U^* = \tilde{U} \cap \mathcal{B}(0; R) \tag{4.121}$$

where $R > 0$ is large enough so that $X \subseteq \mathcal{B}(0, R)$ and $M(e, \ldots, e) \in \mathcal{B}(0, R)$, where M is as in (4.119) and (4.120). Then U^* is a nonempty, bounded, open and convex set in $[C_0^1(\overline{\Omega})]^n$. The essence of the proof that (4.114)–(4.115) has a componentwise positive equilibrium is to show that for any $t > 0$, $\pi(\cdot, t)$ has a fixed point in U^*. It then follows from Bhatia and Szego (1970, Lemma 3.7) that (4.114)–(4.115) has an equilibrium in U^*, which necessarily means the equilibrium is positive in Ω for each component.

To show that $\pi(\cdot, t')$ has a fixed point in U^* for arbitrary $t' > 0$, we appeal to the Asymptotic Schauder Fixed Point Theorem (Zeidler, 1986). The Asymptotic Schauder Fixed Point Theorem will assert that $\pi(\cdot, t')$ has a fixed point in U^* provided that $\pi(\cdot, t')$ is a compact operator (which we know already from (4.38)), and that for some prime number p, $[\pi(\cdot, t')]^p = \pi(\cdot, pt')$ and $[\pi(\cdot, t')]^{p+1} = \pi(\cdot, (p+1)t')$ both map $\overline{U}^*$ into U^*. This last hypothesis will be met if we show that there is a positive integer k_0 such that $\pi(\cdot, kt')$ maps $\overline{U}^*$ into U^* for all $k \geq k_0$. To establish that $\pi(\cdot, kt')$ maps $\overline{U}^*$ into U^* for all $k \geq k_0$ for some positive integer k_0, it will suffice to show that there is a $t^* > 0$ so that $\pi(\cdot, t)$ maps $\overline{U}^*$ into U^* for all $t \geq t^*$. That argument appears in the Appendix to the chapter. With it, we have established the following.

Theorem 4.6. Suppose that (4.114)–(4.115) is permanent. Then (4.114)–(4.115) has an equilibrium solution $(u_1(x), \ldots, u_n(x))$ such that for $i = 1, \ldots, n$, $u_i(x) > 0$ for all $x \in \Omega$.

Appendix

Proof of Theorem 4.1: **Part (i)**: We show first that $\tilde{X} = \overline{\pi(\mathcal{B}(\mathcal{A}, \varepsilon), [t_0, \infty))}$ is compact. Let $x_n \in \mathcal{B}(\mathcal{A}, \varepsilon)$ and $t_n \geq t_0$ for $n = 1, 2, 3, \ldots$. We must show that $\pi(x_n, t_n)$ has a convergent subsequence.

Case 1. $\{t_n\}$ is bounded. Without loss of generality, we may assume that t_n converges to $\bar{t}$, where $\bar{t} \geq t_0$. Then $\pi(x_n, t_n) = \pi(x_n, t_0 + (t_n - t_0)) = \pi(\pi(x_n, t_0), t_n - t_0)$. We know from the compactness of $\pi(\cdot, t_0)$ that $\overline{\pi(\mathcal{B}(\mathcal{A}, \varepsilon), t_0)}$ is compact. Hence there is a subsequence of $\pi(x_n, t_0)$ which converges to $y \in Y$. Again, without loss of generality, we may assume $\pi(x_n, t_0) \to y$. So $\pi(x_n, t_0) \to y$ and $t_n - t_0 \to \bar{t} - t_0$. Continuity of $\pi \Rightarrow \pi(x_n, t_n) \to \pi(y, \bar{t} - t_0)$.

Case 2. $\{t_n\}$ is unbounded. Without loss of generality, we may assume $t_n \to \infty$ as $n \to \infty$. Since $\mathcal{A}$ is a global attractor of π and $\mathcal{B}(\mathcal{A}, \varepsilon)$ is bounded, there is a $t_* > 0$ such that $\pi(\mathcal{B}(\mathcal{A}, \varepsilon), t) \subseteq \mathcal{B}(\mathcal{A}, \varepsilon)$ for all $t \geq t_*$. Taking $t_* \geq t_0$, we have that $\overline{\pi(\mathcal{B}(\mathcal{A}, \varepsilon), t_*)}$ is compact. Let N be large enough so that $t_n > 2t_*$ for $n \geq N$. Then for $n \geq N$, $\pi(x_n, t_n) = \pi(x_n, t_n - t_* + t_*) = \pi(\pi(x_n, t_n - t_*), t_*)$. Since $t_n > 2t_*, t_n - t_* > t_* \Rightarrow \pi(x_n, t_n - t_*) \in \mathcal{B}(\mathcal{A}, \varepsilon)$. So $\pi(x_n, t_n) \in \pi(\mathcal{B}(\mathcal{A}, \varepsilon), t_*)$ for $n \geq N$. Since $\overline{\pi(\mathcal{B}(\mathcal{A}, \epsilon), t_*)}$ is compact, there is a $y \in Y$ so that a subsequence of $\pi(x_n, t_n)$ converges to y.

Next we verify that $\tilde{X}$ is forward invariant. Suppose that $y \in \overline{\pi(\mathcal{B}(\mathcal{A}, \varepsilon), [t_0, \infty))}$ and let $t > 0$. There are $\{x_n\} \subseteq \mathcal{B}(\mathcal{A}, \varepsilon)$ and $\{t_n\}$ with $t_n \geq t_0$ so that $\pi(x_n, t_n) \to y$ as $n \to \infty$.

Then $\pi(x_n, t_n+t) \in \pi(\mathcal{B}(\mathcal{A}, \varepsilon), [t_0, \infty))$ and $\pi(x_n, t_n+t) = \pi(\pi(x_n, t_n), t) \to \pi(y, t)$ by continuity of π. So $\pi(y, t) \in \tilde{X}$ as required.

The continuity of π and the compactness of $\tilde{X}$ implies that $X = \pi(\tilde{X}, t')$ is compact. So suppose $y \in X$ and $t > 0$ is given. Since $y \in X$, there is $x \in \tilde{X}$ so that $y = \pi(x, t')$. Then $\pi(y, t) = \pi(\pi(x, t'), t) = \pi(x, t'+t) = \pi(\pi(x, t), t')$. Since $\tilde{X}$ is forward invariant $\pi(x, t) \in \tilde{X}$. Therefore $\pi(y, t) = \pi(\pi(x, t), t') \in \pi(\tilde{X}, t') = X$. So X is forward invariant.

Part (ii). The definition of X guarantees that if $u = (u_1, \ldots, u_n)$ is a configuration of species densities in X, then for each $i \in \{1, \ldots, n\}$, $u_i \equiv 0$ on the underlying habitat patch or u_i is positive inside the habitat patch. Moreover, $u \in S$ if and only if at least one u_i is identically zero. The result now follows from part (i) and the maximum principle.

Proof of Lemma 4.5: Set $v(x, t) = ke^{(\sigma-\varepsilon)(t-t_*)}\phi(x)$. Then for $t > t_*$, $\frac{\partial v}{\partial t} - \mu\Delta v - [f(x)-\varepsilon]v = (\sigma-\varepsilon)ke^{(\sigma-\varepsilon)(t-t_*)}\phi(x) - \mu ke^{(\sigma-\varepsilon)(t-t_*)}\Delta\phi - [f(x)-\varepsilon]ke^{(\sigma-\varepsilon)(t-t_*)}\phi(x) = ke^{(\sigma-\varepsilon)(t-t_*)}[(\sigma - f(x))\phi(x) - \mu\Delta\phi] = 0$. Since $u(x, t_*) \geq k\phi(x)$ for $x \in \overline{\Omega}$, the result now follows from the method of upper and lower solutions.

Proof of (4.112): As in the text, we let $(z_1(x,t), z_2(x,t)) = \pi((v_1, v_2), t)$ for $(v_1, v_2) \in X \backslash S$. Again, we have that $(z_1(x,t), z_2(x,t))$ converges to $(0, 0)$ uniformly for $x \in \overline{\Omega}$ and $t \in [0, 1]$ as (v_1, v_2) converges in X to $(0, 0) \in \omega(S)$. As a result, we have that for all $s \in [0, 1]$,

$$\sigma_3 + b_1\bar{u}_2(x) - b_1 z_2(x, s) - z_1(x, s) \geq \frac{\sigma_3}{2} \tag{4A.1}$$

and

$$\sigma_4 - b_2\bar{u}_1(x) + b_2 z_1(x, s) - z_2(x, s) \geq \sigma_4 - b_2\bar{u}_1(x) - 1 \tag{4A.2}$$

if (v_1, v_2) is sufficiently close to $(0, 0)$ in X. We use (4A.1) and (4A.2) to get a lower bound on (4.105). In particular, we have that

$$\begin{aligned}
&\exp\left\{\beta_1 \int_0^1 \left[\frac{\int_\Omega w_3(x) z_1(x,s)\{\sigma_3 + b_1\bar{u}_2(x) - b_1 z_2(x,s) - z_1(x,s)\}dx}{\int_\Omega w_3(x) z_1(x,s)dx}\right] ds\right.\\
&\left.\quad + \beta_2 \int_0^1 \left[\frac{\int_\Omega w_4(x) z_2(x,s)\{\sigma_4 - b_2\bar{u}_1(x) + b_2 z_1(x,s) - z_2(x,s)\}dx}{\int_\Omega w_4(x) z_2(x,s)dx}\right] ds\right\}\\
&\geq \exp\left\{\beta_1 \int_0^1 \left[\frac{\int_\Omega w_3(x) z_1(x,s)(\sigma_3/2)dx}{\int_\Omega w_3(x) z_1(x,s)dx}\right] ds\right.\\
&\left.\quad + \beta_2 \int_0^1 \left[\frac{\int_\Omega w_4 z_2(x,s)(\sigma_4 - b_2 \sup_{\overline{\Omega}} \bar{u}_1 - 1)dx}{\int_\Omega w_4(x) z_2(x,s)dx}\right] ds\right\}\\
&\geq \exp\left\{\beta_1 \frac{\sigma_3}{2} + \beta_2(\sigma_4 - b_2 \sup_{\overline{\Omega}} \bar{u}_1 - 1)\right\}.
\end{aligned} \tag{4A.3}$$

Recall from the discussion in the text that the values of β_1 and β_2 need only be positive in order for (4.110) and (4.111) to hold. Consequently, we are free to choose them to our advantage at this point. To this end, we now fix β_2 at some positive value. Once β_2 is fixed, we use the fact that $\sigma_3 > 0$ to choose $\beta_1 > 0$ so that

$$\beta_1 \frac{\sigma_3}{2} + \beta_2 \left(\sigma_4 - b_2 \sup_{\overline{\Omega}} \overline{u}_1 - 1 \right) > 0. \tag{4A.4}$$

It is immediate from (4A.3) and (4A.4) that (4.112) holds and that $a(1,(0,0)) \geq \exp\left\{ \beta_1 \frac{\sigma_3}{2} + \beta_2 \left(\sigma_4 - b_2 \sup_{\overline{\Omega}} \overline{u}_1 - 1 \right) \right\} > 1$.

Completion of the Proof that Abstract Permanence Implies Ecological Permanence in (4.114)–(4.115): The reader may need to recall the discussion of (4.114)–(4.115) in Section 6. We are assuming that π corresponding to (4.114)–(4.115) is permanent with X and S as in Theorem 4.1. In this context, we identify a bounded attracting set $U \subseteq X \backslash S$ so that $\overline{U} \subseteq X \backslash S$. Since X is compact, so is $\overline{U}$. Now choose a smooth function $e(x)$ on the underlying habitat patch $\overline{\Omega}$ so that $e(x) > 0$ in Ω and $\nabla e \cdot \vec{n} < 0$ on $\partial\Omega$. Each $u = (u_1, \ldots, u_n) \in \overline{U}$ has the property that for $i = 1, \ldots, n$, $u_i(x) > 0$ in Ω and $\nabla u_i \cdot \vec{n} < 0$ on $\partial\Omega$. So there is a value $m(u)$ so that for each $i = 1, \ldots, n$,

$$u_i(x) > m(u)e(x) \tag{4A.5}$$

in Ω with

$$\nabla u_1(x) \cdot \vec{n} < m(u) \nabla e(x) \cdot \vec{n} \tag{4A.6}$$

on $\partial\Omega$.

We have that (4.114)–(4.115) is ecologically permanent provided that we can show that $m(u)$ in (4A.5) and (4A.6) can actually be chosen independent of u. To this end, for any fixed $u \in \overline{U}$, there is an $\varepsilon(u) > 0$ so that for all $v = (v_1, \ldots, v_n) \in B(u, \varepsilon(u)) \cap \overline{U}$,

$$v_i(x) > \frac{m(u)}{2} e(x)$$

in Ω with

$$\nabla v_i(x) \cdot \vec{n} < \frac{m(u)}{2} \nabla e(x) \cdot \vec{n}$$

for $i = 1, \ldots, n$. Clearly $\overline{U} \subseteq \bigcup_{u \in \overline{U}} B(u, \varepsilon(u))$. Since $\overline{U}$ is compact, there must be a finite subset $\{u^1, u^2, \ldots, u^r\} \subseteq \overline{U}$ so that

$$\overline{U} \subseteq \bigcup_{i=1}^{r} B(u^i, \varepsilon(u^i)).$$

(Note here that $u^i = (u^i_1, \ldots, u^i_n)$.) If now $m = \min\left\{ \frac{m(u^1)}{2}, \ldots, \frac{m(u^r)}{2} \right\}$, we have for all $v \in \overline{U}$ that

$$v_i(x) > me(x)$$

in Ω with

$$\nabla v_i(x) \cdot \vec{n} < m\nabla e(x) \cdot \vec{n}$$

on $\partial\Omega$, for $i = 1, \ldots, n$, as required.

Proof that if U^* is defined by (4.121), then $\pi(\overline{U}^*, t) \subseteq U^*$ for $t \geq \bar{t}$ for some $\bar{t} > 0$: Let $\mathcal{A}$ be the global attractor for π whose existence is guaranteed by (4.38). Let X and S be as in Theorem 4.1. Since π is permanent, it follows as in Hale and Waltman (1989) and Cantrell et al. (1993a) that there is a compact invariant set $\mathcal{A}_0 \subseteq \mathcal{A} \cap (X \backslash S)$ so that for any bounded set V with $\overline{V} \subseteq \text{int}K$ and any $\delta > 0$, there is a $\bar{t}(V, \delta)$ so that $\pi(V, t) \subseteq \mathcal{B}(\mathcal{A}_0, \delta)$ for $t \geq \bar{t}$. (Hale and Waltman (1989) refer to the set V as being *strongly bounded* and to $\mathcal{A}_0$ as the global attractor relative to strongly bounded sets.) Since (4.114)–(4.115) is ecologically permanent, if $(w_1, \ldots, w_n) \in \mathcal{A}_0$, then $w_i(x)$ satisfies (4.117) and (4.118). Consequently, the definitions of $\tilde{U}$ in (4.119) and (4.120) and U^* in (4.121) imply that $\mathcal{A}_0 \subseteq U^*$. Since U^* is open, we may choose $\delta > 0$ so that $B(\mathcal{A}_0, \delta) \subseteq U^*$. Clearly, $\overline{U}^*$ is a strongly bounded set. So there is a t^* so that if $t \geq t^*$, $\pi(\overline{U}^*, t) \subseteq B(\mathcal{A}_0, \delta)$. Consequently, $\pi(\overline{U}^*, t) \subseteq U^*$ for $t \geq t^*$, as required.

5

Beyond Permanence: More Persistence Theory

5.1 Introduction

In the preceding chapter, we discussed the notion of permanence as a conceptual framework for considering the issue of asymptotic coexistence in models for several interacting biological species. Our aim in Chapters 6 and 7 will be to employ permanence to analyze mathematically and ecologically the predictions of a number of reaction-diffusion models for interacting species in a bounded spatial habitat patch. As a mathematical assertion of the coexistence of several interacting biological species, ecological permanence is very satisfactory on a qualitative level. It applies broadly to different types of models and species interactions and its assertions can accommodate a wide range of specific asymptotic behavior, as we discussed in Chapter 4. Indeed, it corresponds very closely to the notion in the ecological literature (Strong 1986a,b) of having asymptotic "floors" and "ceilings" on the densities of interacting species. On the other hand, there are ways in which permanence *per se* needs some additional elaboration in order to be a really effective tool for the mathematical and ecological analysis of specific models of interacting biological species, and this chapter is primarily devoted to the discussion of various enhancements of the basic concept of permanence.

An assertion that a model for several interacting biological species is permanent means that any componentwise positive initial configuration of species densities evolves over time toward an attracting set of componentwise positive configurations of species densities, where the distance from the attracting set to any configuration of species densities in which at least one density equals 0 exceeds some positive threshold. However, the size of the distance threshold is arbitrary so long as it is positive. Clearly, for some analyses, such quantitative vagueness becomes problematic. We discuss two ways in which this limitation of the notion of permanence *per se* can be addressed. The first is via a special form of permanence which was formally studied first by Hess and Lazer (1991), and which they termed *compressivity*. Compressivity imposes some *a priori* structure on the species interactions involved. Namely there must be a partial ordering on configurations of species densities that is preserved by the model (i.e., the system is monotone). This feature limits the range of models that may be compressive to single species models, models for two species competition and models for any number of mutualists. However, when compressivity may be asserted, the nature of the conclusions that may be drawn is overtly quantitative. There are natural, quantitatively

Spatial Ecology via Reaction-Diffusion Equations R.S. Cantrell and C. Cosner
 ISBN: 0-471-49301-5

useful choices for asymptotic "floors" and "ceilings", which in autonomous models are perturbations of minimal and maximal componentwise positive equilibria. Moreover, there is a certain measure of control over the transient dynamics inherent in compressive systems. We discuss these issues in Section 2.

We discuss a somewhat related means of attaching quantitative information to a finding of permanence in a model for several interacting biological species in Section 3. Frequently, the components (i.e. individual species densities) in such a model may be upper or lower solutions to single species models, the dynamics of which are well understood. This fact sometimes permits a series of "best case" and "worst case" asymptotic estimates on the components of the model leading to asymptotic "floors" and "ceilings" that are explicitly quantitative (often in terms of the model parameters). Such a regime is an example of an approach to *practical persistence*, a notion introduced in Cao and Gard (1993) and Cao et al. (1996). By practical persistence, we mean ecological permanence with quantitatively explicit asymptotic "floors" and "ceilings". The approach to practical persistence that we consider can in some cases be used to obtain predictions of permanence in nonautonomous models, as we illustrate in Section 3, and to bound transient dynamics, as in Section 4. We discuss abstract permanence in nonautonomous systems in Section 5.

Obtaining a prediction of permanence in a model for n interacting biological species via the Average Lyapunov Function Approach or the Acyclicity Theorem requires fairly precise knowledge of the asymptotic dynamics of all the subsystems of the model that result when at least one species is absent. Once $n \geq 3$, a finding of permanence by itself in such a subsystem likely does not provide enough information to employ the abstract theory to assert permanence in the full system. However, this obstacle to obtaining permanence in models for large numbers of species via abstract tools frequently can be overcome by obtaining practical persistence estimates on the asymptotic dynamics of subsystems. This additional use of practical persistence techniques plays an important role in our analyses of nonmonotone systems in Chapter 7.

A model in which at least one species is such that its density declines when it is low cannot be considered permanent. Of course, such a property does not preclude the possibility that the model predicts coexistence for some initial configurations of species densities. We say that a model for several interacting biological species is *conditionally persistent* if the conclusions of a finding of ecological permanence obtain for some proper open set of the initial componentwise positive configurations of species densities associated with the model and we explore this phenomenon in Section 6. Likewise, if a model for several interacting biological species fails to be permanent, it must be the case that at least one species is predicted to go extinct in the longterm if its density becomes too low. Such a prediction is of course less dire than having the model predict that the species is driven to extinction independent of its initial density. We explore conditions for extinction in the context of models for two species competition in Section 7.

5.2 Compressivity

Compressivity is a variant of permanence that is possible when the semi-dynamical system corresponding to (4.22) or (4.29) enjoys certain suitable monotonicity properties. It offers several advantages in comparison to permanence in general. First, compressivity enables one to give a much more precise quantitative description of the asymptotic "floors" and "ceilings" on the population densities involved in a given model for interacting biological species than would generally be available via permanence techniques *per se*. Secondly, compressivity allows one to assert that the *transient* orbits of the system are reasonably well-behaved, whereas permanence offers no information regarding the transient behavior of the system. Finally, the hypotheses that must be imposed on a system to assert that it is compressive coincide with those needed to assert permanence. With such advantages, one may be tempted to ask why permanence has emerged as the "flagship" concept in persistence theory instead of compressivity. The answer to the question lies in the caveat regarding applicability that is implicit in the phrase "suitable monotonicity properties". Indeed, the restrictions to applying the technique of compressivity are such that compressivity applies essentially only to single population models, models for the competition between two species, and models for mutualism.

To appreciate what we mean by the phrase "suitable monotonicity properties", recall that the basic objects we study via permanence are n-tuples of population densities. Mathematically, n-tuples of population densities are n-tuples of nonnegative numbers or nonnegative functions, depending on the modeling context. In either case, the collection of all possible such n-tuples form a positive cone P. The usual (partial) ordering in P is to have $(u_1, \ldots, u_n) < (v_1, \ldots, v_n)$ provided $u_i < v_i$ for each i from 1 to n. That is, the densities for each of the species being considered are comparable, and moreover, the order in the comparison is consistent across all the species. Suppose, for instance, that species 1 through n are mutualists, and that $(u_1, \ldots, u_n)$ and $(v_1, \ldots, v_n)$ represent two possible initial configurations of population densities, which are ordered by $(u_1, \ldots, u_n) < (v_1, \ldots, v_n)$. Let $\pi((u_1, \ldots, u_n), t)$ and $\pi((v_1, \ldots, v_n), t)$ represent the corresponding time evolutions (continuous, say) of the densities, as predicted by the model. Consider species 1, for example. Since $v_2 > u_2, v_3 > u_3, \ldots, v_n > u_n$, and the species are mutualists, we expect more benefit to accrue to species 1 from the mutualism in the configuration which is initially $(v_1, \ldots, v_n)$ than in the configuration which is initially $(u_1, \ldots, u_n)$, so that the species 1 component of $\pi(v_1, \ldots, v_n), t)$ should always exceed that in $\pi((u_1, \ldots, u_n), t)$. The same argument holds for any of the other species. Consequently, in a model for mutualism, two configurations of the species that are initially ordered in P retain the ordering under the semi-dynamical system. In that case we say that π is *order-preserving* or *monotone*. Having the semiflow be order-preserving is the monotonicity property which is necessary for compressivity of π to be possible.

Continuing with our system of mutualists, if in addition $\pi((u_1, \ldots, u_n), t)$ is componentwise increasing with time and $\pi((v_1, \ldots, v_n), t)$ is componentwise decreasing and $(u_1, \ldots, u_n) < (w_1, \ldots, w_n) < (v_1, \ldots, v_n)$, then $\pi(w_1, \ldots, w_n), t)$ is trapped between the increasing and decreasing orbits and its asymptotics are forced to lie between $\lim_{t\to\infty} \pi((u_1, \ldots, u_n), t)$ and $\lim_{t\to\infty} \pi((v_1, \ldots, v_n), t)$. π will be *compressive* provided first, it is dissipative with $\pi((w_1, \ldots, w_n), t) < (v_1, \ldots, v_n)$ for t large for any choice of $(w_1, \ldots, w_n)$, and second, $\pi((u_1, \ldots, u_n), t)$ is componentwise increasing in t for all $(u_1, \ldots, u_n)$ with $u_i > 0$ but near enough to 0 with $\lim_{t\to\infty} \pi((u_1, \ldots, u_n), t)$ independent of $(u_1, \ldots, u_n)$. Compressivity is clearly a form of permanence with any $(u_1^*, \ldots, u_n^*) \in P$ which is componentwise less than $\lim_{t\to\infty} \pi((u_1, \ldots, u_n), t)$ serving as an asymptotic "floor" and any $(v_1^*, \ldots, v_n^*) \in P$ componentwise larger than $\lim_{t\to\infty} \pi((v_1, \ldots, v_n), t)$ serving as an asymptotic "ceiling". Moreover, once an orbit enters the *order interval* $[(u_1, \ldots, u_n), (v_1, \ldots v_n)]$ (i.e., the collection of $(w_1, \ldots, w_n)$ with $u_i \leq w_i \leq v_i$ for all i), it is constrained to evolve for future time between orbits that are componentwise increasing and decreasing in time, although the orbit itself need not be componentwise monotonic.

The use of the term compressive to describe a semi-dynamical system first occurred in an important 1991 paper of Hess and Lazer (1991). (See also the subsequent monograph (Hess, 1991).) There they examined an abstract competition model and applied their results to a reaction-diffusion system for two competitors in a bounded habitat under the assumption that the coefficients in the model are time periodic, say with period T. Of course, no oscillatory orbit can be componentwise monotonic in time. However, it is reasonable to look for componentwise monotonicity in species densities in such models if one only compares at times corresponding to the same phase in the oscillation, i.e., $t, t+T, t+2T, t+3T$, etc. For this reason, Hess and Lazer employed discrete dynamical systems for their analysis. We shall explore these ideas further when we treat nonautonomous systems in Section 5. For now, we will consider continuous time autonomous systems. We focus in particular on the two species diffusive Lotka-Volterra competition model (4.58)–(4.59).

Compressivity in the case of continuous time autonomous semi-dynamical systems can be derived directly from the general theory of monotone dynamic systems, as developed systematically in the work of Morris Hirsch (Hirsch, 1982, 1985, 1988a, 1988b, 1989, 1990, 1991). Our treatment uses the formulations in the monograph of Hal Smith, which is an excellent, readily accessible source on the theory of monotone dynamical systems that also contains an extensive list of references on the subject (Smith, 1995). Recall also our discussion in Section 1.6.7.

In two species competition, increasing the density of Species 1 has the effect of increasing competitive pressure on Species 2, and consequently should lead to a decrease in the density of Species 2. But then the decrease in the density of Species 2 should have the effect of decreasing competitive pressure on Species 1, and consequently, should lead to an increase in the density of Species 1, thus continuing the pattern. With 3 or more competitors, the outcome is not so simple. Indeed, the May-Leonard example (4.17)–(4.18) shows that the densities of three competing species can oscillate in a cyclical manner (Section 4.1). Mathematicians formulate this special feature of 2 species competition in terms of an order-preserving semi-flow. However, it is not the usual ordering in P that $(u_1, u_2) < (v_1, v_2)$ if and only if $u_1 < v_1$ and $u_2 < v_2$ that is preserved. Rather, if Species 1 and 2 are competitors with densities u_1 and u_2, respectively, we say that the configurations of densities (u_1, u_2) and (v_1, v_2) are ordered in the skew sense and write

$$(u_1, u_2) <_{Sk} (v_1, v_2) \quad \text{if and only if} \quad u_1 < v_1 \quad \text{and} \quad v_2 < u_2. \tag{5.1}$$

(When it is clear from context that we mean ordering for two competitors in the skew sense, we drop the subscript Sk.) Then if $\pi((u_1, u_2), t) = (\pi_1((u_1, u_2), t), \pi_2((u_1, u_2), t))$ is the evolution under the semiflow corresponding to initial density (u_1, u_2), the preceding discussion suggests that

$$(\pi_1((u_1, u_2), t), \pi_2((u_1, u_2), t)) <_{Sk} (\pi_1((v_1, v_2), t), \pi_2((v_1, v_2), t))$$

for all subsequent time, or that π preserves the order in the skew sense.

Let us now recall the system (4.58)–(4.59). For convenience we reidentify the system as

$$\begin{aligned} \frac{\partial u_1}{\partial t} &= \Delta u_1 + u_1[a_1 - u_1 - b_1 u_2] \\ \frac{\partial u_2}{\partial t} &= \Delta u_2 + u_2[a_2 - b_2 u_1 - u_2] \end{aligned} \tag{5.2}$$

in $\Omega \times (0, \infty)$ with

$$u_1 = 0 = u_2 \tag{5.3}$$

on $\partial\Omega \times (0, \infty)$. Recall from Chapter 4 that Ω is the bounded habitat in question, a_i and b_i are positive constants representing local growth rates and competition coefficients respectively, and that for an initial configuration of population densities (u_1^0, u_2^0) in the positive cone K of $C_0^1(\overline{\Omega}) \times C_0^1(\overline{\Omega})$, $\pi((u_1^0, u_2^0), t) \in K$ denotes the value (i.e. pair of population densities) of the corresponding solution to (5.2)–(5.3). That π preserves the ordering (5.1) follows from Proposition 1.3 of Chapter 8 of Smith (1995) or from the Remarks following Theorem 1.20 and may be stated as follows.

Theorem 5.1. Suppose the initial configurations (u_1^*, u_2^*) and (u_1^{**}, u_2^{**}) in K satisfy the ordering in (5.1); specifically

$$u_1^*(x) \le u_1^{**}(x) \quad \text{and} \quad u_2^*(x) \ge u_2^{**}(x) \tag{5.4}$$

for all $x \in \Omega$. Let $\pi(u_1^*, u_2^*), t) = (u_1^*(x,t), u_2^*(x,t))$ and $\pi((u_1^{**}, u_2^{**}), t) = (u_1^{**}(x,t), u_2^{**}(x,t))$ denote the corresponding solutions to (5.2)–(5.3). Then for all $t > 0$

$$\begin{aligned} &u_1^*(x,t) \le u_1^{**}(x,t) && \text{in } \Omega \\ &0 > \frac{\partial u_1^*}{\partial \vec{n}}(x,t) \ge \frac{\partial u_1^{**}}{\partial \vec{n}}(x,t) && \text{on } \partial\Omega \\ &u_2^*(x,t) \ge u_2^{**}(x,t) && \text{in } \Omega \\ &\frac{\partial u_2^*}{\partial \vec{n}}(x,t) \le \frac{\partial u_2^{**}}{\partial \vec{n}}(x,t) < 0 && \text{on } \partial\Omega. \end{aligned} \tag{5.5}$$

Moreover, provided that one of the inequalities in (5.4) is strict for some $x_0 \in \Omega$, all the inequalities in (5.5) are strict for all $x \in \Omega$ (or $x \in \partial\Omega$).

Establishing compressivity in (5.2)–(5.3) requires that we find solutions to (5.2)–(5.3) which are monotonic in time with respect to the skew ordering (5.1). A solution to (5.2)–(5.3) is monotonically "increasing" in time with respect to (5.1) if it increases with time in the first component and decreases with time in the second component. Likewise, a solution to (5.2)–(5.3) is monotonically "decreasing" in time with respect to (5.1) if it decreases with time in the first component and increases with time in the second component. Again, we turn to the theory of monotone dynamical systems to see where to look for such solutions. The following result can be derived from Corollary 3.6 of Chapter 7 of Smith (1995). (Recall also the Remarks following Theorem 1.22.)

Theorem 5.2. Suppose that $(\hat{u}_1^0, \hat{u}_2^0) \in K$ is such that

$$\begin{aligned} \Delta\hat{u}_1^0 + \hat{u}_1^0[a_1 - \hat{u}_1^0 - b_1\hat{u}_2^0] \ge 0 \\ \Delta\hat{u}_2^0 + \hat{u}_2^0[a_2 - b_2\hat{u}_1^0 - \hat{u}_2^0] \le 0 \end{aligned} \tag{5.6}$$

for $x \in \Omega$. Then $\pi((\hat{u}_1^0, \hat{u}_2^0), t) = (\hat{u}_1(x,t), \hat{u}_2(x,t))$ satisfies

$$\begin{aligned} \hat{u}_1(x, t_1) \le \hat{u}_1(x, t_2) \\ \hat{u}_2(x, t_1) \ge \hat{u}_2(x, t_2) \end{aligned}$$

for $0 \le t_1 < t_2$ and $x \in \Omega$. Similarly, if $(\tilde{u}_1^0, \tilde{u}_2^0) \in K$ is such that

$$\begin{aligned}\Delta\tilde{u}_1^0 + \tilde{u}_1^0[a_1 - \tilde{u}_1^0 - b_1\tilde{u}_2^0] &\le 0\\ \Delta\tilde{u}_2^0 + \tilde{u}_2^0[a_2 - b_2\tilde{u}_1^0 - \tilde{u}_2^0] &\ge 0\end{aligned} \tag{5.7}$$

for $x \in \Omega$, $\pi((\tilde{u}_1^0, \tilde{u}_2^0), t) = (\tilde{u}_1(x,t), \tilde{u}_2(x,t))$ satisfies

$$\begin{aligned}\tilde{u}_1(x,t_1) &\ge \tilde{u}_1(x,t_2)\\ \tilde{u}_2(x,t_1) &\le \tilde{u}_2(x,t_2)\end{aligned}$$

for $0 \le t_1 < t_2$ and $x \in \Omega$.

It follows from Theorem 4.4 that componentwise positive solutions to (5.2)–(5.3) are asymptotically bounded above by equilibrium solutions to the associated diffusive logistic equations. It follows that the limits $\lim_{t\to\infty}(\hat{u}_1(x,t), \hat{u}_2(x,t))$ and $\lim_{t\to\infty}(\tilde{u}_1(x,t), \tilde{u}_2(x,t))$ of the solutions to (5.2)–(5.3) in Theorem 5.2 exist. Corollary 3.6 of Chapter 7 of Smith (1995) guarantees that these limits are in fact equilibrium solutions to (5.2)–(5.3). It follows from Theorem 5.1 that the first equilibrium is the smallest equilibrium "above" $(\hat{u}_1^0, \hat{u}_2^0)$ in the skew ordering (5.1) and the second is the largest equilibrium "below" $(\tilde{u}_1^0, \tilde{u}_2^0)$ in the skew ordering (5.1).

We saw in Section 4 of Chapter 4 that the positivity of σ_1 and σ_2 in (4.64), σ_3 in (4.73) and σ_4 in (4.74) were sharp conditions for the permanence of (5.2)–(5.3). We assert that these same conditions guarantee that (5.2)–(5.3) is compressive. To this end, assume $\sigma_i > 0$ for $i = 1, 2, 3, 4$ and let $\tilde{w}_1(x) > 0$ and $\tilde{w}_2(x) > 0$ be the unique equilibrium solutions to (4.63) whose existences are guaranteed by $\sigma_1 > 0$ and $\sigma_2 > 0$, respectively. Theorem 4.3 guarantees that any componentwise positive solution $(u_1(x,t), u_2(x,t))$ of (5.2)–(5.3) will satisfy $u_1(x,t) < (1+\delta)\tilde{w}_1(x)$ and $u_2(x,t) < (1+\delta)\tilde{w}_2(x)$ for all $t \ge t^*$ with t^* depending only upon δ and $(u_1(x,0), u_2(x,0))$. Consequently, in establishing compressivity we need only consider skew order interval $[(0, (1+\delta)\tilde{w}_2(x)), ((1+\delta)\tilde{w}_1(x), 0)]$.

We now claim that suitable choices for $(\hat{u}_1^0, \hat{u}_2^0)$ and $(\tilde{u}_1^0, \tilde{u}_2^0)$ in Theorem 5.2 are $(\varepsilon w_4, (1+\delta)\tilde{w}_2)$ and $((1+\delta)\tilde{w}_1, \varepsilon w_3)$, respectively, where w_3 is as (4.73) and w_4 is as in (4.74). (w_3 and w_4 are the positive eigenfunctions corresponding to the eigenvalues σ_3 and σ_4 which we presume to be positive). We check only $(\varepsilon w_4, (1+\delta)\tilde{w}_2)$. We have

$$\begin{aligned}&\Delta(\varepsilon w_4) + (\varepsilon w_4)(a_1 - \varepsilon w_4 - b_1((1+\delta)\tilde{w}_2))\\ &\quad = \varepsilon w_4[\sigma_4 - a_1 + b_1\tilde{w}_2 + a_1 - \varepsilon w_4 - b_1(1+\delta)\tilde{w}_2]\\ &\quad = \varepsilon w_4[\sigma_4 - \varepsilon w_4 - \delta b_1\tilde{w}_2]\\ &\quad > 0\end{aligned}$$

and

$$\begin{aligned}&\Delta((1+\delta)\tilde{w}_2) + ((1+\delta)\tilde{w}_2)(a_2 - b_2(\varepsilon w_4) - (1+\delta)\tilde{w}_2)\\ &\quad = (1+\delta)\tilde{w}_2[-a_2 + \tilde{w}_2 + a_2 - \varepsilon b_2 w_4 - (1+\delta)\tilde{w}_2]\\ &\quad = (1+\delta)\tilde{w}_2[-\varepsilon b_2 w_4 - \delta\tilde{w}_2]\\ &\quad < 0\end{aligned}$$

in Ω for all $\varepsilon > 0$ and $\delta > 0$ which are sufficiently small.

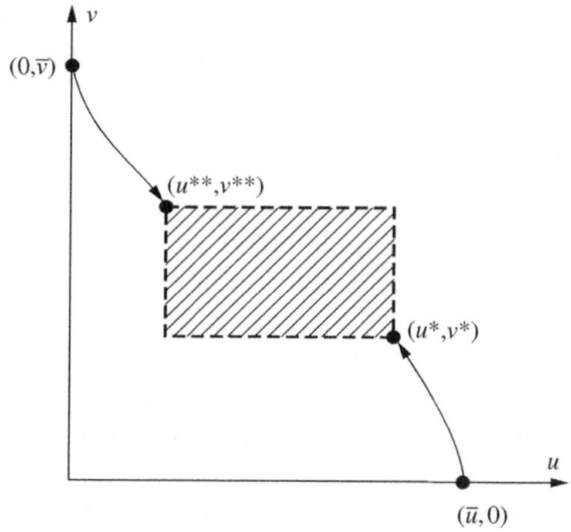

Figure 5.1 This diagram illustrates schematically the notion of compressivity in a competition model. The trajectory starting near the unstable equilibrium ($\bar{u}$,0) will increase in v and decrease in u, and will give a lower bound on v (and an upper bound on u) for trajectories with $u(x,0) < \bar{u}$ and $v(x,0) > 0$. Similarly, the trajectory starting near the unstable equilibrium (0,$\bar{v}$) will increase in u and decrease in v, and will give a lower bound on u (and an upper bound on v) for trajectories with $u(x,0) > 0$ and $v(x,0) < \bar{v}$. As a result, solutions with initial data in the region $0 < u < \bar{u}$, $0 < v < \bar{v}$ are attracted to the order interval where $u^{**} < u < u^{*}$, $v^{*} < v < v^{**}$ as indicated by the shaded region in the diagram. In the typical case where the region $0 < u < \bar{u}$, $0 < v < \bar{v}$, is globally attracting, so is the order interval. Note that in general the equilibria are nonconstant in x, and the trajectories and the shaded order interval are subsets of some Banach space of functions rather than the u, v phase plane.

Certainly we may fix $\delta > 0$ so that $\delta\tilde{w}_2 < \sigma_4$ and $\delta\tilde{w}_1 < \sigma_3$. To conclude that (5.2)–(5.3) is compressive, we need only establish that $[(0, (1+\delta)\tilde{w}_2), ((1+\delta)\tilde{w}_1, 0)]$ contains componentwise positive equilibria $(\underline{u}_1, \underline{u}_2)$ and $(\overline{u}_1, \overline{u}_2)$ to (5.2)–(5.3) so that any componentwise positive equilibrium (u_1, u_2) to (5.2)–(5.3) satisfies

$$(\underline{u}_1, \underline{u}_2) \leq_{Sk} (u_1, u_2) \leq_{Sk} (\overline{u}_1, \overline{u}_2); \tag{5.8}$$

i.e., $\underline{u}_1 \leq u_1 \leq \overline{u}_1$ and $\underline{u}_2 \geq u_2 \geq \overline{u}_2$ in Ω.

It follows from our construction and the maximum principle for elliptic equations (Protter and Weinberger, 1967) that to establish (5.8) we need only eliminate the possibility of having a sequence of equilibria in $[(0, (1+\delta)\tilde{w}_2), ((1+\delta)\tilde{w}_1, 0)]$ which converge in K to $(0, \tilde{w}_2)$ or $(\tilde{w}_1, 0)$. We give a proof of this fact in the Appendix of this chapter.

We have shown that compressivity in (5.2)–(5.3) corresponds exactly to permanence in (5.2)–(5.3). Notice that compressivity clearly implies the existence of a componentwise positive equilibrium to (5.2)–(5.3). Note also that if $(\underline{u}_1, \underline{u}_2) = (\overline{u}_1, \overline{u}_2)$, then (5.2)–(5.3) has $(\underline{u}_1, \underline{u}_2)$ as a globally attracting equilibrium and that uniqueness of a componentwise positive equilibrium to (5.2)–(5.3) necessarily implies that it is globally attracting.

The key to establishing compressivity is the existence of $(\hat{u}_1^0, \hat{u}_2^0)$ and $(\tilde{u}_1^0, \tilde{u}_2^0)$ satisfying (5.6) and (5.7), respectively. In the partial differential equations literature, such pairs of functions are referred to as a subsolution pair and supersolution pair for the corresponding elliptic system (relative to the skew ordering), respectively. The original observation that if the initial data for an orbit in a reaction-diffusion initial-boundary value problem is a subsolution to the corresponding elliptic problem, then the orbit is monotonically increasing in time is due to Aronson and Weinberger (1975), who established the result in the case of a single parabolic equation. Their work strongly influenced subsequent developments.

5.3 Practical Persistence

We saw in Section 4 of Chapter 4, and again in the preceding section, that the positivity of the eigenvalues σ_1 and σ_2 in (4.64), σ_3 in (4.73) and σ_4 in (4.74) were sharp conditions for coexistence via permanence (or compressivity) of two competing species in a bounded isolated habitat patch Ω whose densities are modeled by (5.2)–(5.3). For instance, if σ_1, σ_2 and σ_3 are positive but the value of σ_4 switches from positive to negative, then the prediction of the model (5.2)–(5.3) changes from an assertion that the species coexist to one that has species 1 excluded by species 2, provided that the initial density of species 1 is sufficiently low. The value of σ_4 in (4.74) is determined by three parameters in the original model (5.2)–(5.3): the local intrinsic growth rate a_1 of species 1, the local intrinsic growth rate a_2 of species 2, and the parameter b_1 measuring the competitive impact of species 2 on species 1. We see from Section 4.4.3 that switching from $\sigma_4 > 0$ to $\sigma_4 < 0$ in (4.74) corresponds to crossing the level surface in the three dimensional Euclidean space with coordinates (a_1, a_2, b_1), which is given by

$$\lambda_1^+(a_1 - b_1 \tilde{w}_2(a_2)) = 1 \tag{5.9}$$

and which is equivalent to $\sigma_4 = 0$. (In this instance crossing the level surface results from decreasing a_1, increasing a_2, increasing b_1 or some combination thereof.)

Exploiting such exact knowledge as (5.9) of the boundary in the space of model parameters between regions where the prediction of the model is permanence and regions where the prediction is some form of low density extinction enables us to make meaningful biogeographical analyses. As an example, we note our observation in Section 6.3 that increasing the hostility of the matrix habitat surrounding an isolated refuge can alter the competitive balance between two species sharing the refuge. On the other hand, ecologists frequently consider the notion of a "minimal viable population" for some species (Gilpin and Soulé, 1986; Soulé, 1987), and may require that a model predict that its asymptotic density exceed a suitable *particular* preset threshold in order to regard the model as predicting that the species persists. If such a requirement is placed upon the densities of each species in a model for several interacting species, the region in the Euclidean space of model parameters for which the model predicts coexistence is *necessarily* smaller than the region for which the model predicts permanence. It is simple to understand why such should be the case. Permanence in a model for n interacting biological spaces can be understood as follows. The global attractor for the model (which is basically a collection of n-tuples of species densities) separates into two distinct pieces, one in the "interior" corresponding to having all n species densities be positive and the other in the "boundary" which corresponds to having at least one of the n species extinct, and in addition, all trajectories which originate in the "interior" tend to the part of the global attractor in the "interior". As one moves in the space of model parameters toward the boundary of the region for which the prediction of the model is permanence (e.g., by decreasing a_1 in (5.9)), one expects the portion of the global attractor for the model in the "interior" to move toward the "boundary". Otherwise, permanence should remain the prediction of the model as the parametric boundary threshold is crossed. Consequently, the asymptotic floor (recall (4.26) and (4.27) in the context of continuous time models) for at least one species must tend toward zero, and hence go below the preset density level for parameter values close enough to the parametric boundary threshold.

The notion of *practical persistence* is motivated by considerations such as the preceding. By practical persistence in a model for several interacting biological species, we shall mean *ecological permanence with concrete asymptotic floors*, where in the continuous time case an asymptotic floor is said to be "concrete" if m in (4.26) or (4.27) is given with some specificity, usually as a function of the parameters of the model. The preceding discussion shows that such a function must approach 0 as the model parameters approach the boundary of a region in parameter space corresponding to a model prediction of permanence, as practical persistence implies permanence. However, the process of establishing a "concrete" asymptotic floor may place additional constraints on the parameters of a model, so that the region in parameter space in which one may obtain a prediction of practical persistence for the model may well be smaller than the region in parameter space in which abstract permanence may be detected.

The term "practical persistence" was introduced in Cao and Gard (1993) and Cao et al. (1996) in the context of ordinary delay differential models for several interacting biological species. In ODE models for the n species configuration $u = (u_1, \ldots, u_n)$, the product expression

$$V(u) = \prod_{i=1}^{n} u_i^{\beta_i}, \tag{5.10}$$

where $\beta_1, \ldots, \beta_n$ are constants, has been employed with some success as an average Lyapunov function to establish permanence (Hutson, 1984). In the more mathematically challenging context of delay differential equation models, Cao, Gard and Fan (Cao and Gard, 1993; Cao et al., 1996) modify the average Lyapunov function approach. Following earlier work by Wendi and Ma Zhien (1991), they make use of n different expressions of the form (5.10) (which exhibit suitable interactions among their exponents) to assert permanence in the systems they consider. Moreover, they observe that their derivation produces "concrete" asymptotic floors on the densities of the species in question in terms of the original model parameters; i.e. practical persistence.

In Cantrell and Cosner (1996), we developed an approach for obtaining practical persistence results suitable for treating a rather broad class of reaction-diffusion models for n interacting species in an isolated habitat patch $\Omega \subseteq \boldsymbol{R}^m$, for $m = 1, 2, 3$. The models we considered in Cantrell and Cosner (1996) have the form

$$\begin{aligned} (u_i)_t &= L_i u_i + f_i(x, t, u_1, \ldots, u_n) u_i \quad \text{in} \quad \Omega \times (0, \infty) \\ \rho_i(x) \frac{\partial u_i}{\partial \vec{n}} &+ (1 - \rho_i(x)) u_i = 0 \qquad \text{on} \quad \partial\Omega \times (0, \infty), \end{aligned} \tag{5.11}$$

for $i = 1, 2, \ldots, n$, where $(u_1(x, t), \ldots, u_n(x, t))$ gives the densities of the n species in question at spatial location $x \in \overline{\Omega}$ and time t. The weight $\rho_i(x)$ in the boundary condition for the ith species density may exhibit spatial but not temporal dependence and lies in the interval $[0, 1]$ for all $x \in \partial\Omega$, allowing for a range of standard homogeneous boundary conditions including absorbing ($\rho_i(x) \equiv 0$) and reflecting ($\rho_i(x) \equiv 1$). The elliptic dispersal term L_i allows for a range of isotropic or anisotropic diffusive and advective components to the motion of species i in Ω. The coefficients in L_i may depend on spatial location and/or time, so long as the dependence is appropriately continuous and any temporal dependence is T-periodic for some fixed $T > 0$. The temporal dependence in the local growth rate $f_i(x, t, u_1, \ldots, u_n)$ for species i may be rather more general so long as it remains bounded between appropriate local growth rates which are T-periodic in time. In this way, the local growth rates may reflect the impact of important environmental factors such as temperature or precipitation which fluctuate somewhat randomly from day to day within a known range that varies seasonally. It would be desirable to be able to relax the T-periodicity requirement on the temporal dependence of the coefficients of L_i as well. In all likelihood, to do so is a mathematical technicality, but it is one that we do not know how to remove at present. We are reminded that no matter how much mathematical knowledge advances, Skellam's remark in his landmark 1951 paper (Skellam, 1951) regarding the inadequacy of "orthodox analytical methods" will always be able to find a context in which it is pertinent.

The results of the preceding section indicate that practical persistence results may be obtained when a model for interacting biological species is compressive. However, the results there also show that compressivity is only possible in the context of order preserving systems, which essentially limits one to single-species models, 2-species competition or to mutualism. The results in Cantrell and Cosner (1996) can be applied to these type of ecological interactions. However, they also may be applied in the context of many other forms of ecological interactions, such as predator-prey interactions, multi-species competition or food pyramids. The heart of the utility of the approach to practical persistence in Cantrell and Cosner (1996) is that upper and lower estimates on species densities are made on an equation by equation (or species by species) basis, permitting one to exploit the order-preserving properties of solutions to models for single equations even when there is no order-preserving property attached to the overall interaction. Such an approach to practical persistence is akin to the multi-average Lyapunov function approach of Cao, Fan and Gard (Cao and Gard, 1993; Cao et al., 1996), wherein a collection of average Lyapunov functions equal in number to the number of interacting species in question is exploited. However, in Cantrell and Cosner (1996), each average Lyapunov function would depend on the density of only one species and, as such, would be somewhat special in form. Implementing either approach to practical persistence requires having *asymptotic ceilings* that are sufficiently sharp in a quantitative sense. In practice, obtaining such asymptotic upper bounds appears to require a strong enough self-regulation mechanism. We return to this point later in this section, and also in Section 4, when we discuss how to extend the practical persistence approach of Cantrell and Cosner (1996) to bounding trajectories in reaction-diffusion models for predator-prey interactions.

Our principal aim for the remainder of this section is to execute the practical persistence regime in Cantrell and Cosner (1996) in the context of multi-species diffusive Lotka-Volterra competition models, in order to parallel and inform the discussion of 2-species diffusive Lotka-Volterra competition in the preceding section. However, before proceeding with this example, we give a brief heuristic overview of the regime in the context of a food pyramid. Our reasons are two-fold. First, the results in Cantrell and Cosner (1996) apply to a broad range of types of ecological interactions, and we believe it is beneficial to reinforce this point via a second significant type of interaction without getting overly detailed. Secondly, and perhaps more important, is the fact that the process of obtaining asymptotic ceilings and floors in the multi-species competition context does not require and hence does not illustrate the full power of the entire method so fully and so well as does a food pyramid. Note that, for (5.11) to model a food pyramid, we shall require that species 1 through n are ordered so that species j may be a resource for species i only if $j > i$. (This ordering puts species 1 atop the food pyramid; we could just as easily have arranged the species so that species n is atop the food pyramid. For more on food pyramids in the context of reaction-diffusion systems, see Williams and Chow (1978) and Alikakos (1979).) Assuming that the local intrinsic growth rates $f_i(x, t, u_1, \ldots, u_n)$ in (5.11) are smooth functions of the species densities, the food pyramid condition can be expressed as

$$\frac{\partial f_i}{\partial u_j}(x, t, u_1, \ldots, u_n) \begin{cases} \leq 0 & \text{if} \quad j < i \\ \geq 0 & \text{if} \quad j > i. \end{cases} \tag{5.12}$$

For the sake of specificity, we now take the number of species $n = 4$ and assume that all the inequalities in (5.12) are strict. (We also assume that the L_i and f_i in (5.11) are autonomous, reserving consideration of nonautonomous models to Section 5.) Under these assumptions, species 4 is the base resource in the food pyramid, and as such must have resources not included in the system. Species 3 then utilizes species 4 as a resource, species 2 utilizes species 3 and species 4, and finally, species 1 utilizes all the other species.

The practical persistence regime proceeds as follows. Observe first that the most advantageous situation for species 4 in the context of the food pyramid is the absence of the remaining species. So set the densities of species 1 through 3 equal to 0 in the intrinsic growth rate f_4 in (5.11) and use a self-regulation mechanism on species 4 to obtain an upper bound $\overline{\theta}_4$ on u_4, so that we have

$$u_4 \leq \overline{\theta}_4 \tag{5.13}$$

as an asymptotic ceiling on species 4. In (5.13), in general, $\overline{\theta}_4$ is a function of location x in the isolated habitat Ω. However, we sometimes take $\overline{\theta}_4$ to be a constant. Asymptotically, the remaining species can not expect the density of species 4 to exceed $\overline{\theta}_4$. So for species 3, the most advantageous situation plausible asymptotically is to have species 1 and 2 absent and $u_4 = \overline{\theta}_4$ in f_3 in (5.11). Then use a self-regulation mechanism on species 3 to obtain

$$u_3 \leq \overline{\theta}_3 (= \overline{\theta}_3(\overline{\theta}_4)) \tag{5.14}$$

as an asymptotic ceiling on species 3. Next, set $u_1 = 0$, $u_3 = \overline{\theta}_3$ and $u_4 = \overline{\theta}_4$ in f_2 of (5.11) and use a self-regulation mechanism on species 2 to obtain

$$u_2 \leq \overline{\theta}_2 (= \overline{\theta}_2(\overline{\theta}_3(\overline{\theta}_4), \overline{\theta}_4)) \tag{5.15}$$

as an asymptotic ceiling on species 2. We get the last of our asymptotic ceilings by setting $u_4 = \overline{\theta}_4$, $u_3 = \overline{\theta}_3$ and $u_2 = \overline{\theta}_2$ in f_1 in (5.11) and using a self regulation mechanism on species 1 to obtain

$$u_1 \leq \overline{\theta}_1 (= \overline{\theta}_1(\overline{\theta}_2(\overline{\theta}_3(\overline{\theta}_4)), \overline{\theta}_3(\overline{\theta}_4), \overline{\theta}_4)). \tag{5.16}$$

Inequalities (5.13), (5.14), (5.15) and (5.16) now give asymptotic ceilings on all the species in the food pyramid. Of course, practical persistence requires concrete asymptotic floors. We now return to species 4, the base of the food pyramid. Having the densities of species 1,2 and 3 fixed equal to $\overline{\theta}_1$, $\overline{\theta}_2$ and $\overline{\theta}_3$ respectively, represents an asymptotic upper bound on the predatory or consumer pressure that the food pyramid can place upon species 4; i.e., asymptotically, a worst case scenario. So we set $u_1 = \overline{\theta}_1$, $u_2 = \overline{\theta}_2$ and $u_3 = \overline{\theta}_3$ in f_4 in (5.11) and analyze the equation for the density u_4 of species 4 that results. If possible, we make a comparison that allows us to conclude that if u_1, u_2 and u_3 are fixed at $\overline{\theta}_1$, $\overline{\theta}_2$ and $\overline{\theta}_3$, then

$$u_4 \geq \underline{\theta}_4 \tag{5.17}$$

asymptotically, where $\underline{\theta}_4$ is a "concrete" function which is positive throughout the isolated habitat Ω. Inequality (5.17) indicates that asymptotically the base resource of the food pyramid should be available to the consumer species at a level at least equal to $\underline{\theta}_4$. Next consider species 3. Having the densities of species 1 and 2 fixed equal to $\overline{\theta}_1$ and $\overline{\theta}_2$ represents an asymptotic upper bound on the predatory or consumer pressure that the food pyramid can place on species 3, while having the density of species 4 fixed equal to $\underline{\theta}_4$ represents an asymptotic lower bound on the resources available to species 3. The combination is its asymptotic worst case scenario. So we set $u_1 = \overline{\theta}_1,\ u_2 = \overline{\theta}_2$ and $u_4 = \underline{\theta}_4$ in f_3 in (5.11) and analyze the equation for the density u_3 of species 3 that results. Again, if possible, we make a comparison that allows us to conclude that if $u_1,\ u_2$ and u_4 are fixed at $\overline{\theta}_1,\ \overline{\theta}_2$ and $\underline{\theta}_4$, then asymptotically

$$u_3 \geq \underline{\theta}_3 \tag{5.18}$$

where again $\underline{\theta}_3$ is a "concrete" function which is positive throughout Ω. Turning to species 2, its asymptotic worst case combination of upper estimate on predatory pressure and lower estimate on resource availability comes from fixing $u_1 = \overline{\theta}_1,\ u_2 = \underline{\theta}_3$ and $u_4 = \underline{\theta}_4$ in f_2 of (5.11). As before, we analyze the resulting equation. This time we hope for a comparison that allows us to conclude that the density u_2 of species 2 satisfies

$$u_2 \geq \underline{\theta}_2 \tag{5.19}$$

asymptotically, where $\underline{\theta}_2$ is a "concrete" function that is positive throughout Ω. Finally, we come to the top of the food pyramid. We achieve practical persistence if we can now set $u_2 = \underline{\theta}_2,\ u_3 = \underline{\theta}_3$ and $u_4 = \underline{\theta}_4$ in f_1 of (5.11), analyze the resulting equation for the density u_1 of species 1, and make a comparison that allows us to conclude

$$u_1 \geq \underline{\theta}_1 \tag{5.20}$$

asymptotically, where $\underline{\theta}_1$ is a "concrete" function which is positive throughout Ω.

Let us now consider a diffusive Lotka-Volterra model for n competing species in an isolated bounded habitat patch Ω, where we assume that the boundary $\partial\Omega$ of the patch is lethal to all species under consideration. The model is a direct analogue of (5.2)–(5.3). Specifically, we consider the system

$$u_{it} = \Delta u_i + u_i\left[a_i - \sum_{j=1}^{n} b_{ij}u_i\right] \quad \text{in} \quad \Omega \times (0, \infty) \tag{5.21}$$

$$u_i = 0 \quad \text{on} \quad \partial\Omega \times (0, \infty), \tag{5.22}$$

$i = 1, \ldots, n$. As in (5.2)–(5.3), Ω is a bounded smooth domain in $I\!R^m$, $m = 1, 2$, or 3, and the coefficients represent local intrinsic growth rates (a_i), competition coefficients (b_{ij}, $j \neq i$) and self-regulation coefficients (b_{ii}). We assume throughout this discussion that a_i and b_{ij}, $j \neq i$, $i, j = 1, \ldots, n$, are smooth positive functions of x in $\overline{\Omega}$ and that b_{ii} is normalized by the condition

$$b_{ii} \equiv 1 \tag{5.23}$$

for $i = 1, \ldots, n$.

We initially restrict the local intrinsic growth rates a_i and competition coefficients b_{ij}, $j \neq i$, to be positive constants, in which case (5.21)–(5.22) plus (5.23) reduces to (5.2)–(5.3) in case $n = 2$. Following our description of practical persistence results for the constant coefficient case, we will return to the case of spatially heterogeneous local intrinsic growth rates and competition coefficients. Recall that for the single-species model

$$\begin{aligned} \frac{\partial u}{\partial t} &= \Delta u + u(a-u) \text{ in } && \Omega \times (0,\infty) \\ u &= 0 && \text{on } \partial\Omega \times (0,\infty), \end{aligned} \tag{5.24}$$

where a is a positive constant, Theorem 4.4 asserts the existence of a globally attracting positive equilibrium precisely when $a > \lambda_0^1(\Omega)$, the principal eigenvalue for $-\Delta$ on Ω subject to zero Dirichlet boundary conditions. Let us denote this equilibrium by $\theta(a) = \theta(a)(x)$. Theorem 4.4 further asserts that

$$\theta(a)(x) < a \tag{5.25}$$

for all $x \in \overline{\Omega}$. (More generally, Propositions 3.2 and 3.3 assert that (5.24) admits a unique globally attracting positive equilibrium (again denoted $\theta(a)$) so long as $a = a(x)$ is a continuous function such that

$$a(x_0) > 0 \tag{5.26}$$

for some $x_0 \in \Omega$ and the eigenvalue σ for which

$$\begin{aligned} \Delta\phi + a(x)\phi &= \sigma\phi \text{ in } && \Omega \\ \phi &= 0 && \text{on } \partial\Omega \end{aligned} \tag{5.27}$$

admits a positive eigenfunction is itself positive. Further, if (5.26) holds and σ is positive in (5.27),

$$\theta(a)(x) < \max_{x\in\overline{\Omega}} a(x) \tag{5.28}$$

for all $x \in \overline{\Omega}$).

It is immediate from (5.21) and the positivity of b_{ij} that if $(u_1(x,t), u_2(x,t), \ldots, u_n(x,t))$ is a solution to (5.21)–(5.22) corresponding to componentwise nonnegative initial densities $(u_1^0(x), \ldots, u_n^0(x)) = (u_1(x,0), \ldots, u_n(x,0))$, then $u_i(x,t)$ is a lower solution to (5.24) with $a = a_i$. If we now let $\overline{u}_i(x,t)$ denote the solution to (5.24) with initial density $\overline{u}_i(x,0) = u_i^0(x)$, then $u_i(x,t) \leq \overline{u}_i(x,t)$ for all $t \geq 0$ and $\overline{u}_i(x,t)$ converges in the norm of $C_0^1(\overline{\Omega})$ to $\theta(a_i)(x)$ as $t \to \infty$. Consequently, given any positive number ε, there corresponds a time value $t_0 = t_0(\varepsilon, u_1^0, \ldots, u_n^0)$ so that

$$u_i(x,t) \leq (1+\varepsilon)\theta(a_i)(x) \tag{5.29}$$

for all $x \in \overline{\Omega}$, $t \geq t_0$ and $i = 1, \ldots, n$. If, in addition, ε is small enough so that $(1+\varepsilon)\theta(a_i)(x) \leq a_i$ for all $x \in \overline{\Omega}$, we have

$$u_i(x,t) \leq a_i \tag{5.30}$$

for all $x \in \overline{\Omega}$, $t \geq t_0$ and $i = 1, \ldots, n$. Inequalities (5.29) and (5.30) provide asymptotic ceilings on the density of each of the species in question. Unlike the case of a food pyramid, the asymptotic ceiling on the density of species i given in (5.29) or (5.30) is independent of the density of any of its competitors. Indeed, we may make the same basic estimate for each competitor all at once. For small ε, (5.29) is a sharper asymptotic estimate than is (5.30). In other words, (5.29) provides "lower" asymptotic "ceilings" than does (5.30). However, (5.30) is simpler than (5.29) and will lead to a simpler practical persistence result, albeit one that holds for a more restrictive collection of system parameters, as we will now see.

Let us first employ (5.30) for the asymptotic ceilings on the densities of the species in question. Then we have

$$u_{it} \geq \Delta u_i + u_i \left[a_i - \sum_{\substack{j=1 \\ j \neq i}}^{n} a_j b_{ij} - u_i \right] \quad \text{in} \quad \Omega \times (t_0, \infty) \tag{5.31}$$

for $i = 1, 2, \ldots, n$, where t_0 depends on $(u_1^0, \ldots, u_n^0)$. If w_i denotes the solution of

$$\begin{aligned} w_{it} &= \Delta w_i + w_i \left[a_i - \sum_{\substack{j=1 \\ j \neq i}}^{n} a_j b_{ij} - w_i \right] && \text{in} \quad \Omega \times (t_0, \infty) \\ w_i &= 0 && \text{on} \quad \partial\Omega \times (t_0, \infty) \\ w_i(x, t_0) &= u_i(x, t_0), \end{aligned} \tag{5.32}$$

then it follows from the method of upper and lower solutions that

$$u_i(x, t) \geq w_i(x, t) \tag{5.33}$$

for $t \geq t_0$, $x \in \overline{\Omega}$ and $i = 1, \ldots, n$. So long as

$$a_i - \sum_{\substack{j=1 \\ j \neq i}}^{n} a_j b_{ij} > \lambda_0^1(\Omega), \tag{5.34}$$

it follows from (5.32), (5.33) and Theorem 4.4 that

$$u_i(x, t) \geq (1 - \varepsilon)\theta(a_i - \sum_{\substack{j=1 \\ j \neq i}}^{n} a_j b_{ij})(x) \tag{5.35}$$

for $t \geq t_1$, $x \in \overline{\Omega}$ and $i = 1, \ldots, n$, where t_1 depends on $(u_1^0, \ldots, u_n^0)$. Inequalities (5.35) give "concrete" asymptotic "floors" on the densities of the species in question and hence practical persistence. However, the "floors" involve the solution to a nonlinear partial differential equation, namely (5.24), and one may well ask if the "floors" might be made more "concrete". The answer is yes and is so without any restrictions on the system parameters a_i and b_{ij} beyond those already imposed in (5.34). Indeed, if (5.34) holds and ϕ satisfies

$$\begin{aligned} -\Delta\phi &= \lambda_0^1(\Omega)\phi \text{ in } \Omega \\ \phi &= 0 \quad \text{on} \quad \partial\Omega \\ \phi &> 0 \quad \text{in} \quad \Omega \\ \max_{\overline{\Omega}}\phi(x) &= 1, \end{aligned} \tag{5.36}$$

it is easy to show (via the method of upper and lower solutions) that $\theta\left(a_i - \sum_{\substack{j=1\\j\neq i}}^{n} a_j b_{ij}\right)(x) \geq \left(a_i - \sum_{\substack{j=1\\j\neq i}}^{n} a_j b_{ij} - \lambda_0^1(\Omega)\right)\phi(x)$ so that

$$u_i(x,t) \geq (1-\varepsilon)\left(a_i - \lambda_0^1(\Omega) - \sum_{\substack{j=1\\j\neq i}}^{n} a_j b_{ij}\right)\phi(x) \tag{5.37}$$

for $t \geq t_1$, $x \in \overline{\Omega}$ and $i = 1, 2, \ldots, n$. The eigenfunction ϕ in (5.36) depends only on the geometry of the isolated habitat Ω, and in particular, is independent of the system parameters a_i and b_{ij}. By virtue of these facts alone, the inequalities in (5.37) are more "concrete" than those in (5.35). Moreover, $\phi(x)$ is readily computable numerically and is in fact known analytically in a number of cases. For instance, if Ω is the square $(0,\pi)\times(0,\pi)$, $\phi = \sin x \sin y$.

On the other hand, if we use inequalities (5.29) as our asymptotic ceilings, in place of (5.30), (5.31) becomes

$$u_{it} \geq \Delta u_i + u_i\left[a_i - (1+\varepsilon)\sum_{\substack{j=1\\j\neq i}}^{n} b_{ij}\theta(a_j) - u_i\right] \quad \text{in} \quad \Omega \times (t_0, \infty) \tag{5.38}$$

for $i = 1, 2, \ldots, n$, where again t_0 depends on $(u_1^0, \ldots, u_n^0)$. So long as the principal eigenvalue $\tilde{\sigma}$ of

$$\begin{aligned} \Delta\psi + \left[a_i - (1+\varepsilon)\sum_{\substack{j=1\\j\neq i}}^{n} b_{ij}\theta(a_j)\right]\psi &= \tilde{\sigma}\psi \text{ in } \Omega \\ \psi &= 0 \text{ on } \partial\Omega \end{aligned} \tag{5.39}$$

is positive, we may proceed as before to obtain that for any $\delta \in (0, 1)$,

$$u_i(x,t) \geq (1-\delta)\theta\left(a_i - (1+\varepsilon)\sum_{\substack{j=1\\j\neq i}}^{n} b_{ij}\theta(a_j)\right)(x) \tag{5.40}$$

for $t \geq t_1$, $x \in \overline{\Omega}$ and $i = 1, \ldots, n$, where t_1 depends on $(u_1^0, \ldots, u_n^0)$. The practical persistence estimates in (5.40) are admittedly more complicated in appearance than those in (5.35) (not even to mention (5.37)). However, they also represent higher asymptotic "floors" and hold for a larger range of the system parameters a_i, and b_{ij} as well. Why? To start with, we know from Proposition 3.3 that if for two functions $a(x)$ and $\tilde{a}(x)$ on $\overline{\Omega}$, $\theta(a)$ and $\theta(\tilde{a})$ are defined (i.e. the principal eigenvalues σ and $\tilde{\sigma}$ in (5.27) for a and $\tilde{a}$ are positive, respectively) and $a(x) \overset{<}{\not\equiv} \tilde{a}(x)$, then $\theta(a)(x) < \theta(\tilde{a})(x)$ for all $x \in \Omega$. Since $\theta(a_j) < a_j$ for $j = 1, \ldots, n$, the right-hand side of (5.40) therefore is higher than that of (5.35) for all small enough ε and δ, and hence the asymptotic "floors" given in (5.40) are higher than those given in (5.35). In addition, since (5.34) is equivalent to having the principal eigenvalue σ^* positive in

$$\begin{aligned} \Delta\rho + \left[a_i - \sum_{\substack{j=1 \\ j \neq i}}^{n} a_j b_{ij} \right] \rho &= \sigma^* \rho \text{ in } \Omega \\ \rho &= 0 \text{ on } \partial\Omega \end{aligned} \tag{5.41}$$

and since σ in (5.27) is monotonically increasing with respect to $a(x)$, we have that $\tilde{\sigma}$ in (5.39) is positive for any choice of a_i and b_{ij} for which (5.34) holds (or equivalently, for which σ^* in (5.41) is positive). In other words, for any a_i and b_{ij} for which the practical persistence estimates in (5.35) hold, the practical persistence estimates in (5.40) hold. So the estimates in (5.40) are sharper and hold for a larger collection of system parameters a_i and b_{ij}.

As we noted earlier, the parameters a_i and b_{ij} in (5.21)–(5.22) are in general not merely constants but functions which vary with x in $\overline{\Omega}$. In this case we can still sometimes obtain practical persistence results for (5.21)–(5.22) along the lines of those just obtained. We will require bounds on a_i and b_{ij} as follows. Let $0 < \underline{a}_i \leq a_i(x) \leq \overline{a}_i$ on $\overline{\Omega}$ and $0 < \underline{b}_{ij} \leq b_{ij}(x) \leq \overline{b}_{ij}$ on $\overline{\Omega}$ where $\underline{a}_i, \overline{a}_i, \underline{b}_{ij}$ and $\overline{b}_{ij}$ are constants and the inequalities hold for all $i, j \in \{1, 2, \ldots, n\}$. (By (5.23), we take $\underline{b}_{ii} = \overline{b}_{ii} = 1$.) Then analogues to (5.35), (5.37) and (5.40) hold. As an illustration, if

$$\underline{a}_i - \sum_{\substack{j=1 \\ j \neq i}}^{n} \overline{a}_j \overline{b}_{ij} > \lambda_0^1(\Omega) \tag{5.42}$$

we may obtain for $\epsilon \in (0, 1)$ that

$$u_i(x, t) \geq (1 - \varepsilon) \left(\underline{a}_i - \sum_{\substack{j=1 \\ j \neq i}}^{n} \overline{b}_{ij} \overline{a}_j - \lambda_0^1(\Omega) \right) \phi(x) \tag{5.43}$$

for $t \geq t_1$, $x \in \overline{\Omega}$ and $i = 1, 2, \ldots, n$, where t_1 depends on $(u_1^0, \ldots, u_n^0)$ as well as the bounds on the system parameters. It should be evident that (5.42) is an analogue to (5.34) and that (5.43) is an analogue to (5.37). As in the preceding examples, we again proceed by identifying the asymptotic "worst case scenario"of the system for each of its component species. Note that in the case of (5.21)–(5.22) with spatially varying coefficients, to do so requires lower estimates on intrinsic rates of growth (i.e., $\underline{a}_i$) and upper estimates on competitive impact (i.e., $\overline{a}_j\overline{b}_{ij}$).

The approach to practical persistence we have described for a system of the form (5.11) consists of a succession of best case and worst case comparisons that when successful leads to "concrete" asymptotic "ceilings" and "floors" on the densities being modeled. The comparisons are made via the method of upper and lower solutions on an equation by equation basis against suitable single population models which represent best or worst case scenarios and whose asymptotic dynamical behavior is well-understood. It is reasonable to suspect that the approach should carry over to other classes of models, and indeed such is the case. We direct the interested reader to Cosner (1996), in which a wide array of deterministic models (continuous *versus* discrete time, continuous *versus* discrete space, structured *versus* unstructured) are shown to be amenable to a suitably analogous practical persistence analysis. (See also Cosner (1994).) A rather different but nevertheless fruitful approach to practical persistence is given in Hutson and Mischaikow (1998). This approach applies when one or more of the species being modeled exhibits interactions that are slow when compared to those of the remaining species in the model. Again the approach applies to a wide variety of deterministic models. In the realm of stochastic models, practical persistence results may be found in Gard (1997).

5.4 Bounding Transient Orbits

The approach to practical persistence that we described in the previous section has one additional advantage. Namely, it may sometimes be extended to obtain bounds on densities of interacting biological species which hold for *transient* as well as asymptotic time. We illustrate with a diffusive predator-prey model on an isolated habitat patch. In the situation we envision, the prey species is "well-established" on the patch when the predator species is introduced there at low densities. The model we consider is an extension of the Lotka-Volterra model (4.84), specifically,

$$\begin{aligned} u_{1t} &= \Delta u_1 + [a_1(x,t,u_1,u_2) - b_{11}(x,t,u_1,u_2)u_1 - b_{12}(x,t,u_1,u_2)u_2]u_1 \\ u_{2t} &= \Delta u_2 + [a_2(x,t,u_1,u_2) + b_{21}(x,t,u_1,u_2)u_1 - b_{22}(x,t,u_1,u_2)u_2]u_2 \end{aligned} \tag{5.44}$$

in $\Omega \times (0,\infty)$, subject to

$$\rho_i(x)u_i + (1-\rho_i(x))\nabla u_i \cdot \vec{n} = 0 \tag{5.45}$$

on $\partial\Omega \times (0,\infty)$. As in (5.11), $\rho_i(x)$ in (5.45) is required to lie in the interval $[0,1]$ for all $x \in \partial\Omega$. We allow the system parameters in (5.44) (i.e., growth rates a_i, predation/update rates $b_{ij}, i \neq j$, and self-regulation rates b_{ii}), however, to depend on spatial location x, time t, prey density u_1 and predator density u_2 in a rather general way. The coefficients are subject only to reasonable smoothness requirements and the bounds

$$\underline{a}_i(x) \le a_i(x,t,u_1,u_2) \le \overline{a}_i(x) \tag{5.46}$$

$$0 < b_{ii}^0 < \underline{b}_{ii}(x) \le b_{ii}(x,t,u_1,u_2) \le \overline{b}_{ii}(x) \tag{5.47}$$

$$0 \le \underline{b}_{ij}(x) \le b_{ij}(x,t,u_1,u_2) \le \overline{b}_{ij}(x) \tag{5.48}$$

for $x \in \overline{\Omega}$, $t \geq 0$, $u_1 \geq 0$, $u_2 \geq 0$ and $i, j = 1, 2$, $j \neq i$. In (5.47), b_{11}^0 and b_{22}^0 are positive constants. The upper and lower bounds in (5.46)–(5.48) are smooth functions of x in the isolated habitat $\overline{\Omega}$.

Our first task is to postulate what we mean when we assert that the prey species is "well-established" at the point in time when the predator species is introduced into the isolated habitat patch at a low density, which for the sake of convenience we take to be $t = 0$. To do so we examine (5.44) under the assumption that the density u_2 of the predator species is 0. By (5.46) and (5.47), the density u_1 of the prey species is a subsolution of

$$\begin{aligned} u_t &= \Delta u + (\overline{a}_1(x) - \underline{b}_{11}(x)u)u) \text{ in } \quad \Omega \times (0, \infty) \\ \rho_i(x)u &+ (1 - \rho_1(x))\nabla u \cdot \vec{n} = 0 \text{ on } \quad \partial\Omega \times (0, \infty). \end{aligned} \tag{5.49}$$

If now, the principal eigenvalue $\tilde{\sigma}_1$ in

$$\begin{aligned} \Delta\phi + \overline{a}_1(x)\phi &= \tilde{\sigma}_1\phi \qquad \text{in} \quad \Omega \\ \rho_1(x)\phi + (1 - \rho_1(x))\nabla\phi \cdot \vec{n} &= 0 \text{ on } \quad \partial\Omega \end{aligned} \tag{5.50}$$

is positive, Propositions 3.2 and 3.3 imply there is a unique positive equilibrium u_1^* of (5.49) so that all nonnegative, nontrivial solutions of (5.49) converge in the topology of $C^1(\overline{\Omega})$ to u_1^*. Since u_1 is a subsolution of (5.49), it lies below any solution to (5.49) with $u(x, 0) \geq u_1(x, 0)$. In particular, if $u_1(x, 0) \leq u_1^*(x)$, $u_1(x, t) \leq u_1^*(x)$ for all $t \geq 0$ and $x \in \Omega$. Employing (5.46) and (5.47) again, we see that the prey density u_1 (*in the absence of the predator*) is a supersolution of

$$\begin{aligned} u_t &= \Delta u + (\underline{a}_1(x) - \overline{b}_{11}(x)u)u \quad \text{in} \quad \Omega \times (0, \infty) \\ \rho_1(x)u &+ (1 - \rho_1(x)) \cdot \nabla u \cdot \vec{n} = 0 \text{ on } \quad \partial\Omega \times (0, \infty). \end{aligned} \tag{5.51}$$

If now, the principal eigenvalue $\tilde{\sigma}_1$ in

$$\begin{aligned} \Delta\phi + \underline{a}_1(x)\phi &= \tilde{\sigma}_1\phi \qquad \text{in} \quad \Omega \\ \rho_1(x)\phi + (1 - \rho_1(x))\nabla\phi \cdot \vec{n} &= 0 \text{ on } \quad \partial\Omega \end{aligned} \tag{5.52}$$

is positive, Propositions 3.2 and 3.3 imply there is a unique positive equilibrium $\tilde{u}_1$ of (5.51) so that all nonnegative, nontrivial solutions of (5.51) converge in the topology of $C^1(\overline{\Omega})$ to $\tilde{u}_1$. (Note that if $\tilde{\sigma}_1$ is positive in (5.52), then $\overline{\sigma}_1$ is positive in (5.50) since $\underline{a}_1 \leq \overline{a}_1$.) Since u_1 is a supersolution to (5.51), it lies above any solution to (5.51) with $u(x, 0) \leq u_1(x, 0)$. In particular, if $u_1(x, 0) \geq \tilde{u}_1(x)$, then $u_1(x, t) \geq \tilde{u}_1(x)$ for all $t \geq 0$ and $x \in \overline{\Omega}$. It follows that $u_1^*(x) \geq \tilde{u}_1(x)$ on Ω. So to view the prey species as "well-established" in the absence of the predator from the isolated habitat patch, we assume $\tilde{\sigma}_1$ is positive in (5.52) and postulate

$$\tilde{u}_1(x) \leq u_1(x, 0) \leq u_1^*(x) \quad \text{in} \quad \Omega. \tag{5.53}$$

Then

$$u_1(x, t) \leq u_1^*(x) \quad \text{in} \quad \Omega \times (0, \infty) \tag{5.54}$$

and *in the absence of the predator*

$$\tilde{u}_1(x) \le u_1(x,t) \quad \text{in} \quad \Omega \times (0,\infty). \tag{5.55}$$

We now think of introducing the predator species into the isolated habitat patch Ω at a low density. By (5.46)–(5.48) and (5.53)–(5.54) the predator density u_2 is a subsolution of

$$\begin{aligned} u_t &= \Delta u + [\overline{a}_2(x) + \overline{b}_{21}(x)u_1^*(x) - \underline{b}_{22}(x)u]u \text{ in } && \Omega \times (0,\infty) \\ \rho_2(x)u &+ (1-\rho_2(x))\nabla u \cdot \vec{n} = 0 && \text{on } \partial\Omega \times (0,\infty). \end{aligned} \tag{5.56}$$

If now the principal eigenvalue $\overline{\sigma}_2$ in

$$\begin{aligned} \Delta\phi + [\overline{a}_2(x) + \overline{b}_{21}(x)u_1^*(x)]\phi &= \overline{\sigma}_2\phi \text{ in } && \Omega \\ \rho_2(x)\phi + (1-\rho_2(x))\nabla\phi \cdot \vec{n} &= 0 && \text{on } \partial\Omega \end{aligned} \tag{5.57}$$

is positive, Propositions 3.2 and 3.3 imply there is a unique positive equilibrium u_2^* of (5.56), so that all nonnegative, nontrivial solutions of (5.56) converge in the topology of $C^1(\overline{\Omega})$ to u_2^*. Since u_2 is a subsolution of (5.56), it lies below any solution to (5.56) with $u(x,0) \ge u_2(x,0)$. In particular, if $u_2(x,0) \le u_2^*(x)$, $u_2(x,t) \le u_2^*(x)$ for all $t \ge 0$ and $x \in \overline{\Omega}$.

So if $\tilde{\sigma}_1$ in (5.52) and $\overline{\sigma}_2$ in (5.57) are positive and $\tilde{u}_1(x) \le u_1(x,0) \le u_1^*(x)$ and $u_2(x,0) \le u_2^*(x)$, then $u_1(x,t) \le u_1^*(x)$ and $u_2(x,t) \le u_2^*(x)$ for all $t \ge 0$. It now follows from (5.46)–(5.48) that u_1 is a supersolution of

$$\begin{aligned} u_t &= \Delta u + [\underline{a}_1(x) - \overline{b}_{12}(x)u_2^*(x) - \overline{b}_{11}(x)u]u \text{ in } && \Omega \times (0,\infty) \\ \rho_1(x)u &+ (1-\rho_1(x))\nabla u \cdot \vec{n} = 0 && \text{on } \partial\Omega \times (0,\infty). \end{aligned} \tag{5.58}$$

If now the principal eigenvalue $\hat{\sigma}_1$ of

$$\begin{aligned} \Delta\phi + [\underline{a}_1(x) - \overline{b}_{12}(x)u_2^*(x)]\phi &= \hat{\sigma}_1\phi \text{ in } && \Omega \\ \rho_1(x)\phi + (1-\rho_1(x))\nabla\phi \cdot \vec{n} &= 0 && \text{on } \partial\Omega \end{aligned} \tag{5.59}$$

is positive, Propositions 3.2 and 3.3 imply there is a unique positive equilibrium u_1^{**} of (5.58), so that all nonnegative, nontrivial solutions of (5.58) converge in the topology of $C^1(\overline{\Omega})$ to u_1^{**}. It is easy to see that u_1^{**} is a subsolution of (5.51). It follows that $u_1^{**}(x) \le \tilde{u}_1(x)$. Since u_1 is a supersolution of (5.58), it lies above any solution to (5.58) with $u(x,0) \le u_1(x,0)$. Since $u_1^{**}(x) \le \tilde{u}_1(x) < u_1(x,0)$ by (5.53), we have $u_1(x,t) \ge u_1^{**}(x)$ for all $t \ge 0$ and $x \in \overline{\Omega}$. (Again, notice that if $\hat{\sigma}_1$ is positive in (5.59), $\tilde{\sigma}_1$ is positive in (5.52).)

It now follows from (5.46)–(5.48) that u_2 is a supersolution of

$$\begin{aligned} u_t &= \Delta u + [\underline{a}_2(x) + \underline{b}_{21}(x)u_1^{**}(x) - \overline{b}_{22}(x)u]u \text{ in } && \Omega \times (0,\infty) \\ \rho_2(x)u &+ (1-\rho_2(x))\nabla u \cdot \vec{n} = 0 && \text{on } \partial\Omega \times (0,\infty). \end{aligned} \tag{5.60}$$

If now the principal eigenvalue $\hat{\sigma}_2$ of

$$\begin{aligned} &\Delta\phi + [\underline{a}_2(x) + \underline{b}_{21}(x)u_1^{**}(x)]\phi = \hat{\sigma}_2\phi \text{ in } \Omega \\ &\rho_2(x)\phi + (1 - \rho_2(x))\nabla\phi \cdot \vec{n} = 0 \qquad \text{on } \partial\Omega \end{aligned} \tag{5.61}$$

is positive, Propositions 3.2 and 3.3 imply there is a unique positive equilibrium u_2^{**} of (5.60), so that all nonnegative, nontrivial solutions of (5.60) converge in the topology of $C^1(\overline{\Omega})$ to u_2^{**}. Since the predator density u_2 is an upper solution of (5.60), it lies above any solution to (5.60) with $u(x,0) \le u_2(x,0)$. Suppose now that ϕ is the unique positive solution of (5.61) having the property that $\max_{\overline{\Omega}} \phi(x) = 1$. Then for any sufficiently small $\varepsilon > 0$, $\varepsilon\phi(x)$ is a temporally homogeneous subsolution of (5.60). By Aronson and Weinberger (1975), if u is the solution to (5.60) with $u(x,0) = \varepsilon\phi(x)$, then $u(x,t)$ is strictly increasing in t with $\lim_{t\to\infty} u(x,t) = u_2^{**}(x)$. Since there is an $\varepsilon > 0$ depending on $u_2(x,0)$ so that $u_2(x,0) \ge \varepsilon\phi$, we have that for all $t \ge 0$, $u_2(x,t)$ lies above a solution to (5.60), which monotonically increases to $u_2^{**}(x)$. Since $u_2(x,t) \le u_2^*(x)$ for all $t \ge 0$ and $x \in \overline{\Omega}$, it must follow that $u_2^{**}(x) \le u_2^*(x)$. Moreover, if in fact $u_2(x,0) \ge u_2^{**}(x)$, then $u_2(x,t) \ge u_2^{**}(x)$ for all $t \ge 0$ and $x \in \overline{\Omega}$. (Finally, note again that if $\hat{\sigma}_2$ is positive in (5.61), then $\overline{\sigma}_2$ is positive in (5.57).)

Summarizing, we have the following result:

Theorem 5.3. Assume that the eigenvalues $\overline{\sigma}_1$ in (5.50), $\tilde{\sigma}_1$ in (5.52), $\overline{\sigma}_2$ in (5.57), $\hat{\sigma}_1$ in (5.59) and $\hat{\sigma}_2$ in (5.61) are all positive. Let $u_1^*(x)$, $\tilde{u}_1(x)$, $u_2^*(x)$, $u_1^{**}(x)$ and $u_2^{**}(x)$ be the unique globally attracting positive equilibria to (5.49), (5.51), (5.56), (5.58) and (5.60), respectively, whose existence is guaranteed by the positivity of $\overline{\sigma}_1$, $\tilde{\sigma}_1$, $\overline{\sigma}_2$, $\hat{\sigma}_1$ and $\hat{\sigma}_2$, respectively. Let u_1 and u_2 denote the densities of a prey and predator species whose interaction is modeled by (5.44)–(5.45), assuming (5.46)–(5.48). If

$$\begin{aligned} &\tilde{u}_1(x) \le u_1(x,0) \le u_1^*(x) \\ &0 < u_2(x,0) \le u_2^*(x) \end{aligned} \tag{5.62}$$

in Ω, then

$$\begin{aligned} &u_1^{**}(x) \le u_1(x,t) \le u_1^*(x) \\ &u(x,t) \le u_2(x,t) \le u_2^*(x) \end{aligned} \tag{5.63}$$

in $\Omega \times (0,\infty)$, where $u(x,t)$ is a positive solution to (5.60) which increases monotonically in time with $\lim_{t\to\infty} u(x,t) = u_2^{**}(x)$ in $C^1(\overline{\Omega})$. If, in addition, $u_2^{**}(x) \le u_2(x,0)$ in Ω, then $u_2^{**}(x) \le u_2(x,t)$ in $\Omega \times (0,\infty)$.

Theorem 5.3 may be regarded as an analogue in the context of predator-prey systems to the notion of compressivity in two species competition systems. Of course, the analogy is only practical due to the restrictions that are placed upon the initial prey and predator densities to obtain upper and lower bounds on the densities which hold for all subsequent time. Such restrictions are hardly surprising, since predator-prey systems lack the property (namely a natural ordering of the associated dynamical or semi-dynamical system) that is an essential precondition for compressivity. However, Theorem 5.3 does cover an ecological situation of substantial importance, the introduction of a predator at low densities into a habitat in which a prey species is "well-established". Indeed, it is noteworthy that a result *as strong as Theorem 5.3 is possible at all*, using only comparison techniques and the natural ordering properties for single-species models to compensate for not having an order structure for the system as a whole.

While the predator-prey systems covered by Theorem 5.3 lack the order structure which underlies compressivity results, they do share an important feature with the two-species competition models that we treated via compressivity in Section 2. Namely, in both cases, we require self-regulation on both species' densities. Of course, not all predator-prey models impose explicit self-regulation of the predator density. We considered such a model (namely (4.85)) in Section 4.5, and in fact gave conditions for it to be permanent. However, the self-regulation assumptions on (5.44)–(5.45) can not be relaxed if we are to use comparison techniques *vis-á-vis* single-species models to obtain upper and lower bounds on transient orbits. Consequently, as presently formulated, Theorem 5.3 does not apply to (4.85). Moreover, predator-prey systems without self-regulation on the predator's density may exhibit transient "boom bust" cycles that would be ruled out by Theorem 5.3, even when the systems are permanent. Consequently, it seems likely that the self-regulation requirements on (5.44)–(5.45) are necessary to obtain a result such as Theorem 5.3, and not merely a mathematical artifice.

One should also note that other than requiring self-regulation of both species, the local ecological interaction terms in (5.44) are rather general. For instance, we could treat Holling II functional responses via $b_{12}(x, t, u_1, u_2) = \dfrac{\alpha(x, t)}{\beta(x, t) + \gamma(x, t)u_1}$. Dispersal terms as general as those in (5.11) with time periodic coefficients may also be accommodated, provided we replace the globally attracting positive equilibria in the statement of Theorem 5.3 with globally attracting periodic orbits as appropriate. We shall return to this point in the next section. Finally, just as with our approach to practical persistence in general, we may make comparisons to bound transient orbits of predator-prey interactions in other modeling contexts.

5.5 Persistence in Nonautonomous Systems

As we noted in Section 5.3, practical persistence techniques may be employed to analyze the asymptotic dynamics of certain nonautonomous reaction-diffusion models for several interacting biological species in an isolated bounded habitat patch. The approach to practical persistence we described in Section 5.3 depends on the possibility of making a succession of order comparisons between a component of the model at hand and a reaction-diffusion model for a single biological species in an isolated bounded habitat patch whose long term dynamics are completely understood. We required any explicit temporal dependence in the coefficients of such a single species model to be periodic with a common period. Such models represent the most general class of single species reaction-diffusion models that have so far been used in such comparisons. (Recall the description of the asymptotics of these models from Section 3.6. Such a model may or may not exhibit explicit spatial dependence in its coefficients.) As for the original model for several interacting biological species, we can obtain practical persistence results even when the temporal dependence in its reaction terms is very general, so long as the dependence is appropriately continuous and there are suitable single species models for the order comparisons. Temporal periodicity in the coefficients of the reaction terms in the model is not necessary. However, in the order comparisons, the dispersal mechanism for a component of the system is the same as that in the single species model to which it is compared, and hence we require any explicit temporal dependence in the coefficients of the diffusion terms of the model to be periodic.

There is exciting on-going research into the asymptotics of nonautonomous reaction-diffusion models of the form (5.11) whose coefficients exhibit more general explicit temporal dependence than periodic (Hetzer and Shen, 2002; Zhao, 2002; Shen and Yi, 1998). The form of temporal dependence that has received the most attention is a kind of approximate temporal periodicity, known in the mathematical literature as *almost periodicity*. (For a precise mathematical definition of an almost periodic function as well as some valuable historical context, see Yi (1998).) In the single species case the dynamics of such models are now reasonably well understood. Indeed, one could now employ them in practical persistence comparisons, although to our knowledge no one has done so to date. In the realm of models for interacting species, the most significant work of which we are aware is that of Hetzer and Shen (2002) on coexistence in certain two species diffusive Lotka-Volterra competition models with almost periodic local growth rates, competition rates and self-regulation rates. However, we anticipate that over the next few years there will be many more studies of the asymptotic dynamics of models of the form (5.11) with almost periodic coefficients.

As we noted in Section 4.2 of Chapter 4, solutions to Cauchy problems for systems of nonautonomous ordinary differential equations or reaction-diffusion equations do not automatically form dynamical or semi-dynamical systems, in contrast with their counterparts in systems of autonomous equations. This difference between solutions to nonautonomous systems and solutions to autonomous systems becomes significant if we are to employ abstract permanence techniques such as the Method of Average Lyapunov Functions or the Acyclicity Theorem to obtain ecological permanence results for nonautonomous systems. For the notion of abstract permanence requires a dynamical system or semi-dynamical system context. Fortunately, the theory of dynamical systems is sufficiently developed so that there is a somewhat standardized method of attaching a dynamical or semi-dynamical system to an ODE or reaction-diffusion model for interacting biological species. Moreover, this approach, known as a *skew-product flow* or *semiflow*, is such that, in the cases which have so far been worked out, a finding of abstract permanence in the skew product flow or semiflow implies ecological permanence for the original model.

The study of skew-product dynamical or semi-dynamical systems began with the work of Miller (1965) and Sell (1967). Good introductions to the subject may be found in Sell (1971) and Saperstone (1981). Skew product flow constructions were first used to obtain permanence in nonautonomous ecological models in Burton and Hutson (1991) for the case of ODE models, and in Hutson and Zhao (1994) for the case of reaction-diffusion models with time periodicity in the reaction terms.

To understand more fully why a tool such as a skew product flow construction is needed to convert a nonautonomous system of ODEs or reaction-diffusion equations into a dynamical or semi-dynamical system, recall that a dynamical or semi-dynamical system is required to satisfy three properties. Specifically, if $\pi(u, t)$ denotes the dynamical or semi-dynamical system with state variable u and time variable t, then we have that

$$\begin{aligned}
&\text{(i)}\ \ \pi(u,0) = u \text{ for all } u \\
&\text{(ii)}\ \ \pi(u,t) \text{ is continuous in } u \text{ and } t \\
&\text{(iii)}\ \ \pi(u,t+s) = \pi(\pi(u,t),s) \text{ for all } u, \text{ for all } t \text{ and} \\
&\qquad \text{for all } s \geq 0 \text{ for which } \pi(u,t) \text{ and } \pi(u,t+s) \text{ are defined.}
\end{aligned} \tag{5.64}$$

In the case of an autonomous system of ODEs or reaction-diffusion equations, if $\pi(u,t)$ denotes the solution to the system such that $\pi(u,0) = u$, then (i)–(iii) of (5.64) hold. However, if the system has explicit temporal dependence in one or more of its components and $\pi(u,t)$ is defined in the same way, only (i) and (ii) of (5.64) continue to hold. The third property, the so-called semi-group property, can no longer be verified. The reason is simple. Namely, in a nonautonomous system, the underlying set of equations itself depends on time. Consequently, one must keep track of the "phase" of the system with respect to time or else there will be hopeless ambiguity in (iii) of (5.64). A skew product flow construction keeps track of the "phase" of the system by augmenting the original state variable u with a "phase marker" P_τ so that we consider a function $\pi((u, P_\tau), t)$ instead of a function $\pi(u,t)$.

To describe the construction of $\pi((u, P_\tau), t)$ explicitly, let us focus on a specific class of systems. To that end, consider the system of reaction-diffusion equations

$$(u_i)_t = L_i(t,x)u_i + u_i f_i(t,x,u_1,\ldots,u_n) \quad \text{in} \quad \Omega \times (0,\infty) \tag{5.65}$$

modeling the propagation and interaction of n biological species in a bounded habitat Ω with

$$u_i = 0 \quad \text{on} \quad \partial\Omega \times (0,\infty). \tag{5.66}$$

Here, beyond sufficient smoothness in the dependence of the coefficients of (5.65) upon their arguments (see Henry (1981), for example), we require common temporal periodicity of the components. Specifically, we require that there is a $T > 0$ so that for $i = 1,\ldots,n$,

$$\begin{aligned} L_i(t+T,x) &= L_i(t,x) \\ f_i(t+T,x,u_1,\ldots,u_n) &= f_i(t,x,u_1,\ldots,u_n) \end{aligned} \tag{5.67}$$

for all $t \geq 0$, $x \in \overline{\Omega}$ and $u_i \geq 0$, $i = 1,\ldots,n$.

For an n-tuple $u = (u_1,\ldots,u_n)$ of species densities and times $s \geq t \geq 0$, we let $\phi(u,t,s)$ denote the unique solution to (5.65)–(5.66) so that

$$\phi(u,t,t) = u. \tag{5.68}$$

Now let $\tau \geq 0$ be fixed and consider the system

$$\begin{aligned} (u_i)_s &= L_i(s+\tau,x)u_i + u_i f_i(s+\tau,x,u_1,\ldots,u_n) \text{ in } && \Omega \times (0,\infty) \\ u_i &= 0 && \text{on} \quad \partial\Omega \times (0,\infty). \end{aligned} \tag{5.69}$$

Note that (5.69) represents a phase shift of τ in the system (5.65)–(5.66). For $s \geq 0$, let $\varphi(u,\tau,s)$ denote the unique solution of (5.69) so that

$$\varphi(u,\tau,0) = u. \tag{5.70}$$

It follows from the uniqueness of solutions to Cauchy problems for (5.65)–(5.66) and (5.69) that $\phi(u,\tau,\tau+t) = \varphi(u,\tau,t)$ for $t \geq 0$, or equivalently, that

$$\phi(u,\tau,t) = \varphi(u,\tau,t-\tau) \tag{5.71}$$

for $t \geq \tau \geq 0$.

We may now define a semiflow π on $K \times S^1$ where K is the positive cone of $[C_0^1(\overline{\Omega})]^n$ and S^1 is the unit circle in $I\!R^2$. Recall from Chapter 4 that K consists of all componentwise nonnegative n-tuples $u = (u_1, \ldots, u_n)$ of smooth functions on $\overline{\Omega}$ satisfying absorbing (i.e., homogeneous Dirichlet) boundary conditions on $\partial\Omega$. We represent S^1 as $\{P_\tau : \tau \in I\!R\}$ where $P_\tau = e^{\frac{2\pi i \tau}{T}}$.

Clearly $P_\tau \cdot P_{\tau'} = P_{\tau+\tau'}$ and $P_\tau = P_{\tau+T}$ for any $\tau,\ \tau' \geq 0$. The metric on $K \times S^1$ is

$$d((u, P_\tau), (v, P_\omega)) = ||u - v||_{[C_0^1(\bar{\Omega})]^n} + \sqrt{1 - \cos\left(\frac{2\pi(\tau - \omega)}{T}\right)}. \tag{5.72}$$

Then $\pi : K \times S^1 \times [0, \infty) \to K \times S^1$ is given by

$$\begin{aligned}\pi((u, P_\tau), t) &= (\phi(u, \tau, \tau + t), P_{\tau+t}) \\ &= (\varphi(u, \tau, t), P_{\tau+t})\end{aligned} \tag{5.73}$$

by (5.71). Note that by (5.67) and the uniqueness of solutions to Cauchy problems for (5.65)–(5.66),

$$\phi(u, \tau, t) = \phi(u, \tau + T, t) \tag{5.74}$$

for any $\tau \geq 0$. It follows from (5.74) that π is well-defined by (5.73).

π is a semiflow. The second property of semiflows from (5.64), the continuity of π in u, P_τ and t, follows from the continuous dependence of ϕ (or φ) on u, τ and t. It follows from (5.68) that

$$\pi((u, P_\tau), 0) = (\phi(u, \tau, \tau), P_\tau) = (u, P_\tau) \tag{5.75}$$

so that (i) of (5.64) holds. As for property (iii), it is straightforward to calculate that having the property hold boils down to having it be the case that

$$\varphi(\varphi(u, \tau, t), \tau + t, s) = \varphi(u, \tau, t + s) \tag{5.76}$$

for all $t, s \geq 0$ which are such that both sides of (5.76) are defined. To this end, fix $t \geq 0$. Then if $s = 0$, it follows from (5.70) that $\varphi(\varphi(u, \tau, t), \tau + t, 0) = \varphi(u, \tau, t) = \varphi(u, \tau, t + 0)$ so that (5.76) holds when $s = 0$. One may then calculate, as in Avila (1995), that for $s > 0$, both sides of (5.76) solve (5.69) with τ replaced by $\tau + t$. Consequently, (5.73) defines a semiflow on $K \times S^1$, which is termed a skew product semiflow for (5.65)–(5.66). Additional details regarding the construction may be found in Hutson and Zhao (1994), Avila (1995) and Avila and Cantrell (1995).

Recall from Section 4.3 that either the Average Lyapunov Function or Acyclicity Approach to permanence of a dynamical or semi-dynamical system π requires the existence of a compact global attractor for π. Moreover, recall also that to employ the classical results of Billotti and LaSalle (1971) to guarantee the existence of such a compact global attractor for π requires that π be dissipative and such that $\pi(\cdot, t)$ is compact as a map on the state space for all $t > t_0$ for some $t_0 \geq 0$. For the skew product semiflow π defined in (5.73), the state space is $K \times S^1$. S^1 is itself compact. So the dissipativity and compactness for all large enough fixed positive values of t of the second component $(u, P_\tau, t) \to P_{\tau+t}$ of π from (5.73) is automatic. Hence these properties for π as a map from $K \times S^1 \times [0, \infty)$

into $K \times S^1$ can be expected to follow from the theory of reaction-diffusion equations as described in Section 1.6 so long as suitable assumptions are placed upon the *per capita* growth laws $f_i(t, x, u_1, \ldots, u_n)$ in (5.65)–(5.66). See Avila (1995) and Avila and Cantrell (1995, 1997) for particular examples, and Hutson and Zhao (1994) for examples when (5.66) is replaced by the reflecting boundary condition

$$\nabla u_i(x) \cdot \vec{n} = 0 \tag{5.77}$$

on $\partial\Omega$.

As we mentioned earlier, permanence in (5.73) implies the ecological permanence of (5.65)–(5.66). That such is the case follows as in Section 4.6.1. The chief point is that the theory of reaction-diffusion equations still implies in the context of (5.65)–(5.66) that

$$u_i(x) > 0 \quad \text{for} \quad x \in \Omega$$

and

$$\nabla u_i(x) \cdot \vec{n}(x) < 0 \quad \text{for} \quad x \in \partial\Omega$$

for $i = 1, \ldots, n$ for any point $(u_1, \ldots, u_n, P_\tau) \in K \times S^1$ in the global attractor of π which is also not in $\partial(K \times S^1) = (\partial K) \times S^1$. As a consequence, the result follows as in the autonomous case as given in the Appendix to Chapter 4.

Either of the two main tools for establishing permanence can be brought to bear in the context of (5.73). In Avila (1995) and Avila and Cantrell (1995), the Average Lyapunov Function Approach is used to obtain permanence results for (5.65)–(5.66) in the case of predator-prey models. (The Average Lyapunov Function Approach is also used in Hutson and Zhao (1994) to establish the permanence of the skew product flow which arises from comparison predator-prey models with the absorbing boundary condition (5.66) replaced by (5.77). The reflecting boundary condition (5.77) allows for a simpler average Lyapunov function construction than is possible with a lethal or absorbing boundary condition.) The Acyclicity Approach is employed in Avila (1995) and Avila and Cantrell (1997) to establish permanence in systems of three competing species. The analysis in Avila (1995) and Avila and Cantrell (1997) requires a detailed understanding of the asymptotics of the two species competitive systems that arise when one of the competitors is absent. While it is possible to analyze these two species subsystems with the goal of establishing permanence via the skew product semiflow (5.73), more precise information (which in turn leads to sharper permanence results for the full system) can be obtained if the subsystems are examined from the point of view of compressivity as in Hess and Lazer (1991) and Hess (1991). Using compressivity results to analyze two species competition models of the form (5.65)–(5.66) means that the skew product flow (5.73) on $K \times S^1$ is replaced by a discrete semi-dynamical system $\pi^* : K \times I\!N \to K$, where K is as in (5.73) and $I\!N$ is the nonnegative integers, given by

$$\pi^*(u, n) = \phi(u, 0, nT) \tag{5.78}$$

where $\phi(u, t, s)$ is the unique solution to (5.65)–(5.66) which is defined for $s \geq t$ satisfying (5.68) and $T > 0$ is the time period of the coefficients in (5.67). We alluded to this construction in Section 5.2 prior to our discussion of compressivity in the case of temporally autonomous models, and employed the analogous discrete semi-dynamical system in

Section 3.6 to analyze the dynamics of a species in a bounded isolated habitat patch whose density is modeled by a reaction-diffusion equation with T-periodic coefficients.

We now illustrate how to employ the Average Lyapunov Function Approach to establish permanence in the temporally periodic Lotka-Volterra predator-prey system

$$\begin{aligned} \frac{\partial u_1}{\partial t} &= \Delta u_1 + u_1(a_1(x,t) - u_1 - b_1(x,t)u_2) \\ &\qquad\qquad\qquad \text{in} \quad \Omega \times (0, \infty) \\ \frac{\partial u_2}{\partial t} &= \Delta u_2 + u_2(a_2(x,t) + b_2(x,t)u_1 - u_2) \end{aligned} \tag{5.79}$$

$$u_1 = 0 = u_2 \quad \text{on} \quad \partial\Omega \times (0, \infty). \tag{5.80}$$

Note that (5.79)–(5.80) corresponds to (4.84) from Section 4.5 with spatiotemporal dependence in the intrinsic growth rates a_i and predation coefficients b_i. Recall that in Section 4.5 the Average Lyapunov Function Approach was used to establish conditions for permanence in (4.84). Indeed, our present treatment will parallel that of (4.84) in Section 4.5.

So that the theory of reaction-diffusion equations as described in Section 1.6 apply to (5.79)–(5.80), we shall assume that

$$a_i, b_i \in C^{\alpha,\alpha/2}(\overline{\Omega} \times [0, T]), \tag{5.81}$$

where T is the time period. *To have the possibility* that the prey species persists in the absence of the predator, we must require

$$a_1(x,t) > 0 \tag{5.82}$$

for some x in the habitat Ω and some $t > 0$. If $a_1(x,t) \leq 0$ for all $x \in \overline{\Omega}$ and all $t \geq 0$, the prey species goes extinct as $t \to \infty$, whether the predator is present or not. Our present treatment will parallel that in Section 4.5 for (4.84) in the case when the predator has resources other than the prey and may possibly survive in the absence of the prey. In Section 4.5, this case corresponded to having $a_2 > 0$ in (4.84). Here we make the parallel assumption

$$a_2(x,t) > 0 \tag{5.83}$$

for some $x \in \Omega$ and $t > 0$. If $a_2(x,t) \leq 0$ for $x \in \overline{\Omega}$ and all $t \geq 0$, the predator species goes extinct in the absence of the prey species. In such case, permanence of (5.79)–(5.80) is still possible. However, the average Lyapunov function we would construct for (5.79)–(5.80) would be modified in a way which parallels the modification of the average Lyapunov function for (4.84) when $a_2 \leq 0$. We leave the details to the interested reader. For the predation coefficients, we require

$$b_i(x,t) \geq 0 \tag{5.84}$$

for $x \in \overline{\Omega}$ and $t \geq 0$.

We should note that while having $a_1 > 0$ and $a_2 > 0$ was necessary if each species modeled in (4.84) was to survive in the absence of the other, it was not sufficient. The same is true in (5.79)–(5.80). Conditions (5.82) and (5.83) are necessary for the persistence of each species in the absence of the other, but are not sufficient. We do give necessary and sufficient conditions for the persistence of each species in the absence of the other in the course of identifying conditions for permanence in (5.79)–(5.80). All of the conditions we obtain are expressed in terms of the signs of eigenvalues μ for problems of the form

$$\begin{aligned} \frac{\partial w}{\partial t} - \Delta w - r(x,t)w = \mu w \text{ in } \quad & \Omega \times (0,\infty) \\ w = 0 \text{ on } \quad & \partial\Omega \times (0,\infty) \\ w > 0 \text{ in } \quad & \Omega \times (0,\infty) \end{aligned} \tag{5.85}$$

where $r \in C^{\alpha,\alpha/2}(\overline{\Omega}\times[0,T])$ and is T-periodic in time. (Recall Section 2.5.) Following our identification of conditions for permanence in (5.79)–(5.80), we discuss how to find readily calculable conditions on the growth rates $a_i(x,t)$ and predation coefficients $b_i(x,t)$ which enable us to assert that the conditions for permanence are met. In particular, we address the issue of finding readily calculable conditions on $a_i(x,t)$ under which each species may persist in the absence of the other.

Our assumptions (5.81)–(5.83) allow us to conclude that if (u_1, u_2) represent the densities of a prey-predator pair whose interaction is governed by (5.79)–(5.80), then for $i = 1, 2$

$$\frac{\partial u_i}{\partial t} \leq u_i(\overline{a}_i - u_i) \quad \text{on} \quad \Omega \times (0,\infty) \tag{5.86}$$

where $\overline{a}_i = \max_{(x,t)\in\overline{\Omega}\times[0,\infty)} a_i(x,t)$. Consequently, Theorem 4.1 implies there is a $t_0 > 0$ depending only on $(u_1(x,0), u_2(x,0))$ so that for $i = 1, 2$,

$$u_i(x,t) \leq \overline{a}_i + 1 \tag{5.87}$$

for all $x \in \overline{\Omega}$ and $t \geq t_0$. It now follows as in Section 4.5 that the semiflow π in (5.73) corresponding to (5.79)–(5.80) is dissipative on $K \times S^1$, where K is now the positive cone in $[C_0^1(\overline{\Omega})]^2$. Since the compactness of π on $K \times S^1$ for $t > 0$ fixed follows from the theory of reaction-diffusion equations as described in Section 1.6 and the compactness of S^1, we may apply the classical result of Billotti and LaSalle (1971) to assert the existence of a global attractor $\mathcal{A}$ for π in $K \times S^1$. As in Section 4.3, we define, for some $t_0 > 0$,

$$\tilde{X} = \overline{\pi(\mathcal{B}(\mathcal{A},\varepsilon),[t_0,\infty))} \tag{5.88}$$

and then

$$X = \pi(\tilde{X}, t') \tag{5.89}$$

for some $t' > 0$. By Theorem 4.1, if $S = X \cap [(\partial K) \times S^1]$, X and S are compact and forward invariant under π and we may restrict our attention to X in determining conditions under which π is permanent.

As in Section 4.5, we require detailed information regarding the asymptotics of (5.79)–(5.80) when one of the species is absent in order to formulate a suitable average Lyapunov function. Lemma 2.15 and Propositions 3.1–3.3 allow us to address this issue, as follows.

Consider, for $i = 1, 2$

$$\begin{aligned} \frac{\partial u_i}{\partial t} = \Delta u_i + u_i(a_i(x,t) - u_i) \text{ in } \quad & \Omega \times (0,\infty) \\ u_i = 0 \text{ on } \quad & \partial\Omega \times (0,\infty). \end{aligned} \tag{5.90}$$

Since $a_i(x,t) = a_i(x,t+T)$ for all $t \geq 0$, Lemma 2.15 asserts that there is a unique value $\mu_i \in \mathbb{R}$, so that the eigenvalue problem

$$\begin{aligned}\frac{\partial w_i}{\partial t} - \Delta w_i - a_i(x,t)w_i &= \mu_i w \text{ in } \quad \Omega \times (-\infty,\infty) \\ w_i &= 0 \text{ on } \quad \partial\Omega \times (-\infty,\infty)\end{aligned} \tag{5.91}$$

admits an eigenfunction $w_i(x,t)$ such that

$$\begin{aligned} &w_i(x,t) > 0 \\ &\qquad\qquad\qquad \text{on} \quad \Omega \times (-\infty,\infty). \\ &w_i(x,t+T) = w_i(x,t)\end{aligned} \tag{5.92}$$

Propositions 3.1–3.3 tell us that if $\mu_i \geq 0$, all solutions to (5.90) with $u_i(x,0) \geq 0$ on $\overline{\Omega}$ converge in $C_0^1(\overline{\Omega})$ to 0 as $t \to \infty$, whereas if $\mu_i < 0$, there is a unique T-periodic solution u_i^* of (5.90) such that

$$u_i^*(x,t) > 0 \tag{5.93}$$

for $x \in \Omega$ and $t \geq 0$ and such that all solutions to (5.90) with $u_i(x,0) \overset{>}{\not\equiv} 0$ on $\overline{\Omega}$ converge in $C_0^1(\overline{\Omega})$ to u_i^* as $t \to \infty$, i.e. $||u_i(x,t) - u_i^*(x,t)||_{C_0^1(\overline{\Omega})} \to 0$ as $t \to \infty$. Consequently, if for $i = 1, 2$,

$$\mu_i < 0 \quad \text{in (5.91)} \tag{5.94}$$

we have that the model predicts that each species persists in the absence of the other. Recall from Section 2.5 that having (5.94) hold in the nonautonomous context of (5.79)–(5.80) is the exact counterpart in the autonomous context of (4.84) to having $\sigma_i > 0$, $i = 1, 2$ in (4.96).

Let us now assume that (5.94) holds for $i = 1, 2$, and let $u_i^*(x,t)$ be the T-periodic solution to (5.90) satisfying (5.93). We may apply Lemma 2.15 once again. We conclude first of all that for $i = 1, 2$, there is a unique real value μ_i^* so that the eigenvalue problems

$$\begin{aligned}\frac{\partial \phi_1}{\partial t} - \Delta\phi_1 - \phi_1(a_1(x,t) - b_1(x,t)u_2^*(x,t)) &= \mu_1^*\phi_1 \text{ in } \quad \Omega \times (-\infty,\infty) \\ \phi_1 &= 0 \text{ on } \quad \partial\Omega \times (-\infty,\infty)\end{aligned} \tag{5.95}$$

$$\begin{aligned}\frac{\partial \phi_2}{\partial t} - \Delta\phi_2 - \phi_2(a_2(x,t) + b_2(x,t)u_1^*(x,t)) &= \mu_2^*\phi_2 \text{ in } \quad \Omega \times (-\infty,\infty) \\ \phi_2 &= 0 \text{ on } \quad \partial\Omega \times (-\infty,\infty)\end{aligned} \tag{5.96}$$

both admit eigenfunctions which are T-periodic in time and positive on $\Omega \times (-\infty,\infty)$. Moreover, μ_1^* and μ_2^* are also the unique real values for which the corresponding adjoint eigenvalue problems

$$\begin{aligned}-\frac{\partial \psi_1}{\partial t} - \Delta\psi_1 - \psi_1(a_1(x,t) - b_1(x,t)u_2^*(x,t)) &= \mu_1^*\psi_1 \text{ in } \quad \Omega \times (-\infty,\infty) \\ \psi_1 &= 0 \text{ on } \quad \partial\Omega \times (-\infty,\infty)\end{aligned} \tag{5.97}$$

and

$$-\frac{\partial \psi_2}{\partial t} - \Delta\psi_2 - \psi_2(a_2(x,t) + b_2(x,t)u_1^*(x,t)) = \mu_2^*\psi_2 \text{ in } \quad \Omega \times (-\infty, \infty)$$
$$\psi_2 = 0 \text{ on } \quad \partial\Omega \times (-\infty, \infty) \tag{5.98}$$

admit eigenfunctions which are T-periodic in time and positive on $\Omega \times (-\infty, \infty)$.
We may now identify when (5.79)–(5.80) is permanent, as follows.

Theorem 5.4. Consider the system (5.79)–(5.80), subject to the conditions (5.81)–(5.84). Assume that for $i = 1, 2$, (5.94) holds (i.e. $\mu_i < 0$ in (5.91)) and let u_i^* be the globally attracting positive T-periodic solution to (5.90) whose existence is guaranteed by (5.94). Then if

$$\mu_1^* < 0 \quad \text{in (5.95) (or equivalently (5.97))} \tag{5.99}$$

and

$$\mu_2^* < 0 \quad \text{in (5.96) (or equivalently (5.98))} \tag{5.100}$$

(5.79)–(5.80) is ecologically permanent.

Proof: Let K be the positive cone in $[C_0^1(\overline{\Omega})]^2$ and let $\pi : K \times S^1 \times [0, \infty) \to K \times S^1$ be the skew product semiflow associated with (5.79)–(5.80) by the formula (5.73), under the assumption that (5.81)–(5.84) hold. Let X be as in (5.89) and let $S = X \cap [\partial(K \times S^1)] = X \cap [(\partial K_1) \times S^1]$. (Recall that for $(u_1, u_2, P_\tau) \in X$, $(u_1, u_2, P_\tau) \in S$ if and only if $u_1 \equiv 0$ or $u_2 \equiv 0$.) Define $\rho : X\backslash S \to (0, \infty)$ by

$$\rho((v_1, v_2), P_\tau) = \left(\int_\Omega v_1(x)\psi_1(x,\tau)dx\right)^{\beta_1} \left(\int_\Omega v_2(x)\psi_2(x,\tau)dx\right)^{\beta_2}, \tag{5.101}$$

where $\psi_1(x,t) > 0$ is a normalized T-periodic eigenfunction for (5.97) corresponding to $\mu_1^* < 0$ and $\psi_2(x,t) > 0$ is a normalized T-periodic eigenfunction for (5.98) corresponding to $\mu_2^* < 0$.

Now let $((u_1, u_2), P_{\tau_0}) \in S$, so that $u_1 \equiv 0$ or $u_2 \equiv 0$, and let $t > 0$. Define

$$\alpha(t, ((u_1, u_2), P_{\tau_0})) = \liminf_{\substack{((v_1,v_2),P_\tau)\to(u_1,u_2,P_{\tau_0}) \\ ((v_1,v_2),P_\tau)\in X\backslash S}} \left(\frac{\rho(\pi(((v_1, v_2), P_\tau), t))}{\rho((v_1, v_2), P_\tau)}\right). \tag{5.102}$$

We show in the Appendix to this chapter that we may choose $\beta_1 > 0$ and $\beta_2 > 0$ in (5.101) so that

$$\sup_{t>0} \alpha(t, ((u_1, u_2), P_{\tau_0})) > 0 \tag{5.103}$$

for all $((u_1, u_2), P_{\tau_0}) \in S$ and

$$\sup_{t>0} \alpha(t, ((u_1, u_2), P_{\tau_0})) > 1 \tag{5.104}$$

for all $((u_1, u_2), P_{\tau_0}) \in \omega(S)$. Since (5.103) and (5.104) hold, Theorem 4.2 (Average Lyapunov Function Theorem) applies. Hence π is permanent and thus (5.79)–(5.80) is ecologically permanent.

Under the assumption that the predator species has resources other than the prey which are sufficient for its persistence in the absence of the prey (i.e. $\mu_2 < 0$ in (5.91)), the conditions for permanence in (5.79)–(5.80) are precise in the sense that if $\mu_1 > 0$ in (5.91) or $\mu_1^* > 0$ in (5.95) (or equivalently (5.97)), the system fails to be permanent. (Note that if $\mu_2 < 0$ and $\mu_1 < 0$ in (5.91), $\mu_2^* < 0$ in (5.96) (or equivalently (5.98).) The reason is that in that case $a_2(x,t) + b_2(x,t)u_1^*(x,t) > a_2(x,t)$ on $\Omega \times [0, T]$, and consequently the monotonicity property for eigenvalue problems of the form (5.85), Lemma 2.15, implies that $\mu_2^* < \mu_2 < 0$.) Moreover, the quantities μ_1, μ_1^*, μ_2 and μ_2^* have natural biological interpretations. In particular, $-\mu_i$ represents the average rate of growth for species i at low densities in the absence of species j over the spatial habitat Ω throughout one full time cycle $[0, T]$, while $-\mu_i^*$ represents the average rate of growth for species i when species j is at its (space and time varying) carrying capacity u_j^*, over the spatial habitat Ω throughout one full time cycle $[0, T]$. However, it is also highly desirable that we understand how the quantities μ_i and μ_i^* depend on the coefficients a_i and b_i in (5.79)–(5.80) well-enough to formulate sufficient conditions for permanence in (5.79)–(5.80) in terms of the coefficients a_i and b_i. Such conditions are less precise that (5.94), (5.97) and (5.98), but are considerably more transparent. The following derivations follow Hess (1991, Chapter II, Sections 14–17) and Avila (1995, Chapter 4).

We consider first the periodic-parabolic principal eigenvalue problem (5.85). For each value t in the time interval, let $\hat{r}(t) = \min_{x \in \overline{\Omega}} r(x,t)$, and consider

$$\begin{aligned} \frac{\partial z}{\partial t} - \Delta z - \hat{r}(t)z = \hat{\mu}z \text{ in } \quad & \Omega \times [0, T] \\ z = 0 \text{ on } \quad & \partial\Omega \times [0, T]. \end{aligned} \tag{5.105}$$

Lemma 2.15 (Monotonicity) implies $\mu \leq \hat{\mu}$. Let $\lambda_0^1(\Omega)$ represent the principal eigenvalue of $-\Delta$ on Ω subject to homogeneous Dirichlet boundary data with corresponding eigenfunction $\phi(x) > 0$ in Ω. Since $\hat{\mu}$ is a simple eigenvalue, we find that $z(x,t) = \alpha(t)\phi(x)$ with $\alpha(t+T) = \alpha(t)$ so long as

$$\alpha'(t) + [\lambda_0^1(\Omega) - \hat{r}(t) - \hat{\mu}]\alpha(t) = 0. \tag{5.106}$$

From (5.106), $\alpha(t) = \alpha(0)e^{\int_0^t [\lambda_0^1(\Omega) - \hat{r}(s) - \hat{\mu}]ds}$, so that $\alpha(T) = \alpha(0)$ implies

$$\int_0^T [\lambda_0^1(\Omega) - \hat{r}(t) - \hat{\mu}]dt = 0,$$

or equivalently,

$$\hat{\mu} = \lambda_0^1(\Omega) - \frac{1}{T}\int_0^T \hat{r}(t)dt. \tag{5.107}$$

Consequently, we have from (5.107), as in Hess (1991, Section II), that μ in (5.85) satisfies

$$\mu \leq \lambda_0^1(\Omega) - \frac{1}{T}\int_0^T \left[\min_{x\in\overline{\Omega}} r(x,t)\right] dt. \tag{5.108}$$

Let us now apply (5.108) in the context of (5.79)–(5.80), following Avila (1995, Chapter 4). We get that for $i = 1, 2,$ μ_i satisfies

$$\mu_i \le \lambda_0^1(\Omega) - \frac{1}{T}\int_0^T \left[\min_{x\in\overline{\Omega}} a_i(x,t)\right] dt. \tag{5.109}$$

Observe that if $a_i(x,t)$ has the constant value $a_i > 0$, (5.109) reduces to

$$\mu_i \le \lambda_0^1(\Omega) - a_i. \tag{5.110}$$

From (5.110), it is apparent that $a_i > \lambda_0^1(\Omega)$ implies $\mu_i < 0$. Consequently, the condition for $\mu_i < 0$ in (5.91) in this case is the same as the condition for $\sigma_i > 0$ in (4.96) in the corresponding constant coefficient autonomous model. In both cases, the *per capita* growth rate of the species at low densities (a_i) must be large enough to compensate for the loss via diffusion through a dissipative boundary. In general, (5.109) may be interpreted as saying that μ_i is negative provided that the time average over $[0, T]$ of the (temporally varying) minimums of the *per capita* growth rate over $\overline{\Omega}$ is large enough to compensate for the loss via diffusion through the dissipative boundary. Notice that $a_i(x,t)$ is allowed to be *negative somewhere in* Ω *for some values of t so long as it is positive enough throughout* Ω *for enough other values of t* to compensate appropriately.

As we have noted, $\mu_2^* \le \mu_2$, where μ_2^* is as in (5.96) and μ_2 is as in (5.91), so that $\mu_2^* < 0$ provided $\mu_2 < 0$. Consequently, the condition

$$\frac{1}{T}\int_0^T \left[\min_{x\in\overline{\Omega}} a_2(x,t)\right] > \lambda_0^1(\Omega) \tag{5.111}$$

is sufficient to have $\mu_2 < 0$ in (5.91) and $\mu_2^* < 0$ in (5.96). If we now apply (5.108) to (5.95), we obtain

$$\mu_1^* \le \lambda_0^1(\Omega) - \frac{1}{T}\int_0^T \left[\min_{x\in\overline{\Omega}}(a_1(x,t) - b_1(x,t)u_2^*(x,t))\right] dt,$$

so that $\mu_1^* < 0$ provided

$$\frac{1}{T}\int_0^T \left[\min_{x\in\overline{\Omega}}(a_1(x,t) - b_1(x,t)u_2^*(x,t))\right] dt > \lambda_0^1(\Omega). \tag{5.112}$$

Since

$$\min_{x\in\overline{\Omega}}[a_1(x,t) - b_1(x,t)u_2^*(x,t)] \ge \min_{x\in\overline{\Omega}} a_1(x,t) - \max_{x\in\overline{\Omega}} b_1(x,t)u_2^*(x,t),$$

we have that (5.112) holds provided

$$\frac{1}{T}\int_0^T \left(\min_{x\in\overline{\Omega}} a_1(x,t)\right) dt > \frac{1}{T}\int_0^T \left(\max_{x\in\overline{\Omega}} b_1(x,t)u_2^*(x,t)\right) dt + \lambda_0^1(\Omega). \tag{5.113}$$

Note that $u_2^*(x,t)$ is a lower solution to

$$\begin{aligned} \frac{\partial w}{\partial t} &= \Delta w + w\left(\left[\max_{\substack{x\in\overline{\Omega}\\ t\in[0,T]}} a_2(x,t)\right] - w\right) && \text{in} \quad \Omega\times(0,\infty) \\ w &= 0 && \text{on} \quad \partial\Omega\times(0,\infty). \end{aligned} \tag{5.114}$$

By Theorem 4.4, given $\varepsilon > 0$, any positive solution w of (5.114) satisfies

$$w < \left[\max_{\substack{x\in\overline{\Omega} \\ t\in[0,T]}} a_2(x,t) \right] + \varepsilon$$

for all sufficiently large t. Consequently,

$$u_2^*(x,t) \le \max_{\substack{x\in\overline{\Omega} \\ t\in[0,T]}} a_2(x,t) \tag{5.115}$$

for all $x \in \overline{\Omega}$ and $t \in [0, T]$. From (5.115), (5.113) holds (and hence $\mu_1^* < 0$) provided

$$\begin{aligned} &\frac{1}{T}\int_0^T \left(\min_{x\in\overline{\Omega}} a_1(x,t) \right) dt \\ &> \left[\max_{\substack{x\in\overline{\Omega} \\ t\in[0,T]}} a_2(x,t) \right] \left[\frac{1}{T}\int_0^T \left(\max_{x\in\overline{\Omega}} b_1(x,t) \right) dt \right] + \lambda_0^1(\Omega). \end{aligned} \tag{5.116}$$

Summarizing, we have the following.

Proposition 5.5. The conditions (5.111) and (5.116) are sufficient to guarantee that the model (5.79)–(5.80) is ecologically permanent.

Companion results to Proposition 5.5 when the temporally periodic Lotka-Volterra dynamics in (5.79)–(5.80) are replaced with temporally periodic Holling Type 2 dynamics may also be found in Avila (1995, Chapter 4). In general, estimating μ in (5.85) is a difficult mathematical problem deserving of further study. For some recent results on this topic, see Hutson et al. (2001, 2002).

We conclude this section with some additional remarks on temporal dependence in the coefficients of ODE or reaction-diffusion models. One of our hypotheses throughout our discussion of the reaction-diffusion model (5.65)–(5.66) has been that all of its coefficients which are explicitly time dependent are periodic and share a common period. As a "first take" at including explicit temporal dependence into models such as (5.65)–(5.66), such an assumption is quite reasonable. Moreover, it is more general than it may appear on the surface in the following sense. Periodicity in a coefficient in (5.65)–(5.66) reflects seasonality in some underlying biotic and/or environmental feature(s). Certainly, not all biotic and environmental features of a system of interacting biological species exhibit seasonal cycles of the same length. However, the periodicity assumption (5.67) for (5.65)–(5.66) does not require all seasonal cycles to be of the same length. It merely requires the seasonal cycles to have lengths which are commensurate in the sense of having a common multiple. For example, a quantity which is periodic and cycles through a period in six weeks may be thought of as being periodic and cycling through a period in 12 weeks or 18 weeks or any multiple of six weeks for that matter. There is no requirement in (5.67) that T be the smallest possible period. So, for instance, when considering a model such as (5.65)–(5.66) we could make assumption (5.67) when all underlying seasonal cycles are assumed to be measurable in, say, integer numbers of weeks.

On the other hand, the specific framework for examining (5.65)–(5.66) that we have described will be inadequate if either all its coefficients are periodic but not all the periods involved are commensurate or the coefficients are only approximately temporally periodic. It is at this point that the notion of almost periodicity arises. (The case when all the coefficients in the model are periodic but not all the periods are commensurate is known as *quasiperiodicity*. Quasiperiodicity is an important special case of almost periodicity.) A skew-product flow construction as in Sell (1971) is still possible (Yi, 1998; Shen and Yi, 1998a,b). However, it is no longer possible to employ S^1 as a "phase marker" as in (5.73). In the quasiperiodic situation, the phases of the model may be tracked via a torus T^n in place of the circle S^1, where n is the number of incommensurate periods involved in the coefficients. For instance, if some of the coefficients are 1-periodic in time while others are 2π-periodic in time, the phases of the model could tracked via T^2. It is likely that the program we have described for obtaining permanence via π in (5.73) for (5.65)–(5.66) subject to (5.67) could be adapted to work for a skew product semiflow on $K \times T^n$ in the quasiperiodic case. However, to our knowledge, such a program has not been explicitly carried out, although we believe from an applied viewpoint it may be profitable to do so. There are some reasons why. First of all, the results on the periodic coefficient case themselves are relatively recent, dating only from the mid 1990s. Secondly, there are some genuine mathematical challenges involved, so that an extension to the setting of (5.65)–(5.66) with quasiperiodic coefficients is not merely a routine mathematical exercise once one understands the program in the case of (5.65)–(5.66) with periodic coefficients. However, we believe that the main reason is a matter of mathematical sensibilities. Most of the current research into the topic focuses on models such as (5.65)–(5.66) with almost periodic coefficients. The model (5.65)–(5.66) with either periodic or quasiperiodic coefficients may be regarded as a special case of the model with almost periodic coefficients, and there is often a preference among mathematicians to study the more general question. Moreover, in the context of a single real valued function dependent upon a single time variable, almost periodicity is a widely studied concept and is frequently regarded as the "next step up" in complexity of temporal dependence from periodicity, and hence benefits from familiarity. However, the mathematics involved in analyzing the asymptotics of a model such as (5.65)–(5.66) with almost periodic coefficients in general is substantially more challenging than the results we have presented in this section (Yi, 1998; Shen and Yi, 1998a,b). As in the case of periodic coefficients, one must track the change in the model itself as time progresses. However, it is no longer the case that the model repeats itself. In the case of periodic coefficients the model starting at τ is the same as the model starting at $\tau + T$. Such is no longer the case with almost periodic coefficients in general. One is forced to consider the differential operators and coefficients as objects in a function space which vary with time. When an appropriate topology is placed on the function space, the assumption of almost periodicity allows one to conclude that the closure of the collection of differential operators and/or coefficients in the function space (called the *closed hull* of the collection) is compact. The property of compactness is needed if one is to employ results of dynamical systems theory with an eye towards obtaining permanence results. However, the fact that one is forced to consider the closure of the collection of differential operators and/or coefficients means that in general when a model that we consider arises as a limit in the function space as time tends to infinity, it need not correspond to any time phase of the original model, which is a substantial difficulty. We should note, however, that in the special case when the coefficients are quasiperiodic, the collection of possible phases of the model corresponds to a torus T^n, and hence will be compact, so that any limit in this case will correspond to starting the original model at some finite time.

5.6 Conditional Persistence

We observed in Section 4.1 that a prediction of permanence in a model for several interacting biological species should be independent of the initial densities of the species so long as the initial densities are all positive. This feature eliminates the possibility of permanence *per se* in a model where the population density of at least one of the species being modeled always declines if it is below some threshold. Such a population is said to exhibit an Allee effect. Models incorporating Allee effects have long been prominent in the ecological literature, dating back at least to the seminal work of Allee (1931, 1938), and arise naturally when, for example, species feed less efficiently or are more vulnerable to predation at low densities. (For a fuller ecological discussion of Allee effects, including references on ecological mechanisms that give rise to Allee effects, see Lewis and Kareiva (1993).)

Although models which incorporate Allee effects fail to satisfy the hypotheses necessary for applying abstract permanence, the notion of an asymptotic "floor" or "ceiling" from ecological permanence may nevertheless serve as an organizing principle in the analysis of the longterm dynamics of such models. Consider, for instance, the ODE model

$$\frac{du}{dt} = u(u-a)(K-u) \tag{5.117}$$

where $0 < a < \frac{K}{2}$. In (5.117), the population density u declines if it lies between the equilibrium densities 0 and a or above the equilibrium density K and increases if $a < u < K$. It is easy to establish, either by directly solving (5.117) or by an elementary analysis that if $u(t)$ is a positive solution of (5.117) then

$$\lim_{t\to\infty} u(t) = 0 \tag{5.118}$$

when $0 < u(0) < a$ and

$$\lim_{t\to\infty} u(t) = K \tag{5.119}$$

if $u(0) > a$. (See Figure 5.2.) It follows from (5.118) and (5.119) that any value larger than K (e.g., $2K$) can serve as an asymptotic "ceiling" for a species density modeled by (5.117) while any value between a and K $\left(\text{e.g. } \frac{K+a}{2}\right)$ can serve as an asymptotic "floor" *so long as* $u(0) > a$. Consequently, we may draw the same asymptotic conclusions about (5.117) as we would for a model exhibiting ecological permanence, except that we may only draw the conclusions if the initial density of the species being modeled exceeds the threshold a.

We have opted to employ the term *conditional persistence* to describe the asymptotic behavior exhibited by models such as (5.117). Generally speaking, we will say that a model for several interacting biological species is *conditionally persistent* if the conclusions of ecological permanence hold for an open subset of the set of componentwise positive initial densities. To be more specific, let us consider the general form (4.22) of a model for n interacting biological species whose densities are given by $(u_1, \ldots, u_n)$ when time is viewed as a continuous variable. Such a model is *ecologically permanent* if for any configuration $(u_1^0, \ldots, u_n^0)$ of initial species densities with all $u_i^0 > 0$ (in the sense suitable to the model) there is a $t_0 > 0$ so that for $t \geq t_0$, $u_i(x,t)$ satisfies (4.26) (or (4.27), as appropriate) for $i = 1, \ldots, n$. This phenomenon could be labelled as *deterministic* in the sense that its occurrence only requires that $u_i^0 > 0$ for $i = 1, \ldots, n$; i.e., that each species under

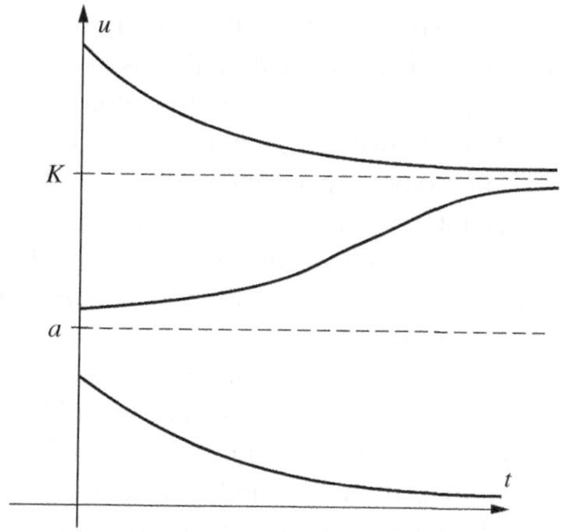

Figure 5.2 Trajectories for the ordinary differential equation $du/dt = u(u - a)(K - u)$ for a population with Allee effect. Trajectories other than the equilibrium $u = a$ either approach $u = K$ or $u = 0$ as t approaches infinity.

consideration is present when the interaction being modeled commences. The model (4.22) will be said to be *conditionally persistent* provided there is a subset H of the collection of n-tuples $(u_1^0, \ldots, u_n^0)$ of initial species densities with $u_i^0 > 0$ for $i = 1, \ldots, n$ so that the interior $H^0 = \text{int}H$ of H is nonempty, and so that if $(u_1^0, \ldots, u_n^0) \in H$, there is a $t_0 > 0$ so that for $t \geq t_0$, $u_i(x, t)$ satisfies (4.26) (or (4.27), as appropriate) for $i = 1, \ldots, n$. In such a situation, if H is a proper subset of the collection of componentwise positive initial densities, requiring that $(u_1^0, \ldots, u_n^0) \in H$ in order to have an asymptotic "floor" and "ceiling" means that the *history* of the interaction is relevant. Such an H exists, for example, provided there is a positive value a (which we call a *threshold value*) so that if for all $i = 1, \ldots, n$, either

$$u_i^0 > a \tag{5.120}$$

or

$$u_i^0 > ae_i(x) \tag{5.121}$$

as appropriate, then there is a $t_0 > 0$ so that if $t \geq t_0$, $u_i(x, t)$ satisfies (4.26) (or (4.27), as appropriate) for $i = 1, 2, \ldots, n$. Clearly, such is the case for (5.117).

Use of the term "conditional persistence" as above is, so far as we are aware, original with us. Consequently, we do not assert it represents definitive nomenclature. Indeed, one might argue that *conditional permanence* is a more accurate term, since the term *persistence* (as introduced in Freedman and Waltman (1984)) does not imply a single asymptotic "floor" independent of initial densities. (See Section 4.1.) However, in terms of vernacular usage of the words, "conditional permanence" is something of an oxymoron, so that we prefer the term *conditional persistence*.

As we discussed in Chapter 4, there is now a substantial body of mathematical literature devoted to the notion of permanence in models for several interacting biological species and to methods for asserting its occurrence. To our knowledge, there does not yet exist a companion development of conditional persistence in models for several interacting biological species. There is a basic reason why. Obviously, the concept *per se* (whatever it is called) is in a sense a relatively new one. Certainly, the notion of a threshold density for population growth (i.e., Allee effect) in ecological models is not new, and similar notions are widely employed in models for nerve conduction in physiology (Fife, 1979) and in genetics (Aronson and Weinberger, 1975). However, viewing conditional persistence as a means of addressing some of the limitations of permanence as a tool for analyzing asymptotic behavior in models for several interacting biological species is a decidedly different and broader "take on things". But from such a point of view, conditional persistence would only arise as a topic for study and consideration after such limitations have been identified and digested, a relatively recent event.

While there is not as yet a systematic treatment of conditional persistence, there are nevertheless several significant observations that should be noted regarding the phenomenon. For definiteness, we confine our remarks to model systems of the form (4.22) where time is a continuous variable. If such a model is dissipative, it will be conditionally persistent if there is a threshold configuration of species densities $(\tilde{u}_1, \ldots, \tilde{u}_n)$ with $\tilde{u}_i > 0$ in a sense appropriate to the model for $i = 1, \ldots, n$, so that for any initial configuration $(u_1^0, \ldots, u_n^0)$ of species densities with

$$u_i^0 \geq \tilde{u}_i \tag{5.122}$$

for $i = 1, 2, \ldots, n$, the corresponding orbit $(u_1(x,t), \ldots, u_n(x,t))$ satisfies

$$u_i(x,t) \geq \tilde{u}_i \tag{5.123}$$

for $t > 0$, $i = 1, \ldots, n$, and $x \in \overline{\Omega}$ (if relevant). In such a situation (5.123) gives an asymptotic "floor" corresponding to the numerically equal threshold in (5.122), a special case of having asymptotic "floors" when initial configurations of species densities exceed (5.120) or (5.121), as appropriate. In a sense, being able to assert the existence of a uniform asymptotic "floor" and "ceiling" so long as initial densities are as in (5.120) or (5.121) is the most desirable way to obtain conditional persistence, assuming ecological permanence is not possible. For, in such a case, the only issue is having sufficient initial abundances of the species in question. *Their relative abundances do not matter in this case.*

Clearly, (5.122)–(5.123) hold with $\tilde{u} = \tilde{u}_1 = a$ in the case of (5.117). Notice that (5.122)–(5.123) is a kind of monotonicity condition, and is more likely in a monotone dynamical or semi-dynamical system than in a nonmonotone system. As an example, consider the simple ODE model for two mutualists given by

$$\begin{aligned}\frac{du_1}{dt} &= u_1(u_1 - a_1)(K_1 - u_1) + \frac{\alpha_1 u_1 u_2}{1 + u_1 + u_2}\\ \frac{du_2}{dt} &= u_2(u_2 - a_2)(K_2 - u_2) + \frac{\alpha_2 u_1 u_2}{1 + u_1 + u_2},\end{aligned} \tag{5.124}$$

where $0 < a_i < K_i$ (as in (5.117) and $\alpha_i > 0$, $i = 1, 2$. The dynamics of each species in the absence of the other are of Allee type, as is clear from a comparison with (5.117), so that each species experiences population decline at low densities in the absence of the other. The symbiotic terms have a Beddington-DeAngelis form. Since for $i, j = 1, 2$, with $i \neq j$,

$$\frac{\partial}{\partial u_j}\left(\frac{\alpha_i u_i u_j}{1 + u_i + u_j}\right) = \frac{\alpha_i u_i (1 + u_i)}{(1 + u_i + u_j)^2} \geq 0$$

for $u_i \geq 0$ and $u_j \geq 0$, (5.124) satisfies a *quasimonotone condition* and hence the dynamical system generated by the solutions to (5.124) corresponding to nonnegative initial species densities is monotone. (See, for example Smith (1995).) Since

$$(u_i - a_i)(K_i - u_i) + \frac{\alpha_i u_j}{1 + u_i + u_j} \leq (u_i - a_i)(K_i - u_i) + \alpha_i \tag{5.125}$$

and the right-hand side of (5.125) is negative for large enough u_i, $i = 1, 2$, the system (5.124) is also dissipative. Since $\dfrac{du_i}{dt} \geq 0$ for $u_i \in [a_i, K_i]$ independent of the density u_j, $j \neq i$, (5.122)–(5.123) holds with $(\tilde{u}_1, \tilde{u}_2) = (a_1, a_2)$. If

$$\alpha_i < a_i K_i \tag{5.126}$$

for $i = 1, 2$, the right-hand side of (5.125) is negative for small values of the density u_i, so that ecological permanence is not a possibility. Consequently, (5.124) is conditionally persistent.

Of course, monotone dynamical and semi-dynamical systems are somewhat special and, as we have noted, in the ecological context are chiefly limited to single-species models, models for mutualism, and models for two-species competition. In the event of conditional persistence in the absence of monotonicity, one should in general anticipate that *the relative abundances of the initial densities of the species in question as well as their absolute abundances* play a role in determining whether the particular initial configuration of species densities evolves over time to configurations of densities which remain between the asymptotic "floor" and asymptotic "ceiling". In fact, in some instances, the collection of initial configurations of species densities which evolve to lie between the asymptotic "floor" and "ceiling" may even be bounded. However, in all cases, *conditional persistence* may be viewed as a *generalization of the notion of a locally attracting equilibrium* among configurations of positive species densities in very much the same way *permanence* is viewed as a *generalization of a globally attracting equilibrium* among configurations of positive species densities. The nature of the long term dynamics which are possible with conditional persistence is really no different from that which is possible in the case of permanence, except that the "basin of attraction" is necessarily a proper subset of the collection of initial configurations of positive species densities. Notice, however, that it is certainly not the case that all models for interacting biological species exhibit conditional persistence. As a famous example, the May-Leonard model for three competing species ((4.17) subject to (4.18)) is not conditionally persistent, since for any componentwise positive initial data other than the equilibrium $\left(\dfrac{1}{1+\alpha+\beta}, \dfrac{1}{1+\alpha+\beta}, \dfrac{1}{1+\alpha+\beta}\right)$, solution trajectories move in an infinitely repeating pattern from being uncomfortably close to extinction for one of the competitors to another to another.

Spatial effects may serve to mediate or even to induce conditional persistence. Suppose, for instance, that the right-hand side of (5.117) represents the local growth rate of a species at every point of some bounded one-dimensional patch that it inhabits. For the sake of specificity, let $(0, \ell)$ denote the location of the patch along some axis. Suppose also that the dispersal of the species in the patch may be described by simple diffusion and that the boundary of patch (i.e., the points 0 and ℓ) is lethal to the species. Then if $\mu > 0$ is the rate of diffusion, the density of the species is described by

$$\begin{aligned} u_t &= \mu u_{xx} + u(u-a)(K-u) \text{ in } && (0,\ell)\times(0,\infty) \\ u &= 0 && \text{on } \{0,\ell\}\times(0,\infty). \end{aligned} \tag{5.127}$$

In this case, the qualitative effect that introducing spatial heterogeneity in this way to (5.117) has on the *conditional persistence* of the model (5.117) is precisely analogous to the qualitative effect that introducing spatial heterogeneity to the model

$$\frac{du}{dt} = u(1-u) \tag{5.128}$$

by considering

$$\begin{aligned} u_t &= \mu u_{xx} + u(1-u) \text{ in } && (0,\ell)\times(0,\infty) \\ u &= 0 && \text{on } \{0,\ell\}\times(0,\infty) \end{aligned} \tag{5.129}$$

has on the *permanence* of (5.128). Namely, in the case of both (5.127) and (5.129), for all patches of length ℓ below a critical threshold, the effect of loss of members of the species through the boundary is strong enough relative to the size of the habitat patch so as to overwhelm local population growth in the patch and lead to extinction of the species in the patch. The threshold value of ℓ, denoted ℓ_*, is usually referred to as a *minimum patch size*. Of course, the value of ℓ_* in (5.127) usually differs from that for (5.129). In (5.129), once ℓ exceeds ℓ_*, the habitat patch is invasible by the species in question and the model predicts permanence. That such is the case follows from Propositions 3.1 and 3.2 once one recognizes that the model (5.129) on the habitat patch $(0,\ell)$ is equivalent to the model

$$\begin{aligned} v_t &= \frac{\mu}{\ell^2} v_{xx} + v(1-v) \text{ in } && (0,1)\times(0,\infty) \\ v &= 0 && \text{on } \{0,1\}\times(0,\infty), \end{aligned} \tag{5.130}$$

where the size of the habitat patch has been normalized to a unit length by adjusting the rate of diffusion from μ to $\frac{\mu}{\ell^2}$. Propositions 3.1 and 3.2 then assert that (5.130) is permanent precisely when $\frac{\ell^2}{\mu} = \left(\frac{1}{\frac{\mu}{\ell^2}}\right) > \pi^2$, the principal eigenvalue of $-v''$ on $(0,1)$ subject to Dirichlet boundary conditions, so that in this instance $\ell_* = \pi\sqrt{\mu}$. (Propositions 3.1 and 3.2 actually assert that (5.129) admits a unique globally attracting positive equilibrium when $\ell > \ell_*$ and that 0 is globally attracting when $0 < \ell \leq \ell_*$. But for our present purposes, the essential feature of the model is a prediction of extinction for $\ell < \ell_*$ and permanence for $\ell > \ell_*$.) In the case of (5.127), the relative local growth rate at $u = 0$ is $-aK < 0$, so that the zero equilibrium solution to (5.127) is at least locally stable no matter how large ℓ becomes, precluding the possibility that the habitat is invasible by the species, and hence the possibility that the model predicts permanence. For $\ell < \ell_*$, the zero equilibrium solution to (5.127) is globally attracting, while for $\ell > \ell_*$, it is only locally stable and the prediction of (5.127) is *conditional persistence*. That such is the case follows from Smoller and Wasserman (1981) or Ouyang and Shi (1998). Specifically, for (5.127) with $\ell > \ell^*$, there are exactly two positive equilibria u_1 and u_2 with $u_1(x) < u_2(x)$ on $(0, \ell)$ and

$$\lim_{t\to\infty} u(x,t) = u_2(x) \tag{5.131}$$

in $C_0^1([0,\ell])$ for all solutions $u(x,t)$ of (5.127) with

$$u(x,0) > u_1(x) \quad \text{on} \quad (0,\ell). \tag{5.132}$$

Consequently, if $\ell > \ell_*$, (5.122) and (5.123) hold for (5.127) with $\tilde{u} = u_1(x)$. (See Figure 5.3.)

The effect of incorporating spatial heterogeneity into the models (5.128) and (5.117) was to mitigate the occurrence of permanence or conditional persistence, respectively, via a dissipative boundary. For habitat patches of size below the minimal patch size, the loss of the species in question through the boundary overwhelms any local population growth and the species tends toward extinction. But once enough of the patch is sufficiently buffered from the dissipative boundary, local population growth leads to permanence or conditional persistence, respectively.

Simple diffusion is clearly far from the only means of incorporating a spatial effect into a population model such as (5.128) or (5.117). A slightly more general formulation might incorporate density dependence into the diffusion leading, in the case of (5.128) to a model such as

$$\begin{aligned} u_t &= (d(u)u_x)_x + u(1-u) \text{ in} \quad && (0,\ell)\times(0,\infty) \\ u &= 0 && \text{on} \quad \{0,\ell\}\times(0,\infty). \end{aligned} \tag{5.133}$$

If $d(u) \equiv d(0) > 0$, then the dispersal mechanism is purely diffusive and (5.133) reduces to (5.129) with $\mu = d(0)$. The dispersive term in (5.133) can be derived as a limit of mechanistic rules for the dispersal of individuals (Turchin, 1989). If

$$d'(0) < 0 \text{ and } d(u) \text{ is increasing for } u \text{ large} \tag{5.134}$$

Turchin (1989) asserts that such a population level description of dispersal is reasonable for insects which use visual cues to congregate near conspecifics at low densities. Such an aggregative process is akin to *area restricted search* (Kareiva and Odell, 1987), but in

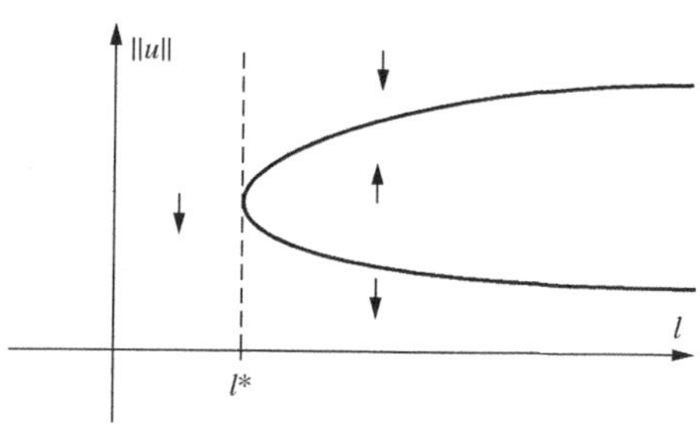

Figure 5.3 This bifurcation diagram shows conditional persistence in a reaction-diffusion model where there is an Allee effect built into the reaction terms. For $l < l^*$ all solutions approach zero asymptotically as t approaches infinity so no population can persist. For $l > l^*$ solutions with large initial data will be bounded below by a positive equilibrium while solutions with small initial data will still decline toward zero, so that there is conditional persistence.

this case, conspecifics rather than prey are cued upon. If $d(u)$ remains positive for all levels of density u, Turchin (1989) classifies the dispersal as *weakly aggregative*. If $d(u)$ is actually negative for some intermediate values of the density, he terms the dispersal *strongly aggregative* and observes that strong aggregation may induce the formation of pattern in homogeneous environments in the absence of population dynamics, but weak aggregation cannot. Mathematically, however, strong aggregation leads to an ill-posed partial differential equation problem. Since we want to be able to recast (5.133) as a semi-dynamical system, we consider (5.133) in the weakly aggregative case.

A detailed analysis of the longterm dynamics for a class of reaction-diffusion models for an aggregative biological species including (5.133) is given in Cantrell and Cosner (2000, 2001a), with further discussion of the ecological ramifications of the analysis in Cantrell and Cosner (2001b). As with (5.129), the patch size $\ell_* = \pi\sqrt{d(0)}$ is a critical threshold for (5.133). For patch sizes ℓ below ℓ_*, the zero equilibrium is (at least) locally stable, meaning the patch is not invasible by the species in question and that (5.133) cannot be permanent. Once ℓ exceeds ℓ_*, the habitat patch is invasible by the species in question and (5.133) is permanent. As with (5.129), enough of the patch is sufficiently buffered from the dissipative boundary once $\ell > \ell_*$ so that local population growth can lead to permanence. However, in the case of (5.129), the minimal positive equilibrium tends to zero as $\ell \to \ell_*$. In (5.133) it does not. This *quantitative* enhancement of asymptotic species abundance in the habitat patch in the case of (5.133) can only be attributed to the tendency to move away from a lethal boundary which results from the aggregative component of movement in (5.133).

A more dramatic, in fact, *qualitative*, change in the predictions of the model (5.133) relative to those of (5.129) occurs when the patch size is only slightly below the threshold ℓ_*. Namely, the model (5.133) predicts that there is a minimal positive equilibrium $\underline{u}_1(x)$ which is unstable and such that any solution trajectory $u(x,t)$ for (5.133) corresponding to an initial population density $u(x,0) \geq \underline{u}_1(x)$ has the property that $u(x,t) \geq \underline{u}_1(x)$ for all $t > 0$. Consequently, $\underline{u}_1(x)$ is a suitable $\tilde{u}_1$ for (5.122) and (5.123), and the model (5.133) predicts *conditional persistence*. Such a prediction holds for all patch sizes ℓ lying between a secondary threshold $\ell_{**}(< \ell_*)$ and ℓ_*. For such ℓ, the zero equilibrium to (5.133) is only locally stable. Moreover, as $\ell \nearrow \ell_*$, the minimal positive equilibrium does tend toward zero, lowering the threshold for a successful invasion toward zero and making conditional persistence less "conditional". However, for patch sizes below ℓ_{**}, the loss of species through the boundary overwhelms both aggregation and local population growth and the zero equilibrium to (5.133) becomes a global attractor, meaning that the model predicts the extinction of the species in question. (Recall the discussion of Allee effects in Section 3.5.2.)

Local population growth of logistic type admits only extinction or permanence when coupled with purely diffusive species movement over a bounded habitat with a dissipative boundary. Consequently, the conditional persistence exhibited in (5.132) must be attributed to the aggregative component of species movement when the habitat is only slightly too small for local population growth in the interior of the patch to compensate for the loss of species through the boundary. Notice that this mechanism for inducing conditional persistence is inherently spatial and moreover, is distinct from simply assuming negative local population growth at low densities (i.e., an Allee effect).

5.7 Extinction Results

As we have noted, when permanence fails to obtain in a model for several interacting biological species, it means that at least one of the model species faces *the risk of extinction from low densities*. It does not mean that one of the species in the community in question is necessarily predicted to go extinct independent of the initial configuration of the community. To make such a prediction requires more information about the model than a finding of "not permanent". In the particular context of models for two-species competition, the additional information needed to predict the extinction of one of the competitors independent of initial densities may be expressed in terms of the signs of principal eigenvalues for appropriate operators, just as were the conditions for persistence that we have described in this and the preceding chapter. Of course, to conclude that one of the competitors goes extinct, we should expect that some of the eigenvalues indicate a decline in the species' density at some configurations of the competitors. Indeed, such is the case, as we shall now demonstrate with the Lotka-Volterra reaction-diffusion model (5.2) for two competitors in an isolated bounded habitat, where the habitat has a boundary which is lethal to both species. The reader should observe, however, that the results we now present for (5.2) not only hold for general two species competition models in the reaction-diffusion context, but also have natural counterparts in other modeling contexts, such as ODE's or matrix models.

Let us relabel (5.2)–(5.3) as

$$\begin{aligned} (u_1)_t &= \Delta u_1 + u_1(a_1 - b_1 u_2 - u_1) \\ & \qquad\qquad\qquad \text{in} \quad \Omega \times (0,\infty) \\ (u_2)_t &= \Delta u_2 + u_2(a_2 - b_2 u_1 - u_2) \\ u_1 &= 0 = u_2 \qquad\qquad \text{on} \quad \partial\Omega \times (0,\infty), \end{aligned} \tag{5.135}$$

where Ω is the isolated bounded habitat in question. We shall presume that both species persist in the absence of the other, which by Theorem 4.4 requires the positivity of the principal eigenvalue σ_i, $i = 1, 2$, in

$$\begin{aligned} \Delta\phi_i + a_i\phi_i &= \sigma_i\phi_i \text{ in } \Omega \\ \phi_i &= 0 \text{ on } \partial\Omega \\ \phi_i &> 0 \text{ in } \Omega. \end{aligned} \tag{5.136}$$

As a consequence, again by Theorem 4.4, recall that for $i = 1, 2$,

$$\begin{aligned} (w_i)_t &= \Delta w_i + w_i(a_i - w_i) \text{ in } \Omega \times (0,\infty) \\ w_i &= 0 \qquad\qquad \text{on } \partial\Omega \times (0,\infty) \end{aligned} \tag{5.137}$$

admits a globally attracting positive equilibrium, which we denote, as in Sections 4.4 and 5.2, as $\tilde{w}_i(x)$. We then let σ_3 and σ_4 be as in (4.73) and (4.74), respectively; i.e., σ_3 is the principal eigenvalue for

$$\begin{aligned} \Delta\phi_3 + (a_2 - b_2\tilde{w}_1)\phi_3 &= \sigma_3\phi_3 \text{ in } \Omega \\ \phi_3 &= 0 \qquad \text{on } \partial\Omega, \end{aligned} \tag{5.138}$$

and σ_4 is the principal eigenvalue for

$$\begin{aligned} \Delta\phi_4 + (a_1 - b_1\tilde{w}_2)\phi_4 &= \sigma_4\phi \text{ in } \Omega \\ \phi_4 &= 0 \qquad \text{on } \partial\Omega. \end{aligned} \tag{5.139}$$

We now have the following result from Cantrell et al. (1993b).

Theorem 5.6. Consider the two species competition model (5.135). Assume for $i = 1, 2$, that $\sigma_i > 0$ in (5.136) and that $\tilde{w}_i$ is the corresponding globally attracting positive equilibrium for (5.137). Then:

(i) If $\sigma_3 > 0$ in (5.138) and there is no componentwise positive equilibrium to (5.135), then any solution to (5.135) corresponding to nonnegative nontrivial initial data (i.e., $u_1(x,0) \stackrel{>}{\not\equiv} 0,\ u_2(x,0) \stackrel{>}{\not\equiv} 0$) converges in $[C_0^1(\Omega)]^2$ to $(0, \tilde{w}_2)$ as $t \to \infty$.

(ii) If $\sigma_4 > 0$ in (5.139) and there is no componentwise positive equilibrium to (5.135), then any solution to (5.135) corresponding to nonnegative nontrivial initial data converges to $(\tilde{w}_1, 0)$ as $t \to \infty$.

Proof: We establish (i) and leave the completely analogous proof of (ii) to the interested reader. Let (u_1, u_2) be a solution to (5.135) with $u_i(x, 0) \stackrel{>}{\not\equiv} 0$ for $i = 1, 2$. Since for $i = 1, 2$,

$$\frac{\partial u_i}{\partial t} \leq \Delta u_i + u_i(a_i - u_i)$$

and $\tilde{w}_i$ is the global attractor for (5.137), the method of upper and lower solutions (Theorem 1.19) guarantees that for any $\varepsilon > 0$ and for $i = 1, 2$,

$$u_i \leq (1+\varepsilon)\tilde{w}_i \tag{5.140}$$

for t sufficiently large, depending only on ε and $u_i(x, 0)$. Since $\sigma_3 > 0$ in (5.138), the continuous dependence of principal eigenvalues upon coefficients and parameters (Theorem 2.1) guarantees that for $\varepsilon > 0$ and sufficiently small the principal eigenvalue $\sigma_{3\varepsilon}$ of

$$\begin{aligned} \Delta\rho + (a_2 - b_2(1+\varepsilon)\tilde{w}_1)\rho &= \sigma_{3\varepsilon}\rho \text{ in } \Omega \\ \rho &= 0 \text{ on } \partial\Omega \\ \rho &> 0 \text{ in } \Omega \end{aligned} \tag{5.141}$$

is positive.

To establish our claim that (u_1, u_2) converges to $(0, \tilde{w}_2)$, we apply Theorem 5.2. To this end we need a lower-upper solution pair and an upper-lower solution pair for the equilibrium system corresponding to (5.135). Let us assume that ρ in (5.141) is normalized by $\max_{x\in\overline{\Omega}} \rho(x) = 1$. Then for any $\delta \in (0, \sigma_{3\varepsilon})$ we have that $(\tilde{u}_1^0, \tilde{u}_2^0) = ((1+\varepsilon)\tilde{w}_1, \delta\rho)$ satisfies (5.7); i.e.,

$$\begin{aligned} \Delta\tilde{u}_1^0 + \tilde{u}_1^0[a_1 - \tilde{u}_1^0 - b_1\tilde{u}_2^0] &\leq 0 \\ \Delta\tilde{u}_2^0 + \tilde{u}_2^0[a_2 - b_2\tilde{u}_1^0 - \tilde{u}_2^0] &\geq 0. \end{aligned}$$

In addition, $(\hat{u}_1^0, \hat{u}_2^0) = (0, (1+\varepsilon)\tilde{w}_2)$ satisfies (5.6), i.e.,

$$\begin{aligned} \Delta\hat{u}_1^0 + \hat{u}_1^0[a_1 - \hat{u}_1^0 - b_1\hat{u}_2^0] &\geq 0 \\ \Delta\hat{u}_2^0 + \hat{u}_2[a_2 - b_2\hat{u}_1^0 - \hat{u}_2^0] &\leq 0. \end{aligned}$$

Now choose t_0 large enough so that (5.140) holds for $i = 1, 2$ for all $t \geq t_0$, and $\delta > 0$ small enough so that $\tilde{u}_2^0 \leq u_2(x, t_0)$. Then we have

$$\begin{aligned} 0 = \hat{u}_1^0 \leq u_1(x, t_0) \leq \tilde{u}_1^0 \\ \tilde{u}_2^0 \leq u_2(x, t_0) \leq \hat{u}_2^0. \end{aligned} \tag{5.142}$$

Theorem 5.1 and (5.142) imply that if (v_1, v_2) is the solution to (5.135) with $(v_1(x, t_0), v_2(x, t_0)) = (\tilde{u}_1^0, \tilde{u}_2^0)$, then

$$\begin{aligned} 0 = \hat{u}_1^0(x) \leq u_1(x, t) \leq v_1(x, t) \\ v_2(x, t) \leq u_2(x, t) \leq \hat{u}_2^0(x) \end{aligned} \tag{5.143}$$

for all $x \in \Omega$ and $t \geq t_0$. Theorem 5.2 guarantees that for all $x \in \Omega$, $v_1(x, t)$ decreases in t for $t \geq t_0$ and $v_2(x, t)$ increases in t for $t \geq t_0$. Moreover, as in Section 5.2, $\lim_{t\to\infty} (v_1(x, t), v_2(x, t)) = (v_1^*, v_2^*)$ exists in $[C_0^1(\overline{\Omega})]^2$ and is an equilibrium solution to (5.135). Since $\tilde{u}_2^0 > 0$ and there is no componentwise positive equilibrium to (5.135), the only possibility is that $v_1^* = 0$ and $v_2^* = \tilde{w}_2$. It follows from (5.143) that $u_1 \to 0$ in $C^0(\overline{\Omega})$ and then from a subsequent regularity argument that $u_1 \to 0$ and $u_2 \to \tilde{w}_2$ in $C_0^1(\overline{\Omega})$.

Theorem 5.6 gives a criterion for one of the model species in (5.134) to go extinct. In (i), the condition $\sigma_3 > 0$ in (5.138) is the condition in Sections 4.4 and 5.2 for the persistence of species 2 in (5.135) in the presence of competition from species 1. Likewise, in (ii), $\sigma_4 > 0$ in (5.139) is the condition for the persistence of species 1 in the presence of competition from species 2. The other condition in (i) and/or (ii), namely that (5.135) have no componentwise positive equilibrium, is simple to understand. However, it is not immediately apparent when there are no componentwise positive equilibria to (5.135). The following result will enable us to establish a sufficient condition in terms of the sign of a principal eigenvalue. (See Cantrell and Cosner (1993, Theorem 2.3).)

Proposition 5.7. Suppose that (u_1, u_2) is a componentwise positive equilibrium to (5.135).

(i) Then, for $i = 1, 2$, $\sigma_i > 0$ in (5.136).

(ii) If $\sigma_3 > 0$ in (5.138) and $\approx{w}_2$ denotes the unique positive globally attracting equilibrium for

$$\begin{aligned} (w)_t &= \Delta w + w(a_2 - b_2\tilde{w}_1 - w) \quad \text{in} \quad \Omega \times (0, \infty) \\ w &= 0 \quad \text{on} \quad \partial\Omega \times (0, \infty), \end{aligned} \tag{5.144}$$

then the principal eigenvalue σ_5 of

$$\begin{aligned} \Delta\phi + (a_1 - b_1 \approx{w}_2)\phi &= \sigma_5\phi \quad \text{in} \quad \Omega \\ \phi &= 0 \quad \text{on} \quad \partial\Omega \end{aligned} \tag{5.145}$$

is positive.

(iii) If $\sigma_4 > 0$ in (5.139) and $\overset{\approx}{w}_1$ denotes the unique positive globally attracting equilibrium for

$$\begin{aligned} (w_t) &= \Delta w + w(a_1 - b_1\tilde{w}_2 - w) \text{ in } && \Omega \times (0,\infty) \\ w &= 0 && \text{on } \partial\Omega \times (0,\infty), \end{aligned} \tag{5.146}$$

then the principal eigenvalue σ_6 of

$$\begin{aligned} \Delta\phi + (a_2 - b_2 \overset{\approx}{w}_1)\phi &= \sigma_6\phi \text{ in } && \Omega \\ \phi &= 0 && \text{on } \partial\Omega \end{aligned} \tag{5.147}$$

is positive.

Proof: If (u_1, u_2) is a componentwise positive equilibrium for (5.135), then for $i = 1, 2$,

$$-\Delta u_i \le u_i(a_i - u_i). \tag{5.148}$$

It follows from (5.148) and the method of upper and lower solutions that (5.137) admits a positive equilibrium solution $\tilde{w}_i$ and that $u_i \le \tilde{w}_i$. Consequently, $\sigma_i > 0$. So (i) holds. We now establish (ii) and leave the proof of (iii) to the interested reader. So suppose that $\sigma_3 > 0$ in (5.138) and that $\overset{\approx}{w}_2$ is the positive globally attracting equilibrium of (5.144). Since $u_1 \le \tilde{w}_1$, we have that

$$-\Delta u_2 \ge u_2(a_2 - b_2\tilde{w}_1 - u_2). \tag{5.149}$$

It follows from (5.149) and the method of upper and lower solutions that $u_2 \ge \overset{\approx}{w}_2$. Consequently,

$$-\Delta u_1 \le u_1(a_1 - b_1 \overset{\approx}{w}_2 - u_1)$$

and the method of upper and lower solutions guarantees that

$$\begin{aligned} -\Delta w &= w(a_1 - b_1 \overset{\approx}{w}_2 - w) \text{ in } && \Omega \\ w &= 0 && \text{on } \partial\Omega \end{aligned} \tag{5.150}$$

admits a positive solution. Theorem 4.4 then guarantees that σ_5 is positive in (5.145), as asserted.

Corollary 5.8. Suppose that for $i = 1, 2$, $\sigma_i > 0$ in (5.136).

(i) If $\sigma_3 > 0$ in (5.138), $\sigma_5 \le 0$ in (5.145) and $(u_1(x,t), u_2(x,t))$ is a solution of (5.135) with $u_i(x, 0) \overset{>}{\not\equiv} 0$ for $i = 1, 2$, then $u_1(x,t) \to 0$ and $u_2(x,t) \to \tilde{w}_2(x)$ in $C_0^1(\overline{\Omega})$ as $t \to \infty$.

(ii) If $\sigma_4 > 0$ in (5.139), $\sigma_6 \le 0$ in (5.147) and $(u_1(x,t), u_2(x,t))$ is a solution of (5.135) with $u_i(x, 0) \overset{>}{\not\equiv} 0$ for $i = 1, 2$, then $u_1(x,t) \to \tilde{w}_1(x)$ and $u_2(x,t) \to 0$ in $C_0^1(\overline{\Omega})$ as $t \to \infty$.

Notice, for example, that $\approx{w}_2$ is the equilibrium solution to the problem (5.144), i.e.,

$$(u_2)_t = \Delta u_2 + u_2(a_2 - b_2\tilde{w}_1 - u_2),$$

which arises when species 1 is held fixed at the carrying capacity $\tilde{w}_1$ it approaches in the absence of species 2. Holding u_1 fixed at $\tilde{w}_1$ asymptotically maximizes the competitive effect of species 1 on species 2. Thus, $\approx{w}_2$ is the lowest possible asymptotic density for species 2. Having $\sigma_5 \leq 0$ in (5.145) means that species 1 fails to be invasive when species 2 is at its lowest possible asymptotic density. Corollary 2.9 asserts that in this case, species 1 will go extinct, independent of the initial densities of species 1 and 2.

Notice also that since $\tilde{w}_2 > \approx{w}_2$, $a_1 - b_1\tilde{w}_2 < a_1 - b_1 \approx{w}_2$, so that by monotonicity σ_4 in (5.139) is less than σ_5 in (5.145). (Likewise, σ_3 in (5.138) is less than σ_6 in (5.147).) Hence if $\sigma_5 < 0$, so is σ_4. Of course, the reverse is not true, since conceivably we may have $\sigma_4 < 0 < \sigma_5$. To this end, let us examine (5.139) and (5.145) more closely. The only parameter among a_1, a_2, b_1 and b_2 that affects the equilibrium $\tilde{w}_i$ in (5.137) is a_i. So $\tilde{w}_2$ depends on a_2 but not a_1, b_1 or b_2. The equilibrium $\approx{w}_2$ in (5.144) depends explicitly on a_2 and b_2 and implicitly on a_1 via $\tilde{w}_1$. However, $\approx{w}_2$ is independent of b_1. So neither $\tilde{w}_2$ nor $\approx{w}_2$ depend on b_1. Consequently, we may track the signs of σ_4 in (5.139) and σ_5 in (5.145) via b_1. Of course, both σ_4 and σ_5 strictly decrease with increasing b_1. When b_1 is small, σ_4 and σ_5 both approach σ_1 and hence are positive. Now think of increasing b_1. We know that $\tilde{w}_1$ satisfies

$$\begin{aligned} &\Delta\phi + (a_1 - \tilde{w}_1)\phi = 0 \cdot \phi \text{ in } \quad \Omega \\ &\phi = 0 \qquad\qquad\qquad\qquad \text{in } \quad \Omega \end{aligned} \tag{5.151}$$

so that 0 is the principal eigenvalue in (5.151). Since $\tilde{w}_1, \tilde{w}_2$ and $\approx{w}_2$ are elements of the interior of the positive cone in $C_0^1(\overline{\Omega})$, the outer normal derivatives $\nabla\tilde{w}_1 \cdot \vec{n}$, $\nabla\tilde{w}_2 \cdot \vec{n}$ and $\nabla \approx{w}_2 \cdot \vec{n}$ are strictly negative on $\partial\Omega$. Consequently, after a sufficient increase in b_1, we have that

$$\tilde{w}_1 < b_1\tilde{w}_2 \quad \text{in} \quad \Omega. \tag{5.152}$$

Comparison of (5.139) and (5.151) shows that if (5.152) holds, then $\sigma_4 < 0$. For still larger values of b_1,

$$\tilde{w}_1 < b_1 \approx{w}_2 \quad \text{in} \quad \Omega, \tag{5.153}$$

and for values of b_1 so that (5.153) holds, $\sigma_5 < 0$ as well. So by increasing b_1, we eventually go from $0 < \sigma_4 < \sigma_5$ to $\sigma_4 < 0 < \sigma_5$ to $\sigma_4 < \sigma_5 < 0$. Consequently, as we increase the competitive impact of species 2 or species 1, we move from a situation where the two species coexist ((5.135) is both permanent and compressive) first to a situation where (5.135) fails to be permanent and we may assert that species 2 can exclude species 1 at low densities and eventually to a situation where we may assert that species 2 excludes species 1 independent of initial densities.

A more general version of a result equivalent to Proposition 5.7 was established in Cantrell and Cosner (1993, Theorem 2.3). In Cantrell and Cosner (1993), the interaction of interspecies competition and environmental heterogeneity is examined via reaction-diffusion models in the context of refuge design, motivated by the observations of Janzen (1983, 1986) in the ecological literature. Specifically, situations in which a pristine refuge is surrounded by partially disrupted secondary successional environments are compared to situations in which the surroundings of the refuge are completely disrupted and hence inhospitable to the species in question. The somewhat surprising conclusion is reached (as is suggested in Janzen's work) that if one of two competing species is better adapted to the partially disrupted environments, in some cases the two competitors may coexist when the buffer zone is completely disrupted, but not when the refuge is surrounded by secondary but not totally inhospitable habitat. Moreover, in the later case, the competitor which is better adapted to the secondary habitat excludes the other species in the refuge itself, independent of initial population densities. We shall consider this model and its implications in detail in the next chapter when we consider edge-mediated effects.

Appendix

Proof that (5.8) holds: By the maximum principle and our construction, it suffices to eliminate the possibility that we have a sequence of componentwise positive equilibria to (5.2)–(5.3), say $\{(u_1^n, u_2^n)\}$ with $(u_1^n, u_2^n) \to (\tilde{w}_1, 0)$ or $(0, \tilde{w}_2)$. Suppose $(u_1^n, u_2^n) \to (\tilde{w}_1, 0)$. We have

$$\Delta u_2^n + u_2^n[a_2 - b_2 u_1^n - u_2^n] = 0$$

in Ω. Let $||u_2^n||$ denote the norm of u_2^n in $C_0^1(\overline{\Omega})$ and $w_2^n = \dfrac{u_2^n}{||u_2^n||}$. Then

$$\Delta w_2^n + w_2^n[a_2 - b_2 u_1^n - u_2^n] = 0$$

in Ω, with

$$w_2^n = 0$$

on $\partial\Omega$. Standard compactness arguments imply that $w_2^n \to w^*$ with $w^* \geq 0$, $||w^*|| = 1$ and

$$\Delta w^* + w^*[a_2 - b_2\tilde{w}_1] = 0$$

in Ω. Now

$$\Delta w_3 + w_3[a_2 - b_2\tilde{w}_1] = \sigma_3 w_3$$

with $\sigma_3 > 0$. Multiplying the top equation by w_3 and the bottom by w^*, integrating over Ω and subtracting the bottom equation from the bottom yields

$$\sigma_3 \int_\Omega w_3 w^* dx = 0.$$

Since $\int_\Omega w_3 w^* dx \neq 0$, $\sigma_3 = 0$, a contradiction. So there can be no such sequence. Similarly for $(0, \tilde{w}_2)$. So (5.8) must be obtained.

Completion of the Proof of Theorem 5.4: We shall need the following auxiliary result.

Lemma 5A.1. Suppose $((u_1, u_2), P_\tau) \in \omega(S)$ and $(u_1, u_2) \not\equiv (0, 0)$. Then either $u_1 \equiv 0$ and $u_2 = u_2^*(\cdot, \tau)$ or $u_1 = u_1^*(\cdot, \tau)$ and $u_2 \equiv 0$.

Proof: We know from the construction of S that either $u_1 \equiv 0$ or $u_2 \equiv 0$. So suppose $u_2 \equiv 0$ and $u_1 \not\equiv 0$. Then there are $u' \in \mathrm{int}C^1_{0+}(\overline{\Omega})$ and $\tau' \in [0, T)$ and a sequence of times $t_n \to \infty$ so that

$$\pi(((u', 0), P_{\tau'}), t_n) \to (u_1, 0, P_\tau)$$

as $n \to \infty$. Since $t_n \to \infty$, there is an N_1 such that $t_n > T$ for $n \geq N_1$. We let $s_n = t_n - (T - \tau')$, and observe that

$$\pi((u', 0), P_{\tau'}), t_n) = \pi(\pi(((u', 0), P_{\tau'}), T - \tau'), s_n).$$

Notice that

$$\begin{aligned}
&\pi(((u', 0), P_{\tau'}), T - \tau') \\
&\quad = ((\phi(u', \tau', T), 0), P_T) \\
&\quad = ((u'', 0), P_0),
\end{aligned}$$

where $u'' \in \mathrm{int}C^1_{0+}(\overline{\Omega})$. Consequently, $s_n \to \infty$ and

$$\pi(((u'', 0), P_0), s_n) \to (u_1, 0, P_\tau)$$

as $n \to \infty$. There is an $N_2 \geq N_1$ so that $s_n > T$ for $n \geq N_2$. For $n \geq N_2$, let k_n be the unique positive integer such that

$$k_n T \leq s_n < (k_n + 1)T.$$

We have

$$\begin{aligned}
\pi(((u'', 0), P_0), s_n) &= \pi(\pi(((u'', 0), P_0), k_n T), s_n - k_n T) \\
&= \pi(((\phi(u'', 0, k_n T), 0), P_0), s_n - k_n T).
\end{aligned}$$

Proposition 3.19 implies that $\phi(u'', 0, k_n T) \to u_1^*(\cdot, 0)$ since $k_n \to \infty$ as $n \to \infty$. But now $0 \leq s_n - k_n T < T$. So there is a subsequence n_i of n so that $s_{n_i} - k_{n_i} T \to r$ for some $r \in [0, T]$.

Continuity of π implies

$$\begin{aligned}
&\lim_{i \to \infty} \pi(((\phi(u'', 0, k_{n_i} T), 0), P_0), s_{n_i} - k_{n_i} T) \\
&\quad = \pi(((u_1^*(\cdot, 0), 0), P_0), r) \\
&\quad = ((u_1^*(\cdot, r), 0), P_r).
\end{aligned}$$

Consequently, $u_1 = u_1^*(\cdot, r)$ and $P_r = P_\tau$. Since $r, \tau \in [0, T]$, we may conclude $r = \tau$.

We may now complete the proof of Theorem 5.4. We must show that we can choose $\beta_1 > 0$ and $\beta_2 > 0$ in (5.101) so that (5.103) and (5.104) hold. Proceeding as in Section 4.5, we have that if $((v_1, v_2), P_\tau) \in X \backslash S$, the quotient $\dfrac{\rho(\pi(((v_1, v_2), P_\tau), t))}{\rho((v_1, v_2), P_\tau)}$ in (5.102) may be expressed as

$$\exp\left\{\sum_{i=1}^{2}\beta_i \int_0^t \left(\frac{\int_\Omega ([(\partial v_i/\partial s)(x, s)]\psi_i(x, \tau + s) + v_i(x, s)((\partial \psi_i/\partial s)(x, \tau + s)))dx}{\int_\Omega v_i(x, s)\psi_i(x, \tau + s)dx}\right) ds\right\} \tag{5A.1}$$

where $\pi(((v_1, v_2), P_\tau), t) = (v_1(x, t), v_2(x, t), P_{\tau+t})$. The definition of π implies that

$$\begin{aligned}\frac{\partial v_1}{\partial s} &= \Delta v_1(x, s) + v_1(x, s)(a_1(x, \tau + s) - v_1(x, s) - b_1(x, \tau + s)v_2(x, s))\\ \frac{\partial v_2}{\partial s} &= \Delta v_2(x, s) + v_2(x, s)(a_2(x, \tau + s) + b_2(x, \tau + s)v_1(x, s) - v_2(x, s)).\end{aligned} \tag{5A.2}$$

From (5A.2), it follows that (5A.1) can be rewritten

$$\begin{aligned}\exp\Bigg\{&\beta_1 \int_0^t \left(\frac{\int_\Omega v_1(x, s)\psi_1(x, \tau + s)[b_1(x, \tau + s)(u_2^*(x, \tau + s) - v_2(x, s)) - v_1(x, s) - \mu_1^*]dx}{\int_\Omega v_1(x, s)\psi_1(x, \tau + s)dx}\right) ds\\ &+ \beta_2 \int_0^t \left(\frac{\int_\Omega v_2(x, s)\psi_2(x, \tau + s)[b_2(x, \tau + s)(v_1(x, s) - u_1^*(x, \tau + s)) - v_2(x, s) - \mu_2^*]dx}{\int_\Omega v_2(x, s)\psi_2(x, \tau + s)dx}\right) ds\Bigg\}.\end{aligned} \tag{5A.3}$$

Since X is compact, the expressions $b_1(x_1, \tau + s)(u_2^*(x, \tau + s) - v_2(x, s)) - v_1(x, s) - \mu_1^*$ and $b_2(x, \tau + s)(v_1(x, s) - u_1^*(x, \tau + s)) - v_2(x, s) - \mu_2^*$ are uniformly bounded below in x, τ and s. As a consequence, (5.103) holds.

We must now establish that (5.104) holds as well. By Lemma 5A.1, $\omega(S) = \{((0, 0), P_{\tau_0}), ((u_1^*(\cdot, \tau_0), 0), P_{\tau_0}), ((0, u_2^*(\cdot, \tau_0)), P_{\tau_0}) : \tau_0 \in [0, T]\}$. Our proof breaks down into cases. Suppose first that $v_1(x, s) \to 0$, $v_2(x, s) \to 0$ and $P_\tau \to P_{\tau_0+s}$. Then

$$\begin{aligned}&b_1(x, \tau + s)(u_2^*(x, \tau + s) - v_2(x, s)) - v_1(x, s) - \mu_1^*\\ &\quad \to b_1(x, \tau_0 + s)u_2^*(x, \tau_0 + s) - \mu_1^* \geq -\mu_1^* \qquad \text{for} \quad s \in [0, T]\end{aligned} \tag{5A.4}$$

and

$$\begin{aligned}&b_2(x, \tau + s)(v_1(x, s) - u_1^*(x, \tau + s)) - v_2(x, s) - \mu_2^*\\ &\quad \to -b_2(x, \tau_0 + s)u_1^*(x, \tau_0 + s) - \mu_2^*\\ &\quad \geq c - \mu_2^* \qquad \text{for} \quad s \in [0, T],\end{aligned} \tag{5A.5}$$

where $c \in (-\infty, \infty)$ may be chosen independent of τ_0. By (5A.4) and (5A.5) it follows from (5A.3) that

$$\alpha(t, ((0,0), P_{\tau_0})) \geq \exp([\beta_1(-\mu_1^*) + \beta_2(c - \mu_2^*)]t).$$

Hence (5.104) holds at $((0,0), P_{\tau_0})$ for any $\tau_0 \in [0, T]$ so long as $\beta_1 > 0$ and $\beta_2 > 0$ are such that

$$\beta_1(-\mu_1^*) + \beta_2(c - \mu_2^*) > 0. \tag{5A.6}$$

Since (5.99) implies $-\mu_1^* > 0$, it is clear that we may choose $\beta_1 > 0$ and $\beta_2 > 0$ so that (5A.6) is satisfied.

Now suppose that $v_1(x,s) \to u_1^*(x, \tau_0 + s)$, $v_2(x,s) \to 0$, and $P_\tau \to P_{\tau_0+s}$. We have from (5A.1) that

$$\begin{aligned}
&\int_\Omega \left[\left(\frac{\partial v_i}{\partial s}\right)(x,s)\right]\psi_i(x,\tau+s) + v_i(x,s)\left[\left(\frac{\partial \psi_i}{\partial s}\right)(x,\tau+s)\right]dx \\
&\quad = \int_\Omega \frac{\partial}{\partial s}[v_i(x,s)\psi_i(x,\tau+s)]dx \\
&\quad = \frac{d}{ds}\int_\Omega v_i(x,s)\psi_i(x,\tau+s)dx
\end{aligned}$$

so that

$$\begin{aligned}
&\int_0^t \left(\frac{\dfrac{d}{ds}\left[\displaystyle\int_\Omega v_i(x,s)\psi_i(x,\tau+s)dx\right]}{\displaystyle\int_\Omega v_i(x,s)\psi_i(x,\tau+s)dx}\right)ds \\
&\quad = \int_0^t \frac{d}{ds}\left[\log\int_\Omega v_i(x,s)\psi_i(x,\tau+s)dx\right]ds \\
&\quad = \log\int_\Omega v_i(x,t)\psi_i(x,\tau+t)dx - \log\int_\Omega v_i(x,0)\psi_i(x,\tau)dx \\
&\quad = \log\left[\frac{\displaystyle\int_\Omega v_i(x,t)\psi_i(x,\tau+t)dx}{\displaystyle\int_\Omega v_i(x,0)\psi_i(x,\tau)dx}\right].
\end{aligned} \tag{5A.7}$$

We have from (5A.1), (5A.3) and (5A.7) that

$$\alpha(t, ((u_1^*(\cdot,\tau_0), 0), P_{\tau_0})) \geq \exp\left\{\beta_1 \log\left(\frac{\displaystyle\int_\Omega u_1^*(x,\tau_0+t)\psi_1(x,\tau_0+t)dx}{\displaystyle\int_\Omega u_1^*(x,\tau_0)\psi_1(x,\tau_0)dx}\right) + \beta_2(-\mu_2^*)t\right\}. \tag{5A.8}$$

By the periodicity of u_1^* and ψ_1 we have that

$$u_1^*(x, \tau_0 + T)\psi_1(x, \tau_0 + T) = u_1^*(x, \tau_0)\psi_1(x, \tau_0).$$

So we have from (5A.8) that

$$\alpha(T, ((u_1^*(\cdot, \tau_0), 0), P_{\tau_0})) \geq \exp \beta_2(-\mu_2^*)T. \tag{5A.9}$$

Since (5.100) implies that $-\mu_2^* > 0$, it follows from (5A.9) that (5.104) holds for $((u_1^*(\cdot, \tau_0), 0), P_{\tau_0})$ for any choice of $\beta_2 > 0$, independent of $\tau_0 \in [0, T]$.

A completely analogous argument can be made when $v_1(x, s) \to 0$, $v_2(x, s) \to u_2^*(x, \tau_0 + s)$ and $P_\tau \to P_{\tau_0+s}$, since $-\mu_1^* > 0$ by (5.99) and u_2^* and ψ_2 are T-periodic in time. Consequently, (5.104) holds for all $(u, v, P_\tau) \in \omega(S)$ and the proof of Theorem 5.4 is complete.

6

Spatial Heterogeneity in Reaction-Diffusion Models for Two Competing Species

6.1 Introduction

The purpose of this chapter is to discuss some of the particular insights into the competition between two biological species simultaneously inhabiting a bounded isolated patch which may be gleaned from spatially heterogeneous reaction-diffusion models. Spatially heterogeneous reaction-diffusion models for interacting biological species in a bounded isolated habitat patch are especially well-suited for capturing biogeographic features of species' interactions and/or movements that explicitly involve the size, shape or heterogeneity (both abiotic and biotic) of the patch or the interface and the interplay between the patch and its surrounding environmental matrix. One reason for this utility is that such a model is spatially explicit (so that the habitat patch itself is given a mathematical coordinate structure, usually as a subset of 1, 2 or 3 dimensional Euclidean space). Another reason is that it also permits the direct incorporation of spatially varying biotic and environmental features of species' interactions and movements into the coefficients of the model equations and boundary conditions. Such encoding makes it natural to examine mathematically how various aspects of the predictions of the model depend on the spatial variation in biotic and environmental parameters, or on the interplay between the habitat patch and the surrounding environmental matrix. The powerful techniques of modern mathematics may then be employed for this purpose. The hope is that the results of the mathematical analysis can be realized in terms that are ecologically meaningful. Frequently, such is the case.

The models under consideration in this chapter can all be expressed in the form

$$\begin{aligned}\frac{\partial u_1}{\partial t} &= \sum_{i,j=1}^{n} a_{ij}^1(x)\frac{\partial^2 u_1}{\partial x_j \partial x_i} + \sum_{i=1}^{n} a_i^1(x)\frac{\partial u_1}{\partial x_i} + u_1 f_1(x, u_1, u_2)\\ \frac{\partial u_2}{\partial t} &= \sum_{i,j=1}^{n} a_{ij}^2(x)\frac{\partial^2 u_2}{\partial x_j \partial x_i} + \sum_{i=1}^{n} a_i^2(x)\frac{\partial u_2}{\partial x_i} + u_2 f_2(x, u_1, u_2)\end{aligned} \tag{6.1}$$

$$\text{in} \quad \Omega \times (0, \infty)$$

Spatial Ecology via Reaction-Diffusion Equations R.S. Cantrell and C. Cosner
 ISBN: 0-471-49301-5

where for $i = 1, 2$,

$$\beta_i(x)u_i + (1 - \beta_i(x))\nabla u_i \cdot \vec{n} = 0 \text{ on } \quad \partial\Omega \times (0, \infty) \tag{6.2}$$

with $0 \leq \beta_i(x) \leq 1$.

Models at this level of generality allow for spatial dependence in both (isotropic and/or anisotropic) diffusion and advection. In most particular examples, only some limited selection of the coefficients is considered to be spatially varying, so as to highlight a specific form of spatial heterogeneity. In (6.1)–(6.2), u_i of course denotes the density of species i, while n is the underlying space dimension of the bounded habitat Ω, which is assumed to have a sufficiently smooth boundary. For the terms $\sum_{i,j=1}^{n} a_{ij}^k(x)\dfrac{\partial^2 u_k}{\partial x_j \partial x_i}$, $k = 1, 2$, which account for spatially dependent isotropic or anisotropic diffusive motion, we require uniform ellipticity, i.e.,

$$\sum_{i,j=1}^{n} a_{ij}^k(x) z_i z_j \geq \gamma \sum_{i=1}^{n} z_i^2 \tag{6.3}$$

for $(z_1, \ldots, z_n)$ where $\gamma > 0$ is independent of k and x in $\overline{\Omega}$. (To this end there is no loss of generality in assuming $a_{ij}^k(x) = a_{ji}^k(x)$. For the purposes of regularity of solutions we assume $a_{ij}^k(x)$ is Hölder continuous in x; i.e., $a_{ij}^k \in C^\alpha(\overline{\Omega})$.) In the terms $\sum_{i=1}^{n} a_i^k(x)\dfrac{\partial u_i}{\partial x_i}$ which account for spatially dependent advection, we only require Hölder continuity (i.e., $a_i^k \in C^\alpha(\overline{\Omega})$) for purposes of regularity of solutions. Since (6.1)–(6.2) is intended to be a model for two competing biological species, we require

$$\frac{\partial f_1}{\partial u_2}(x, u_1, u_2) \leq 0, \quad \frac{\partial f_2}{\partial u_1}(x, u_1, u_2) \leq 0 \tag{6.4}$$

for all $x \in \overline{\Omega}$, $u_1 > 0$, $u_2 > 0$ to account for competition and

$$\frac{\partial f_i}{\partial u_i}(x, u_1, u_2) \leq 0 \tag{6.5}$$

for $i = 1, 2$, $x \in \overline{\Omega}$, $u_1 > 0$, $u_2 > 0$ with

$$f_i(x, u_1, u_2) \leq 0 \tag{6.6}$$

for $u_i \geq K_i$, $x \in \overline{\Omega}$, $u_j \geq 0$, $j \neq i$, to account for self-regulation.

Models of competition have been prominent throughout the development of theoretical ecology, starting with the seminal work of Lotka (1925), Volterra (1928) and Gause (1934), and continuing through to the present (e.g. Lehman and Tilman (1997), Pacala and Levin (1997), among many others). Of course, in any competitive modeling regime, two-species competition plays an important role as a basic building block in the regime. Mathematically, models for two-species competition have an advantage in terms of tractability over models for larger numbers of competing species well beyond the obvious feature of having to consider fewer species. Namely, as we recall from Section 5.2, models for two competing species may be equipped with a natural ordering (cf. (5.1)) in such a way that their asymptotic predictions can essentially be determined via order methods, as in our discussion of compressivity in Section 5.2 and of extinction results in Section 5.7. Of course, the discussions in Section 5.2 and Section 5.7 focused on reaction-diffusion models, but as we noted in those sections, the utility of order methods in analyzing models for two-species competition cuts across modeling frameworks. As an illustration, let us establish the *principle of competitive exclusion* in the context of two consumers competing for one resource. The principle of competitive exclusion is one of the earliest insights of theoretical ecology, going back in its most basic formulation to the work of Lotka (1925) and Volterra (1928). In this work, they demonstrated for mass action ODE models of resource consumption that two or more consumers could not coexist at equilibrium on a single resource. Later work extended the principle to assert that n consumers could not coexist at equilibrium on $k < n$ resources. (See MacArthur and Levins (1964) or Levins (1968) as historical references, and Yodzis (1989), Tilman (1982), Armstrong and McGehee (1980), and Lehman and Tilman (1997), for more recent perspectives.)

To be specific, consider the mass action system of ordinary differential equations

$$\begin{aligned}\frac{du_1}{dt} &= c_1u_1(-d_1 + a_1 v)\\ \frac{du_2}{dt} &= c_2u_2(-d_2 + a_2 v)\\ \frac{dv}{dt} &= rv(1 - \frac{v}{K}) - v(b_1u_1 + b_2u_2)\end{aligned} \tag{6.7}$$

which may be viewed as a model for two consumers (with densities u_1 and u_2) subsisting on a common resource (with density v). (Our formulation follows that in Yodzis (1989, Chapter 5), which in turn follows MacArthur (1972). See Yodzis (1989, p. 119) for a detailed description of the interpretation of the coefficients in (6.7).) Following MacArthur (1972), we assume that resource consumption and resource dynamics occur on a much faster time scale than does consumer population dynamics. As a consequence, we employ a "pseudoequilibrium hypothesis" concerning the resource. Namely, for the purposes of understanding resource dynamics, we may assume that consumer densities are constant relative to the resource and that the resource equilibrates in effect instantly to a level determined by the consumer densities. As a result, the resource density may be expressed in terms of the consumer densities (we say that the resource "tracks" the consumers) and the system (6.7) reduced to the Lotka-Volterra competition system

$$\begin{aligned}\frac{du_1}{dt} &= c_1u_1\left(a_1K - d_1 - a_1\frac{K}{r}b_1u_1 - a_1\frac{K}{r}b_2u_2\right)\\ \frac{du_2}{dt} &= c_2u_2\left(a_2K - d_2 - a_2\frac{K}{r}b_1u_1 - a_2\frac{K}{r}b_2u_2\right).\end{aligned} \tag{6.8}$$

It is immediate from (6.8) that if $a_iK - d_i \le 0$ consumer species i goes extinct independent of any competitive pressure it may feel from species j, $j \neq i$. Following Yodzis (1989, p. 119), the requirement

$$a_iK - d_i > 0 \tag{6.9}$$

can be interpreted as requiring that the overall nutritional intake to consumer species i exceed its starvation rate. Note that (6.9) is equivalent to $a_i/d_i > 1/K$. Generically, the ratios a_1/d_1 and a_2/d_2 are unequal.

Let us assume

$$\frac{a_1}{d_1} > \frac{a_2}{d_2} \tag{6.10}$$

and show that then the model predicts that species 2 goes extinct; i.e., species 1 "competitively excludes" species 2. To this end, we employ the analogue to Theorem 5.6 for ODE models. We see from Section 5.7 that we need only verify two conditions. First, we must show that species 1 can invade the system when species 2 is at its carrying capacity in the absence of species 1, and secondly, we need to show that (6.8) admits no equilibrium with both u_1 and u_2 positive.

It is easy to see that in the absence of species 1, the second equation of (6.8) is logistic and species 2 converges to carrying capacity K_2, with

$$K_2 = \frac{r}{b_2}\left(1 - \frac{d_2}{a_2 K}\right). \tag{6.11}$$

The system (6.8) is invasible by species 1 when $u_2 = K_2$ so long as

$$a_1 K - d_1 - \frac{a_1 K}{r} b_2 K_2 > 0. \tag{6.12}$$

It follows from (6.11) that (6.12) is equivalent to (6.10). Consequently, species 1 may invade the system when $u_2 = K_2$. A componentwise positive equilibrium requires that

$$r\left(1 - \frac{d_1}{a_1 K}\right) = b_1 u_1 + b_2 u_2 = r\left(1 - \frac{d_2}{a_2 K}\right), \tag{6.13}$$

or equivalently, $\frac{a_1}{d_1} = \frac{a_2}{d_2}$, in contradiction to (6.10). Consequently, (6.8) can have no componentwise positive equilibrium if (6.10) holds, and the model predicts that species 2 goes extinct, so that asymptotically, only species 1 remains.

Of course, it has long been thought that a model such as (6.7) must necessarily represent at best a very limited explanation of the outcome of competitive interaction (Hutchinson, 1959, 1961), as there are evidently far more species which appear to coexist in competition with each other than there are limiting resources to support them (e.g. Lehman and Tilman, 1997). This phenomenon is known as "the paradox of the plankton" after the famous 1961 article of Hutchinson (1961), or more generally as "the paradox of diversity" (e.g. Lehman and Tilman, 1997). As Lehman and Tilman (1997) put it, "resolving the paradox became a central issue in theoretical ecology". It is not within the scope of this treatment to attempt anything approaching a complete discussion of the topic, either scientifically or historically. Indeed our initial purpose in discussing the principle of competitive exclusion in the first place has been to reinforce the importance of monotone techniques in analyzing two species competition, regardless of the modeling regime. But we do believe it is worthwhile to examine very briefly some of the limitations inherent in a model such as (6.7)–(6.8) and describe some of the efforts to address these limitations that have been made as part of an on-going effort to offer a fuller theoretical explanation of diversity. Our hope is to point to some of the connections to persistence theory, particularly in the context of spatially heterogeneous reaction-diffusion models for interacting biological species in a bounded isolated habitat. For the reader interested in exploring more fully the efforts to resolve the paradox of diversity, we suggest Armstrong and McGehee (1980), Tilman (1982) and Yodzis (1989) as useful places to begin.

The model (6.7) is limited in several ways. First of all, the competition in the model arises solely from the consumption of a common resource, and resource consumption is quantified in a way so that the "pseudoequilibrium hypothesis" as in MacArthur (1972) is applicable. As a result, v in (6.7) can be expressed in terms of u_1 and u_2, say $v = f(u_1, u_2)$ and (6.7) reduced to the two species competition model (6.8). The facts that resource consumption in (6.7) is modeled by mass action, and that the resource grows logistically in the absence of consumption mean that the pseudoequilibrium $f(u_1, u_2)$ has the form

$$f(u_1, u_2) = \frac{K}{r}(r - b_1 u_1 - b_2 u_2). \tag{6.14}$$

As a result, (6.8) has the property that the *per capita* growth rates $\frac{du_1}{dt}/u_1$ and $\frac{du_2}{dt}/u_2$ in (6.8) are affine linear expressions in the densities u_1 and u_2, so that one gets a Lotka-Volterra model. But, in addition, the ratio of the coefficients of u_1 in $\frac{du_1}{dt}/u_1$ and $\frac{du_2}{dt}/u_2$ is the same as the ratio of coefficients of u_2 in $\frac{du_1}{dt}/u_1$ and $\frac{du_2}{dt}/u_2$, so that the isoclines for the consumers in (6.8) are lines with the same slope. Generically, as when (6.10) holds, these lines are parallel and non-intersecting, so that a component positive equilibrium is ruled out, and competitive exclusion follows from the analogue of Theorem 5.6. Armstrong and McGehee (1980) demonstrate that it is not the mass action form of the growth laws for the consumer species in (6.7), nor the linear dependence of the pseudoequilibrium on u_1 and u_2 in (6.14) *per se* that support the conclusion that one of the consumer species excludes the other. Rather what is essential is having

$$\frac{du_i}{dt}/u_i = g_i(v) \quad \text{with} \quad \frac{dg_i}{dv} > 0 \tag{6.15}$$

and

$$\frac{\partial}{\partial u_i}(f(u_1, u_2)) < 0 \tag{6.16}$$

for $i = 1, 2$. (See Armstrong and McGehee (1980, Appendix D).) The forms of $\frac{du_i}{dt}/u_i$ in (6.7) and $f(u_1, u_2)$ in (6.14) are simply special cases of (6.15) and (6.16).

It follows from the preceding discussion that there are two possible ways to modify a model such as (6.7) so as to have two consumer species coexist on a single resource. The first possibility is to incorporate aspects of intraspecific and/or interspecific competition beyond pure resource consumption in the *per capita* growth rates of the consumers. In the terminology of Yodzis (1989), such would require including some form of interference competition beyond exploitation of a common resource. Basically, this approach is that taken in the famous article of Ayala et al. (1973), which followed the observation of Ayala (1969) that different species of *Drosophila* (i.e. fruit flies) appeared to coexist in an experimental setting on a single resource. In Ayala et al. (1973), the authors tested various ODE competition models to see which best fit the *Drosophila* data. (Inclusion of interference competition in the models tends to produce curved isoclines.) The models themselves were formulated essentially on a phenomenological basis. In other work, Schoener (1973, 1974, 1976, 1978, 1986) constructed ODE competition models on a mechanistic basis and obtained coexistence of two consumer species competing for a single resource. Some of his models are similar to those considered by Ayala et al. (1973). In those models, predictions of coexistence rely on the inclusion of intraspecific interference. Others of his models posit a rather different mechanism for resource consumption than that in MacArthur (1972). Other important contributions to the discussion include those of Gilpin and Justice (1972), Jensen (1987), Sun, Cui and Song (1991) and Belovsky (1984). In a more recent paper, Van Vuuren and Norbury (2000) provide conditions for permanence for some 13 different ODE competition models. Among these are Lotka-Volterra models and many of the models which arose from the ecological discussion of the validity of the principle of competitive exclusion which followed the observations of Ayala (1969) and which are formulated in the papers just referenced. As to the inclusion of Lotka-Volterra models in the list, the reader should note that the Lotka-Volterra competition model (6.8) which arises in the derivation of the principle of competitive exclusion is itself very special within the class of Lotka-Volterra models, and that Lotka-Volterra competition models may certainly be constructed whose prediction is coexistence, as in (4.43). The work of Van Vuuren and Norbury (2000) extends to the case of reaction-diffusion models for several competing species which share an isolated bounded habitat with a reflecting boundary.

The second way to modify (6.7) so as to have a prediction that two consumers can coexist on a single common resource is to alter resource consumption in such a way that the pseudoequilibrium hypothesis no longer holds. In such a case, a reduction from a system of three ODEs to a model for two competing species is not possible, and the long term dynamics of the system become more complicated to analyze. In Armstrong and McGehee (1980), such an example is presented, where the mass action form of consumption is changed to a Holling II form for one of the consumers. Specifically, (6.7) becomes

$$\begin{aligned} \frac{du_1}{dt} &= c_1 u_1 \left(-d_1 + \frac{a_1 v}{v + \Gamma} \right) \\ \frac{du_2}{dt} &= c_2 u_2 (-d_2 + a_2 v) \\ \frac{dv}{dt} &= rv\left(1 - \frac{v}{K}\right) - v\left(\frac{b_1 u_1}{v + \Gamma} + b_2 u_2\right). \end{aligned} \tag{6.17}$$

Notice that generically the system (6.17) *does* fail to have a componentwise positive equilibrium, since such an equilibrium requires that

$$v = \frac{a_2}{d_2} = \frac{d_1 \Gamma}{a_1 - d_1}. \tag{6.18}$$

(If (6.18) held for some choice of the five parameters a, d, Γ, a_2 and d_2, any small perturbation of any one of them causes (6.18) to fail.) Consequently, we cannot expect coexistence at an equilibrium. Moreover, we know from Section 4.6 that if (6.18) does not hold, the system (6.17) cannot be permanent. However, as Armstrong and McGehee (1976a,b, 1977) demonstrate, for appropriate choices of the parameters, the system (6.17) exhibits a compact local attractor in the interior of the positive cone in $\mathbb{R}^3$. Consequently, for a range of componentwise positive initial data, the prediction of the system is coexistence in the sense of asymptotic "floors" and "ceilings". Hence, in the parlance in Section 5.6, the system is conditionally persistent. See Freedman and Waltman (1984) for further discussion.

The model (6.7) is limited in other, more fundamental ways. By virtue of being a system of constant coefficient ODEs, it does not allow for any spatial or temporal variation in the growth rates of the consumers and resource. It also ignores any additional trophic levels which might impinge upon the system. Given the theme of this work, we shall confine our brief comments to issues regarding spatial variation. While it is possible to incorporate spatial variation into ODE models where the dependent variables represent population densities as in (6.7) (e.g. Cantrell, Cosner and Fagan (2002)), generally speaking, if one plans to employ spatial variation as a means of addressing the paradox of diversity, one must move beyond the modeling context of (6.7). There are several ways to proceed. For instance, employing spatially implicit metapopulation like models, Tilman (1994) demonstrated that "any number of [competing] species can stably coexist in an implicitly spatial [but] homogeneous environment, even though the best competitor immediately displaces all others locally" (Lehman and Tilman, 1997). Such results are based upon a supposition that there is a tradeoff between a species' competitive and dispersal abilities (i.e. a strong competitor is a poor disperser and *vice versa*) and a "limit to similarity of these traits" (Lehman and Tilman, 1997) among the competitors. While the underlying metapopulation modeling regime which supports Tilman's conclusions is itself an ODE model with time as independent variable, the interpretation of the dependent variables is radically different from that of (6.7). Specifically, the model envisions an archipelago of possible habitat sites for a collection of competing species and the ith dependent variable represents the portion of possible sites that are occupied by the ith competitor. The spatial location of the habitat sites in relation to one another is not specified. (Recall that Section 1.4 provides further details and references on metapopulation models.)

A general modeling regime which has been used to address the paradox of diversity and which incorporates an explicit representation of the spatial component of the environment is that of interacting particle systems (recall Section 1.5). In such models, space is conceived as discrete patches idealized as points along a grid. Frequently, the grid is envisioned as a two-dimensional lattice with equal horizontal and vertical distances between grid points (i.e. habitat patches). In the current context, each habitat patch is viewed as populated by discrete numbers of individuals of one or more of the competing species. Such models can be very complicated. Consequently, computer simulations have been used extensively as a means of studying them (De Angelis and Gross, 1992). More recently, so-called moment closure methods (e.g. see Pacala and Levin (1997)) have been used to study interacting particle systems analytically. The results of such analyses have led Pacala and Levin (1997) to offer the following scenario as a possible mechanistic explanation to the paradox of diversity. Strong competitive interactions among individuals at small (i.e. micro) spatial scales lead to segregation of species and patchiness in species patterns at these scales. The effect of the spatial segregation is to reduce competition at intermediate (i.e. meso) spatial scales, and this reduction in competition at intermediate spatial scales ultimately leads to coexistence at large (macro) spatial scales (as is frequently observed in nature.)

A third alternative for a modeling regime in which spatial variation may be employed to address the paradox of diversity (and one that should not surprise the reader we might add) is that of reaction-diffusion models for interacting biological species in a bounded habitat. Indeed, recent work of Lopez-Gomez (1997) on "competition with refuges" captures rather explicitly the phenomenon of the coexistence of an arbitrarily large number of competing species. His model envisions M competing species in a bounded habitat Ω in n-dimensional Euclidean space, and takes the explicit form

$$\frac{\partial u_k}{\partial t} = \sum_{i,j=1}^{n} a_{ij}^k(x)\frac{\partial^2 u_k}{\partial x_j \partial x_i} + \sum_{i=1}^{n} a_i^k(x)\frac{\partial u_k}{\partial x_i} - a_0^k(x)u_k$$

$$+ \lambda_k u_k - c_{kk}(x)u_k^2 - \sum_{\substack{j=1 \\ j\neq k}}^{M} s_{kj}c_{kj}(x)u_k u_j \qquad (6.19)$$

$$\text{in} \quad \Omega \times (0, \infty)$$

$$\beta_k(x)u_k + (1 - \beta_k(x))\nabla u_k \cdot \vec{n} = 0$$

$$\text{on} \quad \partial\Omega \times (0, \infty),$$

where for $k = 1, \ldots, M$, u_k denotes the density of the kth competing species. The model (6.19) is an M species analogue to the special case of (6.1)–(6.2) with spatially dependent Lotka-Volterra local interactions given by

$$f^k(x, u_1, \ldots, u_M) = \lambda_k - a_0^k(x) - c_{kk}(x)u_k - \sum_{\substack{j=1 \\ j\neq k}}^{M} s_{kj}c_{kj}(x)u_j. \qquad (6.20)$$

As with (6.1)–(6.2), the coefficients in (6.19) are all of class $C^\alpha(\overline{\Omega})$ and for each species, the diffusive component of dispersal satisfies (6.3). The interpretation of the terms in (6.20) is as expected. Namely, $\lambda_k - a_0^k(x)$ is the *per capita* growth rate for species k, which may be negative in unfavorable portions of the habitat (cf. Cantrell and Cosner (1989), for example), $c_{kk}(x) > 0$ accounts for self-regulation, and $s_{kj}c_{kj}(x) \geq 0$ accounts for the competitive impact of species j upon species k, with $s_{kj} > 0$ as "tunable" scale parameter on the competitive effect.

The novel aspect of the model that captures spatial segregation is Lopez-Gomez's assumption that each species has a "refuge" ($\mathcal{R}_k$ in his notation), i.e., a contiguous (i.e. connected) subhabitat in which it is free from competition from any of the other species. (Note that such means that $c_{kj}(x) = 0$ for all $x \in \mathcal{R}_k$ for all $j \neq k$.) On $\mathcal{R}_k$, the equation for the density of species k reduces to the diffusive-advective logistic form

$$\frac{\partial u_k}{\partial t} = \mathcal{L}_k u_k + \lambda_k u_k - c_{kk}(x)u_k^2, \qquad (6.21)$$

where

$$\mathcal{L}_k u_k = \sum_{i,j=1}^{n} a_{ij}^k(x)\frac{\partial^2 u_k}{\partial x_j \partial x_i} + \sum_{i=1}^{n} a_i^k(x)\frac{\partial u_k}{\partial x_i} - a_0^k(x)u_k. \qquad (6.22)$$

The boundary $\partial\mathcal{R}_k$ of $\mathcal{R}_k$ may or may not have an overlap with the boundary $\partial\Omega$ of Ω. On portions of $\partial\mathcal{R}_k$ that coincide with portions of $\partial\Omega$, the density u_k satisfies the original boundary condition from (6.19). On the remainder of the boundary, u_k satisfies an inhomogeneous spatially and temporally varying Dirichlet boundary condition, and by the method of upper and lower solutions $u_k \geq w_k$, where $w_k > 0$ solves (6.21) subject to

$$\begin{aligned} &\beta_k(x)w_k + (1+\beta_k(x))\nabla w_k \cdot \vec{n} = 0 \text{ on } && (\partial\mathcal{R}_k \cap \partial\Omega) \times (0,\infty) \\ &w_k = 0 && \text{on } \partial\mathcal{R}_k \backslash \partial\Omega \times (0,\infty). \end{aligned} \tag{6.23}$$

Having a homogeneous Dirichlet condition on $\partial\mathcal{R}_k\backslash\partial\Omega$ represents a completely lethal boundary at the interface between the part of Ω where species k faces competition from at least one other species and the refuge $\mathcal{R}_k$. It represents a "worst possible case" scenario that the density for species k necessarily beats on $\mathcal{R}_k$. Lopez-Gomez's result has two parts. First, if for all k the eigenvalue $\sigma_k > 0$ is positive in the eigenvalue problem

$$\begin{aligned} &\mathcal{L}_k\phi + \lambda_k\phi = \sigma_k\phi \text{ in } && \mathcal{R}_k \\ &\phi \quad \text{satisfies (6.23) on } && \partial\mathcal{R}_k \\ &\phi > 0 && \text{in } \mathcal{R}_k, \end{aligned} \tag{6.24}$$

then the original system (6.19) is permanent and all M species are predicted to coexist. In essence, he is saying that if for each species, the refuge is large enough to sustain the species subject to the constraint of a lethal interface between its refuge and the remainder of Ω (where it is subject to competition), then all species are predicted to persist longterm. Notice that the size of s_{kj} (which "tunes" the intensity of the competitive effect of species j or species k) plays no role whatsoever. So competition outside the refuges can be strong. On the other hand, if for some $k \in \{1, \ldots, M\}$, $\sigma_k < 0$ in (6.24), for any initial configuration of the densities, say $(u_1^0, \ldots, u_M^0)$, there will be threshold values of the competitive intensity scale factors s_{kj}, $j \neq k$, say s_{kj}^*, which depend only on $(u_1^0, \ldots, u_M^0)$ so that if $s_{kj} \geq s_{kj^*}$, the u_k component of the solution to (6.19) corresponding to $(u_1^0, \ldots, u_M^0)$ is such that the limit $\lim_{t\to\infty} u_k(x,t) = 0$ uniformly for $x \in \overline{\Omega}$. In other words, if $s_{kj} \geq s_{kj}^*$, species k is driven extinct by the other species.

This example complements very nicely the observations drawn from interacting particle system models by Pacala and Levin (1997). In both cases, coexistence results from a spatially mediated reduction in competition. However, there is a difference in the spatial scales at which competition is mediated. In the interacting particle system models discussed by Pacala and Levin (1997), spatial segregation of species at a microscale reduces competition at a mesoscale. In the work of Lopez-Gomez (1997), spatial mediation in competition is posited at the macroscale in an essentially phenomenological manner. In Section 1.5, we discussed how reaction-diffusion models may be derived mechanistically from interacting particle systems by hydrodynamic limits, based on the work of Durrett and Levin (1994). Our discussion there only treats models for two interacting species. However, since (6.19) only involves pairwise interactions among species, it should be possible to derive it from interacting particle systems via hydrodynamic limits as well, so that (6.19) can be regarded as having a mechanistic basis. However, the patchiness in species patterns at the microscale observed in Pacala and Levin (1997) gets smoothed out and in essence lost when passing to the macroscale via a hydrodynamic limit.

Spatial segregation of competing species at the macroscale is not necessarily the source of the reduction of competition in (6.19). Nevertheless, there are reaction-diffusion models for competing species in an isolated bounded habitat where coexistence is mediated by spatial segregation at a macroscale (e.g., Cantrell, Cosner and Hutson (1996), Cantrell and Cosner (1998), Mimura and Matano (1983)). We discuss these models in detail in the next section.

There is one caveat regarding reaction-diffusion models for competing species in an isolated habitat that we should note. Namely, it is not the mere introduction of explicit spatial coordinates and diffusive dispersal *per se* that serves to mediate coexistence. For instance, consider the Lotka-Volterra system considered by Brown (1980), given by

$$\begin{aligned} \frac{\partial u_1}{\partial t} &= \Delta u_1 + u_1[a - bu_1 - cu_2] \\ \frac{\partial u_2}{\partial t} &= \Delta u_2 + u_2[d - eu_1 - fu_2] \\ &\qquad \text{in} \quad \Omega \times (0, \infty) \end{aligned} \tag{6.25}$$

subject to the boundary condition

$$\nabla u_i \cdot \vec{n} = 0 \quad \text{on} \quad \partial\Omega \times (0, \infty), \tag{6.26}$$

where a, b, c, d, e and f are positive constants. Notice that (6.25)–(6.26) is a special case of (6.1)–(6.2). Brown (1980) showed that if

$$af - cd > 0 \quad \text{and} \quad bd - ae > 0 \tag{6.27}$$

then any solution to (6.25)–(6.26) corresponding to componentwise positive initial data in Ω converges over time to the spatially homogeneous componentwise positive equilibrium

$$\left(\frac{af - cd}{bf - ce}, \frac{bd - ae}{bf - ce} \right), \tag{6.28}$$

uniformly for $x \in \overline{\Omega}$. Under assumption (6.27), the corresponding ODE system

$$\begin{aligned} \frac{du_1}{dt} &= u_1[a - bu_1 - cu_2] \\ \frac{du_2}{dt} &= u_2[d - eu_1 - fu_2] \end{aligned} \tag{6.29}$$

also has the property that any solution corresponding to componentwise positive initial data converges over time to the global equilibrium given in (6.28). (Notice that both results could be made to follow from the results of Section 5.2.) On the other hand, if (6.27) is replaced with

$$af - cd > 0 \quad \text{but} \quad bd - ae < 0 \tag{6.30}$$

then solutions to (6.25)–(6.26) corresponding to componentwise positive initial data converge over time (and uniformly with respect to $x \in \overline{\Omega}$) to the spatially homogeneous equilibrium

$$\left(\frac{a}{b}, 0\right), \tag{6.31}$$

meaning that species 1 competitively excludes species 2. The corresponding result holds for (6.29) (and both cases could be made to follow from Theorem 5.6 or its analogue for ODEs). Consequently, whether the result of competition in (6.25)–(6.26) or (6.29) is coexistence (when (6.27) holds) or competitive exclusion (when (6.30) holds), the introduction of explicit spatial coordinates and diffusive dispersal has essentially no impact on the asymptotic outcome of competition. This phenomenon is sometimes referred to as the "lumped parameter assumption" (Smoller 1982).

The "lumped parameter assumption" does not hold for (6.25) (subject to either (6.27) or (6.30)) if the homogeneous Neumann boundary condition (6.26) is replaced with a homogeneous Dirichlet boundary condition, for example, since in such a case the only spatially homogeneous solution to (6.25) is identically zero. If the conditions on the interaction parameters in (6.25) and (6.29) are altered to

$$af - cd < 0 \quad \text{and} \quad bd - ae < 0 \tag{6.32}$$

so that which competitor dominates asymptotically in (6.29) depends on initial conditions, the geometry of the habitat Ω may also affect whether the "lumped parameter assumption" is valid (Kishimoto, 1981; Kishimoto and Weinberger, 1985; Matano and Mimura, 1983). We shall return to this example in the next section when we discuss the role of spatial segregation in mediating coexistence in two species competition. For further discussion of the "lumped parameter assumption" (including a proof that it is valid for (6.25)–(6.26) subject to (6.32) *regardless of geometry if the diffusion rates in* (6.25) *are large enough*), see Smoller (1982, Section 14.D).

6.2 Spatial Heterogeneity within the Habitat Patch

The formulation in (6.1)–(6.2) allows *all* the coefficients in the model to depend on location within the bounded habitat Ω or its boundary $\partial\Omega$. So doing in some sense enhances the "realism" of the model, but it may also make it difficult to analyze the model so as to isolate and examine specific spatial effects. As we suggested in the Introduction, it may be useful to allow only some limited selection of the coefficients to vary spatially and to view the remaining coefficients as non-spatial parameters. Such a choice for the spatial variation in the coefficients in (6.1)–(6.2) is crudely analogous to selecting control parameters in a laboratory or field experiment.

In this section we focus primarily on the effect that spatial variation within the habitat patch may have on the outcome of a competitive interaction between two species sharing the patch. The examples we consider will incorporate spatial variation in the reaction terms of (6.1) but not in the dispersal terms. More specifically, we focus on

$$\begin{aligned}\frac{\partial u_1}{\partial t} &= \mu_1 \Delta u_1 + u_1(m_1(x) - b_{11}(x)u_1 - b_{12}(x)u_2) \\ & \qquad\qquad\qquad\qquad \text{in} \quad \Omega \times (0, \infty) \\ \frac{\partial u_2}{\partial t} &= \mu_2 \Delta u_2 + u_2(m_2(x) - b_{21}(x)u_1 - b_{22}(x)u_2)\end{aligned} \tag{6.33}$$

subject to

$$\beta_i(x)u_i + (1 - \beta_i(x))\nabla u_i \cdot \vec{n} = 0 \quad \text{on} \quad \partial\Omega \times (0, \infty), \tag{6.34}$$

$i = 1, 2$. In (6.33)–(6.34), μ_i is the diffusion rate of species i and $m_i(x)$ is its *per capita* rate of growth at low densities at location x within the habitat patch Ω in the absence of the other species. The coefficients $b_{ii}(x)$ and $b_{ij}(x)$, $j \neq i$, account for intraspecific and interspecific competition, respectively. We require that $b_{ii}(x) \geq c_i > 0$ for all $x \in \Omega$ and some (positive) constant c_i and that $b_{ij}(x) \geq 0$ for all $x \in \Omega$ when $j \neq i$. However, we allow for the possibility that $m_i(x)$ may be negative for x in some region of the patch, so that the *per capita* growth rate of the species in that region is negative. If $m_i(x) > 0$ at some location x, then at location the *per capita* growth rate in (6.33) can be expressed in the more familiar Lotka-Volterra form

$$r_i(x)\left(1 - \frac{u_i}{K_i(x)} - \frac{\alpha_{ij}(x)u_j}{K_i(x)}\right) \tag{6.35}$$

where $K_i(x)$ is interpreted as carrying capacity. In such a case, $r_i(x) = m_i(x)$, $K_i(x) = \dfrac{m_i(x)}{b_{ii}(x)}$ and $\alpha_{ij}(x) = b_{ij}(x)/b_{ii}(x)$.

The fact that the local *per capita* growth rates in (6.33)–(6.34) are linear in the densities of the competing species (i.e., are of Lotka-Volterra form) is primarily a mathematical convenience that serves to simplify our presentation. Results analogous to those we present in this section would obtain for local *per capita* growth rates $f_i(x, u_1, u_2)$ which satisfy (6.4)–(6.6).

The first result of this section is a sufficient condition for permanence in (6.33)–(6.34) which will facilitate some of our observations regarding spatial effects. This result holds for any boundary condition of the form (6.2). Consequently, we allow (6.34) to coincide with (6.2). In some subsequent results, we need to be more specific with regards to boundary conditions.

Necessary and sufficient conditions for permanence in (6.33)–(6.34) have been given in both Sections 4.4 and 5.2. To have the ith species persist in the absence of its competitor requires having a *positive* principal eigenvalue σ_i of the eigenvalue problem

$$\begin{aligned} &\mu_i \Delta\phi_i + m_i(x)\phi_i = \sigma_i\phi_i && \text{in} \quad \Omega \\ &\beta_i(x)\phi_i + (1 - \beta_i(x)\nabla\phi_i \cdot \vec{n} = 0 \text{ on} && \partial\Omega \\ &\phi_i > 0 && \text{in} \quad \Omega \end{aligned} \tag{6.36}$$

so that the single species model

$$\begin{aligned} &\frac{\partial u_i}{\partial t} = \mu_i \Delta u_i + (m_i(x) - b_{ii}(x)u_i)u_i \text{ in} && \Omega \times (0, \infty) \\ &\beta_i(x)u_i + (1 + \beta_i(x))\nabla u_i \cdot \vec{n} = 0 && \text{on} \quad \partial\Omega \times (0, \infty) \end{aligned} \tag{6.37}$$

admits a globally attracting positive equilibrium solution $\overline{u}_i$. (If $\sigma_i \leq 0$ in (6.36) then 0 is a globally attracting equilibrium for (6.37) and hence permanence is not possible in (6.33)–(6.34).) Given that $\sigma_i > 0$ in (6.36) for $i = 1, 2$, permanence in (6.33)–(6.34) is equivalent to having a positive principal eigenvalue $\tilde{\sigma}_i$, $i = 1, 2$, for the eigenvalue problem

$$\begin{aligned} &\mu_i \Delta \psi_i + (m_i(x) - b_{ij}(x)\overline{u}_j)\psi_i = \tilde{\sigma}_i \psi_i && \text{in} \quad \Omega \\ &\beta_i(x)\psi_i + (1 - \beta_i(x))\nabla \psi_i \cdot \vec{n} = 0 && \text{on} \quad \partial\Omega \\ &\psi_i > 0 && \text{in} \quad \Omega, \end{aligned} \tag{6.38}$$

where $j \neq i$.

Multiplying (6.38) by $\overline{u}_i$ and (6.37) by ψ_i and integrating via Green's Second Identity (e.g. see Gilbarg and Trudinger (1977)) yields that

$$\tilde{\sigma}_i \int_\Omega \psi_i \overline{u}_i dx = \int_\Omega (b_{ii}(x)\overline{u}_i - b_{ij}(x)\overline{u}_j)\overline{u}_i \psi_i dx,$$

so that $\tilde{\sigma}_i$ is given by the ratio of spatial averages

$$\frac{\dfrac{1}{|\Omega|}\displaystyle\int_\Omega (b_{ii}(x)\overline{u}_i - b_{ij}(x)\overline{u}_j)\overline{u}_i \psi_i dx}{\dfrac{1}{|\Omega|}\displaystyle\int_\Omega \overline{u}_i \psi_i dx}. \tag{6.39}$$

Hence, given that $\sigma_i > 0$ in (6.36) for $i = 1, 2$, permanence in (6.33)–(6.34) is equivalent to

$$\int_\Omega (b_{ii}(x)\overline{u}_i - b_{ij}(x)\overline{u}_j)\overline{u}_i \psi_i dx > 0, \tag{6.40}$$

for $i, j = 1, 2$, $i \neq j$, where $\overline{u}_i > 0$ is as in (6.37) and $\psi_i > 0$ is as in (6.38). It is immediate from (6.40) that having

$$b_{ii}(x)\overline{u}_i(x) - b_{ij}(x)\overline{u}_j(x) > 0 \quad \text{on} \quad \Omega \tag{6.41}$$

for $i, j = 1, 2, i \neq j$, is a sufficient condition for permanence in (6.33)–(6.34). It is easy to see that (6.41) is equivalent to the inequality

$$b_{ij}(x) < b_{ii}(x)\frac{\overline{u}_i(x)}{\overline{u}_j(x)} \tag{6.42}$$

for $i, j = 1, 2, i \neq j$ and all $x \in \Omega$. It should be clear that (6.42) holds if throughout the habitat patch the interspecific competition rate b_{ij} measuring the impact of species j upon species i is sufficiently small in comparison to the intraspecific competition or self regulation rate b_{ii} for species i. (Of course, how small depends not only on b_{ii} but also on the ratio $\dfrac{\overline{u}_i}{\overline{u}_j}$, and hence by (6.37) on $\mu_i, \mu_j, m_i(x)$ and $m_j(x)$ as well.) Such a condition is natural, particularly in light of nonspatial theory. However, since from (6.40) the actual condition for permanence in (6.33)–(6.34) is positivity when $(b_{ii}\overline{u}_i - b_{ij}\overline{u}_j)\overline{u}_i\psi_i$ is *averaged* over the entire habitat Ω, it is unlikely that the *pointwise* condition (6.42) can offer further insight into how spatial heterogeneity serves to facilitate the coexistence of two competitors in the habitat patch. It is simply too coarse an approximation. On the other hand, the presence of the factor ψ_i in the integrand in the inequality (6.40) makes it difficult to decide in general when (6.40) holds. (The reason is that ψ_i depends on the (in principle) unknown function $\overline{u}_j$ as well as the coefficients in (6.33). The equilibria $\overline{u}_i$ and $\overline{u}_j$ depend only on μ_i, m_i, b_{ii} and μ_j, m_j, b_{jj}, respectively, and are only one step removed from the (presumably known) coefficients.) With the difficulty presented by having ψ_i in the integrand in (6.40) in mind, Cantrell et al. (1996) derived the following average value condition for positivity in (6.40), the proof of which is given in the Appendix to the chapter.

Theorem 6.1. Assume that for $i = 1, 2,$ σ_i is positive in (6.36) and that $\overline{u}_i$ is the corresponding globally attracting positive equilibrium for (6.37). Then if for $i, j = 1, 2,$ $i \neq j,$

$$\int_\Omega (b_{ii}\overline{u}_i - b_{ij}\overline{u}_j)\overline{u}_i^2 dx > 0, \tag{6.43}$$

(6.40) obtains for $i, j = 1, 2, i \neq j$ and (6.33)–(6.34) is permanent.

6.2.1 How Spatial Segregation May Facilitate Coexistence

We begin with the following result, which is a corollary to Theorem 6.1.

Theorem 6.2. Consider the ecological model (6.33)–(6.34) and assume that the principal eigenvalue σ_i is positive in (6.36) for $i = 1, 2$. Then if

$$\int_\Omega \left((m_i)_+ - \frac{b_{ij}(m_j)_+}{b_{jj}}\right)\left(\frac{(m_i)_+}{b_{ii}}\right)^2 dx > 0 \tag{6.44}$$

for $i, j = 1, 2, i \neq j$, (6.33)–(6.34) is permanent for μ_1 and μ_2 sufficiently small.

Proof: Proposition 3.16 implies that $\lim_{\mu_i \to 0} \overline{u}_i = \dfrac{(m_i)_+}{b_{ii}}$ uniformly on compact subsets of Ω. Such convergence is sufficient to guarantee that

$$\lim_{\substack{\mu_i \to 0 \\ \mu_j \to 0}} \int_\Omega (b_{ii}\overline{u}_i - b_{ij}\overline{u}_j)\overline{u}_i^2 dx$$
$$= \int_\Omega \left((m_i)_+ - \frac{b_{ij}}{b_{jj}}(m_j)_+\right)\left(\frac{(m_i)_+}{b_{ii}}\right)^2 dx$$

which is positive by (6.44). Consequently, (6.43) obtains so long as μ_1 and μ_2 are sufficiently small and (6.33)–(6.34) is permanent by Theorem 6.1.

An immediate consequence of Theorem 6.2 is that the model (6.33)–(6.34) predicts coexistence of two slowly diffusing competing species in the habitat patch Ω so long as there is no overlap between the regions of Ω which are favorable for population growth for each species in the absence of the other. To see that such is the case, let

$$m_i^{-1}((0, \infty)) = \{x \in \Omega : m_i(x) > 0\}, \tag{6.45}$$

for $i = 1, 2$. It is clear from (6.36) that $m_i^{-1}((0, \infty))$ in (6.45) is nonempty. If

$$m_i^{-1}((0, \infty)) \cap m_j^{-1}((0, \infty)) = \emptyset \tag{6.46}$$

for $i, j = 1, 2, i \neq j$, then

$$\int_\Omega \left((m_i)_+ - \frac{b_{ij}}{b_{jj}} (m_j)_+ \right) \left(\frac{(m_i)_+}{b_{ii}} \right)^2 dx = \int_{m_i^{-1}((0,\infty))} \frac{m_i^3}{b_{ii}^2} dx > 0,$$

so that (6.44) holds for $i, j = 1, 2, j \neq j$.

Condition (6.46) is a rather extreme form of spatial segregation at the macroscale level between species 1 and 2, and consequently there is no restriction upon the intensity of the competition between the two species as measured in the coefficients b_{12} and b_{21}. The result does, however, require that species 1 and species 2 diffuse slowly. It is of interest to note that the interacting particle systems results in Pacala and Levin (1997) in which macroscale coexistence of two competitors arises from strong competition and spatial segregation at the local or microscale level also require slow dispersal.

In a sense, we substantially reduce the spatial segregation of the competing species in the bounded habitat Ω indicated by (6.46) if we assume $m_i(x) > 0$ throughout Ω for $i = 1, 2$. Nevertheless, we can still obtain coexistence in (6.33)–(6.34) for slowly diffusing competitors in the face of uniformly strong competition. To this end, assume that

$$b_{ij} \equiv 1 \tag{6.47}$$

in (6.33) for $i, j = 1, 2$. In this case, (6.44) reduces to

$$\int_\Omega (m_1 - m_2) m_1^2 dx > 0 \tag{6.48}$$

and

$$\int_\Omega (m_2 - m_1) m_2^2 dx > 0. \tag{6.49}$$

Conditions (6.48) and (6.49) hold provided that m_2 is small when m_1 is large and *vice versa*. Figure 6.1 illustrates this requirement on m_1 and m_2.

Notice that (6.44) implies that for $i, j = 1, 2, i \neq j$,

$$(m_i)_+ - \frac{b_{ij}}{b_{jj}} (m_j)_+ > 0 \tag{6.50}$$

on an open subset of Ω so long as m_i and b_{ij} are piecewise continuous. However, the converse need not hold, so that (6.50) is a weaker assumption upon the coefficients in

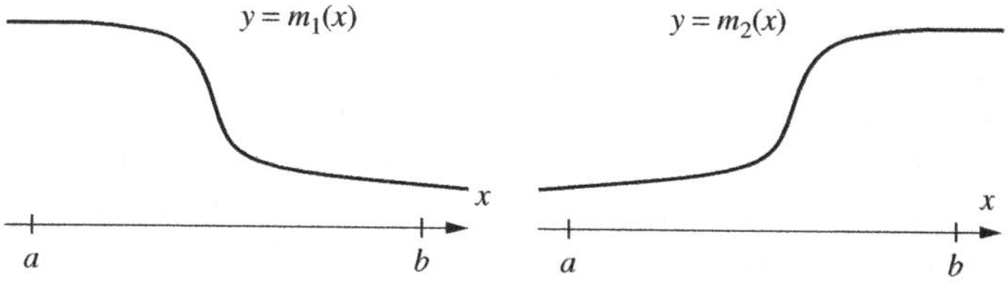

Figure 6.1 The local growth rates in these graphs have a form that could allow two competitors to coexist in the interval [a,b]. See (6.47)–(6.49).

(6.33) than is (6.44). Nevertheless, we have the following result from Cantrell and Cosner (1998), the proof of which is given in the Appendix to the chapter.

Theorem 6.3. Consider (6.33)–(6.34) and assume that for $i, j = 1, 2, i \neq j$, (6.50) holds. Then (6.33)–(6.34) is permanent for μ_1 and μ_2 sufficiently small.

There is given in Cantrell and Cosner (1998) a companion result to Theorem 6.3 which asserts that the outcome of the interaction modeled by (6.33)–(6.34) is competitive exclusion. We next state the result without proof and refer those who seek more mathematical detail to Cantrell and Cosner (1998).

Theorem 6.4. Consider (6.33)–(6.34). Assume that for $i = 1, 2, m_i > 0$ on an open subset of Ω. Suppose that for $\{i, j\} = \{1, 2\}$ or $\{2, 1\}$

$$(m_i)_+ - \frac{b_{ij}}{b_{jj}}(m_j)_+ > 0 \tag{6.51}$$

on an open subset of Ω but that

$$m_j - \frac{b_{ji}}{b_{ii}}\left[m_i - \frac{b_{ij}}{b_{jj}}(m_j)_+\right] < 0 \tag{6.52}$$

on $\overline{\Omega}$. Then if μ_1 and μ_2 are sufficiently small, species i competitively excludes species j.

Notice that in the spatially homogeneous case (where the coefficients are constant) the condition (6.50) is equivalently to (6.27) and is precisely the condition for permanence in (6.29) (or (6.33)–(6.34) with $\beta_1 \equiv 0 \equiv \beta_2$). Consequently, Theorem 6.3 may be interpreted as asserting that if each of the two competing species has some region within the overall habitat Ω in which it can coexist with or exclude the other, then for sufficiently low dispersal rates, the two species can coexist on the habitat Ω. On the other hand, the condition (6.52) for the exclusion of species j by species i when dispersal rates are low must hold everywhere in Ω. Notice that in the case of constant coefficients, (6.52) reduces to

$$m_j - \frac{b_{ji}m_i}{b_{ii}} < -\frac{b_{ji}b_{ij}}{b_{ii}b_{jj}}m_j \tag{6.53}$$

which is stronger than the condition

$$m_j - \frac{b_{ji}m_i}{b_{ii}} < 0 \tag{6.54}$$

for the exclusion of species j by species i in the nonspatial case. Indeed, in the diffusive case, permanence and exclusion are not the only practical possibilities. Even in the spatially uniform case, there is the case of contingent competition, in which both the states $(\overline{u}_1, 0)$ and $(0, \overline{u}_2)$ are locally stable and there is an unstable componentwise positive equilibrium. In the spatially uniform situation almost any small deviation from the coexistence equilibrium results in the extinction of one or the other of the species. So the practical prediction of the model is that coexistence will not be observed. However, in the diffusive case, there may be situations (as observed by Mimura and Matano (1983), for example), wherein the extinction equilibria $(\overline{u}_1, 0)$ and $(0, \overline{u}_2)$ for (6.33)–(6.34) are locally stable but (6.33)–(6.34) admits locally stable componentwise positive equilibria as well. Moreover, this phenomenon can occur even when the coefficients in (6.33) are constants, provided the geometry (i.e.

shape) of the habitat is suitable. We discuss how the shape of the habitat can serve to mediate coexistence of competing species later in this section. We discuss the existence of componentwise positive equilibria to (6.33)–(6.34) in the case wherein the coefficients in (6.33) are constant in Section 4 of this chapter.

There are analogues to Theorem 6.3 and 6.4 for (6.33)–(6.34) when one of the competitors diffuses rapidly and the other diffuses slowly or when both of the competitors diffuse rapidly. Of course, reflecting or Neumann boundary data must be imposed on the boundary $\partial\Omega$ of the habitat Ω if a species which diffuses rapidly throughout Ω is to persist, even in the absence of competition from another species. We next state these results without proof and again refer the interested reader to Cantrell and Cosner (1998) for additional mathematical details. In that which follows, the notation $\overline{m}$ is used to denote $\int_\Omega m dx$ for a function $m : \Omega \to I\!R$.

Theorem 6.5. Consider (6.33)–(6.34). For $i = 1$ or $i = 2$, let $\beta_i \equiv 0$ in (6.34) and assume $\overline{m}_i \geq 0$. Suppose for $j \in \{1, 2\}$ and $j \neq i$,

$$\overline{m}_i - \int_\Omega \frac{b_{ij}(m_j)_+}{b_{jj}} dx > 0 \tag{6.55}$$

and

$$(m_j)_+ - \frac{b_{ji}\overline{m}_i}{\overline{b}_{ii}} > 0 \tag{6.56}$$

on an open subset of Ω. Then for μ_i sufficiently large and μ_j sufficiently small, (6.33)–(6.34) is permanent.

Theorem 6.6. Consider (6.33)–(6.34). For $i = 1$ or $i = 2$, let $\beta_i \equiv 0$ in (6.34) and assume $m_i \geq 0$.

(i) Suppose for $j \in \{1, 2\}$ and $j \neq i$, (6.55) holds but (6.56) is replaced by

$$m_j - \frac{b_{ji}\left[\overline{m}_i - \int_\Omega \frac{b_{ij}(m_j)_+}{b_{jj}} dx\right]}{\overline{b}_{ii}} < 0 \tag{6.57}$$

on $\overline{\Omega}$. Then species i excludes species j if μ_i is sufficiently large and μ_j is sufficiently small.

(ii) Suppose for $j \in \{1, 2\}$ and $j \neq i$, (6.56) holds but (6.55) is replaced by

$$\overline{m}_i - \int_\Omega \left[\frac{b_{ij}\left(m_j - \frac{b_{ji}\overline{m}_i}{\overline{b}_{jj}}\right)_+}{b_{jj}}\right] dx < 0. \tag{6.58}$$

Then species j excludes species i if μ_i is sufficiently large and μ_j is sufficiently small.

Theorem 6.7. Consider (6.33)–(6.34). For $i = 1, 2$, let $\beta_i \equiv 0$ in (6.34) and assume $\overline{m}_i \geq 0$. Then if

$$\overline{m}_i - \frac{\overline{b}_{ij}\overline{m}_j}{\overline{b}_{jj}} > 0 \tag{6.59}$$

for $i, j = 1, 2, i \neq j$, (6.33)–(6.34) is permanent if μ_1 and μ_2 are sufficiently large.

Theorem 6.8. Consider (6.33)–(6.34). For $i = 1, 2$, let $\beta_i \equiv 0$ in (6.34) and assume $\overline{m}_i \geq 0$. Fix $i, j \in \{1, 2\}$ with $j \neq i$. Then if (6.59) holds but

$$\overline{m}_j - \overline{b}_{ji} \frac{\left(\overline{m}_i - \dfrac{\overline{b}_{ji}\overline{m}_j}{\overline{b}_{jj}}\right)}{\overline{b}_{ii}} < 0, \tag{6.60}$$

species i excludes species j if μ_1 and μ_2 are sufficiently large.

As in the case of slowly dispersing competitors, the conditions for invasibility in Theorem 6.5 (i.e., (6.55) and (6.56)) and Theorem 6.7 (i.e., (6.59)) reduce to the standard conditions for invasibility in a nonspatial Lotka-Volterra model when the coefficients are constant. Another similarity to the case of low diffusion rates is that the conditions for each competitor to persist may fail to hold locally on some regions of the habitat Ω but the populations may still persist. Take the situation when one of the competitors diffuses slowly while the other diffuses rapidly. In the case of the slow disperser the essential point of its invasibility is that there must be some region where that competitor has an advantage relative to a certain average of the strength of the rapid disperser. The mechanism is that the slow disperser tends to remain in the region where its population has an advantage and is thus able to recruit effectively. The rapid disperser can also persist with only a local advantage if the advantage is sufficiently great, but the mechanism is different. Notice that (6.55) is an averaged condition, so that if $m_i - \left(\dfrac{b_{ij}(m_j)_+}{b_{jj}}\right)$ is sufficiently large in some places the condition may hold even if $m_i - \left(\dfrac{b_{ij}(m_j)_+}{b_{jj}}\right) < 0$ elsewhere. Such a scenario might occur if the fast dispersers had a sufficiently great advantage in some region and visited that region often enough (because of their rapid dispersal) so that the local advantage was still adequate for persistence after averaging over the environment.

Some of the averages which occur when at least one of the competitors diffuses rapidly are not simple averages over Ω of biologically interpretable quantities such as carrying capacities or competition coefficients, but instead are averages of combinations or functions of these quantities. This fact can lead to some rather striking disparities between the local competitive interaction of the species and the outcome of the competition on the habitat Ω as a whole, as we explore in the next subsection.

6.2.2 Some Disparities Between Local and Global Competition

Numerical experiments in Pacala and Roughgarden (1982) support counterintuitive results regarding coexistence of two competing species in a bounded habitat. The system considered in Pacala and Roughgarden (1982) is a special case of (6.33)–(6.34) in one space dimension when both species are subject to reflecting or Neumann boundary conditions and the growth rates for both species are positive, so that the reaction term in (6.33) can be expressed via (6.35). We write the system for general habitat patches Ω as

$$\begin{aligned} \frac{\partial u_i}{\partial t} &= \mu_i \Delta u_i + r_i \left[1 - \frac{u_1}{K_i(x)} - \frac{\alpha_{ij} u_j}{K_i(x)}\right] u_i \text{ in } \quad \Omega \times (0, \infty) \\ \nabla u_i \cdot \vec{n} &= 0 \qquad\qquad \text{on} \quad \partial\Omega \times (0, \infty) \end{aligned} \tag{6.61}$$

for $i = 1, 2$. As in Pacala and Roughgarden (1982), we take the growth r_i and competition coefficient α_{ij} to be positive constants but allow spatial dependence in carrying capacity $K_i(x)$. Notice that in (6.61) the invasibility conditions (6.55), (6.56) and (6.59) would all reduce to

$$\alpha_{ij} < \frac{K_i}{K_j} \tag{6.62}$$

in the spatially homogeneous case. Our first result regarding (6.61) is:

Theorem 6.9. Consider (6.61). When μ_i is relatively large and μ_j is relatively small, $\tilde{\sigma}_j$ may be positive (and hence species j may invade when species i is at its equilibrium in the absence of competition) even though

$$\alpha_{ji} \geq \frac{K_j(x)}{K_i(x)} \tag{6.63}$$

for all $x \in \Omega$.

Proof: One need only establish that (6.56) holds; i.e., $m_j - b_{ji}\dfrac{\overline{m}_i}{\overline{b}_{ii}} > 0$ on an open subset of Ω. Using $m_j = r_j$, $b_{ii} = \dfrac{r_i}{K_i}$, $b_{ji} = \alpha_{ji} b_{jj}$, it is easy to obtain that (6.56) reduces to

$$\alpha_{ji} < \frac{K_j(x)}{\left(1/\left\{\frac{1}{|\Omega|}\int_\Omega \frac{1}{K_i} dx\right\}\right)} \tag{6.64}$$

on an open subset of Ω.

Following the lead of Pacala and Roughgarden (1982), we show that (6.64) can be obtained for $K_i(x)$, $i = 1, 2$ of the form

$$K_i(x) = \begin{cases} K_{i1} \text{ on } & \Omega_1 \\ K_{i2} \text{ on } & \Omega_2 = \Omega \backslash \Omega_1, \end{cases} \tag{6.65}$$

where Ω_1 is a proper subset of Ω. Strictly speaking, the discontinuity of K_i in (6.65) may create difficulties applying the permanence results as formulated in Chapter 4. However, since (6.61) is a model for two competing species, the compressivity results of Section 5.2 are also available to us, and there is no difficulty in using them. When K_1 and K_2 are as in (6.65) and $i = 1,\ j = 2$, (6.64) reduces to

$$\alpha_{21} < K_{21}\left\{\frac{K_{12}|\Omega_1| + K_{11}|\Omega_2|}{K_{11}K_{12}|\Omega|}\right\} \quad \text{on} \quad \Omega_1$$

or (6.66)

$$\alpha_{21} < K_{22}\left\{\frac{K_{12}|\Omega_1| + K_{11}|\Omega_2|}{K_{11}K_{12}|\Omega|}\right\} \quad \text{on} \quad \Omega_2.$$

Suppose now that K_1 and K_2 satisfy

$$\alpha_{21}K_{11} > K_{21} > \alpha_{21}K_{12} > K_{22}. \tag{6.67}$$

It is clear from (6.67) that

$$\alpha_{21} > \frac{K_2(x)}{K_1(x)} \tag{6.68}$$

for all $x \in \Omega$. Observe that

$$\lim_{|\Omega_1|\to 0} K_{21}\left\{\frac{K_{12}|\Omega_1| + K_{11}|\Omega_2|}{K_{11}K_{12}|\Omega|}\right\} = \frac{K_{21}}{K_{12}} \tag{6.69}$$

and that (6.67) implies

$$\frac{K_{21}}{K_{12}} > \alpha_{21}. \tag{6.70}$$

It now follows from (6.69) and (6.70) that (6.66) holds on Ω_1 provided that $|\Omega_1|$ is relatively small. Since (6.68) holds, the result is established.

The quantity $\left(1/\left\{\frac{1}{|\Omega|}\int_\Omega \frac{1}{K_i}dx\right\}\right)$ in (6.64) which arises from the ratio $\frac{\overline{m}_i}{\overline{b}_{ii}}$ of integral averages is the harmonic mean of K_i on Ω, as opposed to its arithmetic mean $\frac{1}{|\Omega|}\int_\Omega K_i(x)dx$. It is well known that the harmonic mean on Ω is always less than or equal to the arithmetic mean on Ω. Our second result concerning (6.61) is:

Theorem 6.10. Consider (6.61). When μ_j is large, it may be impossible for species j to invade species i (i.e. $\tilde{\sigma}_j < 0$ in (6.38)) even though

$$\alpha_{ji} < \frac{K_j(x)}{K_i(x)} \tag{6.71}$$

for all $x \in \Omega$.

Proof: Since r_i and r_j are positive, for any $\mu_i > 0$ and $\mu_j > 0$, $\overline{u}_i$ and $\overline{u}_j$ in (6.37) exist. Fix $\mu_i > 0$ and $\overline{u}_i$. Consider the eigenvalue problem

$$\mu_j\Delta\phi_j + r_j\left[1 - \frac{\alpha_{ji}\overline{u}_i}{K_j(x)}\right]\phi_j = \tilde{\sigma}_j\phi_j \quad \text{in} \quad \Omega$$

on $\partial\Omega$ (6.72)

0 in Ω.

(Note that (6.72) is simply (6.38) for the special case (6.61) of (6.33)–(6.34). If $1-\dfrac{\alpha_{ji}\overline{u}_i}{K_j(x)} \lneqq 0$ on Ω, $\tilde{\sigma}_j < 0$. So we may assume $1 - \dfrac{\alpha_{ji}\overline{u}_i}{K_j(x)} > 0$ on an open subset of Ω.) We know that if $\int_\Omega \left[1 - \dfrac{\alpha_{ji}}{K_j(x)}\overline{u}_i\right] dx < 0$, then $\tilde{\sigma}_j < 0$ for μ_j satisfying

$$\mu_j > \frac{1}{\lambda_+^1\left(r_j\left[1 - \dfrac{\alpha_{ji}\overline{u}_i}{K_j(x)}\right]\right)}.$$

Since $\mu_i \Delta\overline{u}_i + r_i(1 - \dfrac{\overline{u}_i}{K_i})\overline{u}_i = 0$,

$$\int_\Omega \left(1 - \frac{\overline{u}_i}{K_i}\right) dx = -\frac{\mu_i}{r_i}\int_\Omega \frac{\Delta\overline{u}_i}{\overline{u}_i} dx. \tag{6.73}$$

(Since $\nabla\overline{u}_i \cdot \vec{n} = 0$ on $\partial\Omega$, $\overline{u}_i > 0$ on $\overline{\Omega}$.) It is simple to calculate that

$$\frac{\Delta\overline{u}_i}{\overline{u}_i} = \nabla \cdot \frac{\nabla\overline{u}_i}{\overline{u}_i} + \frac{|\nabla\overline{u}_i|^2}{\overline{u}_i^2}.$$

As a consequence, we obtain from (6.73) that

$$\begin{aligned}\int_\Omega \left(1 - \frac{\overline{u}_i}{K_i}\right) dx &= -\frac{\mu_i}{r_i}\int_\Omega \left[\nabla \cdot \left(\frac{\nabla\overline{u}_i}{\overline{u}_i}\right) + \frac{|\nabla\overline{u}_i|^2}{\overline{u}_i^2}\right] dx \\ &= -\frac{\mu_i}{r_i}\int_{\partial\Omega} \frac{\nabla\overline{u}_i}{\overline{u}_i} \cdot \vec{n}\, dS - \frac{\mu_i}{r_i}\int_\Omega \frac{|\nabla\overline{u}_i|^2}{\overline{u}_i^2} dx \\ &= -\frac{\mu_i}{r_i}\int_\Omega \frac{|\nabla\overline{u}_i|^2}{\overline{u}_i^2} dx < 0\end{aligned} \tag{6.74}$$

so long as $\overline{u}_i$ is not constant. (Such will be the case if K_i is not constant on Ω.)

Notice that for $\gamma \in (0, 1)$, $\int_\Omega \left(1 - \dfrac{\gamma\overline{u}_i}{K_i(x)}\right) dx = \int_\Omega \left[(1-\gamma) + \gamma - \dfrac{\gamma\overline{u}_i}{K_i(x)}\right] dx = (1-\gamma)|\Omega| + \gamma\int_\Omega \left(1 - \dfrac{\overline{u}_i}{K_i(x)}\right) dx$, so that $\lim_{\gamma\to 1}\int_\Omega \left(1 - \dfrac{\gamma\overline{u}_i}{K_i(x)}\right) dx = \int_\Omega \left(1 - \dfrac{\overline{u}_i}{K_i(x)}\right) dx < 0$ by (6.74). So we may choose $\gamma \in (0, 1)$ sufficiently close to 1 so that

$$\int_\Omega \left(1 - \frac{\gamma\overline{u}_i}{K_i(x)}\right) dx < 0. \tag{6.75}$$

Let us now follow the lead of Pacala and Roughgarden (1982), and show that we may establish the result when

$$K_i(x) = \begin{cases} K_{i1} \text{ on } & \Omega_1 \\ K_{i2} \text{ in } & \Omega_2 = \Omega\backslash\Omega_1, \end{cases}$$

where $K_{i1} \neq K_{i2}$. For specificity, take $i = 1$ and $j = 2$. Pick any such K_1 and any $\alpha_{21} > 0$. Choose γ sufficiently close to 1 so that (6.75) holds. Since $\gamma < 1$, the intervals

$$\left(\alpha_{21}K_{11}, \left(\frac{1}{\gamma}\right)\alpha_{21}K_{11}\right)$$

and

$$\left(\alpha_{21}K_{12}, \left(\frac{1}{\gamma}\right)\alpha_{21}K_{12}\right)$$

are nonempty. Choose $K_{21} \in \left(\alpha_{21}K_{11}, \left(\frac{1}{\gamma}\right)\alpha_{21}K_{11}\right)$ and $K_{22} \in \left(\alpha_{21}K_{12}, \left(\frac{1}{\gamma}\right)\alpha_{21}K_{12}\right)$. Then

$$\alpha_{21} < \frac{K_{21}}{K_{11}} < \frac{1}{\gamma}\alpha_{21}$$

and

$$\alpha_{21} < \frac{K_{22}}{K_{12}} < \frac{1}{\gamma}\alpha_{21}$$

so that

$$\gamma\frac{K_2(x)}{K_1(x)} < \alpha_{21} < \frac{K_2(x)}{K_1(x)} \tag{6.76}$$

on Ω. It is immediate from (6.76) that $\dfrac{\gamma}{K_1(x)} < \dfrac{\alpha_{21}}{K_2(x)}$ so that

$$\int_\Omega \left(1 - \frac{\alpha_{21}\bar{u}_1}{K_2(x)}\right) dx < \int_\Omega \left(1 - \frac{\gamma\bar{u}_1}{K_1(x)}\right) dx < 0 \tag{6.77}$$

by (6.75).

6.2.3 Coexistence Mediated by the Shape of the Habitat Patch

The question of "spatial pattern formation" has been a prominent topic of inquiry in mathematical biology at least since the ground breaking work of Turing (1952) on diffusion-driven instabilities. (See Murray (1993, Chapter 14) for a detailed discussion.) The term "spatial pattern formation" usually refers to a steady-state, either an equilibrium or a periodic orbit, for the biological model in question which is spatially inhomogeneous and in some sense stable. For the model (6.25)–(6.26) for two-species competition in a bounded habitat (which is the special case of (6.33)–(6.34) with constant coefficients in (6.33) and $\beta_i \equiv 0$ in (6.34)), we observed in the Introduction that if either the conditions (6.27) under which the system is permanent or the conditions (6.30) under which the system exhibits competitive exclusion are imposed, the outcome of the interaction is convergence to a spatially homogeneous equilibrium. If we impose the conditions (6.32) under which the spatially homogeneous extinction states $(a/b, 0)$ and $(0, d/f)$ are locally asymptotically stable while the coexistence state $\left(\dfrac{af - cd}{bf - ce}, \dfrac{bd - ae}{bf - ce}\right)$ is unstable, then the work of Kishimoto (1981) and Kishimoto and Weinberger (1985) shows that any spatially inhomogeneous equilibrium to (6.25)–(6.26) is necessarily unstable *so long as the bounded habitat is convex*. When this result is combined with the observation of Hirsch (1982, 1985, 1988) that any periodic orbit to (6.25)–(6.26) is necessarily unstable (independent of the geometry of the bounded habitat), we may conclude in essence that "spatial pattern formation" does not occur in (6.25)–(6.26) so long as the bounded habitat is convex.

In essence, convexity of the spatial habitat means that any point in the habitat is connected to any other point in the habitat by a line segment entirely contained in the habitat. In a very real way, all pairs of points in the habitat are then equally accessible to each other. In such a case, if the ecological interaction is homogeneous (that is, doesn't depend on spatial locale), simply explicitly accounting for spatial structure does not serve to mediate the outcome of two-species competition. It is interesting to note that if (6.25)–(6.26) is considered on a nonconvex domain where not all pairs of points are equally accessible and a change of spatial variable is made so that the underlying habitat is convex, spatial heterogeneity arises in the coefficients of the transformed version of equation (6.25). A similar effect has been captured in metapopulation models by Hanski (1997) by having distance between patches reflect barriers or obstacles between them along with physical distance.

There is another important observation of Hirsch (1982, 1985, 1988) that applies to (6.25)–(6.26) when the conditions (6.32) are imposed that allows one to sharpen the preceding conclusion in ecological terms. Namely, under the conditions (6.32) "almost" all solutions to (6.25)–(6.26) corresponding to componentwise positive initial data converge to the set of equilibria for (6.25)–(6.26). The instability of any spatially nonconstant equilibria to (6.25)–(6.26) as in Kishimoto and Weinberger (1985) means that the only outcomes to the competitive interaction that can be expected are convergence to the equilibrium $(a/b, 0)$ and convergence to the equilibrium $(0, d/f)$. Consequently, whether conditions (6.27) for permanence, (6.30) for competitive exclusion, or (6.32) are imposed, the outcome of the competitive interaction is essentially the same as in the corresponding ODE model (6.29). As a result, the explicit introduction of a spatial component into the model (6.29) by means of (6.25)–(6.26) fails to alter the outcome of the interaction so long as the bounded spatial habitat in question is convex and homogeneous.

Matano and Mimura (1983) were the first to establish that "spatial pattern formation" is, in fact, possible in the model (6.25)–(6.26) when the conditions (6.32) are imposed. It should be clear from the preceding discussion that they required the competition to occur in a nonconvex bounded habitat. Indeed, the habitat in Matano and Mimura (1983) is a highly non-convex "dumbbell" shaped region consisting of two large subhabitats joined by a narrow corridor (see Figure 6.2). When the phenomenon of "spatial pattern formation" occurs, each of the competing species has a relatively high population density in one of the subhabitats and a low population density in the other, and the species coexist by virtue of the fact that the two subhabitats are largely isolated from one another. (The subhabitat in which the density of species 1 is high is the one in which the density of species 2 is low, and *vice versa*.) In other words, the shape of the habitat serves to mediate coexistence via spatial segregation. Since we impose the conditions (6.32), the system (6.25)–(6.26) is not permanent. Indeed, we have another example of *conditional persistence* as in Section 5.6, and for a certain rage of initial spatial configurations in the densities of the two species, the outcome of the interaction is markedly different from what would be observed in the corresponding ODE model.

The result of Matano and Mimura (1983) in the context of (6.25)–(6.26) can be expressed as follows:

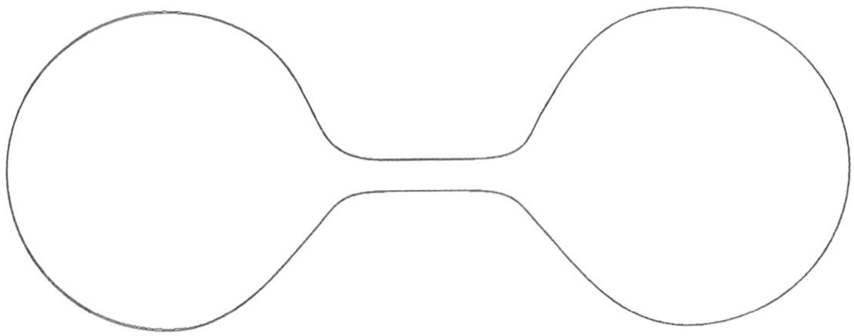

Figure 6.2 This type of nonconvex domain can support stable nonconstant equilibria of reaction-diffusion models with bistable reaction terms under no-flux boundary conditions as in Theorem 6.11.

Theorem 6.11. Consider (6.25)–(6.26). If we impose the conditions (6.32), then there is a choice of the bounded habitat Ω so that (6.25)–(6.26) admits a componentwise positive equilibrium which is spatially inhomogeneous and locally asymptotically stable.

The proof of Theorem 6.11 in Matano and Mimura (1983) is very much along the lines of the proof of compressivity for (5.2)–(5.3) in Section 5.2. Both proofs rely on the same basic fact. Namely, if the initial configuration of species densities for a competitive system of the form (6.33)–(6.34) is a sub-(respectively, super-) solution for the corresponding elliptic system, then the positive orbit of species densities increases (respectively, decreases) with respect to the skew ordering and converges to an equilibrium solution to (6.33)–(6.34). Consequently, if one has a componentwise positive subsolution to the elliptic system below a componentwise positive supersolution to the elliptic system in the skew ordering, the system (6.33)–(6.34) has an attracting order interval whose extremes are componentwise positive equilibria. Of course, in the situation of Theorem 6.11, the basin of attraction for the order interval cannot be all componentwise positive configurations of species densities, for then the system would be permanent, which it is not if we impose the conditions (6.32). Moreover, the construction of the requisite sub and supersolutions in Matano and Mimura (1983) is far more complicated than that for (5.2)–(5.3) in Section 5.2, and relies on making the corridor joining the two aforementioned subhabitats (see Figure 6.2) sufficiently narrow. We refer the interested reader to Matano and Mimura (1983) for the mathematical details of the construction.

Further refinements of the results of Matano and Mimura (1983) on (6.25)–(6.26) may be found in Mimura et al. (1991). See also Ikeda and Mimura (1993). For corresponding results when homogeneous Dirichlet or absorbing boundary conditions are imposed in place of the reflecting or homogeneous Neumann boundary conditions, see the work of Dancer (1985, 1988).

6.3 Edge Mediated Effects

Reaction-diffusion models for interacting biological species in a bounded habitat patch suppose a distinction between the habitat patch Ω and the environs that surround it, which we have called the environmental matrix. The distinction may be one that is readily perceived by humans, such as a patch of forest that is surrounded by grasslands. On the other hand, the difference may not be at all apparent to a casual human observer, such as a change in soil chemical content (Landsberg, 1988). But whether the distinction is perceptible to humans or not, it may serve to alter ecological processes in some way. For instance, Amazonian carrion beetles may readily cross forest-clear cut edges only to die of dessication in the more exposed habitats (Klein, 1989). The beetle's apparent inability to avoid crossing into areas of increased mortality appears to be linked to a compositional shift in the carrion-feeding guild. Ants become increasingly dominant and carrion beetles relatively less important in small fragments and clear cut areas where the effects of edges are most pronounced (Klein, 1989). Even when species are averse to crossing hostile edges, they may suffer from fragmentation via different mechanisms, such as increased predation near edges that decreases *per capita* recruitment locally and results in landscape-wide population decreases that exceed the loss of preferred habitat (e.g., (Gates and Gysel, 1972) for birds). Davies, Margules and Lawrence (2000) detail the individual responses of 69 diverse beetle species to habitat fragmentation in a tropical forest.

We use the term *edge-mediated effects* (Fagan et al., 1999) to describe changes in ecological processes that result in some way from the juxtaposition of disparate habitat types, as in the preceding examples. We are particularly interested in situations in which habitat edges change species interactions, and in collaboration with W.F. Fagan (Fagan et al., 1999) identified four general classes of mechanisms through which ecological edges can alter species interactions in fundamental ways:

1. Edges can change species interactions by altering species' movement patterns.
2. Edges can change community dynamics by differentially inducing species' mortality.
3. Edges can alter species interactions through cross-boundary subsidies.
4. Edges can create new opportunities for species interactions.

The ecological literature contains numerous examples of each of these classes of mechanisms for edge-mediated effects, and an extensive list of references is given in Fagan et al. (1999).

Our aim in this section is to explore edge-mediated effects in the context of two-species competition models of the form (6.1)–(6.2). We consider three models and in so doing highlight three of the four classes of mechanisms for edge-mediated effects (numbers 2, 3 and 4 above, to be specific). However, it is important to recognize that these classes of mechanisms are far from mutually exclusive. Indeed, our first example explores how increasing the hostility of the matrix surrounding a habitat patch may lead to a switch in the competitive advantage between two competitors *inside* the patch. In our analysis, such a "reversal of fortune" is due to a differential in mortality as the hostility of the surrounding matrix increases. However, it would also be possible to view the example in terms of altered movement patterns.

In models such as (6.1)–(6.2), the distinction between the habitat patch and the surrounding matrix environment is captured through the boundary condition (6.2), and the boundary of a habitat patch is frequently viewed as the ecological edge of principal interest. On the other hand, habitat patches are often spatially heterogeneous, and the ecological edge of most significance in a particular example may actually consist of a broad transition zone starting at an interface of some sort in the interior of the patch. In such a case, the ecological edge is captured via spatial heterogeneity in the coefficients in (6.1). As the reader shall see, our examples encompass both sorts of ecological edges.

6.3.1 A Note About Eigenvalues

The reader will notice that the principal eigenvalue of the negative Laplacian relative to the homogeneous boundary condition (6.2) figures prominently in our examples. (Such prominence is perhaps to be expected in light of the discussion in Section 2.2, where we observe that this quantity is closely related to the concept of "core habitat" in the ecological literature.) Our basic notation for this quantity in this section will be $\lambda^1(\Omega, \beta)$; i.e. $\lambda = \lambda^1(\Omega, \beta)$ is the unique nonnegative number for which the eigenvalue problem

$$-\Delta\phi = \lambda\phi \quad \text{in} \quad \Omega \tag{6.78}$$

$$\beta(x)\phi + (1 - \beta(x))\nabla\phi \cdot \vec{n} = 0 \quad \text{on} \quad \partial\Omega \tag{6.79}$$

admits an eigenfunction ϕ which is positive throughout Ω, where $0 \le \beta(x) \le 1$ on $\partial\Omega$. In this notation, $\beta = 1$ corresponds to Dirichlet (or absorbing or lethal) boundary data, while $\beta \equiv 0$ corresponding to Neumann (or reflecting) boundary data. If β is less than 1 throughout $\partial\Omega$, (6.79) can be expressed as

$$\nabla\phi \cdot \vec{n} + \gamma(x)\phi = 0 \quad \text{on} \quad \partial\Omega \tag{6.80}$$

where $\gamma(x) = \dfrac{\beta(x)}{1 - \beta(x)}$. In (6.80), $\gamma(x) \in [0, \infty)$ and, as we shall see, may be regarded as a measure of the level of hostility of the matrix environment near locale x on the boundary $\partial\Omega$. We emphasize this feature of the boundary condition in our first example and use the notation $\lambda_1^\gamma(\Omega)$ to denote the principal eigenvalue of (6.78) and (6.80). Notice that

$$\lambda_1^\gamma(\Omega) = \lambda^1(\Omega, \frac{\gamma}{\gamma + 1}). \tag{6.81}$$

It is natural to extend $\lambda_1^\gamma(\Omega)$ to the case of homogeneous Dirichlet boundary data by setting $\gamma = \infty$, so that in analogy to (6.81) we have

$$\lambda_1^\infty(\Omega) = \lambda^1(\Omega, 1) = \lambda_0^1(\Omega)$$

in the case of a completely lethal boundary, where $\lambda_0^1(\Omega)$ is the symbol we used to represent this quantity in Chapters 2 and 4.

We shall also encounter in our examples in this section the principal eigenvalue μ for the more general eigenvalue problem

$$-\Delta\phi + q(x)\phi = \mu m(x)\phi \quad \text{in} \quad \Omega \tag{6.82}$$

$$\beta(x)\phi + (1 - \beta(x))\nabla\phi \cdot \vec{n} = 0 \quad \text{on} \quad \partial\Omega \tag{6.83}$$

where q and m are in $L^\infty(\Omega)$ and m is positive on an open subset of Ω (see Manes and Micheletti (1973)). We shall use the notation $\mu^1(q, m, \beta)$ to denote this quantity. Observe that

$$\mu^1(0, 1, \beta) = \lambda^1(\Omega, \beta). \tag{6.84}$$

If we consider (6.82)–(6.80) in place of (6.82)–(6.83), we use the notation $\mu_1^\gamma(q, m)$ for the principal eigenvalue, and as in (6.81) we have

$$\mu_1^\gamma(q, m) = \mu^1\left(q, m, \frac{\gamma}{\gamma + 1}\right). \tag{6.85}$$

6.3.2 Competitive Reversals Inside Ecological Reserves Via External Habitat Degradation: Effects of Boundary Conditions

We consider the special case of (6.1)–(6.2) given by

$$\begin{aligned}\frac{\partial u_1}{\partial t} &= D_1\Delta u_1 + [a_1 - u_1 - b_1u_2]u_1 \\ & \qquad\qquad \text{in} \quad \Omega\times(0,\infty) \\ \frac{\partial u_2}{\partial t} &= D_2\Delta u_2 + [a_2 - b_2u_1 - u_2]u_2\end{aligned} \tag{6.86}$$

$$\alpha_i\nabla u_i\cdot\vec{n} + \beta u_i = 0 \qquad \text{on} \quad \partial\Omega\times(0,\infty) \tag{6.87}$$

where u_1 and u_2 represent the densities of two species competing in the bounded habitat patch Ω. The equations in (6.86) can be derived from the more general Lotka-Volterra model

$$\begin{aligned}\frac{\partial U_1}{\partial t} &= D_1\Delta U_1 + r_1\left[1 - \frac{U_1}{K_1} - \frac{B_1}{K_1}U_2\right]U_1 \\ & \qquad\qquad \text{in} \quad \Omega\times(0,\infty) \\ \frac{\partial U_2}{\partial t} &= D_2\Delta U_2 + r_2\left[1 - \frac{B_2}{K_2}U_1 - \frac{U_2}{K_2}\right]U_2\end{aligned} \tag{6.88}$$

via the rescalings $u_1 = \frac{r_1U_1}{K_1}$, $u_2 = \frac{r_2U_2}{K_2}$, $a_i = r_i$, $b_1 = B_1\left(\frac{r_1K_2}{K_1r_2}\right)$ and $b_2 = B_2\left(\frac{r_2K_1}{K_2r_1}\right)$, as in Section 4.4. The results we obtain in this subsection could certainly have been expressed in terms of (6.88) in place of (6.86), but would have a somewhat more cumbersome presentation. Indeed, we have taken Lotka-Volterra interaction terms primarily because they are convenient to talk about in the context of this volume and are widely used. However, our conclusions apply more broadly, in that qualitatively similar results could be obtained from many cases of two species competition as in equation (6.1). We should note that both b_i in (6.86) and B_i in (6.88) reflect the level of competitive pressure species j exerts on species i.

The boundary condition (6.87) can be derived from considerations along the lines of the classic paper of Ludwig et al. (1979). Namely consider a one dimensional habitat, say $\Omega = (a, b)$, viewing the infinite rays $(-\infty, a)$ and (b, ∞) as the surrounding matrix environment. Assume that in (a, b), the density u of a species is governed by a diffusive logistic growth law

$$\frac{\partial u}{\partial t} = D_{\text{in}}\Delta u + r\left(1 - \frac{u}{K}\right)u \quad \text{in} \quad (a,b)\times(0,\infty) \tag{6.89}$$

and that the density is governed in the matrix by the linear model

$$\frac{\partial u}{\partial t} = D_{\text{out}}\Delta u - su \quad \text{in} \quad [(-\infty,a)\cup(b,\infty)]\times(0,\infty), \tag{6.90}$$

where $s > 0$ is the death rate of the species in the matrix environment, D_{in} is the diffusion rate for the species in Ω, and D_{out} is (the possibly different) diffusion rate in the matrix environment. Matching densities and fluxes for an equilibrium to (6.89) with the unique bounded equilibrium to (6.90) leads to the equation

$$\frac{D_{\text{in}}}{\sqrt{D_{\text{out}}}} \nabla u \cdot \vec{n} + \sqrt{s}u = 0 \tag{6.91}$$

which is of the form (6.87) with $\alpha = \dfrac{D_{\text{in}}}{\sqrt{D_{\text{out}}}}$ and $\beta = \sqrt{s}$. Consequently, we regard β as a measure of the hostility of the matrix environment. When two species are involved, their death rates in the matrix environment may differ but be proportional to some overarching level of matrix mortality, induced, for example, via road density or pesticide contamination. In such a case, we could replace s in (6.90) by $c_i s$ for $i = 1, 2$, so that (6.91) becomes

$$\frac{(D_{\text{in}})_i}{\sqrt{c_i (D_{\text{out}})_i}} \nabla u_i \cdot \vec{n} + \sqrt{s}u_i = 0 \tag{6.92}$$

which is of the form (6.87). Since the equation

$$\alpha \nabla u \cdot \vec{n} + \beta u = 0$$

can be rewritten

$$\frac{\alpha}{\beta} \nabla u \cdot \vec{n} + u = 0$$

the case of a completely lethal boundary corresponds to $\beta = +\infty$ in (6.87).

The question of coexistence in (6.86)–(6.87) is addressed in both Sections 4.4 and Section 5.2, and applied to (6.33)–(6.34) in the preceding section, and the conditions for coexistence or elimination at low densities here are essentially identical to those for (6.33)–(6.34) in the preceding section. Namely, we first assume that the principal eigenvalue σ_i is positive in the eigenvalue problem

$$\begin{aligned} &D_i \Delta \phi_i + a_i \phi_i = \sigma_i \phi_i \text{ in } \quad \Omega \\ &\alpha_i \nabla \phi_i \cdot \vec{n} + \beta \phi_i = 0 \quad \text{on} \quad \partial\Omega \\ &\phi_i > 0 \qquad\qquad\qquad \text{in} \quad \Omega, \end{aligned} \tag{6.93}$$

for $i = 1, 2$. Recall that $\sigma_i > 0$ in (6.93) implies that the population model

$$\frac{\partial u_i}{\partial t} = D_i \Delta u_i + (a_i - u_i)u_i \text{ in } \quad \Omega \times (0, \infty) \tag{6.94}$$

$$\alpha_i \nabla u_i \cdot \vec{n} + \beta u_i = 0 \text{ on } \quad \partial\Omega \times (0, \infty) \tag{6.95}$$

for $i = 1, 2$, admits a globally attracting equilibrium $\overline{u}_i = \overline{u}_i(\beta)$, which is positive on the habitat Ω. The density $\overline{u}_i$ may be regarded as the carrying capacity for species i in Ω relative to the loss to the surrounding matrix reflected in (6.95) in the absence of competition from species j. That (6.94)–(6.95) admits a positive carrying capacity means, ecologically speaking, that the bounded habitat Ω is large enough relative to the loss through the boundary to sustain the population in the absence of competition from the other species, i.e., there is enough "core habitat" for this purpose. As we have seen several times, $\sigma_i > 0$ is equivalent to having the ratio of the local *per capita* growth rate at low densities, in this case a_i, to the diffusion rate D_i exceed the principal eigenvalue of $-\Delta$ in the habitat Ω subject to boundary condition (6.95), i.e.,

$$\frac{a_i}{D_i} > \lambda_1^{\beta/\alpha_i}(\Omega). \tag{6.96}$$

Next, under the assumption that (6.96) holds for $i = 1, 2$, we must look at the eigenvalue problems

$$D_i \Delta \psi_i + (a_i - b_i \overline{u}_j)\psi_i = \tilde{\sigma}_i \psi_i \quad \text{in} \quad \Omega \tag{6.97}$$

$$\alpha_i \nabla \psi_i \cdot \vec{n} + \beta \psi_i = 0 \quad \text{on} \quad \partial\Omega \tag{6.98}$$

for $i = 1, 2$, where $j \neq i$ and $\overline{u}_j$ is the globally attracting positive equilibrium for (6.94)–(6.95). The predictions of the model are determined by the sign of the principal eigenvalue $\tilde{\sigma}_i$ in (6.97)–(6.98). Namely, if $\tilde{\sigma}_i > 0$, then species i can invade Ω when species j is present at its carrying capacity and consequently is predicted to persist if it competes with species j in Ω. On the other hand, if $\tilde{\sigma}_i < 0$, species j will eliminate species i from the habitat patch if the density of species i is low. In particular, species i cannot invade Ω when species j is at carrying capacity.

Our principal objective regarding (6.86)–(6.87) is to examine how the predictions of the model depend on the level of hostility in the matrix surrounding the habitat patch, which we track via the parameter β in (6.87). In particular, is it possible to switch from a situation in which $\tilde{\sigma}_i > 0$ and $\tilde{\sigma}_j < 0$ (so that species i has a clear competitive advantage in Ω) to a situation in which $\tilde{\sigma}_i < 0$ and $\tilde{\sigma}_j > 0$ (so that species j has the advantage) by *changing only the parameter* β *in* (6.86)–(6.87)? In seeking to answer this question, we first note that $\tilde{\sigma}_i = 0$ and $\tilde{\sigma}_j = 0$ represent the thresholds between different model predictions for species i and species j, respectively. Geometrically each of $\tilde{\sigma}_i = 0$ and $\tilde{\sigma}_j = 0$ can be realized as a hypersurface in an appropriate Euclidean space. Clearly, the value of $\tilde{\sigma}_i$ in (6.97)–(6.98) depends on the parameters which appear explicitly in the equations, namely $a_i, D_i, \alpha_i, \beta$ and b_i. It is also clear that the value of $\tilde{\sigma}_i$ in (6.97)–(6.98) depends on $\overline{u}_j$ as well. We observe from (6.94)–(6.95) that $\overline{u}_j$ depends on a_j, D_j, α_j and β but not on b_j. Likewise, the value of $\tilde{\sigma}_j$ depends on $a_i, D_i, \alpha_i, a_j, D_j, \alpha_j, \beta$ and b_j but not on b_i. So $\tilde{\sigma}_i = 0$ and $\tilde{\sigma}_j = 0$ may each be regarded as a hypersurface in $\mathbb{R}^8$, with seven of the eight coordinates common to both (i.e., $a_1, a_2, D_1, D_2, \alpha_1, \alpha_2$, and β). We have not attempted anything like a complete analysis of these hypersurfaces. Instead, we have fixed configurations of the parameters $a_1, a_2, D_1, D_2, \alpha_1$ and α_2 and examined the relationship between b_1 and β on the hypersurface given by $\tilde{\sigma}_1 = 0$ and b_2 and β on the hypersurface $\tilde{\sigma}_2 = 0$. So doing has enabled us to recast the model predictions in terms of the competition coefficients b_1 and b_2 as in the following result, whose proof is contained in the Appendix to the chapter.

Theorem 6.12. Consider (6.86)–(6.87). Assume that for $i = 1, 2, a_i, D_i$ and α_i are fixed with

$$\frac{a_i}{D_i} > \lambda_1^{\infty}(\Omega). \tag{6.99}$$

Then for each $\beta \in [0, \infty]$, there are unique values of the competition coefficients b_1 and b_2, denoted $\overline{b}_1(\beta)$ and $\overline{b}_2(\beta)$, respectively, with the following properties:

(i) For $i = 1, 2$ and $j \neq i$, if $b_i < \overline{b}_i(\beta)$, species i can invade the habitat Ω when species j is at its carrying capacity density in the absence of competition (i.e. $\overline{u}_j = \overline{u}_j(\beta)$), and hence may be expected to persist in the face of competition from species j. If $b_i > \overline{b}_i(\beta)$, species j eliminates species i from Ω when species i is at low densities. In particular, species i cannot invade Ω when species j is at its carrying capacity in the absence of competition.

(ii) For $i = 1, 2$, $\overline{b}_i(\beta)$ is a differentiable function from $[0, \infty]$ into $(0, \infty)$.

It follows from (2.14) that $\lambda^{\beta/\alpha_i}(\Omega)$ is an increasing function of β on $[0, \infty]$. In light of (6.96), condition (6.99) can be interpreted as requiring that both species can persist in Ω in the absence of competition whatever the level of matrix hostility. When $b_i = 0$, it should be clear that $\tilde{\sigma}_i = \sigma_i > 0$, where $\tilde{\sigma}_i$ is as in (6.97)–(6.98) and σ_i is as in (6.93). Our proof shows for any β that $\tilde{\sigma}_i$ is a strictly decreasing function of b_i on $[0, \infty)$ which becomes negative for large enough b_i. Consequently, there is a unique $\overline{b}_i(\beta)$ as asserted in Part (i) of Theorem 6.12. The proof of Part (ii) is somewhat complicated, involving two different applications of the Implicit Function Theorem.

Suppose now that for some habitat Ω and some fixed configuration of the six parameters $a_1, a_2, D_1, D_1, \alpha_1$ and α_2 in (6.86)–(6.87) we can establish that

$$\overline{b}_1(0) > 1 > \overline{b}_2(0) \tag{6.100}$$

but

$$\overline{b}_1(\infty) < 1 < \overline{b}_2(\infty). \tag{6.101}$$

Then we can choose competition coefficients b_1 and b_2 so that species 1 has a competitive advantage when the degree of matrix hostility as represented by β is low, but loses the competitive advantage to species 2 when the degree of matrix hostility is very high. For if (6.100) and (6.101) hold, then Part (ii) of Theorem 6.12 implies that

$$\overline{b}_1(\beta) > 1 > \overline{b}_2(\beta)$$

for small β, while

$$\overline{b}_1(\beta) < 1 < \overline{b}_2(\beta)$$

for large β, and hence we can choose competition coefficients b_1, b_2 so that

$$\begin{aligned}
b_1 &< \overline{b}_1(\beta) \text{ for small } \beta\\
b_1 &> \overline{b}_1(\beta) \text{ for large } \beta\\
b_2 &> \overline{b}_2(\beta) \text{ for small } \beta\\
b_2 &< \overline{b}_2(\beta) \text{ for large } \beta
\end{aligned}$$

and the assertion follows directly from Theorem 6.12. As a result, we get a shift in competitive advantage in Ω solely on the basis of the change in the degree of the hostility of the matrix environment. Consequently, in such an instance, degradation of the area

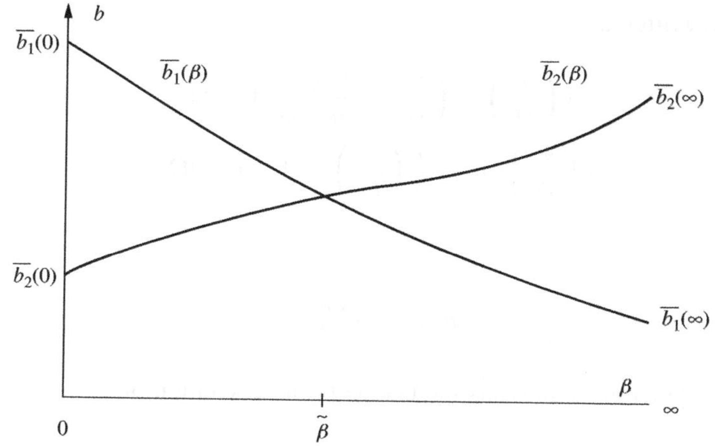

Figure 6.3 In this graph β represents the degree of hostility of the matrix surrounding a habitat patch and $\overline{b}_i(\beta)$ represents the maximum level of competition that species i can sustain from species j and still be predicted to persist for any positive initial density. This particular scenario shows a reversal of competitive advantage as β increases.

surrounding the habitat patch can lead to a profound change in the outcome of an ecological process inside the patch. See Figure 6.3.

So our objective is met if we can identify a habitat Ω and a configuration of the parameters $a_1, a_2, D_1, D_2, \alpha_1$ and α_2 so that (6.100) and (6.101) hold. To this end, note that it follows from the definition of $\overline{b}_i$ given in Theorem 6.12 that

$$\begin{aligned} D_i \Delta\psi_i + (a_i - \overline{b}_i(\beta)\overline{u}_j(\beta))\psi_i &= 0 \text{ in } \quad \Omega \\ \nabla\psi_i \cdot \vec{n} + \frac{\beta}{\alpha_i}\psi_i &= 0 \qquad \text{on} \quad \partial\Omega, \end{aligned}$$

so that

$$-\Delta\psi_i + \frac{\overline{b}_i(\beta)\overline{u}_j(\beta)}{D_i}\psi_i = \frac{a_i}{D_i}\psi_i \quad \text{in} \quad \Omega,$$

and hence that the threshold value $\overline{b}_i(\beta)$ must satisfy

$$\mu_1^{\beta/\alpha_i}\left(\frac{\overline{b}_i\overline{u}_j}{D_i}, 1\right) = \frac{a_i}{D_i}. \tag{6.102}$$

Our aim is to exploit (6.102) to obtain (6.100) and (6.101) for an appropriate selection of the habitat Ω and the parameters $a_1, a_2, D_1, D_2, \alpha_1$ and α_2. To do so, it will be useful to express $\overline{u}_i(\beta), i = 1, 2$, in a form that highlights its dependence on a_i, D_i and α_i. Note that $\overline{u}_i(\beta)$ satisfies

$$\begin{aligned} -D_i\Delta\overline{u}_i &= (a_i - \overline{u}_i)\overline{u}_i \text{ in } \quad \Omega \\ \nabla\overline{u}_i \cdot \vec{n} + \frac{\beta}{\alpha_i}\overline{u}_i &= 0 \qquad \text{on} \quad \partial\Omega \end{aligned}$$

which can be rewritten as

$$-\Delta\left(\frac{\overline{u}_i}{D_i}\right) = \left(\frac{a_i}{D_i} - \frac{\overline{u}_i}{D_i}\right)\frac{\overline{u}_i}{D_i} \quad \text{in} \quad \Omega$$

$$\nabla\left(\frac{\overline{u}_i}{D_i}\right)\cdot\vec{n} + \frac{\beta}{\alpha_i}\left(\frac{\overline{u}_i}{D_i}\right) = 0 \quad \text{on} \quad \partial\Omega.$$

Consequently,

$$\overline{u}_i = D_i\theta_{a_i/D_i}^{\beta/\alpha_i} \tag{6.103}$$

where for $a > \lambda_1^\gamma(\Omega)$ θ_a^γ represents the unique positive solution of

$$-\Delta\theta = (a-\theta)\theta \quad \text{in} \quad \Omega \tag{6.104}$$

$$\nabla\theta\cdot\vec{n} + \gamma\theta = 0 \quad \text{on} \quad \partial\Omega. \tag{6.105}$$

Estimates on the quantity $\mu_1^{\gamma_2}(e\theta_a^{\gamma_1}, 1)$ are the key to analyzing (6.102) closely enough to get the estimates on $\overline{b}_1$ and $\overline{b}_2$ in (6.100) and (6.101). The kinds of estimates we need on $\mu_1^{\gamma_2}(e\theta_a^{\gamma_1}, 1)$ were first derived in Cantrell and Cosner (1987) in the case of homogeneous Dirichlet boundary conditions (i.e., when $\gamma_1 = \gamma_2 = \infty$), and were extended in Cantrell et al. (1998) to the case of arbitrary positive γ_1 and γ_2. We next state the general result from Cantrell et al. (1998). Once we have this result we can identify conditions on Ω and configurations of $a_1, a_2, D_1, D_2, \alpha_1$ and α_2 in (6.86)–(6.87) so that (6.100) and (6.101) hold. We shall discuss the derivation of the estimates in the next section.

Theorem 6.13. (Cantrell et al., 1998)
Suppose that $0 \le \gamma_1 \le \gamma_2$ and that $a > \lambda_1^{\gamma_2}(\Omega)$.

(i) For $0 < e < 1$, $\mu_1^{\gamma_2}(e\theta_a^{\gamma_1}, 1)$ satisfies

$$\begin{aligned} ae + \lambda_1^{\gamma_2}(\Omega)(1-e) &< \mu_1^{\gamma_2}(e\theta_a^{\gamma_1}, 1) \\ &\le \lambda_1^{\gamma_2}(\Omega) + ae \end{aligned} \tag{6.106}$$

while for $e > 1$, $\mu_1^{\gamma_2}(e\theta_a^{\gamma_1}, 1)$ satisfies

$$a < \mu_1^{\gamma_1}(e\theta_a^{\gamma_1}, 1) \le \lambda_1^{\gamma_2}(\Omega) + ae. \tag{6.107}$$

(ii) For $0 < e < 1$, $\mu_1^{\gamma_1}(e\theta_a^{\gamma_2}, 1)$ satisfies

$$\lambda_1^{\gamma_1}(\Omega) < \mu_1^{\gamma_1}(e\theta_a^{\gamma_2}, 1) < a, \tag{6.108}$$

while for $e > 1$, $\mu_1^{\gamma_1}(e\theta_a^{\gamma_2}, 1)$ satisfies

$$\mu_1^{\gamma_1}(e\theta_a^{\gamma_2}, 1) \le ae + (1-e)\lambda_1^{\gamma_1}(\Omega). \tag{6.109}$$

We may now establish a result which enables us to identify conditions on Ω and configurations of $a_1, a_2, D_1, D_2, \alpha_1$ and α_2 so that (6.100) and (6.101) hold.

Theorem 6.14. Suppose that $\dfrac{a_2}{D_2} > \dfrac{a_1}{D_1} > \lambda_1^\infty(\Omega)$ and that $a_1 > a_2$. Then

$$\overline{b}_1(0) > 1 > \overline{b}_2(0) \tag{6.110}$$

while

$$\begin{aligned}\overline{b}_1(\infty) &\le \frac{a_1 - D_1\lambda_1^\infty(\Omega)}{a_2 - D_2\lambda_1^\infty(\Omega)}\\ \text{and } \overline{b}_2(\infty) &\ge \frac{a_2 - D_2\lambda_1^\infty(\Omega)}{a_1 - D_1\lambda_1^\infty(\Omega)}.\end{aligned} \tag{6.111}$$

Proof: When $\beta = 0$, $\sigma_i = a_i$ and $\overline{u}_i = a_i$, so that $\tilde{\sigma}_i = a_i - b_i a_j$. It follows that $\overline{b}_i(0) = \dfrac{a_i}{a_j}$ and hence that (6.110) holds.

We have from (6.102) and (6.103) that

$$\mu_1^\infty\left(\frac{\overline{b}_1(\infty)}{D_1}D_2\theta^\infty_{\frac{a_2}{D_2}}, 1\right) = \frac{a_1}{D_1}. \tag{6.112}$$

It follows as in Section 2.2 that $\mu_1^\gamma(q, 1)$ increases in q. If $\overline{b}_1(\infty)\dfrac{D_2}{D_1} \ge 1$ in (6.112), then $\overline{b}_1\dfrac{D_2}{D_1}\theta^\infty_{\frac{a_2}{D_2}} \ge \theta^\infty_{\frac{a_2}{D_2}}$, so that

$$\begin{aligned}\frac{a_1}{D_1} &= \mu_1^\infty\left(\overline{b}_1\frac{D_2}{D_1}\theta^\infty_{\frac{a_2}{D_2}}, 1\right)\\ &\ge \mu_1^\infty\left(\theta^\infty_{\frac{a_2}{D_2}}, 1\right)\\ &= a_2/D_2.\end{aligned}$$

This last equality follows from the definition of θ_a^γ in (6.104)–(6.105). Since we have hypothesized that $\dfrac{a_2}{D_2} > \dfrac{a_1}{D_1}$, we must conclude that $\overline{b}_1\dfrac{D_2}{D_1} < 1$ in (6.112). Applying (6.106) we get

$$\begin{aligned}\frac{a_1}{D_1} &= \mu_1^\infty\left(\overline{b}_1\frac{D_2}{D_1}\theta^\infty_{\frac{a_2}{D_2}}, 1\right)\\ &\ge \frac{a_2}{D_2}\left(\overline{b}_1\frac{D_2}{D_1}\right) + \lambda_1^\infty(\Omega)\left(1 - \overline{b}_1\frac{D_2}{D_1}\right).\end{aligned} \tag{6.113}$$

It follows from (6.113) that

$$\frac{a_1 - D_1\lambda_1^\infty(\Omega)}{a_2 - D_2\lambda_1^\infty(\Omega)} \ge \overline{b}_1(\infty).$$

It also follows from (6.102) and (6.103) that

$$\mu_1^\infty\left(\overline{b}_2\frac{D_1}{D_2}\theta^\infty_{\frac{a_1}{D_1}}, 1\right) = \frac{a_2}{D_2}. \tag{6.114}$$

If $\overline{b}_2(\infty)\frac{D_1}{D_2} \leq 1$ in (6.114), then

$$\begin{aligned}\frac{a_2}{D_2} &= \mu_1^\infty\left(\overline{b}_2\frac{D_1}{D_2}\theta^\infty_{\frac{a_1}{D_1}}, 1\right)\\ &\leq \mu_1^\infty\left(\theta^\infty_{\frac{a_1}{D_1}}, 1\right)\\ &= a_1/D_1,\end{aligned}$$

again a contradiction to our hypotheses. We conclude that $\overline{b}_2(\infty)\frac{D_1}{D_2} > 1$ in (6.114) and hence by (6.109) that

$$\begin{aligned}\frac{a_2}{D_2} &= \mu_1^\infty\left(\overline{b}_2\frac{D_1}{D_2}\theta^\infty_{\frac{a_1}{D_1}}, 1\right)\\ &\leq \left(\frac{a_1}{D_1}\right)\left(\overline{b}_2\frac{D_1}{D_2}\right) + \left(1 - \overline{b}_2\frac{D_1}{D_2}\right)\lambda_1^\infty(\Omega).\end{aligned} \tag{6.115}$$

We get from (6.115) that

$$\frac{a_2 - D_2\lambda_1^\infty(\Omega)}{a_1 - D_1\lambda_1^\infty(\Omega)} \leq \overline{b}_2(\infty)$$

so that (6.111) holds.

Corollary 6.15. Suppose that the conditions of Theorem 6.14 hold. Then if the habitat patch Ω is such that

$$\lambda_1^\infty(\Omega) > \frac{a_1 - a_2}{D_1 - D_2}, \tag{6.116}$$

(6.101) holds. Consequently, species 1 has a competitive advantage when the hostility in the matrix habitat surrounding the habitat patch Ω is low but loses the advantage to species 2 when the hostility in the matrix environment is very high.

Proof: Note that $\frac{a_2}{D_2} > \frac{a_1}{D_1}$ in the hypotheses of Theorem 6.14 implies $D_1 > \left(\frac{a_1}{a_2}\right) D_2$. Since we also hypothesize that $a_1 > a_2$, we have $D_1 > D_2$. So the quantity on the right-hand side of (6.116) is positive.

Notice from (6.111) that (6.101) holds provided $\frac{a_2 - D_2\lambda_1^\infty(\Omega)}{a_1 - D_1\lambda_1^\infty(\Omega)} > 1$, and it is easy to calculate this condition is equivalent to (6.116).

Let us now illustrate our results with a numerical example, based on parameter values which are plausible for small mammals (Okubo et al., 1989). (See also Cantrell et al. (1998).) Assume the parameters in (6.88) are

$$r_1 = 1.5/yr \; B_1 = 1.0 \; K_1 = 60\frac{\text{individuals}}{km^2}$$
$$r_2 = 1.0/yr \; B_2 = 1.0 \; K_2 = 40\frac{\text{individuals}}{km^2}.$$

so that

$$\begin{aligned} a_1 &= 1.5 \; b_1 = 1.0 \\ a_2 &= 1.0 \; b_2 = 1.0 \end{aligned} \tag{6.117}$$

in (6.86). Since $a_1 - \overline{b}_1(0)a_2 = 0$ and $a_2 - \overline{b}_2(0)a_1 = 0$, we have $\overline{b}_1(0) = 1.5 > 1 > 2/3 = \overline{b}_2(0)$, so that (6.100) holds. Hence $b_1 < \overline{b}_1(0)$ and $b_2 > \overline{b}_2(0)$. Consequently, species 1 has the advantage in any habitat patch Ω when the hostility in the surrounding matrix is low so long as the diffusion rates D_1 and D_2 are such that $\frac{a_2}{D_2} > \frac{a_1}{D_1}$ which in this case requires

$$D_1 > 1.5D_2. \tag{6.118}$$

Suppose for instance that

$$\begin{aligned} D_1 &= 1.0 \; km^2/yr \\ D_2 &= 0.25 \; km^2/yr. \end{aligned} \tag{6.119}$$

Clearly, if D_1 and D_2 are as in (6.119), (6.118) holds. We will get $\overline{b}_1(\infty) < 1 < \overline{b}_2(\infty)$ (and hence a "reversal of fortune" between species 1 and species 2 in Ω when the matrix hostility is high) provided (6.116) holds. In this case this condition becomes

$$\lambda_1^{\infty}(\Omega) > 2/3. \tag{6.120}$$

In a square patch, we know $\lambda_1^{\infty}(\Omega) = \frac{2\pi^2}{|\Omega|}$, where $|\Omega|$ denotes the area of the patch. So (6.120) holds in the case of a square patch provided

$$|\Omega| < 3\pi^2 \; km^2 \approx 29.6 \; km^2. \tag{6.121}$$

In ongoing work (Cantrell et al., 2002), we have shown that it is sometimes possible to have multiple reversals of competitive advantage as exterior matrix hostility increases. Such a phenomenon means that the outcome of a competitive interaction inside an isolated bounded habitat patch may be very sensitive to the level of hostility in the matrix habitat.

6.3.3 Cross-Edge Subsidies and the Balance of Competition in Nature Preserves

We now envision a nature preserve with two competing species. There is a buffer zone between the preserve and a highly hostile exterior matrix environment, and the two species may move freely throughout the preserve and the buffer zone. The buffer zone may be kept cleared or may be allowed to remain as a presumably lower grade habitat. The question we want to address was initially suggested by Janzen (1983, 1986). Namely, does the choice of management regime for the buffer zone impact upon the coexistence of the two species in the reserve if one of them has a significant reproductive advantage over the other in an uncleared buffer zone but neither can reproduce in a cleared area? More specifically, let us assume that when the buffer zone is kept cleared the two species will coexist. If the buffer zone is allowed to remain as a lower grade habitat, one species will have a much higher *per capita* growth rate in the buffer zone than the other. Presumably such a reproductive advantage leads to a higher density for the one species in the buffer zone than can serve to subsidize its density in the preserve. The question is, can this subsidy from the buffer zone lead to the elimination of the other species, something that would not happen if the buffer zone were kept cleared.

In this example the ecological "edge" of most interest is the interface between the preserve and the buffer zone. However, we want to regard the nature preserve plus the buffer zone as the habitat patch in this instance, so we must incorporate the "edge" in our model equations rather than in the boundary conditions as the preceding example.

We model the situation via

$$\frac{\partial U_1}{\partial t} = d_1 \Delta U_1 + \left[r_1 \chi_{\Omega_1} - \frac{r_1}{K_1} U_1 - \frac{C_1 U_2}{K_1} \right] U_1$$

$$\text{in} \quad \Omega \times (0, \infty) \qquad (6.122)$$

$$\frac{\partial U_2}{\partial t} = d_2 \Delta U_2 + \left[r_2 (\chi_{\Omega_1} + \ell \chi_{\Omega - \Omega_1}) - \frac{r_2}{K_2} U_2 - \frac{C_2 U_1}{K_2} \right] U_2$$

$$U_1 = 0 = U_2 \text{ on } \quad \partial\Omega \times (0, \infty), \qquad (6.123)$$

where $\Omega_1 \subseteq \Omega$ is the nature preserve in question and the overall habitat Ω consists of the nature preserve plus the surrounding buffer zone. (See Figure 6.4.) Recall that the "cut-off" function χ_{Ω_1} is given by

$$\chi_{\Omega_1}(x) = \begin{cases} 1, & x \in \Omega_1 \\ 0, & x \in \Omega \backslash \Omega_1. \end{cases}$$

So species 1 has a local *per capita* growth rate of r_1 throughout the nature preserve Ω_1 and zero *per capita* growth rate in the buffer zone $\Omega \backslash \Omega_1$. Species 2 has a *per capita* growth rate of r_2 in the preserve, but its *per capita* growth rate in the buffer zone is $r_2 \ell$, where ℓ is a number in the interval $[0, 1]$. Having $\ell = 0$ corresponds to a cleared buffer zone where neither species can reproduce effectively. If $\ell = 1$, then the *per capita* rate of growth for species 2 is r_2 throughout the overall habitat Ω, so that there is no difference between the preserve and buffer zone in terms of the *per capita* reproductive rate for species 2.

Our analysis in essence is to compare the predictions of the model for these two extreme cases to establish that there are values of r_1, d_1, r_2 and d_2 so that when $\ell = 0$ the model (6.122)–(6.123) predicts coexistence for the two species in Ω via permanence (or compressivity) but that when $\ell = 1$ the model predicts that species 2 competitively excludes species 1 from Ω, and in particular from Ω_1. Our results are robust in the sense that the

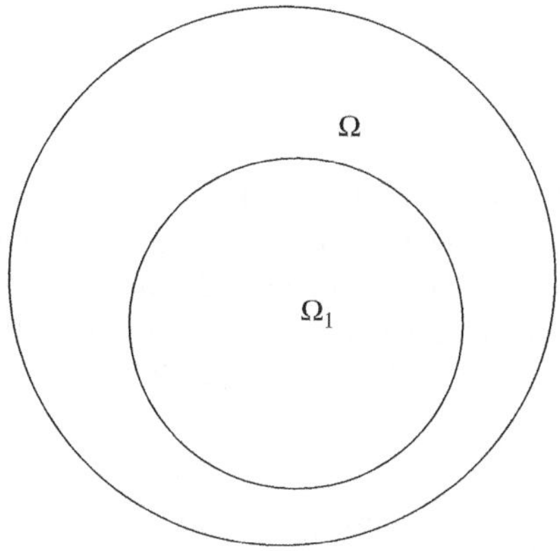

Figure 6.4 The domain Ω_1 corresponds to a nature reserve and the region $\Omega \backslash \Omega_1$ corresponds to a buffer zone surrounding the nature reserve as in the model (6.122), (6.123).

same dichotomy would hold between the predictions of the model when $\ell = 0$ and when $\ell \in (0, 1)$ when the *per capita* growth rate for species 1 is given by

$$r_1(\chi_{\Omega_1} + \varepsilon \ell \chi_{\Omega - \Omega_1})$$

in place of $r_1 \chi_{\Omega_1}$, where ε is positive but very small. Namely, we can identify values of r_1, d_1, r_2 and d_2 so that the model switches from a prediction of coexistence when the buffer zone is kept cleared to a prediction of competitive exclusion of species 1 by species 2 so long as species 2 has a sufficiently large reproductive advantage in the buffer zone. It is not necessary that species 1 be totally unable to reproduce in the buffer zone nor that species 2 have as high a *per capita* growth rate in the buffer zone as it does in the reserve, just that there be a strong enough disparity. However, the mathematical analysis will be cleaner in the case we have chosen to treat.

The model (6.122)–(6.123) can be rescaled to

$$\left(\frac{1}{d}\right)\frac{\partial u_1}{\partial t} = \Delta u_1 + u_1\left[\frac{r_1}{d_1}\chi_{\Omega_1} - u_1 - c_1 u_2\right]$$

$$\text{in} \quad \Omega \times (0, \infty) \qquad (6.124)$$

$$\frac{\partial u_2}{\partial t} = \Delta u_2 + u_2\left[\frac{r_2}{d_2}(\chi_{\Omega_1} + \ell\chi_{\Omega - \Omega_1}) - c_2 u_1 - u_2\right]$$

$$u_1 = 0 = u_2 \quad \text{on} \quad \partial\Omega \times (0, \infty) \qquad (6.125)$$

via the substitutions $r_1(x, t) = \frac{r_1}{d_1}\frac{U_1}{K_1}\left(x, \frac{t}{d_2}\right)$, $u_2(x, t) = \frac{r_2}{d_2}\frac{U_2}{K_2}\left(x, \frac{t}{d_2}\right)$, $c_1 = \frac{C_1}{r_2 d}\frac{K_2}{K_1}$ and $c_2 = \frac{C_2 d}{r_1}\frac{K_1}{K_2}$, where $d = \frac{d_1}{d_2}$. The model (6.124)–(6.125) is permanent and predicts the coexistence of the two species provided

$$\sigma_i = \sigma_i(\ell) > 0 \qquad (6.126)$$

and

$$\tilde{\sigma}_i = \tilde{\sigma}_i(\ell) > 0 \tag{6.127}$$

for $i = 1, 2$, where $\sigma_i(\ell), \overline{u}_i(\ell)$ and $\tilde{\sigma}_i(\ell)$ are the obvious analogues for (6.124)–(6.125) to their counterparts for (6.86)–(6.87) given in (6.93), (6.94)–(6.95), and (6.97)–(6.98), respectively. On the other hand, if we assume (6.126) and $\tilde{\sigma}_2(\ell) > 0$, species 2 can invade Ω when species 1 is at its carrying capacity density in Ω in the absence of competition from species 2, and moreover, there will be a unique globally attracting positive equilibrium solution $\tilde{u}_2(\ell)$ for

$$\frac{\partial u_2}{\partial t} = \Delta u_2 + u_2\left[\frac{r_2}{d_2}(\chi_{\Omega_1} + \ell\chi_{\Omega-\Omega_1}) - c_2\overline{u}_1(\ell) - u_2\right] \quad \text{in} \quad \Omega \times (0, \infty) \tag{6.128}$$

$$u_2 = 0 \quad \text{on} \quad \partial\Omega \times (0, \infty). \tag{6.129}$$

If the principal eigenvalue $\approx{\sigma}_1(\ell)$ in

$$\begin{aligned} &\Delta \approx{\phi}_1 + \left(\frac{r_1}{d_1}\chi_{\Omega_1} - c_1\tilde{u}_2(\ell)\right)\approx{\phi}_1 = \approx{\sigma}_1 \approx{\phi}_1 \quad \text{in} \quad \Omega \\ &\approx{\phi}_1 = 0 \qquad\qquad\qquad\qquad\qquad\qquad \text{on} \quad \partial\Omega \end{aligned} \tag{6.130}$$

is nonpositive, it follows as in Proposition 5.7 and Corollary 5.8 that there is no componentwise positive equilibrium to (6.124)–(6.125), and hence that species 2 excludes species 1 in Ω.

Our aim is to demonstrate that there is an open region of ordered pairs of ratios of growth to diffusion rates $\left(\frac{r_1}{d_1}, \frac{r_2}{d_2}\right)$ so that when $\ell = 0$, (6.126) and (6.127) hold (and hence the model (6.124)–(6.125) predicts coexistence), but when $\ell = 1$, (6.126) holds, $\tilde{\sigma}_2(1) > 0$ and $\approx{\sigma}_1 \leq 0$ in (6.130) (so that the model predicts that species 2 eliminates species 1 from Ω). We employ an extension of Theorem 6.13 as in Cantrell and Cosner (1993) for this purpose. Proceeding as in the previous example, we obtain that (6.126)–(6.127) is equivalent to

$$\frac{r_1}{d_1} > \mu_1^\infty(c_1\overline{u}_2(0), \chi_{\Omega_1}) \tag{6.131}$$

and

$$\frac{r_2}{d_2} > \mu_1^\infty(c_2\overline{u}_1(0), \chi_{\Omega_1}). \tag{6.132}$$

when $\ell = 0$. When $\ell = 1$, having $\tilde{\sigma}_2 > 0$ and $\approx{\sigma}_1 \leq 0$ requires

$$\frac{r_1}{d_1} \leq \mu_1^\infty(c_1\tilde{u}_2(1), \chi_{\Omega_1}). \tag{6.133}$$

In direct analogy to (6.103), we may identify $\overline{u}_1(0), \overline{u}_2(0)$ and $\tilde{u}_2(1)$ via

$$\overline{u}_i(0) = \theta\left(\frac{r_i}{d_i}\chi_{\Omega_1}\right) \tag{6.134}$$

and

$$\tilde{u}_2(1) = \theta\left(\frac{r_2}{d_2} - c_2\theta\left(\frac{r_1}{d_1}\chi_{\Omega_1}\right)\right) \tag{6.135}$$

where for $r > \mu_1^\infty(q, m)$, $\theta(rm - q)$ denotes the globally attracting positive equilibrium for

$$\frac{\partial u}{\partial t} = \Delta u + (rm - q - u)u \quad \text{in} \quad \Omega \times (0, \infty) \tag{6.136}$$

$$u = 0 \quad \text{on} \quad \partial\Omega \times (0, \infty). \tag{6.137}$$

Substituting (6.134) into (6.131) and (6.132) and (6.135) into (6.133), one may estimate the locus of points satisfying (6.131), (6.132) and (6.133) simultaneously using extensions to Theorem 6.13, as in Cantrell and Cosner (1993). To this end, assume that the competition coefficients c_1 and c_2 in (6.124)–(6.125) lie in the interval (0, 1). Then if (6.131)–(6.132) hold,

$$c_2\left(\frac{r_1}{d_1} - \mu_1^\infty(0, \chi_{\Omega_1})\right) \le \frac{r_2}{d_2} - \mu_1^\infty(0, \chi_{\Omega_1}) \le \frac{1}{c_1}\left(\frac{r_1}{d_1} - \mu_1^\infty(0, \chi_{\Omega_1})\right) \tag{6.138}$$

whereas if

$$\frac{r_2}{d_2} - \lambda_1^\infty(\Omega) \ge K\left(\frac{r_1}{d_1} - \mu_1^\infty(0, \chi_{\Omega_1})\right) \tag{6.139}$$

for a $K > \dfrac{1}{c_1} > 1$ which depends on the geometric relationship between Ω_1 and Ω, then (6.133) holds. Since $\mu_1^\infty(0, \chi_{\Omega_1}) > \lambda_1^\infty(\Omega)$, we have the picture shown in Figure 6.5.

In Figure 6.5, the region inside the curves D and E represents the collection of pairs of growth to diffusion rate ratios $\left(\dfrac{r_1}{d_1}, \dfrac{r_2}{d_2}\right)$ for which the model (6.124)–(6.125) predicts coexistence of species 1 and species 2 when $\ell = 0$. The lower and upper linear approximations to this region given in (6.138) are represented graphically by Lines A and B, respectively. We know from (6.139) that for any pairs of growth to diffusion rate ratios $\left(\dfrac{r_1}{d_1}, \dfrac{r_2}{d_2}\right)$ that lie above Line C, the model predicts the exclusion of species 1 when $\ell = 1$. Since K in (6.139) exceeds $1/c_1$ and $\mu_1^\infty(0, \chi_{\Omega_1}) > \lambda_1^\infty(\Omega)$, the picture must be as we have shown. Consequently, the shaded region represents values of $\left(\dfrac{r_1}{d_1}, \dfrac{r_2}{d_2}\right)$ for which the model predicts coexistence if the buffer zone is kept cleared but exclusion of species 1 by species 2 when it is not. Notice that these values are those near point $(\mu_1^\infty(0, \chi_{\Omega_1}), \mu_1^\infty(0, \chi_{\Omega_1}))$ and that having $\dfrac{r_i}{d_i} > \mu_1^\infty(0, \chi_{\Omega_1})$ is equivalent to having $\sigma_i > 0$ when $\ell = 0$. So the growth to diffusion rate ratios for which we can assert that a cross-boundary subsidy can affect a switch in the model predictions are those which are only slightly above what is needed for either of the species to persist in the absence of competition from the other when the buffer is kept cleared.

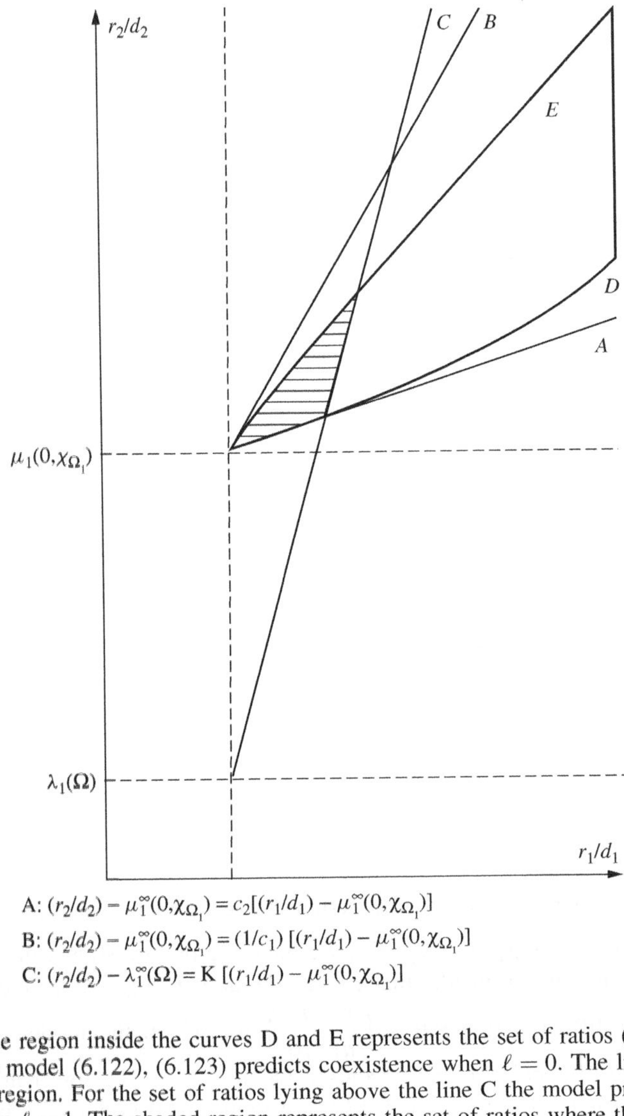

A: $(r_2/d_2) - \mu_1^\infty(0,\chi_{\Omega_1}) = c_2[(r_1/d_1) - \mu_1^\infty(0,\chi_{\Omega_1})]$

B: $(r_2/d_2) - \mu_1^\infty(0,\chi_{\Omega_1}) = (1/c_1)\,[(r_1/d_1) - \mu_1^\infty(0,\chi_{\Omega_1})]$

C: $(r_2/d_2) - \lambda_1^\infty(\Omega) = \mathrm{K}\,[(r_1/d_1) - \mu_1^\infty(0,\chi_{\Omega_1})]$

Figure 6.5 The region inside the curves D and E represents the set of ratios (r_1/d_1) and (r_2/d_2) for which the model (6.122), (6.123) predicts coexistence when $\ell = 0$. The lines A and B are bounds on that region. For the set of ratios lying above the line C the model predicts exclusion of species 1 when $\ell = 1$. The shaded region represents the set of ratios where the model predicts coexistence when the buffer zone is kept cleared ($\ell = 0$) but predicts the exclusion of species 1 when it is not ($\ell = 1$).

The analysis suggests that the concerns raised by Janzen (1983, 1986) about the effects of cross boundary subsidies on nature preserves may well be justified, at least in cases where the preserves are just large enough to sustain a population of some focal species. In such situations the nature of the habitat immediately surrounding the preserve can have profound effects on the outcome of competitive interactions within the preserve. The lesson for management of preserves is that conditions immediately outside a preserve should be considered as a factor that may influence the effectiveness of the preserve. Furthermore, modifying or maintaining the surroundings of a preserve in ways that are locally favorable to the focal species may actually be harmful to that species within the preserve *if* the surroundings are even more favorable to a competing species.

6.3.4 Competition Mediated by Pathogen Transmission

In conservation biology, scientists are particularly concerned with the influences of diseases on the persistence or extinction of threatened species. Emphasis is often directed at understanding the roles of "reservoir species", those that by virtue of abundance and/or immunity to a disease, facilitate the persistence of the disease within a region of habitat. The role of reservoir species is often critical, because populations of rare or threatened species, which might otherwise be too small to allow establishment of diseases on their own, may face added extinction risks via infection from contacts with reservoir species. Such contacts are sometimes increased by habitat fragmentation, as increased "edginess" in the terrain fosters encounters between species that would not happen otherwise. For example, increased edginess of forests in the Northeastern United States and Canada has been implicated as a key factor influencing the transmission of brainworm infections from the common white-tailed deer to other rarer ungulates such as moose, woodland caribou and elk. The effects of brainworm infections are relatively benign in the white-tailed deer but may be lethal to other ungulates (Anderson, 1972; Holmes, 1996). Similarly, forest-matrix edge encounters between the pearly-eyed thrasher and the Puerto Rican parrot in remnant Puerto Rican rainforests can lead to parasitic botfly infections in parrot populations (Snyder et al., 1987; Loye and Carroll, 1995).

A useful way to examine the impact that edge-mediated pathogen transmission from a reservoir species to a rare or threatened species may have upon the persistence or extinction of the rare or threatened species in some bounded habitat is through a model for competition between the reservoir and threatened species. Such a model is akin to models for apparent competition mediated by a mutual predator (Holt and Lawton, 1993, 1994), but with an edge-mediated spatial complicating factor. In Cantrell et al. (2001), we derive such a reaction-diffusion model starting with a four component (susceptibles and infectives for both the reservoir and threatened species) ODE epidemiological model for pathogen transmission between the species. We reduce the four component ODE model to a two component ODE competition model via the following working assumptions:

(i) the pathogen is quickly lethal to the threatened species;

(ii) the effects of the pathogen upon the reservoir species are mild;

(iii) the rates of transmission (and when possible) recovery from infection are very large compared to the ecological birth and death rates and competition coefficients in the system;

(iv) in the absence of the pathogen the competitive effect of the threatened species on the reservoir species is mild.

We then incorporate the spatial aspects into the model via diffusion by envisioning overlapping bounded habitats for the two species (see Figure 6.6). In this scenario the "edge" in question is the overlap in habitat patches where pathogen transmission from the reservoir species to the threatened species takes place. This feature requires that as in the previous example we account for the edge mediated effect in the model equations rather than in boundary conditions.

The model we arrive at is

$$\frac{\partial u_i}{\partial t} = D_i \Delta u_i + (r_i - u_i - \alpha_{ij} U_j(u_j))u_i \quad \text{in} \quad \Omega_i \times (0, \infty) \tag{6.140}$$

$$\beta_i(x)u_i + (1 - \beta_i(x))\nabla u_i \cdot \vec{n} = 0 \quad \text{on} \quad \partial\Omega_i \times (0, \infty), \tag{6.141}$$

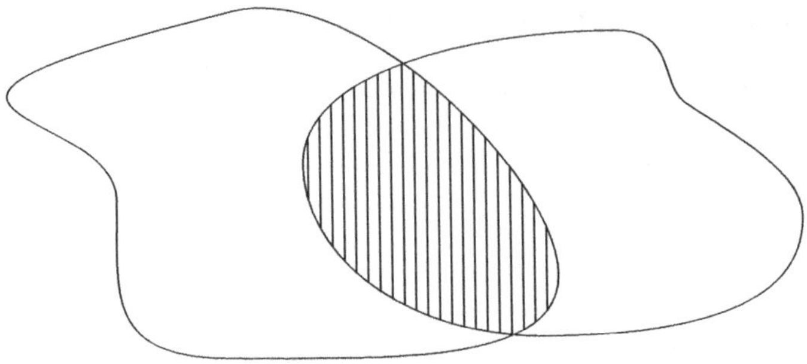

Figure 6.6 The shaded area shows the region where the ranges of two interacting species, that is, the spatial domains for the two equations in the model (6.140), (6.141) overlap.

$i = 1, 2,\ j \neq i$. In (6.140)–(6.141), u_i denotes the density of species i on its habitat Ω_i. For the sake of specificity we take species 1 as the focal threatened species and species 2 as the reservoir species. The species only come into contact with each other in the overlap $\Omega_E = \Omega_1 \cap \Omega_2$ of their habitat patches. We use the term $U_j(u_j)$ to account for this feature of the model, where $U_j(u_j)$ is the function *defined on* Ω_i *where* $i \neq j$ by

$$U_j(u_j(x,t)) = \begin{cases} u_j(x,t) & \text{on} \quad \Omega_E \\ 0 & \text{on} \quad \Omega_i \backslash \Omega_E. \end{cases} \tag{6.142}$$

The fact that the two equations in (6.140) have different underlying spatial domains is a complicating factor in the modeling. Nevertheless, (6.140)–(6.141) can be recast as a semi-dynamical system π on a function space X in such a way that π is dissipative and $\pi(\cdot, t) : X \to X$ is compact for $t > 0$. Consequently, the permanence results of Chapter 4 are applicable. The space X in question embeds in $C^{1+\alpha}(\overline{\Omega}_1) \times C^{1+\alpha}(\overline{\Omega}_2)$. Additionally, as should be expected in models for two-species competition, the semi-dynamical system π preserves the skew ordering $(u_1, u_2) \leq_{Sk} (v_1, v_2)$ given by $u_1 \leq v_1$ and $u_2 \geq v_2$, and the compressivity results of Section 5.2 are also applicable in the analysis of (6.140)–(6.41). See Cantrell et al. (2001) for further details.

The system (6.140)–(6.141) is permanent (and compressive) provided $\sigma_i > 0$ and $\tilde{\sigma}_i > 0,\ i = 1, 2$, where σ_i, $\overline{u}_i$ and $\tilde{\sigma}_i$ have the same meanings as in the two preceding examples. We have presupposed that species 2 does not face an extinction risk from the competition with species 1, independent of the extent of overlap between their habitats or the degree of boundary loss species 1 is subject to on $\partial\Omega_1$. We can quantify this feature as follows. Species 2 is predicted to persist in the face of competition from species 1 so long as $\tilde{\sigma}_2 > 0$ in

$$D_2 \Delta \psi + [r_2 - \alpha_{21} U_1(\overline{u}_1)]\psi = \tilde{\sigma}_2 \psi \quad \text{in} \quad \Omega_2 \tag{6.143}$$

$$\beta_2 \psi + (1 - \beta_2) \nabla \psi \cdot \vec{n} = 0 \quad \text{on} \quad \partial\Omega. \tag{6.144}$$

In the absence of species 2, the density of species 1 is governed by a diffusive logistic equation in Ω_1. We know from the maximum principle that $\overline{u}_1 \leq r_1$ independent of $\beta_1(x) \in [0, 1]$. Moreover, if $\beta_1(x) \equiv 0$, $\overline{u}_1 \equiv r_1$ and $\overline{u}_1$ converges to r_1 as $\beta_1 \to 0$ on $\partial\Omega$. By (6.142) we have

$$U_1(\overline{u}_1) \leq r_1 \chi_{\Omega_E}.$$

Consequently, if $\tilde{\sigma}_2$ is to be positive in (6.143)–(6.144) independent of the extent of Ω_E, we need to have the principal eigenvalue ρ be positive in

$$D_2\Delta w + [r_2 - \alpha_{21}r_1]w = \rho w \text{ in } \quad \Omega_2$$
$$\beta_2 w + (1-\beta_2)\nabla w \cdot \vec{n} = 0 \quad \text{ on } \quad \partial\Omega_2.$$

It is easy to calculate that

$$\rho = r_2 - \alpha_{21}r_1 - D_2\lambda^1(\Omega_2, \beta_2)$$

so that $\tilde{\sigma}_2 > 0$ in (6.143)–(6.144) independent of Ω_E and β_1 provided

$$r_2 > \alpha_{21}r_1 + D_2\lambda^1(\Omega_2, \beta_2). \tag{6.145}$$

Since $\sigma_2 > 0$ is equivalent to

$$r_2 > D_2\lambda^1(\Omega_2, \beta_2),$$

(6.145) holds so long α_{21} is sufficiently small relative to the local *per capita* growth rate for species 1, which is consistent with our modeling assumptions. We shall assume (6.145) for the remainder of this discussion.

Let us now examine the principal eigenvalue $\tilde{\sigma}_1$ in

$$D_1\Delta\phi + [r_1 - \alpha_{12}U_2(\overline{u}_2)]\phi = \tilde{\sigma}_1\phi \quad \text{in} \quad \Omega_1 \tag{6.146}$$
$$\beta_1\phi + (1-\beta_1)\nabla\phi \cdot \vec{n} = 0 \text{ on } \quad \partial\Omega_1 \tag{6.147}$$

more closely. To this end, let

$$M_2 = \max_{\overline{\Omega}_E} \overline{u}_2.$$

We know that $M_2 \le r_2$. So if $m \in [M_2, r_2]$, $\tilde{\sigma}_1 \ge \rho$, where ρ is the principal eigenvalue for

$$D_1\Delta w + [r_1 - \alpha_{12}m\chi_{\Omega_E}]w = \rho w \quad \text{in} \quad \Omega_1 \tag{6.148}$$
$$\beta_1 w + (1-\beta_1)\nabla w \cdot \vec{n} = 0 \quad \text{on} \quad \partial\Omega_1. \tag{6.149}$$

If we want to have permanence independent of the extent of Ω_E, we would need the principal eigenvalue $\tilde{\rho} > 0$ in

$$D_1\Delta\tilde{w} + [r_1 - \alpha_{12}m]\tilde{w} = \tilde{\rho}\tilde{w} \quad \text{in} \quad \Omega_1 \tag{6.150}$$
$$\beta_1\tilde{w} + (1-\beta_1)\nabla\tilde{w} \cdot \vec{n} = 0 \text{ on } \quad \partial\Omega_1, \tag{6.151}$$

where $m \in \left[\max_{\overline{\Omega}_2} \overline{u}_2, r_2\right]$. It follows from (6.150)–(6.151) that (6.140)–(6.141) is permanent so long as

$$r_1 > \alpha_{12}m + D_1\lambda^1(\Omega_1, \beta_1) \tag{6.152}$$

holds in addition to (6.144), where again $m \in \left[\max_{\overline{\Omega}_2} \overline{u}_2, r_2\right]$. Since $\max_{\overline{\Omega}_2} \overline{u}_2$ decreases as β_2 increases, increasing β_2 allows us to take smaller values of m in (6.152).

Let us return to (6.148)–(6.149), with $m \in [M_2, r_2]$, and assume that $\beta_1(x) < 1$ on $\partial\Omega_1$, so that $w \neq 0$ anywhere on $\partial\Omega_1$. Consequently, (6.148) can be rewritten

$$D_1 \frac{\Delta w}{w} + [r_1 - \alpha_{12} m \chi_{\Omega_E}] = \rho \quad \text{in} \quad \Omega_1. \tag{6.153}$$

Integrating (6.153) on Ω_1 and using the identity

$$\frac{\Delta w}{w} = \nabla \cdot \left(\frac{\nabla w}{w}\right) + \frac{|\nabla w|^2}{w^2}$$

we obtain

$$\rho = r_1 + \frac{D_1}{|\Omega_1|}\left[\int_{\Omega_1} \frac{|\nabla w|^2}{w^2} dx - \int_{\partial\Omega_1} \frac{\beta_1}{1-\beta_1} dS\right] - \alpha_{12} m \frac{|\Omega_E|}{|\Omega_1|}. \tag{6.154}$$

From (6.154), we have that (6.140)–(6.141) is permanent if, in addition to (6.145),

$$r_1 > \frac{D_1}{|\Omega_1|}\int_{\partial\Omega_1} \frac{\beta_1}{1-\beta_1} dS + \alpha_{12} m \frac{|\Omega_E|}{|\Omega_1|} \tag{6.155}$$

holds. Note that if $\beta_1 < 1$ is constant, (6.155) becomes

$$r_1 > \frac{D_1 \beta_1}{1-\beta_1} \cdot \frac{|\partial\Omega_1|}{|\Omega_1|} + \alpha_{12} m \frac{|\Omega_E|}{|\Omega_1|}. \tag{6.156}$$

Comparisons between (6.152) and (6.155) or (6.156) are in order. Formulas (6.155) and (6.156) are more explicit in showing the relationships among the system parameters that lead to a prediction of persistence for the threatened species than is (6.152). They explicitly take the size of the overlapping region Ω_E relative to Ω_1 into account. If Ω_E is small relative to Ω_1, the contact between the species is low and $\alpha_{12} m \frac{|\Omega_E|}{|\Omega_1|}$ is much smaller than the corresponding term in (6.152). On the other hand, suppose that

$$-\Delta z = \lambda^1(\Omega_1, \beta_1) z \quad \text{in} \quad \Omega_1 \tag{6.157}$$

$$\beta_1 z + (1-\beta_1)\nabla z \cdot \vec{n} = 0 \quad \text{on} \quad \partial\Omega, \tag{6.158}$$

with $\beta_1 < 1$. If we divide both sides of the equation (6.157) by z and integrate, we can conclude that

$$\frac{D_1}{|\Omega_1|}\int_{\partial\Omega_1} \frac{\beta_1}{1-\beta_1} dS \geq D_1 \lambda^1(\Omega_1, \beta_1). \tag{6.159}$$

As $\beta_1 \to 0$, both sides of (6.159) tend to zero, so that for small values of β_1, (6.155) or (6.156) is a less stringent condition for persistence of species 1 than is (6.152), since $\alpha_{12} m \frac{|\Omega_E|}{|\Omega_1|} < \alpha_{12} m$. However, as β_1 approaches 1 on $\partial\Omega$, $\int_{\partial\Omega_1} \frac{\beta_1}{1-\beta_1} dS$ approaches $+\infty$, while $\lambda^1(\Omega_1, \beta_1)$ approaches $\lambda^1(\Omega_1, 1) < \infty$. Consequently, as the dissipative effect along the boundary of Ω_1 becomes stronger, (6.152) becomes a less stringent condition for the persistence of species 1 than (6.155) or (6.156). Moreover, (6.152) does not require any additional assumption on β_1. Finally, note that in the case in which both Ω_1 and Ω_2 are closed, i.e., $\beta_1 \equiv 0 \equiv \beta_2$, $\overline{u}_2 = r_2$ and (6.156) reduces to

$$r_1 > \alpha_{12} r_2 \frac{|\Omega_E|}{|\Omega_1|}.$$

We may also extract conditions for the exclusion of species 1 in (6.140)–(6.141) by employing practical persistence techniques, as in Section 5.3. Let us assume $\sigma_1 > 0$ and that (6.145) holds. Suppose $(u_1(x,t), u_2(x,t))$ is the solution to (6.140)–(6.141) corresponding to the initial densities $(u_1(x,0), u_2(x,0))$. Since $\overline{u}_1 \leq r_1$ on $\overline{\Omega}_1$, a comparison of $u_1(x,t)$ with $\overline{u}_1$ allows us to conclude that for any given $\varepsilon > 0$

$$u_1(x,t) < r_1 + \varepsilon$$

for $t \geq T$, where the value of T depends on $u_1(x,0)$ and ε. So $u_2(x,t)$ is an upper solution for the logistic model

$$\frac{\partial w}{\partial t} = D_2 \Delta w + (r_2 - \alpha_{21}(r_1 + \varepsilon) - w)w \quad \text{in} \quad \Omega_2 \times (T, \infty) \tag{6.160}$$

$$\beta_2 w + (1 - \beta_2)\nabla w \cdot \vec{n} = 0 \quad \text{on} \quad \partial\Omega_2 \times (T, \infty). \tag{6.161}$$

Since (6.145) implies that

$$r_2 > \alpha_{21}(r_1 + \varepsilon) - D_2 \lambda^1(\Omega_2, \beta_2)$$

for small enough ε, (6.160)–(6.161) admits a globally attracting positive solution $\underline{u}_2(\varepsilon)$ and another comparison allows us to conclude that for $t > T' > T$ and $x \in \Omega$

$$u_2(x,t) > (1 - \varepsilon)\underline{u}_2(\varepsilon).$$

Hence $u_1(x,t)$ is a lower solution to

$$\frac{\partial y}{\partial t} = D_1 \Delta y + [r_1 - \alpha_{12} U_2((1 - \varepsilon)\underline{u}_2) - y]y \quad \text{in} \quad \Omega_1 \times (T', \infty) \tag{6.162}$$

$$\beta_1 y + (1 - \beta_1)\nabla y \cdot \vec{n} = 0 \quad \text{on} \quad \partial\Omega_1 \times (T', \infty). \tag{6.163}$$

If the principal eigenvalue $\underline{\rho}$ for

$$D_1 \Delta \phi + [r_1 - \alpha_{12} U_2((1 - \varepsilon)\underline{u}_2(\varepsilon))]\phi = \underline{\rho}\phi \quad \text{in} \quad \Omega_1 \tag{6.164}$$

$$\beta_1 \phi + (1 - \beta_1)\nabla \phi \cdot \vec{n} = 0 \quad \text{on} \quad \partial\Omega_1 \tag{6.165}$$

is negative, 0 is the global attractor for positive solutions to (6.162)–(6.163). By comparison, $u_1(x,t) \to 0$ on $\overline{\Omega}_1$ on $\overline{\Omega}_1$ as $t \to \infty$, so that (6.140)–(6.141) predicts exclusion of species 1.

We can make the condition $\underline{\rho} \leq 0$ in (6.164)–(6.165) more explicit when D_2 is small. To this end, note that Proposition 3.16 implies that $\lim_{D_2 \to 0} \underline{u}_2(\varepsilon) = r_2 - \alpha_{21}(r_1 + \varepsilon)$ uniformly on any open subdomain Ω of Ω_2 with $\overline{\Omega} \subset \Omega_2$. As a consequence, $\underline{\rho} \leq 0$ in (6.164)–(6.165) when D_2 is small so long as the principal eigenvalue $\rho(\Omega_E)$ in

$$D_1 \Delta \psi + [r_1 - \alpha_{12}(r_2 - \alpha_{21} r_1)\chi_{\Omega_E}]\psi = \rho(\Omega_E)\psi \quad \text{in} \quad \Omega_1 \tag{6.166}$$

$$\beta_1 \psi + (1 - \beta_1)\nabla\psi \cdot \vec{n} = 0 \quad \text{on} \quad \partial\Omega \tag{6.167}$$

is negative. We know from Section 2.2 that $\rho(\Omega_E)$ converges to $\rho(\Omega_1)$ as $|\Omega_E|$ approaches $|\Omega_1|$, where $\rho(\Omega_1)$ is the principal eigenvalue of

$$D_1 \Delta z + [r_1 - \alpha_{12}(r_2 - \alpha_{21} r_1)]z = \rho(\Omega_1)z \quad \text{in} \quad \Omega_1. \tag{6.168}$$

$$\beta_1 z + (1 - \beta_1)\nabla z \cdot \vec{n} = 0 \quad \text{on} \quad \partial\Omega_1. \tag{6.169}$$

A simple calculation gives

$$\rho(\Omega_1) = r_1 - \alpha_{12}(r_2 - \alpha_{21} r_1) - D_1 \lambda^1(\Omega_1, \beta_1).$$

Consequently, we have that (6.140)–(6.141) predicts that species 2 eliminates species 1 from Ω_1 via pathogen transmission in Ω_E so long as the diffusion rate of species 2 is small, the region of overlap Ω_E constitutes a large portion of Ω_1 and

$$r_1 < \frac{1}{(1 + \alpha_{12}\alpha_{21})}(\alpha_{12} r_2 + D_1 \lambda^1(\Omega_1, \beta_1)). \tag{6.170}$$

Note that (6.170) places a constraint on the value of r_1 beyond the reversal of the condition (6.152) for the persistence of species 1, consistent with some of our earlier findings.

6.4 Estimates and Consequences

We have seen in Section 4.4 that when $D_i = 1$, $\beta = \infty$ and $0 < b_i < 1$ in (6.86)–(6.87) the locus of pairs of *per capita* growth rates (a_1, a_2) for which $\tilde{\sigma}_i = 0$ in (6.97)–(6.98) for $i = 1, 2$, are curves in $\{(a_1, a_2) : a_i \geq \lambda_1^\infty(\Omega), i = 1, 2\}$ which meet only at the point $(\lambda_1^\infty(\Omega), \lambda_1^\infty(\Omega))$ and lie on opposite sides of the diagonal ray $\{a_1 = a_2 : a_1 \geq \lambda_1^\infty(\Omega)\}$. Indeed $\tilde{\sigma}_1 = 0$ can be parameterized by a_2 for $a_2 \geq \lambda_1^\infty(\Omega)$ and lies to the left of the diagonal ray, while $\tilde{\sigma}_2 = 0$ can be parameterized by a_1 for $a_1 \geq \lambda_1^\infty(\Omega)$ and lies below the diagonal ray. Since $\tilde{\sigma}_i$ increases with a_i, it follows that the collection of pairs (a_1, a_2) which lie between the curves $\tilde{\sigma}_1 = 0$ and $\tilde{\sigma}_2 = 0$ in this case are precisely those for which the prediction of (6.86)–(6.87) is permanence. This collection of points forms a "wedge" in $\{(a_1, a_2) : a_i \geq \lambda_1^\infty(\Omega)\}$.

Mathematically speaking, it is quite natural to ask for a more detailed description of the aforementioned set of points in terms of the parameters a_i, b_i and $\lambda_1^\infty(\Omega)$. Such was the motivation for the original formulation of Theorem 6.13 in Cantrell and Cosner (1987), which implies that if $D_i = 1$, $\beta = \infty$ and $0 < b_i < 1$, then $\{(a_1, a_2) : \tilde{\sigma}_i > 0 \text{ in (6.97)–(6.98) for } i = 1, 2\}$ is contained in the set of ordered pairs (a_1, a_2) for which

$$b_2(a_1 - \lambda_1^\infty(\Omega)) \leq a_2 - \lambda_1^\infty(\Omega) \leq \frac{1}{b_1}(a_1 - \lambda_1^\infty(\Omega)).$$

However, as we saw in the preceding section, estimates resulting from Theorem 6.13 and its extensions have proved highly useful to the study of edge-mediated effects. This phenomenon is but one more example of how a mathematical result may arise for purely mathematical reasons but then find a meaningful application in a different context. We next opt to present a proof of Theorem 6.13, followed by some remarks, which are somewhat historical in nature, that aim to make the reader more aware of a body of mathematical work concerning (6.86)–(6.87) when $\beta = \infty$ beyond Theorem 6.13.

To establish Theorem 6.13, we begin with the equations defining $\mu_1^{\gamma_2}(e\theta_a^{\gamma_1}, 1)$, where $0 \le \gamma_1 \le \gamma_2$ and $a > \lambda_1^{\gamma_2}(\Omega)$, namely

$$-\Delta\psi_e + e\theta_a^{\gamma_1}\psi_e = \mu_1^{\gamma_2}(e\theta_a^{\gamma_1}, 1)\psi_e \tag{6.171}$$

$$\vec{n}\cdot\nabla\psi_e + \gamma_2\psi_e = 0 \quad \text{on} \quad \partial\Omega, \tag{6.172}$$

where $\psi_e > 0$ in Ω. If we normalize ψ_e by the requirement

$$\int_\Omega \psi_e^2 dx = 1,$$

an Implicit Function Theorem argument akin to that which establishes part (ii) of Theorem 6.12 guarantees that ψ_e is a differentiable function of e.

Choose e_1 and e_2 nonnegative with $e_1 \neq e_2$. Multiply (6.171) when $e = e_1$ by ψ_{e_2} and employ Green's Second Identity to obtain

$$[\mu_1^{\gamma_2}(e_1\theta_a^{\gamma_1}, 1) - \mu_1^{\gamma_2}(e_2\theta_a^{\gamma_1}, 1)]\int_\Omega \psi_{e_1}\psi_{e_2}dx = (e_1 - e_2)\int_\Omega \theta_a^{\gamma_1}\psi_{e_1}\psi_{e_2}dx. \tag{6.173}$$

If we set $e_1 = e$ and $e_2 = 0$ in (6.173) and use the fact that $\theta_a^{\gamma_1} \le a$ in $\overline{\Omega}$ we obtain

$$\mu_1^{\gamma_2}(e\theta_a^{\gamma_1}, 1) \le ae + \lambda_1^{\gamma_2}(\Omega), \tag{6.174}$$

which establishes the right-hand sides of both (6.106) and (6.107).

Now return to (6.173), set $\gamma_2 = \gamma_1 = \gamma$ and divide by $(e_1 - e_2)\int_\Omega \psi_{e_1}\psi_{e_2}dx$ to obtain

$$\frac{[\mu_1^{\gamma}(e_1\theta_a^{\gamma}, 1) - \mu_1^{\gamma}(e_2\theta_a^{\gamma}, 1)]}{e_1 - e_2} = \frac{\int_\Omega \theta_a^{\gamma}\psi_{e_1}\psi_{e_2}dx}{\int_\Omega \psi_{e_1}\psi_{e_2}dx}. \tag{6.175}$$

Passing to the limit in (6.175) as $e_2 \to e_1$ yields

$$\frac{d}{de_1}\{\mu_1^{\gamma}(e_1\theta_a^{\gamma}, 1)\} = \int_\Omega \theta_a^{\gamma}\psi_{e_1}^2 dx. \tag{6.176}$$

Rewrite (6.171) as

$$\theta_a^{\gamma}\psi_e = \frac{1}{e}[\Delta\psi_e + \mu_1^{\gamma}(e\theta_a^{\gamma}, 1)\psi_e] \quad \text{in} \quad \Omega,$$

multiply by ψ_e and integrate. Green's First Identity and (6.175) imply

$$\frac{d}{de}\{\mu_1^{\gamma}(e\theta_a^{\gamma}, 1)\} = \frac{1}{e}\left[\mu_1^{\gamma}(e\theta_a^{\gamma}, 1) - \gamma\int_{\partial\Omega}\psi_e^2 dS - \int_\Omega |\nabla\psi_e|^2 dx\right].$$

Theorem 2.1 implies that $\int_\Omega |\nabla\psi_e|^2 dx + \gamma \int_{\partial\Omega} \psi_e^2 dS \geq \lambda_1^\gamma(\Omega)$, so that

$$\frac{e\frac{d}{de}\{\mu_1^\gamma(e\theta_a^\gamma, 1)\} - \mu_1^\gamma(e\theta_a^\gamma, 1)}{e^2} \leq -\frac{\lambda_1^\gamma(\Omega)}{e^2},$$

or equivalently,

$$\frac{d}{de}\left\{\frac{\mu_1^\gamma(e\theta_a^\gamma, 1)}{e}\right\} \leq -\frac{\lambda_1^\gamma(\Omega)}{e^2} \tag{6.177}$$

for all $e > 0$. If we think of fixing an $e \in (0, 1)$ and integrate (6.177) between e and 1 and use the fact that $\mu_1^\gamma(\theta_a^\gamma, 1) = a$ we obtain

$$ae + (1-e)\lambda_1^\gamma(\Omega) \leq \mu_1^\gamma(e\theta_a^\gamma, 1). \tag{6.178}$$

If now $\gamma_1 \leq \gamma_2$, an upper and lower solution argument as in Cantrell et al. (1998) shows that $\theta_a^{\gamma_2} \leq \theta_a^{\gamma_1}$. So from (6.178) we get

$$\begin{aligned} ae + (1-e)\lambda_1^{\gamma_2}(\Omega) &\leq \mu_1^{\gamma_2}(e\theta_a^{\gamma_2}, 1) \\ &\leq \mu_1^{\gamma_2}(e\theta_a^{\gamma_1}, 1), \end{aligned}$$

establishing the left-hand side of (6.106) and hence Part (i) of Theorem 6.13.

To complete the proof of Theorem 6.13, we only need to establish (6.109). To that end, fix an $e > 1$ and integrate (6.177) between 1 and e. Again, using the fact that $\mu_1^\gamma(\theta_a^\gamma, 1) = a$, we obtain

$$\mu_1^\gamma(e\theta_a^\gamma, 1) \leq ae + (1-e)\lambda_1^\gamma(\Omega). \tag{6.179}$$

Consequently,

$$\begin{aligned} \mu_1^{\gamma_1}(e\theta_a^{\gamma_2}, 1) &\leq \mu_1^{\gamma_1}(e\theta_a^{\gamma_1}, 1) \\ &\leq ae + (1-e)\lambda_1^{\gamma_1}(\Omega), \end{aligned}$$

establishing (6.109) as required.

So now let us reconsider the collection of pairs of *per capita* growth rates (a_1, a_2) with $a_i > \lambda_1^\infty(\Omega)$ for which the prediction of (6.86)–(6.87) is permanence when $D_i = 1$, $\beta = \infty$ and $0 < b_i < 1$. We know from Section 4.6 that for all such points (6.86)–(6.87) admits an equilibrium (u_1, u_2) with $u_i > 0$ in the habitat patch Ω for $i = 1, 2$. In fact, we know from Section 5.2 that the system is compressive. So (6.86)–(6.87) admits a globally attracting order interval whose "end points" are componentwise positive equilibria to (6.86)–(6.87), say $(\overline{u}_1, \overline{u}_2)$ and $(\underline{u}_1, \underline{u}_2)$, with $\underline{u}_1 \leq \overline{u}_1$ and $\underline{u}_2 \geq \overline{u}_2$ in $\overline{\Omega}$. When $\underline{u}_i = \overline{u}_i$ for $i = 1, 2$, there is a unique such equilibrium and species densities corresponding to nonnegative nontrivial initial configurations necessarily evolve to the equilibrium configuration over time.

In general, the question of uniqueness of a componentwise positive equilibrium for (6.86)–(6.87) when $D_i = 1$, $\beta = \infty$, $0 < b_i < 1$ and (a_1, a_2) are such that $\tilde{\sigma}_i > 0$ for $i = 1, 2$, is an open mathematical problem. It is known that such an equilibrium is unique when $a_1 = a_2$ (Cosner and Lazer, 1984). In fact, Cosner and Lazer (1984) give an explicit formula in terms of $\theta_{a_i}^{\infty}$ and b_i. Exploiting the uniqueness property when $a_1 = a_2$, Cantrell and Cosner (1989) show that there is a unique componentwise positive equilibrium for (a_1, a_2) with a_1 and a_2 close enough to each other. Otherwise, not much about the uniqueness of componentwise positive equilibria is known for sure in this case, although Eilbeck et al. (1994) do offer some bifurcation theoretic evidence which suggests that uniqueness obtains.

We have observed in Section 5.7 that having a model for several interacting biological species such as (6.86)–(6.87) fail to be permanent does not mean that the model always predicts that at least one of the species is driven to extinction. In particular, the model (6.86)–(6.87), with $D_i = 1$, $\beta = \infty$ and $0 < b_i < 1$, may well admit componentwise positive equilibria for (a_1, a_2) outside the set with $\tilde{\sigma} > 0$ for $i = 1, 2$. To be sure, to our knowledge no one has shown the existence of such equilibria, and indeed, the evidence in Eilbeck et al. (1994) suggesting uniqueness of componentwise equilibria in this case also suggests that such equilibria do not exist. But the possibility has not been ruled out, again to the best of our knowledge. What is true is that Proposition 5.7 holds. Namely, there is a "wedge" in $\{(a_1, a_2) : a_i > \lambda_1^{\infty}(\Omega)\}$ containing $\{(a_1, a_2) : \tilde{\sigma}_i > 0 \text{ for } i = 1, 2\}$ outside of which exist no componentwise positive equilibria to (6.86)–(6.87) when $D_i = 1$, $\beta = \infty$ and $0 < b_i < 1$. (We should note that all the aforementioned results have analogues in the model (6.124)–(6.125) for cross-edge subsidies in nature preserves. Indeed, the essence of our argument there is that part of "inner wedge" when $\ell = 0$ is exterior to the "outer wedge" when $\ell = 1$. See Section 6.3.3.)

Given the deceptively simple appearance of (6.86)–(6.87) when $D_i = 1$ and $\beta = \infty$ plus the fact that the corresponding system of ODEs is completely understood, the reader may well be surprised at just how much remains to be discovered about the system when $0 < b_i < 1$ for $i = 1, 2$ and may be curious about what is known when at least one of the b_i's exceeds 1. To address this issue, recall that $\tilde{\sigma}_i = 0$ in (6.97)–(6.98) corresponds to $a_i = \mu_1^{\infty}(b_i\theta_{a_j}^{\infty}, 1)$ for $i = 1, 2$, $j \neq i$. Note that if $a_j > \lambda_1^{\infty}(\Omega)$ is fixed, increasing b_i forces a_i to increase. Consequently, if we examine the locus of $\tilde{\sigma}_i = 0$ by looking at its projection in $\{(a_1, a_2) : a_i > \lambda_1^{\infty}(\Omega)\}$ for different choices of b_1 and b_2, increasing b_1 causes the locus of $\tilde{\sigma}_1 = 0$ to move to the right and increasing b_2 causes the locus of $\tilde{\sigma}_2 = 0$ to move up. Recall also that when $b_i = 1$, $\tilde{\sigma}_i = 0$ corresponds to $a_1 = a_2$. As long as b_1 and b_2 are both less than 1, the two curves only intersect at the point $(\lambda_1^{\infty}(\Omega), \lambda_1^{\infty}(\Omega))$. However, if, for example, $b_1 < 1$ but $b_2 > 1$, the curves $\tilde{\sigma}_1 = 0$ and $\tilde{\sigma}_2 = 0$ could intersect at other points as well. Indeed, it is known that such is sometimes the case. The most comprehensive discussion of the possibilities to our knowledge may be found in Eilbeck et al. (1994), where a number of additional references are given. Notice that in such an instance, one would have for fixed b_1 and b_2 alternating open regions in $\{(a_1, a_2) : a_i > \lambda_1^{\infty}(\Omega)\}$, first with say $\tilde{\sigma}_i > 0$ for $i = 1, 2$, and then with $\tilde{\sigma}_i < 0$ for $i = 1, 2$. When $\tilde{\sigma}_i > 0$ for $i = 1, 2$, the model is permanent, while when $\tilde{\sigma}_i < 0$ for $i = 1, 2$, both $(\theta_{a_i}^{\infty}, 0)$ and $(0, \theta_{a_2}^{\infty})$ are stable. When both $\tilde{\sigma}_1 < 0$ and $\tilde{\sigma}_2 < 0$, Dancer (1985) shows the existence of an unstable componentwise positive equilibrium. Dancer (1988, 1991) also shows that pattern formation is possible in the model in the sense of Section 6.2.3, depending on the geometry of the bounded habitat Ω. When both b_1 and b_2 exceed 1, the locus of $\tilde{\sigma}_1 = 0$ is now to the right of $a_1 = a_2$ while the locus of $\tilde{\sigma}_2 = 0$ is above $a_1 = a_2$. Consequently, there are no (a_1, a_2) with $\tilde{\sigma}_1 > 0$ and $\tilde{\sigma}_2 > 0$ in this case. For some additional insights into the equilibria for the model in this case when $a_1 = a_2$, see Gui and Lou (1994).

Now seems to be an appropriate juncture to focus briefly and particularly on an important feature of reaction-diffusion models for several interacting biological species in a bounded habitat that is recurring theme in this volume. Namely, such models encode an enormous amount of information via their coefficients. Consequently, one is confronted with models that are frequently *very* multiparameter in nature. The complete mathematical analysis of models with many parameters is difficult. Not many people can visualize effectively in dimensions higher than three. What we have done throughout this chapter is to freeze parameter configurations in all but two or three components and analyze the problem at hand as those components vary. In other words, we look at the problem through a particular slice of parameter space. By varying the location of the "frozen" parameters, a more complete picture of the phenomena that is possible with the model emerges. What parameters to consider as varying for the purpose of mathematical analysis will depend on the investigator and the purpose of the investigation. The preceding discussion regarding (6.86)–(6.87) when $D_i = 1$ and $\beta = \infty$ "froze" those variables and b_1 and b_2 as well and thought of the *per capita* growth rates a_1 and a_2 as the varying parameters. However, some of the investigations into the problem (e.g. Dancer (1985, 1988)) have focused on varying competition coefficients b_1 and b_2 instead, with equally fruitful results.

Appendix

Proof of Theorem 6.1: An Implicit Function Theorem argument along the lines of that in Cantrell and Cosner (1987) guarantees that the principal eigenvalue $\alpha_i(s)$ and positive eigenfunction $\psi_i(s)$ for the eigenvalue problem

$$\mu_i \Delta\psi_i(s) + [m_i - (1-s)b_{ii}\overline{u}_i - sb_{ij}\overline{u}_j]\psi_i(s) = \alpha_i(s)\psi_i(s) \quad \text{in} \quad \Omega \tag{6A.1}$$

$$\beta_i\psi_i + (1-\beta_i)\nabla\psi_i \cdot \vec{n} = 0 \quad \text{on} \quad \partial\Omega \tag{6A.2}$$

depend differentiably as functions of s (from $[0, 1]$ into $I\!R$ and $C^{1+\alpha}(\overline{\Omega})$, respectively) so long as $\psi_i(s)$ is required to satisfy

$$\int_\Omega \psi_i^2(s)dx = 1. \tag{6A.3}$$

Note that (6.37) implies that $\alpha_i(0) = 0$ and $\psi_i(0) = \dfrac{\overline{u}_i}{||\overline{u}_i||_{L^2}}$ and that (6.38) implies that $\alpha_i(1) = \tilde{\sigma}_i$. If we differentiate both sides of (6A.1) with respect to s, letting $\psi_i' = \psi_i'(s)$ and $\alpha_i' = \alpha_i'(s)$, we obtain

$$\mu_i\Delta\psi_i' + [m_i - (1-s)b_{ii}\overline{u}_i - sb_{ij}\overline{u}_j]\psi_i' + (b_{ii}\overline{u}_i - b_{ij}\overline{u}_j)\psi_i = \alpha_i\psi_i' + \alpha_i'\psi_i \quad \text{in} \quad \Omega \tag{6A.4}$$

with ψ_i' satisfying (6A.2). Multiplying (6A.1) by ψ_i' and (6A.4) by ψ_i, integrating and subtracting the resulting equations via Green's Second Identity, and employing (6A.3) leads to

$$\alpha_i' = \int_\Omega (b_{ii}\overline{u}_i - b_{ij}\overline{u}_j)\psi_i^2 dx. \tag{6A.5}$$

Next, rewrite (6A.1) in the alternative form

$$\mu_i\Delta\psi_i + (m_i - b_{ii}\overline{u}_i)\psi_i + s(b_{ii}\overline{u}_i - b_{ij}\overline{u}_j)\psi_i = \alpha_i\psi \tag{6A.6}$$

in Ω. Multiply both sides of (6A.6) by $-\psi_i$, integrate via the Divergence Theorem and employ (6A.3) to obtain

$$\begin{aligned}\mu_i \int_\Omega |\nabla\psi_i|^2 dx - \int_\Omega (m_i - b_{ii}\bar{u}_i)\psi_i^2 dx - s\int_\Omega (b_{ii}\bar{u}_i - b_{ij}\bar{u}_j)\psi_i^2 dx \\ + \mu_i \int_{\{x\in\partial\Omega:\beta_i(x)<1\}} \left(\frac{\beta_i(x)}{1-\beta_i(x)}\right)\psi_i^2 dS = -\alpha_i.\end{aligned} \tag{6A.7}$$

By (6.37) and the variational characterization of eigenvalues (e.g., see Courant and Hilbert (1953))

$$\mu_i \int_\Omega |\nabla\psi_i|^2 dx - \int_\Omega (m_i - b_{ii}\bar{u}_i)\psi_i^2 dx + \mu_i \int_{\{x\in\partial\Omega:\beta_i(x)<1\}} \frac{\beta_i}{1-\beta_i}\psi_i^2 dS \geq 0.$$

Consequently, (6A.6) and (6A.7) imply that

$$-s\alpha_i'(s) = -s\int_\Omega (b_{ii}\bar{u}_i - b_{ij}\bar{u}_j)\psi_i^2 \leq -\alpha_i(s),$$

so that for $s \in (0, 1]$,

$$\alpha_i'(s) \geq \frac{\alpha_i(s)}{s} \tag{6A.8}$$

Note that $\alpha_i'(0) = \displaystyle\int_\Omega (b_{ii}\bar{u}_i - b_{ij}\bar{u}_j)\left(\frac{\bar{u}_i}{||\bar{u}||}\right)^2 dx$ so that (6.43) implies that $\alpha_i'(0) > 0$. Since $\alpha_i'(0) > 0$ and $\alpha_i(0) = 0$, there is a $\delta \in (0, 1)$ so that $\alpha_i(s) > 0$ for $s \in (0, \delta)$. From (6A.8), we see that $\alpha_i'(s) > 0$ for $s \in (0, \delta)$, so that $\alpha_i(s)$ is positive *and* increasing on $(0, \delta)$. Iterating the argument allows us to conclude that $\tilde{\sigma}_i = \alpha_i(1) > 0$, as required.

Proof of Theorem 6.3: (Cantrell and Cosner 1998) Recall from Theorem 2.6 that if $m > 0$ on an open subset of Ω, then $\sigma > 0$ is equivalent to $\mu < \dfrac{1}{\lambda_+^1(m)}$, where σ is the principal eigenvalue for

$$\begin{aligned}\mu\Delta\phi + m\phi &= \sigma\phi \quad &\text{in} \quad \Omega \\ \beta\phi + (1-\beta)\nabla\phi\cdot\vec{n} &= 0 \text{ on} \quad &\partial\Omega\end{aligned} \tag{6A.9}$$

and $\lambda_+^1(m) \geq 0$ is the principal eigenvalue for

$$\begin{aligned}-\Delta\psi &= \lambda m\psi \quad &\text{in} \quad \Omega \\ \beta\psi + (1-\beta)\nabla\psi\cdot\vec{n} &= 0 \text{ on} \quad &\partial\Omega.\end{aligned} \tag{6A.10}$$

(Note that $\lambda_+^1(m) = 0$ only if $\beta \equiv 0$ and $\displaystyle\int_\Omega m \geq 0$. In this case, $1/\lambda_+^1(m)$ is interpreted as $+\infty$, and $\sigma > 0$ for any $\mu > 0$.)

Since $(m_i)_+ - \dfrac{b_{ij}}{b_{jj}}(m_j)_+ > 0$ on an open subset of Ω, so are m_i and $m_i - \dfrac{b_{ij}}{b_{jj}}(m_j)_+$ so that $\lambda_+^1(m_i) \geq 0$ and $\lambda_+^1\left(m_i - \dfrac{b_{ij}}{b_{jj}}(m_j)_+\right) \geq 0$ exist. Consequently, if $\mu_i < \dfrac{1}{\lambda_+^1(m_i)}$, $\sigma_i > 0$

in (6.36) and the globally attracting equilibrium $\overline{u}_i$ of (6.37) exists. We shall establish the result first in the case when $\lambda_+^1\left(m_i - \frac{b_{ij}}{b_{jj}}(m_j)_+\right) > 0$ and then in the case when $\lambda_+^1\left(m_i - \frac{b_{ij}}{b_{jj}}(m_j)_+\right) = 0$.

We have by Proposition 3.16 that $\overline{u}_j \to \frac{(m_j)_+}{b_{jj}}$ uniformly on compact subsets of Ω as $\mu_j \to 0$, so that $m_i - b_{ij}\overline{u}_j \to m_i - b_{ij}\frac{(m_j)_+}{b_{jj}}$ uniformly on compact subsets of Ω as $\mu_j \to 0$. Consequently, $m_i - b_{ij}\overline{u}_j > 0$ on an open subset of Ω for μ_j sufficiently small, which implies that $\lambda_+^1(m_i - b_{ij}\overline{u}_j) \geq 0$ exists for all sufficiently small $\mu_j > 0$. Moreover, $m_i - b_{ij}\overline{u}_j \to m_i - \frac{b_{ij}}{b_{jj}}(m_j)_+$ in $L^p(\Omega)$ as $\mu_j \to 0$ for any $p \in (1, \infty)$. Theorem 2.4 implies as a result that $\lambda_+^1(m_i - b_{ij}\overline{u}_j) \to \lambda_+^1\left(m_i - \frac{b_{ij}}{b_{jj}}(m_j)_+\right) > 0$ as $\mu_j \to 0$. Hence there is a $\mu_j^* > 0$ so that if $0 < \mu_j < \mu_j^*$,

$$0 < \lambda_+^1(m_i - b_{ij}\overline{u}_j) < 2\lambda_+^1\left(m_i - \frac{b_{ij}}{b_{ii}}(m_j)_+\right). \tag{6A.11}$$

It follows from (6A.11) that if

$$\mu_i < \frac{1}{2\lambda_+^1(m_i - \frac{b_{ij}}{b_{jj}}(m_j)_+)}, \tag{6A.12}$$

then $\mu_i < \frac{1}{\lambda_+^1(m_i - b_{ij}\overline{u}_j)}$ and hence $\tilde{\sigma}_i > 0$ in (6.38), as required.

If $\lambda_+^1\left(m_i - \frac{b_{ij}(m_j)_+}{b_{ij}}\right) = 0$ it must be the case that $\beta_1 \equiv 0$ and $\int_\Omega m_i - \frac{b_{ij}(m_j)_+}{b_{jj}}dx \geq 0$. Suppose $\int_\Omega m_i - \frac{b_{ij}(m_j)_+}{b_{jj}}dx > 0$. Since $m_i - b_{ij}\overline{u}_j \to m_i - \frac{b_{ij}}{b_{jj}}(m_j)_+$ uniformly on compact subsets of Ω as $\mu_j \to 0$, there is a $\mu_j^* > 0$ so that $\int_\Omega m_i - b_{ij}\overline{u}_j dx > 0$ for $0 < \mu_j < \mu_j^*$. Since $\beta_i \equiv 0$, $\int_\Omega m_i - b_{ij}\overline{u}_j dx > 0$ implies that $\tilde{\sigma}_i$ in (6.38) is positive.

Finally, suppose $\int_\Omega m_i - \frac{b_{ij}}{b_{jj}}(m_j)_+ dx = 0$. We know that $m_i - \frac{b_{ij}(m_j)_+}{b_{jj}} > 0$ on an open subset of Ω. Since $\left(m_i - \frac{b_{ij}}{b_{jj}}(m_j)_+\right)^{-1}((0,\infty)) = \cup_{n=1}^\infty (m_i - \frac{b_{ij}(m_j)_+}{b_{jj}})^{-1}((1/n, \infty))$, there is an $n \geq 1$ so that $(m_i(x) - \frac{b_{ij}(m_j)_+}{b_{jj}})^{-1}((1/n, \infty))$ has nonempty interior. Consequently, there is an open subset of Ω so that $m_i(x) - \frac{b_{ij}(x)(m_j)_+(x)}{b_{jj}(x)} - \delta > 0$ for any $\delta \in (0, 1/n)$. For such a δ, $\int_\Omega \left(m_i - \frac{b_{ij}(m_j)_+}{b_{jj}} - \delta\right) dx = -\delta|\Omega| < 0$ and hence $\lambda_+^1\left(m_i - \frac{b_{ij}(m_j)_+}{b_{jj}} - \delta\right) > 0$. So there is a $\mu_j^* > 0$ so that $0 <$

$\lambda_+^1(m_i - b_{ij}\overline{u}_j - \delta) < 2\lambda_+^1\left(m_i - \dfrac{b_{ij}(m_j)_+}{b_{jj}} - \delta\right)$ if $0 < \mu_j < \mu_j^*$. If now $\mu_i < \dfrac{1}{2\lambda_+^1(m_i - \frac{b_{ij}}{b_{jj}}(m_j)_+ - \delta)}$, $\mu_i < \dfrac{1}{\lambda_+^1(m_i - b_{ij}\overline{u}_j - \delta)}$ and there is $\alpha_i > 0$ so that α_i is the principal eigenvalue of

$$\begin{aligned} &\mu_i\Delta\phi + (m_i - b_{ij}\overline{u}_j - \delta)\phi = \alpha_i\phi \text{ in } \Omega \\ &\nabla\phi \cdot \vec{n} = 0 \qquad \text{on } \partial\Omega. \end{aligned}$$

Hence $\tilde{\sigma}_i$ in (6.38) satisfies $\tilde{\sigma}_i = \alpha_i + \delta > 0$.

Proof of Theorem 6.12: (Cantrell et al., 1998) As noted, Part (i) of Theorem 6.12 is tantamount to showing that for all $\beta \in [0, \infty]$ there is a unique $b_i(\beta)$ so that $\tilde{\sigma}_i > 0$ for $b_i < b_i(\beta)$, $\tilde{\sigma}_i = 0$ when $b_i = b_i(\beta)$ and $\tilde{\sigma}_i < 0$ when $b_i > b_i(\beta)$. We establish these facts by showing that $\tilde{\sigma}_i$ decreases as b_i increases, $\tilde{\sigma}_i > 0$ when $b_i = 0$ and $\tilde{\sigma}_i < 0$ for b_i large enough. To see that $\tilde{\sigma}_i$ decreases as b_i increases, fix a $\beta \in [0, \infty]$. Let $0 \leq b_i < b_i'$ and let $\tilde{\sigma}_i$, $\tilde{\sigma}_i'$, ψ_i, ψ_i' denote the corresponding principal eigenvalues and eigenfunctions in (6.97)–(6.98), so that

$$\begin{aligned} &D_i\Delta\psi_i + (a_i - b_i\overline{u}_j)\psi_i = \tilde{\sigma}_i\psi_i \text{ in } \Omega \\ &\alpha_i\nabla\psi_i \cdot \vec{n} + \beta\psi_i = 0 \qquad \text{on } \partial\Omega \end{aligned} \tag{6A.13}$$

and

$$\begin{aligned} &D_i\Delta\psi_i' + (a_i - b_i'\overline{u}_j)\psi_i' = \tilde{\sigma}_i'\psi_i' \text{ in } \Omega \\ &\alpha_i\nabla\psi_i' \cdot \vec{n} + \beta\psi' = 0 \qquad \text{on } \partial\Omega. \end{aligned} \tag{6A.14}$$

If we multiply (6A.13) by ψ_i' and (6A.14) by ψ_i, integrate over Ω and subtract we obtain by Green's Second Identity that

$$(b_i' - b_i)\int_\Omega \overline{u}_j\psi_i\psi_i' = (\tilde{\sigma}_i - \tilde{\sigma}_i')\int_\Omega \psi_i\psi_i'. \tag{6A.15}$$

It is immediate from (6A.15) that $b_i < b_i'$ implies $\tilde{\sigma}_i > \tilde{\sigma}_i'$, so that $\tilde{\sigma}_i$ decreases as b_i increases.

Observe that if $b_i = 0$, $\tilde{\sigma}_i = \sigma_i > 0$, where σ_i is the principal eigenvalue in (6.93). To see that $\tilde{\sigma}_i < 0$ for b_i large, we consider two cases, $\beta < \infty$ and $\beta = \infty$. Suppose first that $0 \leq \beta < \infty$. Then $\overline{u}_j = \overline{u}_j(\beta) > 0$ on $\overline{\Omega}$. Multiplying (6A.13) by ψ_i and integrating we obtain

$$D_i\int_\Omega \psi_i\Delta\psi_i dx + \int_\Omega (a_i - b_i\overline{u}_j)\psi_i^2 dx = \tilde{\sigma}_i\int_\Omega \psi_i^2 dx \tag{6A.16}$$

The Divergence Theorem implies that $D_i\int_\Omega \psi_i\Delta\psi_i dx = -D_i\int_\Omega |\nabla\psi_i|^2 dx - \dfrac{\beta D_i}{\alpha_i}\int_{\partial\Omega}\psi_i^2 dS$, so that (6A.16) can be rewritten as

$$-D_i\int_\Omega |\nabla\psi_i|^2 dx - \frac{\beta D_i}{\alpha_i}\int_{\partial\Omega}\psi_i^2 dS - b_i\int_\Omega \overline{u}_j\psi_i^2 dx = (\tilde{\sigma}_i - a_i)\int_\Omega \psi_i^2 dx. \tag{6A.17}$$

It follows from (6A.17) that $\tilde{\sigma}_i < a_i - b_i\left(\min_{\overline{\Omega}} \overline{u}_j\right)$, which implies that $\tilde{\sigma}_i < 0$ for sufficiently large b_i.

Notice that if $\beta = \infty$, $\min_{\overline{\Omega}} \overline{u}_j = 0$, so the preceding argument will not work. To circumvent this difficulty, we proceed as follows. Recall from Theorem 2.1 that the principal eigenvalue σ in

$$\begin{aligned} D_i \Delta\phi + m(x)\phi &= \sigma\phi \quad \text{in} \quad \Omega \\ \phi &= 0 \quad \text{on} \quad \partial\Omega \end{aligned} \tag{6A.18}$$

depends continuously on $m \in L^\infty(\Omega)$ thought of as a subspace of $L^p(\Omega)$ for p large. Consequently, there is an open subset Ω' of Ω with $\overline{\Omega}' \subset \Omega$ so that if $m_{\Omega'}(x) = \begin{cases} a_i & \text{in} \quad \Omega\backslash\Omega' \\ -1 & \text{in} \quad \Omega' \end{cases}$, the principal eigenvalue in (6A.18) with $m = m_{\Omega'}$ is negative. Since $\min_{\overline{\Omega}'} \overline{u}_j > 0$, $a_i - b_i\overline{u}_j < m_{\Omega'}$ for b_i sufficiently large. Since σ in (6A.18) increases with m, $\tilde{\sigma}_i < 0$ for b_i sufficiently large.

To establish Part (ii) of Theorem 6.12, we first employ the Implicit Function Theorem as in Section 3.4 to establish that the mapping $\beta \to \overline{u}_i(\beta)$ is differentiable from $[0, \infty]$ into $C^{2+\alpha}(\overline{\Omega})$. We next normalize $\psi_i(\beta)$ in (6.97) by

$$\int_\Omega \psi_i^2 dx = 1 \tag{6A.19}$$

and define a differentiable map $\rho : C^{2+\alpha}(\overline{\Omega}) \times [0, \infty) \times [0, \infty) \to C^\alpha(\overline{\Omega}) \times I\!R \times C^{1+\alpha}(\partial\Omega)$ by

$$\rho(\psi, b, \beta) = \left(D_i\Delta\psi + (a_i - b\overline{u}_j(\beta))\psi, \int_\Omega \psi^2 dx - 1, \alpha_i\nabla\psi\cdot\vec{n} + \beta\psi\right). \tag{6A.20}$$

It is not difficult to calculate from (6A.20) that

$$\begin{aligned} \left[\frac{\partial\rho}{\partial(\psi, b)}(\psi, b, \beta)\right](z, c) = \Big(& D_i\Delta z + (a_i - b\overline{u}_j(\beta))z \\ & -c\overline{u}_j(\beta)\psi, \ 2\int_\Omega \psi z dx, \ \alpha_i\nabla z\cdot\vec{n} + \beta z\Big). \end{aligned} \tag{6A.21}$$

If we choose $(\psi, b, \beta) = (\psi_i(\beta), \overline{b}_i(\beta), \beta)$, then $\rho(\psi, b, \beta) = (0, 0, 0)$. Suppose for this choice of (ψ, b, β) that $\left[\frac{\partial\rho}{\partial(\psi, b)}(\psi, b, \beta)\right](z, c) = (0, 0, 0)$. Then

$$\begin{aligned} & D_i\Delta z + (a_i - \overline{b}_i(\beta)\overline{u}_j(\beta))z - c\overline{u}_j(\beta)\psi_i = 0 \ \text{in} \quad \Omega \\ & \int_\Omega z\psi_i = 0 \\ & \alpha_i\nabla z\cdot\vec{n} + \beta z = 0 \qquad \text{on} \quad \partial\Omega. \end{aligned} \tag{6A.22}$$

The Fredholm Alternative, Theorem 1.10 (Gilbarg and Trudinger, 1977), guarantees that there can be a $z \in C^{2+\alpha}(\overline{\Omega})$ solving the first and third equations in (6A.22) if and only if

$c\int_\Omega \overline{u}_j(\beta)\psi_1^2 = 0$ which in turn is equivalent to $c = 0$. Consequently, $z = k\psi_i$ for some constant k. But from the second equation in (6A.22), $k\int_\Omega \psi_i^2$ must be 0, and hence k is zero. Consequently, the mapping $\left[\frac{\partial p}{\partial(\psi, b)}(\psi, b, \beta)\right]$ is injective.

To see that the mapping is also surjective, consider the system

$$\begin{aligned} &D_i\Delta z + (a_i - \overline{b}_i(\beta)\overline{u}_j(\beta))z - c\overline{u}_j(\beta)\psi_i = f \text{ in } \quad \Omega \\ &\int_\Omega z\psi_i dx = g \\ &\alpha_i\nabla z\cdot\vec{n} + \beta z = h \qquad\qquad \text{on} \quad \partial\Omega \end{aligned} \tag{6A.23}$$

with $f \in C^\alpha(\overline{\Omega})$, $g \in I\!R$ and $h \in C^{1+\alpha}(\partial\Omega)$. The Fredholm Alternative implies that there will be a z solving the first and third equations of (6A.23) if and only if

$$\frac{D_i}{\alpha_i}\int_{\partial\Omega} h\psi_i dS = \int_\Omega (f + c\overline{u}_j\psi_i)\psi_i dx,$$

or equivalently,

$$c = \frac{-\int_\Omega f\psi_i dx + \dfrac{D_i}{\alpha_i}\int_{\partial\Omega} h\psi_i dS}{\int_\Omega \overline{u}_j\psi_i^2 dx}. \tag{6A.24}$$

Since $\overline{u}_j$ and ψ_i are positive in Ω, c in (6A.24) is well defined. So choose c as in (6A.24) and let z_0 be some solution to the first and third equations in (6A.23). It is not difficult to observe that $z_0 + k\psi_i$ also solves the first and third equations in (6A.23) for any choice of $k \in I\!R$. To satisfy the second equation in (6A.23), we need only have

$$\int_\Omega (z_0 + k\psi_i)\psi_i dx = g$$

or equivalently

$$k = g - \int_\Omega z_0\psi_i dx.$$

Consequently, $\left[\frac{\partial \rho}{\partial(\psi, b)}(\psi, b, \beta)\right]$ is surjective, and hence is a linear homeomorphism.

Part (ii) of Theorem 6.12 now follows from the Implicit Function Theorem for $\beta < \infty$. To extend the result to $\beta = \infty$, we rewrite the boundary condition on Ω as

$$\alpha_i\gamma\nabla u\cdot\vec{n} + u = 0 \tag{6A.25}$$

with $\gamma = \dfrac{1}{\beta}$ and argue as before.

7

Nonmonotone Systems

7.1 Introduction

This chapter brings our present discussion of modeling interacting biological species in bounded habitat patches via reaction-diffusion equations to an end. However, so very much remains to be discovered mathematically and interpreted and communicated ecologically about the impact of spatial heterogeneity in such models that this volume is better taken not as definitive treatise on the subject, but rather as a primer leading (hopefully) to further inquiry. As we shall see, such is especially the case with respect to the topic of this chapter, *nonmonotone systems*. We recall that a dynamical or semi-dynamical system associated with a model for several interacting biological species is said to be *monotone* if there is a partial ordering of the underlying configurations (i.e., tuples) of species densities that is preserved by the dynamical or semi-dynamical system. Put another way, the system is monotone if whenever two configurations of initial densities are ordered, their temporal evolutions under the system are ordered in the same way for all subsequent time. When no such ordering is known, the system is referred to as *nonmonotone*.

We observed in Section 5.2 that in the context of a monotone system associated with a model for several interacting biological species the conclusions that can be drawn from a prediction of permanence in the system are substantially sharpened. Following Hess and Lazer (1991), we have used the term *compressive* to describe a permanent monotone system. In a compressive system, the asymptotic "floor" and "ceiling" come from perturbing the "endpoints" of a globally attracting order interval of componentwise positive configurations of spaces densities. With permanence, in general, this kind of quantitative information about an asymptotic "floor" or "ceiling" is lost. One knows the existence of an asymptotic "floor" and "ceiling" but *a priori* nothing about their "heights". Moreover, the *transient* dynamics in a compressive system are also well behaved. In particular, the trajectory corresponding to any particular initial configuration is trapped between a trajectory that increases toward the lower "endpoint" of the globally attracting order interval and a trajectory that decreases toward the upper "endpoint". In general, just knowing that a system is permanent provides no information regarding its transient behavior at all. (At this point it is perhaps useful to emphasize some subtle features of nonautonomous reaction-diffusion models regarding monotonicity and increasing or decreasing orbits. Recall from Section 5.5 that in contrast with their counterparts in autonomous models, the solution trajectories for nonautonomous models do not by themselves constitute continuous time semi-dynamical systems. However, having the solution trajectories for a reaction-diffusion system preserve the ordering between

Spatial Ecology via Reaction-Diffusion Equations R.S. Cantrell and C. Cosner
 ISBN: 0-471-49301-5

two configurations of initial densities does not depend on whether the system is autonomous or nonautonomous, but rather depends only on having the appropriate form of species interactions (two species competition, mutualism among any number of species). So, for instance, in a nonautonomous reaction-diffusion model for two competitors in a bounded habitat, if two configurations of initial densities are ordered, the orbits corresponding to these initial density configurations maintain the same ordering for all subsequent time. However, this feature does not mean either orbit is monotonic in continuous time, nor should it. Recall that the original work of Hess and Lazer (1991) on compressivity was concerned with periodic-parabolic models of two-species competition in a bounded habitat. Such models are, of course, continuous time models, although their solution trajectories do not by themselves constitute a continuous time dynamical system. But, to study them, Hess and Lazer employed a discrete time dynamical system framework via a period or Poincaré map. This approach amounts to looking at a "snapshot" of the time evolution of species densities every T units of time, where T is the underlying period. Relative to this dynamical system framework, an orbit is say, increasing, when points corresponding to the same "phase" relative to T are increasing from one discrete time measurement to the next. In contrast, in the temporally autonomous counterpart, when a trajectory is increasing, the configuration of species densities at some instant in time is below the configuration at all subsequent times. [Recall that such orbits arise when the initial configuration of densities is a lower solution to the corresponding elliptic system.] However, in the periodic-parabolic setting, the "end points" of the globally attracting order interval for the discrete dynamical system correspond to points on periodic orbits of the continuous time model. Consequently, the sense of increasing we have just described is all that can or should be expected.)

The preceding discussion indicates what nonmonotone systems associated with models for several interacting biological species in general *lack* in comparison with monotone systems. It is natural to ask if there are commonalities regarding nonmonotone systems or mathematical methods for analyzing them significant enough or widespread enough to warrant noting them at this point in our discussion. We believe there are several such. The first is a recurring theme in this volume. Namely, flows or semi-flows for nonmonotone systems can exhibit a broad range of dynamic behavior (both at transient and asymptotic time scales), and permanence is a mathematical construct describing asymptotic coexistence in such systems that can accommodate a wide array of particular dynamical behavior. In Section 4.1, we illustrated some of the range of possible dynamic outcomes in nonmonotone systems via three ODE examples ((4.2), a Beddington-DeAngelis predator-prey model in which componentwise positive trajectories spiral toward a componentwise positive limit cycle; (4.16), a Hastings-Powell three species food chain model which is permanent but exhibits chaotic dynamics; and (4.17), a May-Leonard model for three competing species in which componentwise positive trajectories approach a heteroclinic cycle located among the configurations of species densities in which at least one competitor is absent, so that the system fails to be permanent). The extreme sensitivity of the outcome of the dynamics of a system such as (4.16) to initial data brings home once more the point from Section 4.1 that asymptotic coexistence is less about the complexity of the dynamics of a system of interacting species than it is about the location of the dynamics relative to the extinction states of the system. Indeed, in a system with chaotic dynamics a finding of permanence may be as strong a statement about asymptotic coexistence as is possible.

A second widespread feature in nonmonotone systems for interacting biological species is that the range of asymptotic behaviors that are possible may depend strongly on the qualitative form of local species interactions. As an example, consider the model for two species competing in a bounded habitat Ω given by

$$\frac{\partial u_i}{\partial t} = \mu_i \Delta_i + u_i(r_i - u_i + g_i(x, u_1, u_2)) \quad \text{in} \quad \Omega \times (0, \infty) \tag{7.1}$$

$$\beta_i(x)u_i + (1 - \beta_i(x))\nabla u_i \cdot \vec{n} = 0 \quad \text{on} \quad \partial\Omega \times (0, \infty), \tag{7.2}$$

$i = 1, 2$, where $\dfrac{\partial g_i}{\partial u_j} \leq 0$ for $i = 1, 2$ and $g_1(x, u_1, 0) \equiv 0 \equiv g_2(x, 0, u_2)$. We know that (7.1)–(7.2) is permanent provided that each species persists in the absence of the other and that for $i, j = 1, 2 \quad i \neq j$, species i can invade the habitat when species j is at its carrying capacity $\overline{u}_j$ in the absence of species i. Moreover, we also know that mathematically these conditions translate into the positivity of the principal eigenvalues σ_i and $\tilde{\sigma}_i$, $i = 1, 2$, for appropriate analogues to the elliptic operators in (6.36) and (6.38). Here, for instance,

$$\tilde{\sigma}_1 = \max_{\substack{\phi \in V \\ \phi \neq 0}} \left\{ \frac{-\mu_1 \int_\Omega |\nabla \phi_1|^2 dx + \int_{\Omega_1} (r_1 + g_1(x, 0, \overline{u}_2))\phi_1^2 dx - \mu_1 \int_{\{x:\beta_1 \neq 1\}} \frac{\beta_1}{1 - \beta_1} \phi_1^2 dS}{\int_\Omega \phi_1^2 dx} \right\} \tag{7.3}$$

for an appropriate subspace V of $W^{1,2}(\Omega)$ (which will depend on β_1 (cf. (2.11))). It is evident that whether species 1 is predicted to persist in (7.1)–(7.2) depends only on whether its *per capita* growth rate r_1 is large enough relative to the competitive effect of species 2 (reflected in the term $g_1(x, 0, \overline{u}_2)$ in (7.3)) and its loss through the boundary $\partial\Omega$ of the habitat patch.

The form of g_1 will definitely impact on whether or not species 1 does invade the habitat when the density of species 2 is $\overline{u}_2$, and hence may alter the conditions under which the model (7.1)–(7.2) predicts persistence of species 1. However, the form of g_1 cannot eliminate the possibility that species 1 invades the habitat when the density of species 2 is $\overline{u}_2$. Put another way, the form of g_1 influences the outcome of the interaction between the two competing species quantitatively, and such may have significant ecological ramifications. However, the form of g_1 does not qualitatively alter the possible outcomes of the interaction. On the other hand, in three species competition models, the form of the local competitive interactions may in some instances determine whether or not permanence is a possible outcome for the system. We treat such a situation in Section 7.3 when we discuss the phenomenon of competition mediated coexistence (Cantrell and Ward, 1997; May and Leonard, 1975). By the term "competition mediated coexistence," we refer to a situation in which a species coexists with two other competitors when both are present but fails to persist when one of the other competitors is absent. Such a phenomenon has significant implications for community structure. We shall see that whether competition mediated coexistence is possible at all may depend in a delicate way on the nature of the local interactions.

As another example, we showed in Section 5.4 that the inclusion of predator self-regulation in a two-species predator-prey model may allow one to control the transient as well as asymptotic phases of trajectories when the predator is introduced at low densities into a habitat patch where its prey is well-established. Another context in which predator self-regulation can play a pivotal role is in facilitating the persistence of two consumers on a single resource. To this end, let us modify the example (6.17) (Armstrong and McGehee, 1980) from Section 6.1, and consider

$$\begin{aligned}\frac{du_1}{dt} &= c_1u_1\left(-d_1+\frac{a_1v}{v+e_1u_1+\Gamma_1}\right)\\ \frac{du_2}{dt} &= c_2u_2\left(-d_2+\frac{a_2v}{v+e_2u_2+\Gamma_2}\right)\\ \frac{dv}{dt} &= rv\left(1-\frac{v}{K}\right)-v\left(\frac{b_1u_1}{v+e_1u_1+\Gamma_1}+\frac{b_2u_2}{v+e_2u_2+\Gamma_2}\right).\end{aligned}\tag{7.4}$$

Notice that as was the case with (6.17) we cannot employ a pseudoequilibrium hypothesis to reduce (7.4) to a system of two species competition, and instead must analyze the dynamics of the full three component system. Hence, in the context of our discussion in Section 6.1, the basic means by which we are allowing for the possibility of two consumers coexisting on a single resource in (7.4) is the same as was the case with (6.17). However, the inclusion of predator self-limitation via Beddington-DeAngelis functional and numerical response terms in (7.4) is a major difference between the models, and one that has a profound impact, as we shall see.

In analyzing (7.4), observe that the resource equation in (7.4) is logistic in form in the absence of the consumers, so that for any $\epsilon > 0$, the resource density $v(t) \le K + \varepsilon$ for large enough values of t, independent of its initial density $v(0)$. Consequently, it follows from the observations that $\dfrac{a_iv}{v+e_iu+\Gamma_i} \le \dfrac{a_iv}{v+\Gamma_i}$ and that $\dfrac{a_iv}{v+\Gamma_i} \le \dfrac{a_i(K+\varepsilon)}{(K+\varepsilon)+\Gamma_i}$ for all large values of t that if

$$-d_i+\frac{a_iK}{K+\Gamma_i}<0,$$

$\dfrac{1}{u_i}\dfrac{du_i}{dt} < 0$ for all large values of t and consequently $\lim_{t\to\infty} u_i(t) = 0$, independent of $u_i(0)$. As a result, we must posit that

$$\frac{a_iK}{K+\Gamma_i} > d_i \tag{7.5}$$

in order for there to be a realistic possibility of species i persisting. So we assume (7.5) holds for $i = 1, 2$.

For large values of t, we have

$$\frac{du_i}{dt} \le c_iu_i\left(-d_i+\frac{a_i(K+\varepsilon)}{K+\varepsilon+e_iu_i+\Gamma_i}\right) \tag{7.6}$$

for $i = 1, 2$. Set $F_i(w) = -d_i + \dfrac{a_i(K+\varepsilon)}{K+\varepsilon+e_iw+\Gamma_i}$. Then $F_i(0) > 0$ by (7.5), $F_i'(w) = \dfrac{-a_i(K+\varepsilon)e_i}{(K+\varepsilon+e_iw+\Gamma_i)^2} < 0$ for all $w \ge 0$, and $F_i(w) = 0 \Leftrightarrow w = \dfrac{a_i(K+\varepsilon)-d_i(K+\varepsilon+\Gamma_i)}{d_ie_i}$, which is greater than 0, again by (7.5). Consequently, all positive solutions to

$$\frac{dw}{dt} = c_iwF_i(w)$$

converge to $w = \dfrac{a_i(K+\varepsilon) - d_i(K+\varepsilon+\Gamma_i)}{d_i e_i}$. By (7.6), we may assert that

$$u_i(t) < (1+\varepsilon)\frac{(a_i(K+\varepsilon) - d_i(K+\varepsilon+\Gamma_i))}{d_i e_i} = \gamma_i(\varepsilon) \tag{7.7}$$

for all large values of t. Consequently, we have asymptotic ceilings on all three components of (7.4). Employing (7.7) we have

$$\frac{dv}{dt} \geq rv\left(1 - \frac{v}{K}\right) - v\left(\frac{b_1\gamma_1(\varepsilon)}{v + e_1\gamma_1(\varepsilon) + \Gamma_1} + \frac{b_2\gamma_2(\varepsilon)}{v + e_2\gamma_2(\varepsilon) + \Gamma_2}\right) \tag{7.8}$$

for all large values of t. Setting $G(y) = r\left(1 - \frac{y}{K}\right) - \dfrac{b_1\gamma_1(\varepsilon)}{y + e_1\gamma_1(\varepsilon) + \Gamma_1} - \dfrac{b_2\gamma_2(\varepsilon)}{y + e_2\gamma_2(\varepsilon) + \Gamma_2}$, we have $G'(y) < 0$ for $y \geq 0$ (provided e_1 and e_2 are large enough) and $G(K) < 0$. Note that $G(0) = r - \dfrac{b_1\gamma_1(\varepsilon)}{e_1\gamma_1(\varepsilon) + \Gamma_1} - \dfrac{b_2\gamma_2(\varepsilon)}{e_2\gamma_2(\varepsilon) + \Gamma_2} > r - \dfrac{b_1}{e_1} - \dfrac{b_2}{e_2}$ so $G(0) > 0$ if e_1 and e_2 are large enough. Consequently, in such case, positive solutions of

$$\frac{dy}{dt} = yG(y)$$

converge to the unique positive root of $G(y) = 0$. If we label this value by $\overline{y} = \overline{y}(\varepsilon)$, we may assert from (7.8) that $v \geq \overline{y}(\varepsilon)(1-\varepsilon)$ for all large values of t, providing an asymptotic floor on v. Thus for large values of t,

$$\frac{du_i}{dt} \geq c_i u_i\left(-d_i + \frac{a_i\overline{y}(\varepsilon)(1-\varepsilon)}{\overline{y}(\varepsilon)(1-\varepsilon) + e_i u_i + \Gamma_i}\right) \tag{7.9}$$

Setting $\overline{F}_i(w) = -d_i + \dfrac{a_i\overline{y}(\varepsilon)(1-\varepsilon)}{\overline{y}(\varepsilon)(1-\varepsilon) + e_i w + \Gamma_i}$, we may show as in the preceding that for $i = 1, 2$, positive solutions to $\dfrac{dw}{dt} = w\overline{F}_i(w)$ converge to the unique positive solution of $\overline{F}_i(w) = 0$, provided that $\overline{F}_i(0) > 0$, which is equivalent to

$$\frac{a_i\overline{y}(\varepsilon)(1-\varepsilon)}{\overline{y}(\varepsilon)(1-\varepsilon) + \Gamma_i} > d_i. \tag{7.10}$$

It is not too difficult to show that $\lim\limits_{\substack{e_1\to\infty \\ e_2\to\infty}} \overline{y}(\varepsilon) = K$, so that (7.10) holds by (7.5) for large enough values of e_1 and e_2. One may then employ (7.9) to obtain asymptotic floors on u_1 and u_2.

Recalling the interpretation of e_1 and e_2 from Section 4.1, we have that the model (7.4) predicts that two consumers can coexist on a single resource if the mutual interference among predators is a sufficiently strong self-regulatory effect. Indeed, we have shown that for large enough e_1 and e_2, (7.4) is permanent. In particular, it follows from Section 4.6 that then there is a componentwise positive equilibrium. (Recall that the prediction of (6.17) was conditional persistence and that generically (6.17) did not admit a componentwise positive equilibrium.)

Our discussion here and in Section 6.1 of ODE models for two consumers sharing an essential resource shows that such models can predict dynamical behavior that ranges from competitive exclusion of one consumer by the other to permanence of the system, depending upon the assumptions that are made upon the functional and numerical response terms in the model. In particular, inclusion of the predator self-limitation in the functional and numerical response terms of the model allows for the possibility of permanence in the system, in very much the same way that inclusion of predator self-limitation allowed for bounding transient orbits for predator-prey models in Section 5.4. Indeed, these examples offer very strong evidence that predator-self limitation has a stabilizing impact in nonmonotone systems. In the case of ODE systems, the work of Butler et al. (1990) provides theoretical support for this assertion as well.

The preceding discussion of whether two consumer species can coexist on a single essential resource species not only serves to illustrate that the possible outcomes in a nonmonotone system are quite dependent upon the assumptions that are placed upon local species interactions but also demonstrates that mathematical techniques based on monotonicity or order may play an important role in the mathematical analysis of nonmonotone systems. Indeed, our finding of permanence in (7.4) was based on a practical persistence argument of the type described in Section 5.3. We will take such an approach again in Section 7.4 when we discuss the impact of habitat size on the persistence of food chains. In other nonmonotone systems for several interacting biological species, practical persistence techniques or estimates may need to be combined with abstract permanence results (such as the use of Average Lyapunov Functions or the Acyclicity Theorem) to demonstrate that the system predicts the coexistence of the species involved. We illustrate such an approach in Section 7.3 with two different examples involving three competitors.

7.2 Predator Mediated Coexistence

An important notion in community ecology is that of a "keystone species" (Paine, 1969). A "keystone species" may be defined as "one on whose presence the whole structure and dynamics of [an ecological] community depends" (Horn et al., 1989). In Paine's own words, "[t]hese individual populations are the keystone of the community's structure, and the integrity of the community and its unaltered persistence through time, that is, stability, are determined by their activities and abundances" (Paine, 1969).

Paine's conceptualization was motivated by his study (Paine, 1966) of ecological communities within the rocky intertidal pools that exist along the Pacific coast of North America. In such pools, the adult stages of many organisms are sessile, and competition among them is primarily competition for space. Preceding studies (Connell, 1961a, b) had demonstrated a hierarchy in terms of competitive dominance among some such organisms, and that competition was ameliorated by predation. Paine (1966) showed experimentally that systematically removing a certain predator, the starfish species *Pisaster ochraceus*, from an intertidal pool "patch" caused the remaining ecological community to become increasingly less diverse, tending in a successional manner over time toward what was essentially a monoculture of the bi-valve *Mytilus californianus*. In Paine's words, "in the absence of a complicating factor (predation), there is a "winner" in the competition for space, and the local system tends toward simplicity" (Paine, 1966).

The phenomenon illustrated in Paine's work, wherein the top predator acts as a keystone species, is frequently referred to as *predator-mediated coexistence* (Butler and Wolkowicz, 1987; Hsu, 1981; Abrams, 1999; Takeuchi, 1985). In the paper in which they developed the notion of *strong persistence* (see Section 4.1), Freedman and Waltman (1984) established predator-mediated coexistence (in the sense of strong persistence) in a Lotka-Volterra ODE model for two competing species that are preyed upon by a common predator. In the absence of the predator, one of the prey species competitively excludes the other. The predator is presumed to be able to persist on either of the prey species by itself. Freedman and Waltman (1984) show that if the impact of predation is felt more strongly by the dominant prey species, the inclusion of the predator may result in coexistence of the three species community. These results were extended in Dunbar et al. (1986) to the spatially explicit case via a reaction-diffusion model in which species interactions take place in a bounded habitat with a closed or reflecting boundary. Again coexistence is in the sense of strong persistence. (See also Caristi et al. (1992).) Hutson and Schmitt (1992) showed that in fact the coexistence of the three species in the models of Freedman and Waltman (1984) and Dunbar et al. (1986) is in fact in the sense of permanence. Cantrell et al. (1993b) established permanence in the reaction-diffusion Lotka-Volterra setting via the Acyclicity Theorem when the underlying bounded habitat is taken to have an absorbing or lethal boundary. Our treatment here follows that in Cantrell et al. (1993b).

Consider the model

$$\begin{aligned}\frac{\partial u_1}{\partial t} &= \Delta u_1 + u_1(a - u_1 - c_{12}u_2 - c_{13}u_3)\\ \frac{\partial u_2}{\partial t} &= \Delta u_2 + u_2(a - c_{21}u_1 - u_2 - c_{23}u_3)\\ \frac{\partial u_3}{\partial t} &= \Delta u_3 + u_3(a + c_{31}u_1 + c_{32}u_2 - u_3)\end{aligned} \tag{7.11}$$

$$\text{in} \quad \Omega \times (0, \infty)$$

$$u_i = 0 \tag{7.12}$$

$$\text{on} \quad \partial\Omega \times (0, \infty)$$

for $i = 1, 2, 3$. Here Ω, as usual, denotes the isolated bounded habitat patch. The densities of the two prey species are given by u_1 and u_2, while u_3 denotes the density of the predator species. The local *per capita* growth rates, competition coefficients and predation coefficients are all taken to be the positive constants.

The model (7.11)–(7.12) is a special case of the model (4.1) in Cantrell et al. (1993b). Specifically, we take the local *per capita* growth and diffusion rates to be the same for each of the species in question. Such an assumption allows us to simplify our presentation substantially and consequently to explore more fully the relations among the coefficients of (7.11)–(7.12) that are required for permanence. In the more general (and likely more realistic) case in Cantrell et al. (1993b), the local *per capita* growth for species i is a_i. There a_3 may be either positive or negative. When $a_3 > 0$, the predator must be a generalist with resources other than species 1 and 2. The model imposes self-regulation upon the predator species. As we demonstrated in Section 4.5, such a requirement is not strictly speaking needed in order to obtain permanence in the two two-species predator-prey subsystems which arise in analyzing (7.11)–(7.12). However, in order to obtain permanence in the full three-species system, we require quantitative estimates for the asymptotic "ceilings" and "floors" of the two species subsystems and hence have imposed self-regulation on the predator. With this requirement of self-regulation, we may proceed in a manner analogous to that in Section 4.4 to convert (7.11)–(7.12) into a dissipative semi-dynamical system on the positive cone K of $[C_0^1(\overline{\Omega})]^3$.

To have each species persist in the bounded habitat in the absence of the other two, we must consider the logistic equation

$$\frac{\partial u_i}{\partial t} = \Delta u_i + u_i(a - u_i) \quad \text{in} \quad \Omega \times (0, \infty) \tag{7.13}$$

subject to the boundary condition

$$u_i = 0 \quad \text{on} \quad \partial\Omega \times (0, \infty). \tag{7.14}$$

As is now familiar (see, for example, Section 4.4), we require that the principal eigenvalue σ_1 be positive in

$$\begin{aligned} &\Delta\phi_1 + a\phi_1 = \sigma_1\phi \text{ in } \quad \Omega \\ &\phi_1 = 0 \qquad\qquad\qquad \text{on} \quad \partial\Omega. \end{aligned} \tag{7.15}$$

Recall that $\sigma_1 > 0$ in (7.15) is equivalent to $a > \lambda_0^1(\Omega)$, whereas in Section 4.4, $\lambda_0^1(\Omega)$ denotes the principal eigenvalue of $-\Delta$ on the habitat Ω subject to homogeneous Dirichlet boundary conditions. We know in such case that θ_a is the global attractor for positive solutions to (7.13)–(7.14) where θ_a is the unique positive solution of

$$\begin{aligned} &\Delta w + w(a - w) = 0 \text{ in } \quad \Omega \\ &w = 0 \qquad\qquad\qquad \text{on} \quad \partial\Omega. \end{aligned} \tag{7.16}$$

Next let us look at the subsystem of (7.11)–(7.12)

$$\begin{aligned} \frac{\partial u_1}{\partial t} &= \Delta u_1 + u_1(a - u_1 - c_{12}u_2) \\ \frac{\partial u_2}{\partial t} &= \Delta u_2 + u_2(a - c_{21}u - u_2) \\ &\qquad\qquad\qquad \text{in} \quad \Omega \times (0, \infty) \end{aligned} \tag{7.17}$$

$$\begin{aligned} u_1 &= 0 = u_2 \\ &\qquad\qquad\qquad \text{on} \quad \partial\Omega \times (0, \infty), \end{aligned} \tag{7.18}$$

that arises under the assumption that the predator is absent. As in Section 4.4, we know that species 1 can invade the habitat when species 2 is at its equilibrium density θ_a precisely when the principal eigenvalue σ_2 is positive in

$$\begin{aligned} &\Delta\phi_2 + \phi_2(a - c_{12}\theta_a) = \sigma_2\phi_2 \text{ in } \quad \Omega \\ &\phi_2 = 0 \qquad\qquad\qquad \text{on} \quad \partial\Omega. \end{aligned} \tag{7.19}$$

Moreover, we know that σ_2 is positive in (7.19) if and only if

$$c_{12} < 1. \tag{7.20}$$

It follows from Theorem 5.7 that all componentwise positive solutions to (7.17)–(7.18) converge to $(\theta_a, 0)$ provided that (7.17)–(7.18) fails to have a componentwise positive equilibrium. This issue was addressed initially in Cosner and Lazer (1984). Supposing that (7.17)–(7.18) admits a componentwise positive equilibrium, multiply the first equation in (7.17) by the density u_2 and the second by u_1 and integrate. Green's Second Identity implies

$$\int_\Omega [(c_{21} - 1)u_1 + (1 - c_{12})u_2]u_1 u_2 dx = 0. \tag{7.21}$$

It is immediate from (7.20) and (7.21) that no such equilibrium to (7.17)–(7.18) is possible if

$$c_{21} > 1. \tag{7.22}$$

Now let us examine the two subsystems of (7.11)–(7.12) that arise under the assumption that one of the prey species is absent. For specificity, let us assume that $u_2 = 0$ and consider

$$\begin{aligned} \frac{\partial u_1}{\partial t} &= \Delta u_1 + u_1(a - u_1 - c_{13}u_3) \\ \frac{\partial u_3}{\partial t} &= \Delta u_3 + u_3(a + c_{31}u - u_3) \\ &\qquad\qquad \text{in} \quad \Omega \times (0, \infty) \end{aligned} \tag{7.23}$$

$$\begin{aligned} u_1 &= 0 = u_3 \\ &\qquad \text{on} \quad \partial\Omega \times (0, \infty). \end{aligned} \tag{7.24}$$

We know from Section 4.5 that (7.23)–(7.24) is permanent provided the principal eigenvalues σ_3 and σ_4 are positive in

$$\begin{aligned} &\Delta\phi_3 + \phi_3(a - c_{13}\theta_a) = \sigma_3\phi_3 \text{ in} \quad \Omega \\ &\phi_3 = 0 \qquad\qquad \text{on} \quad \partial\Omega \end{aligned} \tag{7.25}$$

and

$$\begin{aligned} &\Delta\phi_4 + \phi_4(a + c_{31}\theta_a) = \sigma_4\phi_4 \text{ in} \quad \Omega \\ &\phi_4 = 0 \qquad\qquad \text{on} \quad \partial\Omega, \end{aligned} \tag{7.26}$$

respectively. Consequently we must require

$$c_{13} < 1 \tag{7.27}$$

to have $\sigma_3 > 0$, but we only need to have

$$c_{31} > 0$$

to have $\sigma_4 > 0$. So long as

$$c_{13}(1 + c_{31}) < 1 \tag{7.28}$$

we may employ a somewhat involved practical persistence argument as in Cantrell et al. (1993b) to conclude that componentwise positive solutions to (7.23)–(7.24) converge to

$$\left(\frac{1-c_{13}}{1+c_{13}c_{31}}\theta_a, \frac{1+c_{31}}{1+c_{13}c_{31}}\theta_a\right). \tag{7.29}$$

(For the sake of exposition, we choose to present the argument in the Appendix to the chapter.)

Analogous reasoning applies for the subsystem

$$\begin{aligned}\frac{\partial u_2}{\partial t} &= \Delta u_2 + u_2(a - u_2 - c_{23}u_3)\\ \frac{\partial u_3}{\partial t} &= \Delta u_3 + u_3(a + c_{32}u_2 - u_3)\\ &\qquad\qquad \text{in} \quad \Omega \times (0, \infty)\end{aligned} \tag{7.30}$$

$$\begin{aligned} u_2 = 0 = u_3\\ \text{on} \quad \partial\Omega \times (0, \infty)\end{aligned} \tag{7.31}$$

that arises under the assumption that $u_1 = 0$. As a result, if

$$c_{23} < 1, \tag{7.32}$$

$c_{32} > 0$ and

$$c_{23}(1 + c_{32}) < 1 \tag{7.33}$$

componentwise positive solutions to (7.30)–(7.31) converge to

$$\left(\frac{1-c_{23}}{1+c_{23}c_{32}}\theta_a, \frac{1+c_{32}}{1+c_{23}c_{32}}\theta_a\right). \tag{7.34}$$

Proceeding as in Sections 4.3 and 4.4, we may now establish that if $X \subseteq K$ and $S \subseteq \partial K$ are as in Theorem 4.1, then $\omega(S)$ has the covering

$$\{M_1, M_2, M_3, M_4, M_5, M_6\}$$

where

$$\begin{aligned} M_1 &= \{(0, 0, 0)\}\\ M_2 &= \{(\theta_a, 0, 0)\}\\ M_3 &= \{(0, \theta_a, 0)\}\\ M_4 &= \{(0, 0, \theta_a)\}\\ M_5 &= \left\{\left(\frac{1-c_{13}}{1+c_{13}c_{31}}\theta_a, 0, \frac{1+c_{31}}{1+c_{13}c_{31}}\theta_a\right)\right\}\\ M_6 &= \left\{\left(0, \frac{1-c_{23}}{1+x_{23}c_{32}}\theta_a, \frac{1+c_{32}}{1+c_{23}c_{32}}\theta_a\right)\right\}. \end{aligned}$$

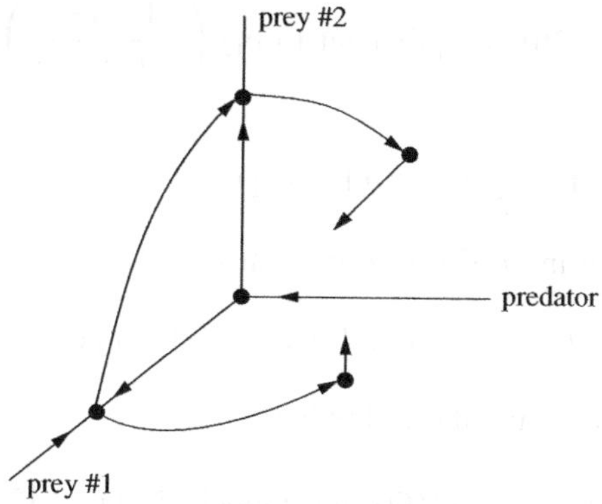

Figure 7.1 Schematic diagram of the equilibria and connecting orbits for a one-predator two-prey system corresponding to predator-mediated coexistence. The geometry of the diagram indicates that the flow in the boundary of the positive cone is acyclic and that no equilibrium has a stable manifold which intersects the interior of the positive cone. This structure can be used to establish permanence, which then implies that the system predicts predator-mediated coexistence.

It follows from (7.15), (7.20), (7.22), (7.27), (7.28), (7.32) and (7.33) that the covering of $\omega(S)$ is acyclic (see Figure 7.1), and that for $i = 1, 2, 3, 4$, M_i is isolated with respect to the semiflow corresponding to (7.11)–(7.12) on both X and S and also that

$$W^s(M_i) \cap \operatorname{int} K = \emptyset. \tag{7.35}$$

To invoke Theorem 4.3 to assert that (7.11)–(7.12) is permanent, we need for it to be the case that species 1 can invade Ω when (u_2, u_3) are as in (7.34) and that species 2 can invade Ω when (u_1, u_3) are as in (7.29). In that case M_5 and M_6 are isolated with respect to both X and S and (7.35) holds for $i = 5, 6$, and hence Theorem 4.3 can be applied. Invasibility of species 1 when $(u_2, u_3) = \left(\frac{1 - c_{23}}{1 + c_{23}c_{32}}\theta_a, \frac{1 + c_{32}}{1 + c_{23}c_{32}}\theta_a\right)$ is equivalent to having the principal eigenvalue σ_5 be positive in

$$\begin{aligned} &\Delta\phi_5 + \phi_5\left(a - c_{12}\left(\frac{1 - c_{23}}{1 + c_{23}c_{32}}\theta_a\right) - c_{13}\left(\frac{1 + c_{32}}{1 + c_{23}c_{32}}\theta_a\right)\right) = \sigma_5\phi_5 && \text{in } \Omega \\ &\phi_5 = 0 && \text{on } \partial\Omega. \end{aligned} \tag{7.36}$$

Likewise, invasibility of species 2 when $(u_1, u_3) = \left(\frac{1 - c_{13}}{1 + c_{13}c_{31}}\theta_a, \frac{1 + c_{31}}{1 + c_{13}c_{31}}\theta_a\right)$ is equivalent to having the principal eigenvalue σ_6 be positive in

$$\begin{aligned} &\Delta\phi_6 + \phi_6\left(a - c_{21}\left(\frac{1 - c_{13}}{1 + c_{13}c_{31}}\theta a\right) - c_{23}\left(\frac{1 + c_{31}}{1 + c_{13}c_{31}}\theta_a\right)\right) = \sigma_6\phi_6 && \text{in } \Omega \\ &\phi_6 = 0 && \text{on } \partial\Omega. \end{aligned} \tag{7.37}$$

Having σ_5 be positive in (7.36) is equivalent to $c_{12}\left(\frac{1-c_{23}}{1+c_{23}c_{32}}\right)+c_{13}\left(\frac{1+c_{32}}{1+c_{22}c_{32}}\right)<$ 1, or

$$c_{12}(1-c_{23})+c_{13}(1+c_{32})<1+c_{23}c_{32}, \tag{7.38}$$

while having σ_6 be positive in (7.37) is equivalent to

$$c_{21}(1-c_{13})+c_{23}(1+c_{31})<1+c_{13}c_{31}. \tag{7.39}$$

Summarizing our results, we have the following.

Theorem 7.1. Assuming that $a>\lambda_0^1(\Omega)$, the model (7.11)–(7.12) is permanent provided that (7.20), (7.22), (7.27), (7.28), (7.32), (7.33), (7.38) and (7.39) hold.

The requirement $a>\lambda_0^1(\Omega)$ is by now familiar. Each species must have enough "core habitat" in the absence of the other two species to increase its population density in Ω when introduced at low densities. The inequalities $c_{12}<1$, $c_{21}>1$, $c_{13}<1$ and $c_{23}<1$ then "set the stage" for us to consider the possibility of having "predator mediated coexistence." Given these preconditions, (7.28), (7.33), (7.38), and (7.39) represent an interplay among predation and competition rates which produces the phenomenon. Let us examine them more closely. It is natural to expect that in the case of predator mediated coexistence the predator should impact the dominant competitor more than the weaker competitor. Indeed, it follows from (7.39) that without such an assumption, species 2 cannot invade Ω when $(u_1,u_3)=\left(\frac{1-c_{13}}{1+c_{13}c_{31}}\theta_a,\frac{1+c_{31}}{1+c_{13}c_{31}}\theta_a\right)$ and hence the predator does not mediate the coexistence of the two species. To see that such is the case, notice that (7.39) can be rewritten as

$$c_{21}(1-c_{13})+c_{23}+(c_{23}-c_{13})c_{31}<1.$$

If $c_{23}\geq c_{13}$, the left-hand side of the inequality exceeds $c_{21}(1-c_{23})+c_{23}>1-c_{23}+c_{23}=1$. In this event, (7.39) cannot hold. Consequently, we must assume

$$c_{23}<c_{13}. \tag{7.40}$$

We can then rewrite (7.38) and (7.39) as

$$(c_{13}-c_{23})c_{32}<1-c_{12}(1-c_{23})-c_{13} \tag{7.41}$$

and

$$c_{21}(1-c_{13})+c_{23}-1<(c_{13}-c_{23})c_{31}. \tag{7.42}$$

Observe that (7.41) first requires that

$$c_{12}(1-c_{23})+c_{13}<1 \tag{7.43}$$

and then that

$$c_{32}<\frac{1-c_{12}(1-c_{23})-c_{13}}{c_{13}-c_{23}}. \tag{7.44}$$

The left-hand side of (7.43) represents the sum of the predation pressure on species 1 from species 3 and the predation ameliorated competition pressure on species 1 from species 2. Given that the sum is not too high (as measured by having (7.43) hold), we see from (7.44) that to have "predator mediated coexistence" there must be an "upper bound" on the benefit species 3 derives from species 2. Note that our argument that the joint densities of species 2 and species 3 in the absence of species 1 tend toward the globally attracting equilibrium given in (7.34) also required an upper bound on c_{32} in terms of c_{23}, namely $\frac{1}{c_{23}} - 1$. Note also that $\frac{1}{c_{23}} - 1$ decreases as c_{23} increases. However, if we set

$$f(c_{23}) = \frac{1 - c_{12} + c_{12}c_{23} - c_{13}}{c_{13} - c_{23}}$$

for $0 < c_{23} < c_{13} < 1$,

$$f'(c_{23}) = \frac{(1 - c_{12})(1 - c_{13})}{(c_{13} - c_{23})^2} > 0,$$

meaning that the upper limit on the benefit species 3 can draw from species 2 given by (7.44) does increase with predation pressure on species 2. So there is some tension between the two requirements (7.33) and (7.44). Nevertheless, there will always be a range of values of c_{32} satisfying both requirements for any suitable choice of c_{12}, c_{13} and c_{23}.

As for (7.42), assuming (7.40), any $c_{31} > 0$ will satisfy (7.42) so long as

$$c_{21}(1 - c_{13}) + c_{23} - 1 \leq 0,$$

but once the competitive pressure c_{21} on species 2 from species 1 is great enough so that

$$c_{21}(1 - c_{13}) + c_{23} - 1 > 0,$$

(7.42) imposes a lower bound on how much benefit species 3 must derive from species 1, namely

$$c_{31} > \frac{c_{21}(1 - c_{13}) + c_{23} - 1}{c_{13} - c_{23}}. \tag{7.45}$$

On the other hand, (7.28) imposes an upper limit on c_{31} in terms of c_{13}. Notice that if $c_{21} = 1$, $c_{21}(1 - c_{13}) + c_{23} - 1 = c_{23} - c_{13} < 0$ by (7.40). So if $c_{21} - 1$ is not too large there will be a range of c_{31} satisfying both (7.28) and (7.45) and the model predicts "predator mediated coexistence." However, if c_{21} becomes too large, there will be no such value of c_{31}. So assuming (7.28) and (7.33), the effect of the predator on the dominant competitor will not be sufficient for the persistence of the weaker competitor if the competitive advantage of the dominant competitor is too great.

Suppose now that we take $c_{12} = 1/2$, $c_{13} = 1/2$ and $c_{23} = 1/4$. Then $c_{12}(1-c_{23})+c_{13} = 7/8$, so (7.43) holds and (7.44) reduces to $c_{32} < 1/2$. (In this case (7.33) just requires $c_{32} < 3$.) The requirement on c_{31} in (7.45) becomes

$$c_{31} > 2c_{21} - 3$$

whereas the requirement from (7.28) is

$$c_{31} < 1.$$

If $c_{21} \in (1, 3/2]$, (7.45) holds for any $c_{31} > 0$, so that the range of c_{31} for which the model predicts "predator mediated coexistence" is $(0, 1)$. Once $c_{21} > 3/2$, we need $c_{31} \in (2c_{21} - 3, 1)$ for the model to predict "predator mediated coexistence." If $c_{21} \in (3/2, 2)$, the interval $(2c_{21} - 3, 1)$ is not empty, but if $c_{21} \geq 2$, there are no values of c_{31} satisfying both (7.28) and (7.45).

7.3 Three Species Competition

In light of the preceding discussion of predator-mediated coexistence, it seems natural to ask if a community of competing species in which at least one competitor tends toward extinction can be "stabilized" in some sense by the introduction of additional competitors into the community. In particular, can it be the case that the larger community is in fact permanent? The simplest context in which such an outcome is conceivable is when a single additional competitor is introduced into a pre-existing two species community in which a dominant competitor can be expected to exclude an inferior competitor over time. Of course, within such a scenario, there are a number of possibilities for a model, depending on the outcomes of the newly introduced species' interactions with each of the pre-existing competitors. In this section, we consider two of the possible cases. The first case we consider is the situation in which the newly introduced species coexists with the dominant pre-existing competitor and excludes the other competitor over time. The second is the situation in which the newly introduced species excludes the dominant pre-existing competitor over time yet is itself excluded over time by the other competitor. The reader will recognize this last case as that considered in May and Leonard (1975), which we discussed in Section 4.1. Of course, in Section 4.1, the coefficients in the May-Leonard example were such that the three species competition model failed to be permanent. Indeed, as we noted in Section 4.1, the example was a prominent consideration in the development of the Acyclicity Approach for ascertaining permanence in a model for several interacting biological species. Clearly, we shall need to modify the conditions on the coefficients in the May-Leonard example if we expect to obtain a prediction of permanence. Moreover, we will not be able to employ the Acyclicity Theorem for this purpose, but instead turn to its main alternative, the Average Lyapunov Function Approach. On the other hand, we shall see that the Acyclicity Theorem is a suitable mathematical tool in the first case above.

7.3.1 How Two Dominant Competitors May Mediate the Persistence of an Inferior Competitor

In principle, it is easy to understand what would underlie such a phenomenon should it occur. Namely, the competition between the two dominant competitors (which we designate by species 1 and species 2 for the sake of specificity) would need to be sufficiently intense so as to reduce the combined densities of species 1 and 2 enough to allow the inferior competitor (which we designate by species 3) to coexist. In particular, we know from Section 4.1 that if the three species model is to be permanent, it must be the case that species 3 can invade any asymptotic density configuration in the global attractor for the interaction of species 1 and 2 in its absence. We know from Section 5.2 that in the modeling regimes of primary interest in this volume the global attractor for the interaction of species 1 and 2 (assuming it is permanent) is contained in an order interval in the skew ordering (where $(u_1, u_2) \leq (v_1, v_2) \Leftrightarrow u_1 \leq v_1$ but $u_2 \geq v_2$) whose "end points" are componentwise positive equilibrium densities for the interaction. Where such equilibria are located relative to the configurations $(K_1, 0)$ and $(0, K_2)$, where K_i denotes the carrying capacity of species i in the absence of any competition, depends on the specific forms of the local competitive interactions between species 1 and 2. This observation was a key component in the analysis in Ayala et al. (1973). As noted in our discussion of the principle of competitive exclusion in Section 6.1, Ayala in 1969 observed in an experimental setting that different species of *Drosophila* appeared to coexist on a single resource, contrary to the orthodox ecological theory of the time. In Ayala et al. (1973), various ODE competition models were tested to see how they fit the *Drosophila* data. The data indicated that a componentwise positive equilibrium for two such competitors should lie beneath the line segment joining $(K_1, 0)$ and $(0, K_2)$. Not all two species competition models have this property. In particular, if a two species ODE Lotka-Volterra model is permanent, the intersection of its isoclines (which is the location of the componentwise positive equilibrium) must be above the line joining $(K_1, 0)$ and $(0, K_2)$. Consequently, the relative growth rates $\frac{du_i}{dt}/u_i$ for a model for *Drosophila* competition were deemed in Ayala et al. (1973) to require nonlinear interference competition terms. In our case, permanence will require that the global attractor for the interaction of species 1 and 2 be near $(0, 0)$. Consequently, we should anticipate that the local *per capita* growth rate terms for species 1 and 2 in our model would also contain nonlinear interference competition terms.

In Cantrell and Ward (1997) we explored the possibility of two dominant competitors mediating the coexistence of a third inferior competitor in the modeling regime of reaction-diffusion models for interacting biological species in an isolated bounded habitat. However, before we turn to those results, for the sake of illustration, let us first explore an ODE formulation for the interaction of species 1 and 2 incorporating a particular choice of nonlinear interference competition. To this end, consider the two species model

$$\begin{aligned} \frac{du_1}{dt} &= u_1(a - u_1 - \alpha_{12}u_2 - \beta_{12}u_2^2) \\ \frac{du_2}{dt} &= u_2(a - \alpha_{21}u_1 - \beta_{21}u_1^2 - u_2) \end{aligned} \tag{7.46}$$

with $a, \alpha_{12}, \alpha_{21}, \beta_{12}$ and β_{21} positive constants. (This model is one of those considered in Ayala et al. (1973).) We shall regard a, α_{12} and α_{21} as fixed and allow the coefficients of the interference terms $-\beta_{12}u_2^2u_1$ and $-\beta_{21}u_1^2u_2$ to vary.) In the absence of species j, species i converges to a for $i, j = 1, 2, i \neq j$. Consequently, it is easy to determine that permanence in (7.46) is equivalent to

$$\beta_{12} < \frac{1 - \alpha_{12}}{a} \quad \text{and} \quad \beta_{21} < \frac{1 - \alpha_{21}}{a}. \tag{7.47}$$

Of course, (7.47) requires that we assume $\alpha_{12} < 1$ and $\alpha_{21} < 1$ (meaning that the part of the competitive interaction that is captured by the Lotka-Volterra interaction terms $-\alpha_{12}u_1u_2$ and $-\alpha_{21}u_1u_2$ denotes not by itself preclude coexistence). Now observe that a componentwise positive equilibrium (u_1, u_2) for (7.46) satisfies

$$\begin{aligned} u_1 + \alpha_{21}u_2 &= a - \beta_{12}u_2^2 \\ \alpha_{21}u_1 + u_2 &= a - \beta_{21}u_1^2. \end{aligned}$$

Hence if such an equilibrium is to approach $(0, 0)$, it must be the case that β_{12} and β_{21} become arbitrarily large, which is incompatible with (7.47). The basic problem is that the terms $-\beta_{12}u_2^2u_1$ and $-\beta_{21}u_1^2u_2$ are *linear* in u_1 and u_2 respectively so that β_{12} and β_{21} end up factoring into whether or not (7.46) is permanent. Consequently, we see that not only should the local *per capita* growth rates for species 1 and 2 contain nonlinear terms, but also that *particular choice* of nonlinearities does matter.

With these considerations in mind, let us consider the model (Cantrell and Ward, 1997)

$$\begin{aligned}
\frac{\partial u_1}{\partial t} &= \Delta u_1 + u_1(a - u_1 - \alpha_{12}u_2 - \beta_{12}u_1u_2 - \alpha_{13}u_3)\\
\frac{\partial u_2}{\partial t} &= \Delta u_2 + u_2(a - \alpha_{21}u_1 - \beta_{21}u_1u_2 - u_2 - \alpha_{23}u_3)\\
\frac{\partial u_3}{\partial t} &= \Delta u_3 + u_3(a' - \alpha_{31}u_1 - \alpha_{32}u_2 - u_3) && \text{in } \Omega \times (0, \infty)\\
u_1 &= u_2 = u_3 = 0 && \text{on } \partial\Omega \times (0, \infty)
\end{aligned} \tag{7.48}$$

where Ω is the bounded habitat in question. The parameters in (7.48) are constants satisfying

$$\begin{aligned}
&a \geq a' > \lambda_0^1(\Omega)\\
&\alpha_{12}, \alpha_{13}, \alpha_{21}, \alpha_{23} \in (0, 1)\\
&\alpha_{31} > 1, \ \alpha_{32} > 1\\
&\beta_1 \geq 0, \ \beta_2 \geq 0,
\end{aligned} \tag{7.49}$$

where again $\lambda_0^1(\Omega)$ denotes the principal eigenvalue of $-\Delta$ on the habitat Ω subject to homogeneous Dirichlet boundary conditions. Consequently, as in the preceding section, we have that $(\theta_a, 0, 0)$, $(0, \theta_a, 0)$, and $(0, 0, \theta_{a'})$ are the global attractors for (7.48) restricted to $(u_1, 0, 0)$, $(0, u_2, 0)$ and $(0, 0, u_3)$, respectively, where θ_a is the unique positive solution of (7.16). The two two-species subsystems involving species 3 may be analyzed in a manner strictly analogous to the analysis of (7.17)–(7.18) in the preceding section, allowing us to conclude that species 1 excludes species 3 in the absence of species 2 and that species 2 excludes species 3 in the absence of species 1.

In the absence of species 3, (7.48) reduces to

$$\begin{aligned}
\frac{\partial u_1}{\partial t} &= \Delta u_1 + u_1(a - u_1 - \alpha_{12}u_2 - \beta_{12}u_1u_2)\\
\frac{\partial u_2}{\partial t} &= \Delta u_2 + u_2(a - \alpha_{21}u_1 - \beta_{21}u_1u_2 - u_2) && \text{in } \Omega \times (0, \infty)\\
u_1 &= 0 = u_2 && \text{on } \partial\Omega \times (0, \infty).
\end{aligned} \tag{7.50}$$

By (7.49), $\alpha_{12} < 1$ and $\alpha_{21} < 1$. Consequently, the principal eigenvalues σ_1 and σ_2 are positive in

$$\begin{aligned}
\Delta\phi_1 + \phi_1(a - \alpha_{12}\theta_a) &= \sigma_1\phi_1 && \text{in } \Omega\\
\phi_1 &= 0 && \text{on } \partial\Omega
\end{aligned} \tag{7.51}$$

and

$$\begin{aligned} \Delta\phi_2 + \phi_2(a - \alpha_{21}\theta_a) &= \sigma_2\theta_2 \text{ in } \quad \Omega \\ \phi_2 &= 0 \qquad\qquad \text{on} \quad \partial\Omega. \end{aligned} \tag{7.52}$$

Because the terms $-\beta_{12}u_1^2u_2$ and $-\beta_{21}u_1u_2^2$ are higher order in u_1 and u_2, respectively, the values of β_{12} and β_{21} do not play a role in deciding whether species 1 can invade the habitat Ω when species 2 is at its carrying capacity density, or *vice versa*. As a result, (7.50) is permanent for all values of $\beta_{12} \geq 0$ and $\beta_{21} \geq 0$.

Whether the full three species competition model (7.48) is permanent (that is, whether the competition between the two determinant competitors mediates the coexistence of the inferior competitor) does depend on the values of β_{12} and β_{21}. Note that if K once again denotes the positive cone in $[C_0^1(\overline{\Omega})]^3$ and $X \subseteq K$ and $S \subseteq \partial K$ are as in Theorem 4.1,

$$\omega(S) = \{(0, 0, 0), (\theta_a, 0, 0), (0, \theta_a, 0), (0, 0, \theta_a), M_5\}$$

where M_5 denotes the global attractor for solutions to (7.50) corresponding to componentwise positive initial data. As in the preceding section, the covering of $\omega(S)$ is acyclic (see Figure 7.2), and it follows as in Avila (1995) and Cantrell and Ward (1997) that the model (7.48) is permanent provided that there is a value $c > 0$ so that the principal eigenvalue $\sigma_3 \geq c$ in

$$\begin{aligned} \Delta\phi_3 + (a' - \alpha_{31}\tilde{u}_1 - \alpha_{32}\tilde{u}_2)\phi_3 &= \sigma_3\phi_3 \text{ in } \quad \Omega \\ \phi_3 &= 0 \qquad\qquad \text{on} \quad \partial\Omega \end{aligned} \tag{7.53}$$

for any $(\tilde{u}_1, \tilde{u}_2) \in M_5$. As we noted earlier, M_5 is contained in an order interval in the skew ordering whose "endpoints" are componentwise positive equilibria. If we denote these

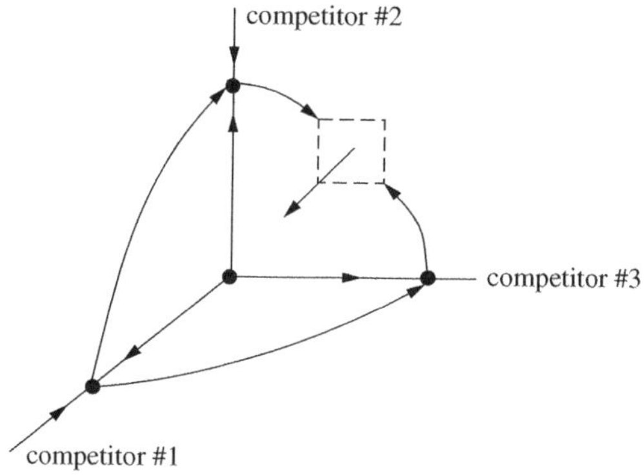

Figure 7.2 Schematic diagram showing the stability types and connecting orbits of the equilibria and attracting sets in the boundary of the positive cone for a three-species competition system corresponding to competition-mediated coexistence. The geometry of the diagram indicates that the flow in the boundary of the positive cone is acyclic and that no equilibrium or attracting set has a stable manifold which intersects the interior of the positive cone. This structure can be used to establish permanence, which then implies that the system predicts competition-mediated coexistence.

equilibria by (u_{1*}, u_{2*}) and (u_1^*, u_2^*) with $u_{1*} \le u_1^*$ and $u_2^* \le u_{2*}$, σ_3 in (7.53) will be as required so long as the principal eigenvalue σ_4 is positive in

$$\begin{aligned} \Delta\phi_4 + (a' - \alpha_{31}u_1^* - \alpha_{32}u_{2*})\phi_4 &= \sigma_4\phi_4 \text{ in } \quad \Omega \\ \phi_4 &= 0 \qquad\qquad \text{on} \quad \partial\Omega. \end{aligned} \tag{7.54}$$

Notice that (u_1^*, u_{2_*}) is an equilibrium for (7.50) only in the case when $(u_1^*, u_2^*) = (u_{1*}, u_{2*})$ and hence M_5 reduces to $\{(u_1^*, u_2^*)\}$. However, if $(\tilde{u}_1, \tilde{u}_2) \in M_5$, then $\tilde{u}_1 \le u_1^*$ while $\tilde{u}_2 \le u_{2*}$ and u_1^* and u_{2_*} are the smallest functions for which these inequalities hold separately. Setting $\tilde{u}_1 = u_1^*$ and $\tilde{u}_2 = u_{2*}$ in (7.53) over-estimates the competitive pressure on species 3 when the densities of species 1 and 2 lie in M_5, in essence giving "competitive overkill" or a "worst case scenario" *vis-á-vis* the possibility of species 3 invading the habitat Ω when species 1 and 2 are well-established. The value σ_4 in (7.54) that arises from taking $\tilde{u}_1 = u_1^*$ and $\tilde{u}_2 = u_{2*}$ in (7.53) will in general be less than the c connected with (7.53), since the quantity in (7.53) to be approximated is the combination $\alpha_{31}\tilde{u}_1 + \alpha_{32}\tilde{u}_2$ and $\alpha_{31}u_1^* + \alpha_{32}u_{2*}$ is not a sharp upper bound for this quantity. However, it is a natural choice. Moreover, our goal at this point is to establish that competition mediated coexistence is possible for appropriate large values of β_{12} and β_{21}, and showing that $\sigma_4 > 0$ in (7.54) for some range of large values of β_{12} and β_{21} is sufficient for this purpose. If we use $\tilde{u}_1 = u_1^*$ and $\tilde{u}_2 = u_{2*}$, we shall be able to employ the maximum principle to this end.

Let us first dispose of the case when $\beta_{12} = 0$ and $\beta_{21} = 0$, in which (7.48) becomes a three species Lotka-Volterra competition model. In this case, Cosner and Lazer (1984) have shown that all componentwise positive solutions to (7.50) converge over time to $\left(\dfrac{1-\alpha_{12}}{1-\alpha_{12}\alpha_{21}}\theta_a, \dfrac{1-\alpha_{21}}{1-\alpha_{12}\alpha_{21}}\theta_a\right)$, so that $M_5 = \left\{\left(\dfrac{1-\alpha_{12}}{1-\alpha_{12}\alpha_{21}}\theta_a, \dfrac{1-\alpha_{21}}{1-\alpha_{12}\alpha_{21}}\theta_a\right)\right\}$. In this case, there is only one choice of $(\tilde{u}_1, \tilde{u}_2)$ in (7.53) and σ_3 in (7.53) equals σ_4 in (7.54). As in the preceding section, we have that $\sigma_3 < 0$ in (7.53) if

$$\alpha_{31}\left(\frac{1-\alpha_{12}}{1-\alpha_{12}\alpha_{21}}\right) + \alpha_{32}\left(\frac{1-\alpha_{21}}{1-\alpha_{12}\alpha_{21}}\right) > 1. \tag{7.55}$$

In such case, species 3 cannot invade Ω when species 1 and 2 are well-established and the system (7.48) fails to be permanent. To see that (7.55) holds, notice that by (7.49)

$$\alpha_{31}\left(\frac{1-\alpha_{12}}{1-\alpha_{12}\alpha_{21}}\right) + \alpha_{32}\left(\frac{1-\alpha_{21}}{1-\alpha_{12}\alpha_{21}}\right) > \frac{1-\alpha_{12}}{1-\alpha_{12}\alpha_{21}} + \frac{1-\alpha_{21}}{1-\alpha_{12}\alpha_{21}} = \frac{2-\alpha_{12}-\alpha_{21}}{1-\alpha_{12}\alpha_{21}}.$$

Now since $(1-\alpha_{21})(1-\alpha_{21}) > 0$, we have that

$$2 - \alpha_{12} - \alpha_{21} > 1 - \alpha_{12}\alpha_{21} > 0$$

so that (7.55) holds and (7.48) is not permanent.

So now let us suppose that β_{12} and β_{21} are positive. We will first treat the case when $\beta_{12} = \beta_{21}$ and then consider the case when they are not. So let $\beta_{12} = \beta_{21} = \beta$. As in Cantrell and Ward (1997), if (u_1, u_2) is a componentwise positive equilibrium to (7.50), then $w = u_2 - \left(\dfrac{1-\alpha_{21}}{1-\alpha_{12}}\right)u_1$ is a solution of the eigenvalue problem

$$\begin{aligned} -\Delta w + (\beta u_1 u_2 + u_1 + u_2)w &= \gamma w \text{ in } \quad \Omega \\ w &= 0 \qquad\qquad \text{on} \quad \partial\Omega \end{aligned} \tag{7.56}$$

with $\gamma = a$. We know from Section 2.2 that the smallest eigenvalue γ_1 of (7.56) corresponds to a positive eigenfunction, say z. If we multiply (7.56) (with w replaced by z) by u_1, integrate both sides and employ Green's Second Identity, we obtain

$$\int_\Omega (a + (1-\alpha_{12})u_2)u_1 z dx = \gamma_1 \int_\Omega u_1 z dx. \tag{7.57}$$

Since $\alpha_{21} \in (0, 1)$ and u_1 and u_2 are positive in Ω, we may conclude from (7.57) that $\gamma_1 > a$. Consequently, (7.56) with $\gamma = a$ has only the zero solution and $u_2 \equiv \dfrac{1-\alpha_{21}}{1-\alpha_{12}} u_1$.

So u_1 satisfies

$$\begin{aligned} -\Delta u_1 &= u_1 \left(a - u_1 - \alpha_{12}\left(\frac{1-\alpha_{21}}{1-\alpha_{12}}\right) u_1 - \beta\left(\frac{1-\alpha_{21}}{1-\alpha_{12}}\right) u_1^2 \right) \\ &= u_1 \left(a - \left(\frac{1-\alpha_{12}\alpha_{21}}{1-\alpha_{12}}\right) u_1 - \beta\left(\frac{1-\alpha_{21}}{1-\alpha_{12}}\right) u_1^2 \right) \end{aligned} \tag{7.58}$$

in Ω. If we set $f(u) = a - \left(\dfrac{1-\alpha_{12}\alpha_{21}}{1-\alpha_{12}}\right) u - \beta\left(\dfrac{1-\alpha_{21}}{1-\alpha_{12}}\right) u^2$, then $f'(u) < 0$ for $u \geq 0$ and $f(u) \leq 0$ when $u \geq K^*(\beta)$, where

$$K^*(\beta) = \frac{-(1-\alpha_{12}\alpha_{21}) + \sqrt{(1-\alpha_{12}\alpha_{21})^2 + 4a\beta(1-\alpha_{12})(1-\alpha_{21})}}{2\beta(1-\alpha_{21})}. \tag{7.59}$$

It follows from Propositions 3.2 and 3.3 that (7.58) has a unique positive solution. Consequently there can be only one positive function u_1 solving (7.58) and hence only one componentwise positive equilibrium to (7.50), which we denote $\left(u_1^*, \left(\dfrac{1-\alpha_{21}}{1-\alpha_{12}}\right) u_1^*\right)$. We can employ the maximum principle or the method of upper and lower solutions to conclude that $u_1^* \leq K^*(\beta)$ where $K^*(\beta)$ is as in (7.59).

In this case, $M_5 = \left\{\left(u_1^*, \left(\dfrac{1-\alpha_{21}}{1-\alpha_{12}}\right) u_1^*\right)\right\}$ so (7.48) is permanent provided the principal eigenvalue σ_3 is positive in

$$\begin{aligned} \Delta\phi_3 + \left(a' - \alpha_{31}u_1^* - \alpha_{32}\left(\frac{1-\alpha_{21}}{1-\alpha_{12}}\right) u_1^*\right)\phi_3 &= \sigma_3\phi_3 \quad \text{in} \quad \Omega \\ \phi_3 &= 0 \quad \text{on} \quad \partial\Omega. \end{aligned} \tag{7.60}$$

Since $u_1^* \leq K^*(\beta)$, $\sigma_3 \geq \sigma_5$ where σ_5 is the principal eigenvalue in

$$\begin{aligned} \Delta\phi_5 + \left(a' - \alpha_{31}K^*(\beta) - \alpha_{32}\left(\frac{1-\alpha_{21}}{1-\alpha_{12}}\right) K^*(\beta)\right)\phi_5 &= \sigma_5\phi_5 \quad \text{in} \quad \Omega \\ \phi_5 &= 0 \quad \text{on} \quad \partial\Omega. \end{aligned}$$

It is easy to calculate that ϕ_5 satisfies

$$\sigma_5 = a' - \left(\alpha_{31} + \alpha_{32}\left(\frac{1-\alpha_{21}}{1-\alpha_{12}}\right)\right) K^*(\beta) - \lambda_0^1(\Omega).$$

It is evident from inspection of (7.59) that $\lim_{\beta \to \infty} K^*(\beta) = 0$. Consequently, σ_5 is positive since $a' > \lambda_0^1(\Omega)$ and thus σ_3 is positive in (7.60). We conclude that if $\beta = \beta_{12} = \beta_{21}$ in (7.48) and β is sufficiently large, then (7.48) is permanent. In this case, the competition between species 1 and species 2 mediates the coexistence of species 3.

So now let us assume that β_{12} and β_{21} are positive and unequal. For the sake of specificity, we will assume $\beta_{12} < \beta_{21}$. (Analogous results hold when $\beta_{12} > \beta_{21}$. See Cantrell and Ward (1997).) In this case, to the best of our knowledge, M_5 does not reduce to a single componentwise positive equilibrium for (7.50). As a result, we need to show that σ_4 in (7.54) is positive to conclude that (7.48) is permanent. To do so, we first obtain upper bounds on both components of an arbitrary componentwise positive equilibrium (u_1, u_2) to (7.50) via the method of upper and lower solutions. Because of the nature of the skew ordering, two applications of the method of upper and lower solutions will be needed. So suppose

$$\begin{aligned} -\Delta u_1 &= u_1(a - u_1 - \alpha_{21}u_2 - \beta_{12}u_1u_2) \\ & \qquad\qquad\qquad\qquad \text{in} \quad \Omega, \\ -\Delta u_2 &= u_2(a - \alpha_{21}u_1 - \beta_{21}u_1u_2 - u_2) \\ \text{with} \quad & u_1 = u_2 = 0 \qquad\qquad \text{on} \quad \partial\Omega, \end{aligned}$$

and let ϕ be a positive eigenfunction for

$$\begin{aligned} -\Delta\phi &= \lambda_0^1(\Omega)\phi \text{ in } \quad \Omega \\ \phi &= 0 \qquad\qquad \text{on} \quad \partial\Omega. \end{aligned}$$

Then we have

$$\begin{aligned} -\Delta u_1 &\geq u_1(a - u_1 - \alpha_{12}u_2 - \beta_{21}u_1u_2) \\ & \qquad\qquad\qquad\qquad \text{in} \quad \Omega \\ -\Delta u_2 &\leq u_2(a - \alpha_{21}u_1 - \beta_{21}u_1u_2 - u_2) \end{aligned}$$

since $\beta_{12} < \beta_{21}$. Moreover, if $(\overline{u}_1, \overline{u}_2) = (\varepsilon\phi, a)$, then

$$\begin{aligned} -\Delta\overline{u}_1 &\leq \overline{u}_1(a - \overline{u}_1 - \alpha_{12}\overline{u}_2 - \beta_{21}\overline{u}_1\overline{u}_2) \\ -\Delta\overline{u}_2 &\geq \overline{u}_2(a - \alpha_{21}\overline{u}_1 - \beta_{21}\overline{u}_1\overline{u}_2 - \overline{u}_2) \end{aligned}$$

for small enough positive values of ε provided

$$(1 - \alpha_{12})a > \lambda_0^1(\Omega). \tag{7.61}$$

Consequently, provided that (7.61) holds, the Remarks following Theorem 1.22 imply that the unique componentwise positive equilibrium to (7.50) when $\beta_{12} =$

β_{21}, $(u_1^*(\beta_{21}), \left(\dfrac{1-\alpha_{21}}{1-\alpha_{12}}\right) u_1^*(\beta_{21}))$, satisfies

$$\overline{u}_1 \le u_1^*(\beta_{21}) \le u_1$$
$$u_2 \le \left(\frac{1-\alpha_{21}}{1-\alpha_{12}}\right) u_1^*(\beta_{21}) \le \overline{u}_2$$

so that in particular

$$u_2 \le \left(\frac{1-\alpha_{21}}{1-\alpha_{12}}\right) u_1^*(\beta_{21}). \tag{7.62}$$

To obtain an upper bound on u_1, we make an additional assumption regarding β_{12} and β_{21}. Namely, we require that

$$\beta_{21} - \beta_{12} < \frac{1-\alpha_{21}}{a}. \tag{7.63}$$

If (7.63) holds, we have that

$$a(\beta_{21} - \beta_{12}) < 1 - \alpha_{21}$$

so that we can obtain a value $\tilde{\alpha}_{21} \in (\alpha_{21}, 1)$ so that

$$a(\beta_{21} - \beta_{12}) < \tilde{\alpha}_{21} - \alpha_{21}. \tag{7.64}$$

Since $u_2 < a$, it follows from (7.64) that

$$(\beta_{21} - \beta_{12})u_2 < \tilde{\alpha}_{21} - \alpha_{21},$$

which in turn implies

$$\alpha_{21}u_1 + \beta_{21}u_1u_2 < \tilde{\alpha}_{21}u_1 + \beta_{12}u_1u_2. \tag{7.65}$$

From (7.65), we get

$$\begin{aligned} -\Delta u_1 &\le u_1(a - u_1 - \alpha_{12}u_2 - \beta_{12}u_1u_1) \\ & \qquad\qquad \text{in} \quad \Omega. \\ -\Delta u_2 &\ge u_2(a - \tilde{\alpha}_{21}u_1 - \beta_{12}u_1u_2 - u_2) \end{aligned}$$

If we now set $(\hat{u}_1, \hat{u}_2) = (a, \varepsilon\phi)$, we have

$$\begin{aligned} -\Delta\hat{u}_1 &\ge \hat{u}_1(a - \hat{u}_1 - \alpha_{12}\hat{u}_2 - \beta_{12}\hat{u}_1\hat{u}_2) \\ -\Delta\hat{u}_2 &\le \hat{u}_2(a - \tilde{\alpha}_{21}\hat{u}_1 - \beta_{12}\hat{u}_1\hat{u}_2 - \hat{u}_2) \text{ in } \quad \Omega, \end{aligned}$$

provided

$$(1 - \tilde{\alpha}_{21})a > \lambda_0^1(\Omega). \tag{7.66}$$

In this case, the Remarks following Theorem 1.22 imply that the unique componentwise equilibrium to (7.50) with $\beta_{21} = \beta_{12}$ and $\alpha_{21} = \tilde{\alpha}_{21}$, $\left(u_1^*(\beta_{12}), \left(\frac{1-\tilde{\alpha}_{21}}{1-\alpha_{12}}\right) u_1^*(\beta_{12})\right)$, satisfies

$$u_1 \leq u_1^*(\beta_{12}) \leq \hat{u}_1$$

$$\hat{u}_2 \leq \left(\frac{1-\tilde{\alpha}_{21}}{1-\alpha_{21}}\right) u_1^*(\beta_{12}) \leq u_2$$

so that in particular

$$u_1 \leq u_1^*(\beta_{12}). \tag{7.67}$$

Consequently, (7.62) and (7.67) imply that σ_4 in (7.54) exceeds the principal eigenvalue σ_6 in

$$\begin{aligned} \Delta\phi_6 + \left(a' - \alpha_{31} u_1^*(\beta_{12}) - \alpha_{32}\left(\frac{1-\alpha_{21}}{1-\alpha_{12}}\right) u_1^*(\beta_{21})\right)\phi_6 &= \sigma_6\phi_6 \quad \text{in} \quad \Omega \\ \phi_6 = 0 \qquad\qquad\qquad\qquad\qquad\qquad &\qquad\quad \text{on} \quad \partial\Omega \end{aligned} \tag{7.68}$$

provided that (7.61), (7.64) and (7.66) hold. So (7.48) is permanent if these constraints hold and β_{12} and β_{21} are large enough.

Two comments are in order regarding the extra constraints (7.61), (7.64) and (7.66) we have placed upon the parameters in (7.48) in order to have the interaction between the dominant competitors facilitate the persistence of the inferior competitor. First, we have not "pulled a fast one". There are parameter values in (7.48) with (7.61), (7.64) and (7.66) holding and with β_{12} and β_{21} large enough so that σ_6 in (7.68) is positive. To this end, let α_{12} and α_{21} take arbitrary but fixed values in (0, 1) and let $\tilde{\alpha}_{21}$ be any fixed value in $(\alpha_{21}, 1)$. Then (7.61) and (7.66) require

$$a > \lambda_0^1(\Omega) \cdot \max\left\{\frac{1}{1-\alpha_{12}}, \frac{1}{1-\tilde{\alpha}_{21}}\right\}. \tag{7.69}$$

Notice that the larger α_{12} and α_{21} are, the larger a must be in (7.69). Once such an a is identified, (7.64) holds if

$$\beta_{21} - \beta_{12} < \frac{\tilde{\alpha}_{21} - \alpha_{21}}{a}. \tag{7.70}$$

Note that if the pair of parameters (β_{12}, β_{21}) satisfies (7.70) so does $(\beta_{12}+s, \beta_{21}+s)$ for any $s > 0$. Consequently, (7.70) holds for (β_{12}, β_{21}) lying in a thin strip along the main diagonal $\beta_{12} = \beta_{21}$ in $\beta_{12} - \beta_{21}$ parameter space, and we can find β_{12} and β_{21} which are arbitrarily large but satisfy (7.70).

Our second comment regarding the side conditions (7.61), (7.64) and (7.66) is that they may be mathematical conveniences that are not required to obtain competition mediated coexistence in (7.48), but we don't know that for sure. The conditions are needed to obtain the estimates in (7.62) and (7.67) via the method of upper and lower solutions given our choices of $(\overline{u}_1, \overline{u}_2)$ and $(\hat{u}_1, \hat{u}_2)$. Nevertheless, we cannot rule out the possibility that a different means of obtaining estimates might relax the constraints (7.61), (7.66) and especially (7.64). On the other hand, requiring β_{12} and β_{21} to exceed some size threshold in order to obtain competition mediated coexistence in (7.48) is essential, since the system fails to have the property when $\beta_{12} = 0 = \beta_{21}$, the Lotka-Volterra case.

7.3.2 The May-Leonard Example Revisited

In the preceding example, either of the two dominant competitors could be considered as the "keystone species". In the May-Leonard example that follows, any of the three species may be regarded as "keystone", since removal of any one of the three species leads to a collapse of the ecological community. The version we consider is

$$\begin{aligned}\frac{\partial u_1}{\partial t} &= \mu\Delta u_1 + u_1(1 - u_1 - \alpha u_2 - \beta u_3)\\ \frac{\partial u_2}{\partial t} &= \mu\Delta u_2 + u_2(1 - \beta u_1 - u_2 - \alpha u_3)\\ \frac{\partial u_3}{\partial t} &= \mu\Delta u_3 + u_3(1 - \alpha u_1 - \beta u_2 - u_3)\\ &\qquad\qquad\text{in} \quad \Omega \times (0, \infty)\end{aligned} \tag{7.71}$$

$$u_1 = u_2 = u_3 = 0 \qquad \text{on} \quad \partial\Omega \times (0, \infty), \tag{7.72}$$

where the diffusion rate μ and the competition coefficients α and β are positive constants and $0 < \alpha < 1 < \beta$.

Notice that the reaction terms in (7.71)–(7.72) are precisely the same as in the ODE case (4.17) that we discussed in Section 4.1. In Section 4.1, we noted that the assumption

$$\alpha + \beta > 2 \tag{7.73}$$

ruled out the possibility of permanence in (4.17). Moreover, we noted the significance of (4.17) in the development of one of the principal tools for detecting permanence in a model for several interacting biological species, namely the Acyclicity Theorem. For, in (4.17), if (7.73) holds, the boundary of the set of componentwise positive triples (u_1, u_2, u_3) of species densities contains a cycle consisting of the equilibria $(1, 0, 0)$, $(0, 1, 0)$ and $(0, 0, 1)$ along with heteroclinic orbits connecting them. This cycle viewed as a whole attracts componentwise positive orbits for (4.17) even though the equilibria by themselves do not. The Acyclicity Theorem addresses the possibility of such an outcome in a model for several interacting biological species by making the explicit exclusion of a boundary cycle one of its hypotheses. In very many cases (such as the preceding example (7.48) or the treatment of predator-mediated coexistence in Section 7.2), a direct examination of limit sets for the flow or semi-flow when one or more species under consideration are absent shows that no such cycle is possible. When there may be such a cycle, another means of detecting permanence must be employed. In that which follows we employ the Average Lyapunov Function Test, i.e., Theorem 4.2.

The condition for each of the competitors modeled in (7.71)–(7.72) to persist in the absence of the other two is the by now familiar requirement that the principal eigenvalue σ_1 be positive in

$$\begin{aligned}\mu\Delta\phi_1 + \phi_1 &= \sigma_1\phi_1 \text{ in } \quad \Omega\\ \phi_1 &= 0 \qquad\quad \text{on} \quad \partial\Omega.\end{aligned} \tag{7.74}$$

When σ_1 is positive in (7.74) (which we remind the reader is equivalent to $1/\mu > \lambda_0^1(\Omega)$) and two of the competitors are absent, the diffusive single species model

$$\begin{aligned} \frac{\partial u}{\partial t} &= \mu \Delta u + u(1-u) \text{ in } \quad \Omega \times (0, \infty) \\ u &= 0 \qquad\qquad\qquad \text{on} \quad \partial\Omega \times (0, \infty) \end{aligned} \tag{7.75}$$

admits a unique positive equilibrium, which we denote by $\overline{u}(\mu)$ and which is the global attractor for orbits of the remaining species corresponding to positive initial data. Consequently, when (7.71)–(7.72) is converted to a semi-dynamical system on K, the positive cone in $[C_0^1(\overline{\Omega})]^3$, and $X \subseteq K$ and $S \subseteq \partial K$ are as in Theorem 4.1, the collection

$$\{(0, 0, 0), (\overline{u}(\mu), 0, 0), (0, \overline{u}(\mu), 0), (0, 0, \overline{u}(\mu))\} \tag{7.76}$$

is contained in $\omega(S)$. Since $0 < \alpha < 1 < \beta$, the three two-species subsystems for (7.71)–(7.72) all exhibit competitive exclusion as in the system (7.17)–(7.18) so that $\omega(S)$ is given by (7.76).

If we aim to exploit Theorem 4.2 to determine conditions under which (7.71)–(7.72) is permanent, we need a candidate for an average Lyapunov function. Recall that we constructed the average Lyapunov function in a diffusive predator-prey model in Section 4.5 using the positive eigenfunctions which arose in the invasibility conditions (4.97) and (4.98) (or (4.96) and (4.98)). In the current situation, each of the single species equilibria is invasible by one of the other species. Specifically, species 1 can invade Ω when species 2 has its density $u_2 = \overline{u}(\mu)$, species 2 can invade Ω when species 3 has its density $u_3 = \overline{u}(\mu)$ and species 3 can invade Ω when species 1 has its density $\mu_1 = \overline{u}(\mu)$. Mathematically, these phenomena are captured by the positivity of the principal eigenvalue σ_2 in

$$\begin{aligned} \mu \Delta \phi_2 + \phi_2(1 - \alpha \overline{u}(\mu)) &= \sigma_2 \phi_2 \text{ in } \quad \Omega \\ \phi_2 &= 0 \qquad\qquad\qquad \text{on} \quad \partial\Omega. \end{aligned} \tag{7.77}$$

(Notice that if one multiplies (7.77) by $\overline{u}(\mu)$, integrates and employs Green's Second Identity, (7.75) implies that

$$\sigma_2 = (1 - \alpha) \frac{\displaystyle\int_\Omega \overline{u}^2 \phi_2 dx}{\displaystyle\int_\Omega \overline{u} \phi_2 dx}, \tag{7.78}$$

which is positive since $\alpha < 1$.) Consequently, if we simply mimic the construction in Section 4.5, we should expect the average Lyapunov function $P : X \to [0, \infty)$ to have the form

$$P(v_1, v_2, v_3) = \prod_{i=1}^{3} \left(\int_\Omega \phi_2 v_i dx \right)^{\beta_i}.$$

The symmetry in (7.71)–(7.72) allows us to take $\beta_i = 1$ for $i = 1, 2$, and 3, so that

$$P(v_1, v_2, v_3) = \prod_{i=1}^{3} \left(\int_\Omega \phi_2 v_i dx \right). \tag{7.79}$$

Recall from Theorem 4.2 that (7.71)–(7.72) is permanent if for $(u_1, u_2, u_3) \in S$ (which means at least one of u_1, u_2 and u_3 is identically zero) the quantity

$$a(t, (u_1, u_2, u_3)) = \liminf_{\substack{(v_1,v_2,v_3)\to(u_1,u_2,u_3)\\(v_1,v_2,v_3)\in X\setminus S}} \left(\frac{P(\pi((v_1, v_2, v_3), t))}{P(v_1, v_2, v_2)}\right) \tag{7.80}$$

satisfies

$$\sup_{t>0} a(t, (u_1, u_2, u_3)) > \begin{cases} 0 & \text{for} \quad (u_1, u_2, u_3) \in S \\ 1 & \text{for} \quad (u_1, u_2, u_3) \in \omega(S). \end{cases} \tag{7.81}$$

Using (7.79) and calculating as in (4.104), we obtain that

$$\frac{P(\pi((v_1, v_2, v_3), t))}{P(v_1, v_2, v_3)} = \exp\left\{\int_0^t \left[\sum_{i=1}^{3}\left(\frac{\displaystyle\int_\Omega \phi_2 \frac{\partial z_i}{\partial s}(x, s)dx}{\displaystyle\int_\Omega \phi_2 z_i(x, s)dx}\right)\right] ds\right\}, \tag{7.82}$$

where $P(\pi((v_1, v_2, v_3), t)) = (z_1(x, t), z_2(x, t), z_3(x, t))$. Proceeding as in Section 4.5, it is straightforward to establish that $\sup_{t>0} a(t, (u_1, u_2, u_3)) > 0$ for $(u_1, u_2, u_3) \in S$.

To establish the remainder of (7.81) (that is, to show $\sup_{t>0} a(t, (u_1, u_2, u_3)) > 1$ when $(u_1, u_2, u_3) \in \omega(S)$), we need only consider the equilibria $(0, 0, 0)$ and $(\overline{u}(\mu), 0, 0)$, since the symmetry in the model guarantees that

$$\sup_{t>0} a(t, (\overline{u}(\mu), 0, 0)) = \sup_{t>0} a(t, (0, \overline{u}(\mu), 0)) = \sup_{t>0} a(t, (0, 0, \overline{u}(\mu))).$$

Let us first consider $\sup_{t>0} a(t, (0, 0, 0))$. It will suffice to show that $a(1, (0, 0, 0)) > 1$. As in Section 4.5, the compactness of the set X guarantees $\pi : X \times [0, 1] \to X$ is uniformly continuous. Consequently, if two initial configurations (v_1, v_2, v_3), (v_1^*, v_2^*, v_3^*) of species densities in X are close, so are $\pi((v_1, v_2, v_3), t)$ and $\pi(v_1^*, v_2^*, v_3^*), t)$ for all $t \in [0, 1]$. In particular, we may assert that given $\varepsilon > 0$, there will correspond a $\delta > 0$ so that if $||(v_1, v_2, v_3) - (0, 0, 0)||_{[C^1(\overline{\Omega})]^3} < \delta$, then $||(z_1, (x, t), z_2(x, t), z_3(x, t)) - (0, 0, 0)||_{[C(\overline{\Omega})]^3} < \varepsilon$ for all $t \in [0, 1]$. So now examine (7.82) for $(u_1, u_2, u_3) = (0, 0, 0)$. Note, for instance, that by Green's Second Identity and (7.77)

$$\int_0^1 \left[\frac{\displaystyle\int_\Omega \phi_2 \frac{\partial z_1}{\partial s}(x, s)dx}{\displaystyle\int_\Omega \phi_2 z_1(x, s)dx}\right] ds$$

$$= \int_0^1 \left[\frac{\displaystyle\int_\Omega \phi_2(\mu\Delta z_1(x, s) + z_1(x, s)(1 - z_1(x, s) - \alpha z_2(x, s) - \beta z_3(x, s)))dx}{\displaystyle\int_\Omega \phi_2 z_1(x, s)dx}\right] ds$$

$$= \int_0^1 \left[\frac{\displaystyle\int_\Omega (\mu\Delta\phi_2) z_1(x, s)dx + \int_\Omega \phi_2 z_1(x, s)(1 - z_1(x, s) - \alpha z_2(x, s) - \beta z_3(x, s))dx}{\displaystyle\int_\Omega \phi_2 z_1(x, s)dx}\right] ds$$

$$= \int_0^1 \left[\frac{\int_\Omega \phi_2 z_1(x,s)(\sigma_2 + \alpha\overline{u}(\mu) - z_1(x,s) - \alpha z_2(x,s) - \beta z_3(x,s))dx}{\int_\Omega \phi_2 z_1(x,s)dx} \right] ds$$

$$> \sigma_2 - \int_0^1 ||z_1(x,s) + \alpha z_2(x,s) - \beta z_3(x,s)||_{C(\overline{\Omega})} ds$$

$$> \frac{\sigma_2}{2}$$

provided $||(v_1, v_2, v_3)||_{[C^1(\overline{\Omega})]^3}$ is sufficiently small. Similar calculations hold for $\int_0^1 \left[\frac{\int_\Omega \phi_2 \frac{\partial z_2}{\partial s}(x,s)dx}{\int_\Omega \phi_2 z_2(x,s)dx} \right] ds$ and $\int_0^1 \left[\frac{\int_\Omega \phi_2 \frac{\partial z_3}{\partial s}(x,s)dx}{\int_\Omega \phi_2 z_3(x,s)dx} \right] ds$. Consequently, we have that $a(1,(0,0,0)) > e^{3\sigma_2/2} > 1$.

Suppose now that $(u_1, u_2, u_3) = (\overline{u}(\mu), 0, 0)$ and examine (7.82). We have

$$\int_0^1 \left[\frac{\int_\Omega \phi_2 \frac{\partial z_1}{\partial s}(x,s)dx}{\int_\Omega \phi_2 z_1(x,s)dx} \right] ds$$

$$= \int_0^1 \left[\frac{\int_\Omega \phi_2(\mu\Delta z_1(x,s) + z_1(x,s)(1 - z_1(x,s) - \alpha z_2(x,s) - \beta z_3(x,s)))dx}{\int_\Omega \phi_2 z_1(x,s)dx} \right] ds$$

$$\to \int_0^1 \left[\frac{\int_\Omega \phi_2(\mu\Delta\overline{u}(\mu) + \overline{u}(\mu)(1 - \overline{u}(\mu)))dx}{\int_\Omega \phi_2 \overline{u}(\mu)dx} \right] ds = 0$$

as $(v_1, v_2, v_3) \to (\overline{u}(\mu), 0, 0)$. So $\exp \int_0^1 \left[\frac{\int_\Omega \phi_2 \frac{\partial z_1}{\partial s}(x,s)dx}{\int_\Omega \phi_2 z_1(x,s)dx} \right] ds \to 1$ as $(v_1, v_2, v_3) \to (\overline{u}(\mu), 0, 0)$.

Next,

$$\int_0^1 \left[\frac{\int_\Omega \phi_2 \frac{\partial z_2}{\partial s}(x,s)dx}{\int_\Omega \phi_2 z_2(x,s)dx} \right] ds$$

$$= \int_0^1 \left[\frac{\int_\Omega (\sigma_2 + \alpha\overline{u}(\mu) - \beta z_1(x,s) - z_2(x,s) - \alpha z_3(x,s))\phi_2 z_2(x,s)dx}{\int_\Omega \phi_2 z_2(x,s)dx} \right] ds$$

$$= \int_0^1 \left[\frac{\int_\Omega \phi_2 z_2(x,s)(\sigma_2 + (\alpha - \beta)\overline{u}(\mu) - \beta(z_1(x,s) - \overline{u}(\mu)) - z_2(x,s) - \alpha z_3(x,s))dx}{\int_\Omega \phi_2 z_2(x,s)dx} \right] ds$$

$$> \sigma_2 + (\alpha - \beta)||\overline{u}(\mu)||_{C(\overline{\Omega})} - \int_0^1 ||\beta(z_1(x,s) - \overline{u}(\mu)) + z_2(x,s) + \alpha z_3(x,s)||_{C(\overline{\Omega})} ds$$

$$> \sigma_2 + (\alpha - \beta)||\overline{u}(\mu)||_{C(\overline{\Omega})} - \varepsilon$$

for an arbitrary $\varepsilon > 0$ provided $||(v_1, v_2, v_3) - (\overline{u}(\mu), 0, 0)||_{[C^1(\overline{\Omega})]^3}$ is sufficiently small.
Likewise

$$\int_0^1 \left[\frac{\int_\Omega \phi_2 \frac{\partial z_3}{\partial s}(x,s)dx}{\int_\Omega \phi_2 z_3(x,s)dx} \right] ds$$

$$= \int_0^1 \left[\frac{\int_\Omega (\sigma_2 + \alpha\overline{u}(\mu) - \alpha z_1(x,s) - \beta z_2(x,s) - z_3(x,s))\phi_2 z_3(x,s)dx}{\int_\Omega \phi_2 z_3(x,s)dx} \right] ds$$

$$> \sigma_2 - \int_0^1 ||\alpha(z_1(x,s) - \overline{u}(\mu)) + \beta z_2(x,s) + z_3(x,s)||_{C(\overline{\Omega})} ds$$

$$> \sigma_2 - \varepsilon$$

for any arbitrary $\varepsilon > 0$ provided $||(v_1, v_2, v_3) - (\overline{u}(\mu), 0, 0)||_{[C^1(\overline{\Omega})]^3}$ is sufficiently small.
As a consequence we have that

$$a(1, (\overline{u}(\mu), 0, 0)) \geq e^{2\sigma_2 + (\alpha - \beta)||\overline{u}(\mu)||_{C(\overline{\Omega})}}. \tag{7.83}$$

Since $a(1, (0, 0, 0)) > e^{3\sigma_2/2} > 1$, we have from (7.83) and Theorem 4.2 that the 3-species model (7.71)–(7.72) is permanent so long as

$$2\sigma_2 + (\alpha - \beta)||\overline{u}(\mu)||_{C(\overline{\Omega})} > 0. \tag{7.84}$$

Let us now examine the condition (7.84) for permanence in (7.71)–(7.72). The condition represents an interplay between the spatial and local interactional aspects of the model. Purely spatial aspects of the model are explicitly accounted for in (7.84) through the quantity $||\overline{u}(\mu)||_{C(\overline{\Omega})}$. The existence of a positive equilibrium $\overline{u}(\mu)$ to (7.75) is predicated upon having sufficient "core area" in the habitat after taking boundary dissipation into account to support each species in the absence of competition from the other two. Mathematically, this requirement is captured in the positivity of the principal eigenvalue σ_1 in (7.74), which as we noted is equivalent to the inequality

$$\frac{1}{\mu} > \lambda_0^1(\Omega). \tag{7.85}$$

The values α and β determine the intensity of the local competitive interactions. In the corresponding ODE model (4.17) they must satisfy

$$\alpha + \beta < 2 \tag{7.86}$$

for (4.17) to be permanent. Since $\int_\Omega \overline{u}^2(\mu)\phi_2 dx < ||\overline{u}(\mu)||_{C(\overline{\Omega})} \int_\Omega \overline{u}(\mu)\phi_2 dx$, it follows from (7.78) that

$$\sigma_2 < (1 - \alpha)||\overline{u}(\mu)||_{C(\overline{\Omega})}. \tag{7.87}$$

If (7.84) holds, (7.87) implies

$$||\overline{u}(\mu)||_{C(\overline{\Omega})}(2 - \alpha - \beta) > 0,$$

so that (7.86) holds. Consequently, if our criteria for permanence (7.84) is met, then we must meet the requirement (7.85) on the geometry of the habitat and the requirement (7.86) on the local competitive interactions. However, since the inequality in (7.87) is strict for all $\mu \in (0, 1/\lambda_0^1(\Omega))$, (7.85) and (7.86) can hold even though (7.84) does not. So the geometric condition (7.85) and the condition (7.86) on local competitive interactions are not in and of themselves sufficient to guarantee (7.84). However, as the rate of diffusion diminishes the conditions (7.84) and (7.86) coalesce. To this end, we know from Proposition 3.15 that as $\mu \to 0^+$, $u(\mu)$ converges to the constant 1 uniformly on interior subsets of the habitat Ω, so that $||\overline{u}(\mu)||_{C(\overline{\Omega})} \to 1$ as $\mu \to 0^+$ and hence the left hand side of (7.84) tends toward $2 - \alpha - \beta$.

7.4 Three Trophic Level Models

Food webs are central to the understanding of community structure in ecology (Cohen, 1989). In this section, we explore a very basic example of a food web, namely, a three species food chain in which a resource species is preyed upon by an intermediate predator which in turn is preyed upon by a dominant predator. If we envision these interactions as taking place in an isolated habitat patch, it is natural to ask what role the size of the patch plays in the sustainability of the three trophic level community. To this end, we consider the model (Cantrell and Cosner, 2001)

$$\begin{aligned}
\frac{\partial u_1}{\partial t} &= d_1 \Delta u_1 + u_1(1 - u_1) - \frac{A_1 u_1 u_2}{1 + B_1 u_1 + C_1 u_2} \\
\frac{\partial u_2}{\partial t} &= d_2 \Delta u_2 + \frac{E_1 u_1 u_2}{1 + B_1 u_1 + C_1 u_2} - \frac{A_2 u_2 u_3}{1 + B_2 u_2 + C_2 u_3} - D_2 u_2 \\
\frac{\partial u_3}{\partial t} &= d_3 \Delta u_3 + \frac{E_2 u_2 u_3}{1 + B_2 u_2 + C_2 u_3} - D_3 u_3
\end{aligned} \qquad (7.88)$$

$$\text{in} \quad \Omega \times (0, \infty)$$

$$u_1 = u_2 = u_3 = 0 \quad \text{on} \quad \partial\Omega \times (0, \infty), \qquad (7.89)$$

where Ω is the habitat patch in question, u_1 is the density of the resource species, u_2 is the density of the intermediate predator, and u_3 is the density of the dominant predator. To examine the effect that the size of Ω has upon the sustainability of the ecological community in Ω, we introduce a new parameter ℓ into (7.88) in essence by "freezing" the shape of Ω. Specifically, we envision some base habitat patch, say $\Omega_0 \subseteq I\!R^2$ or R^3, and consider habitat patches Ω of the form

$$\Omega = \ell\Omega_0 \qquad (7.90)$$

for $\ell > 0$, where $\ell\Omega_0 = \{(\ell x, \ell y) : (x, y) \in \Omega_0\}$ when $\Omega_0 \subseteq I\!R^2$ and $\ell\Omega_0 = \{(\ell x, \ell y, \ell z) \in \Omega_0\}$ when $\Omega_0 \subseteq I\!R^3$. In (7.90), Ω and Ω_0 have the same basic shape. We arrive at Ω from Ω_0 by expanding or contracting the linear dimensions of Ω_0 by a scale factor ℓ, with $\ell > 1$ when Ω_0 is expanded and $0 < \ell < 1$ when Ω_0 is contracted.

By making a change of the independent variables, (7.88)–(7.89) can be recast as a model on the base habitat Ω_0, namely

$$\begin{aligned}
\frac{\partial u_1}{\partial t} &= \frac{d_1}{\ell^2}\Delta u_1 + u_1(1-u_1) - \frac{A_1u_1u_2}{1+B_1u_1+C_1u_2}\\
\frac{\partial u_2}{\partial t} &= \frac{d_2}{\ell^2}\Delta u_2 + \frac{E_1u_1u_2}{1+B_1u_1+C_1u_2} - \frac{A_2u_2u_3}{1+B_2u_2+C_2u_3} - D_2u_2 \qquad (7.91)\\
\frac{\partial u_3}{\partial t} &= \frac{d_3}{\ell^2}\Delta u_3 + \frac{E_2u_2u_3}{1+B_2u_2+C_2u_3} - D_3u_3 \qquad \text{in} \quad \Omega_0\times(0,\infty),
\end{aligned}$$

with

$$u_1 = u_2 = u_3 \qquad \text{on} \quad \partial\Omega_0 \times (0,\infty). \qquad (7.92)$$

Then for a habitat patch Ω satisfying (7.90), the predictions of (7.88)–(7.89) for the three species in question coincide with those of (7.91)–(7.92) for the appropriate choice of ℓ. Notice that the $\frac{1}{\ell^2}$ term in each equation in (7.91) arises from the two differentiations with respect to each space variable in the Laplacian. Whether the base habitat Ω_0 is thought of as 2 or 3 (or for the matter n) dimensional is irrelevant. The Dirichlet or absorbing boundary condition (7.89) is unaltered by the change of variables. Of course, boundary dissipation is necessary in order for the size of the habitat patch to play a role in the outcome of the interaction described in (7.88)–(7.89). To that end, we could have imposed a Robin boundary condition in (7.89) in one or more of the species in question instead of the Dirichlet condition. However, such a choice would require us to have a factor $\frac{1}{\ell}$ multiplying $\nabla u_i \cdot \vec{n}$ in one or more of the equations in (7.92), and would further complicate the mathematical analysis of the model (7.91)–(7.92) without necessarily providing additional ecological insight. As a consequence, we have chosen to treat the case in which absorbing boundary conditions are placed upon each species in (7.89).

The functional and numerical response terms in (7.91)–(7.92) have the Beddington-DeAngelis form. We discussed ODE predator-prey models with Beddington-DeAngelis functional and numerical response terms in some detail in Section 4.1, and in Section 7.1, showed that Beddington-DeAngelis functional and numerical response terms can mitigate the coexistence of two consumer species upon a single resource. Our discussion in Section 7.1 made use of a key feature of Beddington-DeAngelis ODE predator-prey models that carries over to the reaction-diffusion setting, and will be important in our analysis of (7.91)–(7.92). Namely, the dynamics of the predator equations

$$\frac{dv}{dt} = \left(\frac{Eu}{1+Bu+Cv} - D\right)v \qquad (7.93)$$

in the ODE case, and

$$\begin{aligned}
\frac{\partial v}{\partial t} &= d\Delta v + \left(\frac{Eu}{1+Bu+Cv} - D\right)v \quad \text{on} \quad \Omega\times(0,\infty)\\
v &= 0 \qquad \text{on} \quad \partial\Omega\times(0,\infty)
\end{aligned} \qquad (7.94)$$

in the reaction-diffusion setting, are, for a fixed prey density u (or $u(x)$), essentially the same as the corresponding logistic equations. So, for either (7.93) or (7.94), solutions corresponding to positive initial densities tend over time toward the zero density or a positive equilibrium density, depending upon the prey density and the model parameters. That such is the case follows from Propositions 3.1–3.3, since $\frac{d}{dv}\left(\frac{Eu}{1+Bu+Cv}-D\right)=\frac{-ECu}{(1+Bu+Cv)^2}<0$ for all $v\geq 0$ provided $u>0$ and $\frac{Eu}{1+Bu+Cv}-D\leq 0$ for $v\geq\frac{(E-BD)u-D}{CD}$.

The objective of our analysis of (7.91)–(7.92) is to demonstrate that the prediction of (7.91)–(7.92) is permanence for all large enough values of ℓ. Indeed, we shall be able to show that such is the case provided the corresponding ODE model is permanent, which will require some "reasonable assumptions" on the remaining parameters in the model. Let us begin our discussion with the simple but important observations that just as in the ODE case if species 1 is absent from Ω, species 2 and 3 are necessarily driven toward extinction in Ω_0 with increasing time, and likewise, if species 2 is absent from Ω_0, species 3 is driven toward extinction in Ω_0 with increasing time. As a result, if (7.91)–(7.92) is converted to a semiflow on the positive cone K in $[C_0^1(\overline{\Omega}_0)]^3$ and $X\subseteq K$ and $S\subseteq\partial K$ are as in Theorem 4.1, then the set $\omega(S)$, under the assumption that species 1 and 2 persist in Ω, is given by

$$\omega(S)=\{(0,0,0),(\overline{u}_1,0,0),M_2\}. \tag{7.95}$$

In (7.95), $\overline{u}_1$ is the globally attracting equilibrium for

$$\begin{aligned}\frac{\partial u_1}{\partial t}&=\frac{d_1}{\ell^2}\Delta u_1+u_1(1-u_1) &&\text{in}\quad \Omega_0\times(0,\infty)\\ u_1&=0 &&\text{on}\quad \partial\Omega_0\times(0,\infty)\end{aligned} \tag{7.96}$$

and M_2 is the global attractor for solutions to (7.91)–(7.92) corresponding to initial population density configurations $(u_1^0,u_2^0,0)$ in which $u_1^0>0$ and $u_2^0>0$ on Ω_0. The structure of $\omega(S)$ in (7.95) means that in principle the Acyclicity Theorem (Theorem 4.3) is available as a means for asserting permanence in (7.91)–(7.92).

As we have now seen numerous times, species 1 persists in the absence of species 2 and 3 provided the principal eigenvalue σ_1 is positive in

$$\begin{aligned}\frac{d_1}{\ell^2}\Delta\phi_1+\phi_1&=\sigma_1\phi_1 &&\text{in}\quad\Omega_0\\ \phi_1&=0 &&\text{on}\quad\partial\Omega_0.\end{aligned} \tag{7.97}$$

Since our present focus is to demonstrate permanence in (7.91)–(7.92) when ℓ is large, it will be beneficial to recast requiring the positivity of principal eigenvalues such as σ_1 in (7.97) in terms of a size requirement on the scale parameter ℓ. (Recall that the positivity of a principal eigenvalue such as σ_1 can be interpreted as a condition for the invasibility of $\Omega=\ell\Omega_1$ by the species in question. Consequently, recasting a positivity condition on a principal eigenvalue as a size requirement on ℓ can be interpreted as identifying a "minimal patch size" for Ω for invasibility of the species in question.) In the case of σ_1 in (7.97), a direct calculation shows that

$$\ell^2-d_1\lambda_0^1(\Omega)=\sigma_1\ell^2$$

so having σ_1 positive in (7.97) is equivalent to

$$\ell^2 > d_1 \lambda_0^1(\Omega). \tag{7.98}$$

When (7.98) holds, $\Omega = \ell\Omega_0$ is large enough so that species 1 can persist in spite of its dissipation through the boundary of Ω. More generally, if r is a smooth function on Ω_0 (representing the local *per capita* growth rate at low densities for some species), which is such that $r(x_0) > 0$ for some $x_0 \in \Omega_0$ and the principal eigenvalue σ in

$$\begin{aligned} \frac{d}{\ell^2}\Delta\phi + r(x)\phi &= \sigma\phi \quad \text{in} \quad \Omega_0 \\ \phi &= 0 \qquad\quad \text{on} \quad \partial\Omega_0 \end{aligned} \tag{7.99}$$

denotes its average growth rate at low densities over Ω_0, then having $\sigma > 0$ in (7.99) is equivalent to the patch size requirement

$$\ell^2 > d\lambda_+^1(r(x)), \tag{7.100}$$

where $\lambda_+^1(r(x)) > 0$ is as in Section 4.4. The variability of $r(x)$ in (7.100) reflects the impact of the presence of other species upon the local *per capita* growth rate of the species at low densities. (Note that if $\mu_1^\infty(q, m)$ is as in Section 6.3, then $\lambda_+^1(r(x)) = \mu_1^\infty(0, r(x))$.)

Assuming that ℓ satisfies (7.98), (7.96) admits the globally attracting equilibrium $\overline{u}_1 = \overline{u}_1(\ell)$. The system (7.91)–(7.92) when $u_3 = 0$ reduces to

$$\begin{aligned} \frac{\partial u_1}{\partial t} &= \frac{d_1}{\ell^2}\Delta u_1 + u_1(1 - u_1) - \frac{A_1 u_1 u_2}{1 + B_1 u_1 + C_1 u_2} \\ \frac{\partial u_2}{\partial t} &= \frac{d_2}{\ell^2}\Delta u_2 + \left(\frac{E_1 u_1}{1 + B_1 u_1 + C_1 u_2} - D_2\right) u_2 \\ &\qquad\qquad\qquad\qquad \text{in} \quad \Omega_0 \times (0, \infty) \end{aligned} \tag{7.101}$$

$$u_1 = u_2 = 0 \quad \text{on} \quad \partial\Omega_0 \times (0, \infty). \tag{7.102}$$

Notice that the second equation in (7.101) is of the form (7.94). It follows from our consideration of (7.94) that when (7.101)–(7.102) is converted to a semi-dynamical system on $[C_0^1(\Omega_0)]^2$, the system is dissipative. As a result, permanence in (7.101)–(7.102) may be established in a manner strictly analogous to that employed for the Lotka-Volterra predator-prey model (4.84) in Section 4.5. So (7.101)–(7.102) is permanent provided (7.98) holds,

$$\frac{E_1 \overline{u}_1(x_0)}{1 + B_1 \overline{u}_1(x_0)} - D_2 > 0 \tag{7.103}$$

for some $x_0 \in \Omega_0$ and

$$\ell^2 > d_2 \lambda_+^1 \left(\frac{E_1 \overline{u}_1(x)}{1 + B_1 \overline{u}(x)} - D_2\right). \tag{7.104}$$

Since $\overline{u}_1 < 1$ by the maximum principle, having (7.103) hold requires

$$\frac{E_1}{1 + B_1} > D_2. \tag{7.105}$$

Notice that (7.105) is the same as the requirement (4.7) for permanence in the ODE Beddington-DeAngelis predator-prey model (4.2). We may conclude that permanence in the corresponding ODE is a necessary condition for permanence in (7.101)–(7.102), for if (7.105) fails the local *per capita* growth rate for the species 2 at low densities when species 1 is at its carrying capacity is negative. Hence species 1 is an inadequate resource to sustain species 2 regardless of domain size. Inequality (7.105) is then the first of several "reasonable assumptions" we place upon the parameters of (7.91)–(7.92) to obtain permanence.

Let us focus on condition (7.103) at this point. As noted, if (7.103) holds, so does (7.105). However, since $\overline{u}_1(x) < 1$ on Ω_0, the converse is not true, i.e., it may be the case that (7.105) holds but (7.103) fails to hold. Indeed, we know that such is the case if ℓ is only slightly larger in size than $\sqrt{d_1\lambda_0^1(\Omega_0)}$, since Propositions 3.9–3.11 and Corollary 3.14 tell us that

$$\lim_{\ell \to \sqrt{d_1\lambda_0^1(\Omega_0)}^+} \overline{u}(\ell) = 0 \quad \text{in} \quad C_0^1(\overline{\Omega}_0),$$

and hence

$$\frac{E_1\overline{u}_1(x)}{1 + B_1\overline{u}_1(x)} < D_2/2 \tag{7.106}$$

on $\overline{\Omega}_0$ if ℓ is close enough to $\sqrt{d_1\lambda_0^1(\Omega_0)}$. We know that the u_1 component of a solution to (7.101)–(7.102) is a lower solution to (7.96) and that all positive solutions to (7.96) converge uniformly to $\overline{u}_1(x)$ as $t \to \infty$. So suppose ℓ is such that (7.106) holds. Then if (u_1, u_2) is a componentwise positive solution to (7.101)–(7.102), and $\varepsilon > 0$ is small enough so that

$$\frac{E_1(1+\varepsilon)\overline{u}_1(x)}{1 + B_1(1+\varepsilon)\overline{u}_1(x)} < \frac{D_2}{2}$$

on $\overline{\Omega}_0$, then for large enough t (depending on $(u_1(x,0), u_2(x,0))$) we have that u_2 satisfies

$$\begin{aligned}\frac{\partial u_2}{\partial t} &\le \frac{d_2}{\ell^2}\Delta u_2 + \left(\frac{E_1(1+\varepsilon)\overline{u}_1(x)}{1 + B_1(1+\varepsilon)\overline{u}_1(x) + C_1 u_2} - D_2\right) u_2 \\ &\le \frac{d_2}{\ell^2}\Delta u_2 - \frac{D_2}{2}u_2,\end{aligned} \tag{7.107}$$

so that $u_2 \to 0$ on $\overline{\Omega}_0$ as $t \to \infty$. We may conclude that the size threshold on ℓ for permanence in (7.101)–(7.102) is *strictly higher* than the threshold for the persistence of species 1 in the absence of species 2. A similar result holds for (7.101)–(7.102) in comparison with the three species system (7.91)–(7.92), i.e., the size threshold for ℓ to have permanence in (7.91)–(7.92) is strictly higher than the needed for permanence in (7.101)–(7.102).

Next consider (7.103) in relation to (7.104). If (7.103) holds, the eigenvalue $\lambda_+^1\left(\frac{E_1\overline{u}_1(x)}{1 + B_1\overline{u}_1(x)} - D_2\right) > 0$ exists, but its existence *per se* does not necessarily force (7.104) to hold. However, as ℓ increases the effect of diffusion decreases and the equilibrium $\overline{u}_1(x)$ approaches the equilibrium of the corresponding ODE on the interior of Ω_0. Proposition 3.15 captures this phenomenon with mathematical precision. Namely Proposition 3.15 asserts that

$$\lim_{\ell\to\infty} \overline{u}_1(x) = 1 \tag{7.108}$$

uniformly on any subset Ω^* of Ω_0 which is such that $\overline{\Omega}^* \subseteq \Omega_0$. The convergence in (7.108) is mathematically sufficient to guarantee that as $\ell \to \infty$ the functions $\dfrac{E_1\overline{u}_1(x)}{1+B_1\overline{u}_1(x)} - D_2$ converge to $\dfrac{E_1}{1+B_1} - D_2$ in an integral average or $L^p(\Omega_0)$ sense for any $p \in [1,\infty)$. Theorem 2.4 then implies that

$$\lim_{\ell\to\infty} \lambda_+^1\left(\frac{E_1\overline{u}_1(x)}{1+B_1\overline{u}_1(x)} - D_2\right) = \lambda_+^1\left(\frac{E_1}{1+B_1} - D_2\right). \tag{7.109}$$

(Note that $\lambda_+^1\left(\dfrac{E_1}{1+B_1} - D_2\right) > 0$ exists by virtue of (7.105).) By (7.109), (7.104) holds for sufficiently large values of ℓ, since clearly $\ell^2 > \lambda_+^1\left(\dfrac{E_1}{1+B_1} - D_2\right)$ for large enough ℓ.

We can actually say even more. Note that $\overline{u}_1(\ell)$ satisfies

$$\begin{aligned} -\Delta\overline{u}_1 &= \frac{\ell^2}{d_1}\overline{u}_1(1-\overline{u}_1) \\ &\le \frac{(\ell')^2}{d_1}\overline{u}_1(1-\overline{u}_1) \end{aligned}$$

for any $\ell' > \ell$, so that $\overline{u}_1(\ell) < \overline{u}_1(\ell')$ on Ω_0 by the method of upper and lower solutions. So if (7.103) holds for some value of ℓ, it holds for all large values of ℓ. Likewise, the variational formula (2.15) for λ_1^+ implies that

$$\lambda_+^1\left(\frac{E_1\overline{u}_1(\ell)}{1+B_1\overline{u}_1(\ell)} - D_2\right) > \lambda_+^1\left(\frac{E_1\overline{u}_1(\ell')}{1+B_1\overline{u}_1(\ell')} - D_2\right)$$

if $\ell < \ell'$. Consequently, if ℓ is such that (7.104) holds, then (7.104) holds for all larger values of ℓ.

Summarizing our discussion, we may assert that (7.101)–(7.102) is permanent provided the corresponding ODE model is permanent (i.e., (7.105) holds) and the domain is large enough to compensate for dissipation of the species through the boundary (i.e., ℓ is large enough in (7.101)). Observe, though, that the spatial aspect of the requirement for permanence in (7.101)–(7.102) is more subtle than just having Ω large enough. There is an additional increase in the size of domain that is necessary each time a new (i.e., higher) trophic level is to be maintained long term. Put mathematically, species 1 persists once ℓ exceeds $\sqrt{d_1\lambda_0^1(\Omega_0)}$, and assuming (7.105) holds, species 2 persists once ℓ exceeds ℓ^* for some ℓ^* which is strictly larger than $\sqrt{d_1\lambda_0^1(\Omega_0)}$.

The story for the full three trophic level model (7.91)–(7.92) is essentially the same. Namely, to add and sustain the highest trophic level (i.e., species 3), we need a still larger domain (so that the effect of dissipation through the boundary via diffusion is muted even further), and the corresponding ODE model will need to be permanent. However, establishing permanence in (7.91)–(7.92) with mathematical rigor is a somewhat intricate task. To understand why such is the case, assume (7.105) and that the scale parameter ℓ exceeds ℓ^*. Then $\omega(S)$ is as in (7.95), and we are set up to employ the Acyclicity Theorem (Theorem 4.3) to assert that (7.91)–(7.92) is permanent. The conditions we have imposed so far, specifically (7.98), (7.105) and (7.104) enable us to conclude that the requirements of the Acyclicity Theorem are met at $(0,0,0)$ and $(\overline{u}_1,0,0)$. So we will be able to assert that (7.91)–(7.92) is permanent precisely so long as the instability criterion is met for M_2, namely

$$W^s(M_2) \cap \operatorname{int} K = \emptyset. \tag{7.110}$$

Since M_2 is the global attractor for the system when species 1 and species 2 are both present but the dominant predator (species 3) is absent, (7.110) boils down to having species 3 uniformly repelled from M_2, which happens if there is a $c > 0$ so that for all $(u_1, u_2, 0) \in M_2$, the principal eigenvalue σ_3 exceeds c in

$$\begin{aligned} &\frac{d_3}{\ell^2}\Delta\phi_3 + \left(\frac{E_2u_2}{1+B_2u_2} - D_3\right)\phi_3 = \sigma_3\phi_3 \text{ in } \quad \Omega_0 \\ &\phi_3 = 0 \qquad\qquad\qquad\qquad\qquad\qquad\qquad \text{on} \quad \partial\Omega_0. \end{aligned} \tag{7.111}$$

It is at this point that a dilemma arises. We don't know just from having (7.101)–(7.102) be permanent exactly what the structure of M_2 is, just that it is compact and uniformly bounded away from states with either $u_1 \equiv 0$ or $u_2 \equiv 0$. So we employ practical persistence techniques to give a sharp enough estimate on the location of M_2 to assert that $\sigma_3 \geq c > 0$ in (7.111) for all $(u_1, u_2) \in M_2$. Once we can make such a claim, we then apply the abstract result Theorem 4.3 to guarantee that (7.91)–(7.92) is permanent. This example illustrates how it sometimes becomes essential to exploit the monotonicity properties of the equations of a nonmonotone system when taken one at a time to establish that the system as a whole is permanent.

We can see from (7.111) and the fact that σ_3 increases as $\dfrac{E_2u_2}{1+B_2u_2} - D_3$ increases that we need a lower estimate, namely a $\underline{u}_2$ so that if $(u_1,u_2,0) \in M_2$, then $u_2 \geq \underline{u}_2$. (At this point, the reader may want to review our discussion of food chain models in Section 5.3.) To get our $\underline{u}_2$, we start by exploiting the fact that for any initial data of the form $(u_1^0, u_2^0, 0)$ with $u_1^0 > 0$ and $u_2^0 > 0$ in Ω_0, the u_1 component of the corresponding solution to (7.91)–(7.92) must be less than $1+\varepsilon$ for all large enough values of t so that for such values of t

$$\frac{\partial u_2}{\partial t} \leq \frac{d_2}{\ell^2}\Delta u_2 + u_2\left(\frac{E_1(1+\varepsilon)}{1+B_1(1+\varepsilon)+C_1u_2} - D_2\right). \tag{7.112}$$

The corresponding equation

$$\frac{\partial w}{\partial t} = \frac{d_2}{\ell^2}\Delta w + w\left(\frac{E_1(1+\varepsilon)}{1+B_1(1+\varepsilon)+C_1w} - D_2\right)$$

is of the form (7.94). It follows from (7.105) that all positive solutions to the equation converge to a positive equilibrium. Moreover, as ℓ increases the equilibrium approaches the equilibrium of the corresponding ODE, uniformly on strictly interior portions of Ω_0. We use these facts to parlay (7.112) into a *trackable* asymptotic upper bound on the u_2

component of any such solution. In particular, we get an upper bound on the u_2 component of any point $(u_1, u_2, 0) \in M_2$ for all large enough values of ℓ. We next use this asymptotic upper bound on species 2 in the first equation of (7.91)–(7.92) to obtain an asymptotic lower bound on species 1. Again, we use the fact that solutions to the comparison reaction-diffusion equation converge to a unique equilibrium which itself tracks the equilibrium to the corresponding ODE as $\ell \to \infty$. To know that the globally attracting equilibrium in the comparison reaction-diffusion equation is actually positive, we require

$$A_1 < C_1 \tag{7.113}$$

or equivalently $1 - \dfrac{A_1}{C_1} > 0$, the second of our "reasonable assumptions" on the local species interactions in the model. (The condition (7.113) may well be an artifact of the practical persistence approach, but it arises in the ODE setting as well.) Finally, we may then obtain the desired asymptotic lower bound on species 2 from the second equation in (7.91)–(7.92), assuming (7.105), (7.113) and large enough values of ℓ. So we get $\underline{u}_2$ so that if $(u_1, u_2, 0) \in M_2$, then $u_2 \geq \underline{u}_2$. Then σ_3 in (7.111) exceeds $c = \underline{\sigma}_3 > 0$ for all $(u_1, u_2, 0) \in M_2$ as required provided that the principal eigenvalue $\underline{\sigma}_3$ is positive in the eigenvalue problem

$$\begin{aligned} &\frac{d_3}{\ell^2}\Delta\phi + \left(\frac{E_2\underline{u}_2}{1 + B_2\underline{u}_2} - D_3\right)\phi = \underline{\sigma}_3\phi \quad \text{in} \quad \Omega_0 \\ &\phi = 0 \qquad\qquad\qquad\qquad\qquad\qquad\quad \text{on} \quad \partial\Omega_0. \end{aligned} \tag{7.114}$$

We will be able to assert that $\underline{\sigma}_3 > 0$ in (7.114) (and hence that (7.91)–(7.92) is permanent) for all large enough values of ℓ provided that $\underline{u}_2$ is such that the local *per capita* growth rate for species 3 at low densities when $u_2 = \underline{u}_2$ satisfies

$$\frac{E_2\underline{u}_2}{1 + B_2\underline{u}_1} - D_3 \geq r(x) \tag{7.115}$$

for all large enough values of ℓ, where r is a piecewise continuous function on Ω_0 so that $r(x_0) > 0$ for some $x_0 \in \Omega_0$. If (7.115) holds, then $\underline{\sigma}_3 > 0$ in (7.114) for all values of ℓ so that (7.115) holds and

$$\ell > \sqrt{d_3\lambda_+^1(r(x))}. \tag{7.116}$$

It clearly must be the case that if (7.115) holds then

$$E_2 - B_2 D_3 > 0, \tag{7.117}$$

since otherwise $\dfrac{E_2\underline{u}_2}{1 + B_2\underline{u}_2} - D_3 \leq 0$ regardless of the values of $\underline{u}_2$ on Ω_0. Indeed, the corresponding system of ODE's fails to be permanent, and (7.117) is our third "reasonable assumption" on the local species interactions in the model. However, it is clear from (7.115) that (7.117) is not enough. The values of $\underline{u}_2$ on Ω_0 matter. What we need is that the equilibrium to the pertinent ODE, say u_2^*, satisfy

$$\frac{E_2 u_2^*}{1 + B_2 u_2^*} - D_3 > 0. \tag{7.118}$$

The condition for (7.118) to hold is complicated but can be written in a way that shows how much beyond (7.105), (7.113) and (7.117) is needed. Namely, we require

$$(E_1 - B_1 D_2 - D_2)\left(1 - \frac{A_1}{C_1}\right)(E_2 - B_2 D_3) > C_1 D_2 D_3. \tag{7.119}$$

If (7.105), (7.113) and (7.117) hold, then each of the quantities in parentheses on the left-hand side is positive, so that having (7.119) hold is a possibility. However, it may be the case that (7.105), (7.113) and (7.117) hold but (7.119) fails to hold. There are various scenarios under which (7.119) will hold, assuming that (7.105), (7.113) and (7.117) hold, for instance, having E_2 large or D_3 small. It is instructive to note that the rate at which members of species 3 interfere with each other (which translates into C_2) plays no role in the permanence of (7.91)–(7.92).

So we have that (7.91)–(7.92) is permanent provided the conditions (7.105), (7.113), (7.117) and (7.119) on the local species interactions hold and the scale parameter ℓ is sufficiently large. The mathematical details (principally consisting of obtaining and estimating $\underline{u}_2$) are presented in the Appendix to the chapter.

Appendix

Proof of dynamics for (7.23)–(7.24): (Cantrell et al., 1993b, Lemma 3.6) It will suffice to show that for $u_1(x,0)$ and $u_3(x,0)$ with $u_1(x,0) \overset{>}{\not\equiv} 0$ and $u_3(x,0) \overset{>}{\not\equiv} 0$, if $(v_1, v_3) \in \omega((u_1(x,0), u_3(x,0)))$, then $(v_1, v_3) = \left(\dfrac{1 - c_{13}}{1 + c_{13}c_{31}}\theta a, \dfrac{1 + c_{31}}{1 + c_{13}c_{31}}\theta a\right)$. To this end, let $(u_1(x,t), u_3(x,t))$ denote the solution to (7.23)–(7.24) corresponding to $(u_1(x,0), u_3(x,0))$. We know first of all that u_1 is a lower solution to

$$\begin{aligned} \frac{\partial w}{\partial t} &= \Delta w + w(a - w) \text{ in } && \Omega \times (0, \infty) \\ w &= 0 && \text{on } \partial\Omega \times (0, \infty). \end{aligned} \tag{7A.1}$$

The assumption $\sigma_1 > 0$ in (7.15) (which is equivalent to $a > \lambda_0^1(\Omega)$) means that positive solutions to (7A.1) converge in $C^1(\overline{\Omega})$ to θ_a as $t \to \infty$. In particular, the solution w to (7A.1) which has the value $u_1(x,0)$ when $t = 0$ has this property. As a result, if $\varepsilon > 0$ is given, then $w(x,t) < (1+\varepsilon)\theta_a(x)$ on Ω for all $t \geq t_1$, where t_1 depends on $u_1(x,0)$. So $u_1(x,t) < (1+\varepsilon)\theta_a(x)$ on Ω for $t \geq t_1$. Since ε is arbitrary, we may conclude that $v_1 \leq \theta_a$.

Now let $t > t_1$. Then u_3 is a lower solution to

$$\begin{aligned} \frac{\partial z}{\partial t} &= \Delta z + z(a + c_{31}(1+\varepsilon)\theta_a - y) \text{ in } && \Omega \times (t_1, \infty) \\ z &= 0 && \text{on } \partial\Omega \times (t_1, \infty). \end{aligned} \tag{7A.2}$$

If z is the solution to (7A.2) with $z(x, t_1) = u_3(x, t_1)$, $u_3(x,t) \leq z(x,t)$ for $t > t_1$ and z converges to the unique positive equilibrium to (7A.2). (That such is the case follows from the fact that the principal eigenvalue σ is positive in

$$\begin{aligned} \Delta\phi + \phi(a + c_{31}(1+\varepsilon)\theta_a) &= \sigma\phi \text{ in } && \Omega \\ \phi &= 0 && \text{on } \partial\Omega, \end{aligned}$$

since $a + c_{31}(1+\varepsilon)\theta_a > a$ and $\sigma_1 > 0$ in (7.15).) A simple calculation will show the equilibrium to (7A.2) is $(1 + c_{31}(1+\varepsilon))\theta_a$. (See also Cosner and Lazer (1984).) Consequently, for some $t_2 > t_1$, it will be the case that $u_3(x,t) < z(x,t) < (1+\varepsilon)(1 + c_{31}(1+\varepsilon))\theta_a(x)$ on Ω for $t > t_2$. Since ε is arbitrary, we may conclude $v_3 \leq (1 + c_{31})\theta_a$.

Now let $\delta > 0$ be given and choose $\varepsilon > 0$ so that $(1+\varepsilon)(1 + c_{31}(1+\varepsilon))\theta_a(x) < (1+\delta)(1+c_{31})\theta_a(x)$. Then for the value of t_2 corresponding to ε we will have that u_1 is an upper solution to

$$\begin{aligned} \frac{\partial w}{\partial t} &= \Delta w + w(a - c_{13}(1+\delta)(1+c_{31})\theta_a - w) \text{ in } \quad \Omega \times (t_2, \infty) \\ w &= 0 \qquad\qquad\qquad\qquad\qquad\qquad\qquad \text{on} \quad \partial\Omega \times (t_2, \infty). \end{aligned} \tag{7A.3}$$

If now the principal eigenvalue σ is positive in

$$\begin{aligned} \Delta\phi + \phi(a - c_{13}(1+\delta)(1+c_{31})\theta_a) &= \sigma\phi \text{ in } \quad \Omega \\ \phi &= 0 \qquad\qquad \text{on} \quad \partial\Omega, \end{aligned} \tag{7A.4}$$

we will have that the solution w to (7A.3) with $w(x,t_2) = u_1(x,t_2)$ converges in $C^1(\overline{\Omega})$ to the unique positive equilibrium of (7A.3). It is at this point that the restriction (7.28) on c_{13} and c_{31} comes into play. Since

$$\Delta\phi_a + \phi_a(a - \theta_a) = 0 \quad \text{in} \quad \Omega$$

we need $c_{13}(1 + c_{31}) < 1$ in order to have $\sigma > 0$ in (7A.4) for small enough δ. (Note that (7.23)–(7.24) is still permanent for $c_{13} < 1$ and $c_{31} > 0$ arbitrary. We need the extra assumption to carry out this practical persistence regime.) Since the positive equilibrium to (7A.3) is $(1 - c_{13}(1+\delta)(1+c_{31}))\theta_a$, we may conclude for some $t_3 > t_2$ that $u_1(x,t) > (1-\delta)(1 - c_{13}(1+\delta)(1+c_{31}))\theta_a(x)$ on Ω for $t > t_3$. As a result, since δ is arbitrary

$$v_1 \geq (1 - c_{13} - c_{13}c_{31})\theta_a.$$

Now for $t > t_3$, u_3 is an upper solution to

$$\begin{aligned} \frac{\partial z}{\partial t} &= \Delta z + z(a + c_{31}(1-\delta)(1 - c_{13}(1+\delta)(1+c_{31}))\theta_a - z) \text{ in } \quad \Omega \times (t_3, \infty) \\ z &= 0 \qquad\qquad\qquad\qquad\qquad\qquad\qquad\qquad\qquad \text{on} \quad \partial\Omega \times (t_3, \infty). \end{aligned} \tag{7A.5}$$

As before, the solution z of (7A.5) with $z(x,t_3) = u_3(x,t_3)$ converges in $C^1(\overline{\Omega})$ to $(1 + c_{31}(1-\delta)(1 - c_{13}(1+\delta)(1+c_{31})))\theta_a$. Hence for some $t_4 > t_3$, $u_3(x,t) > (1-\delta)(1 + c_{31}(1-\delta)(1 - c_{13}(1+\delta)(1+c_{31})))\theta_a(x)$ in Ω for $t > t_4$. Since $\delta > 0$ is arbitrary,

$$\begin{aligned} v_3 &\geq (1 + c_{31}(1 - c_{13}(1 + c_{31})))\theta_a \\ &= (1 + c_{31} - c_{13}c_{31} - c_{13}c_{31}^2)\theta_a. \end{aligned}$$

We now have the estimates

$$\begin{aligned}(1-c_{13}-c_{13}c_{31})\theta_a \le v_1 \le \theta_a\\(1+c_{31})(1-c_{13}c_{31})\theta_a \le v_3 \le (1+c_{31})\theta_a.\end{aligned} \tag{7A.6}$$

It should be clear that the procedure can be continued indefinitely. The next step is to use the preceding lower bound on u_3 to obtain an upper bound on u_1 (always for large enough values of t). The upper bound on u_1 that we obtain leads to an upper bound on u_3 which in turn produces a lower bound on u_1. Then the lower bound on u_1 give us a new lower bound on u_3 and we start all over. After the next round we may update the estimates in (7A.6) to

$$\begin{aligned}&[(1-c_{13})(1-c_{13}c_{31}+c_{13}^2c_{31}^2)-c_{13}^3c_{31}^3]\theta_a \le v_1 \le [(1-c_{13})(1-c_{13}c_{31})+c_{13}^2c_{31}^2]\theta_a\\&(1+c_{31})(1-c_{13}c_{31}+c_{13}^2c_{31}^2-c_{13}^3c_{31}^3)\theta_a \le v_3 \le (1+c_{31})(1-c_{13}c_{31}+c_{13}^2c_{31}^2)\theta_a.\end{aligned} \tag{7A.7}$$

Proceeding from (7A.7) we get in general that v_1 satisfies

$$\begin{aligned}&\left[(1-c_{13})\sum_{n=0}^{2k}(-1)^n(c_{13}c_{31})^n - c_{13}^{2k+1}c_{31}^{2k+1}\right]\theta_a\\&\le v_1\\&\le \left[(1-c_{13})\sum_{n=0}^{2k-1}(-1)^n(c_{13}c_{31})^n + c_{13}^{2k}c_{31}^{2k}\right]\theta_a\end{aligned} \tag{7A.8}$$

while v_3 satisfies

$$\begin{aligned}&\left[(1+c_{31})\sum_{n=0}^{2k+1}(-1)^n(c_{13}c_{31})^n\right]\theta_a\\&\le v_3\\&\le \left[(1+c_{31})\sum_{n=0}^{2k}(-1)^n(c_{13}c_{31})^n\right]\theta_a.\end{aligned} \tag{7A.9}$$

The result now follows from (7A.8) and (7A.9) using the fact that $c_{13}c_{31} < c_{13}(1+c_{31}) < 1$ implies that $\sum_{n=0}^{\infty}(-1)^n(c_{13}c_{31})^n = \dfrac{1}{1+c_{13}c_{31}}$ and $\lim_{n\to\infty} c_{13}^n c_{31}^n = 0$.

Completion of the proof of permanence in (7.91)–(7.92): (Cantrell and Cosner, 2001) As discussed in Section 7.4, we need to show that under the assumption that (7.105), (7.113), (7.117) and (7.119) hold, then for sufficiently large values of ℓ, there will be a $c > 0$ so that if $(u_1, u_2, 0) \in M_2 = M_2(\ell)$ (the global attractor for (7.101)–(7.102)), then the principal eigenvalue σ_3 in (7.111) exceeds c. If such is the case, permanence in (7.91)–(7.92) follows from Theorem 4.3. To establish the existence of such a value of c, we make practical persistence estimates on M_2 in order to obtain a $\underline{u}_2 = \underline{u}_2(\ell)$ so that if $(u_1, u_2, 0) \in M_2$, then $u_2 \ge \underline{u}_2$. In such a case, the principal eigenvalue σ_3 in (7.111) exceeds the principal eigenvalue $\underline{\sigma}_3$ in (7.114). We show that $\underline{\sigma}_3 > 0$ by finding a piecewise continuous function $r(x)$ on Ω_0 with $r(x_0) > 0$ for some $x_0 \in \Omega_0$ so that (7.115) holds for large enough values of ℓ (i.e. $\dfrac{E_2\underline{u}_2}{1+B_2\underline{u}_2} - D_3 \geq r(x)$). Finding such an $r(x)$ completes the proof. Why? Since $r(x_0) > 0$ for some $x_0 \in \Omega_0$, $\lambda_+^1(r(x)) > 0$ exists. Consequently, the principal eigenvalue σ_r is positive in

$$\begin{aligned} &\frac{d_3}{\ell^2}\Delta\phi + r(x)\phi = \sigma_r\phi \text{ in } \quad \Omega_0 \\ &\phi = 0 \qquad\qquad\qquad\qquad \text{on} \quad \partial\Omega_0 \end{aligned} \tag{7A.10}$$

for sufficiently large ℓ, and $\underline{\sigma}_3$ in (7.114) exceeds σ_r in (7A.10).

To get the estimate we require on M_2, we begin with the solution $(u_1(x,t), u_2(x,t))$ of (7.101)–(7.102) corresponding to initial data $u_1(x,0) \overset{>}{\not\equiv} 0$ and $u_2(x,0) \overset{>}{\not\equiv} 0$. Any asymptotic bounds we obtain for $u_1(x,t)$ or $u_2(x,t)$ hold for any u_1 or u_2 in M_2 by definition of M_2. We assume to start that $\ell > \ell^*$ so that (7.101)–(7.102) is permanent. As we have now seen a number of times, u_1 is a lower solution to

$$\begin{aligned} &\frac{\partial w}{\partial t} = \frac{d_1}{\ell^2}\Delta w + w(1-w) \text{ in } \quad \Omega_0 \times (0,\infty) \\ &w = 0 \qquad\qquad\qquad\qquad \text{on} \quad \partial\Omega_0 \times (0,\infty). \end{aligned} \tag{7A.11}$$

All solutions to (7A.11) converge over time in $C_0^1(\overline{\Omega}_0)$ to the unique globally attracting positive equilibrium $\overline{w}$ of (7A.11). By the maximum principle, $\overline{w} < 1$ on Ω_0. Consequently, given any $\varepsilon > 0$, there is a $t_1 > 0$ so that if w is the solution to (7A.11) with $w(x,0) = u_1(x,0)$, then $u_1(x,t) < w(x,t) < 1+\varepsilon$ on $\overline{\Omega}_0$ for $t > t_1$.

As a result, $u_2(x,t)$ is a lower solution to

$$\begin{aligned} &\frac{\partial z}{\partial t} = \frac{d_2}{\ell^2}\Delta z + \left(\frac{E_1(1+\varepsilon)}{1+B_1(1+\varepsilon)+C_1 z} - D_2\right) z \text{ in } \quad \Omega_0 \times (t_1,\infty) \\ &z = 0 \qquad\qquad\qquad\qquad\qquad\qquad \text{on} \quad \partial\Omega_0 \times (t_1,\infty). \end{aligned} \tag{7A.12}$$

Assumption (7.105) implies that $\lambda_+^1\left(\dfrac{E_1(1+\varepsilon)}{1+B_1(1+\varepsilon)} - D_2\right) > 0$ exists. So if $\ell > \sqrt{d_2\lambda_+^1\left(\dfrac{E_1(1+\varepsilon)}{1+B_1(1+\varepsilon)} - D_2\right)}$, all positive solutions to (7A.12) converge in $C_0^1(\overline{\Omega}_0)$ to a globally attracting equilibrium $\overline{z} = \overline{z}(\ell)$ for (7A.12). By the maximum principle, $\overline{z}(\ell)$ on $\overline{\Omega}_0$ is less than the root of the equation

$$\frac{E_1(1+\varepsilon)}{1+B_1(1+\varepsilon)+C_1 z} - D_2 = 0,$$

namely

$$\frac{(E_1 - B_1 D_2)(1+\varepsilon) - D_2}{D_2 C_1}.$$

If z is the solution to (7A.12) so that $z(x,t_1) = u_2(x,t_1)$ there is a $t_2 > t_1$ so that

$$u_2(x,t) < z(x,t) < \frac{(1+\varepsilon)[(E_1 - B_1 D_2)(1+\varepsilon) - D_2]}{D_2 C_1}$$

for $t > t_2$. So if $\delta > 0$ is given, by an appropriate choice of ε in the preceding discussion, we may guarantee that for some $\ell^{**} > \ell^*$

$$u_2(x,t) < (1+\delta)\left(\frac{E_1 - B_1 D_2 - D_2}{D_2 C_1}\right)$$

for $\ell > \ell^{**}$ and $t > t_2 = t_2(\ell)$.

For $\ell > \ell^{**}$ and $t > t_2 = t_2(\ell)$, u_1 is an upper solution to

$$\frac{\partial w}{\partial t} = \frac{d_1}{\ell^2}\Delta w + w(1-w) - \frac{A_1 w(1+\delta)[(E_1 - B_1 D_2 - D_2)/D_2 C_1]}{1 + C_1(1+\delta)[(E_1 - B_1 D_2 - D_2)/D_2 C_1]} \quad \text{in} \quad \Omega_0 \times (t_2, \infty)$$

$$w = 0 \quad \text{on} \quad \partial\Omega_0 \times (t_2, \infty)$$

which simplifies to

$$\begin{aligned} \frac{\partial w}{\partial t} &= \frac{d_1}{\ell^2}\Delta w + w\left(1 - \frac{(1+\delta)A_1(E_1 - B_1 D_2 - D_2)}{C_1(D_2 + (1+\delta)(E_1 - B_1 D_2 - D_2))} - w\right) \quad \text{in} \quad \Omega_0 \times (t_1, \infty) \\ w &= 0 \quad \text{on} \quad \partial\Omega_0 \times (t_2, \infty). \end{aligned} \tag{7A.13}$$

So long as $1 - \dfrac{(1+\delta)A_1(E_1 - B_1 D_2 - D_2)}{C_1(D_2 + (1+\delta)(E_1 - B_1 D_2 - D_2))} > 0$, there will be a value $\ell^{***} > \ell^{**}$ so that if $\ell > \ell^{***}$ all solutions to (7A.13) converge over time in $C_0^1(\overline{\Omega}_0)$ to a unique globally attracting positive equilibrium solution $\underline{u}_1 = \underline{u}_1(\ell, \delta)$. Note that $1 - \dfrac{(1+\delta)A_1(E_1 - B_1 D_2 - D_1)}{C_1(D_2 + (1+\delta)(E_1 - B_1 D_2 - D_2))} > 0$ for all sufficiently small $\delta > 0$ precisely so long as $\dfrac{A_1}{C_1}\dfrac{(E_1 - B_1 D_2 - D_2)}{(E_1 - B_1 D_2)} < 1$. It should be evident that the assumption (7.113) (i.e. $A_1 < C_1$) is more than sufficient for this purpose. So now if $\delta > 0$ is such that $1 - \dfrac{(1+\delta)A_1(E_1 - B_1 D_2 - D_2)}{C_1(D_2 + (1+\delta)(E_1 - B_1 D_2 - D_2))} > 0$ and $\gamma \in (0,1)$, then

$$u_1(x,t) > (1-\gamma)\underline{u}_1(\ell, \delta)(x)$$

on Ω_0 for $\ell > \ell^{***}$ and $t > t_3 = t_3(\ell, \delta, \gamma)$.

Now for $\ell > \ell^{***}$ and $t > t_3 = t_3(\ell, \delta, \gamma)$, u_2 is an upper solution to

$$\begin{aligned} \frac{\partial z}{\partial t} &= \frac{d_2}{\ell^2}\Delta z + z\left(\frac{E_1(1-\gamma)\underline{u}_1(\ell,\delta)}{1 + B_1(1-\gamma)\underline{u}_1(\ell,\delta) + C_1 z} - D_2\right) \quad \text{in} \quad \Omega_0 \times (t_3, \infty) \\ z &= 0 \quad \text{on} \quad \partial\Omega_0 \times (t_3, \infty). \end{aligned} \tag{7A.14}$$

To have positive solutions to (7A.14) converge in $C_0^1(\overline{\Omega}_0)$ to a unique positive equilibrium $\underline{u}_2(\ell, \delta, \gamma)$ as $t \to \infty$, we require first that

$$\frac{E_1(1-\gamma)\max\limits_{\overline{\Omega}_0}\underline{u}_1(\ell,\delta)}{1+B_1(1-\gamma)\max\limits_{\overline{\Omega}_0}\underline{u}_1(\ell,\delta)} > D_2. \tag{7A.15}$$

If (7A.15) holds, the convergence of positive solutions to (7A.14) in $C_0^1(\overline{\Omega}_0)$ is as asserted provided that

$$\ell > \sqrt{d_2\lambda_+^1\left(\frac{E_1(1-\gamma)\underline{u}_1(\ell,\delta)}{1+B_1(1-\gamma)\underline{u}_1(\ell,\delta)} - D_2\right)}. \tag{7A.16}$$

To establish (7A.15) and (7A.16) note that Proposition 3.15 implies that if Ω' is an open subdomain of Ω_0 with $\overline{\Omega}' \subseteq \Omega_0$, $\underline{u}_1(\ell, \delta)$ converges uniformly on $\overline{\Omega}'$ to

$$1 - \frac{(1+\delta)A_1(E_1 - B_1D_2 - D_2)}{C_1(D_2 + (1+\delta)(E_1 - B_1D_2 - D_2))}$$

as $\ell \to \infty$. Consequently, we may assert that (7A.15) holds for large enough values of ℓ and small enough values of δ and γ provided that

$$\frac{E_1[1 - (A_1(E_1 - D_1D_2 - D_2)/C_1(E_1 - B_1D_2))]}{1 + B_1[1 - (A_1(E_1 - B_1D_2 - D_2)/C_1(E_1 - B_1D_2))]} > D_2. \tag{7A.17}$$

An elementary, albeit tedious, calculation will show that (7A.17) is equivalent to

$$(E_1 - B_1D_2 - D_2)(C_1 - A_1) > 0. \tag{7A.18}$$

In light of (7.105), (7A.18) is equivalent to (7.113) and so (7A.15) holds if $\ell > \ell^{***}$ is large enough and δ and γ are small enough.

We need to establish (7A.16). To this end choose $\rho \in (0, 1)$ so that

$$\frac{E_1\rho[1 - (A_1(E_1 - B_1D_2 - D_2)/C_1(E_1 - B_1/D_2))]}{1 + B_1\rho[1 - (A_1(E_1 - B_1D_2 - D_2)/C_1(E_1 - B_1D_2))]} > D_2. \tag{7A.19}$$

Next choose $\rho' \in (0, 1)$ so that $(\rho')^3 > \rho$. Choose $\delta' > 0$ so that for $0 \le \delta \le \delta'$,

$$1 - \frac{(1+\delta)A_1(E_1 - B_1D_2 - D_2)}{C_1(D_2 + (1+\delta)(E_1 - B_1D_2 - D_2))} > \rho'\left(1 - \frac{A_1(E_1 - B_1D_2 - D_2)}{C_1(E_1 - B_1D_2)}\right).$$

Now choose $\ell' > \ell^{***}$ large enough so that

$$\underline{u}_1(\ell, \delta') > \rho'\left(1 - \frac{(1+\delta')A_1(E_1 - B_1D_2 - D_2)}{C_1(D_2 + (1+\delta')(E_1 - B_1D_2 - D_2))}\right)$$

on $\overline{\Omega}'$ if $\ell > \ell'$. It follows from (7A.13) and the method of upper and lower solutions that $\underline{u}_1(\ell, \delta)$ is decreasing in δ for $\delta > 0$ and small. So now if $\gamma \in (0, 1 - \rho')$, $\delta \in (0, \delta')$ and

$\ell > \ell'$,

$$
\begin{aligned}
(1-\gamma)\underline{u}_1(\ell,\delta) &> \rho'\underline{u}_1(\ell,\delta) \\
&> \rho'\underline{u}_1(\ell,\delta') \\
&> (\rho')^2\left(1-\frac{(1+\delta')A_1(E_1-B_1D_2-D_2)}{C_1(D_2+(1+\delta')(E_1-B_1D_2-D_2))}\right) \\
&> (\rho')^3\left(1-\frac{A_1(E_1-B_1D_2-D_2)}{C_1(E_1-B_1D_2)}\right) \\
&= \rho\left(1-\frac{A_1(E_1-B_1D_2-D_2)}{C_1(E_1-B_1D_2)}\right)
\end{aligned}
$$

on $\overline{\Omega}'$. We have now that if $\gamma \in (0, 1-\rho')$, $\delta \in (0, \delta')$ and $\ell > \ell'$, $\dfrac{E_1(1-\gamma)\underline{u}_1(\ell,\delta)}{1+B_1(1-\gamma)\underline{u}_1(\ell,\delta)} - D_2 > m(x)$, where

$$
m(x) = \begin{cases} \dfrac{E_1\rho\left(1-\dfrac{A_1(E_1-B_1D_2-D_2)}{C_1(E_1-B_1D_2)}\right)}{1+B_1\rho\left(1-\dfrac{A_1(E_1-B_1D_2-D_2)}{C_1(E_1-B_1D_2)}\right)} - D_2 & \text{on } \overline{\Omega}' \\ -D_2 & \text{on } \Omega_0\backslash\overline{\Omega}'. \end{cases}
$$

By (7A.19), $m(x) > 0$ for $x \in \overline{\Omega}'$, and hence $\lambda_+^1(m(x)) > 0$ exists. Since the formula (2.15) implies that

$$
\lambda_+^1(m(x)) > \lambda_+^1\left(\frac{E_1(1-\gamma)\underline{u}_1(\ell,\delta)}{1+B_1(1-\gamma)\underline{u}_1(\ell,\delta)} - D_2\right),
$$

we have that (7A.16) holds if $\ell > \ell'$ and

$$
\ell > \sqrt{d_2\lambda_+^1(m(x))}.
$$

We may now assert that if $\ell > \ell^{****} > \ell^{***}$, $\delta < \delta'$ and $\gamma < 1-\rho'$, all positive solutions to (7A.14) converge in $C_0^1(\overline{\Omega}_0)$ to a unique positive equilibrium $\underline{u}_2(\ell,\delta,\gamma)$. Multiparameter bifurcation theory (Alexander and Antman, 1981), combined with the uniqueness of positive equilibria for (7A.13) and (7A.14), implies that $\underline{u}_2(\ell,\delta,\gamma)$ is continuous as a map from $(\ell^{****},\infty)\times[0,\delta']\times[0,1-\rho']$ into $C_0^1(\overline{\Omega}_0)$. Since $u_2(x,t)$ is an upper solution to (7A.16) for large enough values of t, we may conclude that if $(u_1,u_2,0) \in M_2 = M_2(\ell)$, where $\ell > \ell^{****}$, then

$$
u_2 \geq \underline{u}_2(\ell,\delta,\gamma) \tag{7A.20}
$$

for $\delta \leq \delta'$ and $\gamma \leq 1-\rho'$.

We now aim to show that $\underline{\sigma}_3$ in (7.114) is positive. To this end, we note that $\underline{u}_2(\ell, \delta, \gamma)(x)$ is an upper solution of the elliptic boundary value problem

$$
\begin{aligned}
0 &= \frac{d_2}{\ell^2}\Delta y + \frac{E_1\rho\left[1-\left(\dfrac{A_1(E_1-B_1D_2-D_2)}{C_1(E_1-B_1D_2)}\right)\right]y}{1+B_1\rho\left[1-\left(\dfrac{A_1(E_1-B_1D_2-D_2)}{C_1(E_1-B_2D_2)}\right)\right]+C_1y} - D_2y \text{ in } \quad \Omega' \\
y &= 0 \qquad\qquad \text{on} \quad \partial\Omega'.
\end{aligned}
\tag{7A.21}
$$

It follows from (7A.19) that for some $\ell^{\#} > \ell^{****}$, (7A.21) admits a unique positive solution $y = y(\ell)$ for $\ell > \ell^{\#}$. Moreover, Proposition 3.15 implies that if Ω'' is an open subdomain of Ω' with $\overline{\Omega}'' \subseteq \Omega'$, $y = y(\ell)$ converges uniformly on $\overline{\Omega}''$ to the root of the equation

$$
\frac{E_1\rho\left[1-\left(\dfrac{A_1(E_1-B_1D_2-D_2)}{C_1(E_1-B_1D_2)}\right)\right]}{1+B_1\rho\left[1-\left(\dfrac{A_1(E_1-B_1D_2-D_2)}{C_1(E_1-B_1D_2)}\right)\right]+C_1y} - D_2 = 0
$$

as $\ell \to \infty$. A calculation shows that this root is given by

$$
K = \frac{\rho(E_1-B_1D_2)\left(1-\dfrac{A_1(E_1-B_1D_2-D_2)}{C_1(E_1-B_1D_2)}\right) - D_2}{C_1D_2}. \tag{7A.22}
$$

Consequently, for any $\beta \in (0, 1)$ we may conclude that if $\ell > \ell^{\#}$, $\delta < \delta'$ and $\gamma < \rho'$,

$$
\underline{u}_2(\ell, \delta, \gamma)(x) > \beta K \tag{7A.23}
$$

on $\overline{\Omega}''$ with K as in (7A.22). Now define $h = h(\beta)$ on Ω_0 by

$$
h(x) = \begin{cases} \beta K \text{ on } & \overline{\Omega}'' \\ 0 \quad \text{ on } & \Omega_0 \backslash \overline{\Omega}'' \end{cases}
$$

It follows from (7A.23) that if $\ell > \ell^{\#}$, $\delta < \delta'$ and $\gamma < \rho'$

$$
\frac{E_2\underline{u}_2(\ell, \delta, \gamma)}{1+B_2\underline{u}_2(\ell, \delta, \gamma)} - D_3 > \frac{E_2h(\beta)}{1+B_2h(\beta)} - D_3 \tag{7A.24}
$$

on Ω_0. Let $r(x) = \dfrac{E_2h(\beta)(x)}{1+B_2h(\beta)(x)} - D_3$, where $\beta \in (0, 1)$ is yet to be determined. The function $r(x)$ is the function to which we alluded at the beginning of the proof. As a result, we have from (7A.24) that $\underline{\sigma}_3$ in (7.114) is positive (and hence (7.91)–(7.92) is permanent by Theorem 4.3) so long as the principal eigenvalue σ_r in (7A.10) is positive.

Recall from the beginning of the proof that we observed that, to establish that σ_r in (7A.10) is positive for sufficiently large values of ℓ, it suffices to establish that $r(x_0) > 0$ for some $x_0 \in \Omega_0$. To that end, we need to show that

$$\frac{E_2 \beta K}{1 + B_2 \beta K} - D_3 > 0 \tag{7A.25}$$

where $K = K(\rho)$ is as in (7A.22). To have (7A.25) hold for ρ and β sufficiently close to 1, we need

$$\frac{\dfrac{E_2}{C_1 D_2}\left[(E_1 - B_1 D_2)\left(1 - \dfrac{A_1(E_1 - B_1 D_2 - D_2)}{C_1(E_1 - B_1 D_2)}\right) - D_2\right]}{1 + \dfrac{B_2}{C_1 D_2}\left[(E_1 - B_1 D_2)\left(1 - \dfrac{A_1(E_1 - B_1 D_2 - D_2)}{C_1(E_1 - B_1 D_2)}\right) - D_2\right]} > D_3,$$

which after a calculation simplifies to

$$(E_1 - B_1 D_2 - D_2)(C_1 - A_1)(E_2 - B_2 D_3) > C_1^2 D_2 D_3. \tag{7A.26}$$

Since (7A.26) holds by virtue of (7.105), (7.113), (7.117) and (7.119), we have that r is as required for β and ρ sufficiently near 1 and the proof is complete.

References

Abrams, P.A. (1999). Is predator-mediated coexistence possible in unstable systems? *Ecology* **80**, 608–621.

Abrams, P.A. and L.R. Ginzburg. (2000). The nature of predation: prey dependent, ratio dependent, or neither? *Trends in Ecology and Evolution* **15**, 337–341.

Adams, R.A. (1975). *Sobolev Spaces*. Academic Press, New York.

Agmon, S., A. Douglis, and L. Nirenberg. (1959). Estimates near the boundary for solutions of elliptic partial differential equations satisfying general boundary conditions I. *Comm. Pure and Appl. Math.* **12**, 623–727.

Alexander, J.C. and S.S. Antman. (1981). Global and local behavior of bifurcating multi-dimensional continua of solutions for multiparameter nonlinear eigenvalue problems. *Arch. Rat. Mech. Anal.* **76**, 339–354.

Alikakos, N.D. (1979). An application of the invariance principle to reaction-diffusion equations, *J. Diff. Equations* **33**, 201–225.

Allee, W.C. (1931). *Animal Aggregations, a Study on General Sociology*. University of Chicago Press, Chicago, IL.

Allee, W.C. (1938). *The Social Life of Animals*. W.W. Norton and Co., New York.

Amann, H. (1976). Fixed point equations and nonlinear eigenvalue problems in ordered Banach spaces. *SIAM Review* **18**, 620–709.

Amann, H. (1985). Global existence for semilinear parabolic systems. *J. Reine Angew. Math.* **360**, 47–84.

Amann, H. (1986). Quasilinear evolution equations and parabolic systems. *Trans. Amer. Math. Soc.* **293**, 191–227.

Amann, H. (1988). Dynamic theory of quasilinear parabolic equations I: Abstract evolution equations. *Nonlinear Analysis TMA* **12**, 895–919.

Amann, H. (1989). Dynamic theory of quasilinear parabolic systems III: Global existence. *Math. Z.* **202**, 219–250.

Amann, H. (1990). Dynamic theory of quasilinear parabolic equations II: Reaction-diffusion systems. *Differential and Integral Equations* **3**, 13–75.

Anderson, R.C. (1972). The ecological relationships of meningeal worm (*Pneumostrongylus tenuis Dougherty*) and native cervids in North America. *J. Wildlife Disease* **7**, 304–310.

Andow, D.A., P. Kareiva, S.A. Levin, and A. Okubo. (1990). Spread of invading organisms. *Landscape Ecology* **4**, 177–188.

Angelstam, P. (1986). Predation on ground-nesting birds' nests in relation to predator densities and habitat edge. *Oikos* **47**, 365–373.

Apostol, T.M. (1975). *Mathematical Analysis*. Addison-Wesley, Reading, MA.

Arditi, R. and A.A. Berryman. (1991). The biological control paradox. *Trends in Ecology and Evolution* **6**, 32.

Armstrong, R.A. and R. McGehee. (1976a). Coexistence of two competitors on one resource. *J. Theor. Biol.* **56**, 499–502.

Armstrong, R.A. and R. McGehee. (1976b). Coexistence of species competing for shared resources. *Theor. Pop. Biol.* **9**, 317–328.

Armstrong, R.A. and R. McGehee. (1980). Competitive exclusion. *American Naturalist* **115**, 151–170.

Spatial Ecology via Reaction-Diffusion Equations R.S. Cantrell and C. Cosner

ISBN: 0-471-49301-5

Aronson, D.G. and H.F. Weinberger. (1975). Nonlinear diffusion in population genetics, combustion, and nerve pulse propagation. *In*: *Partial Differential Equations and Related Topics*. Lecture Notes in Mathematics **446**, 5–49. Springer-Verlag, New York.

Aronson, D.G. and H.F. Weinberger. (1978). Multidimensional nonlinear diffusions arising in population genetics. *Advances in Math.* **30**, 33–76.

Avila-Vales, E.J. (1995). *Permanence in Seasonal Ecological Models with Spatial Heterogeneity*. Ph.D. Thesis, University of Miami.

Avila-Vales, E.J. and R.S. Cantrell. (1995). Permanence in periodic-parabolic ecological systems with spatial heterogeneity. *In* R.P. Agarwal, ed. *Dynamical Systems and Applications*. Series in Applicable Analysis **4**, 63–76. World Scientific, Singapore.

Avila-Vales, E.J. and R.S. Cantrell. (1997). Permanence of three competitors in seasonal ecological models with spatial heterogeneity. *Canadian Appl. Math. Quar.* **5**, 1–24.

Ayala, F.J. (1969). Experimental invalidation of the principle of competitive exclusion. *Nature* **224**, 1076–1079.

Ayala, F.J., M.E. Gilpin, and J.G. Ehrenfeld. (1973). Competition between species: Theoretical models and experimental tests. *Theor. Pop. Biol.* **4**, 331–356.

Bach, C.E. (1984). Plant spatial pattern and herbivore population dynamics: Plant factors affecting the movement patterns of a tropical cucurbit specialist (Acallymma Innubum). *Ecology* **65**, 175–190.

Bandle, C. (1980). *Isoperimetric Inequalities and Applications*. *Monographs and Studies in Mathematics* 7. Pitman, Boston.

Beddington, J.R. (1975). Mutual interference between parasites or predators and its effect on searching efficiency. *J. Animal Ecology* **44**, 331–340.

Belgacem, F. (1997). *Elliptic Boundary Value Problems with Indefinite Weights: Variational Formulations of the Principal Eigenvalue and Applications*. *Pitman Research Notes in Mathematics* **368**. Longman, Harlow, Essex, UK.

Belgacem, F. and C. Cosner. (1995). The effects of dispersal along environmental gradients on the dynamics of populations in heterogeneous environments. *Canadian Applied Math. Quarterly* **4**, 379–397.

Belovsky, G.E. (1984). Moose and snowshoe hare competition and a mechanistic explanation from foraging theory. *Oecologia* **61**, 150–159.

Berman, A. and J. Plemmons. (1979). *Nonnegative Matrices in the Mathematical Sciences*. Academic Press, New York.

Berryman, A.A. (1992). The origins and evolution of predator-prey theory. *Ecology* **73**, 1530–1535.

Berysticki, H. and P.-L. Lions. (1980). Some applications of the method of super and subsolutions, *In* C. Bardos, J.M. Lasry and M. Schatzman, eds. *Bifurcation and Nonlinear Eigenvalue Problems: Proceedings, Paris, October 24, 1978*. Lecture Notes in Mathematics **782**, 16–41, Springer-Verlag, New York.

Beverton, R.J.H. and S.J. Holt. (1957). *On the Dynamics of Exploited Fish Populations*. Ministry of Agriculture, Fisheries and Food. Fishery Investigations Series **II**. **19**. London.

Bhatia, N.P. and G.P. Szegö. (1970). *Stability Theory of Dynamical Systems*. Springer-Verlag, New York.

Bilotti, J. and J.P. LaSalle. (1971). Periodic dissipative processes. *Bull. Amer. Math. Soc.* **6**, 1082–1089.

Blat, J. and K.J. Brown. (1984). Bifurcation of steady-state solutions in predator-prey and competition systems. *Proc. Royal Soc. Edingburgh* **97A**, 21–34.

Brand, L.A., and T.L. George. (2000). Predation risks for nesting birds in fragmented coast redwood forest. *J. Wildlife Management* **64**, 42–51.

Brauer, F. and J. Nohel. (1969). *The Qualitative Theory of Ordinary Differential Equations*. W.A. Benjamin, Menlo Park, CA.

Brown, J.H. (1971). Mammals on mountaintops: Nonequilibrium insular biogeography. *American Naturalist* **105**, 467–478.

Brown, K.J. and S.S. Lin. (1980). On the existence of positive eigenfunctions for an eigenvalue problem with indefinite weight function. *J. Math. Anal. Appl.* **75**, 112–120.

Brown, P.N. (1980). Decay to uniform states in ecological interactions. *SIAM J. Appl. Math.* **38**, 22–37.

Burton, T.A. and V. Hutson. (1991). Permanence for non-autonomous predator-prey systems. *Diff. Int. Equations* **4**, 1269–1280.

Butler, G.J., H.I. Freedman, and P. Waltman. (1986). Uniformly persistent systems. *Proc. Amer. Math. Soc.* **96**, 425–430.

Butler, G.J., R. Schmid, and P. Waltman. (1990). Limiting the complexity of limit sets in self-regulating systems. *J. Math. Anal. Appl.* **47**, 63–68.

Butler, G.J. and G.S.K. Wolkowicz. (1987). Predator-mediated coexistence in a chemostat: coexistence and competition reversal. *Math. Modelling* **8**, 781–785.

Cantrell, R.S. and C. Cosner. (1987). On the steady-state problem for the Volterra-Lotka competition model with diffusion. *Houston J. Math.* **13**, 337–352.

Cantrell, R.S. and C. Cosner. (1989a). Diffusive logistic equations with indefinite weights: Population models in disrupted environments. *Proc. Royal Soc. Edinburgh* **112A**, 293–318.

Cantrell, R.S. and C. Cosner. (1989b). On the uniqueness and stability of positive solutions in the Lotka-Volterra competition model with diffusion. *Houston J. Math.* **15**, 341–361.

Cantrell, R.S. and C. Cosner. (1991a). Diffusive logistic equations with indefinite weights: Population models in disrupted environments II. *SIAM J. Math. Anal.* **22**, 1043–1064.

Cantrell, R.S. and C. Cosner. (1991b). The effects of spatial heterogeneity in population dynamics. *J. Math. Biol.* **29**, 315–338.

Cantrell, R.S. and C. Cosner. (1993). Should a park be an island? *SIAM J. Appl. Math.* **53**, 219–252.

Cantrell, R.S. and C. Cosner. (1994). Insular biogeographic theory and diffusion models in population dynamics. *Theor. Pop. Biol.* **45**, 177–202.

Cantrell, R.S. and C. Cosner. (1996a). Models for predator-prey systems at multiple scales. *SIAM Review* **38**, 256–286.

Cantrell, R.S. and C. Cosner. (1996b). Practical persistence in ecological models via comparison methods. *Proc. Royal Soc. Edinburgh* **126A**, 247–272.

Cantrell, R.S. and C. Cosner. (1997). Bounds on trajectories in diffusive predator-prey models. *In*: *Proceedings of the 4th International Conference on Mathematical Population Dynamics. Advances in Mathematical Population Dynamics: Molecules, Cells and Man*, 733–746. World Scientific, Singapore.

Cantrell, R.S. and C. Cosner. (1998). On the effects of spatial heterogeneity on the persistence of interacting species. *J. Math. Biol.* **37**, 103–145.

Cantrell, R.S. and C. Cosner. (1999). Diffusion models for population dynamics incorporating individual behavior at boundaries: Applications to refuge design. *Theor. Pop. Biol.* **55**, 189–207.

Cantrell, R.S. and C. Cosner. (2000). Upper and lower solutions for a homogeneous Dirichlet problem with nonlinear diffusion and the principle of linearized stability. *Rocky Mth. J. Math.* **30**, 1229–1236.

Cantrell, R.S. and C. Cosner. (2001a). Effects of domain size on the persistence of populations in a diffusive food chain model with DeAngelis-Beddington functional response. *Natural Resource Modeling* **14**, 335–367.

Cantrell, R.S. and C. Cosner. (2001b). On the dynamics of predator-prey models with the Beddington-DeAngelis functional response. *J. Math. Anal. Appl.* **257**, 206–222.

Cantrell, R.S. and C. Cosner. (2001c). Spatial heterogeneity and critical patch size: Area effects via diffusion in closed environments. *J. Theor. Biol.* **209**, 161–171.

Cantrell, R.S. and C. Cosner. (2002a). Conditional persistence in logistic models via nonlinear diffusion. *Proc. Royal Soc. Edinburgh* **132A**, 267–281.

Cantrell, R.S. and C. Cosner. (2002b). Deriving reaction-diffusion models in ecology from interacting particle systems. Preprint.

Cantrell, R.S. and C. Cosner. (2002c). Effects of aggregative movement on population dynamics and critical patch size. *J. Math. Biol., to appear.*

Cantrell, R.S., C. Cosner, and W.F. Fagan. (1998). Competitive reversals inside ecological reserves: the role of external habitat degradation. *J. Math. Biol.* **37**, 491–533.

Cantrell, R.S., C. Cosner, and W.F. Fagan. (2001). Brucellosis, botflies and brainworms: The impact of edge habitats on pathogen transmission and species extinction. *J. Math. Biol.* **42**, 95–119.

Cantrell, R.S., C. Cosner, and W. Fagan. (2002). Habitat edges and predator-prey interactions: Effects on critical patch size. *Math. Biosci.* **175**, 31–55.

Cantrell, R.S., C. Cosner, and V. Hutson. (1993a). Permanence in ecological systems with diffusion. *Proc. Royal Soc. Edinburgh* **123A**, 533–559.

Cantrell, R.S., C. Cosner, and V. Hutson. (1993b). Permanence in some diffusive Lotka-Volterra models for three interacting species. *Dynamic Syst. Appl.* **2**, 505–530.

Cantrell, R.S., C. Cosner, and V. Hutson. (1996). Ecological models, permanence and spatial heterogeneity. *Rocky Mth. J. Math.* **26**, 1–35.

Cantrell, R.S., C. Cosner, and Y. Lou. (2002). Preprint.

Cantrell, R.S. and J.R. Ward, Jr. (1997). On competition mediated coexistence. *SIAM J. Appl. Math.* **57**, 1311–1327.

Cao, Y. and T.C. Gard. (1993). Uniform persistence for population models with time delay using multiple Lyapunov functions. *Diff. Int. Equations* **6**, 883–898.

Cao, Y., J. Fan, and T.C. Gard. (1996). Peristence in Kolmogorov population models with time delay. *In*: V. Lakshmikantham, ed. *Proceedings of the First World Congress of Nonlinear Analysts*, 3741–3751. Walter de Gruyter, Berlin.

Caristi, G., K.P. Rybakowski, and T. Wessolek. (1992). Persistence and spatial patterns in a one predator-two prey Lotka-Volterra model with diffusion. *Annali di Mat. Pur. Appl.* **46**, 345–377.

Castro, A. and A.C. Lazer. (1982). Results on periodic solutions of parabolic equations suggested by elliptic theory. *Bollettino Della Unione Matematica Italina Section B, Series 6* **1**, 1089–1104.

Caswell, H. (1989). *Matrix Population Models: Construction, Analysis, Interpretation*. Sinauer, Sunderland, MA.

Chang, K.W. and F. Howes. (1984). *Nonlinear Singular Perturbation Phenomena: Theory and Applications*. Applied Mathematical Sciences **56**. Springer-Verlag, New York.

Chesson, P. (1997). Making sense of spatial models in ecology. *In* J. Bascompte and R. Solé, eds. *Modelling Spatiotemporal Dynamics in Ecology*, 155–166. Landes Bioscience, Georgetown, Texas.

Cohen, J.E. (1989). Food webs and community structure. *In* J. Roughgarden, R.M. May and S.A. Levin, eds. *Perspective in Ecological Theory*, 181–202. Princeton University Press, Princeton, NJ.

Comins, H.N., M.P. Hassell, and R. May. (1992). The spatial dynamics of host-parasitoid systems. *J. Animal Ecology* **61**, 735–748.

Connell, J.H. (1961a). Effect of competition, predation by *Thaislapillus* and other factors on natural populations of the barnacle *Balanus balanoides*. *Ecol. Monogr.* **31**, 61–104.

Connell, J.H. (1961b). The influence of interspecific competition and other factors on the distribution of the barnacle *Chthamalus stallatus*. *Ecology* **42**, 710–723.

Cosner, C. (1994) Persistence (permanence), compressivity, and practical persistence in some reaction-diffusion models from ecology. *In* X. Liu and D. Siegel eds. *Comparison Methods and Stability Theory*, 101–115. Marcel Dekker, New York.

Cosner, C. (1996). Variability, vagueness, and comparison methods for ecological models. *Bull. Math. Biol.* **58**, 207–246.

Cosner, C. (1997). Comparison principles for systems that embed in cooperative systems, with applications to Lokta-Volterra models. *Dynamics of Discrete, Continuous, and Impulsive Systems* **3**, 283–303.

Cosner, C., D.L. DeAngelis, J.S. Ault, and D.B. Olson. (1999). Effects of spatial grouping on the functional response of predators. *Theor. Pop. Biol.* **56**, 65–75.

Cosner, C. and A.C. Lazer. (1984). Stable coexistence in the Volterra-Lotka competition model with diffusion. *SIAM J. Appl. Math.* **44**, 1112–1132.

Courant, R. and D. Hilbert. (1953). *Methods of Mathematical Physics, Vol. I*. Interscience, New York.

Crandall, M.G. and P.H. Rabinowitz (1971). Bifurcation from simple eigenvalues. *J. Functional Analysis* **8**, 321–340.

Crandall, M.G. and P.H. Rabinowitz (1973). Bifurcation, perturbation of simple eigenvalues, and linearized stability. *Arch. Rat. Mech. Anal.* **52**, 161–180.

Cui, G. and H. Freedman. (1999). A model for the diffusion of populations in annular patchy environments, *Quarterly Appl. Math.* **57**, 339–354.

Dancer, E.N. (1984). On positive solutions of some pairs of differential equations. *Trans. Amer. Math. Soc.* **284**, 729–743.

Dancer, E.N. (1985). On positive solutions of some pairs of differential equations II. *J. Diff. Equations* **60**, 236–258.

Dancer, E.N. (1988). The effect of domain shape on the number of positive solutions of certain nonlinear equations. *J. Diff. Equations* **74**, 120–156.

Dancer, E.N. (1991). On the existence and uniqueness of positive solutions for competing species models with diffusion. *Trans. Amer. Math. Soc.* **326**, 829–859.

Davies, K.F., C.R. Margules, and K.F. Lawrence. (2000). Which traits of species predict population declines in experimental forest fragments? *Ecology* **81**, 1450–1461.

DeAngelis, D.L., R.A. Goldstein, and R.V. O'Neill. (1975). A model for trophic interaction. *Ecology* **56**, 881–892.

DeAngelis, D.L. and L.J. Gross. (1992). *Individual-Based Models and Approaches in Ecology: Populations, Communities and Ecosystems.* Chapman & Hall, New York.

DeAngelis, D.L., K.A. Rose, and M.A. Huston. (1994). Individual-oriented approaches to modeling ecological populations and communities. *In* S.A. Levin, ed. *Frontiers in Mathematical Biology.* Lecture Notes in Biomathematics **100**. Springer-Verlag, New York.

de Figueiredo, D.G. (1982). Positive solutions of semilinear elliptic problems. *In* D.G. deFigueiredo, ed. *Differential Equations.* Lecture Notes in Mathematics, **957**. Springer-Verlag, Berlin.

Deimling, K. (1985). *Nonlinear Functional Analysis.* Springer-Verlag, New York.

DeMasi, A. and E. Presutti. (1991). *Mathematical Methods for Hydrodynamic Limits.* Springer-Verlag, New York.

DeSanti, A.J. (1986). Boundary and interior layer behavior of solutions of some singularly perturbed semilinear elliptic boundary value problems. *J. Mathes. Pures Appl.* **65**, 227–262.

DiGiorgi, E. (1957). Sulla differenziabilità e lànaliticità della estremali degli integrali multipli regolari. *Mem. Accad. Sci. Torino Cl. Sci. Fis. Mat. Natur.* **3**, 25–43.

Dunbar, S.R., K.P. Rybakowski, and K. Schmitt. (1986). Persistence in models of predator-prey populations with diffusion. *J. Diff. Equations* **65**, 117–138.

Durrett, R. and S.A. Levin. (1994). The importance of being discrete (and spatial). *Theor. Pop. Biol.* **46**, 363–394.

Durrett, R. and S.A. Levin. (1996). Spatial models for species area curves. *J. Theor. Biol.* **179**, 119–127.

Dwass, M. (1970). *Probability and Statistics.* W.A. Benjamin, Menlo Park, CA.

Eilbeck, J.C., J.E. Furter, and J. Lopez-Gomez. (1994). Coexistence in the competition model with diffusion. *J. Diff. Equations* **107**, 96–139.

Ellner, S.P., A. Sasaki, Y. Haraguchi, and H. Matsuda (1998). Speed of invasion in lattice population models: Pair-edge approximation. *J. Math. Biol.* **36**, 469–484.

Elton, C. (1933). *The Ecology of Animals.* (Reprinted 1966 by Science Paperbacks and Methuen, London.)

Enright, J.T. (1976). Climate and population regulation: The biogeographer's dilemma. *Oecologia* **24**, 295–310.

Fagan, W.F., R.S. Cantrell, and C. Cosner. (1999). How habitat edges change species interactions. *American Naturalist* **153**, 165–182.

Fife, P. (1979). *Mathematical Aspects of Reacting and Diffusing Systems.* Lecture Notes in Biomathematics **28**. Springer-Verlag, New York.

Fisher, R.A. (1937). The wave of advance of advantageous genes. *Annals of Eugenics* **7**, 355–369.

Fitzgibbon, W.E., J.J. Morgan, and R. Sanders. (1992). Global existence and boundedness for a class of inhomogeneous semilinear parabolic systems. *Nonlinear Analysis TMA* **19**, 885–899.

Fleming, W.H. (1975). A selection-migration model in population genetics. *J. Math. Biol.* **2**, 219–233.

Flierl, G., D. Grünbaum, S. Levin, and D. Olson. (1999). From individuals to aggregations: the interplay between behavior and physics. *J. Theor. Biol.* **196**, 397–454.

Freedman, H.I. (1980). *Deterministic Mathematical Models in Population Ecology.* Marcel Dekker, New York.

Freedman, H. and T. Kriztin. (1992). Global stability in models of population dynamics with diffusion I. Patchy environments. *Proc. Royal Soc. Edinburgh* **122A**, 69–84.

Freedman, H., J.B. Shukla, and Y. Takeuchi. (1989). Population diffusion in a two-patch environment. *Math. Biosci.* **95**, 111–123.

Freedman, H.I. and P. Moson. (1990). Persistence definitions and their connections. *Proc. Amer. Math. Soc.* **109**, 1025–1033.

Freedman, H.I. and P. Waltman. (1977). Mathematical analysis of some three-species food-chain models. *Math. Biosci.* **33**, 257–276.

Freedman, H.I. and P. Waltman. (1984). Persistence in three interacting predator-prey populations. *Math. Biosci.* **68**, 213–231.

Fretwell, S.D. (1972). *Populations in a Seasonal Environment.* Princeton University Press, Princeton, NJ.

Fretwell, S.D. and H.L. Lucas Jr. (1970). On territorial behavior and other factors influencing habitat distribution in birds. I. Theoretical development. *Acta Biotheoretica* **19**, 16–36.

Friedman, A. (1964). *Partial Differential Equations of Parabolic Type.* Prentice Hall, Englewood Cliffs, NJ.

Friedman, A. (1976). *Partial Differential Equations.* Krieger, New York.

Furter, J.E. and J. Lopez-Gomez. (1995). On the existence and uniqueness of coexistence states for the Lotka-Volterra competition model with diffusion and spatially dependent coefficients. *Nonlinear Anal.* **25**, 363–398.

Furter, J.E. and J. Lopez-Gomez. (1997). Diffusion-mediated permanence problem for a heterogeneous Lotka-Volterra competition model. *Proc. Royal Soc. Edinburgh* **127A**, 281–336.

Gard, T. (1997). *Practical persistence in stochastic population models.* In: S. Ruan, G.S.K. Wolkowicz, and J. Wu, eds. *Differential Equations with Applications to Biology. Fields Institute Communications* 21, 205–215. American Mathematical Society, Providence, RI.

Gardiner, C.W. (1985). *Handbook of Stochastic Methods* (2nd edition). Springer-Verlag, New York.

Gates, J.E. and L.W. Gysel. (1972). Avian nest dispersion and fledgling success in field-forest ecotones. *Ecology* **59**, 871–883.

Gause, G.F. (1935). *The Struggle for Existence.* Williams and Wilkins, Baltimore, MD.

Getz, W.M. (1994). A metaphysiological approach to modeling populations and communities. *In*: S.A. Levin, ed., *Frontiers in Mathematical Biology.* Lecture Notes in Biomathematics **100**, 411–442. Springer-Verlag, New York.

Gilbarg, D. and N.S. Trudinger. (1977). *Elliptic Partial Differential Equations of Second Order.* Springer-Verlag, New York.

Gilpin, M.E. and K.E. Justice. (1972). Reinterpretation of the invalidation of the principle of competitive exclusion. *Nature* **236**, 273–274.

Gilpin, M.E. and M.E. Soulé. (1986). Minimal viable populations: Processes of species extinction. *In* M.E. Soulé, ed. *Conservation Biology: The Science of Scarcity and Diversity,* 19–34, Sinauer, Sunderland, MA.

Ginzburg, L.R. (1998). Assuming reproduction to be a function of consumption raises doubts about some popular predator-prey models. *J. Animal Ecology* **67**, 325–327.

Gleason, H. (1926). The individualistic concept of the plant association. *Bull. Torrey Bot. Club.* **53**, 1–20.

Glendinning, P. (1994). Island chain models and gradient systems. *J. Math. Biol.* **32**, 171–178.

Goh, B.S. (1976). Global stability in two species interactions. *J. Math. Biol.* **3**, 313–318.

Goldwasser, L., J. Cook, and E.D. Silverman. (1994). The effects of variability on metapopulation dynamics and rates of invasion. *Ecology* **75**, 40–47.

Goszczynski, J. (1979a). Density estimation for an urban population of the field mouse. *Acta Theriologica* **24**, 411–414.

Goszczynski, J. (1979b). Penetration of mammals over urban green spaces in Warsaw. *Acta Theriologica* **24**, 419–423.

Grindrod, P. (1996). *The Theory and Applications of Reaction-Diffusion Equations: Patterns and Waves*, 2nd ed. Oxford University Press, New York.

Groom, M.J. and N. Schumaker. (1990). Evaluating landscape change: Patterns of worldwide deforestation and local fragmentation. *In*: P. Kareiva, J.G. Kingsolver, and R.B. Huey, eds. *Biotic Interactions and Global Change*, 24–44. Sinauer, Sunderland, MA.

Grünbaum, D. (1994). Translating stochastic density-dependent individual behavior with sensory constraints into an Eulerian model of animal swimming. *J. Math. Biol.* **33**, 139–161.

Grünbaum, D. (1999). Advection-diffusion equations for generalized tactic searching behaviors. *J. Math. Biol.* **38**, 169–194.

Gui, C. and Y. Lou. (1994). Uniqueness and nonuniqueness of coexistence states in the Lotka-Volterra competition model. *Comm. Pure. Appl. Math.* **47**, 1571–1594.

Hale, J.K. and H. Kocak. (1991). *Dynamics and Bifurcations*. Springer-Verlag, New York.

Hale, J.K. and P. Waltman. (1989). Persistence in infinite dimensional systems. *SIAM J. Math. Anal.* **29**, 388–395.

Hanski, I. (1997). Predictive and practical metapopulation models: The incidence function approach. *In*: D. Tilman and P. Kareiva, eds. *Spatial Ecology*, 21–45. Princeton University Press, Princeton, NJ.

Hanski, I. (1999). *Metapopulation Ecology*. Oxford University Press, Oxford, UK.

Hanski, I. and O. Ovaskainen (2000). The metapopulation capacity of a fragmented landscape. *Nature* **404**, 755–758.

Hanski, I. and O. Ovaskainen (2001). Spatially structured metapopulation models: Global and local assessment of metapopulation capacity. *Theor. Pop. Biol.* **60**, 281–302.

Hardin, D.P., P. Takac, and G.F. Webb. (1988a). Asymptotic properties of a continuous-space discrete-time population model in a random environment. *J. Math. Biol.* **26**, 361–374.

Hardin, D.P., P. Takac, and G.F. Webb. (1988b). A comparison of dispersal strategies for survival of spatially heterogeneous populations. *SIAM J. Appl. Math* **48**, 1396–1423.

Hardin, D.P., P. Takac, and G.F. Webb. (1990). Dispersion population models discrete in time and continuous in space. *J. Math. Biol.* **28**, 1–20.

Harrison, J.M. and L.A. Shepp. (1981). On skew Brownian motion. *Ann. Probability* **9**, 309–313.

Hassell, M.P., H.N. Comins, and R.M. May. (1994). Spatial structure and chaos in insect populations. *Nature* **353**, 255–258.

Hastings, A. and T. Powell. (1991). Chaos in a three species food chain. *Ecology* **72**, 896–903.

Henry, D. (1981) *Geometric Theory of Semilinear Parabolic Equations*. Lecture Notes in Math. **840**. Springer-Verlag, New York.

Hess, P. (1977). On uniqueness of positive solutions of nonlinear elliptic boundary value problems. *Math. Zeitschrift* **154**, 17–18.

Hess, P. (1991). *Periodic-Parabolic Boundary Value Problems and Positivity*. Pitman Research Notes in Mathematics **247**. Longman, Scientific and Technical, Harlow, Essex, UK.

Hess, P. and Kato, T. (1980). On some linear and nonlinear eigenvalue problems with an indefinite weight function. *Comm. in Partial Differential Equations* **5**, 999–1030.

Hess, P. and A.C. Lazer. (1991). On an abstract competition model and applications. *Nonlinear Anal.* **16**, 917–940.

Hetzer, G. and W. Shen. (2002). Uniform persistence, coexistence, and extinction in almost periodic/nonautonomous competition diffusion systems. *SIAM J. Math. Anal.* **34**, 204–227.

Hirsch, M. (1982). Systems of differential equations which are competitive or cooperative I: Limit sets. *SIAM J. Appl. Math.* **13**, 167–179.

Hirsch, M. (1985). Systems of differential equations which are competitive or cooperative II: Convergence almost everywhere. *SIAM J. Math. Anal.* **16**, 423–439.

Hirsch, M. (1988a). Systems of differential equations which are competitive or cooperative III: Competing species. *Nonlinearity* **1**, 51–71.

Hirsch, M. (1988b). Stability and convergence in strongly monotone dynamical systems. *J. Reine. Angew. Math.* **383**, 1–53.

Hirsch, M. (1989). Systems of differential equations which are competitive or cooperative V: Convergence in 3-dimensional systems. *J. Diff. Equations* **80**, 94–106.

Hirsch, M. (1990). Systems of differential equations which are competitive or cooperative IV: Structural stability in three dimensional systems. *SIAM J. Math. Anal.* **21**, 1225–1234.

Hirsch, M. (1991). Systems of differential equations which are competitive or cooperative, VI: A local C^r closing lemma for 3-dimensional systems. *Ergod. Th. Dynamical Sys.* **11**, 443–454.

Hirsch, M.W. and S. Smale. (1974). *Differential Equations, Dynamical Systems, and Linear Algebra*. Academic Press, New York.

Hofbauer, J. (1981). A general cooperation theorem for hypercycles. *Monatsh. Math.* **91**, 233–240.

Hofbauer, J. and K. Sigmund. (1988). *Dynamical Systems and the Theory of Evolution*. Cambridge University Press, Cambridge, UK.

Hofbauer, J. and K. Sigmund. (1998). *Evolutionary Games and Population Dynamics*. Cambridge University Press, Cambridge, UK.

Holland, C.J. (1977). A new energy characterization of the smallest eigenvalue of the Schrodinger equation. *Comm. Pure and Applied Math.* **30**, 755–765.

Holland, C.J. (1978). A minimum principle for the principal eigenvalue for second order linear elliptic equations with natural boundary conditions. *Comm. Pure and Applied Math.* **31**, 509–519.

Holling, C.S. (1959). Some characteristics of simple types of predation and parasitism, *Canadian Entomologist* **91**, 385–398.

Holling, C.S. (1973). Resilience and stability of ecological systems. *Ann. Rev. Ecol. Syst.* **4**, 1–23.

Hollis, S.L. and J.J. Morgan. (1991). Global existence and asymptotic decay for systems of convective reaction-diffusion equations. *Nonlinear Analysis TMA* **17**, 725–739.

Holmes, E.E. (1993). Are diffusion models too simple? A comparison with telegraph models of invasion. *American Naturalist* **142**, 779–795.

Holmes, E.E., M.A. Lewis, J.E. Banks, and R.R. Veit. (1994). Partial differential equations in ecology: Spatial interactions and population dynamics. *Ecology* **75**, 17–29.

Holmes, J.C. (1996). Parasites as threats to biodiversity in shrinking ecosystems. *Biodiversity and Conservation* **5**, 975–983.

Holt, R.D. (1992). A neglected facet of island biogeography: The role of internal spatial dynamics in area affects. *Theor. Pop. Biol.* **41**, 354–371.

Holt, R.D. and J.H. Lawton. (1993). Apparent competition and enemy-free space in insect host-parasitoid communities. *American Naturalist* **142**, 623–645.

Holt, R.D. and J.H. Lawton. (1994). The ecological consequences of shared natural enemies. *Ann. Rev. Ecol. Syst.* **25**, 495–520.

Horn, H.S., H.H. Shugart, and D.L. Urban. (1989). Simulators as models of forest dynamics. *In* J. Roughgarden, R.M. May and S.A. Levin, eds. *Perspectives in Ecological Theory*, 256–267. Princeton University Press, Princeton, NJ.

Hsu, S.B. (1981). Predator-mediated coexistence and extinction. *Math. Biosci.* **54**, 231–248.

Huffaker, C.B. (1958). Experimental studies on predation: Dispersion factors and predator-prey oscillations. *Hilgardia* **27**, 343–383.

Hutchinson, G.E. (1959). Homage to Santa Rosalia, or why are there so many kind or animals? *American Naturalist* **93**, 145–159.

Hutchinson, G.E. (1961). The paradox of the plankton. *American Naturalist* **95**, 137–145.

Hutson, V. (1984). A theorem on average Liapunov functions. *Monatsh. Math.* **98**, 267–275.

Hutson, V. and K. Mischaikow. (1998). An approach to practical persistence. *J. Math. Biol.* **37**, 447–466.

Hutson, V., K. Mischaikow, and P. Polácik. (2001). The evolution of dispersal rates in a heterogeneous time-periodic environment. *J. Math. Biol.* **43**, 501–533.

Hutson, V. and K. Schmitt. (1992). Permanence and the dynamics of biological systems. *Math. Biosci.* **111**, 1–71.

Hutson, V., W. Shen, and G.T. Vickers. (2000). Estimates for the principal spectrum point for certain time-dependent parabolic operators. *Proc. Amer. Math. Soc.* **129**, 1669–1679.

Hutson, V. and X.Q. Zhao. (1994). Permanence in Kolmogorov periodic predator-prey models with diffusion. *Nonlinear Anal.* **23**, 651–668.

Ikeda, T. and M. Mimura. (1993). An interfacial approach to regional segregation of two competing species mediated by a predator. *J. Math. Biol.* **31**, 215–240.

Janzen, D.H. (1983). No park is an island: Increase in interference from outside as park size decreases. *Oikos* **41**, 402–410.

Janzen, D.H. (1986). The eternal external threat. *In* M.E. Soulé, ed. *Conservation Biology: The Science of Scarcity and Diversity*, 286–303. Sinauer, Sunderland, MA.

Janzen, V.A.A. and K. Sigmund. (1998). Shaken not stirred: On permanence in ecological communities. *Theor. Pop. Biol.* **54** (1998), 195–201.

Jensen, A.L. (1987). Simple models for exploitative and interference competition. *Ecological Modelling* **35**, 113–121.

John, F. (1982). *Partial Differential Equations*, 4th. ed. Springer-Verlag, New York.

Kareiva, P.M. (1985). Finding and losing host plants by Phyllotreta: Patch size and surrounding habitat. *Ecology* **66**, 1809–1816.
Kareiva, P.M. (1989). Renewing the dialogue between theory and experiments in population ecology. *In* J. Roughgarden, R.M. May and S.A. Levin, eds. *Perspectives in Ecology Theory*, 68–88. Princeton University Press, Princeton, NJ.
Kareiva, P.M. and G. Odell. (1987). Swarms of predators exhibit "prey taxis" if individual predators use area-restricted search. *American Naturalist* **130**, 233–270.
Kareiva, P.M., J.G. Kingsolver, and R.B. Huey, eds. (1993). *Biotic Interactions and Global Change.* Sinauer, Sunderland, MA.
Kato, T. (1966). *Perturbation Theory for Linear Operators.* Springer, New York.
Kawasaki, K. and N. Shigesada. (1997). *Biological Invasions.* Oxford University Press, New York.
Kevorkian, J. and J.D. Cole. (1996). Multiple Scale and Singular Perturbation Methods. *Applied Mathematical Sciences* **114**. Springer-Verlag, New York.
Kierstead, H. and L.B. Slobodkin. (1953). The size of water masses containing plankton bloom. *J. Marine Research* **12**, 141–147.
Kipnis, C. and C. Landin. (1999). *Scaling Limits of Interacting Particle Systems.* Springer-Verlag, New York.
Kishimoto, K. (1981). Instability of non-constant equilibrium solutions of a system of competition-diffusion equations. *J. Math. Biol.* **13**, 105–114.
Kishimoto, K. and H.F. Weinberger. (1985). The spatial homogeneity of stable equilibrium of some reaction-diffusion systems on convex domains. *J. Diff. Equations* **58**, 15–21.
Klebanoff, A. and A. Hastings. (1994a). Chaos in one predator, two-prey models: General results from bifurcation theory. *Math. Biosci.* **122**, 221–233.
Klebanoff, A. and A. Hastings (1994b). Chaos in three species food chains. *J. Math. Biol.* **32**, 427–451.
Klein, B.C. (1989). Effects of forest fragmentation on dung and carrion beetle communities in central Amazonia. *Ecology* **70**, 1715–1725.
Korman, P. and A. Leung. (1986). A general monotone scheme for elliptic systems with applications to ecological models. *Proc. Royal Soc. Edinburgh* **102A**, 315–325.
Krasnosel'skii, M.A. (1964). *Positive Solutions of Operator Equations.* Noordhoff, Groningen, The Netherlands.
Kriztin, T. and H. Freedman. (1993). Global stability in models of population dynamics with diffusion II. Continuously varying environments. *Rocky Mountain J. Math.* **24**, 155–163.
Kshatriya, M. and C. Cosner. (2002). A continuum formulation of the ideal free distribution and its implications for population dynamics. *Theor. Pop. Biol* **61**, 277–284.
Kuang, Y. and E. Beretta. (1998). Global qualitative analysis of a ratio-dependent predator-prey system. *J. Math. Biol.* **36**, 1172–1189.
Ladyzhenskaya, O., V. Solonikov, and N. Ural'ceva. (1968). *Linear and Quasilinear Equations of Parabolic Type*, American Math. Soc., Providence, RI.
Landsberg, J. (1988). Dieback of rural eucalyptus: Tree phenology and damage caused by leaf-feeding insects. *Australian J. Ecol.* **13**, 251–267.
Law, R. and R.D. Morton. (1996). Permanence and the assembly of ecological communities. *Ecology* **77**, 762–775.
Lazer, A.C. (1982). Some remarks on periodic solutions of parabolic differential equations, *In*: A. Bednarek and L. Cesari, eds. *Dynamical Systems* II, 227–246. Academic Press, New York.
Lehman, C.L. and D. Tilman. (1997). Competition in spatial habitats. *In* D. Tilman and P. Kareiva, eds. *Spatial Ecology*, 185–203. Princeton University Press, Princeton, NJ.
Leslie, P.H. (1948). Some further notes on the uses of matrices in population mathematics. *Biometrika* **35**, 213–245.
Leung, A. (1980). Equilibria and stabilities for competing species, reaction-diffusion equations with Dirichlet boundary data. *J. Math. Anal. Appl.* **73**, 204–218.
Leung, A.W. (1989). *Systems of Nonlinear Partial Differential Equations.* Kluwer Academic, MA, Boston.
Levin, S.A. ed. (1994). *Frontiers in Mathematical Biology. Lecture Notes in Biomathematics* 100. Springer-Verlag, New York.
Levins, R. (1968). *Evolution in Changing Environments.* Princeton University Press, Princeton, NJ.

Levins, R. (1969). Some demographic and genetic consequences of environmental heterogeneity for biological control. *Bull. Entomological Soc. of America* **15**, 237–240.

Lewis, M.A. (1997). Variability, patchiness, and jump dispersal in the spread of an invading population *In*: D. Tilman and P. Kareiva, eds. *Spatial Ecology*, 46–69. Princeton University Press, Princeton, NJ.

Lewis, M.A. and P. Kareiva. (1993). Allee dynamics and the spread of living organisms. *Theor. Pop. Biol.* **43**, 141–158.

Liang, X.T., Z.D. Wu, and X.X. Liang. (1983). The weak maximum principle and uniqueness theorem for generalized solutions of uniformly parabolic equations. *J. Math. Research and Exposition* **3**, 67–71.

Liggett, T.M. (1985). *Interacting Particle Systems*. Springer-Verlag, New York.

Lions, P.-L. (1982). On the existence of positive solutions to semilinear elliptic equations. *SIAM Review* **24**, 441–467.

Lopez-Gomez, J. (1997). Strong competition with refuges. *Nonlinear Anal.* **30**, 5167–5178.

Lotka, A. (1925). *Elements of Physical Biology*. Williams and Wilkins, Baltimore. (Reprinted 1956 by Dover Press, New York.)

Lovejoy, T.E., A.O. Bierregaard, Jr., A.B. Rylands, J.R. Malcolm, C.E. Quintela, L.H. Harper, K.S. Brown, Jr., A.A. Powel, G.V. Powell, H.O.R. Schubart, and M.B. Hays (1986). Edge and other effects of isolation on Amazon forest fragments. *In*: M.E. Soulé, ed. *Conservation Biology: The Science of Scarcity and Diversity*, 257–285, Sinauer, Sunderland, MA.

Loye, J. and Carroll. (1995). Birds, bugs and blood: Avian parasitism and conservation. *Trends Ecol. Evol.* **10**, 232–235.

Ludwig, D., D.G. Aronson, and H.F. Weinberger. (1979). Spatial patterning of the spruce budworm. *J. Math. Biol.* **8**, 217–258.

Ludwig, D., D.D. Jones, and C.S. Holling (1978). Qualitative analysis of insect outbreak systems: The spruce budworm and the forest. *J. Animal Ecol.* **47**, 315–332.

Lui, R. (1982). Existence and stability of traveling wave solutions of a nonlinear integral operator. *J. Math. Biol.* **16**, 199–220.

MacArthur, R.H. (1972). *Geographical Ecology*. Harper and Row, New York. (Reprinted 1984 by Princeton University Press, Princeton, NJ.)

MacArthur, R.H. and R. Levins. (1964). Competition, habitat selection, and character displacement in a patchy environment. *Proc. Nat. Acad. Sci. U.S.A.* **51**, 1207–1210.

MacArthur, R.H. and E.O. Wilson (1967). *The Theory of Island Biogeography*. Princeton University Press, Princeton, NJ.

Malthus, T.R. (1798). An essay on the principle of population. (Reprinted in *The World of Mathematics*, v. 2, James R. Newman, Simon and Schuster, New York, 1956.)

Manes, A. and A.M. Micheletti. (1973). Un' estensione della teoria variazionale classica degli autovalori per operatori ellitici del secondo ordine. *Boll. Un. Mat. Ital.* **7**, 285–301.

Mangel, M. and C. Tier. (1993). A simple direct method for finding persistence times of populations and applications to conservation problems. *Proc. Natl. Acad. Sci. U.S.A.* **90**, 1083–1086.

Matano, H. and M. Mimura. (1983). Pattern formation in competition-diffusion systems in non convex domains. *Publ. RIMS, Kyoto Univ.* **19**, 1049–1079.

May, R.M. and W.J. Leonard. (1975). Nonlinear aspects of competition between three species. *SIAM J. Appl. Math* **29**, 243–253.

Maynard Smith, J. (1982). *Evolution and the Theory of Games*. Cambridge University Press, Cambridge, UK.

McCann, K. and P. Yodzis. (1994). Biological conditions for chaos in a three-species food chain. *Ecology* **75**, 561–564.

McGehee, R. and R.A. Armstrong. (1977). Some mathematical problems concerning the ecological principle of competitive exclusion. *J. Diff. Equations* **23**, 30–52.

McKelvey, K., B.R. Noon, and R.H. Lamberson. (1993). Conservation planning for species occupying fragmented landscapes: The case of the Northern Spotted Owl. *In*: P.M. Kareiva, J.G. Kingsolver and R.B. Huey, eds. *Biotic Interactions and Global Change*, 424–450. Sinauer, Sunderland, MA.

Mikhalov, V.P. (1978). *Partial Differential Equations*. Mir, Moscow.

Miller, R.K. (1965). Almost periodic differential equations as dynamical systems with applications to the existence of almost periodic solutions. *J. Diff. Equation* **1**, 337–345.

Mimura, M., S.-I. Ei, and Q. Fang. (1991). Effect of domain-shape on coexistence problems in a competition-diffusion system. *J. Math. Biol.* **29**, 219–237.

Molofsky, J. (1994). Population dynamics and pattern formation in theoretical populations. *Ecology* **75**, 30–39.

Mora, X. (1983). Semilinear parabolic problems define semiflows on C^k spaces. *Trans. Amer. Math. Soc.* **278**, 21–55.

Morgan, J.J. (1990). Boundedness and decay results for reaction-diffusion systems. *SIAM J. Math. Anal.* **21**, 1172–1189.

Moser, J. (1960). A new proof of DeGiorgi's theorem concerning the regularity problem for elliptic differential equations. *Comm. Pure Appl. Math.* **13**, 457–468.

Murray, J.D. (1993). *Mathematical Biology* (2nd edition). Springer-Verlag, New York.

Murray, J.D. and R.P. Sperb. (1983). Minimum domains for spatial patterns in a class of reaction-diffusion equations. *J. Math. Biol.* **18**, 169–184.

Nash, J. (1958). Continuity of solutions of parabolic and elliptic equations. *Amer. J. Math.* **80**, 931–954.

Noss, R.F., E. Dinerstein, B. Gilbert, M. Gilpin, B.J. Miller, J. Terborgh, and S. Trombulak. (1989). Core areas: Where nature reigns. *In*: M.E. Soulé and J. Terborgh, eds. *Continental Conservation: Foundations of Regional Reserve Networks*, 99–128. Island Press, Washington, DC.

Nussbaum, R. (1984). Positive operators and elliptic eigenvalue problems. *Math. Zeitschrift* **186**, 247–264.

Okubo, A. (1980). *Diffusion and Ecological Problems: Mathematical Models*. Springer-Verlag, New York.

Okubo, A., P.K. Maini, M.H. Williamson, and J.D. Murray. (1989). On the spatial spread of the grey squirrel in Britain. *Proc. Royal Soc. London* **B238**, 113–125.

Olson, D.B. and R.R. Hood (1994). Modelling pelagic biogeography. *Prog. Oceanogr.* **34**, 161–205.

Ouyang, T. and J. Shi. (1998). Exact multiplicity of positive solutions for a class of semilinear problems. *J. Diff. Equations* **146**, 121–156.

Pacala, S.W. and S.A. Levin. (1997). Biologically generated spatial pattern and the coexistence of competing species. *In*: D. Tilman and P. Kareiva, eds. *Spatial Ecology*, 204–232. Princeton University Press, Princeton, NJ.

Pacala, S.W. and J. Roughgarden. (1982). Spatial heterogeneity and interspecific competition. *Theor. Pop. Biol* **21**, 92–113.

Paine, R.T. (1966). Food web complexity and species diversity. *American Naturalist* **100**, 65–75.

Paine, R.T. (1969). A note on trophic complexity and community stability. *American Naturalist* **103**, 91–93.

Pao, C.V. (1969). Coexistence and stability of a competition-diffusion system in population dynamics. *J. Math. Anal. Appl.* **83**, 54–76.

Pao, C.V. (1992). *Nonlinear Parabolic and Elliptic Equations*. Plenum Press, New York.

Patterson, B.D. and W. Atmar (1986). Nested subsets and the structure of insular mammalian faunas and archipelagoes. *Biological J. Linnean Soc.* **28**, 83–98.

Pazy, A. (1983). *Semigroups of Linear Operators and Applications to Partial Differential Equations*. *Applied Mathematical Sciences* 44, Springer-Verlag, New York.

Pimm, S.L. and M.E. Gilpin. (1989). Theoretical issues in conservation ecology. *In*: J. Roughgarden, R.M. May, and S.A. Levin, eds. *Perspectives in Ecological Theory*, 287–305. Princeton University Press, Princeton, NJ.

Protter, M.H. and H.F. Weinberger (1966). On the spectrum of general second order operators. *Bull. Amer. Math. Soc.* **72**, 251–255.

Protter, M.H. and H.F. Weinberger. (1967). *Maximum Principles in Differential Equations*. Prentice-Hall, Englewood Cliffs, NJ.

Pulliam, H.R. (1988). Sources, sinks, and population regulation. *American Naturalist* **132**, 652–661.

Quinn, J.F. and J.R. Karr. (1993). Habital fragmentation and global change. *In*: P.M. Kareiva, J.G. Kingsolver and R.B. Huey, eds. *Biotic Interactions and Global Change*, 451–463. Sinauer, Sunderland, MA.

Quinn, T.J. and R.B. Deriso. (1999). *Quantitative Fish Dynamics.* Oxford University Press. New York.

Rabinowitz, P.H. (1971). Some global results for nonlinear eigenvalue problems. *J. Functional Analysis* **7**, 487–413.

Rabinowitz, P.H. (1973). Some aspects of nonlinear eigenvalue problems. *Rocky Mountain J. Math.* **3**, 161–202.

Ricker, W.E. (1954). Stock and recruitment. *Fisheries Research Board of Canada* **11**, 559–623.

Ricker, W.E. (1975). Computation and Interpretation of Biological Statistics of Fish Populations. Bulletin of the Fisheries Research Board of Canada **191**, Ottawa.

Rosenzweig, M.L. (1971). Paradox of enrichment: Destabilization of exploitation ecosystems in ecological time. *Science* **171**, 385–387.

Roughgarden, J. (1989). The structure and assembly of communities. *In*: J. Roughgarden, R.M. May and S.A. Levin, eds. *Perspectives in Ecology Theory*, 203–226. Princeton University Press, Princeton, NJ.

Roughgarden, J., S.A. Levin, and R.M. May, eds. (1989). *Perspectives in Ecological Theory*. Princeton University Press, Princeton, NJ.

Royden, H. (1968). *Real Analysis*, 2nd ed. Collier-Macmillan, London.

Rudin. W. (1976). *Principles of Mathematical Analysis.* McGraw-Hill, New York.

Rudin, W. (1966). *Real and Complex Analysis.* McGraw-Hill, New York.

Ruxton, G., W.S.C. Gurney, and A. De Roos. (1992). Interference and generation cycles. *Theor. Pop. Biol.* **42**, 235–253.

Saperstone, S.H. (1981). Semidynamical Systems in Infinite Dimensional Spaces. *Appl. Math. Sci.* 37. Springer-Verlag, New York.

Schaefer, A.H. (1971). *Topological Vector Spaces.* Springer, New York.

Schaefer, P.W. ed. (1988). *Maximum Principles and Eigenvalue Problems in Partial Differential Equations*, Pitman Research Notes in Mathematics **175**. Longman, Harlow, Essex, UK.

Schaffer, W.M., S. Ellner, and M. Kot. (1986). Effects of noise on some dynamical models in ecology. *J. Math. Biol.* **24**, 479–523.

Schechter, M. (2002). *Principles of Functional Analysis.* Graduate Studies in Math. **36**. American Mathematical Society, Providence, Rhode Island.

Schiaffino, A. and A. Tesei. (1982). Competition systems with Dirichlet boundary conditions. *J. Math. Biol.* **15**, 93–105.

Schoener, T.W. (1973). Population growth regulated by intraspecific competition for energy or time: some simple representations. *Theor. Pop. Biol.* **4**, 56–84.

Schoener, T.W. (1974). Competition and the form of habitat shift. *Theor. Pop. Biol.* **6**, 265–307.

Schoener, T.W. (1976). Alternatives to Lotka-Volterra competition: Models of intermediate complexity. *Theor. Pop. Biol.* **10**, 309–333.

Schoener, T.W. (1978). Effects of density-restricted food encounter on some single-level competition models. *Theor. Pop. Biol.* **13**, 365–381.

Schoener, T.W. (1986). Mechanistic approaches to community ecology: A new reductionism? *Amer. Zool.* **26**, 81–106.

Schuster, P., K. Sigmund, and R. Wolff. (1979a). On ω-limits for competition between three species. *SIAM J. Appl. Math.* **37**, 49–54.

Schuster, P., K. Sigmund, and R. Wolff. (1979b). Dynamical systems under constant organization III: Cooperative and competitive behavior of hypercycles. *J. Diff. Equations* **32**, 357–368.

Sell, G.R. (1967). Nonautonomous differential equations and topological dynamics, I, II. *Trans. Amer. Math. Soc.* **127**, 241–283.

Sell, G.R. (1971). *Topological Dynamics and Ordinary Differential Equations.* Van Nostrand, London.

Senn. S. and P. Hess. (1982). On positive solutions of a linear elliptic eigenvalue problem with Neumann boundary conditions. *Math. Ann.* **258**, 459–470.

Seno, H. (1988). Effect of a singular patch on population persistence in a multi-patch system. *Ecol. Modelling.* **43**, 271–286.

Shen, W. and Y. Yi. (1998a). Almost automorphic and almost periodic dynamics in skew-product semiflows, II: Skew-product semiflows. *Memoirs Amer. Math. Soc.* **136**, 23–52.

Shen, W. and Y. Yi. (1998b). Almost automorphic and almost periodic dynamics in skew-product semiflows, III: Applications to differential equations. *Memoirs Amer. Math. Soc.* **136**, 53–93.

Shen, W. and Y. Yi. (1998c). Convergence in almost periodic Fisher and Kolmogorov models *J. Math. Biol.* **37**, 84–102.

Skalski, G.T. and J.F. Gilliam. (2001). Functional responses with predator interference: Viable alternatives to the Holling type II model. *Ecology* **82**, 3083–3092.

Skellam, J. (1951). Random dispersal in theoretical populations. *Biometrika* **38**, 196–218.

Smith, H. (1995). *Monotone Dynamical Systems.* Mathematical Surveys and Monographs **41**. American Mathematical Society, Providence, RI.

Smoller, J. (1982). *Shock Wave and Reaction-Diffusion Equations.* Springer-Verlag, New York.

Smoller, J. and A. Wasserman. (1981). Global bifurcation of steady-state solutions. *J. Diff. Equations* **39**, 269–290.

Snyder, N.F.R., J.W. Wiley, and C.B. Kepler. (1987). *The Parrots of Luquillo: Natural History and Conservation of the Puerto Rican Parrot.* Western Foundation of Vertebrate Zoology, Los Angeles, CA.

Soderstrom, B., T. Part, and J. Ryden. (1998). Different nest predator faunas and nest predation risk on ground and shrub nests at forest ecotones: An experiment and a review. *Oecologia* (Berlin) **117**, 108–118.

Soulé, M.E., ed. (1986). *Conservation Biology: The Science of Scarcity and Diversity.* Sinauer, Sunderland, MA.

Soulé, M.E., ed. (1987). *Viable Populations for Conservation.* Cambridge University Press, Cambridge, UK.

Soulé, M.E. and J. Terborgh, eds. (1989). *Continental Conservation: Foundations of Regional Reserve Networks.* Island Press, Washington DC.

Sperb, R.P. (1981). *Maximum Principles and Their Applications.* Academic Press, New York.

Spohn, H. (1991). *Large Scale Dynamics of Interacting Particles.* Springer-Verlag, New York.

Stamps, J.A., M. Buechner, and V.V. Krishnan. (1987). The effects of edge permeability and habitat geometry on emigration from patches of habitat. *American Naturalist* **179**, 533–552.

Strauss, W.A. (1992). *Partial Differential Equations.* Wiley, New York.

Strong, D.R. (1986a). Density-vague population change. *Trends in Ecology and Evolution* **1**, 39–42.

Strong, D.R. (1986b). Density vagueness: Abiding the variance in the demography of real populations. *In* J. Diamond and T.J. Case, eds. *Community Ecology*, 257–268. Harper and Row, New York.

Sun, G., Q. Cui, and B. Song. (1991). A new mathematical model of interspecific competition—an expansion of the classical Lotka-Volterra competition equations. *Ecological Modelling* **58**, 273–284.

Takeuchi, Y. (1985). Predator-mediated coexistence in a Volterra model. In *Mathematics in Biology and Medicine (Bari 1983).* Lecture Notes Biomath. **57**, 92–97. Springer-Verlag, New York.

Tanner, J.T. (1975). The stability and the intrinsic growth rates of prey and predator populations. *Ecology* **56**, 855–867.

Tilman, D. (1982). *Resource Competition and Community Structure.* Princeton University Press, Princeton, NJ.

Tilman, D. (1994). Competition and biodiversity in spatially structured habitats. *Ecology* **75**, 2–16.

Tilman, D., C.L. Lehman, and P. Kareiva. (1997). Population dynamics in spatial habitats. In: D. Tilman and P. Kareiva, eds. *Spatial Ecology*, 3–20. Princeton University Press, Princeton, NJ.

Turchin, P. (1989). Population consequences of aggregative movement. *J. Animal Ecol.* **58**, 75–100.

Turchin, P. (1998). *Quantitative Analysis of Movement.* Sinauer, Sunderland Massachusetts.

Turing, A.M. (1952). The chemical basis of morphogenesis. *Phil. Trans. Royal Soc. London.* **B237**, 37–72.

Van Kirk, R. and M. Lewis (1997). Integrodifference models for persistence in fragmented habitats. *Bull. Math. Biol.* **59**, 107–137.

Van Kirk, R.W. and M.A. Lewis (1999). Edge permeability and population persistence in isolated habitat patches. *Nat. Resource Modeling* **12**, 37–64.

Van Vuuren, J.H. and J. Norbury. (2000). Conditions for permanence in well-known biological competition models. *ANZIAM J.* **42**, 195–223.

Verhulst, P.F. (1838). Notice sur la loi que la population suite dans son accroissement. Correspondence Mathematique et. *Physique* **10**, 113–121.

Volterra, V. (1928). Variations and fluctuations of the number of individuals in animal species living together. *Journal du Conseil, Conseil International pour L'Exploration de la Mer* **3**, 2–51.

Volterra, V. (1931). *Lecons sur la Théorie Mathématique de la Lutte pour la Vie*. Marcel Brelot, Paris.

Walsh, J.B. (1978). A diffusion with discontinuous local time. *Astérisque* **52–53**, 37–45.

Walter, W. (1970). *Differential and Integral Inequalities*. Springer-Verlag, New York.

Waltman, P. (1991). A brief survey of persistence in dynamical systems. In: S. Busenberg and M. Martelli, eds., *Delay-Differential Equations and Dynamical Systems*, 31–40. Lecture Notes in Mathematics **1475**, Springer-Verlag, New York.

Webb, G.F. (1985). *Theory of Nonlinear Age-Dependent Population Dynamics*. Marcel Dekker, New York.

Wegner, J.F. and G. Merriam. (1979). Movement by birds and small animals between a wood and adjoining farmland habitats. *J. of Appl. Ecology* **16**, 349–357.

Weinberger, H.F. (1974). *Variational Methods for Eigenvalue Approximation*. CBMS Regional Conference Series in Applied Mathematics **15**. Society for Industrial and Applied Mathematics, Philadelphia.

Weinberger, H.F. (1982). Long-time behavior of a class of biological models, *SIAM J. Math. Anal.* **13**, 353–396.

Wendi, W. and M. Zhien. (1991). Harmless delays for uniform persistence. *J. Math. Anal. Appl.* **158**, 256–268.

Wilcove, D.S. (1985). Nest predation in forest tracks and the decline of migratory songbirds. *Ecology* **66**, 1211–1214.

Williams, S.A. and P.L. Chow. (1978). Nonlinear reaction-diffusion models for interacting populations. *J. Math. Anal. Appl.* **62**, 157–169.

Williamson, M. (1981). *Island Populations*. Oxford University Press, New York.

Yahner, R.H. (1983). Population dynamics of small mammals in farmstead shelterbelts. *J. Mammalogy* **64**, 380–386.

Yi, Y. (1998). Almost automorphic and almost periodic dynamics in skew-product semiflows, I: Almost automorphy and almost periodicity. *Memoirs Amer. Math. Soc.* **136**, 1–22.

Yodzis, P. (1989). *Introduction to Theoretical Ecology*. Harper and Row, New York.

Zeidler, E. (1985). *Nonlinear Functional Analysis and Its Applications I*. Fixed-Point Theorems. Springer-Verlag, New York.

Zeidler, E. (1986). *Nonlinear Functional Analysis and Its Applications II*. Springer-Verlag, New York.

Zhao, X.-Q. (2002). Global attractivity in monotone and subhomogeneous almost periodic systems. *J. Diff. Equations*, In press.

Zinner, B. (1991). Stability of traveling wavefronts for the discrete Nagumo equation. *SIAM J. Math. Analysis* **22**, 1016–1020.

Zinner, B. (1992). Existence of traveling wavefront solutions for the discrete Nagumo equation. *J. Differential Equations* **96**, 1–27.

Zinner, B., G. Harris, and W. Hudson. (1993). Traveling wavefronts for the discrete Fisher's equation. *J. Differential Equations* **105**, 46–62.

Index

Spatial Ecology via Reaction-Diffusion Equations R.S. Cantrell and C. Cosner

ISBN: 0-471-49301-5